COMPUTER INTEGRATED MANUFACTURING

Proceedings of the 3rd International Conference

COMPUTER INTEGRATED MANUFACTURING

11 – 14 July 1995
Singapore International Convention & Exhibition Centre

Volume 1

Editors

J Winsor, A I Sivakumar and R Gay

GINTIC Institute of Manufacturing Technology
Nanyang Technological University
Nanyang Avenue, Singapore

World Scientific
Singapore • New Jersey • London • Hong Kong

Published by

World Scientific Publishing Co. Pte. Ltd.
P O Box 128, Farrer Road, Singapore 9128
USA office: Suite 1B, 1060 Main Street, River Edge, NJ 07661
UK office: 57 Shelton Street, Covent Garden, London WC2H 9HE

COMPUTER INTEGRATED MANUFACTURING
Proceedings of the 3rd International Conference
Copyright © 1995 by World Scientific Publishing Co. Pte. Ltd.

All rights reserved. This book, or parts thereof, may not be reproduced in any form or by any means, electronic or mechanical, including photocopying, recording or any information storage and retrieval system now known or to be invented, without written permission from the Publisher.

For photocopying of material in this volume, please pay a copying fee through the Copyright Clearance Center, Inc., 222 Rosewood Drive, Danvers, Massachusetts 01923, USA.

ISBN 981-02-2376-5 (Set)
      981-02-2973-9 (Vol. 1)
      981-02-2974-7 (Vol. 2)

Printed in Singapore.

ORGANISING COMMITTEE

Chairman		Prof Robert Gay, *Gintic Institute of Manufacturing Technology, Singapore*

Members:		Mrs Foo-Lim Hooi Ling, *National Computer Board, Singapore*
		Dr Hui Siu Cheung, *Nanyang Technological University, Singapore*
		Ms Lim Lai Lee, *Gintic Institute of Manufacturing Technology, Singapore*
		Prof Lennie Lim, *Nanyang Technological University, Singapore*
		Mr James Ling, *Gintic Institute of Manufacturing Technology, Singapore*
		Dr Lye Sun Woh, *Nanyang Technological University, Singapore*
		Prof Dinesh P Mittal, *Nanyang Technological University, Singapore*
		Mr Saw Ken Wye, *National Computer Board, Singapore*
		Mr Teo Ming Yeong, *Nanyang Technological University,Singapore*
		Dr Julian Winsor, *Gintic Institute of Manufacturing Technology, Singapore*
		A/Prof Yeong Hin Yuen, *Nanyang Technological University, Singapore*

TECHNICAL COMMITTEE

Chairman		Dr Julian Winsor, *Gintic Institute of Manufacturing Technology, Singapore*

Co-Chairman	Dr Appa Iyer SivaKumar, *Gintic Institute of Manufacturing Technology, Singapore*
		Ms Lim Lai Lee, *Gintic Institute of Manufacturing Technology, Singapore*
		A/Prof Ward McClure, *Gintic Institute of Manufacturing Technology, Singapore*
		A/Prof Soh Ai Kah, *Gintic Institute of Manufacturing Technology, Singapore*
		Dr Robert de Souza, *Nanyang Technological Universit, Singapore*
		Mr Simon Tsang, *Gintic Institute of Manufacturing Technology, Singapore*
		Prof VC Venkatesh, *Nanyang Technological University,Singapore*

INTERNATIONAL ADVISORY COMMITTEE

Prof L Alting, *Technical University of Denmark, Denmark*
Prof J Banks, *Georgia Institute of Technology, USA*
Prof M Bonney, *The University of Nottingham, United Kingdom*
Prof H J Bullinger, *Fraunhofer-Institute for Industrial Engineering & Organisation, Germany*
Prof A Carrie, *University of Strathclyde, United Kingdom*
Ms R Connolly, *Autodesk Asia, Singapore*
Prof L A Gerhardt, *Rensselaer Polytechnic Institute, USA*
Prof A Ghosh, *Indian Institute of Technology, Kanpur, India*
Prof P O 'Grady, *North Carolina State University, USA*
Prof T N Goh, *National University of Singapore, Singapore*
Dr J P van Griethuysen, *Swiss Federal Institute of Technology Laussane, Switzerland*
Mr R Gunawan, *PT Astra Tech, Indonesia*
Prof Y Hasegawa, *Waseda University, Japan*
Prof S Hiraki, *Hiroshima University, Japan*
Prof W Hoheisel, *Steinbeis-Stiftung Wirtschaftforderung, Germany*
Dr S C Hui, *Structural Dynamics Research Corp, USA*
Prof T Ichimura, *Nihon University, Japan*
Prof K Ishii, *Kanazawa Institute of Technology, Japan*
Dr S Inaba, *Fanuc, Japan*
Prof B Kaftanoglu, *Middle East Technical University, Turkey*
A/Prof A G B Kamarrudin, *University Malaya, Malaysia*
Prof S Kang, *Seoul National University, Korea*
Prof H Katayama, *Waseda University, Japan*
Prof F Kimura, *University of Tokyo, Japan*
Prof A Kochhar, *University of Manchester Institute of Science & Technology, United Kingdom*
Dr R Krishnamurthy, *Indian Institute of Technology, Madras, India*
Prof S V Kumar, *Central Manufacturing Technology Institute , India*

Prof M Kuroda, *Aoyama Gakuin University, Japan*
Prof W P Lewis, *University of Melbourne, Australia*
Prof Li M, *University of Science & Technology, China*
A/Prof M K Lim, *Nanyang Technological University*
Mr C T Lin, *Nanyang Polytechnic, Singapore*
Prof G C I Lin, *University of South Australia, Australia*
Prof L Lindgren, *Luleå Univ of Technology, Sweden*
Prof S C-Y Lu, *University of Illinois at Urbana-Champaign, USA*
Prof K J MacCallum, *Universtiy of Strathclyde, United Kingdom*
Prof N Martensson, *Chalmers University of Technology, Sweden*
Dr K I Mori, *Kansai University, Japan*
Prof A Nee, *National University of Singapore, Singapore*
Dr P J Nolan, *University College, Ireland*
Prof V Orpana, *Lappeenranta University of Technology, Finland*
Prof D Patterson, *Institute of Systems Science, Singapore*
Dr I Persson, *Lund Institute of Technology, Sweden*
Prof D T Pham, *University of Wales, United Kingdom*
Prof F C Pruvot, *Ecole Polytechnique Federale de Laussane, Switzerland*
Mr E Quah, *Singapore Polytechnic, Singapore*
Prof N Ramaswamy, *Indian Institute of Technology, Bombay, India*
Dr P Ranky, *University of East London, United Kingdom*
Dr U R K Rao, *Indian Institute of Technology, New Delhi, India*
Prof F O Rasch, *University of Trondheim, Norway*
Prof J Reissner, *ETH-Zurich Institut-Fur Umformtechnik, Switzerland*
Dr P Sackett, *Cranfield University, United Kingdom*
Prof H Sato, *Chuo University, Japan*
Prof M Susomo, *Waseda University, Japan*
Prof M T Tabucanon, *Asian Institute of Technology, Thailand*
Mr A S Tan, *Ngee Ann Polytechnic, Singapore*
Prof J Triouleyre, *Ecole Nationale Superieure d'Arts et Metiers, France*
Prof T Vamos, *Hungarian Academy of Sciences, Hungary*
Dr N Varaprasad, *Temasek Polytechnic, Singapore*
Prof N Viswanadham, *Indian Institute of Science, India*
Prof K K Wang, *Cornell University, USA*
Prof H Warnecke, *Fraunhofer Institute of Manufacturing Engineering & Automation, Germany*
Dr B D Warthen, *Warthen Technology Information Services, USA*
Prof R Weston, *Loughborough University of Technology, United Kingdom*
Prof D J Williams, *Loughborough University of Technology, United Kingdom*
Prof T Woo, *University of Michigan, USA*
Prof B M Worrall, *Technical University of Nova Scotia, Canada*
Prof M Wozny, *National Institute of Standards & Technology, USA*

PREFACE

The International Conference on Computer Integrated Manufacturing (ICCIM '95) is now in its third session. Once again, it has been held in Singapore and has notably attracted tremendous interest from the professional community world-wide. We have seen a marked increase in activity in the strategic area of Computer Integrated Manufacturing (CIM). This is demonstrated by the substantial response to the call for papers. We received over 300 high quality manuscripts from which we selected those most representative of the latest developments.

It is intended that the proceedings provide the reader with a good overall coverage of the field's diverse subject matter. We have strived to give a good indication of how CIM practices are converging across sectors and throughout the world. CIM professionals are focusing on developing and utilising common standards and architectures. The new themes that people are striving for, be it mass-customisation, globalisation, empowerment or agility, are all bound by a common thread - CIM. The proceedings attest to this.

When setting out this text we intended it to take a timeless role - one which goes beyond the conference. The papers are arranged in order to take the reader through its contents in a logical manner starting with the higher level, pre-manufacturing issues of CIM and progressing to the down-stream, production issues. This, the first of two volumes deals with CIM Management, Product Design, CAD/CAM/CAE, CAD Data Exchange, FEM/FEA and Manufacturing Planning & Control.

ICCIM '95 is being held concurrently with the Society of Manufacturing Engineers' first AUTOFACT-ASIA exhibition for CAD/CAM, CAE & CIM. This has allowed the conference to take on a dual role by providing a forum for the presentation of current CIM research as well as information of the latest industrial applications.

250 referees from over 40 countries completed more than 800 reviews of the received manuscripts. The list of all those who have made a valuable contribution is extensive. Without the help of the learned community it would not have been possible to maintain a high standard. Our sincere thanks go to all the authors, IAC members, Organising Committee, Technical Committee, reviewers and participants in ICCIM '95.

July 1995

J Winsor
A Sivakumar
R Gay

CONTENTS

Preface   vii

Volume 1

CIM MANAGEMENT   1

Management

Re-Engineering Through Natural Structures: The Fractal Factory   5
    W. Sihn

Multidatabase Agents for CIM Systems   13
    M. Wunderli, M. C. Norrie and W. Schaad

Business Process Reengineering: Basis for Successful Information System Planning   21
    K. Mertins and R. Jochem

Hypermedia in CIM: Issues and Opportunities   29
    R. F. Leung, H. C. Leung and J. F. Hill

A Model-driven Toolset for Supporting the Life-cycle of CIM Systems   37
    J. M. Edwards, S. Murgatroyd and R. H. Weston

Business Process Management: The European Approach   45
    H. H. Hinterhuber

An International Perspective of Manufacturing   55
    T. A. Spedding, L. H. S. Luong and R. F. O'Connor

Evaluation of Advanced Manufacturing Technology — The Modes of Thinking   62
    I. Persson

Concurrent Engineering

Organizational Growth Demands Concurrent Engineering   73
    N. K. H. Tang and P. L. Forrester

Developing an Integrated Workgroup Environment for CIM Enterprise   82
    P. W. Lek, C. C. Wong, R. K. L. Gay and R. Y. G. Lim

Models and Architecture

The Development and Application of a Generic "Order Fulfilment"
Process Model — 93
    *A. M. Weaver, S. J. Childe, R. S. Maull, P. A. Smart
and J. Bennett*

Modelling Manufacturing Resources and Activities: An Ontology — 101
    F. Bonfatti, P. D. Monari and P. Paganelli

A Generic Information Infrastructure for Enterprise Integration — 109
    J. Jonker and E. M. Ehlers

Engineering Data and Process Integration in the SUKITS Environment — 117
    B. Westfechtel

Systems Integration

A Case Analysis of the Limits to Integration in Organizations Adopting
Computer Integrated Manufacturing — 127
    P. L. Forrester and J. S. Hassard

The Integration of Document Archives in a Manufacturing Automation
Protocol (MAP) Environment — 135
    A. Prosser

Communication Networks for Computer Integrated Manufacturing — 143
    W. P. Lu and W. F. Lu

Manufacturing Information Enhancement through Data Integration — 151
    A. E. James

Systems Integration — Coping with Legacy Systems — 159
    V. Singh

Implementation

Avoiding Human Error in CIM Implementations — 169
    R. G. Hannam

Integrating CAPM and CAMM: A Bottom-up Approach — 177
    K. F. Pun and J. A. Bull

A Step Towards CIM Using Intelligent Documents — 184
    V. Raja

CIMS in Shenyang Blower Works 192
    R. L. Mu, K. M. Lin, J. S. Xue and H. B. Yu

Computer Aided Logistics

The Effects of Transportation Interval in International Global
Complementary Production Systems 201
    S. Hiraki, K. Ishii and H. Katayama

A Decision Support Tool for Designing Software Interfaces Between
Logistics Centres and Associated Systems 209
    E. J. Fletcher and R. Brunner

PRODUCT DESIGN 217

Geometrical Design

A Prototype System For Early Geometric Configuration Design 221
    X. H. Guan, D. A. Stevenson and K. J. MacCallum

Feature Based Design of Free Form Products 229
    S. R. Mitchell and R. Jones

Geometry-Oriented Information Modeling for Mechanical Products 236
    W. Z. Zhang, F. Y. Wang, H. Ding and S. Z. Yang

Product Development

Solid Freeform Manufacturing as Part of the Integrated Product
Development Process 245
    K. H. Grote, J. L. Miller and T. K. Pflug

Interactive Multiple Criteria Decision Making for Product Development 253
    A. Jeang and D. R. Falkenburg

An Integrated Computer-Aided Gear Design System 261
    J. Y. Lai and W. F. Lu

Development of a Framework System for Tool Intergration in a
Product Information Archive 269
    S. Sum, D. Cheng, D. Koch, D. Domazet and S. S. Lim

Design Support for Configuration Management in Product Development 278
    B. Yu and K. J. MacCallum

Reverse Engineering — Modern Tool for Product-Planning 287
 B. Schumacher and P. Pscheid

Moulded Parts Design

A HyperCAD Expert System for Plastic Product Design 295
 J. Borg and K. J. MacCallum

A Framework to Develop an Expert Injection Mold Planning System
for Early Product Design Decisions 303
 K. S. Chin and T. N. Wong

CAD/CAM & CAE 311

CAD/CAM Technology

A Computational Tolerancing Scheme for CAD/CAM 315
 K. M. Yu

Application of an Environment for International Collaborative
CAD/CAM 323
 G. C. I. Lin, Y. C. Kao, H. C. Liaw, R. S. Lee, L. S. Chen
 and D. Y. Zhang

3D Graphic Collision Control as a Function of CAD/CAM Systems 331
 Z. J. Bao

CAE

Computer Aided Geometric Design of Ship Hull form Using
Bi-Cubic B-Spline Surface 343
 M. R. Bin Mainal and Y. Bin Samian

Virtual Tryout of Die and Molds Using Numerical Simulation
with PAM-STAM™ 351
 L. T. Kisielewicz

The Best Casting Plane of Simple Polyherdron 359
 W. Wang, J. Y. Wang, W. P. Wang and X. X. Meng

Simulation of Conjugate Profiles in Gear Shaping Using
Computer Graphics 366
 S. V. R. Surya Narayana, V. Jayaprakash and M. S. Shunmugam

CAD Tools Development

An Algorithm For Nesting Patterns In Apparel J. Y. Wang, D. Y. Liu, E. W. Lee, T. H. Koh and M. B. Maswan	377
A 3D-Clipping Algorithm for Form Feature Volume Extraction S. R. P. Rao Nalluri, V. Vani and B. Gurumoorthy	385
An Experiment in Integrating CAD Tools M. Bounab and C. Godart	394
Parametric CAD Based on Relation Model Z. Y. Ou, J. Liu and B. Yuan	402
Evaluation of Four CAD Systems Using Analytic Hierarchy Process S. Agarwal, Y. T. Lee and S. B. Tor	409
Joint Application of B-Spline and Bezier Methods for Surface Modeling X. G. Ye, D. Q. Li and F. Y. Wang	419
Tool Paths for Face Milling Considering Cutter Tooth Exit/Entry Conditions Y. S. Ma	427

CAD/CAM Applications

CAD-Based Intelligent Robot Workcell G. C. I. Lin and T. F. Lu	437
CAD-Based Robotic Welding System with Enhanced Intelligence T. Kangsanant and R. G. Wang	445
Efficient Toolpath Geometry for CNC Turning Operations R. N. Ibrahim, P. K. Kee and S. Cabarkapa	453
Determination of Analytic Features from Laser Scan Data S. E. Ebenstein	461

Computer Aided Process Planning

Catch — A Practical Computer Aided Tolerance Charting System G. A. Britton and K. Whybrew	471
Model Based Planning and Calculation for the Manufacture of Dies and Moulds H. K. Toenshoff and J. Trampler	479

Computer Aided Tolerance Control in Process Planning   487
   Y. K. Chow and W. M. Chiu

Computer Aided Facilities Planning

An Evaluation of a PC CAD-Based Facilities Planning Package   497
   C. E. H. Teo

Systolic Algorithm for Improving Efficiency of Facility Layout
Algorithms   505
   Y. L. Qi, B. Sirinaovakul and K. Narue-domkul

An Analysis of Computer Aided Facility Layout Techniques   513
   B. Sirinaovakul and P. Thajchayapong

CAD DATA EXCHANGE   521

Visualizing STEP/EXPRESS Models Based on
Aggregation/Inheritance Hierarchies   525
   S. C. Hui, A. Goh and B. Song

Support for Overlapping Models in an Enterprise Data Model   533
   S. C. F. Chan, P. K. S. Tong and J. W. T. Lee

On the Implementation of EXPRESS Information Models onto
Versant OODB   541
   Q. Z. Yang and B. Song

A Reference Architecture for Information Sharing in Collaborative
Engineering Environments   549
   D. Domazet, D. Sng, F. N. Choong and S. Sum

PROSTEP- An Initiative of the Automotive Industry for Introducing
STEP into Industrial Applications   557
   M. Holland and D. Trippner

FEM/FEA   565

Numerical Simulation of Melt Flow Behavior in Injection Molding   569
   J. F. Bao, P. F. Shao and Y. M. Jin

The Influence of Tool Geometry on Blanking Characteristics   576
   C. M. Choy and R. Balendra

FE Analysis of Pre-Stressed Press Frames 583
 R. Balendra, H. Ou, X. Lu and K. Chodnikiewicz

MANUFACTURING PLANNING & CONTROL 591

Cellular Group Technology

An Improved Assignment Model for Group Technology Application 595
 A. K. Agrawal and Abhinav

An Agile Line Balancing Procedure for Versatile Market 603
 H. Katayama and M. Tanaka

Simulated Annealing Approach to Group Technology 611
 S. M. Sharma and N. Viswanadham

Fuzzy Set Based Machine-Cell Formation in Cellular Manufacturing 619
 C. W. Leem and J. J. G. Chen

Planning

An Object-Oriented Bill-of-Materials System for Dynamic
Product Management 629
 A. J. C. Trappey, T. K. Peng and H. D. Lin

Application of MRP in a Local Manufacturing Company — A Case Study 637
 N. Bin Mohd. Yusof and C. L. Ngeow

MRP II as a Base of Different Production Management Techniques 645
 E. L. J. Bohez and M. A. A. Hasin

A Decision Support Framework for PWB Assembly 653
 Y. Y. Su and K. Srihari

Scheduling

A Scheduling Strategy For Efficient Operation of PCB Assembly Line 663
 Y. H. Lee and D. H. Kim

Modeling the Effect of Hot Lots in Semiconductor
Manufacturing Systems 671
 Y. Narahari and L. M. Khan

Integrating Intelligent Job-Scheduling into a Real-World
Production-Scheduling System 679
    K. Kurbel and A. Ruppel

A Scheduling Algorithm for Dynamic Job Scheduling 687
    Y. X. Zhang, S. Di, H. Cheng and K. F. Cheng

Scheduling Utilizing Market Models 695
    H. H. Adelsberger, W. Conen and R. Krukis

Developing Knowledge-Based System for Calibration Scheduling 703
    F. T. S. Chan

Disjunctive Constraints for Manufacturing Scheduling: Principles
and Extensions 711
    P. Baptiste and C. Le Pape

Intelligent Simulation-Based Scheduling of Work Cells 719
    H. S. Tan and R. de Souza

Control

Intelligent Object Networks — The Solution for Tomorrow Manufacturing
Control Systems 729
    J. Gausemeier, G. Gehnen and K. H. Gerdes

Object-Oriented Integration of Distributed Flexible Manufacturing Systems 737
    S. K. Cha

FMS

A Method for Describing Operations Sequences in Flexible
Manufacturing Systems 749
    G. Jones and M. S. Sodhi

Planning for Modular Fixtures in Flexible Manufacturing Systems 757
    P. C. Pandey and P. Ngamvinijsakul

Inventory

Ordering Alternatives in JIT Production Systems 767
    K. Takahashi and N. Nakamura

Manufacturing System Performance Evaluation with Stock Profiles 775
    M. Aldanondo and B. Archimede

An Integrative Model for Automatic Warehousing Systems 783
    E. Eben-Chaime and N. Pliskin

Tools & Fixture Planning

Planning and Control of Tool and Cutter Grinding in
Manufacturing Systems 795
    G. Petuelli and U. Muller

A Computerized Tool Planning and Scheduling System 802
    K. N. Krishnaswamy, B. G. Raghavendra and D. Sampath

Application of Neural Network in Tool Selection Problem 812
    P. C. Pandey and S. Pal

Petri-Net Applications

An Integrated Object-Oriented Petri Net Approach for Developing the
Manufacturing Control Systems 823
    L. C. Wang

An Integrated MMS and Colour Petri-Net Model for the Distributed
Control of Flexible Assembly Systems 831
    W. H. R. Yeung and P. R. Moore

Volume 2

MODELLING AND SIMULATION 839

Distributed Information Systems

An Application of Manufacturing Systems Theory (MST) to
Production Planning 843
    K. S. Wang

Integration and Simulation for Mastering Design Risks by
Concurrent Engineering 851
    O. Senechal and C. Tahon

Scheduling and Performance Analysis

Simulation Based Performability Analysis of Flexible
Manufacturing Systems 861
    V. Gopalakrishna and N. Viswanadham

Optimizing Production Schedules in Small Scale Industries by
Using PC-Simulation Package 869
    F. T. S. Chan and N. K. H. Tang

Simulation of Scheduling Options in a Flexible
Manufacturing Cell 877
    A. Tredgold and E. Fielding

Simulation Applications

Reducing Manufacturing Cycle Time of Wafer FAB with
Simulation 889
    K. T. Giam, W. T. Ui, A. M. Lie, G. Sun, M. Wang
    and K. Kok

Management of an Assembly Line: An Investigation of Product
Mix Using Simulation 897
    T. A. Spedding, R. de Souza and W. B. Tan

Information Management for the Design of Flexible
Machining Cells 903
    S. T. Newman, S. Rahimifard and R. B. R. de Souza

Witness Helps Process Re-Engineering (BPR) in a
Manufacturing Context 912
    P. Tung, M. Wang and G. Sun

Integration of Simulation and Export System through
Intervention Modeling 920
    W. H. Ip and K. L. Choy

Process Monitoring in Manufacturing Systems — Modeling in
View of Simulation 925
    G. Petuelli and G. Blum

Physical Simulator for Flexible Manufacturing Systems 933
    A. Totu

SimEnvir++: An Object-Based Simulation Environment 939
    Y. H. Zhang and R. de Souza

FACTORY AUTOMATION 947

FMS

Strategies for Planning and Implementation of Flexible Fixturing
Systems in a Computer Integrated Manufacturing Environment 951
    B. Shirinzadeh, G. C. I. Lin and K. C. Chan

Automated In-Line Assembly System in Disc Drive Manufacturing 959
    L. M. Saw and C. L. Chuah

World Class Manufacturing 967
    S. Bansal

Machine and Process Control

An Application of H_∞ Control And Iterative Learning Control
to a Scanner Driving System 981
    Z. W. Zhong

Inverse Theory of Discrete Time System 989
    C. W. Li, Y. Miao and Q. H. Miao

Power Monitoring with Memory Function for Automatic
Machining Systems 997
    F. Liu, B. Dan and Z. J. Xu

Robotics, AGVs and ASRS Applications

Development of an Automated Micro-Pellet Counter and
Feeder System 1007
    K. Baines

Configuration Space Representation for Robot Work Cells 1016
    S. Boopathy, K. Venkatraman and V. Radhakrishnan

Window-Based Teach Unit for Cartesian Robots 1025
    M. J. Er, M. Zribi, C. M. Chang and Y. K. Yeo

Development of an Automated Storage and Retrieval System
for Use in Production Environment 1033
    S. Biswas, K. V. Rao, T. Sriram and B. Ahmed

The Complex Device Integration Quality Design and Operation
Monitoring Approach 1041
    J. Szpytko and S. K. Ghosh

ADVANCED MANUFACTURING PROCESSES   1049

CNC & NC

Developing Three-Co-ordinate CNC Milling Machine as
Economical Multi-Use CNC Machine   1053
    Y. S. Yin and X. C. Liang

5-Axes CNC Programming for 3D-Impellers — Important Aspects
and Optimization Technique for Minimizing Undercuts   1059
    G. Madhavulu, S. V. N. A. Sundar, B. Ahmed and S. K. Bhave

Cutting Technology

Kinematic Model Representation of Machine Tool for Virtual
Machining   1069
    F. Tanaka, M. Suzuki, M. Yamada and T. Kishinami

A New Model for Prediction of Acoustic Emissions in
Orthogonal Cutting Operations   1077
    Y. J. Park and D. Saini

The Mathematical Foundations of Free-Form Surface Machining
with Flat-End-Cutters   1089
    A. Szende and K. W. Lee

Effectiveness of Various Coatings on Cermet Cutting Tools   1100
    M. Rahman, T. N. Goh, K. H. W. Seah and C. H. Lee

A Model for Prediction of Ground Surface Generated with Laser
Dressed Grinding Wheels   1107
    V. Phanindranath and N. R. Babu

An Analysis of Cutting Tools with Negative Side Cutting
Edge Angles   1115
    V. C. Venkatesh, I. A. Kattan, D. Hoy, C. T. Ye and
    J. S. VanKirk

Injection Moulding

A Proposed Automatic Pin Ejector System Design For Plastic
Injection Moulding   1125
    S. Jin, S. B. Tor and G. A. Britton

Automated Machining For Plastic Injection Moulds   1132
    W. L. Li, G. A. Britton and S. B. Tor

Innovative Processes

A Simplified Post-Processor For Wire-Cut EDM K. P. Rao, T. K. Au, Y. K. D. V. Prasad and M. Hua	1141
Process Improvement in Volumetric Roll-Stretch Bending R. G. Weippert and J. Reissner	1149
Micro-Precision by Wire EDM I. Beltrami, A. Bertholds and D. Dauw	1157
Analysis and Evaluation of Errors in the Nett-Form Manufacturing of Components R. Balendra and Y. Qin	1165

Tool Wear and Work Piece Holding

A Study of the Built-up Edge in Drilling with Indexable Coated Carbide Inserts V. C. Venkatesh and W. Xue	1175
A General Frictional Model for Fixturing Verification N. H. Wu, K. C. Chan and S. S. Leong	1183
On-Line Identification and Control of Dynamic Characteristics of Slender Workpieces in Turning C. M. Nicolescu	1191
Crater Wear Measurement Using Computer Vision and Automatic Focusing M. Y. Yang and O. D. Kwon	1200

COMPUTER AIDED INSPECTION AND TESTING 1209

Algorithms and Techniques

A Polynomial Algorithm for the Correspondence Problem K. S. Al-Sultan	1213
A Novel Method for Accurate Evaluation of Minimum Zone for Flatness S. Damodarasamy and S. Anand	1218

Machine Vision and Sensors

Computer-Aided Visual Inspection for Integrated Quality Control   1229
    J. B. Zhang

AVIPS: A Knowledge-Based System for Automatic Industrial Visual
Inspection Planning   1239
    M. G. Jeong and S. I. Yoo

Study on Vision Inspection Technique for Meshes   1249
    S. H. Ye, F. J. Duan, C. H. Wang and J. X. Zhang

A Flexible Inspection Cell for Machined Parts   1255
    M. K. Lee and K. Y. Chan

Process Reliability

Reliability Engineering Through the Concept of Model Accuracy   1263
    J. Y. T. Ang and N. A. J. Hastings

Knowledge-Based Analysis and Optimization of Reliability
Shortened Testing Plan Choice for Manufacturing System   1271
    L. Papic and J. Aronov

Joint Inspection and Replacement Policy for Randomly
Failing Equipment   1281
    A. Chelbi and D. A. Kadi

Applications

An Automated, Computer-Controlled Photo-Flashtube Testing System   1291
    T. K. Lim, S. Swaminathan and K. W. Cheung

A Study on the Evaluation of Related Features with
Geomatrical Constraints   1299
    H. Okamoto, T. Sasaki, F. Tanaka and T. Kisihinami

Sampling of Inspection Points for Free-Form Surface   1307
    Y. J. Kim and K. W. Lee

TOTAL QUALITY MANAGEMENT   1317

Reducing Defects by SPC   1321
    Z. Wu

OLSPCS Application in Malaysian Based Semiconductor Manufacturing Companies *N. Bin Mohd Yusof and L. T. Yew*	1330
Total Quality Management Philosophy Within the Group Technology Cells *M. Gundogan and J. M. Kay*	1337
Application of Fuzzy Sets Theory in System Failure Analysis *B. Vasić, J. Aronov and L. Papić*	1345
Experience of Hypermedia Applications in Total Quality Management *H. C. Leung, R. F. Leung and J. F. Hill*	1352

AI IN MANUFACTURING — 1361

Diagnostics

Increasing the Availability of Flexible Manufacturing Systems (FMS) with Multimedia Supported Diagnostic Expert System *S. Torvinen and R. Milne*	1365
Implementation of a Fault Tree Based Diagnosis System for Incipient Faults in Machine Tool Components *P. J. Nolan, P. Muldoon, M. G. Madden and D. M. Lennon*	1373

Intelligent Support

An Intelligent System for Manufacturing Quality Control in the Food Industry *O. Castillo and P. Melin*	1383
Automated System for Jigs and Fixtures Design *G. Dai and M. M. F. Yuen*	1391
An AI Approach to Process Modelling and Optimum Selection of Grinding Conditions *C. B. Zhu*	1400
Automated Knowledge Acquisition for the Surface Mount PWB Assembly Domain *C. H. Wu and K. Srihari*	1408
A Selection Expert System for Solid Freeform Manufacturing *V. Narayanan, C. K. Chua and B. Ang*	1416

Monitoring and Control

A Fuzzy Control Strategy for Flexible Machines Operating Under
Information Delay — 1427
    R. A. Caprihan, S. Kumar and S. Wadhwa

Monitoring the Accuracy Characteristics of the Machinery by Using
a Dynamic Measurement Approach — 1435
    S. Torvinen, P. H. Anderson, J. Vihinen and R. Milne

Mould and Die Applications

Development of Initial Machine Setting Program for IC Transfer
Molding Process — 1445
    T. S. Yeung, M. M. F. Yuen, W. M. Kwok, T. H. Kuah
    and P. H. Yeung

An Expert System for Injection Molding Defect Correction — 1453
    K. H. Tan and M. M. F. Yuen

An Improved BP Algorithm and its Application in Manufacturing
of Stamping Dies of Car Bodies — 1461
    Y. N. Hu, J. Xia, Y. P. Chen and Z. D. Zhou

Neural Networks

Neural Network for 3D Force/Torque Sensor Calibration and
Robot Control — 1471
    G. C. I. Lin and T. F. Lu

Enhancing Flexibility of Vision-Based Robots Using Artificial
Neural Network Approach — 1479
    S. K. Sim and M. Y. Teo

Neural Simulation of a Prototype Keyboard Assembly Cell with
Adaptive Control — 1491
    W. L. Lee, T. A. Spedding, R. de Souza and S. S. G. Lee

Artificial Neural Network for Sensor Fusion and Tool Wear
Estimation in Face Milling — 1499
    S. Kakade, L. Vijayaraghavan and R. Krishnamurthy

Measurement of Performance Degradation of a Machine and System
Using a CMAC Based Neural Networks Model — 1507
    J. Lee

Planning and Scheduling

Recursive Type Learning Method for Knowledge-Based Planning System   1517
    Y. Ikkai, T. Ohkawa and N. Komoda

Learning in a Mechanical Engineering Planning Domain   1525
    H. Durr and W. Ewert

Author Index   1533

CIM MANAGEMENT

CIM MANAGEMENT

Management

Re-Engineering Through Natural Structures: The Fractal Factory
    W. Sihn

Multidatabase Agents for CIM Systems
    M. Wunderli, M. C. Norrie and W. Schaad

Business Process Reengineering: Basis for Successful Information System Planning
    K. Mertins and R. Jochem

Hypermedia in CIM: Issues and Opportunities
    R. F. Leung, H. C. Leung and J. F. Hill

A Model-driven Toolset for Supporting the Life-cycle of CIM Systems
    J. M. Edwards, S. Murgatroyd and R. H. Weston

Business Process Management: The European Approach
    H. H. Hinterhuber

An International Perspective of Manufacturing
    T. A. Spedding, L. H. S. Luong and R. F. O'Connor

Evaluation of Advanced Manufacturing Technology — The Modes of Thinking
    I. Persson

RE-ENGINEERING THROUGH NATURAL STRUCTURES: THE FRACTAL FACTORY

DR.-ING. WILFRIED SIHN
Fraunhofer Institute for Manufacturing Engineering and Automation (IPA)
Silberburgstr. 119a, 70176 Stuttgart, Germany
E-mail: whs@ipa.fhg.de

ABSTRACT

Many branches of European industry have had to relinquish their lead in the world market, particularly through Asian competition. The reasons are rigid company structures which prevent flexible reaction to constantly changing environmental conditions.

The article illustrates the methods of the "fractal company," necessary to solve the structure crisis. It distinguishes itself through its dynamics and vital, independent ability to react to changing circumstances. The procedures and basic conditions such as company structuring, networking and hierarchy formation, as well remuneration and working time models are explained, based on examples from IPA's work with the automobile industry, their suppliers and the engineering industry.

KEY WORDS: fractal factory, organizational structure, re-engineering

1. The Shift of Goals

There has been a clear shift concerning the objectives of industrial production. Goals of production which have been held up for decades are giving way in favor of speed objectives. This is mainly due to a shift from purchaser markets to seller markets. The cause for this shift is the saturation, or even the over-saturation, of markets. According to surveys, profit losses during commercialization are highest if there is any delay in placing the product on the market. Other factors are less important. There are fewer products whose production is complicated in the conventional sense. On the one hand, fully matured components whose assembly results in final products contribute to this. On the other, manufacturing technologies, becoming increasingly reliable, ensure consistent quality and allow the production of complicated parts by normally trained staff. This means there will be more suppliers serving the market with similar products. The result is an over-saturation of the market.

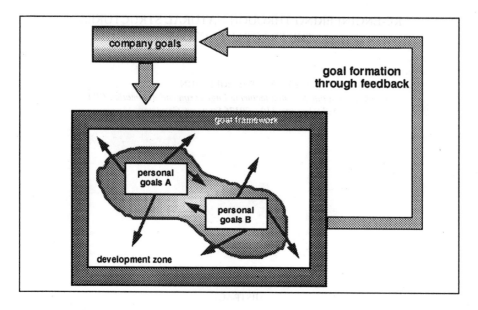

Fig. 1. Goal formation process in the fractal factory

2. Historical Change

Taking a historical look at production development, the number of employed factors and their availability has been steadily increasing. This is remarkable, especially concerning the availability of mechanical energy and information which is crucial for industrial production, allowing the effective and timely usage of resources which results in reduction of the stock of resources. Because of this reduction it is not always possible to have a stable situation, that is, to always run production according to events calculated in advance. Rather, the dynamics of the events will increase. One will have to abandon certain processing stages without prior planning and leave them to their own devices, which means entrusting them to higher control circuits. As a consequence, it is of utmost importance to incorporate into the company and manufacturing strategies the greatest closeness to customers and an extreme ability to react promptly. The speed objective must be pursued vigorously, even regarding the development of operating technologies.

3. Production Factor Information

The decline in prices for electronic components, improvements in efficiency of data processing units and progress concerning communication technologies make any desired quantity of information available at any place at any time. Despite this fact, the processes in our industrial companies are still organized according to the old model, according to which the division of labor has also been used for minimizing the effort for drawing up information. Although each worker could be included in the fetch-principle, as far as the required information is concerned, our organizational structures work according to the bring-principle for information by staff functions and through hierarchy levels. Overcoming this discrepancy is the task of the coming years for each industrial company. These developments could enable the organizational structure to become flatter and will result in more scope for decisions and more responsibility for each individual in the organization.

In the past, one believed the secret of effective performance specification consisted in Taylorism, applying appropriate coordination. Conditions have now radically changed. The summing up of contents of labor is required. Tasks must be interlinked and turned into processes. Also, flow design is called for. The organizational structures will increasingly turn into project organizations.

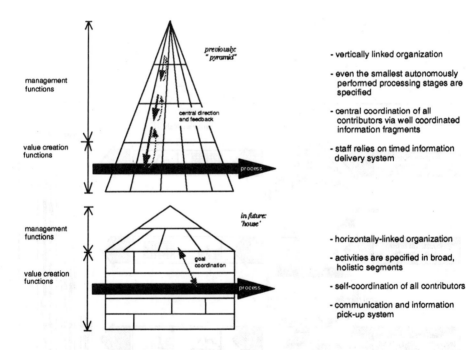

Fig. 2. Interplay of Organization, Information and Value Creation

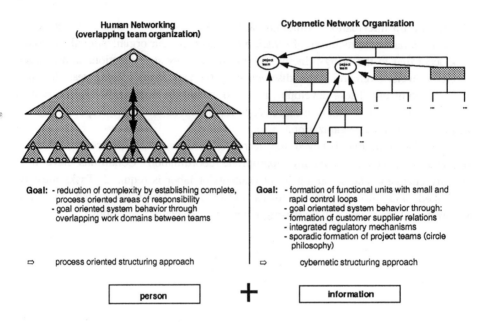

Fig. 3. Organization structures in the fractal factory

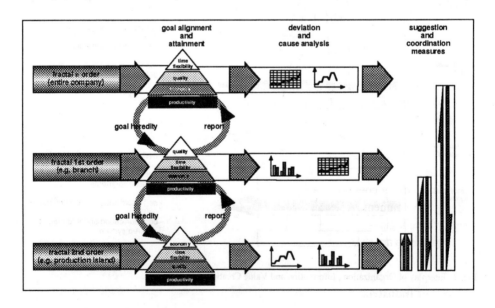

Fig. 4. Interconnected cybernetic structures

4. Limitations of Current System Considerations

The faster moving markets, technologies and information systems available everywhere lead to higher pressure on the companies. There is an increased demand for innovation and speed concerning processes and adaptation. As a reaction to the environmental dynamics and fast-moving events, strong self-dynamics and the ability to evolve are required. However, our current view of a manufacturing company is static. We assume there are definite, clear-cut tasks and that a need exists for authorities who issue instructions on a rigid structure scale. A company develops in a linear fashion, and the economic and arithmetic models are static.

5. Characteristics of Dynamic Change

With such above-described changes concerning important parameters for a manufacturing company, conventionally structured companies are no longer able to fulfill the required self-dynamics with the necessary speed of evolution. The results of this rigidity become increasingly perceptible to everyone in the company. Phenomena such as rising pressure concerning costs, a rapidly worsening profit situation, a waning ability to adapt to market situations and an increasing rush regarding the sequence of events should be mentioned. In such cases additional tasks, overtime and reorganization cannot put things right. They represent a specific consideration of single phenomena like operating times, wages or productivity which can only have a limited and temporary effect. A paradigm change is required — a new conceptional way of thinking which is the basis of the manufacturing company. It must consider that these changes take place with enormous dynamics. In order to survive, a company must learn to secure market shares and profitability in a turbulent environment by developing adaptability. In such an environment it does not develop linearly but with leaps in development and with transformations according to the laws of probability which, although they can be controlled, cannot be accurately predetermined. It is therefore a matter of establishing dynamic systems whose development can become complicated. The dynamics which a company has to develop under these conditions is called *evolution*. Only a continual development secures an appropriate adaptation to the changing environment. As a key statement: a company is an open, complex, dynamic system whose logic — based on the chaotic system of thinking — is one of fractals, and whose dynamics are shaped according to laws of evolution.

6. Design of Industrial Performance Planning

In order to master the future, ways of thinking and points of view must be changed radically. The considerations must move away from static systems toward dynamic systems, away from mechanistic toward organic models of explanation and away from monocausal toward multidimensional explanations. We must learn to understand a manufacturing company as an integrated system with its own processes and structures, a system which does not develop in a linear way, which is not accurately predictable and whose interior and exterior limitations are fuzzy and permeable. The model of the *fractal company* is suggested.

The fractal company is an open system which consists of independently acting self-similar units — the fractals — and is a vital organism due to its dynamic organizational structure. This approach does not describe the world anew, but emphasizes the dynamic and multicausal relations of the real world. The question is, how can you get units into the factory which, regarding their objectives, are self-similar, self-organizing, and able to act independently? This is a structuring task of an integrated kind. Structures must be created which support and develop the above-mentioned abilities.

7. The Basis for the Formation of Fractals: the Level Concept

The methodology must on the one hand contain a sequence of steps, and on the other, enable integrated views. A continual development process must result. In order to reduce the complexity in a company, and with the goal of creating a "fractal company", only horizontal divisions into levels are imaginable. The intermediate results of all projects currently dealt with at IPA show that the basic pattern is a division into six levels which can be treated as a whole, but are separate: 1. Cultural, 2. Strategic, 3. Socio-psychological, 4. Financial, 5. Informational (Information Flow) and 6. Technological.

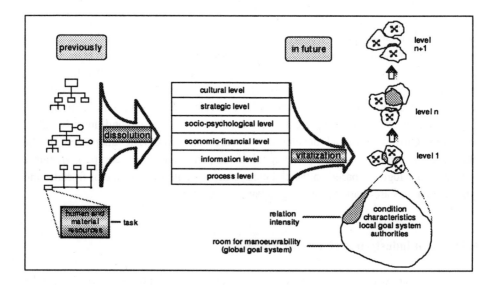

Fig. 5. Dynamic organization structures

Organization culture is the generic term which includes the company culture. Values are an essential part of organization cultures. They express desires and contain the whole field of human preferences. Behavior is value oriented. It becomes clear that a goal oriented company cannot get along without values or cultural elements. Ideals are developed, as are common views of values and principles concerning internal and external harmony.

Important *strategies* can be strategies of innovation, "me too" strategies, specialization and diversification strategies, etc. However, the system of objectives must first be defined. A goal oriented approach is the basis for successful structuring which requires a minimum of time and cost. One must generate a system of objectives based on the philosophy, culture and strategic orientation of the company. According to the goal structures which are evaluated in this way, concrete planning can be established in order to compare the alternative principle solutions which have been worked out.

The *social-informal level* includes all kinds of psychic, social and psychological factors which determine and influence the structure of relations of the entire staff. Organizational structure, communication and ability for teamwork can be identified as central variables on this level. The corresponding methods which must be adapted to the context of the fractal company would be: development of organization, formation of teams and working groups, guidance of the teams, information and communication management and coaching.

The *financial level* of the fractal company deals with methods concerning account settlement of performance. Business management data must be judged with regard to their economic and efficiency related viabilities. This means that sales/cost considerations must be connected with the process design.

The *informational level* deals with the design of technical information flow. The central term is therefore process organization. The main problem exists in maintaining continuity and integrating information systems without hindering the dynamics of the structures. There are many possibilities for a realization apart from computer-integrated manufacturing. The use of information must be adapted to the processes and not vice-versa.

The *technological level* of the fractal company is responsible for the technical design of material flow equipment. The whole complex of logistics and material management, including all kinds of parts of components, belongs to this.

Aimed variables, such as the increase of productivity and flexibility and maintaining schedules concern the entire order processing as well as the decrease of throughput time. They are becoming increasingly important and are the focus of attention. Pilot projects at IPA have already brought desired results such as transparent manufacturing processes, reduction of throughput times and reduction of stock value with simultaneous tripling of output.

8. Level Model and Vitality

The word *vitality* was coined as a superordinate term to measure the viability and efficiency of a fractal. The variables to be taken into account should have the ability to develop a dynamic system behavior. Therefore, vitality must record and evaluate essentially those variables which are included in the individual characteristics and which can be used as a measure for the change or changeability of the individual characteristics of the levels.

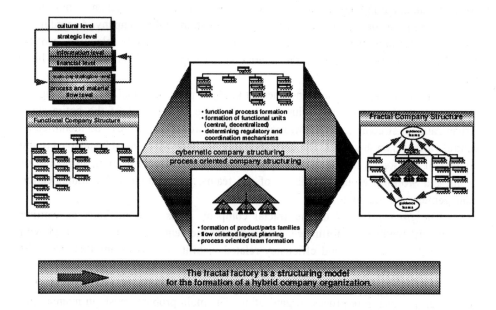

Fig. 6. Company structuring in the fractal factory

9. Summary

The concept of the fractal company, based on western structures, mentality and strengths, is a European answer to Japanese "lean production." With reference to natural organisms, vital and dynamic structures reacting flexibly to environmental conditions were created.

In addition to the fundamental elements and structures of a fractal company, the practical relevance gained from various companies was elucidated.

MULTIDATABASE AGENTS FOR CIM SYSTEMS

Martin Wunderli, Moira C. Norrie, Werner Schaad
Institute for Information Systems, ETH Zentrum, CH-8092 Zurich, Switzerland
Email: {wunderli,norrie,schaad}@inf.ethz.ch

Abstract

We propose a general coordination architecture for CIM systems based on multidatabase technologies. Each component system has an agent which notifies a global coordinator of local operations pertinent to global consistency and, in turn, reacts to operation requests from the global coordinator for the maintenance of global consistency. In order to exploit multidatabase technologies, an agent must provide its component system with the minimum functionality required to apply these technologies. In this paper, we describe how these so-called multidatabase agents are realised in the context of our CIM prototype system, CIM/Z.

1 Introduction

Approaches to Computer Integrated Manufacturing (CIM) have tended to be based on either a partial or total integration of component subsystems. We propose rather a coordination approach which provides a looser coupling of subsystems to minimise loss of autonomy and maximise flexibility. Our approach is data-oriented in that coordination is achieved through the maintenance of global data consistency.

Generally, we adopt a "database approach" to CIM and exploit multidatabase technologies for cooperative working. Each subsystem provides a description of its local data in terms of a global data model which forms the basis for communication between these subsystems and a logically central, global coordinator. The coordinator bases its activity on inter-system dependencies expressed as global consistency constraints together with a set of reaction rules which specify actions to be taken to maintain global consistency. The coordination process operates within a global transaction scheme.

The approach requires that each subsystem be augmented by an agent which provides the necessary coordination interface. Each agent is responsible for notifying the coordinator of local operations which may affect global consistency and, in turn, acting upon operation requests issued by the coordinator for the maintenance of global consistency. Since the agent, in effect, must raise the functionality of the subsystem to that required to apply multidatabase technologies, we refer to it as a multidatabase agent. It is the purpose of this paper, to examine the requirements of such agents and present a general agent architecture which minimises those parts which are subsystem specific.

We begin in section 2, with an overview of our coordination approach. Section 3 analyses the general requirements of an agent and discusses how these are specialised to the various categories of CIM subsystems. The global coordination process is described in section 4. In section 5, we present the general architecture of the agents. Concluding remarks are given in section 6.

2 General Architecture

Figure 1 shows the general coordination architecture; for simplification we show only two subsystems.

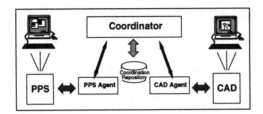

Figure 1: General Coordination Architecture

As an example of coordination, we consider a CAD system and a PPS system. The CAD system supports the design activity and stores information about the various CAD drawings. One of the PPS' tasks is to manage information about the structure of parts. There is a dependency between the two systems; part data referred to in the CAD drawings must exist in the PPS. If a part is discontinued, we must ensure that designers cannot reference this part in future designs – and must somehow inform designers that existing designs using this part are no longer valid. This is an example of an inter-system dependency which is the basis of the coordination process and is stored in the Coordination Repository.

The main task of the coordinator is to ensure, with the help of the agents, the consistent state of the CIM system. The coordinator does this by coordinating activities of the various subsystems according to various inter-system dependencies [NSSW94, NW94]. Each subsystem has an agent which monitors local activity and notifies the coordinator of any actions that are pertinent to the inter-system dependencies. Based on these dependencies, the coordinator will delegate necessary actions to one or more subsystems via their agents.

An agent provides the coordination interface for a subsystem; it must be able to communicate with its subsystem and also with the coordinator. One of the major functions of an agent is to map between local representations of operations and data to the global representation scheme based on the NIAM data model [NH89]. Local system dependency is therefore isolated in part of the local agent and the rest of the coordination components are general and not specific to particular subsystems.

An agent has its own repository to record information about which local operations and data are relevant to coordination, logs of local activities and any other information required to perform its part of the coordination task.

Global transaction management is a prerequisite to ensure that all information about changes and the corresponding coordination activities reach the relevant parties and are acted upon. Since a subsystem may not be a database application system and, therefore, may be without transaction and recovery support, it may be necessary that its agent takes over the role of providing some form of logging and atomicity control.

3 Classification of CIM Subsystems

In this section, a classification scheme for subsystems is proposed which identifies the level of enhancements an agent has to provide to a particular subsystem. To do this, the desirable subsystem has to be characterised; that is, the subsystem which would make a separate agent obsolete and could be easily integrated into a coordination system.

Given the general architecture of a coordination system, we can identify the following main characteristics of a desirable subsystem:

1. the subsystem uses the global data model as its own, local data model;
2. user operations are expressed in terms of the local data model and a request for confirmation of every data change can *easily* be made mandatory;
3. the subsystem can execute actions required by the coordinator in terms of the global data model;
4. the subsystem is able to return to a consistent state by itself after a crash or an abort of a coordination action.

As a consequence of the high degree of heterogeneity found in the CIM area, the first characteristic is unlikely to be exhibited by a subsystem. So it is necessary for an agent to have a description of the local *data and operations* in terms of the global data model and to be able to map between the local and the global description.

The remaining three characteristics now depend on three almost orthogonal properties of subsystems – namely

1. the possible form of communication from the subsystem to the agent,
2. the possibility to induce operations on the subsystem data, and,
3. the support for atomicity and durability of subsystem operations.

The first property concerns how knowledge about user operations is obtained and how the agent may take over control of these operations. The second property concerns how the agent may initiate operations requested by the coordinator. The third property is concerned with how the database properties of the subsystem, especially the recovery and persistence of data, can be ensured. Figure 2 illustrates these properties and the corresponding classifications of three subsystems.

The worst scenario, when trying to get knowledge about the 'what' and the 'when' of user operations and reacting on them, is the case where the subsystem does not provide any information on the operations performed and does not allow any foreign control over its data and operations. This situation may occur with an established parts list management system which uses operating system files to store its data. In such a case, we have to resort to the monitoring of file sizes and timestamps, possibly combined with subsystem version dependent knowledge of internal file structures.

A better situation is that of being able to connect to the subsystem on an intermediate data level; that is, on a higher level than sequences of bytes but below the subsystem's data model. In this case, the data observed is already in a structured form which is easier to interpret. An example of such a situation is a CAD system which manipulates objects such as cylinders and boxes but stores them as sequences of circles and lines in a relational database such as Sybase. In that situation, we could build an agent which allows us to monitor the stream of SQL operations, interpret it and react on it.

Nevertheless, these solutions are less than satisfactory as we would like the connection between the agent and the subsystem to be on the same logical level

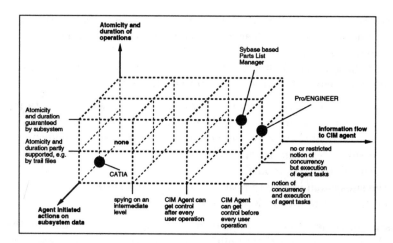

Figure 2: Classification scheme for subsystems

as that of user operation. For example, if a CAD system performs a deletion of a cylinder, we want the agent to be informed of the deletion of a cylinder and not the deletion of the circles and lines that it comprises. If a subsystem provides us with information on the user level, we can further distinguish between two general cases:

1. There is *no* possibility to influence the success of a user operation. For example, information is received *after* the operation has completed and no way exists to interfere.
2. There *is* a possibility to influence the success of a user operation either before it is performed with the possibility to reject it, or, after it is performed with sufficient information available to undo by some form of compensating action.

To summarise, the possibility to influence the success of a user operation and initiate operations, combined with being informed on the same logical level as that of user operation, are preferred subsystem properties for supporting an agent.

The second aspect of supporting an agent has to do with failures. In an ideal world with no system failures, the factors discussed above would be sufficient. Unfortunately, a subsystem failure which results in a loss of data integrity can occur. In the case of a stand-alone system, the worst case would be a (partial) loss of data which, hopefully, would be reduced by restoring data from a backup tape. However, in case of a coordinated system, the state of a subsystem is at least partially known outside the system, and we must guarantee that all state information made persistent in the coordinator really reflects the current situation in the subsystem. This can be achieved if the subsystem guarantees data durability. If a subsystem provides no such notion, then the agent, together with the subsystem, has to provide it. If the subsystem has some form of log files or trail files, the agent can use these to guarantee durability by redoing failed operations. Otherwise the agent will have to perform all logging itself.

Thirdly, an agent may be requested by the coordinator to initiate an operation in its subsystem in order to reestablish system wide consistency. If there is no way to interact with its subsystem, this has to be done by including a user of the subsystem into that process. It is preferred that a subsystem allows external triggering of operations.

4 Global Operations

In this section, we provide a brief overview of global transactions in a coordinated CIM environment.

Assume the situation described in section 2 where a global constraint between the object type 'CAD Drawings' of a CAD system and the object type 'Parts Lists' of a PPS has to be guaranteed: For every CAD drawing there should be exactly one parts list and vice versa.

If the CAD agent detects the creation of a new CAD drawing (for example, by notification through the CAD system's programming interface), it maps this event to the global data model and starts a subtransaction to request confirmation from the coordinator. The coordinator acknowledges this and returns the ID of the global transaction; this transaction will comprise the subtransaction submitted by the agent and any subtransactions on other subsystems necessary to maintain global consistency. The request of the initiating agent is transferred by sending all the operations on globally important data and an end marker. The agent then remains in a state where it is able to abort/compensate or commit its subtransaction.

The coordinator now checks whether the received subtransaction violates any global constraints and, if so, initiates the necessary subtransactions on the relevant agents. Success reports of the agents concerning such subtransactions are done on basis of single operations. This allows the coordinator to take alternative reactions into consideration if some operations of a subtransaction fail while others do not. Alternative reactions can help to avoid discarding successfully completed operations which have been expensive in terms of time and CPU consumption, e.g. a Finite Elements calculation.

If all operations of all subtransactions are completed successfully, the coordinator completes the global transaction by sending commits to its agents and updating its repository.

5 Agent Modules

In this section, we present the general structure of agents and describe the basic principles and modules common to all agents.

We can identify two main tasks of an agent. The first is to monitor a subsystem's actions performed on globally important objects and request confirmation for these actions from the coordinator. The second is to execute actions requested by the coordinator necessary for system wide consistency. Both tasks require a mapping from the local to the global data model, in terms of both data and operations.

In figure 3, we show a general agent architecture comprising four modules, which is intended to isolate system specific dependencies and maximise the reusability of agent components.

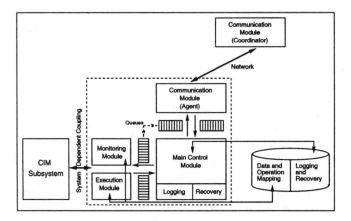

Figure 3: The basic modules of an agent

It is important to note that, apart from the monitoring and execution module, everything is expressed in terms of the global data model. Only the monitoring and execution modules deal with the local data model and the conversion to and from the global data model. Therefore, we usually talk of the insertion, deletion and updating of objects belonging to a certain object type without specifying the form of these objects.

5.1 The Communication Module

The communication module connects the agent to the coordinator. In our project (CIM/Z), we assumed the availability of a socket library on all architectures which are part of the coordination system and that this module can therefore be considered independent of the nature of the subsystem and the hardware on which it runs. The communication module is an independent process communicating with the agent's control module using queues for in- and out-going messages.

The communication modules of the agents and the communication module of the coordinator exchange information by means of a network language that specifies the format of messages.

5.2 The Control Module

The second module of an agent is the control module. The size and task of this module is dependent on the nature of the corresponding subsystem.

In general, we can distinguish two fundamentally different kinds of CIM subsystems. First, there are those with low concurrency, such as CAD systems, which have long running transactions and some sort of check in/check out mechanism. Second, there are multi-user systems, such as PPS, which are based on a database system and typically have short transactions. The difference manifests itself in terms of whether or not a fully fledged and efficient scheduler has to be built. Such a scheduler would have to manage the correct concurrent execution of multiple local (user) transactions on the subsystem's data together with global transactions initiated by the coordinator. In the case of a CAD system, we can almost always accept a simple scheduler which serialises incoming requests from the coordinator. Possible conflicts of global requests with objects currently accessed by the user of the subsystem could be managed by a request to the user to release them temporarily. As we will see in the next subsection 5.3, the execution of coordinator requests will, in any case, frequently involve user interaction. However, in the case where a multi-user system with high concurrency has to be coordinated, an efficient scheduler is a prerequisite. Investigations on this topic have been made in the area of Federated Database Systems [WDSS93] and the proposed solutions can be applied here.

5.3 The Monitoring and Execution Module

The monitoring of subsystem actions and execution of coordinator requests requires transformations between two data models, that of the subsystem and the global data model. In this mapping, we allow that multiple object types of the subsystem are combined into one global object type in order that unnecessary details be hidden. For example, it may be that we are only globally interested in the existence of parts but their representation in a CAD system is spread over several files or relational tables (e.g. used to separate geometric and non-geometric data); then the global object type **part** may map into a number of local object types of the CAD system. However, in our experience, the usual case is a one-to-one mapping between local and global object types.

For each object of global interest, its global id and the ids of the corresponding local parts (if more than one) are known. Each global object belongs to one or more global object types. For each global object type, its global id along with the id of the corresponding local object types (e.g. relation names) are stored. Since the mapping between global and local operations is too system specific to be implemented in a generic way, these are directly implemented using functions in the monitoring and execution module.

We not only have a function for every global operation type (insert, delete and update), but also for the inverse one. The necessity for these stems from the fact that the monitoring and execution module, under the instruction of the global coordinator, may have to compensate actions which have already been performed locally; this depends on the form of control flow between the agent and the subsystem. These compensating actions may save a user's work, rejected by the coordinator for reasons of global consistency, in order that any results may be accessed later and possibly incorporated into future globally important data.

The actual monitoring and execution of coordinator requests is very much subsystem specific. But all variants have in common the fact that they have to perform insertion, deletion and updates of objects and monitor those actions performed by the subsystem and its users. In order that rejected operations can be undone, all have to undertake certain preparatory tasks, such as the generation of compensating actions. Furthermore, they have to inform the user, or administrator, of the subsystem if an action cannot be performed or had to be compensated.

During the construction of the agent for the CAD system Pro/ENGINEER, we noted that insertions, deletions and updates of globally important objects frequently involved user interaction. This means that users and/or an adminis-

trator have to be consulted about some actions such as creating a new document
and the agent may have to wait for the user to perform the operation rather
than performing the action on its own. Of course, there are situations where
an agent can perform actions independently, but this is not always the case. It
may even be desirable that a user is informed about the consequences of his
actions. But this is currently not supported and it would require extensions to
our system wide communication protocol.

6 Conclusions

Our proposed coordination architecture provides a flexible means of integrating
autonomous subsystems to form CIM systems that support cooperative working. Each subsystem is provided with the necessary functionality and interface required for the coordination process through its own multidatabase agent.
Agents are subsystem specific in that their operation depends on the extent to
which the functionality required for coordination is supported by the subsystem. Further, certain features are dependent on the general means of control
flow possible between the subsystem and the agent.

The agent monitoring and execution modules are the most system specific
and exhibit the greatest variation across the categories of subsystem. The size
and complexity of these modules reflect the amount of work required to provide
the subsystem with the necessary coordination capability. Although these components may not be directly reusable, certain general modes of operation can
be reapplied. Further, it is the trend that evolving systems exhibit more of the
features required for coordination, thereby simplifying the task of these modules.

Acknowledgements

The work described is part of the project "Integration of Databases and CIM
subsystems" (CIM/Z) funded by KWF (Swiss Federal Commission for the Advancement of Scientific Research). We thank our partners for their contributions; they are IKB, ETH Zurich, ABB Informatik and Sulzer Informatik.

References

[NH89]   G. M. Nijssen and T. A. Halpin. *Conceptual schema and relational database design: A fact oriented approach.* Prentice Hall, New York, 1989.

[NSSW94] M. C. Norrie, W. Schaad, H.-J. Schek, and M. Wunderli. CIM through Database Coordination. In *Proc. of the Int. Conf. on Data and Knowledge Systems for Manufacturing and Engineering*, Hongkong, May 1994.

[NW94]   M. C. Norrie and M. Wunderli. Coordination System Modelling. In *Proc. of the 13th International Conference on the Entity Relationship Approach*, December 1994.

[WDSS93] G. Weikum, A. Deacon, W. Schaad, and H.-J. Schek. Open Nested Transactions in Federated Database Systems. *IEEE Bulletin on Data Engineering*, 16(2), 1993.

Business Process Reengineering:
Basis for Successful Information System Planning

K. Mertins, R. Jochem
Fraunhofer Institute of Production Systems
and Design Technology (IPK) Berlin
Pascalstraße 8-9, D-10587 Berlin, Germany
Roland.Jochem@ipk.fhg.de

Abstract

Many problems within enterprises appear as a consequence of both organizational and technological issues. The integration of processes regarding aspects of dynamics and concurrency during decision making is a key element for achieving flexibility. Changed tasks and timeframes have to be reflected by restructured process chains.

The authors describe a methodology for integrated modelling of business processes, related organzational structures and information system support based on an object-oriented approach which is under discussion in ISO TC184/SC5/WG1 and CEN TC310/WG1 for standardisation. Examples of industrial application for different areas and a supporting modelling tool prototype are presented.

Keywords: Business Process Modelling, Object-Oriented Approach, Modelling Tool

1. Introduction

In the nineties enterprises face a higher pressure of time and pressure to succeed at global markets, increasing competition, shorter product life cycles and, related to this, a higher flexibility in all areas. The market demands additional product differentiation exactly meeting customer needs, advance of technological level of products, shortening of both product development and order troughput times, advance of delivery time and advance of product quality.

To improve competitiveness, all efforts are traditionally concentrated on optimization of single functions – the enterprise is subdivided into a number of separate functions, which are easier to overview and control. This introduces a number of "interface" problems in organization and optimization of single functions at the expense of the manufacturing process and the organization as a whole.

The integration of separated functions and the optimization of business processes require a higher degree of transparency within the organization. In consideration of the complex relationships – looking on the manufacturing enterprise as a network of functions – modelling methods have to be applied, to support, to ease and to systematize planning and integration of functions to business processes and to describe the related organizational structure. Suitable methods secure a common understanding of business processes and provide mechanisms for structuring the required information about processes and organization.

In the following, the method of Integrated Enterprise Modelling (IEM) is presented. IEM uses the object-oriented modelling technique for modelling business processes, related organizational structures and required information systems as well. It provides a model for planning and optimizing the processes and organizational structures within the enterprise.

Models developed according to the IEM method give a transparent representation of planning information and therefore, are the basis for discussion between project participants. For evaluating the variety of planning information and descripiction requierements, it allows different views on one consistent model.

IEM models provide the means to precisely assign the value of planning goals, like improvements in time, cost, or quality, to each business process and resource and, by that, to optimize the process organization.

2. Object-oriented Modelling

2.1. The Approach

Object-oriented techniques are broadly used for the development of applications in various areas. The main advantage of this approach is the entirety of data and functions operating on these data. Provided with the powerful inheritance mechanism it yields models which are more stable and easier to maintain than those based on other modelling approaches [1].

In order to utilize it's advantages and to provide a comprehensive and extendable enterprise model, the IEM method uses the object-oriented modelling approach, thus allowing the integration of different views on an enterprise in one consistent model and the easy adaptation of the model to changes within the enterprise [1,8,9].

Generic classes of objects: The generic classes Product, Resource and Order form the basis of Integrated Enterprise Modelling for developing models from the user`s point of view. They will be specialized according to the specifics of an individual enterprise [4,7,11]. Each generic class prescribes a specific generic attribute structure, thus defining a frame for describing the structure and behaviour of objects of it's subclasses (cf. Figure 2). Real enterprise objects will be modelled as objects of these subclasses.

Required enterprise data and the business processes, i.e. the tasks referring to objects, are structured in accordance to the object classes (see below). Furthermore, the relations between objects are determined. The result is a complete description of tasks, business processes, enterprise data, production equipment and information systems of the enterprise at any level of detail [5,6].

The model kernel comprises two main views. The tasks, which are to be executed on objects, and the business processes are the focal point of the Business Process Model View, whereas the Information Model View primarily regards the object describing data (**Figure** 1). Thus, the kernel of the enterprise model consists of the data and process representations of classes of objects. The views are interlinked by referring to the same objects and activities, although they represent them in different ways, levels of detail and context. Any view on the model can be derived from this standardized model kernel. Additional features can be tied to the kernel if necessary.

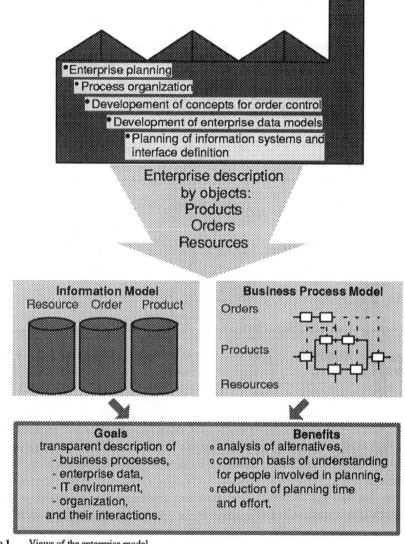

Figure 1. Views of the enterprise model

2.2. Business Processes as Interactions of Objects

Everything that happens in a manufacturing enterprise as part of the manufacturing process can be described by activities. In general, activities process and modify objects which were classified above as Products, Orders and Resources. The execution of any activity requires direct or indirect planning and scheduling, it is executed by Resources owing the needed capability. The modelling elements are described in **Figure 2**. Product classes represent the main results of the whole enterprise process – the products. Resource classes represent all means, including organizational units, being necessary for carrying out any activity in the enterprise. Order classes represent planning and control information. **Figure 2** represents the attribute schemes of the generic classes [5,6,12].

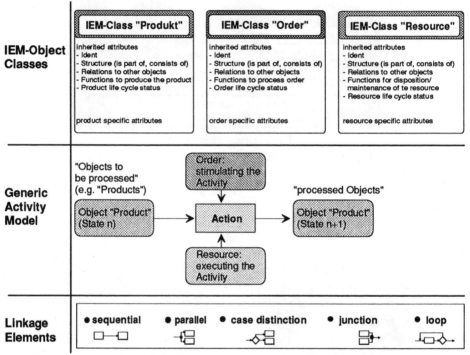

Figure 2: Basic Constructions of the IEM

The Generic Activity Model represents the processing of objects of Product or Order or Resource classes respectively indicating the object interactions at processing. The related organizational structure is described by specific Resource classes along with their interrelations.

Using the linkage elements Activities are combined for representing business processes. The decomposition and aggregation of processes is supported as well.

2.3. Business Process Modelling

The description of enterprise processes starts with the analysis of the actual situation, normally applying a top-down procedure. Business processes along with the respective classes of objects to be processed are the starting point for modelling a certain part of the enterprise. This part is delimited in a first step concerning e.g. Products and respective "ordering" with regard to the main task, required resources and the interfaces to the environment as well. The main task and objects of the application area are described by that [8,9,10]. The products have to be identified as subclass instances of the IEM class Product and the business processes have to be modelled according to these products, independently of organizational structures. Next, resources and controlling orders have to be identified for each function of the defined enterprise processes.

To obtain more detail, business processes of generation, processing and supply of orders and resources as well as of processing of sub-objects of the products should be modelled. The network and the interdependencies of the business processes are described with the concatenating constructs of the IEM [11,12].

The development of particular business process models has to be extended by the order and resource flow, the analysis of concurrency of business processes and their mutual influence. For this purpose simulation and other methods should be applied. The right choice of the level of detail is important for the modelling effort and benefit. For the task of information system planning, an overall, not too detailed modelling of a number of enterprise areas should be preferred [9,10].

2.4. Information Modeling

The collection and structuring of the data of all objects which were identified at modelling lead to a particular enterprise information model. For this purpose, a structuring frame for representing the relevant data is required.

The differentiation of the generic classes of objects and their internal structure define the structure of the enterprise information model. Three interconnected submodels, the Product, Control and Resource models are defined. The internal structure of the submodels is represented by layers which enable a grouping of object data by different criteria. The independence of data of a specific information processing system secures the extendability and interchangeability of data between several systems. A particular enterprise information model will provide the preconditions for a general use of data bases and support the recognition of priorities at data exchange [8].

The first step for the development of a particular information model of an enterprise is the cataloguing of objects, their descriptive data and object relations in data dictionaries. Different kinds of lists were defined in the layers of the enterprise information model.

2.5. Integration of Business Processes and Information System Support by Additional Views

Further aspects of modelling related to special purposes can be integrated as additional views on the model. Examples of such views are special representations of control mechanisms, organizational units and costs. The relevant properties of the additional views can be represented by deriving specific subclasses of the generic classes:
- Determination of class specific attributes and
- determination of attribute values (**Figure 3**) [9].

3. Modelling Tool

The use of the described method for enterprise modelling is only efficient in connection with a modelling tool. A PC-based verison, which supports the method, is available (**Figure 4**).

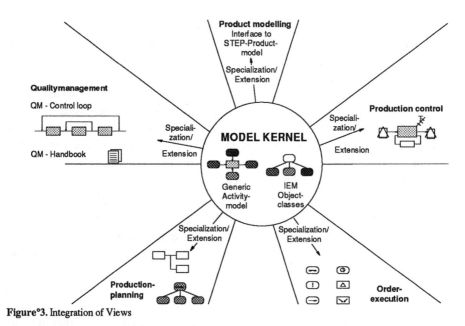

Figure 3. Integration of Views

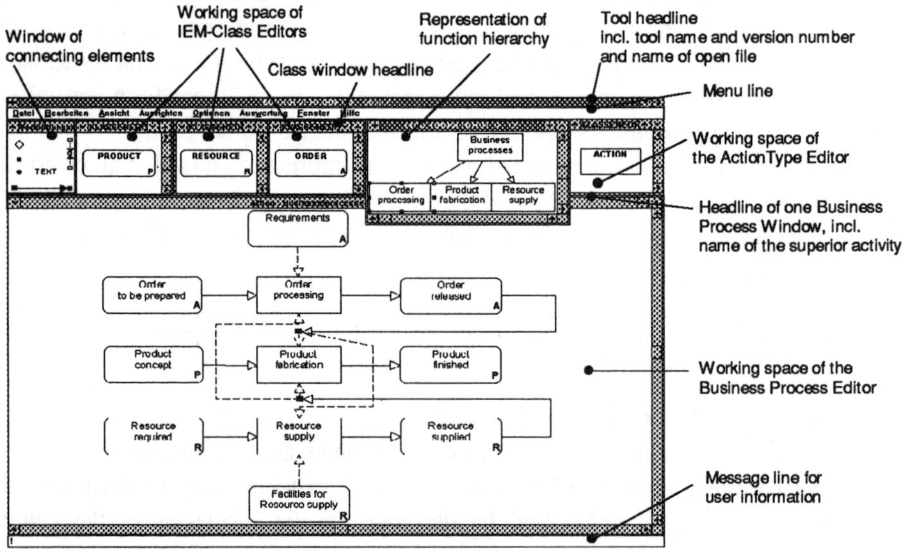

Figure 4. Tool User Interface

It is based on Windows 3.1 platform and uses Windows standard. The development of the graphical notations (diagram technique) of the user-interface was supported by a special commercial software package. The logics and the application rules of the method are implemented in C++ independently of the supporting software via a special programming interface [5]. The following modelling tasks are supported by the functionality of the tool:

- Classification of products, orders and resources of the enterprise.
- Hierarchy of business processes.
- Description of product, order and resource structures.
- Generation of documents.
- Rapid model construction by predefined model structures and navigation facilities.
- Support of different user views.

The Tool is used in industrial projects of IPK as well as in internal projects of industrial and consulting companies.

4. Conclusion

There exist a number of methodologies for enterprise modelling with their own purposes and advantages. In [6] a detailed analysis of several modelling methods was conducted. Here we mention only ICAM/IDEF, SADT and CIM-OSA as examples of function-oriented methods. Generally speaking, the main topic of these methods are data processing functions, data and functions are represented in separate models loosely tied via the names of data entities.

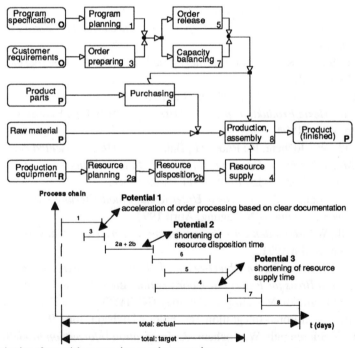

Figure 5. Derivation of potentials to save time at order processing

Besides the well known difficulties during information systems design, implementation and maintenance, the application of these methods for business process reengineering faces the problem that, starting with function analysis, one rather "copies" the existing organizational structure than to analyze the (object-related) nature of processes. Thus, for the task of process reengineering these methods are useful but not optimal.

The aim of IEM is to build a comprehensive (conceptual) enterprise model as an aid for analyzing and reengineering business processes, and as a platform for developing and integrating various applications independently of a particular implementation environment.

An Example for the application of the IEM as an accepted description method in industrial projects is shown in **Figure 5**. The project has had the goal to work out potentials for saving time and costs by improved information systems in order processing.

5. References

1. P. Coad, E. Yourdon: *Object Oriented Analysis*.
   Yourdon Press/Prentice Hall, Englewood Cliffs, NJ, 1990.
2. Flatau, U.: *Digital's CIM-Architecture*, Rev. 1.1.
   Digital Equipment Corporation, Marlboro, MA U.S.A., April 1986
3. Harrington, J.R.: *Understanding the Manufacturing Process. Key to Succesful CAD/CAM Implementation.* Marcel Dekker, inc: 1984.
4. ISO TC 184/SC5 N 148, *Technical Report: Reference Model for Shop Floor Production, Part 1*.
5. Spur, G.; Mertins, K.; Jochem, R.: *Integrierte Unternehmensmodellierung*. Beuth Verlag. Berlin. 1993.
6. Mertins, K.; Süssenguth, W.; Jochem, R.: *Modellierungsmethoden für rechnerintegrierte Produktionsprozesse* (Hrsg.: G. Spur). Carl Hanser Verlag. München, Wien. 1994.
7. Süssenguth, W.; Jochem, R.; Rabe, M.; Bals, B.: An *Object-oriented Analysis and Design Methodology for Computer Integrated Manufacturing Systems*. Proceedings Tools '89, November 13-15, 1989, CNIT Paris/France.
8. Mertins, K.; Jochem, R.: *Integrierte Unternehmensmodellierung – Basis für die Unternehmensplanung*. DIN-Tagung. April 1993.
9. Süssenguth, W.: *Methoden zur Planung rechnerintegrierter Produktionsprozesse*. Dissertation. Berlin 1991.
10. Mertins, K.; Jochem, R.: *An Object-oriented Method for Integrated Enterprise modelling as a Basis for Enterprise Coordination*. International Conference on Enterprise Integration Modeling Technology (ICEIMT). Hilton Head (South Carolina), US Air Force-Integration Technology Division, June '92.
11. Mertins, K.; Süssenguth, W.; Jochem, R.: *Integration Information Modelling. In: Proceedings of Fourth IFIP Conference on Computer Applications in Production and Engineering (CAPE '91)*. Bordeaux, France. Elsevier Science Publisher B.V. (North Holland).
12. Mertins, K.; Jochem, R.. *Planning of Enterprise-Related CIM-Structures*. In: Proceedings of 8th International Conference CARS and FOF. Metz; France, 17.-19. August 1992.

HYPERMEDIA IN CIM: ISSUES AND OPPORTUNITIES

HORRIS C. LEUNG[‡]
E-mail: esryj@csv.warwick.ac.uk

and

RUTH F. LEUNG[‡], JOHN F. HILL[‡]
[‡]*Department of Engineering, University of Warwick*
Coventry CV4 7AL, United Kingdom

ABSTRACT

Hypermedia as a methodology of delivering static information such as on-line documentation and computer based training has provided with the manufacturing industry a number of successful solutions. At a time when the industry is searching for resolutions on information integration, it is generally believed that hypermedia could have been a candidate for some long lasting information integration problems. It has however failed to respond in the way as expected. The paper attempts to identify the issues which have to be resolved for hypermedia to gain wider acceptance as an integration tool in the manufacturing industry. Considering the existing problems in terms of information integration in Computer Integrated Manufacturing (CIM), the paper discusses also the areas which can be addressed by the hypermedia.

Keywords: CIM; Hypertext; Multimedia;

1. Introduction

Multimedia computing is becoming more a reality following the ever increasing desktop computing power, cheaper mass storage and advances in telecommunication technologies. It is now a technology accessible not only to the scientists in research laboratories, but even to the home computer users in every household. Hypermedia, a concept in extension to hypertext (Conklin 1987), is the most commonly adopted technique in structuring multimedia information to facilitate the information retrieval and navigation. Having an undisputed position in addressing the problems associated with CIM information requirements, hypermedia have been adopted in applications which include on-line documentation and operation manuals; Computer Based Training (CBT) and Computer Aided Instruction (CAI); information management and information retrieval; enhancements on CAD systems; and enhancements on the collaborative product development process.

While the use of hypermedia has provided effective solutions in the areas as mentioned, it is suggested that there are potentials to be deployed when we contrast the characteristics of hypermedia to the problems concerning the information management of manufacturing. The aim of this paper is to further explore these opportunities and to discuss the related issues that have to be addressed.

2. Opportunities of Hypermedia in Manufacturing

Considering manufacturing as an information driven process, we have identified the following areas in which the concept of hypermedia will have significant contribution.

2.1. Information/Knowledge Management

This is the trend of using hypermedia in managing and retrieving static information such as on-line documentation and multimedia data entities residing in certain database management systems. Currently, we can see that the trend is supported by two streams of development, the Computer-Aided Acquisition Logistic Support (CALS) Initiative (Malcolm et al. 1991, Devlin 1992), and the increasing popularity of multimedia information systems (Leung et al. 1994, Grosky 1994). A large part of the CALS initiatives by the US Department of Defence (DoD) is a electronic documentation interchange standard that all the DoD contractors should gear up for its use. This standard provides a way of storing all information content, be it part list, documentation, drawing and so on. Already, some hypermedia developers feel that CALS provides all the basic building blocks required by any hyperdocument. Multimedia information system is a new class of information system developed to support the human's natural perception on information. Originally developed for supporting scientific research in the fields such as medical, physics, the usefulness has attracted already certain amount of business interest (Frolick and Ramarapu 1993). Undoubtedly, the research in multimedia information system for CIM should not be too far away if such effort does not exist already. Hypermedia is probably the most natural user interface to such an information system.

2.2. Combining the Technology with Artificial Intelligence

It is generally agreed that artificial intelligence (AI) - expert systems in particular, and hypermedia are two complementary technologies. The way that hypermedia structures fragments of information provides expert systems with a promising knowledge representation scheme for the construction of a knowledge base. AI techniques on the other hands have been found suitable to solve problems associated with the extreme flexibility available from hypermedia for the users to browse through the information space. Lost in the hyperspace is probably the most embracing description on most of these problems.

Littleford (1991) has suggested in particular that AI can be used to embed organisational knowledge and domain knowledge of a hyperdocument. Organisational knowledge is comprised of statements about the organisation of a document. By tracking the users' access to the document, for example, if a particular user has been found reading the introductory and tutorial sessions all the time, then it is a strong evidence that s/he is a novice and probably the system has to skip through the difficult bit for a little while. Domain knowledge deals with the actual content of the document. For example in a user manual of a complex system, if the user has been accessing information mainly concerning the control sub-system, then s/he is most likely a electronic technician and the system can safely skip through the mechanical sub-system to provide the quickest access to the required information.

2.3. Hypermedia as an Integration Tool

It is agreed in general that a greater acceptance of hypermedia within organisations will occur only when the technology is better integrated with other organisational systems (Malcolm et al. 1991, Gertley and Magee 1991, Isakowitz 1993). This is of particular importance in manufacturing. It is a unique environment in a way that there exists a large number of software systems, user manuals, databases and standards documents. Considering the design engineers as an example, they need routine consultation to material databases, component databases/catalogues, CAD, CAE, simulation systems among the others. A scenario described by Parunak (1991) as a manufacturing information maze. Getting together all the required information and compiling them into a form useful to the engineers is already time consuming, additionally, they have to learn to use all the systems involved.

There exists some form of solution for the problem just mentioned, for example, many CAE systems can accept design models directly from CAD to proceed with the engineering analysis. These are considered rather passive measures from the information integration point of view. Theoretically, the concept of hypermedia can provide a higher level of integration in two aspects. Firstly to logically associate information residing in these legacy systems to save time in hunting for the information (Gertley and Magee 1991). Secondly, the hypermedia system can exchange dynamic information with the legacy systems (Isakowitz 1991, Bieber 1993), to provide the users with a unified computer user interface. A possible application of this in manufacturing is that a simulation system can accept parameters from the hypermedia system, proceed with the calculations and have the results displayed visually in the hyperdocument. A rudimentary example of this in the business community is that sending through a hypermedia system the name of a company, the information about its stock market performance can be retrieved from the database and the corresponding graph is plotted and displayed in the hyperdocument. Implications in manufacturing is of course a lot more complicated but its usefulness is apparently cannot be under estimated.

2.4. Collaborative Manufacturing

By collaborative manufacturing, we imply a situation where a group of geographically dispersed people working together on certain product development or manufacturing operations (Leung 1995). Computer Supported Co-operative Work (CSCW) paradigms, in particular desk-top video conferencing (Gale 1991) and CAD with shared workspace are the dominating approaches. Practical experience is that it takes much longer time to arrange people across country boundaries and time zones to be able to sit together than the actual time spending together, and solving problems productively. Networked hypertext systems remove this restriction by allowing participating individuals working at their own convenient time, yet aware of what the rest of the group are progressing by referring to the shared hyperdocument. By adding their own nodes and links to the document, they can inform the rest of the group on their contributions and progress. The concept, called collaborative authoring in hypertext terms, was originally developed for intellects in different locations co-authoring a literature. Applying to manufacturing for collaborations, it is a potential alternative in addition to what we normally perceive in terms of CSCW. However, as Benton and Devlin (1991) have remarked, there exists quite a few issues to be resolved due to the complexity of engineering systems.

2.5. Inter-organisational Information Sharing

The need of inter-organisational information integration in manufacturing is accelerated by the increasing popularity of Total Quality Management (TQM) and the emerging concept of virtual companies (Goldman and Nagel 1993). As more companies are working closer together, the information exchange between companies on the same or between different knowledge domains will see exponential growths. General protocols from normal post, facsimile to the lately electronic mail will continue to support routine operational procedures. They contribute very little however to resolve the information tangleness as highlighted by Engelbart (1991) in the US aerospace industry. In a particular fighter aircraft development project, he had proposed an Open Hyperdocument System (OHS) to facilitate the on-line information sharing between firstly, the different knowledge domains in his own company, and secondly, between the 6000 sub-contractors involved in the project. The OHS is seen useful not only to support the project development but also the re-use of information considering that the life cycle expectancy of such an aircraft is likely to be 25 years. Although the system or standard is still evolving, it is the authors' belief that the impact on manufacturing, in particular on inter-organisational integration will be quite substantial.

This trend of development is further supported by the new dimensions supplemented recently to the CALS initiatives announced by the DoD in January 1994 (Pistenon 1994). CALS under this supplement, has a new acronym named Continuous Acquisition and Life-Cycle Support. Instead of just a standard for digital document exchange as mentioned, the new CALS initiatives will go as far as a standard for global information sharing between organisations through out the complete product life cycle. The hypermedia provides with this new initiative a valuable alternative for the implementation of the man-machine interface for information navigation. It is of course worth noticing that both CALS and OHS are proposed for defence or weapon industry whereas the real value for civilian manufacturing industry is sometimes questionable. However, with so many successful examples in the past on the technology transfer from the defence industry to civilian industry, it should be fairly optimistic that if the concept is proved to be useful, the civilian industry will adapt and make use of the technology very quickly.

3. Issues in Manufacturing

It is rather obvious from the discussion in the previous section that the potential of hypermedia in manufacturing has yet to be deployed. This section serves as remarks on the issues to be resolved if hypermedia is to gain such acceptance as perceived.

3.1. Development Support

There exist in the market a few hypermedia application development systems which have demonstrated in many cases the usefulness in delivering static information such as on-line manuals and CBT materials. Various systems demonstrate different combinations of desirable features, but no one system currently has the combination of features to support the needs for engineering applications (Parunak 1991). Current hypermedia systems on the one hand have provided successful solutions to a specific range of manufacturing problems, on the other hand have prevented hypermedia to gain wider acceptance.

A number of requirements have been suggested by various authors in this respect and we believe that interoperability (Malcolm et al. 1991) and dynamic information exchange (Isakowitz 1993, Bieber 1993) are of particular significance to manufacturing. Interoperability specifies a hypermedia system to be able to operate under different hardware platforms and Operating Systems (OS), a characteristic which only a few hypermedia systems are capable of. Dynamic information exchange is a property by which hypermedia systems can access information residing in other applications such as database management systems (DBMS), CAD or engineering analysis/simulation packages. Except for Structured Query Language (SQL) which is agreed upon relational DBMS, dynamic information exchange of hypermedia systems and other legacy systems is still largely unaddressed.

3.2. Information Security

It is generally felt that storing information by electronic means is less secure than its paper-based counterpart. This in a way is a misconception because providing reliable security control over printed materials is generally more difficult, requires additional resources, and in many cases entail considerable inconvenience (Woodhead 1991). Imposing distribution control and recording the history of access on sensitive document is relatively easy for its electronic counterpart.

The concept of hypermedia naturally encourages information openness, it is not surprising that information security has been an issue under the heaviest scrutiny when the concept comes to the terms of industrial applications. Unfortunately, there is still no satisfactory answer from the community in response to the past record of network intrusions. Passive measures such as firewalls are commonly used in the industry serving as a form of barrier between the public and private information. The drawbacks are, sometimes it is difficult to draw a line between private and public in information terms. Further, it is a counter-productive measure on the ideal towards information integration. Until the issues of information security are satisfactorily resolved, the use of hypermedia in manufacturing industry is likely to be restricted.

3.3. Standards

Standards has always been a difficult problem to address in the scientific community because we tend to develop derivatives on a concept or technology to cater for specific needs. This is arguably a natural scientific development process, it creates however in many cases the need for additional resources in order that derivatives can communicate or exchange information. The problem is particularly acute and difficult in manufacturing due to its extensive involvement of technologies. The Manufacturing Automation Protocols (MAP) serves a very good example to illustrate the situation, very few people could possibly dispute the ideals behind the concept, yet it is still struggling to find its way into majority of the manufacturing establishments.

We have suggested in the previous section that hypermedia can potentially serve as an integration tool in CIM and that a hypermedia system should be able to exchange information dynamically with the other factory systems. Also it is considered as an enabling technology for inter-organisational information sharing. Among all the pre-requisites for these to happen, standards to allow information exchange is probably the most important.

While the standards for most of the data types like still photographic images, motion pictures, sounds and so on are being addressed and constantly reviewed by the multimedia community (Devlin 1991), we concentrate ourselves to the data types specific to manufacturing. Established standards like IGES (Initial Graphic Exchange Specification) appear to be the immediate solution but being physical files themselves, the limitations in supporting different applications running side by side and exchange information dynamically have long been recognised. For hypermedia system to be a practical tool for manufacturing integration, we believe therefore that conformance to the emerging product data exchange standard STEP (Bezos 1994) is an important criteria. Its very concept on the layer structure, support of neutral file format for dynamic information exchange and flexibility has already attracted the commitment of most application developers. However the manufacturing industry who will be the eventual end user considers the issues somewhat differently, after 10 years of effort, the standard is still striving itself into factory systems, much like the dilemma we have described for MAP. There are however signs for optimism because firstly there exists no comparable competing effort and, secondly it is probably the most promising standard that can support the needs of manufacturing integration in the future.

3.4. Database Technology

Until recently, relational data model is probably the de facto standard database model adopted in the industry. Although it provides a successful platform for business information processing, its inherent limitations in dealing with engineering and manufacturing environment are rather obvious and are well documented. The situation is even more acute when multimedia data entities are to be considered. Recent development allows relational data model to store Binary Large Object (BLOB) (Shetler 1992) such as photographic images and digitised video clips. However, until there are corresponding extension to SQL for supporting database query on these new data entities, the use of this storage mechanism in relational data model is bound to be limited.

Object oriented model is seen and demonstrated in parts by many authors the answer to the relational database limitations. The very fact that the model is still lack of a standardised language that can be compared to SQL for relational data model severely damages its attractiveness. With some measures of database mapping (Ahmadi et al. 1991), it is possible to support the interoperation of different data models. This is however counter-productive to the principle of having a database management system. There exist two reasons to believe that object oriented model is the data model to support hypermedia in manufacturing. Firstly, much of the concept of STEP is based on the object oriented data model (Bezos 1994). Secondly, it is considered by the hypermedia research community that object oriented is the most appropriate data model to support the development purposes. There are of course problems to be resolved. Query language as mentioned is the most important and following that is the ways in which multimedia data entities to be retrieved from the database. An example of this is the question concerning the retrieval of photographic images based on their visual contents.

4. Conclusion

A number of potential areas in which hypermedia can have significant contributions have been identified and the related issues in realising the ideas have been discussed. It is the authors' wish that the paper has encouraged some industrial interest, and subsequently to provide the necessary support for the development of hypermedia applications in the information intensive environment as manufacturing. While hypermedia is an exciting technology, vitally important is that it is not considered as a panacea as we have once thought of AI expert systems, a proper task identification is required to ensure that the use is not being abused and the community end up with disappointments.

5. Acknowledgements

The authors wish to thank the Warwick Manufacturing Group, University of Warwick for the support on this research.

6. References

Ahmadi, J., Nayak, N., Ow, P. S. and Wald, J., (1991) "Key Technologies for CIM in the USA," *Computer Applications in Production and Engineering: Integration Aspects*, pp 41-48.

Benton, P. M. and Devlin, J., (1991) "Work Group Automation and Hypertext" in *Hypertext Hypermedia Handbook*, (Ed) Berk, E. and Devlin, J., McGraw Hill, pp 415-434.

Bezos, A., (1994) "STEP Technology for Product Data Representation," *Proceedings of the 14th International CODATA Conference, Chambery France, 18-22 September 1994*.

Bieber, M., (1993) "Providing Information Systems with Full Hypermedia Functionality," *Proceedings of the 26th Hawaii International Conference on System Sciences*, pp 390-400.

Conklin, J., (1987) "Hypertext - A Survey and Introduction," *IEEE Computer*, **Vol 20** No 9 pp 17-41.

Devlin, J., (1991) "Standards for Hypertext," in *Hypertext/Hypermedia Handbook*, (Ed) Berk, E. and Devlin, J., McGraw Hill, pp 437-444.

Engelbart, D. C., (1991) "Knowledge-Domain Interoperability and an Open Hyperdocument System" in *Hypertext Hypermedia Handbook*, (Ed) Berk, E. and Devlin, J., McGraw Hill, pp 415-434.

Frolick, M. and Ramarapu, N. K., (1993) "Hypermedia: The Future of EIS," *Journal of Systems Management*, July, pp 32-36.

Gale, S., (1991) "Recent Advances in Network and Video Communications," *Computer Supported Co-operative Work, the Multimedia and Networking Paradigm*, pp 80-100.

Gertley, G. G. and Magee, B. R., (1991) "Hypermedia Applied to Manufacturing Environment," *Proceedings of Hypertext'91*, pp 419-424.

Goldman, S. L. and Nagel, R. N., (1993) "Management, Technology and Agility: The Emergence of a New Era in Manufacturing," *International Journal of Technology Management*, **Vol 8** No 1/2 pp 18-38.

Grosky, W. I., (1994) "Multimedia Information Systems" *IEEE Multimedia*, Spring, pp 12-24.

Isakowitz, T., (1993) "Hypermedia, Information Systems and Organisations: A Research Agenda," *Proceedings of the 26th Hawaii International Conferences of System Sciences 1993*, pp 361-369.

Leung, H., Pashby, I. and Tsui, W., (1994) "Multimedia Information System for the Machining of Advanced Aerospace Composites," *Proceedings of the 14th International CODATA Conference, Chambery France, 18-22 September 1994.*

Leung, H., (1995) "A Collaborative Manufacturing Environment with the use of Hypermedia" *Software Systems in Engineering 1995 - Proceedings of the Energy-sources Technology Conference and Exhibition ETCE, 29 January - 1 February 1995 Houston, USA*, (Ed) Cooke, D., Hurley, W. D., Mittermeir, R., and Rossak, W., ASME, **PD-Vol. 67** pp 121-129.

Littleford, A., (1991) "Artificial Intelligence and Hypermedia," in *Hypertext/Hypermedia Handbook*, (Ed) Berk, E. and Devlin, J., McGraw Hill, pp 357-378.

Malcolm, K. C., Poltrock, S. E., and Schuler, D., (1991) "Industrial Strength Hypermedia: Requirements for a Large Engineering Enterprise," *Proceedings of Hypertext'91*, pp 13-25.

Parunak, V. D. M., (1991) "Toward Industrial Strength Hypermedia," in *Hypertext Hypermedia Handbook*, (Ed) Berk, E and Devlin, J, McGraw Hill, pp 381-395.

Pistenon, R. J., (1994) "The CALS Initiative: A Global Model for Acquisition," *Proceedings of the 14th International CODATA Conference, Chambery France, 18-22 September 1994.*

Shetler, T., (1992) "Birth of the BLOB" *BYTE*, February, pp 221-226

Woodhead, N., (1991) *Hypertext and Hypermedia: Theory and Applications*, Sigma Press pp 60.

A MODEL-DRIVEN TOOLSET FOR SUPPORTING THE LIFE-CYCLE OF CIM SYSTEMS

Authors - Dr. JM Edwards, S. Murgatroyd and Prof. RH Weston
MSI Research Institute, Dept. of Manufacturing Engineering,
Loughborough University of Technology, Loughborough,
Leicestershire, LE11 3TU, U.K.
Tel.: +44 (0)509 222919 Fax: +44 (0)509 267725
Email: johne@mansun.lut.ac.uk.

Abstract

The paper provides an overview of a manufacturing systems engineering toolset, which embraces a number of different methods, created to provide computer assistance to the personnel responsible for designing, implementing and running manufacturing systems. The approach is based on:

- a unification and extension of process oriented and object oriented design methods;
- the creation of CASE tools combining design methodologies with simulation and rapid prototyping facilities to enable the iterative generation of manufacturing systems. The rapid prototyping phase incorporates a series of system building tools which provide two separate approaches to structuring and automating implementation processes;
- a toolset to support information modelling, information resource structuring and run time information access;

1.0 Model-Driven CIM

The aim of the Model-Driven CIM research programme is to help manufacturing system designers and constructors to produce more effective and flexible integrated manufacturing systems.

The research has focused on producing a set of software tools, linked via models within a framework, which can be used to systemise design and build processes. The approach brings together tools which can be used for system description, visualisation, simulation and rapid prototyping including support for system construction and the provision of infrastructural utilities. The toolset is used to create, simulate, and rapid prototype or enact the models (which collectively describe the integrated manufacturing system), in an iterative cycle until an optimum solution is found.

The goal has been to extend existing modelling methods in a way which:

(i) structures and controls the behavior of manufacturing functions;

(ii) structures and creates much of the manufacturing integration software;

(iii) defines information entities and their interrelationships on a system-wide basis.

(iv) defines the mapping, via an integrating infrastructure, of logical device independent descriptions of functions and entities onto a set of physical resources available within a specific end system; and

(v) provides means of dealing with non-conformant legacy components of a system.

The approach is designed to impose a degree of prescription (or formalism) on the activities of system designers and constructors and support them in reducing the time required to produce optimum system designs.

Figure 1 shows a simple representation of the principal elements of the complete toolset with reference to a life-cycle axis, from conceptual design to operation and maintenance. It can be seen from Figure 1 that the principal tool subsets, or workbenches incorporate design and implementation.

During the course of the work it has been established that the design phase of the life-cycle is underpinned by a requirement for exercising designs. This has been achieved by both simulation and at a later stage through rapid prototyping or system emulation. Rapid prototyping has been achieved through a system of "model enactment". Here the models created during the design phase are combined with all other "soft" elements of the system to create a working system using emulated manufacturing resources. Thus to move from a rapid prototyped system to the real world requires only the replacement of emulated resources with real functional elements. The cycle of design, simulation, redesign, rapid prototype, redesign, etc. to optimise the final system solution incorporates the implementation phase of all soft integration elements of the final system.

FIGURE 1. Principal elements of the toolset

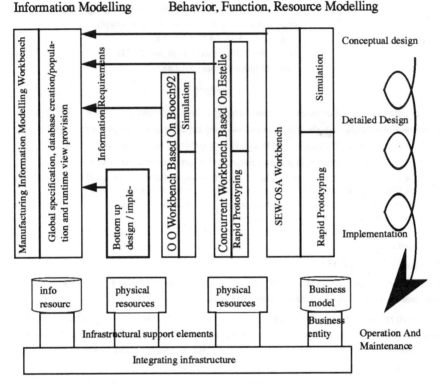

Figure 1 also provides a clear picture of the way the toolset tackles information system design and implementation as a separate discipline from that of modelling the system behaviour and the function of its constituent elements.

The constituent parts of each workbench or tool subset can be classified as one of three elemental types as follows:

- Tools - typically for design, model exercising and information translation;
- Models - reference models (typically product models or factory models), and models created through the use of design tools (typically models of system function or system information);
- Infrastructural Support Elements (ISE) - typically integration services, application libraries, configurable drivers and execution engines.

These types are combined within an overall framework which defines the relationships between the three element types above and forms the structure of the complete toolset.

The following sections provide more detail of the main elements of the toolset identified in Figure 1.

2.0 Modelling The Structure And Behaviour Of Manufacturing Systems

2.1 The SEW-OSA Workbench

SEW-OSA has been designed and consolidated as a combination of CIM-OSA, Generalised Stochastic Time Petri-Nets, Predicate-Transition Petri-Nets and Object-Oriented Design tools coupled to run-time utilities of the CIM-BIOSYS Integrating Infrastructure.

In certain respects this is the most advanced of the Model-Driven CIM workbenches, it is largely complete having been implemented as a formal method using the IPSYS Meta-CASE tool [2]. The life-cycle support capability of SEW-OSA extends from the Conceptual Design phase of systems integration projects (at which stage IT requirements are related to the principal Business Processes of a company) through the Detailed Design phase (where time-based behaviour, functional and resource descriptions of a system are identified), System Build (which is supported via the enactment of SEW-OSA models) onto System Operation. Much of the SEW-OSA modelling capability has been provided by formally implementing the CIM-OSA specification [3]; this has been found to provide a fairly comprehensive methodology when defining the requirements of integrated manufacturing systems but has been significantly extended within SEW-OSA through the provision of its system build and model enactment capabilities.

FIGURE 2. Principal elements of the SEW-OSA Workbench

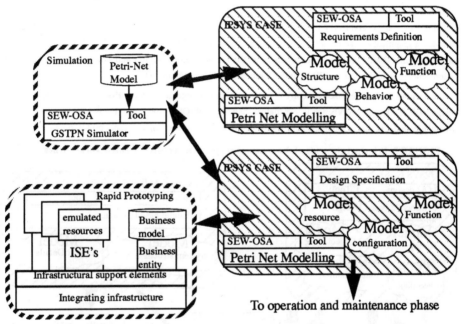

A major strength of SEW-OSA lies in its ability to support much of the life-cycle of integrated manufacturing systems, providing facilities which support the definition and mapping of system functions and behaviour onto a set of physical resources. To date, no information modelling capability is included in SEW-OSA, but in the context of Model-Driven CIM this does not represent a problem. As can be seen from Figure 1 SEW-OSA creates a set of information requirements which are passed to the Manufacturing Information Systems workbench.

Figure 2 shows the principal elements of the SEW-OSA workbench showing support for modelling requirements definition, design specification and facilities for simulation and rapid prototyping.

The complete SEW-OSA workbench is a CASE toolbox that integrates a number of separate Model-Driven CIM elements:
- a CASE tool for the Function View, at the Requirements Definition Modelling Level of CIM-OSA;
- a CASE tool for modelling Function/Resource Views, of the Design Specification Modelling Level of CIM-OSA, and generation of interpreted code for the business entity of SEW-OSA;
- a CASE tool for Predicate-Transition Petri-Net modelling and generation of code for the Predicate-Transition Execution Environment;
- the Business Entity of SEW-OSA;

- Support for the Generalised-Stochastic Time Petri-net (GSTPN) simulator by generation of a description derived from SEW-OSA models.

A detailed description of the SEW-OSA workbench is given by Aguiar [1].

2.2 The Concurrent Systems Workbench based on Estelle

The use of the SEW-OSA workbench has proved to be very successful in formally defining the structure of software applications, particularly during Conceptual Design from the perspective of specifying the functionality and behaviour of integrated manufacturing systems. However, in its current form it does not provide sufficient constructs and formalisms to facilitate the detailed modelling of parallel systems, e.g. where there is a need for formally structured and optimised interaction protocol between distributed processes and processors. Example CIM applications which are parallel in nature include distributed planning, scheduling, costing, and bidding systems and a wide variety of realtime machine and process control systems. To meet this requirement a Concurrent Systems (CS) Workbench has been produced based on an extension of the Estelle formal specification language; which was originally produced for specifying low level communication protocols [6]. The Estelle workbench includes: a graphical design environment (which aids system designers in producing Estelle specifications); an Estelle compiler (based on a compiler produced at NIST); an Estelle generator and simulator; a link to the application generation tools; and the CIM-BIOSYS integrating infrastructure [7].

Figure 3 shows the Estelle based concurrent system design tool together with the Booch91 based object oriented design tool. The Estelle workbench is less mature than the SEW-OSA workbench but is based on the same constituent elements, in that it provides facilities for detailed design modelling and rapid prototyping.

FIGURE 3. Principal elements of the concurrent system and object oriented workbenches

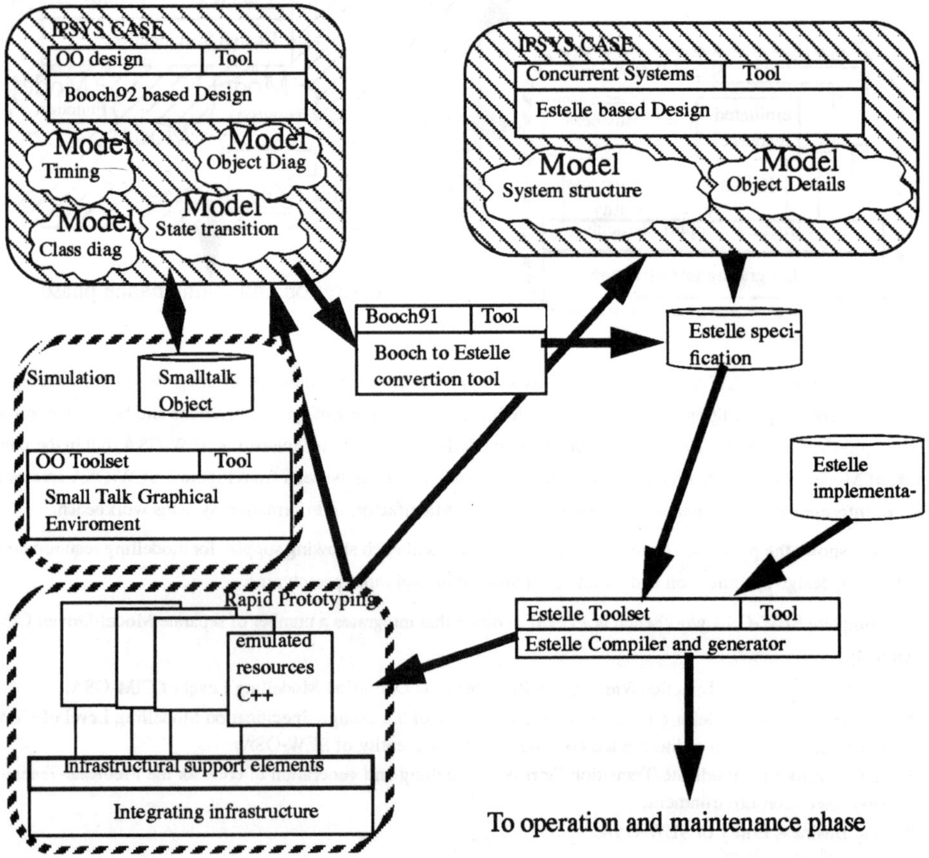

2.2.1 The Estelle Generation Tool

The Estelle Generation CASE tool provides a mechanism for entering a graphical representation of a system by describing its structure and interactions and generating a representation of the system in the Estelle language. Estelle specifications generated by the tool are then available to be used for subsequent manufacturing system generation.

The tool user specifies the system design using graphical objects which correspond with components of the Estelle syntax. The core objects (modules, channels and body definitions) provide the overall system structure and the user is able to enter system specifics through templates associated with each graphical object. The generated Estelle source is available for viewing at any time during the design. The main advantage of the tool is that it provides for more intuitive and incremental specification of the design and also tests for completeness.

2.2.2 The Estelle Implementation Libraries

The Estelle implementation library provides a facility for executing systems (generated from the NIST Estelle compiler) over the CIM-BIOSYS integrating infrastructure. The library provides appropriate communication functions and control such that the applications are able to conform to the protocol as specified in the Estelle specification.

The library provides a link between the C++ source files generated from the NIST Estelle compiler and the C calls provided by the Application Service Interface (ASI) library. Communications functionality and some lower-level handshaking functionality are included in the library for successful execution of the Estelle specification. The library provides a clean link between a high level interaction design model and a specific run-time implementation of distributed systems over common infrastructural services.

2.3 The Object oriented workbench based on Booch91

An important aim of the Model-Driven CIM research has been to enable use of a combination of process (or function) oriented and object-oriented methods, this is to allow different aspects and phases of the life-cycle to be supported in the most appropriate way. Object-Oriented (OO) methods have been widely used within the Model-Driven CIM programme. The SEW-OSA, Information Systems and Concurrent Systems workbenches employ the principles of object orientation as well as incorporating OO as an integral part of some of the methods they support. It was recognised early in the programme that OO could provide an appropriate starting point for supporting the mapping of distributed manufacturing software solutions onto physical resources; it has for example resulted in reduced design to implementation cycle-times through the re-use of software components and had given rise to systems which are easier to maintain. Hence a separate OO workbench was produced through building on and extending the constructs defined by an existing, popular and widely accepted OO software design method (namely Booch91 [8]) to enable it to be used for manufacturing system design. This led to the creation of a self standing OO design tool, which was implemented using the IPSYS meta CASE tool. The OO workbench comprises an OO system design capability and a Smalltalk simulation environment. Because it is derived from a general purpose software design method it can support modelling from the conceptual level to the detailed level. However its general purpose nature means it does not impose as much formalism specific to manufacturing systems design. The Booch91 method has been adapted with the addition of specific messaging constructs suitable for integrated manufacturing and libraries of manufacturing objects classes to help guide the designer.

Figure 3 shows the Booch91 based tool and its close relation to the concurrent systems tool. Simulation facilities are provided via a graphical Smalltalk environment while rapid prototyping can be achieved through the use of a conversion tool which produces an Estelle description, this enables the OO tool to utilise the Estelle model enactment facilities.

The OO tool supports system design through the following textural and diagrammatic constructs:
- Class diagrams and templates
- Object diagrams and templates (with message passing and types)

- State transition diagrams
- Timing, Module and Process diagrams

FIGURE 4. Information System Design Toolset

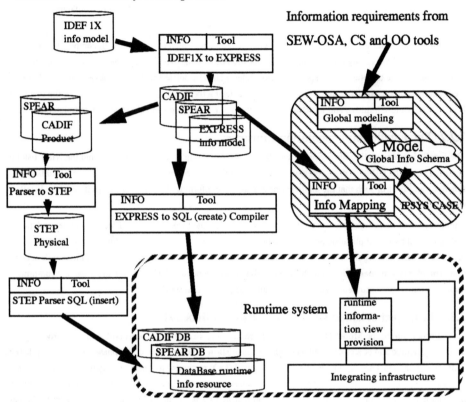

3.0 Manufacturing Information System Design And Implementation

Figure 1 identified information system design and implementation as a separate discipline to the more function based workbenches which have been described. The Manufacturing Information Systems (MIS) workbench comprises a sub-set of tools which in a consistent manner: (A) support various forms of manufacturing information modelling, including IDEF 1X, EXPRESS G, EXPRESS and STEP [4]; (B) creates consistent run-time information resources (database table creation including INGRES, Oracle and PROGRES [5]; (C) populate different forms of data storage device (i.e. data files and databases,) and (D) generate different (but consistent) views of that information to provide information support during system run-time.

The MIS workbench has been designed to support the most popular industrial choices of information modelling and representation techniques and information storage and access standards; this with a view to promoting exploitation of the approach. Many of the tools within the workbench are used for information translation. The most appropriate way of creating this type of tool is through the use of the UNIX utilities Lex and Yacc, together with the "C" programming language. This approach has also helped to facilitate take up by third parties.

The MIS workbench is based on the Three Schema Architecture [9] (comprising external, conceptual and physical views). The aim of this architecture is to provide manufacturing systems with an information architecture which hides much of the underlying system from the user. Runtime access of information will be through automatic generation of SQL statements within the information mapping CASE tool which is currently under development.

As the MIS workbench has been conceived and developed, it has been used in a number of industrial and academic case study applications. In these applications, use of the MIS workbench has proved to be highly beneficial with respect to:
- providing global and local model descriptions of information,
- enabling management of and access to heterogeneous information stores,
- providing object-oriented views of information,
- buffering users from a need for knowledge of the structure and location of information,
- providing structured means of modifying the underlying information models.

Figure 4 shows a further decomposition within the Manufacturing Information Systems (MIS) workbench which identifies the principal tools available.

4.0 Conclusions

The information mapping tool within the MIS workbench provides a unification of the MIS workbench and the SEW-OSA, Estelle and OO workbenches. This unification is an integral part of the final case study of the Model Driven CIM project which will be completed in December 94. The study involves the use of the SEW-OSA, Estelle, and OO workbenches in modelling the SMT Printed Circuit Board assembly line at a Large UK electronics manufacturer. Within this exercise the role of each workbench and their integration to form an overall framework will be critically assessed.

To date the following points have become clear.

The Object Oriented workbench and the Estelle workbench are highly complementary, the non-prescriptive nature of the OO workbench allows the modelling process to start at what ever stage the designer feels is appropriate to a particular application domain. Support for the designer (in terms of domain knowledge) comes primarily from the availability of a set of reference object classes describing elements appropriate to the domain. Capabilities of the OO workbench can be combined with particular abilities of Estelle to describe system interaction. This combination can be used to create systems based on a set of distributed interoperating objects which holistically encapsulate system behaviour. Such systems require an "early binding" of the manufacturing elements to form a working system

The SEW-OSA workbench offers a richer set of diagramming tools and constructs which can structure the design process by prescribing a requirements definition stage prior to detailed design. Requirements definition captures the business processes in an enterprise which form the basis of the business model which drives the manufacturing system at run-time. This concept allows the separation of overall system behaviour (i.e. the business model) from the individual behaviour of the separate application elements of the system. This separation implies a possible "late binding" which can provide a high level of flexibility and support ease of maintenance of the system when replacement of application elements is required.

Thus SEW-OSA (on the one hand) and the OO and Estelle workbenches (on the other hand) represent two distinct types of approach, which create different styles of run-time system requiring different infrastructural software support elements. Further more they are made up of different types of executable code elements (see Figure 2 and 3). On one hand SEW-OSA makes use of Petri-Net execution engines and management and control elements within the business entity, together with an integration infrastructure. These infrastructural elements are driven by Petri-Net model based code which is almost wholly created by the SEW-OSA CASE tool. Here a degree of system change can be accommodated through modification to the driving models. On the other hand the combined use of the OO and Estelle workbenches creates application object code which operates via a thin layer of infrastructural software, which provides an interface between Estelle and the integration infrastructure. Once built, the Estelle based system must be rebuilt to incorporate change.

The lack of prescription imposed by the OO methodology provides scope for wide applicability but requires support from reference models of good practice and a library of domain specific object classes in order to provide the constraints which can help structure the creation of manufacturing systems.

The SEW-OSA methodology encourages the user to begin with a definition of business requirements which imposes a requirement to start the modelling process at a high level of abstraction. Whether this approach is appropriate for all areas of an enterprise and for all types of integration project needs to be assessed. Again support for the modelling exercise through the provision of reference models is already identified as an essential requirement.

To date, application of the three workbenches has revealed that particular parts of each approach can very usefully be applied within the other workbenches. Thus, in addition to a comparison of the methods, the study will allow a more complete and consolidated view to be extracted. Anticipated examples of such a consolidation include: the integration of the resource modelling capability, from the object oriented workbench, within the SEW-OSA workbench, and the utilisation of the implementation path of the Estelle workbench by the OO workbench. Indeed it has become clear that the object oriented workbench can provide much of the information required for an Estelle specification in terms of state transitions (for function descriptions) and detailed object descriptions (interpretable for interface definitions).

The most appropriate choice of method and workbench will depend upon characteristics of the problem domain; such as its scope, complexity, parallelism, legacy and propensity towards information or functional interaction issues. Also best choice will depend on the capabilities, attitudes and previous experience of would-be users, such as do they need or will they accept and cope with high-level conceptual system design issues and tools or do they simply want something that is a bit better than present bottom-up practices, satisfying their immediate needs at minimal cost and effort. Although not described in this paper, Figure 1 identifies tools to support bottom up system build as inevitably many systems will incorporate bottom-up elements.

In summary the research program has demonstrated that a model driven approach, supported by automated tools can be used to facilitate the creation, co-ordination and control of integrated manufacturing systems in a formalised and semi-automated manner. This is expected to have particular benefits in helping to overcome existing problems in respect of specifying, building, managing, maintaining and changing manufacturing systems.

5.0 Acknowledgements

The authors wishes to fully acknowledge the contributions of the MSI researchers, and particularly Ian Coutts (in providing general purpose model enactment capabilities), Shaun Murgatroyd (in creating the OO workbench), Paul Clements (in creating the MIS workbench), Marcos Aguiar (in creating the SEW-OSA workbench) and Paul Gilders (in creating the CS workbench). Particular thanks are also due to the ACME Directorate of the UK's EPSRC.

6.0 References

1. M W Aguiar, I A Coutts and R H Weston. "Rapid Prototyping of Open Software Systems", European Simulation Multi-Conference, Spain, 1-3 June, 1994.

2. A Alderson, "Meta-CASE Technology". Lecture Notes in Comp. Science Software Dev. Env. and CASE Technology. Proc. of Euro. Symp. p81-91. Springer-Verlag. Germany. 1991.

3. ESPRIT/AMICE. CIM-OSA Architecture Description, AD 1.0. 2. ed 1993.

4. P Clements, I A Coutts and R H Weston. "A Means of Achieving Model Driven CIM", 1994.

5. SERC/ACME Final Review Report on the grant entitled, "Application and Information Support Systems for Planning and Control in CIM", April 1993.

6. Estelle: A Formal Description Technique Based on Extended State Transition Model. ISO 9074, 1988.

7. P Gilders and R H Weston. "A Mechanism for the Rapid Generation of Interaction Functionality for CIM Systems based on Estelle". To be submitted to the Computing and Control Engineering Journal.

8. G B Booch, "Object Oriented Design with Applications" Benjamin-Cummings, New York. 1991.

9. Tsichritzis D.C., and Klug, A, 1978, The ANSI/X-3/SPARC DBMS Framework report on the study group on Database Management Systems. Information Systems, 1,173-191.

BUSINESS PROCESS MANAGEMENT: THE EUROPEAN APPROACH

HANS H. HINTERHUBER
Department of Management,
University of Innsbruck, Austria
E-mail: management-c407@uibk.ac.at

ABSTRACT

The purpose of the paper is to show how European companies are approaching the problem of changing traditional organizational structures and eliminating barriers between individuals, functional areas, business units, and hierarchical levels. Business Process Management is becoming an increasingly important element in any change programme in European companies. The paper outlines a six-step approach to implementing Business Process Management and draws some preliminary lessons from the experience in progress of European companies.

KEY WORDS: Strategic planning, stakeholder satisfaction, implementation

1. Introduction

In the past few years, European firms have made intensive and comprehensive efforts to simplify organisational and decision processes, to increase productivity and to improve decisively the international competitiveness of their business units. They have recognised that they will not survive until the end of the millennium if they are not in future able to do (1) the same things better and faster than in the past and (2) to "invent" new markets.

The key success factor which is proving decisive especially in the struggle against US and Japanese competition is the mastery of the transition from functional to process oriented organisational structures. This transition can be compared to a shift in paradigms. Process management means that executives and employees think and act in a way that radically transcends functional and hierarchical boundaries. The aim is to reorganize key business processes in order to reduce costs significantly, improve service considerably and/or shorten cycle times appreciably.

2. The Strategic Planning of Business Processes

Process management is the latest trend in the strategic management of European companies (Figure 1).

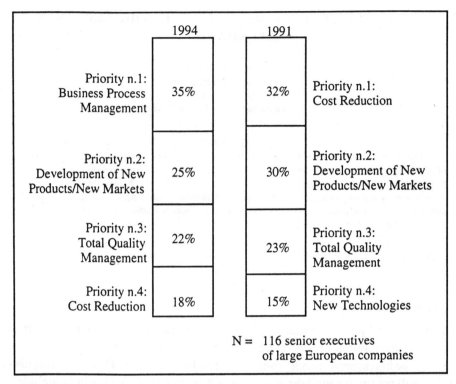

Figure 1: Changes in management priorities from the standpoint of European senior executives (Source: Dept. of Management, University of Innsbruck)

By process management we mean a customer oriented management endeavour to achieve exceptional performance in those business processes which transcend functional boundaries (Figure 2). A business process represents a set of integrated and coordinated activities required for producing products or offering services which:
- increase the satisfaction and competitivity of external customers,
- make the work of internal customers easier and more efficient,
- have a measurable input and output,
- add value,
- may be replicated, and
- fall within the area of responsibility of an executive who manages and coordinates an interdisciplinary and empowered team.

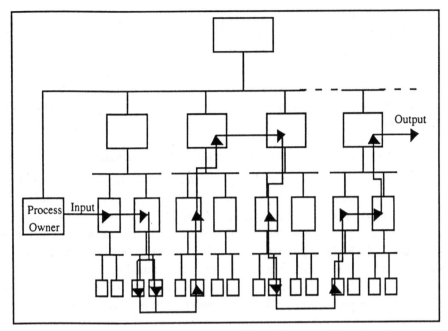

Figure 2: A business process cuts across departmental boundaries and optimizes the performance of the organization as a whole

The objective of the strategy developed for every strategic business unit, is to satisfy all key stakeholders better and quicker than the competitors or other leading reference firms ("best practices") (Figure 3) (Nayak, Drazen and Kastner 1992).

Stakeholders	Satisfaction Criteria	Importance for the Stakeholders	Worse than the competitors or "best practices"					Weighted Number of Points
			-2	-1	+0	+1	+2	
Customers	Products Made Exactly to Customers' Requirements	10					●	+20
Employees	Working Environment	8		●				-8
Owners and Financial Community	Economic Value Added	10	●					-20
Society	Environmental Protection	10				●		+10
Suppliers	Stable Outlets	10			●			–
Partners in Joint Ventures	Open Communication	9					●	+18
1 ... 10: not important ... very important			Total Weighted Number of Points					+20

Figure 3: Satisfying all key stakeholders (illustrative example)

Every strategic business unit increases its ability to compete to the extent that it coordinates and manages its business processes in such a way as to make its customers more competitive and to add value for the other stakeholders. In other words, the impetus of process management is towards making the customers the best in their market segment. The fundamental principle of process management has been stated by a German senior executive the following way: "Not we are the best but we help the customers to become the best".

3. The Implementation of Business Process Management

Process management is not a programme for continuous improvement. It is embedded in a SBU strategy, attempts to achieve a number of breakthroughs in the area of (1) internal and external customer satisfaction and (2) of adding value to all other stakeholders. The important question is: "If we were a new firm or business unit, how would we design the business processes so as to increase the external customers ability to compete or enhance their satisfaction to an extent which would put us all way ahead of our competitors?" The starting point for process management is, therefore, a precise definition of the business processes. These business processes must then be incorporated within the strategy of the firm or within the strategy of particular business units.

1. Defining the business process
2. Installing a process owner
3. Measuring and mastering the business process
4. Coordinating the business process in question with other business processes
5. Concentrating an critical business processes
6. Continous process improvement

Figure 4: Six steps for implementing business process management

Only one manager can be responsible for a particular business process. The task of the process owner is to define the process in detail. He should also be responsible for the composition of the team which he should guide and coordinate in such a way as to lead to continuous improvement in the business processes, which should then run with zero defects. The process owner must have the authority to take all measures necessary to coordinate and improve the business process. The members of the team must be in a position to commit their particular areas of responsibility to the decisions made in the team.

A business process can only be mastered if it can be measured. Only those things which can be measured can be controlled and what can be controlled can also be improved. One of the process facilitator's tasks is to develop suitable methods of measurement. The flow chart is an important tool for initiating and controlling a business pocess.

Depending on the number of business units, more than one business process may be running in a single firm at any one time. SBUmanagers and process owners have to coordinate the business processes in such a way as to use human, physical, financial, and informational resources effectively, taking advantage of the effects of synergy and improving the qualitative structure of the firm as a whole. In German firms, the task of coordination is often undertaken by a "Steering Committee".

By using Pareto analysis, managers, working closely with the process owners, can determine their priorities. They can discover which 20% of the causes account for 80% of the problems and concentrate on key business processes. It is these key processes which will make the biggest contribution to increasing the value of the firm in the middle and long term.

Total Cycle Time is a measure of the firm's ability to react in the market and is, therefore, one of the most important indicators of a firm's ability to compete. Total Cycle Time is the length of time from the moment when the customer requests something to the moment when he receives it and, fully satisfied, pays the bill. Empirical studies have shown that reducing Total Cycle Time is the key factor in increasing productivity (Figure 5). Productivity, defined as the relationship between turnover and total resource input, is the most important key in ensuring the firm's ability to compete (Fromm 1992).

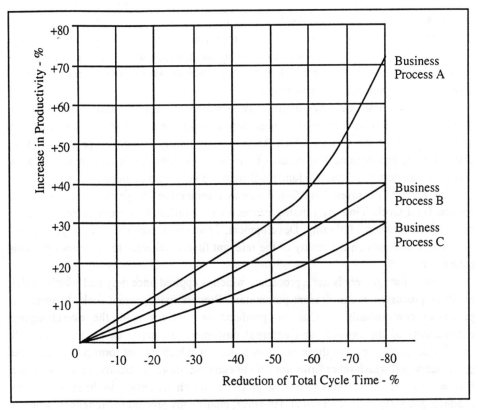

Figure 5: Increase in productivity and reduction of total cycle time in an Austrian electronic firm (Source: Department of Management, University of Innsbruck

Within the Total Cycle Time, a number of activities take place, which, when all added together determine its length. It often appears that those activities which do not create value are exactly those which account for an undesirably high proportion of Total Cycle Time. Examples of such activities are unnecessary waiting time and repair and reworking occasioned by defects.

In analyzing Total Cycle Time, all those problems become apparant which occur in a firm as a result of a too rigid structuring into functional areas. We should like to mention just two.

The first problem concerns the way in which we can most accurately and meaningfully measure Total Cycle Time. If time related yardsticks are used as performance indicators in the firm at all, they are generally applied first to the bottom of the organisational structure. They are then used to assess the performance of each department in turn until the top of the organisational structure is reached. The individual departments are unwilling to record in their statistics times which, "they can't do anything about". The result of this is that much time remains unrecorded although it may play a considerable role in determining how long it takes for a particular operation to run.

A stakeholder oriented firm must, therefore, move away from a system which records time vertically to one which measures Cycle Time horizontally. Only in this way can we achieve real success in reducing the time taken for processes to run. Reducing Total Cycle Time often makes a considerable contribution towards clearing up misunderstandings between departments. It also improves communication, avoids a duplication of effort and reduces the loss of information which may occur when messages move form one level of responsibility or functional area to another.

The second problem concerns the way in which one activity may follow another although it may not, in fact, be at all necessary to order the activities in this way. As an example, think of the way in which R&D is traditionally thought to precede and be distinct from production. The result of this way of thinking is that in many large organisations, between these two functional areas are very rigid barriers.

The name "Concurrent" or "Simultaneous Engineering" is given to the attempt to reduce Total Cycle Time by running processes in parallel. This requires considerable cooperation between Marketing, Development, Production, Service and Recycling and places new demands on the ability of the relevant functional areas to communicate with each other.

Process management is not a procedure which is applied once only and which guides business processes within present parameters. Process management seeks to improve processes continuously so that the products or services meet the everchanging expectations of the internal and external customers. In many cases, the firm will subcontract to third parties those business processes which do not form part of its core competence and which other firms are able to carry out more efficiently. In this way, the firm will be free to concentrate on those processes which are critical to the success of its products and services in the market. However, continuous process improvement alone is not enough: A firm must learn faster and better and innovate more quickly than the competition is able to do if it wants to rank among the winners of the nineties. Employees are expected not only to do their work efficiently but also to be "process improvers". A Japanese entrepreneur is convinced that in his firm 90% of the employees are "process improvers"; in European firms, he thinks, the situation is reversed: 10% are "process improvers" and 90% are concerned only with improving the quality of their own wellbeing.

To sum up we should like to emphasize that process management should benefit all those involved in it. This means not only the internal and external customers but also the process owner, the members of the teams and the employees - that is anybody who in some way contributes to process management. Nasreddin, the Turkish Socrates, is supposed to have said, "Always and in everything, try to combine what is useful for other people with what is pleasant for yourself".

4. Lessons to be Learned from Organisations which are already using Business Process Management

As far as the application of process management is concerned, it appears that those firms which have had to come to terms with great changes in the nature of competition are the ones which are in the lead. Firms such as IBM and NCR in computers, SMH and IWC in the watch industry, Daimler-Benz, VW and Fiat in automobiles as well as the suppliers of major enterprises are examples. These firms, which have had to adapt to the changed nature of competition, are the ones which have implemented business process management.

The following lessons are to be learned from firms which have implemented process management:

1. Process management is part of tactics; Process management can certainly lead to cost savings, reductions in Total Cycle Time of operations or improvements in quality even if it is not embedded in the strategy of a firm. However, unless process management becomes a real part of strategy, it cannot achieve a really radical increase in the value of the firm (Figure 6).

2. Process management is the responsibility of the top management of the firm or the responsibility of the executives in charge of the relevant business units. The managers of business processes must have sufficient authority to coordinate a business process across functional areas and regional units or down through the various levels of responsibility which make up the structure of the firm. Those in charge of staff functions are, generally speaking, not likely to become process owners.

3. Emphasizing the team. The main task of the process owner is the creation of a multidisciplinary team. The members of this team should not be the sort of people who are inclined to have second thoughts. They should be unconventional individuals capable of thinking creatively. They should be drawn from important functional areas and regional units. The team should also include innovative people from outside the firm who, according to Nicolas G. Hayek, have kept "the phantacy of a six year old child"(quoted from Bentivogli et al. 1994). The process owner should not make compromises with regard to the quality of the employees or the way in which the process should run.

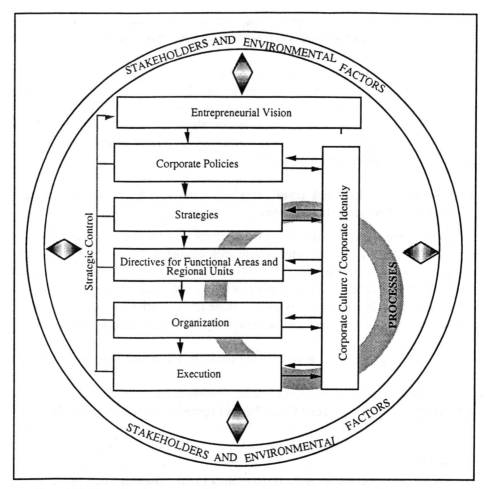

Figure 6: The European approach to strategic management

4. A sense of urgency. The key question in process management is: If we could start again from scratch how would we approach the problem? Firms which have achieved radical breakthroughs attribute their success to a sense of urgency. This provided an "intrinsic motivation" for all those involved. Some firms go so far as to implant in their managers and employees the deeply rooted conviction that their enterprise is on the verge of a crisis. They hope in this way to use time pressure to induce change.

5. Concentrating the best resources on a definite number of business processes and strategic goals. Successful firms start with a few business processes for the most important strategic business units. The starting point is customer satisfaction and the question which is closely related to it, "What does the customer expect from us?" The second question is: "How can we add value for all other stakeholders?"

6. Agreeing on clear measurable aims the achievement of which will lead to managers and employees enjoying a feeling of success even in the short term. Process management is essentially a top-down initiative, since it involves breaking down organisatonal barriers and introducing new ways of thinking. In successful firms, the

conflict between top-down and bottom-up is resolved as follows. Executives and employees are all involved in decisions. In this way clear and measurable aims are agreed upon which means that all those involved enjoy a feeling of success even in the short term.

7. Immediate reward for exceptional achievement and no punishment in cases where the effort towards innovation was well-conceived but failed Exceptional performance either by the teams or by particular individuals should be rewarded immediately. Rewards can be either financial or psychological. Efforts towards innovation which were wellconceived but which, nevertheless failed, should not be punished since the firm wants to learn from its mistakes.

8. Leaders should set a good example. Process management is team management. One team member needs to be able to rely on the other if the aim of satisfying the internal and external customer is to be realised. In all those firms which successfully employ process management, those at the top live according to a mental model which is based upon a holistic way of thinking, transcends individual areas of responsibility and is more oriented towards serving than towards doing (Hinterhuber, 1994).

9. The firm should be seen as a learning system which is continually extending its learning abilities. In successful firms, process management is to be seen as a means of altering the mentality and attitudes of management and employees towards the firm, the customers and other stakeholders and towards the competition. Everyone in the firm is required to demonstrate the ability and readiness to take on responsibilities which are greater than those which have been formally assigned to them.

10. Never rest on your laurels. In all areas of successful firms the prevalent attitude is that the present conditions are the result of far-reaching change and that those attitudes and methods which ensured success in the past do not represent the methods which guarantee success in the market in future. The attitude that, "nothing fails like success" is, in fact, a precondition for the successful implementation of process management.

5. Conclusions

The transition from function to process orientation resembles a change in paradigms. Crucial are the changes in attitudes which take place in the minds of executives and employees at all levels of responsibility, in all functional areas and in all regional units. They are all expected to demonstrate a holistic way of thinking which transcends functional boundaries and is oriented towards the satisfaction of the internal and external customer and, in addition, to add value for all other stakeholders. There are firms which make their customers more competitive by offering them valuable and useful systems, products and services which provide them with a springboard for growth. These are the firms which will be successful in the face of the increasingly fierce competition of the nineties.

In conclusion, I would like to make a few remarks about the future development of process management. I am convinced that the following factor is crucial if firms are to survive and increase their value in the future. The crucial factor is that each and every employee should display the creativity and the will to put themselves in the position of the internal and external customers. They should be able to offer the latter a more useful product or service and offer it more quickly than the competition. Process management has tremendous potential but only if the firm approaches the ideal of an organization which facilitates open communication. This "openly communicating community" provides everyone with open access to information and involves everyone with the decisions made in the firm. In the openly communicating community, each supplier receives immediate feedback from his internal or external customer and this feedback lets him know how satisfied the customer was with his performance.

The real problems which occur when one tries to implement and apply process management are not problems of a technical or business nature. They are cultural problems. The problem of giving the firm a process orientation would be easy to solve if all managers and all employees really understood that the success of the firm depends on satisfying the internal and external customers and the other stakeholders better and quicker than the competitors or other reference firms ("best practices") and that, therefore, everyone must concern him or herself with matters which extend beyond the formal sphere of responsibility assigned to them. Concrete plans should exist to bring the attitudes and behaviour of all managers and employees closer to the ideal of holistic thinking and shared responsibility. Only then will it be possible to talk about "corporate culture". The iniative for this should come from the employees and managers themselves if the corporate culture is to shift in the direction of process management. In the final analysis, process management means that managers and workers must work on their own attitudes. Moving managers and employees in the direction of process management means that the firm must shoulder a heavy burden. Similarly onerous is the task of motivating employees and teams to take strategic initiatives. However, the greater the preparedness of the firm to shoulder this burden the more successful and competitive it will be.

6. References

Ch. Bentivogli H.H. Hinterhuber and S. Trento, *The Watch Industry: a Strategic Analysis*, in *International Review of Strategic Management*, ed.D.E. Hussey, **vol. 5** (1994), pp. 133-170.

Ch. Berger et al., *Kano's Methods for Understanding Customer-defined Quality*, Center for Quality Management Journal **Fall 1993**, pp. 3-27.

H. Fromm, *Management von Zeit und Variabilität in Geschäftsprozessen*, CIM Management, n. 5 (1992), pp. 7-14.

H.H. Hinterhuber, *The European Way to Lean Management*, The International Executive 36, n. 3 (1994), pp. 275-290.

H.H. Hinterhuber and B. Levin, *Strategic Networks - The Organization of the Future*, Long Range Planning 27, **n. 3** (1994), pp. 43-53.

H.H. Hinterhuber and E. Krauthammer, *Flat Hierarchies and Individual Job Security in Times of Recession*, Journal of Strategic Change 3 (1994), pp. 125-131.

P.R. Nayak, E. Drazen and G. Kastner, *The High Performance Business: Accelerating Performance Improvement*, Prism, **First Quarter** (1992), pp. 5-30.

AN INTERNATIONAL PERSPECTIVE OF MANUFACTURING

T. A. SPEDDING
School of Mechanical and Production Engineering, Nanyang Technological University
Singapore 2263
e:mail mtspedding@ntuvax.ntu.ac.sg

L.H.S. LUONG
School of Mechanical Engineering, University of South Australia
The Levels, S.A. 5095, Australia

and

R.F. O'CONNOR
School of Manufacturing and Mechanical Engineering, The University of Birmingham
Edgbaston Birmingham, United Kingdom

ABSTRACT

Manufacturing technologies are advancing globally, but the rate and characteristics of their evolution is uniquely diverse. This advancement is driven by the requirements and aspirations as well as the constraints of the environment which are endemic in a particular region. The manufacturing technologies are often well known but their advancement and the characteristics of their implementation are influenced by cultural and economic forces.

By reviewing the implementation of manufacturing technologies in Europe, Australia and South East Asia, this paper develops a perspective of current enabling technologies for the manufacturing environment. These findings are then used to provide insight into the requirements necessary to support and enhance an integrated global manufacturing environment in the future.

1. Australia

It has long been recognised that successful manufacturing companies in Australia need to be more competitive on an international level (Quantz 1984). Although the Australian manufacturing environment has several positive characteristics such as a significant raw materials, intellectual capacity and an advantageous geographical location it has still been unable to penetrate the world manufacturing market. The problems of Australian Manufacturing can be summarised in the following terms (Small and Baines 1984):

Marketing - Australia companies concentrate on local markets

Product Development - Most companies only manufacture products with no in-house design or development.

Planning - Most Australian companies operate on short term planning cycles making it difficult to respond in the long term to international requirements.

Technology - The international manufacturing community has updated its technological base whilst Australia has lagged behind.

Australian manufactures have been reluctant to position themselves to take advantage of manufacturing technologies. The application of these technologies may take substantial investment which can only be justified if the market is sufficiently large. The viability of flexible manufacturing is still dependent on total throughput even if batch sizes are very small. Technology shift and the ad-hoc introduction of unsuitable technology has had detrimental effects. It has caused confusion and led to failures in industry and distrust in technology. However, with the recent reduction in tariff, many manufacturers have to adopt AMT in order to compete in the world market (Grigg 1993).

Although adopting technologies such as CAD/CAM have shown benefits CIM and FMS are more difficult to implement and register lower payoffs. Technology driven programmes require substantial time and an appropriate set of complementary actions to give positive results.

Technology implemented as an instant solution may result in a negative effect, especially in the short term. Also it is easy to lose sight of the organisation requirements of CIM and FMS.

The key to effective CIM implementation is flexibility. Australian manufacturing is diverse and sensitive so that CIM systems need flexibility to adapt to changes in market demand. If Australian manufacturing is to compete on the world market, as well as in its own market it must incorporate lower costs, high quality levels and short lead time into its manufacturing systems. These are no longer advantages in the world market but basic requirements necessary to compete. The relationship between the CIM structure and the company strategy is therefore of vital importance. Introducing CIM into a company should be carefully controlled exercise so that it is blended into the company structure easily and with the minimum of disturbance.

Several writers have observed that it is not unusual for a well organised company to be successful but with minimum or no technology but it is very rare foe a company with a high degree of technology and hardly any management structure to be successful.

2. Singapore And South East Asia

After the second world war Asian countries such as Korea, Taiwan, Hong Kong and Singapore were amongst the first to realise the importance of establishing a highly trained workforce (Drucker 1992). This low paid but efficient, highly productive and competitive workforce contributed significantly to the success of these countries in the post war period, resulting in many companies in the West moving the manufacturing operations to Asia.

Economic development in South East Asia has lead to several economic regions such as Malaysia, Singapore, Indonesia, the Philippines and Thailand being proposed. Although these countries are culturally diverse, such a large free trade area would have the potential to develop a unified response to international trade (Drucker 1993). The future of the Asia Pacific region will also be heavily influenced by the future development of giants such as China and India (already reflected in the long term investments of countries such as Singapore and Malaysia in these countries), and the continuing success of Japan.

Several commentators have suggested that growth in one Asian country may be at the expense of others. For example there are many fears concerning the continued success of

Singapore's manufacturing sector as it's highly motivated workforce begins to price itself out of the market because of higher labour costs due to an improved standard of living. Many manufacturing companies are moving (partially or completely) their operations to surrounding countries such as Malaysia, Thailand and Indonesia. This can however be misleading. In a recent study Natarajan and Tan (Natarajan and Tan 1993) discovered that many MNC's originally chose Singapore as part of a long term plan in which it is used as a "spring board" to neighbouring countries because of it's political stability. Once they had gained experience and confidence in the in the region the companies then relocate to Malaysia or Thailand. Well aware of this situation the Singapore government have emphasised the need to improve the technology, productivity and quality of their manufacturing sector so that it can compete on a different level to the more competitively priced labour intensive alternatives of neighbouring countries. Several Singapore companies who have recently relocated part of their operations to Thailand or Malaysia have taken the opportunity to upgrade their own operations to new product lines and higher technology activities. The Singapore government in fact offer incentives to relocate labour intensive operations to Malaysia (Singapore Investment News 1993).

In an article in the Singapore Straits Times (Straights Times 1993) the Chief executive Officer of Singapore Institute of Standards and Research pointed out that since ISO 9000 was introduced as a national standard in Singapore in 1988 over 300 companies had been certified by SISIR a figure that was expected to grow by 50% during the next 5 years. Although some of Singapore's manufacturing companies regard ISO 9000 as a strategic key to their International marketing potential, many introduce ISO 9000 on the contractual request of overseas customers and do not realise the full potential of implementing a quality management system (Chua 1992). It is recognised that ISO 9000 is an ideal platform to implement total quality management. Only if the concepts of quality assurance are absorbed into the company at all levels will quality improvement will take place and the benefits of TQM be fully appreciated.

The Singapore government fully realise the importance of a quality driven manufacturing sector Singapore. The introduction of a National Quality Initiative and Quality Award in 1993 underlines Singapore's government commitment and realisation of the importance of quality to their manufacturing sector. A recent study by Williams (Williams 1993) emphasises the success of this strategy; reporting work by Mody (Mody, Suri, Saunders, and Tatikonda 1991) he discusses how products produced by newly industrialised automated economies are more competitive when considered in terms of a quality and cost etc than less developed or newly industrialised manual economies and developed automated countries. The potential of this niche in the market is well understood by countries such as Singapore and Korea. A recently published survey (Il-woon and Jin 1993) concluded that whilst their technology lagged behind Japan and the United States Korea was aggressively restructuring its manufacturing and emphasising product quality and customer service. Aggressive competition in the global market forces countries to progressively improve their technology and move up the value chain (Porter 1990).

Dubois and Olif (DuBois and Oloff 1992) noted that during the early part of it's life cycle a product would manufactured in it's country of origin where it could have closer access to its design and research facilities. It was only towards the end of its life cycle when designs and manufacturing techniques where well proven that multinational companies would make full use of a global infrastructure. Indeed Singapore has deliberately improving its research, development, design and service base so that it can meet future challengers not just production environment but as a complete manufacturing entity offering the complete range of facilities.

3. United Kingdom

Many commentators have chartered the differences between UK manufacturing industry and the rest of the world. A favoured comparison of Western industry is to the achievements of Far Eastern competitors especially the Japanese. As a result many UK and other European countries have strived to improve manufacturing operations attempting to equal productivity levels and cost reductions as those claimed by Japanese counterparts. The realisation has been however that Western industry has a different history and culture as well as a different political environment in which it must operate The main problems in implementing "large" scale projects such as CIM in UK manufacturing can be classified into two broad areas of culture and investment.

UK manufacturing companies have been particularly dogged by the "resistance to change" attitude and industrial relation factors that have held back important investment in new technology and operational methods. This cultural factor has been a significant barrier to investment in CIM. UK manufacturing industry is steeped in tradition and practices some of which were borne out of the strong trade union movement that once existed. During the 1970s and early 1980s advanced technology and/or methods to rationalise business operations were not readily accepted at a time when UK manufacturing should have been investing.

The cost of capital for long term projects has been a significant factor in the limited large scale implementation of CIM. This is especially relevant for UK manufacturing where the cost of capital is far greater than that experienced by leading competitor countries such as the USA, Germany and Japan. Yates (Yates 1990) has suggested that nearly all investment in innovation and growth comes from retained profits. Furthermore, UK industrial companies are strongly linked to pension funds and banks who demand short-term profit. Investment in long term strategic development is therefore not "encouraged". This is in contrast to competitors in countries such as Germany and Japan who can obtain capital from sources who are willing to invest for the long term. Polices of the current UK government do not encourage that type of investment that is required to underpin initiatives such as CIM.

Despite well documented benefits of CIM it is true to say that any major applications of CIM in the UK are limited to large manufacturing companies. Probably the major technological contributors for manufacturing business improvement has been the advent of CAD, CAPP and CAM. Many companies in the UK will typically have invested in one or more of these areas. However, the number of companies known to have truly integrated these business functions within an all embracing Management Information System is very limited. More over such integration is restricted to large multi-national companies who have some scope for investment. Investigations have shown that the approaches to CIM implementation vary from an ad-hoc basis of applying individual systems over time that are often incompatible with other computerised systems to full CIM implementation with well phased project phases (Hargreaves 1991, Goldhar and Sclhie 1991).

Often the specification and installation of a CIM system is carried out by consultant companies that specialise in CIM implantation. A major reason for this and one which has been observed by the authors within several small and large UK companies is the low level of skill or competence associated with the many issues of CIM. This could be a reflection on the importance that many companies pay to CIM since many operate relatively successfully with their existing manufacturing systems and see little benefit in CIM investment. Another factor is the relatively low number of educated people entering manufacturing and an even smaller proportion then specialising in CIM oriented degree courses (Institution of Electrical Engineers 1992, Smithers and Robinson 1991).

There is no doubt that to maintain world class manufacturing then UK companies must be committed to a high level of work force education and training. Again this realisation appears to have prompted larger companies such as Ford, Peugeot, Jaguar, ICL, Mars etc to carry out extensive training programmes which usually deal with general operational philosophies such as SPC and TQM. Little time and effort is devoted to CIM and possibly this is a reflection of the failing of higher management to understand the true scope of what CIM could achieve for the business.

To really succeed, improvement technologies must be established throughout the supply chain. The ideal would be to establish a full CIM infrastructure involving small and large sized companies. This sort of approach sometimes termed "partnership" has been successful with the application of Just-In-Time systems between UK vendors and customer companies. This has shown that improvement in operations can be achieved but time and commitment are required to be successful.

Similar commitment to the development of CIM in UK manufacturing industry must be expended. The outcome would be a leaner manufacturing base comprising companies that have high productivity achieved through the use of integrated computerised methods and systems. A characteristic of CIM oriented companies will be a reduction in the number of people required to effect business operations. This will have significant impact on a society which has relied heavily on manufacturing for employment and wealth creation. The adoption of CIM must be viewed as a major strategy for UK manufacturing industry. It is unlikely that smaller companies can achieve "full" CIM but they must be guided as to what is a realistic strategy for their business. This drive for increased use of CIM within the UK will only be possible however if UK government policies and company ownership are changed to support long term business planning.

4. Discussion

Several characteristics of manufacturing have become apparent from the study of the three regions presented in this paper. Features of global manufacturing include:

Government support - This is perhaps the most significant factor effecting national growth. Foreign and local investments in manufacturing and their future development need to be supported by appropriate incentives and legislation at a national level. This is very apparent when the relative development of manufacturing in countries such as the United Kingdom and Singapore are compared.

Regionalisation - Governments which support their manufacturing enterprise must also look beyond to neighbouring countries. It is increasingly evident that a nation can form strong regional relationships without sacrificing its individual success and future development. It is also apparent that cultural and economic differences between countries need not be a barrier. Long term development, in a local and regional arena could take advantage of regional cross fertilization whilst building on a strong home manufacturing base. The technological development of South East Asia provides a base for its constituent countries to benefit from each others advancements in manufacturing. Singapore is the antithesis of Australia, a very small country with little natural resources in a strategically central geographic location; compared to a vast, relatively isolated country with abundant natural resources and a population mainly distributed around its perimeter.

Training and education at all levels - The importance of education and training for the workforce is again apparent from a comparison of the UK and South East Asia. The recent success of the South East Asian economy is founded on the early recognition of the potential benefits of a well trained workforce with the versatility to adapt to advances is technology.

Integration of CIM throughout the manufacturing enterprise - It is apparent that one of the biggest advantages in modern technology brought about by CIM and its components is the flexibility it brings to the manufacturing environment. It is essential that manufacturing industry is receptive to change and adaptable in a dynamic global arena. Integration of modern technology should not be adopted uniformly throughout an organisation but must take into account and be sensitive to the appropriate characteristics of the manufacturing enterprise, the features of the product and future development so that that the most suitable manufacturing strategy can be developed.

Quality assurance as a platform for TQM - The global market is becoming increasingly competitive and so manufacturers must ensure the "value" of their goods. Quality of goods and services has become a prerequisite for a manufacturing enterprise to compete at the global level. International standards such as ISO 9000 are important for global recognition, but will have little effect on the quality output of a company unless they are adopted for the right reasons. This has not only been recognised as a problem in South East Asia, where many companies are under pressure to adopt quality standards by overseas customers, but also in the UK (and Europe) where the potential of many quality assurance systems have not been fully realised (Barker 1994). The quality principles intrinsic in these standards will not be evident in a company unless the objectives of quality assurance are fully appreciated. Only then can a quality assurance system be used as a base to implement a total quality management philosophy.

5. Conclusion

The successful advancement of manufacturing companies into the twenty-first century requires a customised integration of AMT and CIM structured on an established foundation of quality and value added services and products. Despite cultural and political differences many of the problems facing manufacturing are similar throughout the regions studied in this paper. Most of the problems are generic and need to be offset and supported by proactive government policies. It is increasingly apparent that the principals of TQM, which are founded on management commitment, should be extended beyond the manufacturing enterprise to the national level so that governments are totally committed to supporting and developing their manufacturing resources in the local and international arena.

6. References

Barker, B., *Quality Promotion in Europe* Gower, UK 1994
Chua Stephen, ISO 9000 Certification in Singapore, *Quality Forum*, 1992
Drucker, P.F., *Managing for the Future*, Truman Tally Books / Dutton USA 1992
Drucker P.F., *Post Capitalist Society*, Butterworth Heinmann, Oxford 1993
DuBois, F.L and Oloff, M.D., "International Manufacturing Configuration and Competitive Priorities" in *Manufacturing Strategies Process and Content*, Voss C.A. ed., pp 239-257 Chapman and Hall, London, 1992
Goldhar, J.D. and Sclhie, T.W., "Computer Technology and International Competition - Part 2" *Integrated Manufacturing Systems*, Vol 2, No 22-30, 1991
Grigg, R.G. "Critical Manuafcturing Engineering Issues to be addressed by the Australian Industry to be Internationally Competitive", *Proc Aust. Conf. on Manufacturing*, Adelade, Nov 1993, Insititution of Engineers Australia, pp 7-16
Hargreaves, J., "CIM - A People Centres Approach" *Sandvik Coromant Internal Publication*, December 1991
Il-woon Kim and Jin H Im, "Manufacturing Environemtns in Korea, Japan And the United States" *Asia Pacific Journal of Management*, April 1993
"UK Manufacturing - A Survey of Surveys and a Compendium of Remedies, Public Affairs Board", *The Institution of Electrical Engineers* (IEE), May 1992
Mody, A. Suri R, Saunders J., and Tatikonda, M, "International Competition in Printed Circuit Board Asembly: Keeping Pace with Technological Change", *World Bank Industry and Energy Department Working Paper* No 53, December 1991
Natarajan, S. and Tan, J.M., *The impact of MNC Investments in Malaysia, Singapore and Thailand*, Institute of Southeast Asian Studies, Singapore 1993
Porter, M.E., *The Competitive Anvantage of Nations*, Free Press 1990
Quantz, P.A. (ed), *"Computer Aided Design in Manufacturing: The Role of CAD in Australia's Factories with a Future"*, University of Sidney, Sidney, 1984
Singapore Investment News, Singapore Economic Development Board, May 1993
Small, B.W. and Baines, E.N. "Australian Manufacturing Industry: The Next Decade", *Proc AUTOMACH Australia, Melbourne*, Jul 1985, SME, Dearborn, USA, pp 1-10 to 1-22
Smithers, A. and Robinson, P., "Beyond Compulsory Schooling - A Numerical Picture" *Council for Industry and Higher Education*, November, 1991
Straights Times, Singapore, October 1, 1993
Williams D.J., "The Impact of Quality and Continuous Improvement Phisophies on the Manufacture of Electronics Products", *Proceedings of the Second International Conference on Computer Integrated Manufactruing* September 1993 Singapore
Yates, I., "Improving Britain's Industrial Performance" *Employment Institution Conference*, July 1990

EVALUATION OF ADVANCED MANUFACTURING TECHNOLOGY

THE MODES OF THINKING

INGVAR PERSSON
Dept of Industrial Engineering
Lund University
Box 118, 22100 Lund
Sweden
E-mail: Ingvar.Persson@ie.lth.se

ABSTRACT

Since the introduction of AMT there has been an obvious need for increased knowledge about the implications of investments in the area. Several articles have been published which suggest different concepts, models and techniques. The suggestions in the articles are based on the researcher's mode of thinking. So far, however, no one has tried to structure the suggested models and techniques based on the researcher's underlying mode of thinking and the manager's decision making. This article has that mission. The modes of thinking are here classified as an analytic and a system theoretical mode of thinking.

The basic difference between the appraisal of an AMT investment in a system theoretical and an analytic mode of thinking is analysed. Analytical models are used within both modes of thinking. The interpretation and use of a model, however, is different in the different modes. In the analytic mode, the models are developed for normative use. Within the system theoretical mode the models are used in order to increase the manager's understanding of the situation.

Key words: Investment Process; Appraisal; Advanced Manufacturing Technology

1. Introduction

During the last decade, many articles have been published to help managers evaluate projects in advanced manufacturing technology (AMT). By 1986 Canada already had a list of 113 bibliographies on AMT with short comments. Since that time, the number of articles within the area has increased. Obviously the evaluation of AMT was found to be more difficult than before when evaluating investments in manufacturing.

When the number of articles increases there is a need to structure the knowledge. Suresh and Meredith (1985), and Meredith and Hill (1987) have structured the different justification approaches in a "continuum", from economic to strategic justification. Partly based on the structure above, Swann and O´Keefe (1990:1, 1990:2) go further and structure the data needed. In a review, Proctor & Canada (1992) begin with the traditional financially oriented techniques, analyse the criticism of the capital budgeting process and continue with the less tangible attributes. At the top level, this means strategy linkage considerations and considerations to influencing the ability to compete. Articles emphasising quality and flexibility are also explicitly mentioned. Finally more advanced accounting based measures, as well as, the use of non-traditional manufacturing evaluation techniques are presented.

All of the articles are based on the researcher's assumptions on the manager's decision making process, though it is usually not mentioned. Is the manager's decision making process a rational process or is it an irrational process (Brunsson, 1985). Strategic decision making is an ill-structured task and the manager's diagnosis is an opinion (Mintzberg et al, 1976; Yadav and Khazanchi, 1992). In a study of top executive decision making, Donaldsson and Lorsch (1983) concluded that among the corporate managers there existed distinctive systems of beliefs. " .. these systems of distinctive beliefs act as filter, through which management perceives the realities facing the firm"(p280). Actually, the manager's decision making is based on the manager's interpretation of the situation.

So far, however, no one has tried to structure the suggested models and techniques based on the researcher's underlying mode of thinking on the manager's decision making process. This article has that mission and the modes of thinking are here classified as an analytic and a system theoretical mode of thinking.

2. The analytic and system theoretical mode of thinking on reality

The analytic approach is based on the assumption of an objective reality and a rational decision maker. The preconditions is the existing theory which is possible to verify and falsify by hypothesis. The explanation is focused on cause-effect relationships and the result is logic models and cases which are representative. The result from one study can be used by anyone else in other situations, if the rules are followed. The knowledge is independent of the individuals.

The system theoretical approach is based on system theory. The reality in the system theoretical approach is arranged in such a way that the sum of the parts is different from the whole. Depending on the relationships, the whole can be more or less than the sum of the parts. These effects are called *synergism.*. Strategic decision making is a complex process (Mintzberg,1989; Yadav and Khazanchi, 1992). In research, focus is placed on the relationship between the parts in the system, explaining the driving forces. However it is not a cause-effect relationship. The result from system theoretical research, based on the synergism effects, is merely a mechanism for classifying and creating taxonomies. The knowledge is dependant on the system, and the exchange of knowledge between different systems is made by analogies. Cases can be unique.

3. Appraisal of the AMT-investment from an analytic approach

An analytic approach means researching the cause-effect relationships which AMT has or can have(and which was not known earlier), explaining the difference between the present and past when using the traditional models, and to developing other models which can be used when evaluating in AMT.

Much of the research is focused on the so called intangibles but the suggested solutions are different. One group of articles is based on the traditional capital budgeting techniques (CBT). These techniques transform cash-flow to a profitability measure, taking the time value of money into account. One pre-condition is an estimated cash flow.

Primrose and Leonard (1985, 1987) and Primrose (1991, 1992, 1993) focus on estimating "intangible benefits". In their mode of thinking, there are no intangibles. Since their research (1985), "no one has described a benefit that could not be re-defined into quantifiable terms"(p49, 1991). It is important to quantify every benefit since it overcomes the main excuse managers have for making investments without evaluation.

Everything that can be identified has to be quantified. "No benefits should ever be excluded on the grounds that it is intangible". To make the estimations possible Primrose(1991) provides the managers with a detailed description of parameters that have to be taken into account when making investments in CNC, FMS, MRP, CAD/CAM, robots and CIM. Troxler (1990) suggests the use of cost functions.

Flexibility change, quality change, and change in lead time and inventory are all effects which have been of interest in research. Primrose(1991) does not use the flexibility concept in the evaluation. Instead, flexibility has to be transformed into a parameter which can be measured and quantified in money, for instance as set-up time. Set-up time is, in Azzone and Bertele (1989), used to measure process flexibility. Other flexibility measures used in Azzone and Bertele would probably not be accepted by Primrose, for instance production flexibility, which is "the probability that the manufacturing system will be able to process a new product"(p738)

Azzone and Bertele (1989) focus on flexibility and the firm's strategic position. "The only way to define the most suitable manufacturing system for the firm's strategic position is (to chose) the most profitable system." which "depends not only on the starting mix, as theorists of the economic approach suggest, but also on its dynamics, i.e. on the firm's strategic position"(p736). The strategic position is defined by four numerical values (response time, level of service, probability of new products introduction and probability of technological compatibility). To apply the model, an additional six measures concerning flexibility must be known, apart from costs and revenues which are used in the "traditional economic approach".

In Park and Son (1988), the objective is "to expand current capital budgeting procedures to consider improvements in productivity, quality and flexibility that would result from implementing new manufacturing technology"(p2). Park and Son define current capital budgeting procedures as being based on a "conventional accounting income and expense classification"(p3) which only includes revenues and expenses from a traditional cost accounting system. They define "new" costs as quality costs (preventive maintenance, failure, inspection), and set-up costs; they change the content if inventory costs from work in process and finished goods to holding and backorder and they define opportunity costs. A model of a multi-period production performance includes a linear program based on the NPV model.

Instead of quantifying the effects in money many of the articles published suggest scoring. Scoring means that relevant attributes are selected and the project is scored against each attribute. Each attribute is given a weighted factor. The score and the weighted factor are then multiplied for each attribute and the desirability of a project is calculated as the sum of the weighted scores.

The analytic hierarchy process (AHP) is a scoring method where the process is made in a hierarchy and the weighted factors are developed through the method (Saaty, 1980). The basic assumption is, that reality can be decomposed, analysed and evaluated. The whole is then calculated from the parts.

The AHP model is applicable at different levels. In O'Brien and Smith(1993), the overall strategy is decomposed in several steps. Strategic and tactical considerations are incorporated in a multi-level model. "The model has been constructed so that companies can adopt a top down approach in entering strategies as currently determined, and then at the bottom level the degree to which particular investments match the sub criteria they are being compared against can be entered. The model will then provide a score that indicates how well any particular investment matches the company's overall strategic criteria."(p312)

In Ghosh et al (1991), the AHP model is preferred at a lower level.

"The basic factors to the production system are more realistically estimated than strategic factors such as financial position, competitive position or sales increase. Any attempt to place a score on these items is ill-advised and mere guess-work. There are too many variables which relate to these potential benefits and many are uncontrollable."(p5). The chosen levels were, for instance, workforce, methods and material, equipment, etc.

Capital investment in AMT from a system theoretical approach

In the system theoretical approach the problem is to find out how the AMT investment will change the system and what will influence this change. The new machine or machine system is a new sub-system which will influence the whole system. The system which will be created does not only depend on the AMT investment itself. It also depends on the relationships that will be established between AMT and other parts of the system, relationships which are not created automatically. Hence, it is the manager's responsibility to engage himself in the necessary organisational adoptions in order to achieve the full benefits from AMT (Boer et al ,1990). Moreover, these relationships can change from time to time. The changes are dynamic and very much dependent on other parts of the system (Leonard-Barton & Kraus, 1985). Employees and managers who can directly or indirectly influence the use of the AMT will form the role of AMT in the system. Some researchers suggest a champion as a pre-requisition for success (Ward, 1990)

If the analytic perspective deals mostly with how a manager should make his evaluation, then the focus in the system theoretical based research is to understand what the manager wants to know in order to acquire and manage AMT, and which criteria are important in the successful implementations.

A company is an open system and the effects on competition in the long run can be affected. Hence the strategic implications are important in many system theoretical based articles. Changing market and competition is focused on in Goldhar and Jelinek (1985) . Increased internationalisation, from mass production to tailor-made or customised products and services, shorter and more truncated product life cycles, and increasing complexity in the products are putting new demands on manufacturing.

Slagmulder and Bruggeman (1992) studied nine AMT projects in six companies. In each project the characteristics of the investment decision making process were analysed from strategic, technical, financial, and organisational aspects and the perceived impact on business unit performance was analysed. One conclusion from the study was that the quality of the strategic analysis was the main critical success factor. Financial justification and use of CBT were of secondary importance. If necessary the data were manipulated, that is, the financial justification was an obstacle to overcome (Jansson, 1992). If attention to technical and organisational aspects was not satisfactory, then it created problems in the implementation phase. A champion however was not a sufficient condition for success.

According to Grant et al (1991), the primary objective of strategic investments should be to achieve sustainable competitive advantage. In a dynamic environment this will require continuos innovation. The choice of manufacturing technology hence is not only concerned with static optimisation but also with dynamic optimisation. Dynamic considerations mean that more advanced process technology is preferred. However, learning capabilities is a limitation. If the size of the technological step is too big, the learning and implementation costs will increase too much. "Technological increments improve performance up to the limit of the firm's learning capacity"(p52). Thus, an incremental approach is to be preferred instead of a single technological leap.

Platform investments as "options on the future" are discussed by Kogut and Kulatilaka (1994). The perspective is dynamic. The level is above the strategic business unit since "SBU frequently leads to an under investment in projects with long-term growth"(p58). However the value of a platform investment depends on four conditions: uncertainty, opportunity, time dependence and discretion. To benefit from a platform investment there has to be an information system that helps managers to understand actual opportunities and a management incentive system that forces them to take actions. The most important platform investments are organisational capabilities. FMS is a physical asset that can be bought on the market. "The more important platform value is the development of the capability to know how to run a plant flexible, and then know how to expand this organising heuristic to other operations"(p65). Here we notice the correspondence to the learning capabilities in Grant et al.(Ibid).

Perssons (1990, 1994) conceptual models focus on the dynamic of change, and the relationship between the investment of today and the investments and management of the future. The AMT investment is seen as a tool for managers in remaining competitive. The investment will create both options and pre-options. Pre-options are opportunities in the future which are unknown today but can be expected in a dynamic environment. The suggestion is that the evaluation is made both in a static and in a dynamic mode. The static mode is the expectations which are based on the actual situation. In the dynamic mode some effects can not be foreseen. Instead, attention is placed on the investment as a manager's business tool when managing the dynamic of change.

4.1 Appraisal in the system theoretical approach

In the articles representing the system theoretical mode of thinking above, the tools for practical use are delimited. Strategic analysis, business tool, learning capabilities, options and pre-options are all valuable concepts but how do they make sense when evaluating a project?

In the system theoretical mode of thinking an investment in AMT is not an isolated decision. It is one of many decisions in an on-going business. The investment decision links not only backwards but also forwards, influencing forthcoming decisions in different ways. The approach is dynamic and the effects from an AMT investment are complex. So complex that it is difficult to catch the effects in one general model. Models and quantitative measures can, however, be of value within the system theoretical mode of thinking. In a study on decision making Miller (1956, p91) found that an individual could usually "distinguish about seven categories and that there is a span of attention that will encompass about six object on glance". He also "..suspect that there is ...a span of perceptual dimensionality and that this span is somewhere in the neighbourhood of ten". Applied to the appraisal of AMT investments, the manager's ability to make better judgements increases with an increased number of criteria, up to an average "seven plus minus two". Based on these findings, it is difficult to understand why so much effort is made to reduce the number of criteria to one.

Before going further, the researcher's and manager's roles will be briefly discussed. The manager's role is to make the decision. The most important question is what the manager wants to know before the decision is made. The hypothesis is that he wants to understand the situation - understand the actual situation in the system, understand the implications of the AMT investment and thus the main difference between the system without and with the AMT investment. This understanding is not "objective". It is reality perceived by the managers "subjective lens" (Yadav and Khazanchi, 1992). As Slagmulder and Bruggeman (1992) found, the manager's strategic decision making in a complex environment is not primarily based on evaluation from a cash-flow model.

The researcher's role is to help managers gain an understanding which is based on relevant perspective and has as much "objective facts" as possible. At the system level, researchers can guide those models that would be of value in special situations. I.e. Meredith and Hill (1987) made guide-lines based on a system level approach. Four levels of integration of AMT were suggested: stand alone, cells, linked islands and full integration. Justification techniques were classified as economic, portfolio, analytic and strategic. The relevance of the techniques for each level was then discussed in the article.

A model does not tell the truth but has strengths and weaknesses. At the model level researchers have to describe what the model really does, and what its weaknesses and delimitation's are. For instance, the CBT does nothing but transform cash-flow to a profitability measure. In practice the most difficult is to estimate cash-flow.

Estimations are not facts. In evaluation models we have at least two dimensions of estimation. One dimension is the actual change which is more or less an estimation of what the actual situation would have been if the AMT investment was on hand. The other dimension is an estimation of what will happen during the life-time (Persson, Ibid.). In practice these two dimensions should be separated but are usually not.

Thus several models and criteria can be of use in a special case where each model increases manager's understanding. It is the researcher's role to provide the models, the models constraints, possible contributions, strengths and weaknesses. It is the managers responsibility to find out which models are to be used in the actual case.

5. Articles and the researcher's mode of thinking.

Usually it is possible to notice whether the researcher has a system theoretical or an analytic mode of thinking. The suggestions or conclusions are unmasking. The analytical researcher prefers numeric models. The system oriented researcher is more interested in the main relationships that can influence the outcome.

Analytical models are used within the system theoretical mode but the researcher may emphasise other aspects when using the model or prefer other models. Concerning traditional CBT technologies, the system theoretical researcher suggests sensitivity analysis in order to increase understanding.

The system theoretical researcher would prefer profile scheme before scoring. If the guide-lines in scoring are quite clear, the system theoretical researcher would suggest that scoring is only a roundabout. If the guide-lines are not clear the value of scoring is low or non-existent. There probably is an agreement between the schools concerning the use of risk calculation to judge technological system reliability and natural phenomenon. Probability analysis on parameters that are influenced by management's actions, however, does not seem very attractive to a system theoretical researcher. In this case, the use of these techniques is not neutral; they are worse than nothing. Prices and volumes for instance are parameters which often, directly or indirectly, can be influenced by management actions. The use of risk calculation may then decrease the manager's experience of responsibility. In the system theoretical mode of thinking, actions to reduce uncertainty are preferred above risk calculation.

In studying the articles within the field, some articles are ambivalent. Many articles begin with a system theoretical mode of thinking but end with analytic tools

6. Summary and Conclusion

There is a basic difference between the appraisal of AMT investment in a system theoretical and analytic mode of thinking. The system theoretical mode focuses on the expected influences on business in the management of change. The goal is the manager's understanding of the implication, and the demand for management when the investment is implemented. In the analytic mode, the issue is to isolate and calculate the assumed effects of the AMT investment.

Analytic models are used within both modes of thinking. The interpretation and use of a model however is different in the different modes. In the analytic mode, the models are developed for normative use. Within the system theoretical mode the models are used in order to increase the manager's understanding of the situation. The difference between the modes implies preferences for different models. For instance profile scheme is preferred in a system theoretical mode of thinking while score methods are preferred in the analytic. In the analytic mode of thinking, the CBT techniques create the answer to whether the investment is profitable or not; in the system theoretical mode, the CBT techniques are one of several heuristics which create a pattern. In the system mode sensitivity analysis may be used with the goal to increase managers understanding, which in turn can affect the content of the investment, or influence manager's further on when managing the investment.

7. References

G. Azzone & U. Bertele, Measuring the Economic Effectiveness of Flexible Automation: A New Approach, *Int. Journal of Production Research*, no5, 1989, pp735-746.

H. Boer & W.E. During, Management of Process Innovation - the Case of FMS: A Systems Approach, *Int. Journal of Production Research*, no11, 1987, pp1671-1682.

N. Brunsson, *The Irrational Organization,* John Wiley & Sons, 1985.

J.R. Canada, Non Traditional Method for Evaluation CIM Opportunities Assigns Weights to Intangibles, *Industrial Engineering,* March, 1986, pp 66-71.

G. Donaldsson & J.W. Lorsch, *Decision Making at the Top: The Shaping of Strategic Directions,* Basic Books, 1983.

B.K. Ghosh & R.N. Wabalickis, A Comparative Analysis for the Justification of Future Manufacturing Systems, *Int. Journal of Operations & Prod. Mgmnt,* no9, 1991, pp4-23.

J.D. Goldhar & M. Jalenik, Computer Integrated Flexible Manufacturing Organizational Economic and Strategic Implications, *Interfaces,* no3, 1985, pp94-105.

R.M. Grant, R. Krishnan, A.B.Shani & R. Baet, Appropriate Manufacturing Technology: A strategic Approach, *Sloan Management Review,* Fall, 1991, pp43-54.

D. Jansson, *Spelet kring investeringskalkyler,* Norstedts, 1992.

B.Kogut & N.Kulatilaka, Options Thinking and Platform Investments: Investing in Opportunity, *California Management Review,* no2,1994,pp52-71.

D. Leonard-Barton & W.A.Kraus, Implementing New Technology, *Harvard Business Review,* Nov-dec, 1985, pp104.

J.R. Meredith & M.M.Hill, Justifying New Manufacturing Systems: A Managerial Approach, *Sloan Management Review,* Summer, 1987, pp46-61.

Miller, The Magical Number Seven Plus Minus Two: Some Limits on our Capacity for Processing Information, *Psychological Review,* no2, 1956, pp81-97.

H. Mintzberg , *Mintzberg on Management,* The Free Press, MacMillan, 1989.

H. Mintzberg, D.Raisinghani & , The Structure of Unstructured Decision Processes, *Administrative Science Quarterly,* 1976, pp246-275.

C.O'Brien & S.J.E.Smith, Design of the Decision Process for Strategic Investment in Advanced Manufacturing System, *Int. Journal of Production Economics,* 1993, pp309-322.

C.S.Park & Y.K.Son, An Economic Evaluation Model for Advanced Manufacturing Systems, *The Engineering Economist,* no1, 1988, pp 1-25.

I. Persson, Analysis of Capital Investments - A Conceptual Model. *Engineering Costs and Production Economics* No3 , 1990),pp277-284.

I. Persson, *The Capital Investment Process and New Technology,* WP 1994:60, Institute for Management of Innovation and Technology, 1994.

P.L.Primrose, *Investment in Manufacturing Technology,* Chapman & Hall, 1991.

P.L.Primrose, The Investment Principle, *Manufacturing Eng,* april, 1993, pp63-65.

P.L.Primrose, Is Anything Really Wrong with Cost Mgmnt, *Cost Mgmnt,* Spring, 1992, pp48-57.

P.L.Primrose & R.Leonard, The Use of a Conceptual Model to Evaluate Financially Flexible Manufacturing System Projects, *Proc Inst Mechanical Eng,* No b1, 1985, pp15-21.

P.L.Primrose & R.Leonard, The Financial Evaluation and Economic Application of Advanced Manufacturing Technology, *Proc Inst Mechanical Eng,* March, 1986, pp27-31.

M.D.Proctor & J.R.Canada, Past and Present Methods of Manufacturing Investment Evaluation: A review of the Empirical and Theoretical Literature, *The Engineering Economist,* no1, 1992 pp45-58.

T.L.Saaty, *The Analytical Hierarchy Process,* McGraw Hill, 1980.

R.Slagmulder and W.Bruggeman, Investment Justification of Flexible Manufacturing Technologies: Inferences from Field Research, *International Journal of Operations & Prod. Mgmnt,*no7/8,1992, pp168-186.

N.C.Suresh & J.R.Meredith, Justifying Multimachine Systems: An Integrated Strategic Approach, *Journal of Manufacturing Systems,* no2, 1985, pp117-134.

K. Swann & W.D.O'Keefe, Advanced Manufacturing Technology: Investment Decision Process. Part 1, *Management Decision,* no1, 1990, pp20-31.

K. Swann & W.D.O'Keefe, Advanced Manufacturing Technology: Investment Decision Process. Part 2, *Management Decision,* no3, 1990, pp27-34.

J.W.Troxler, Estimating the Cost Impact of Flexible Manufacturing, *Journal of Cost Management for the Manufacturing Industry,* Summer, 1990, pp26-32.

T.L.Ward, Role of Champions in Justification of Computer Integrated Manufacturing *Systems,* In H.R. Parsai, T.L.Ward and W. Karowski (eds.) *Planning, Design, Justification and Costing* (Elseiver Science Publishers B.V., 1990) pp123-131.

Yadav, S.B.& D.Kahazanchi, Subjective Understanding in Strategic Decision Making - an Information System Perspective, *Decision Support Systems,* no8, 1992, pp55-71.

Concurrent Engineering

Organizational Growth Demands Concurrent Engineering
 N. K. H. Tang and P. L. Forrester

Developing an Integrated Workgroup Environment for CIM Enterprise
 P. W. Lek, C. C. Wong, R. K. L. Gay and R. Y. G. Lim

Organizational Growth Demands Concurrent Engineering

Nelson K H Tang
Management Centre, University of Leicester
Leicester, LE1 7RH, UK
Email:NKHT1@LEICESTER.AC.UK

Paul L Forrester
Department of Management, Keele University
Staffordshire, ST5 5BG, UK
Email: MNA03@KEELE.AC.UK

ABSTRACT

During the past few years a considerable amount of research knowledge regarding Organizational Change and Concurrent Engineering (CE) has been accumulated and this paper suggests that organizational growth provides an environment for the emergence of CE techniques. The paper discusses the view that organizational growth demands CE, the barriers to the implementation of CE and it considers how an organizational structure can best enhance the implementation of CE principles.

1. Introduction

A toy manufacturer retains an organizational structure long after it has served its purpose because management power is derived from its structure. The company eventually goes into bankruptcy. A printing company chooses an organizational structure that is too complicated and formalized for its state of development and limited size. It flounders due to rigidity and bureaucracy for several years and is finally acquired by a larger company. A manufacturing company is dissatisfied with a project manager who is blamed for the lack of progress of systems development, when the underlying cause is really the organizational culture (vision, values, beliefs and goals) that is holding back the development project.

Many of the problems encountered in these actual company case studies are rooted in past decisions rather than in present events or outside market dynamics. Historical forces do indeed shape the future growth of organizations. The management desire for rapid growth of a company often overlooks such critical development questions as: Where have their organizations been? Where are they now? What do the answers to these questions mean for the future? Companies may fail to see that many clues to their future success already lie within their own organizations and their evolving states of development.

2. Phases of Growth

Greiner[1] claims that growing organizations move through five distinguishable phases of development, each of which contains a relatively calm period of growth that ends with a management crisis. Figure 1 shows these five phases of growth; each **evolutionary period** is characterized by the **dominant management style** used to achieve growth, while each **revolutionary period** is characterized by the **dominant management problem** that must be solved before growth can continue. The term "evolution" is used to describe prolonged periods of growth where no major upheaval occurs in organizational practices. The term "revolution" is used to describe those periods of substantial chaos in the life of an organization.

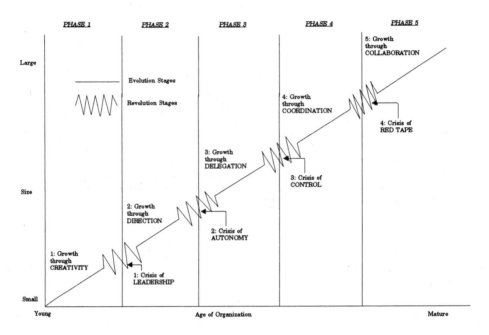

Figure 1 The Five Phases of Growth. (Adopted from Greiner[1])

The pattern presented in Figure 1 seems to be typical for companies in industries with moderate growth over a long time period; companies in faster growing industries tend to experience all five phases more rapidly, while those in slower growing industries encounter only two or three phases over many years.

It is important to note that each phase is both the outcome of the previous phase and a cause for the next phase. For example, the evolution management style in Phase 3 of Figure 1 is "delegation"; this grows out of, and becomes the solution to, demands for greater "autonomy" in the preceding Phase 2 revolution. The principal implication of each phase is that management actions are narrowly prescribed if growth is to occur. For example, a company experiencing an autonomy crisis in Phase 2 will find it difficult to return to directive management for a solution - it needs to adopt a new style of delegation in order to move ahead.

2.1 Phase 1: Creativity To Direction

In the birth stage of an organization, the emphasis is on creating both a product and a market. The characteristics of the period of creative evolution are usually technically or entreprepreneurially oriented; physical and mental energies are absorbed entirely in selling a new product or winning orders. Communication among employees is frequent and informal, and control of activities comes from immediate marketplace feedback; the management acts as the customers react.

All of the forgoing individualistic and creative activities are essential for the company to get off the ground. But therein lies the problem. As the company grows, larger production runs require greater knowledge about manufacturing efficiency than before. An increased number of employees cannot be managed exclusively through informal communication; new employees may not be motivated by the same intense dedication to the product or organization as the pioneers.

At this point a crisis of leadership occurs; this is the onset of the first revolution. Who is to lead the company out of the confusion and solve the managerial problems confronting it? Quite obviously, a strong manager is needed who has the necessary knowledge and skill to introduce new business techniques. But this is easier said than done. The founders of a company often hate to step aside, even though they are probably temperamentally unsuited to be managers. So the first critical developmental choice occurs - to locate and install a strong business manager who is acceptable to the founders and who can pull the organization together.

2.2 Phase 2 : Direction To Delegation

Those companies that survive the first phase by installing a capable business manager usually embark on a period of sustained growth under able and directive leadership. The characteristics of this evolutionary stage are that a functional organizational structure is introduced and job assignments become more specialized. Communication becomes more formal and impersonal as a hierarchy of titles and positions develops. The new manager and his/her "empire" takes most of the responsibility for instituting direction, while middle managers and low-level supervisors are treated more as functional specialists than as autonomous decision-making managers.

Although the new directive techniques channel employee energy more efficiently into growth, eventually they become inappropriate for controlling a larger, more diverse and complex organization. Lower-level employees find themselves restricted by a cumbersome and centralized hierarchy.

A second revolution is imminent as a crisis develops following demands for greater autonomy on the part of lower-level managers. The solution adopted by most companies is to move toward greater delegation, yet it is difficult for top managers who were previously successful at being directive to give up such responsibility. Moreover, lower-level managers are not accustomed to making decisions for themselves. As a result, numerous companies flounder during this revolutionary period, adhering to centralized methods while lower-level employees grow more disappointed and leave the organization.

2.3 Phase 3 : Delegation To Coordination

The next era of growth evolves from the successful application of a decentralized organizational structure. This type of structure gives much greater responsibility to lower-level managers with profit centres and bonuses being used to stimulate motivation. The top executives at headquarters restrain themselves to managing by exception, based on periodic reports from the field. Management often concentrates on making new acquisitions that can be operated alongside other decentralized units.

The delegation stage proves useful for gaining expansion through heightened motivation at lower levels, but a serious problem eventually evolves as top executives sense that they are losing control over a highly diversified field operation. Autonomous field managers prefer to run their own activities without needing to coordinate plans, money, technology and personnel with the rest of the organization.

The Phase 3 revolution is under way when top management seeks to regain control over the total company. Some top managements attempt to return to centralized management; this usually fails because of the vast scope of operations, and most companies that do move ahead find a new solution in the use of special coordination techniques.

2.4 Phase 4 : Coordination to Collaboration

During this phase, the evolutionary period is characterized by the use of formal systems for achieving greater coordination and by the top management team taking responsibility for the initiation and administration of these new systems. Typical characteristics are when decentralized units are merged into product groups (where each group is treated as a profit centre) and where return on invested capital is an important criterion used in allocating funds. Formal planning procedures are established and intensively reviewed and certain technical functions, such as data processing are centralized at headquarters, while at the same time the daily operating decisions remain decentralized.

All these new coordination systems prove useful for achieving growth through more efficient allocation of the limited resources of a company and they prompt field managers to look beyond the needs of their local units. While these managers still have much decision-making responsibility, they learn to justify their actions more carefully to a "watchdog" at headquarters.

But, in time, a lack of confidence gradually develops between line and staff, and between headquarters and the field. Line managers, for example, increasingly resent heavy staff direction from those who are not familiar with local conditions. Staff people, on the other hand, complain about uncooperative and uninformed line managers. Together, both groups criticize the bureaucratic paper system that has evolved. Working as a profit centre may lead to a temptation to reduce investment in operations to maximize short term gain. In short, the organization has become too large and complex to be managed through formal programmes and rigid systems. The Phase 4 revolution is under way.

2.5 Phase 5 : Collaboration To Concurrent Engineering

Phase 4 was managed more through formal systems and procedures; Phase 5 emphasizes greater spontaneity in management action through team work and the skilful confrontation of interpersonel differences. Social control and self-discipline take over from formal control. This transition is especially difficult for those line managers who created the old systems as well as for those line managers who relied on formal methods for answers.

The Phase 5 evolution develops around a more flexible and behavioural approach to management with the focus on solving problems quickly through team action. Teams are combined across functions for task-group activity and a matrix-type structure is frequently used to assemble the right teams for appropriate problems. Previous formal systems are simplified and combined into single multipurpose systems. Educational programmes are utilized to train managers in behavioural skills for achieving better teamwork and conflict resolution.

This background provides the environment for the emergence of Concurrent Engineering (CE). However, some important implications for managers to characterize each growth phase within their organizations will be described before we explore this new management style.

2.6 Implications of the Organization Growth Phases

Every organization and its component parts are at different stages of development. The task of top management is to be aware of these stages; otherwise, it may not recognize when the time for change has come, or it may act in such a way as to impose the wrong solution. In each revolutionary stage it becomes evident that this stage can be ended only by certain specific solutions; moreover, these solutions might be different from those which were applied to the problems of the preceding revolution. Too often it is tempting to choose solutions that have been tried before; this makes it impossible for a new phase of growth to evolve.

At times managers may fail to realize that organizational solutions create problems for the future (i.e. a decision to delegate eventually causes a problem of control). Historical actions are very much determinants of what happens to the company at a much later date. An awareness of this effect should help managers to evaluate company problems with greater historical understanding instead of "pinning the blame" on to a current development.

However, management that is aware of the problems ahead could well decide not to grow. Top managers may, for instance, prefer to retain the informal practices of a small company, knowing that this way of life is inherent in the limited size of the organization. If they choose to grow, they may do themselves out of a job and a way of life they enjoy.

And what about the managements of very large organizations? Can they find new solutions for continued phases of evolution? Or, alternatively, are they reaching a stage where a government will act to break them up because they are too large? Clearly, there is still much to learn about processes of development in organizations. The phases outlined by Greiner are only five in number and are still only approximations. A detailed discussion about the process of development in organizations may be found in other publications[2,3].

3. Concurrent Engineering

Concurrent Engineering (CE) has been called by many different names, including Simultaneous Engineering (SE)[4], Life-Cycle Engineering, Concurrent Product and Process Design, Design for Production, Design for Manufacture and Assembly (DFMA)[5], Integrated and Cooperative Design, Design Fusion, Producibility Engineering and System Engineering. Whatever it is called, CE is not a physical process or a set of procedures; it is a board level responsibility to integrate the teams into effective, cohesive business units.

CE has also been defined in another way by Winner et al [6]:
"Concurrent engineering is a systematic approach to the integrated, concurrent design of products and their related processes, including manufacture of support. This approach is intended to cause the developers, from the outset, to consider all elements of the product life cycle from concept through disposal, including quality, cost, schedule and user

requirements."

A definition by Rolls-Royce of simultaneous engineering[7] (which really defines the objective rather than the process) is:
"Simultaneous engineering attempts to optimise the design of the product and manufacturing process to achieve reduced lead times and improved quality and cost by the integration of design and manufacturing activities and by maximising parallelism in working practices."

These two definitions indicate that the lead time should be significantly reduced as a result of CE, because the design is not passed from group to group or from department to department. All disciplines required are members of the same team, reporting to the same management. At first sight, the changes needed to introduce CE seem to be intimidating, and suitable only for very large companies. Fortunately, CE lends itself to a gradual introduction, and it follows that small companies can pick up the elements they need in the same way as large organizations.

An organization cannot have an improved output if all it does is to bolt a new business philosophy onto the side of its existing organizational system. To implement CE, the first essential is the lead must be taken at board level. It should not be left to the director of product design to implement; once the plan is accepted, the chief executive should make it clear that he/she is the ultimate leader. Thus, his/her role is to push the plan forward and to ensure that it does not become the preserve of one department, such as product engineering, even though it is usual for a product engineer to lead the Task Force.

In some instances it may be sufficient to instruct product design, quality and manufacturing engineering sections that they must rethink how they develop products. However, it may be necessary to change also the structure of the organization. For example, Adam Opel[8] made one director responsible for both product design and production engineering, and then set up an Advanced Product Study Department consisting of both product designers and production engineers working together. An alteration such as this makes it easier for people to be ready for change.

The key to making CE happen is to change the mechanistic culture of efficiency, volume, cost-control and planning to value teamwork, improvement, market-competitiveness, entrepreneurialism and customer satisfaction. Organizations must change their cultures to encompass a long-term vision for CE. The next section relates some industrial applications of implementing CE[9] and discusses the barriers encountered.

4. Applications and Barriers

A minority of UK-owned companies like Rover, Lucas and JCB are already on the right track. They have flattened their organizations, devolved responsibility and eliminated wasteful processes in offices and factories. They are now developing new products much faster than before and introducing products that they believe the customer wants.

Rolls-Royce instituted a CE programme in May 1991, having seen how other business in the automotive industry were taking these principles on board. The Chief Engineer was responsible for implementing the move from development by sequential steps to a multi-disciplined team effort which required different departments to work in parallel. Roll-Royce launched the Silver Spirit model in 1988 and the design and manufacture of the car took seven years. "We want to reduce that", said Mr. Harding the Chief Engineer when speaking of new cars. "In our view, the concurrent principle equals a 20% minimum time saving and some people estimate an improvement in productivity of up to two." Harding admits that, at present, Rolls-Royce is not meeting European practice in product life cycles. He will be striving to meet Japanese standards of three years for the face-lift life cycle - the time taken for a model to be modified - which is between six and 12 months ahead of Western Europe.

Harding thinks that the greatest challenge to introducing the CE lies in transforming the company culture. It may be a challenge for people to free themselves to join in a team. Because they are starting out on a journey, people have to be convinced of the benefits at the outset; there are no tangible benefits to point to at that stage. Potentially more threatening to employees is the fact that the new macro-team style of development means that individuals will not be judged according to their status within a well-established hierarchy but, instead, according to their contribution to specific projects. He confessed that CE practices mean that many managers are overstressed and overworked. But the rewards for success will be considerable and, "By putting concurrent teams together we can incorporate market concerns about the choice of materials, as well as the manufacturing feasibility of that choice, way up front."

JCB, a UK manufacturer of earth moving equipment, was working on a major new product. It was not saying what it was or when it would be on the market, but it was expected that the product would be developed in record time, JCB was using CE. Even the first CE programme was reasonably haphazard because they were learning as they went along. But JCB has left little to chance. Senior managers have visited Japanese factories - where CE is more a way of life. Over the past few years the company has also been laying a solid foundation for the introduction of CE. The idea of team working is not new. JCB has used teams of people from different disciplines on previous development projects (but at the tail-end of those projects) - when major changes are more difficult to make and more costly. It has involved suppliers in new product work but, again, late in the day. On the new JCB development, the team - including people from design, engineering, manufacturing and major suppliers - will work together from product to final production.

Since CE results in quality being guaranteed by design, it fits in completely with the concept of Total Quality Management (TQM). Just as in TQM quality becomes everyone's business, so in CE the design is the concern of all departments. However, many of the UK manufacturers are disillusioned by quality programmes.

The failure of total quality programmes cannot be attributed to the ideas that forging closer working relationships and understanding between people in a company, as well as with suppliers and customers, or by fostering things like teamwork and team problem-solving are at fault. One of the major problems has been a lack of training to instil such techniques into the workforce. Another problem of the quality problem is organizational structure. Companies that are top heavy - not only in terms of numbers but in the amount of power the hierarchy wields, and that are organized functionally (again, each department protecting its own territory) - will fail. Herein also lies the barrier to successful implementation of CE and to UK manufacturing becoming world class.

UK manufacturing used to be restrained by restrictive labour practices. Now it may be management attitudes that are restrictive. Typically UK managers have been measured and rewarded for their ability to sweat their assets, both people and machines. Efficiency was the key. While this is important, it also can lead to companies efficiently manufacturing their way to bankruptcy. Managers who see inventories as assets and

100% machine utilization as a sign that all is well in the factory still exist.

Lucas Engineering & Systems, (which originally acted as consultant to Lucas operating companies but now offers its services worldwide,) has compared the performance of leading Japanese companies with those in the West. It found that, on average, product costs were 30% lower in Japan - despite higher labour costs, that new product lead times were 50% shorter - due to the use of CE - and that stock turns were three times as high - due to the use of just-in-time.

These gaps are large but not unbridgeable. If the average UK manufacturer can accept the fact that its performance is poor compared with the world's best, then it can begin to progress. Unfortunately, many find this admission inadmissable. Many companies feel that they are ahead of the competition in areas such as price, delivery, quality, design, and investment and use of technology. The scale of change is daunting but some companies are achieving improvements year after year. Assuming management is willing to take risks then the biggest but most necessary of these changes is the organizational changes.

5. Organizational Changes

The process of changing the culture of an organization must start at the top. The "top" means any manager with enough autonomy and power to "buck the existing culture". Individuals at the top absolutely must be convinced that changes are required for long-term survival. The impetus for change can come from one of the three sources described by Rigg[10] : education/vision, customer mandate or economic crisis. Management can pursue either the first proactively, or await the arrival of the latter two reactively. However it is initiated, the process begins with senior management commitment. Senior management must establish a comprehensive vision of the organizational changes. The process of creating a vision, however, should not turn into a paper-work drill. The vision should establish detail so that everyone can have a general understanding of the changes.

Once the vision is created, it must be filtered throughout the organization. This educational process is the beginning of the cultural change. Management should deliver the message and receive feedback in order to demonstrate commitment. Communication of the vision should include the general concept of how the various organizational functions will be changing. This allows people to see the all-encompassing nature of the changes, as well as what the impact personally will be. Changes of this magnitude will generate a tremendous amount of anxiety. Based on the feedback, the change leaders can manage the anxiety and modify the plan as necessary.

The entire process of change will take a long period of time because the incremental steps are small. Meaningful change, however, will not come from talk. Meaningful changes will only come about through changes in the organizational functions themselves. What specific changes are necessary in those functions? What values and organizational changes should management articulate as the vision? To a large extent, the answers depend on how far senior management wants to implement organizational changes and how well senior management can understand the changes and need for competitiveness.

6. Conclusion

In this paper, the view that organizational growth demands CE has been discussed. If an organization wants to adopt CE wholeheartedly, its organizational structure and culture must and will change. However, unless the attitude of managers in the organization toward design are changed, the full benefits will not be realised. This paper may provide some fundamental theories for application of the concept of CE in manufacturing companies. However, the the research literatures on the interrelationships between organizational structure, corporate strategy, manufacturing strategy, human resource strategy and IT strategy in concurrent engineering is far away from sufficiency. A paper investigating the technique of employing CE theories to imply the process of integrating strategies will follow to compensate for this shortage.

References
1. Larry E Greiner, Evolution and Revolution as Organizations Grow, Management of Change, Harvard Business Review, Harvard University, 1991.

2. Rigg, Michael, Organizations Must Change Their Cultures With A Long-Term Vision, Journal of Industrial Engineering, April 1992.

3. Hales, Colin, Managing Through Organisation (The Management Process, Forms of Organisation and the Work of Managers), Routledge, 1993.

4. Zhang H C and Atling L, An Exploration of Simultaneous Engineering for Manufacturing Enterprises, Int. J. Adv. Manuf. Technology, Vol 7, 1992, pp101-108.

5. Hartley, J. and Mortimer, J., Simultaneous Engineering: The Management Guide, Industrial Newsletters Ltd in Association with the Department of Trade and Industry, 1990.

6. Winner R T et al, The Role of Concurrent Engineering in Weapon System Acquisition (U), Report R-388, Institute for Defense Analyses, December 1988.

7. De Meyer, A., An Empirical Investigation of Manufacturing Strategies in European Industry, Proceedings of the 5th Int. Conf. of the UK Operations Management Association - Manufacturing Strategy - Theory and Practice, MCB June 1990, pp 555-579.

8. Hartley, J, Simultaneous Engineering, Industrial Newsletters Ltd, 1990.

9. Bleak House", Computing, 6 February 1992, pp22-24.

10. Rigg, M., Organizations Must Change Their Cultures with a Long-Term Vision, Industrial Engineering, April 1992, p10.

DEVELOPING AN INTEGRATED WORKGROUP ENVIRONMENT FOR A CIM ENTERPRISE

LEK PHEOW HWA, WONG CHUN CHONG
Ngee Ann Polytechnic
535, Clementi Road, Singapore 2159
E-mail: LPH1@nova.np.ac.sg

and

PROFESSOR ROBERT K L GAY, ROLAND LIM YAN GUAN
Gintic Institute of Manufacturing Technology, Nanyang Technological Unversity
Nanyang Avenue, Singapore 2263

ABSTRACT

The workgroup concept has been introduced into a CIM-based facility for the implementation of concurrent engineering practices. A structured functional analysis of the enterprise, using the $IDEF_0$ methodology, has identified the minimum number of work groups required to operate in a cohesive and integrated environment. The integrated workgroup environment has been implemented through a client/server network architecture, and a software tool, called the Engineering Data Manager (EDM). A methodology was also used in its design and implementation to provide effective communication, control and management of the data transfer for the integrated workgroup environment.

1. Introduction

In today's competitive manufacturing environment, companies must be able to produce more innovative and quality products, quicker, in greater varieties, and at low cost. The computer-integrated manufacturing (CIM) concept has evolved in response to this situation. Concurrent engineering practices, which form an essential part of CIM, entail organisations to structure their operations in order to maintain a highly responsive, lean, and flexible manufacturing environment.

A medium size manufacturing enterprise, known as Ngee Ann Computer-Integrated Manufacturing (CIM) Kongsi, is being set up in a tertiary institution for training purposes. The enterprise must have the ability to produce engineering components and products quickly and in great variety. To achieve this, a computer-integrated manufacturing (CIM) facility is required and concurrent engineering practices introduced in the business cycle of manufacturing. Marketing, product design and development, production engineering, production planning and control, and production, all these activities must proceed in parallel.

An important part of the concurrent engineering practices being introduced in the CIM facility is the formation of work groups. In the workgroup environment, it is essential that the work groups are integrated in a cohesive manner, so that effective communication, control and management of data transfer can take place, within and between the work groups.

The first phase of the installation was to perform a structured functional analysis and modeling of the enterprise, in order to establish the major activity cycles in the entire business cycle of manufacturing and the minimum number of work groups necessary for effective, integrated and cohesive concurrent engineering practices to be implemented. The second phase was to define the system requirements in terms of hardware and software requirements. The third was the design and implementation phase to provide effective communication, control and management of data transfer for the integrated workgroup environment.

1.1 The Work Group Concept

Enterprises are looking towards organisational structures based on work groups, now recognised as the foundation for the modern enterprise (Datapro, August 1993). Work groups consist of knowledge workers from several interdependent functions. The work groups are empowered to accomplish their objectives with minimal direction, and to report information and results directly to top management.

The Work Group Concept not only increase productivity, but they also support enterprise activities. The new enterprise must enable these self-managed multifunctional teams to coordinate their operations with other teams throughout the enterprise. Some of the tools to enable these work groups to function in a coordinated manner include:-

- Group-specific applications tailored to the requirements of the work group.
- Decision support tools geared to support individual specialists within the work group.
- Scheduling and project management tools that coordinate the activities of the work group.
- Word processing, electronic mail, and multimedia to provide communications within the work group and between work groups.
- Interprocess messaging to allow a specific work group to interact with other work groups.

2. Functional Analysis of the Enterprise

2.1 Ngee Ann Computer-Integrated Manufacturing (CIM) Kongsi

In this manufacturing facility, concepts of CIM and key technologies, such as CAD, CAM, CAE, CAPP, MRP, LAN, etc., are being incorporated in the set-up. Concurrent engineering practices, which form an essential part of CIM, is introduced into the business cycle of manufacturing.

The system architecture is essentially a layered architecture, containing two levels and separately identified as facility or enterprise level and shop floor level, as illustrated in Figure 1. A local area network (LAN), operating on the Ethernet TCP/IP protocol, provides the communication backbone for the enterprise.

The enterprise level consists of various functional departments. Each department uses UNIX workstations loaded with appropriate software for performing the functional activities. A corporate server provides a central vault for storing and managing engineering data and applications for sharing among the functional departments, and other essential network services.

At the shop floor level, an UNIX based workstation acts as the shop floor controller that coordinates and monitors a number of standalone PC-based cell controllers which run on the QNX operating system. The TCP/IP-based shop floor network is linked to another network consisting of a Macintosh server and clients. The latter network provides the material handling system for the entire shop floor facility. The shop floor controller communicates with the enterprise level via the the corporate server.

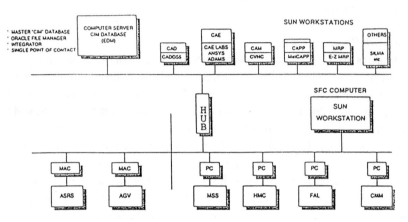

Figure 1: The CIM Architecture of Ngee Ann CIM Kongsi

2.2 $IDEF_0$ Modeling

Figure 2 shows the $IDEF_0$ model of the enterprise from the viewpoint of a system designer. The purpose of this model is to describe the enterprise's entire manufacturing cycle which can be grouped into five major activity cycles.

The market input cycle involves gathering of market information, for example, product specifications from the customer, essentially by the marketing department. Other inputs to this cycle include corporate directives for new products to be developed, and forecasts of sales of existing products. A review process by key personnel of the enterprise marks the end of this cycle.

The Product Design and Development Cycle involves activities related to the design of new products which is usually iterative in nature, starting from concept to preliminary design. The preliminary design stage may also involve analysis work depending on the type of product that is to be developed. Usually the results of each analysis is sent for further analysis, modification, or acceptance of the design. At this stage, key personnel are pulled in from the various departments to consider various aspects of manufacturability, typical of the concurrent engineering approach.

The Detail Design Cycle is where details are added to the product design, documented and released. If the design has to be modified after this, a formal engineering change notification (ECN) and procedure must be used. Once the proposed changes are approved, this cycle is repeated and the new version is released to the next activity cycle.

The Production Engineering Cycle takes the design of the product and determines how to build it, either as designed or with changes to improve its manufacturability. This cycle determines the manufacturing processes that should be used, the desired sequence of the manufacturing, and the specific machine tools and materials to be used. The other function of this cycle involves the generation of programs for NC machine tools, robots, inspection (CMM) machine tool and instructions for work setting and fixturing.

The Production Processing Cycle involves the production planning function where the master production schedule, work orders, and purchase orders are generated. The production planning activity is followed by production execution and purchasing activities where approved purchase orders are processed and work orders released to the shop floor. During the production execution activity, status information is collected by shop floor controller and fed back to the enterprise level to complete the entire loop of manufacturing.

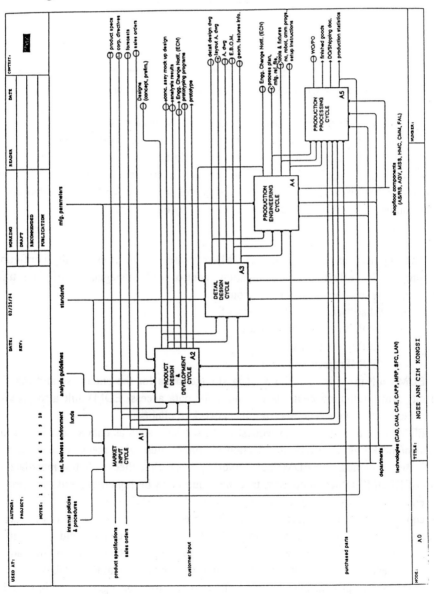

Figure 2: [Node A0] Major Activity Cycles of Ngee Ann CIM Kongsi

3. Design and Implementation

3.1 Client/Server Network Architecture

A client/server network architecture is one of the essential requirements for providing the openness, services, and communication backbone between the various work groups. Figure 3 shows the client/server network set up for the enterprise.

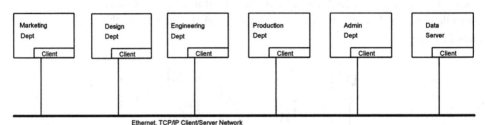

Figure 3: Client/Server Network Architecture Set Up for the Enterprise

The server stores the enterprise's databases and allows data sharing among the various clients. The clients, in this case, are made up of various work groups or functional departments - corporate management, product design and development work group, production engineering work group, and production processing work group. Individual work group performs various activities and stores information into the corporate vault which can be accessed by the same work group or by other work groups.

3.2 EDM Software

The Engineering Data Manager (EDM) software is used to essentially configure the environment for effective communication, control and management of data transfer between and within the various work groups.

Core EDM software modules consist of:-
- EDMVault server software
- EDMClient software
- EDMProjects software

EDMVault provides distributed data storage areas for the enterprise and maximises capacity while retaining centralised control for data access. EDMVault also provides a server facility to handle requests from its network of clients.

EDMClient allows a user to request EDM services, such as data access, data sharing, check-in/check-out, electronic notification via either messaging mechanism or automatic notification schemes. The EDMVault server's process first checks control metadata in the form of RDBMS tables to confirm that the request can be honoured, and then processes the request. If the user has the proper authority, the EDMVault server delivers a copy of the requested EDM data object (including the object's metadata which describes each data object's characteristics such as storage location, owner, version) and locks it from modification by other users. Other users can obtain a copy of the data object to read it, but no one else can modify the data object until it is checked back into the database.

EDMProjects provides the system designer with a tool for organising people and their work into projects for on-line tracking and control. EDMProjects allows for the definitions of project tasks and milestone specifications. Associated with these tasks definitions, EDMProjects allows for the appropriate security access to be defined. With EDMProjects, the system designer can manage and control project or work group data flow.

3.3 Design and Implementation Methodology

The essence of the design and implementation methodology is illustrated by the model as shown in Figure 4.

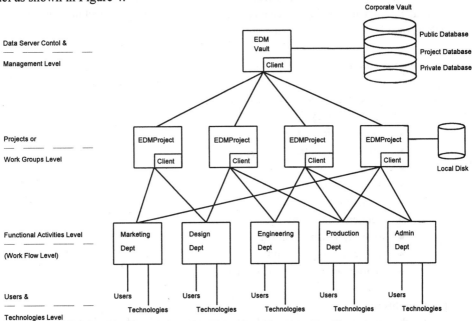

Figure 4: Design and Implementation Model of the Integrated Workgroup Environment

Accessing information from the corporate vault can be effected by various services:
- read - where a copy of the data object is made available to the user;
- get - where the original data object is made available to the user for modifications;
- update - where the data object is updated into the corporate vault after modifications;
- store - where the data object is stored into the corporate vault; the server prevents storing of the same data object again.

3.4 The Integrated Workgroup Environment

The EDMProjects software has been used to essentially create four projects where these projects may each contain one or more phases, with each phase being identified by a specific status-code. Data objects created in a certain phase of a project is stamped with a sequence number, status-code and authority number associated with that phase when it is stored into the corporate vault. The working version of a data object changes to a stable version only after it has been approved through a review/release control process. The approved version is stamped the successive status-code, sequence number and authority number, and becomes the working version in the next phase of the project.

A user or a specific work group can be associated with a project or projects. These users or work groups can be given full or limited access authority through authorisation schemes. Users having authority numbers equal or higher can access the working version or stable version data objects. Formal communication within or between work groups is effected through automatic notification schemes whereas informal communication is effected through messaging mechanism. Each work group uses technologies associated with its own information processing activities.

3.5 Communication, Control and Management of Data Flow

Concurrent communication, control and management of data flow in the enterprise has been implemented through various schemes:
- project life cycle
- authorisation scheme
- authority groups
- status levels
- status codes
- system partition code
- sequence numbers
- release/revision control
- revision sequence codes
- automated review procedures

4. Conclusion

In this paper, the design and development work of an integrated workgroup environment for a CIM-based facility has been described. Concurrent engineering practices has been recognised as being an essential part of the business cycle of manufacturing. The integrated workgroup environment is essential for concurrent engineering activities to be successfully carried out. It also provides the flexibility in its manufacturing operations. System requirements as well as essential tools and methodologies are required for the design and implementation of the integrated workgroup environment.

5. Acknowledgements

The authors wish to thank Professor Robert Gay and Roland Lim for their guidance in the research work related to this project.

6. References

1. Air Force Systems Command, *Integrated Computer-Aided Manufacturing (ICAM) Dynamics Modeling manual (IDEF$_2$); Integrated Computer-Aided Manufacturing (ICAM) Function Modeling Manual (IDEF$_0$)* (Wright-Patterson Air Force Base, Ohio, 1981).

2. C. L. Ang, *System Modeling Using Diagramming Techniques* (Lecture Notes, 1992).
3. Computervision Corporation, *EDM Technical Summary Revision 5.0* (1992).
4. Prof Robert K. L. Gay, P. K. Seet, B. S. Lee, *A Network-based Concurrent Design Environment for Micro-based CAD* (1993).
5. J. R. Hartley, *Concurrent Engineering: Shortening Lead Times, Raising Quality, Lowering Costs* (Productivity Press, Cambridge, Massachusetts, 1990).
6. F. H. Mitchell, Jr., *CIM Systems - An Introduction to Computer-Integrated Manufacturing* (Prentice-Hall, 1991).
7. T. A. Nolle, *Communication Series: LAN Architectures - The LAN and the Workgroup: Foundation of the Future* (Datapro Information Services Group, McGraw-Hill, August 1993), p. 1 - 4.
8. P. R. Warndorf, M.E. Merchant, *Development and Future Trends in Computer-Integrated Manufacturing in the USA* (International Journal of Technology Management 1, 1986), p. 1 - 2, 162.

Models and Architecture

The Development and Application of a Generic "Order Fulfilment" Process Model
    A. M. Weaver, S. J. Childe, R. S. Maull, P. A. Smart and J. Bennett

Modelling Manufacturing Resources and Activities: An Ontology
    F. Bonfatti, P. D. Monari and P. Paganelli

A Generic Information Infrastructure for Enterprise Integration
    J. Jonker and E. M. Ehlers

Engineering Data and Process Integration in the SUKITS Environment
    B. Westfechtel

THE DEVELOPMENT AND APPLICATION OF A GENERIC "ORDER FULFILMENT" PROCESS MODEL

A M WEAVER, S J CHILDE, R S MAULL, P A SMART
School of Computing, University of Plymouth, Drake Circus, Plymouth, PL4 8AA, UK
E-mail : adamw@soc.plym.ac.uk

and

J BENNETT
Teaching Company Centre, University of Plymouth

Abstract

The objective of this paper is to describe how generic process models can be used as a technique within an approach to Business Process Re-engineering that will encourage companies and participants carrying out BPR projects to take a business process perspective. The definition of core processes within organisations is identified as common practice in BPR projects. The paper proposes the development and validation of a set of generic process models to provide a process framework and intervention tool for BPR projects within Small and Medium Sized Manufacturing enterprises.

The development and validation of a generic process model representing the "order fulfilment" process of a manufacturing company is used as an example. IDEFo is used to develop the model. A method of using the generic process models as intervention tools is outlined and its advantages are described.

1. Introduction

"If you want to understand the way work gets done, to improve the way work gets done, and to manage the way work gets done, processes should be the focus of your attention and actions. Viewing issues from a process perspective often reveals a need to make radical changes in goals, in the design of business systems and the management practices." Rummler & Brache (1990)

Rummler and Brache describe the essence of Business Process Re-engineering (BPR) i.e. viewing the issues in an organisation from a business process perspective and changing the design of business systems and management practices.

In recent surveys between 65% and 77% of respondents were carrying out or considering Business Process Re-engineering (BPR) projects. The popularity of BPR suggested by these figures is backed up by an abundance of literature, seminars, conferences and software tools that have emerged over the past few years.

In their survey (Skinner and Pearson 1993), Highams Systems Services Group Ltd , found that the respondents to the survey were implementing BPR projects for a number of reasons. The list of reasons given included "the need for continuous improvement", "increased customer expectations", "increased competition" and "changing market needs". The benefits that these companies hoped to achieve were again various, "increased customer focus", "improved profitability" and "improved corporate flexibility" all featured high on the list of benefits.

A Business Intelligence survey found that their respondents had mixed experiences of using BPR, in fact, "few have succeeded in transforming their total operations" (Harvey 1994). Business Intelligence refer to a number of possible reasons for the mixed experiences including weaknesses at any stage of the methodology resulting in partial or complete failure, corporate cultural barriers and lack of a sound business strategy.

The objective of this paper is to describe how generic process models can be used as a technique within an approach to BPR that will encourage companies and participants carrying out BPR projects to take a business process perspective. The authors believe that the issue of how to encourage individuals at all levels within a company to think in terms of business processes is critical to the success of a BPR project. This is reinforced by Rummler and Brache who have found that;

> "When we ask a manager to draw a picture of his or her business (be it an entire company, a business unit or department), we typically get something that looks like the traditional organisation chart." (Rummler and Brache 1990)

2. Definition of Business Processes

In the majority of documented BPR methodologies, including those developed by Coopers & Lybrand (Johansson et al 1993), IBM (Kane 1986), British Telcom (Harvey 1994), Xerox and Lucas (Parnaby 1993), one of the initial activities is to identify the core business processes. In identifying the core processes the participants in the BPR project are defining boundaries within their organisation using a process perspective.

It is useful to compare a business process to a system. A system embodies four basic ideas; which paraphrased from Checkland (1981) are emergent properties, a hierarchical structure, communication between entities within the system and a process of control. A business process embodies the same four basic ideas. For example products and information are emergent properties of a business process since they are a result of the overall interaction of the entities within the process; a business process can be decomposed into a hierarchy of sub-processes; there are flows of information and physical entities within a business process connecting the entities and the process is managed.

The basic idea of a hierarchy of processes is important when considering generic processes for an industry type and the core processes within a company. The hierarchy of processes provides the framework within which the analysis and redesign will take place. The number of core processes within a company is very much dependent on the level of abstraction at which the organisation decides the core process definition will be meaningful. Business Intelligence's report (Harvey 1994) provides a table of the core process taxonomies of a number of consultants, the numbers ranging from 7 to 20 core processes.

Examples of breaking down businesses into varying numbers of processes include Arthur Andersen (200) and Xerox Nordic (48). The difference in numbers of processes can be explained by the level of analysis, the lower number being more abstract. Other companies where the definition of core processes was evident and was being used at senior management level in the initial stages of a BPR project include IBM which is currently organising its world-wide operations around ten generic internal and external customer facing processes and Lucas Engineering and Systems Ltd. who have developed a model containing 16 generic processes. There are further examples of core process definition by companies in both Davenport (1993) and the Business Intelligence report (Harvey 1994) on BPR and many case studies in journal articles (Shapiro et al 1992, Davenport and Snort 1990).

The prospect of managing businesses in a process organisation may lead to the evolution of standard processes, in the same way that a roughly standardised set of functional divisions (manufacturing, design, sales and marketing, finance, personnel, etc.) developed.

The definition of core processes has required extensive investment by the multi-national organisations discussed above. The purpose of developing the generic process models described in this paper is to provide small and medium-sized manufacturing enterprises with a similar framework without requiring the use of their limited resources. The generic process models described provide a framework and the detail of the generic process models provides the ability for the models to be used as an intervention tools in a BPR approach.

3. The development and validation of the generic "order fulfilment" process model

3.1. The level of analysis

The first objective is to establish the level of analysis or "bound" the model. For the generic model to be of any use it must contain elements which are at a level of abstraction that allows meaningful discussion. Breaking a major business process into 5-10 generic activities and flows would not provide a catalyst for comparison with a company's existing process. Conversely the generic model should not be at a level of abstraction where much of the model is irrelevant to any particular company.

For a generic model to act as an intervention tool to encourage participants in a BPR project to take a process perspective and work with the model as a framework for improvement, it must model a process that is key to the success of the business. The generic process model described in this paper is a model of the "order fulfilment" process within a manufacturing company. The model has been developed in discussions with a number of manufacturing companies varying in size from Times 1000 companies to Small and Medium Manufacturing Enterprises (SMEs) with under 500 employees.

The generic process model of the "order fulfilment" process will cover all four types of manufacturing companies defined by Wortmann (1990); Make-to-stock, Assemble-to-order, Make-to-order and Engineer-to-order. During discussions with companies it was evident that different companies place different emphasis on parts of the "order fulfilment" process. For example a local company that can be classified as engineer-to-order places considerable emphasis on the preliminary stages of the order fulfilment process where the company works closely with the customer to specify the product and plan the manufacture of the product. Another local company that can be classified as make-to-stock considers the activities immediately before shipping to be of particular importance.

3.2. The Modelling Technique

The generic model of the "order fulfilment" process has been developed using IDEFo (CAM-I 1980). IDEFo is widely used in the manufacturing sector for modelling processes. IDEFo comprises:
- A set of methods that assist in understanding a complex subject;
- A graphical language for communicating that understanding;
- A set of management and human-factor considerations for guiding and controlling the use of the technique.

IDEFo uses top-down decomposition to break-up complex topics into small pieces which can be more readily understood. An IDEFo model is an ordered collection of diagrams. The diagrams are related in a precise manner to form a coherent model of the subject. The number of diagrams in a model is determined by the breadth and depth of analysis required for the purpose of that particular model. At all times the relationship of any part of the whole remains graphically visible.

In summary IDEFo provides the ability to show what is being done within a process, what connects the activities and what constrains activities. It uses a structured set of guidelines based around hierarchical decomposition, with excellent guidance on abstraction at higher levels. If used well this ensures good communication and a systemic perspective.

3.3. Information used to develop the generic model

The information used to develop the generic process model has been extracted and assimilated from a number of sources. The activities that are carried out within the "order fulfilment" process were adapted from a generic task model developed by Childe (1991). Childe's task model was based on the proposition that there are a key set of tasks or activities which are consistent throughout manufacturing companies (all manufacturing companies order materials, take orders from customers etc.). The task model does not show any information or physical flows and hence it does not show how the activities within a manufacturing company may be integrated horizontally to produce an output. However it did provide a validated model of activities from which to develop a generic process model of the "order fulfilment" process.

The physical and information flows that integrate the activities to form the "order fulfilment" process were identified by using IDEFo models of manufacturing companies that had been produced by the authors in the course of their research work. Information was also distilled from other models produced in a number of different modelling techniques, from literature and from the experiences of the authors while working with manufacturing companies.

3.4. Validation of the generic process model

The validation of the generic process model is on-going as more knowledge is gained through using the model within manufacturing companies. Validation includes the criticism and comment by academic colleagues, a comparison by third parties to their own generic models of the "order fulfilment" process and experience gained by applying the generic process model as an intervention tool within manufacturing companies interested in BPR

The generic process model of the "order fulfilment" process currently includes over 110 activities integrated by the flows of physical and information entities. Fig 1. shows the second highest level of abstraction of the "order fulfilment" process. The model extends to 5 lower levels of activities and flows. The complete model also includes a glossary of terms.

4. The application of generic process models

In the introduction the critical issue of getting employees to think in terms of business processes was identified. The generic process models are intended to be used as an intervention tool to encourage the participants of a BPR project within a manufacturing company to take a business process perspective. The participants in a BPR project would generally be individuals from the functions who currently perform activities within the process guided the objectives set by senior management.

In the initial stages of the BPR project following the identification of a core process to be redesigned, the participants would be presented with the generic process model and glossary of terms and asked to compare the generic process model against the process within the company that the model is intended to represent. In carrying out a comparison it encourages the participants to;

1. Take a business process perspective as the generic model provides an existing process framework.

2. Develop a consensus view of their own company's process by debating the differences between the generic model and each participants perceive view of the company's process.

3. Identify and change the generic model to represent their company's process.

4. Identify immediate changes that could be made to the company's process as differences between the model and reality are found.

5. Consider the systemic relationship of all parts of the process as IDEFo provides a structured medium where inconsistencies in the changed model can be identified easily.

In comparison with current BPR approaches where the participants are encouraged to develop a process model of the existing business process, it reduces the danger of participants reverting to tradition functional thinking by providing a process focused framework. It also provides greater momentum to the project than a "blank sheet of paper" and the generic process model is non-political having being produced externally. The non-political nature of the generic process model should enable participants to more freely criticise the generic model and in doing so generate debate and understanding amongst the group.

5. Conclusion

The objective of the paper was to describe the development of generic process models and their application within small and medium sized manufacturing enterprises. A generic process model of the "order fulfilment" process was chosen as an example following considerable interest expressed by companies visited during the research project.

The definition of core processes is dependent on the level of abstraction that the organisation finds meaningful. There has been substantial investment by many multinationals in defining core processes within their organisations. The identification of an "order fulfilment" process is a commonly defined across many different sectors of industry. A set of generic process models of core processes within manufacturing companies would provide a framework to encourage a process perspective in companies less able to invest resources in the definition of core processes and could be used as an intervention tool.

The generic process model was developed using IDEFo. IDEFo provided structured approach, hierarchical decomposition and medium to enable easy of communication of the model.

The application of generic process models in SMEs as part of a BPR project encourages a process perspective to be taken by participants, provides an additional momentum to the project and encourages debate and understanding of the existing process within the company.

Acknowledgements

The research work described in this paper has been jointly funded by the Engineering and Physical Science Research Council under a grant (GR/J/95010) entitled "A specification of a Business Process Re-engineering methodology for Small and Medium Sized Manufacturing Enterprises" and British Aerospace (Systems and Engineering) Ltd.

The generic process model was developed using DESIGN/IDEF supplied by IDEFine Ltd.

References

CAM-I (1980), *Architect's Manual ICAM Definition Method "IDEF-0"*, Arlington, Texas

Checkland P (1981), Systems thinking, systems practice, J Wiley & Sons

Childe S J (1991), *The design and implementation of manufacturing infrastructure*, PhD Thesis, Polytechnic South West, UK

Davenport T H and Short J E (1990), The new industrial engineering: information technology and Business Process Redesign, *Sloane Management Review*, Summer

Davenport H J (1993), *Process Innovation*, Harvard Business School Press

Harvey D (1994), *Re-engineering: The Critical Success Factors.* , Business Intelligence

Johansson H J, McHugh P, Pendlebury A J, Wheeler W A (1993), Business Process Re-engineering- BreakPoint Strategies for market dominance, J Wiley & Sons

Kane E J (1986), IBM's Quality Focus on the Business Process, *Quality Progress*, April

Parnaby J (1993), Business Process Systems Engineering, Lucas Industries plc, November

Rummler G A & A P Brache A P (1990), *Improving Performance - How to manage the white space on the organisation chart*, Jossey-Bass Publishers

Shapiro B P, Rangan V K, Sviokla J J (1992), Staple yourself to an order, *Harvard Business Review*, July-August

Skinner C & Pearson J R W (1993), *Business Process Re-engineering in the UK Financial Services Industry*, Highams Systems Services Group Ltd

Wortmann J C (1990), Towards One-of-A-Kind Production: The Future of European Industry, *Advances in Production Management Systems*, Elsevier Science Publishers P.V. Holland

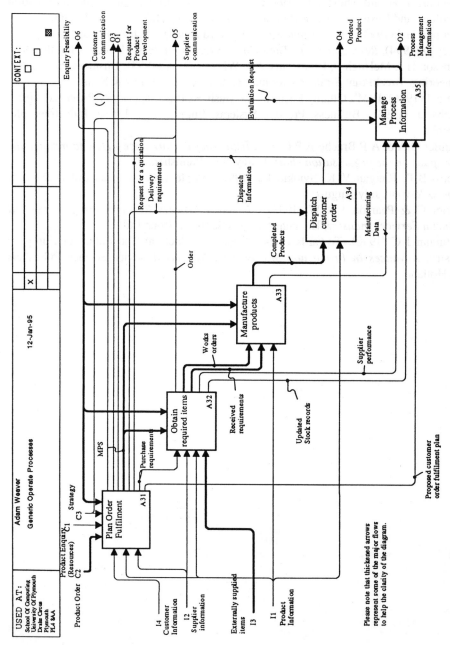

Figure 1. IDEF0 Diagram of the 2nd Level of abstraction of the Generic "Order Fulfilment" Process

Modelling Manufacturing Resources and Activities: an Ontology

FLAVIO BONFATTI, PAOLA DANIELA MONARI
Faculty of Engineering, University of Modena
c/o CICAIA, via Campi 213/B, 41100 Modena (Italy)
tel. +39 59 378514, fax +39 59 378515, E-mail bonfatti@c220.unimo.it

and

PAOLO PAGANELLI
Democenter s.cons.r.l.
Viale Virgilio 55, 41100 Modena (Italy)
tel. +39 59 848810, fax +39 59 848630, E-mail democe@c220.unimo.it

ABSTRACT

Process representation constitutes a critical point in enterprise modelling, due to its role of bridge between product representation and planning functions. The paper proposes the ontology underlying a new process modelling approach; this approach characterises an ESPRIT project aimed at providing small-medium enterprises with advanced computer-based tools supporting product and process design and planning activities. This ontology is based on a clear separation between abstract definitions of product routings and the involved resources, and between general structural aspects of resources and product-dependent behaviours. Resource activities are modelled taking into account occupation over time and configuration constraints expressed as pre- and post-conditions.

KEYWORDS: *Process modelling, Rules, ESPRIT project.*

Introduction

The ability of manufacturing enterprises to use effectively their productive resources strongly depends on how deeply resource behaviours and occupations are known. Idle periods, bottle-necks and wastes can be reduced only having available accurate representations of resource activities, their subdivision into sub-activities, durations, temporal relationships and so forth. On the other hand, a finer representation granularity usually requires larger volumes and varieties of data to acquire and process, which implies a higher complexity of the manufacturing information system. This is particularly burdensome for small-medium enterprises (SMEs) due to their limited resources and the difficulty of subtracting skilled personnel from directly productive activities. Thus, significant efforts have been spent in the last few years in studying advanced enterprise models, often based on rules, to improve management and control capabilities [7, 8, 9, 12].

This is also the aim of the ESPRIT project nr. 8224 - RUMS (a Rule-based Manufacturing Modelling System) that started in April 1994 and will end in august 1996. Objective of the project is to provide SMEs with advanced computer-based product and process modelling techniques supporting a number of fundamental activities: (i) verification and assessment of the current manufacturing organisation; (ii) evaluation of technological enhancement hypotheses concerning production resources; (iii) definition of realistic manufacturing and purchasing plans holding uncertainty of forecasts in due consideration; (iv) production planning with respect to different time horizons; (v) training of human resources involved in manufacturing.

A rule-based product model has already been defined in [3]; it is based on the concepts of product family, structural schema and functional schema, and on the relationships that are established between them. On the other side, planning will involve definition of criteria for resource allocation and flow management, aimed at creating realistic purchasing and production plans on different time horizons. The bridge between these two aspects is constituted by process modelling, that is, representation of manufacturing operations related to resource configurations and behaviours; here we are particularly concerned with this theme.

The definition of a new model for representing manufacturing resources and activities requires that the underlying ontology is clearly and explicitly stated [2, 10]. To this purpose we split the problem into three parts:

-- Transformations and assemblies, required to realise the products belonging to a family, are ordered to form an operational schema which results independent of the resources that could be employed to perform them.

-- Resources are considered at different levels of definition (machines, workcenters, lines); at the finest level, a resource is seen as a complex structure with components, such as tools and fixtures, that contribute to define the lawful working configurations.

-- Working and set-up activities, respectively performed by and on each resource, are identified and understood in terms of time spent, priorities, consumes, involved manpower and so forth; these activities will be mapped onto the operational schemas of the required products in the planning phase.

The paper is subdivided into four sections. Section 1 presents the modelling requirements whose satisfaction is pursued. Section 2 is spent to describe characteristics and properties of operational schemas. In Section 3 the representation of manufacturing resource configurations is addressed at the finest level of detail. Finally, in Section 4 the activities that are performed by and on every resource are modelled taking into account temporal relationships. The concepts introduced in the paper are expressed in a rule-based formalism that results simple and self-explanatory and is suitable to support a possible prototypal development.

1. Process Modelling Requirements

The first aspect a process ontology should address is the variety of information needs the observer perceives when analysing manufacturing management and control functions. If compared with current process modelling practice, they give rise to a number of requirements the representation should meet. Significant limitations of current practice are:

-- At the long- and medium-term planning level, process models are mostly product oriented: bill-of-materials represent complex production phases, in

general concluded by an assembly operation. Routings split phases into simpler operations, providing basic information on resource occupation [1, 6].
-- At the scheduling and shop-floor control level, process models are mostly resource oriented: they include current resource states, operation dispatching policies and materials flow control parameters [1,13].
-- Operations are only defined in association with the machines or workcentres that are in charge of them. This means that it is not possible to obtain a resource-independent process description, while independence is normal in the real world: during the product life-cycle the same operation can be assigned to different machines, or possibly performed by a sub-contractor.

From these and other problems, some requirements are identified:
-- *Family-based modelling*. The production process must be associated to a whole family of products, that is, to a class of products sharing some basic functional and structural features. As for products [3], genericity should be included in the model by means of alternatives, parameters and conditions.
-- *Resource independence*. Process representation must refer to abstract operations, that is, operations defined independently of actual shop-floor resources so as to be virtually executable by means of alternative resources.
-- *Variable granularity*. In general, production process representation should include operations defined at different levels of detail. Depending basically on the planning time horizons, the same operation may be split into simpler ones, or be considered part of a more complex operation.
-- *Mix of technologies*. Due to complexity and cost of automation, most enterprises typically present a mix of automated and traditional working units, and this situation can probably last for many years. The process model should ensure an effective representation of all possible production means.
-- *Finite vs. infinite capacity*. Resource allocation should be planned by expressing different constraints on capacity, criticity and possible partial use of the resource itself and of its components.
-- *Temporal relationships*. The execution of an activity involves primary and auxiliary resources as well as manpower. The actions they perform are temporally related and these inter-dependencies should be clearly expressed.

In order to meet these requirements we distinguish two separate aspects of process modelling : (i) the operational schema, where abstract operations describe the process required by a given family of products, (ii) the factory model, where manufacturing resources are described and basic activities are introduced in order to express how operations are actually performed.

2. The Operational Schema

In general, an operation is any phase of the production process that produces a known output part starting from a given set of input parts [5]. The input and output parts are clearly identified within the product composition structure. We recognise two basic categories of manufacturing operations: *assembly* operations, whose input is constituted by two or more parts, and *transformation* operations, whose input is constituted by a single part. In both cases we can adopt the following *operation* rule:

```
P   <--   Op (r1:T1, r2:T2, ..., rn:Tn)
```

where r_i indicates the role played by the input component type T_i within the output object. Note that a single-input operation does not necessarily imply a physical transformation, but even a logical change in product condition due, for example, to purchasing, control, transportation or handling activities.

In addition, in order to express the differences that characterise the products of a family, *alternative* rules are required:

```
A   <--   T1 / T2 / ... / Tn
```

where A is an alternative type, associated to a certain role, and T_i is the i-th option of A. For example, the following rules:

```
EngFeed  <-- Connect(device: FeedDvc, control: CtrlType)
FeedDvc  <-- Carburetor / Pump / Turbine
CtrlType <-- StandardCtrl / SafetyCntrl
```

introduce six alternative configurations. In this case we assume that the different versions are realised through the same "Connect" operation, but alternative operations could also be considered.

It may occur that some combinations of options are not allowed or do not correspond to actual products; in other terms, consistency problems may arise when options are mutually dependent or exclusive [4]. This calls for *condition* rules that assume two basic forms: *implication* (A => B) and *mutual implication*(A <=> B), where A and B are boolean expressions formed by restriction predicates on options. Implication asserts that if A is true then B must be true. Mutual implication asserts implication in both directions. With reference to "EngFeed", suppose that only four of the six combinations are actually produced; the following conditions express this restriction:

```
FeedDvc eq Carburetor => CtrlType eq StandardCtrl
FeedDvc eq Turbine    => CtrlType eq SafetyCtrl
```

The operational schema associated to a whole family of products is called family operational schema. A family schema contains all the operation definitions, alternatives and constraints required to represent the production process of every product in the family. Consider a possible complete schema associated to the family "Engine":

```
Engine   <-- Testing(untested: UntEngine)
UntEngine <-- FinAssembly(block: EngBlock, feed: EngFeed)
EngFeed  <-- Connect(device: FeedDvc, control: CtrlType)
FeedDvc  <-- Carburetor / Pump / Turbine
CtrlType <-- StandardCtrl / SafetyCtrl
EngBlock <-- PreAssembly(head:EngHead, body:EngBody, pistons:EngPiston#4)
EngHead  <-- Finishing(unfinished: UnfHead)
```

```
UnfHead   <-- Tourning(rough: RoughHead)
RoughHead <-- Acquiring(basepart: ExtHead)
ExtHead   <-- H1 / H2 / H3 / H4
EngBody   <-- Finishing(unfinished: UnfBody)
UnfBody   <-- Acquiring(basepart: ExtBody)
ExtBody   <-- B1 / B2 / B3 / B4
EngPiston <-- Acquiring(basepart: ExtPiston)
ExtPiston <-- P1 / P2 / P3 / P4

ExtHead <=> ExtBody <=> ExtPiston
FeedDvc eq Carburetor => ExtHead in [H1, H2] and CtrlType eq SafetyCtrl
FeedDvc eq Pump => ExtHead in [H2, H3, H4]
FeedDvc eq Turbine => ExtHead eq H3 and CtrlType eq SafetyCtrl
```

In the "EngBlock" definition rule we can observe the presence of a component constituted by a group of four (#4) objects of the type "EngPiston": this is an example of cumulative role, that is, a role where no actual distinction can be found in behaviours of the participating objects. Moreover, the first compatibility condition establishes a correspondence between the options of "ExtHead", "ExtBody" and "ExtPiston". Finally, given a family schema, one obtains the operational schema of each family member by simply choosing the corresponding option within every alternative. If choices leave some open alternatives, the shared schema of a subset of family products is obtained.

3. Resources

As abstract operations can be introduced independently of the real factory status, so resource can be described, to some extent, without explicit reference to the activities they perform. More precisely, we find convenient to separate structural aspects from product-dependent behaviours: in this section we examine the former while the following section is devoted to the latter aspect.

From the structural point of view, we recognise a wide spectrum of resources depending on their nature (machines, persons, transportation units, etc.) and the observation detail (machines, workcentres, lines). In general, resource is any autonomous entity participating in the performance of manufacturing activities. At the chosen detail level we distinguish primary resources, those that are in charge of basic operations, and auxiliary resources, whose roles are defined within primary ones (for instance, fixtures and tools). A person behaves as a primary resource when carrying out a manual assembly, and as an auxiliary resource when fixing a tool on a spindle. Both primary and auxiliary resources may be characterised by complex structures, that we describe by *composition* rules:

```
R   <--   p1:C1, p2:C2, ..., pn:Cn
```

where each position p_i has associated a component type C_i. The rules:

```
Lathe   <-- pallet:Fixture * spindle:Tool * operator:Controller
Fixture <-- coupling: Flange * operator:Mounter
```

introduce three components for "Lathe" and two for "Fixture". Note that all the components are auxiliary resources. Resource complexity is related to the level of detail chosen for activity representation: as we shall see, resource behaviour includes the behaviours of all its components.

By effect of the activities carried out during the production process, a resource modifies its configuration by replacing auxiliary components with others. More precisely, we recognise some resource positions as configurable in that they have associated different components during resource working time. The options are expresses with *alternative* rules similar to those introduced above; for example, depending on circumstances, the tool mounted on lathe can be of type TL1, TL2 or TL3:

```
Tool <-- TL1 / TL2 / TL3.
```

In conclusion, a primary resource is described by an proper set of composition and alternative rules, called the *resource schema*. Consider a possible complete schema of the "Lathe" resource:

```
Lathe   <-- pallet:Fixture * spindle:Tool * operator:Controller
Fixture <-- coupling: Flange * operator:Mounter
Tool    <-- TL1 / TL2 / TL3
Flange  <-- F1 / F2 / F3 / F4
```

The schema describes all the potential configurations for the instances of the corresponding resource type. The component types associated to all the configurable positions define a *resource configuration*. Mutual dependencies or exclusions between options are sometimes identified. They are usually activity-dependent and thus will be considered in the next Section.

4. Activities

Even activities can be seen at different levels of detail. At the finest level they include every product transformation and assembly, as well as single set-up, transportation, picking and storing operation. We consider activity an autonomous step of the production process which is not useful splitting into simpler activities. An activity involves a primary resource indicated as the activity *site*. Different occurrences of the same activity are possible in time, according to the production schedule: identification and sequencing of the required occurrences over the time horizon are tasks of the planning model.

An activity is characterised by three main aspects: (i) utilisation of auxiliary resources within the activity site; (ii) pre-conditions on resource configuration; (iii) effects, including changes of resource configuration. Concerning the first aspect, the involvement of each resource component can be seen as a succession of time intervals where the status of the corresponding position changes alternatively from free to occupied and vice versa. We express the occupied time segments by associating to the resource positions the start and end instants in relative terms with respect to the whole activity start time [11]. The activity definition results straightforwardly (see also Figure 1):

```
activity Task01 on Lathe
        operator                from 0 to 7
        spindle                 from 5 to 20
        pallet                  from 2 to 22
        operator                from 18 to 25
```

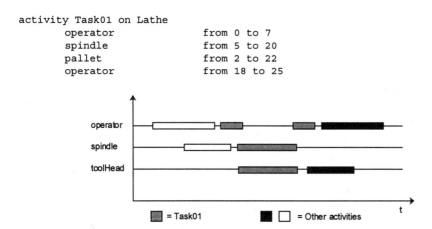

Figure 1: Activities distribution during working time of a certain site.

Chronological dependencies between different segments of the same activity may be recognised and expressed by labelling extremes of time segments and referring to these labels in other segment definitions:

```
activity Task01 on Lathe
        operator                from 0 to t1 + 2
        spindle                 from t1 = 5 to t2 = 20
        pallet                  from t1 - 3 to t2 + 2
        operator                from t2 - 2 to t2 + 5
```

Occupation segments express necessary conditions on activity executability. Further conditions concern proper configurations of the resource positions. For example, in order to perform "Task01" the tool component must be of type TL1. We represent each of these pre-conditions as a *start configuration* rule associated to the corresponding segment definition:

```
activity Task01 on Lathe
        spindle from t1 - 3 to t2 + 2: TL1
```

Note that for all the duration of the segment we are assured that no change will occur on the tool type.

Finally, the execution of an activity sometimes affects the site configuration, in that one or more components can result modified. Activity effects are expressed by segment post-conditions, that is, conditions that are verified after the segment ends. A post-condition is expressed by an *end configuration* rule and where present, completes the definition of the corresponding segment. Consider the activity "ChangeTool21" representing a tool transformation on the "Lathe" resource: in the second sentence, the "-->" operator is used to separate start and end configurations. Note the dynamic dependence established between spindle and pallet configurations, since the presence of flange F1 is expected before tool transformation starts.

```
activity ChangeTool21 on Lathe
        operator from 0 to 25: Controller
        spindle from t1 = 5 to 20: TL2 --> TL1
        pallet from 0 to t1 + 5: coupling = F1
```

Start and end configurations are used to determine if two activities on the same resource can be correctly sequenced or if they require an intermediate activity of configuration transformation. This is a basic information for the following planning algorithms.

Conclusions

The model derived from the ontological approach proposed in this paper is presently under verification by a number of experiments at manufacturing SMEs. Particular attention is paid to ascertain the model capability of capturing the behaviours of both traditional and technologically advanced resources. In addition, the model is checked with respect to different levels of representation details, aiming at establishing a correspondence between the information acquired at the finest level and that used at the other levels. A software prototype to support the modeller activities is under development: its employment will be the basis for disseminating the proposed approach.

References

1. APICS "Dictionary" by T. F. Wallace and J. R.Dougherty, Sixth Edition, 1987.
2. F. Bonfatti, L. Pazzi, *Modeling object complexity and behaviour: towards an ontological paradigm*, COMPEURO 91 Int. Conf., Bologna, 1991.
3. F. Bonfatti, P. D. Monari, P.Paganelli, *Towards a rule-based unified product modelling*, DKSME 94 Int. Conf., Hong Kong, 1994.
4. F. Bonfatti, P. D. Monari, P.Paganelli, *Object-oriented constraint analysis in complex applications*, DEXA 94 Int. Conf., Athens, 1994.
5. F. Bonfatti, P. D. Monari, P.Paganelli, *Modelling process by rules: regularities, alternatives and constraints*, IMSE 94 Int. Conf., Grenoble, 1994.
6. F. Da Villa, A. De Toni, *An innovative production planning*, International Meeting on Production Management, Varenna, Italy, 1992.
7. J. B. Evans, *Description of the production process using temporal expressions*, DKSME 94 Int. Conf., Hong Kong, 1994.
8. F. G. Fadel, M. S. Fox, M. Gruninger, *A resource ontology for enterprise modelling*, WET ICE 94 Int. Workshop, Morgantown, 1994.
9. M. S. Fox, M. Gruninger, *An activity ontology for enterprise modelling*, WET ICE 94 Int. Workshop, Morgantown, 1994.
10. S Greenspan., J. Mylopoulos, A. Borgida, *On formal requirements modeling languages: RML revisited*, ICSE 16 Int. Conf., Sorrento, 1994.
11. J. F. Allen, *Maintaining knowledge about temporal intervals*, Comm. of ACM, 26, 11, 1983.
12. A. Di Leva, P. Giolito, F. Vernadat, *The M*-object methodology for information system design in CIM environments*, INRIA Tech. Report 1918, 1993.
13. T. E. Vollmann, W. L. Berry, D. C. Whybark, *Manufacturing Planning and Control System*, Second Edition, 1988.

A GENERIC INFORMATION INFRASTRUCTURE FOR ENTERPRISE INTEGRATION

J. JONKER, E.M. EHLERS
Department of Computer Science, Rand Afrikaans University, P.O. Box 524
Aucklandpark, 2006, South Africa
E-mail: jjo@rkw.rau.ac.za

ABSTRACT.

The end product of any enterprise integration project, independent of the methodology used, is always an information infrastructure that supports integration. In this paper we will describe a generic model consisting of cooperating information systems to achieve enterprise integration, using hybrid knowledge representation structures. This model can be used with any integration architecture (CIM-OSA, GIM, etc.) it only represents an alternate information view

Keywords: Information Infrastructures, Computer Integrated Manufacturing, Enterprise Integration

1. Introduction

According to Williams et al. 1993, enterprise integration strives to achieve a pro-active, aware enterprise that is able to act in real-time adaptive mode to a changing environment. This is a very idealistic and complex goal and it is clear that solutions to the enterprise integration problem will be equally complex. The purpose of this paper is to present a generic model of an information infrastructure for manufacturing enterprise integration that will facilitate the goal of achieving a pro-active aware enterprise.

One of the major issues in achieving an intelligent, aware enterprise is whether to centralize or decentralize control. We however feel that this is not really the issue. Rather than advocating one or the other, we recommend moving each decision to the appropriate level within the decision making process, in such a way as to minimize unnecessary communication.

2. Problems with Enterprise Integration

At this time there exist a large number of enterprise integration architectures like CIM-OSA, GRAI-GIM and PERA, to name a few (Williams et al. 1993, Katzy et al. 1993, Doumeingts, et al. 1993). The problem with any enterprise integration architecture is that a large part of the success of the integration project is not determined by the architecture used, but the manner in which it is implemented. The success of an integration project can be said to be determined by the internal organizational philosophies rather than a specific architecture. In particular for an integration project to succeed, it is necessary to manage the total information resource of the organization as a single entity. It is for this reason that we defined a model that specifically demonstrates the relationship and interaction between all the information sources within an organization.

Some other problems with enterprise integration, include the following (Williams et al. 1993, Katzy et al. 1993):

- Systems for manufacturing and enterprise integration cannot be bought off the shelf; each organization needs to develop its own integrated information system.

- The necessary expertise to develop an infrastructure for enterprise integration requires that experts of different disciplines cooperate closely.

- Enterprise integration projects are usually initiated in existing organizations in which substantial amounts are already invested in systems that might not support integration.

- The risk involved in an enterprise integration project can be high.

In examining the problems associated with integration projects, it is noteworthy that none of these problems are caused by lack of technology. In fact the problems are caused because technology developed faster than the ability of large organizations to use it optimally. In other words, the problem is not lack of technology, but the way in which we apply it.

3. A Generic Information Infrastructure for Enterprise Integration

In figure 1, we present a high level model of the interaction between the different components of our generic model of the information infrastructure necessary for enterprise integration. This model can be used as an alternate information view within any of the existing enterprise reference architectures, as a high level generic view of the total information system required to achieve enterprise integration.

3.1. General description of the model

Looking at the model of the information infrastructure of an integrated enterprise as depicted in figure 1, it is important to take note of the following:

- The information infrastructure consists of a number of cooperating systems, that are arranged in a number of levels that broadly corresponds to the different management levels encountered within an organization

- In general the model consists of:
    - a meta-system (top-level system),
    - n levels of information subsystems, where n>=1.
    - operational level software (bottom-level systems).

- The number of levels of information subsystems (*n*), is determined by the size and complexity of the specific enterprise.

- Each level of information subsystems can contain any number of systems, but the number of systems in a level will generally increase towards the operational level of the enterprise.

- Each level of information subsystems, acts as a filter for information passing through the enterprise.

In the following paragraphs we will now proceed to discuss the different components of the model in more detail.

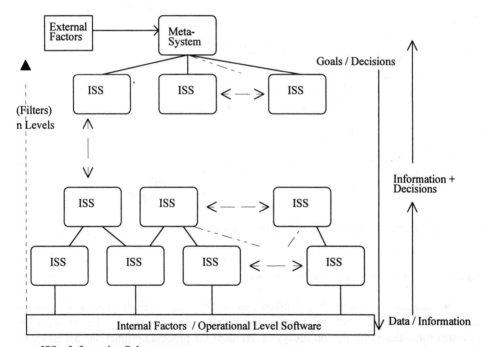

ISS = Information Subsystem

Fig 1. The relationships and data flow between information subsystems in an integrated enterprise.

3.2   Flow of information in the model

In the generic information infrastructure, as indicated in figure 1, there is the following flow of information:

- Goals to achieve, as well as decisions taken are passed down from higher, to lower levels.

- Information and decisions taken are passed from lower levels, upwards.

- The meta-system (top level) processes information about external factors (i.e., external to the enterprise).

- The bottom level information subsystems processes information and data received from operational level software.

3.3 Information Subsystem

The general structure of the information subsystems, is shown in figure 2.

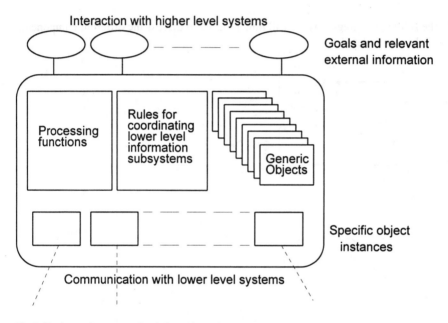

Fig 2. The internal structure of an information subsystem.

As indicated in figure 2, an information subsystem, receives goals and relevant external information from information subsystems, higher up in the information infrastructure. Each information subsystem contains the processing functions necessary to fulfill its local function. In addition to this it contains a number of rules, used to modify its own action according to the information received from the higher levels of information subsystems as well as for coordinating the subsystems it is responsible for. Each information subsystem, also contains a number of generic objects used as generic descriptions of the type of systems it can coordinate, as well as specific object instances of the objects it is coordinating at any time. In other words, each information subsystem contains a representation of its environment that is detailed enough to enable it to act intelligently on information received from the subsystems surrounding it and for it to be able to coordinate the action of the subsystems it is responsible for.

The function of each information subsystem can be summarized as follows:

- Each information subsystem has a local processing function.

- Each information modifies its own actions to accommodate any changes in its environment, and filters information to the appropriate subsystems.

- Each information subsystem needs to inform the subsystems it is responsible for coordinating, of any information that is relevant to that specific subsystem.

The next component, the meta-system is used to achieve intelligent action in a global context.

3.4 The Meta-system

Figure 3 shows the general structure of the meta-system.

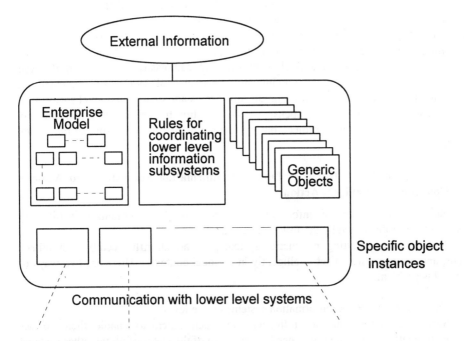

Fig 3. The internal structure of the meta-system.

The meta-system contains a dynamic model of the complete enterprise and its environment. Using this model, the meta-system is able to simulate different scenario's, showing the effect of changes in the environment or in any subsystem on the enterprise as a whole. The important difference between the meta-system and traditional simulations, used to model organizations, is that the meta-system is 'aware' of the current state of all the information subsystems within the organization. This awareness is maintained by means of communication between the meta-system and the information subsystems. This means that any changes within the subsystems are reflected in the meta-system.

The meta-system is the only system that is aware of the current state of the external environment, as well as the complete internal environment. Because of this, the meta-system can be used to coordinate the actions of all the other systems to allow intelligent action in a global context. The function of the meta-system can be compared to that of the conscious mind of a human being, whereas the functionality of the rest of the model can be compared to reflex reactions and the unconscious control of aspects such as breathing and heart beat.

3.5 Operational Level Software

The operational level software represents the root information sources within the organization. The main difference between the operational level software, and the information subsystems, are that the software systems at the operational level are not necessarily 'aware' of their parent systems. This usually means that the information subsystems immediately above the operational level software will need special methods of extracting information from the operational level software, and for transferring information back to the operational level software. This communication is critical for the functioning of the generic information infrastructure for enterprise integration as a whole. The communication with operational level software can easily be achieved if these systems were developed using open architectures (Barker et al. 1993). By implementing this communication between the information subsystems and operational level software, it is possible to continue using the expensive systems within the organization that does not directly support the integration project, but that cannot be discarded because of the capital investment they represent.

4. Reasons for Using Different Levels of Information Subsystems, to Achieve Coordinated Intelligent Action.

Our reasons for using an information infrastructure that is organized in different levels can be best illustrated by the following example.

Consider N, cooperating information systems, that are all influenced by each other's actions, and are supposed to act intelligently in a global context. There are two ways of solving this problem:

1. The first, organizes the information systems in one level:
   Since the systems are all influenced by each other, to enable them to act intelligently, each system will need to be aware of the states of all the other systems at any one time. They will further more need to have a protocol for making joint decisions. Clearly this is a complex and not very effective solution to the problem. This solution also poses serious problems, when adding a new system, as all of the existing systems will need to be changed to accommodate the additional information system.

2. The second, solution adds a second level information system to coordinate the cooperating systems:
   Add an additional system with the primary function of coordinating the actions of the cooperating information systems. This one system can then pass relevant information to each of the information systems that will assure intelligent cooperation between them. This solution needs no complex protocols. It requires less communication, as each information needs only to communicate with the

coordinating system, and the addition of new information systems requires change in only the coordinating system.

From this discussion it should be cleared why we organized the information systems into different levels.

5. Conclusion

The use of this model has a number of advantages:

- Using this generic information infrastructure for manufacturing enterprise integration makes it possible to achieve an 'aware' enterprise by managing the information resource of the organization as a single logical entity.

- This arrangement of information systems makes it possible for the total information system of an organization to react intelligently in a local, as well as a global context.

- Using this model, it is possible to include the existing software systems within the integrated framework, even if they do not directly support integration.

The use of this model also has the following implication from a software engineering perspective:

- Within an integrated organization, each software component should be developed taking into account its final position (in reference to all other software components) within the organization.

- When buying a new software component, the main criteria for deciding which product to choose, should be integration with existing systems.

- In an organization, once the information infrastructure, has been fixed, it is possible to develop the different components in simultaneously, as long as their interfaces with each other remain unchanged.

6. References

1. Barker, H.A. et al. April 1993. Open architecture for computer-aided control engineering. *IEEE Control Systems* : p17 - 27.
2. Doumeingts, G. et al. Towards virtual manufacturing systems. *JSPE-IFIP WG 5.3 Workshop on the design of information infrastructure systems for manufacturing*: p389 - 403. University of Tokyo. Japan.
3. Katzy, B.R. et al. November 1993. CIMOSA pilot implementation for technology transfer. *JSPE-IFIP WG 5.3 Workshop on the design of information infrastructure systems for manufacturing* : 51 - 62. University of Tokyo. Japan.

4. O' Hare, G.M.P. 1991 Designing Intelligent Manufacturing Systems: A distributed artificial intelligence approach. *Intelligent Manufacturing Systems III* : 17 -25. Amsterdam Elsevier Press.
5. Williams, T.J. et al. November 1993. Architectures for integrating manufacturing activities and enterprises. *JSPE-IFIP WG 5.3 Workshop on the design of information infrastructure systems for manufacturing* : 1 - 17. University of Tokyo. Japan.

ENGINEERING DATA AND PROCESS INTEGRATION IN THE SUKITS ENVIRONMENT

BERNHARD WESTFECHTEL
Lehrstuhl für Informatik III, RWTH Aachen, Ahornstr. 55
D–52074 Aachen, Germany
E–mail: bernhard@i3.informatik.rwth–aachen.de

ABSTRACT

SUKITS is a joint project of computer scientists and mechanical engineers which is devoted to the design and implementation of an integrated infrastructure for product development. This infrastructure is called CIM Manager and provides services for managing descriptions of products and manufacturing processes, managing engineering processes, and for communication in a heterogeneous network. The CIM Manager provides a framework for a posteriori integration of heterogeneous application systems. Management of engineering products and processes is tightly integrated, and process management supports concurrent engineering as well as dynamic evolution of process nets during execution.

1. Introduction

A posteriori integration of heterogeneous engineering design applications is a challenging task. *CAD frameworks* [2] address this challenge by providing an infrastructure for data, process, communication, or user interface integration. In this paper, we present such an infrastructure which is called *CIM Manager* and has been developed in the *SUKITS project* [1]. SUKITS is a joint project of computer scientists and mechanical engineers which was launched by the Technical University of Aachen in fall 1991. Within this project, we have developed an infrastructure for a posteriori integration of heterogeneous engineering design applications which have been developed by different vendors, run under different operating systems, and rely on different data management systems.

In contrast to more comprehensive projects like e.g. CIM–OSA [6], we do not intend to cover the whole area of Computer Integrated Manufacturing. Rather, we focus on *product development*, i.e. on all engineering activities which are concerned with developing descriptions of products and manufacturing processes. The results of such activities are captured in *documents* such as CAD designs, manufacturing plans, or NC programs. Documents may contain both product and (manufacturing) process data. So far, we have not investigated those phases of the product life cycle which follow product development, namely production planning and production. On the other hand, our work exceeds the CIM area because we reuse our integrating infrastructure in other domains, specifically in software engineering [7].

This paper describes data and process integration provided by the CIM Manager [9, 10]. For communication integration, the reader is referred to [3, 4]. Only small efforts have been devoted to user interface integration.

2. Data integration

Data integration is concerned with managing the products of engineering design processes. We call these products *engineering design documents* (briefly denoted as documents in the sequel). A document is a logical unit of reasonable size typically manipulated by a single engineer (e.g. a CAD design, a manufacturing plan, an NC program). According to the constraints of a posteriori integration, we do not make any assumptions regarding the internal structure of documents. This approach is called *coarse-grained* because a document is considered an atomic unit. Physically, a document is typically represented as a file or a complex object in an engineering database.

Logically related documents are collected in *document groups*. For example, a document group may comprise all documents describing a single part and its manufacturing process. Although all documents of a group might be created by the same engineer, a document group usually serves as a work context for integrating documents produced by multiple engineers. In general, document groups may be nested such that they form a composition hierarchy of arbitrary depth.

Components of a document group are related by various kinds of *dependencies*. Such dependencies may either connect components belonging to the same work area (e.g. dependencies between descriptions of components of an assembly part), or they may cross work area boundaries (e.g. dependencies between designs and manufacturing plans). Accurate management of dependencies is an essential prerequisite for *consistency control*, i.e. for keeping interdependent components consistent with each other. To this end, tools are needed which operate on the fine-grained level, i.e. on the contents of documents. Since product management follows a coarse-grained approach, the CIM Manager itself cannot perform consistency control on the fine-grained level. However, it provides a reusable framework for embedding domain-specific tools operating on the fine-grained level.

During its evolution history, both documents and document groups evolve into multiple *versions*. Versions may be regarded as snapshots recorded at appropriate points in time. The reasons for managing multiple versions of an object (instead of just its current state) are manifold: reuse of versions, maintenance of old versions having already been delivered to customers, storage of back-up versions, or support of change management (e.g. through a `diff` analysis figuring out what has changed with respect to an old version).

The CIM Manager supports uniform versioning of both documents and document groups. Therefore, we subsume both notions under the generic term *object*. Each object has an evolution history represented by a graph of versions interconnected by *history relations*. Versions of documents and document groups are denoted as *revisions* and *configurations*, respectively. Due to uniform versioning, all benefits gained from version control (e.g. reuse, change management) may be exploited on arbitrary levels of the composition hierarchy.

The CIM Manager database is composed of a set of interrelated *graphs* which are called version graphs, configuration graphs, and document group graphs, respectively. In the sequel, these graphs are explained in turn.

A *version graph* (fig. 1) consists of versions which are connected by history relations. A history relation from v_1 to v_2 indicates that v_2 was derived from v_1 (usually by modifying a copy of v_1). In simple cases, versions are arranged in a sequence reflecting the order in which they were created. Concurrent development of multiple versions causes branches in the evolution history (version tree). Merging of changes performed on different branches results in directed acyclic graphs (dags), where a version may have multiple predecessors. Finally, a version graph may even be separated if multiple branches have been developed in parallel from the very beginning.

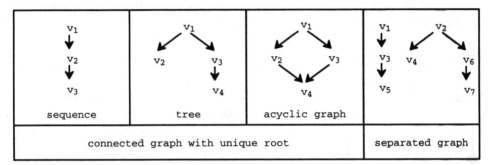

Fig. 1 Version graphs

A *configuration graph* contains version components and their mutual dependencies. As we have already mentioned above, configurations are versions, too. Therefore, configurations of document groups may be employed as units of reuse in the same way as revisions of documents. As a simple example, the top half of fig. 2 shows three configurations of a document group gathering some documents for a certain single part (e.g. a shaft). c_1 contains initial revisions of a CAD design, a manufacturing plan, and two NC programs. The manufacturing plan depends on the design, and the NC programs depend on both the manufacturing plan and the design. In c_2, the manufacturing plan has been changed, resulting in corresponding modifications to both NC programs. The transition from c_2 to c_3 involves a structural change introducing another NC program such that the single part is manufactured in three consecutive steps.

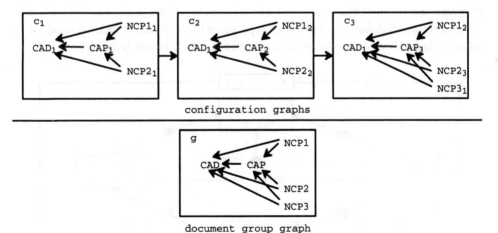

Fig. 2 Configuration graphs and document group graph

A *document group graph* abstracts version-independent information from all configurations of some document group. Its purpose is to express properties common to all members of a family of configurations. A document group graph is constructed by performing a union of all configuration graphs and abstracting from the version numbers. An example is given in the lower half of fig. 2. Note that the CIM Manager offers analysis operations which mark all varying parts, i.e. all elements of the document group graph which do not belong to the intersection of all configuration graphs.

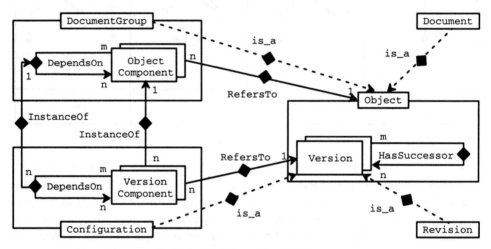

Fig. 3 Meta model for product management

Fig. 3 presents an ER-like schema for product management. This schema is called *meta model* because it is independent of a certain application domain. Labeled boxes represents complex entity types (DocumentGroup, Configuration, and Object, respectively). Component types with cardinality n are shown as double-shaped boxes (ObjectComponent, VersionComponent, and Version, respectively). Relationship types are drawn as black diamonds with adjacent solid lines (e.g. HasSuccessor is an m:n relationship between entities of type Version). Finally, black diamonds with adjacent dashed lines denote inheritance relations (e.g. Document is an Object).

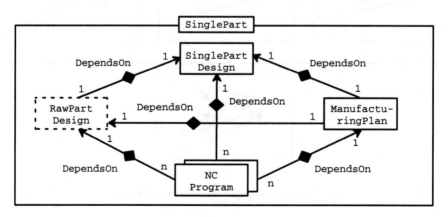

Fig. 4 Concrete model for product management

The CIM Manager is adapted to a certain application domain by defining a *concrete model* which specializes the meta model. An example is given in fig. 4. The sample model refers to the description of single parts by means of designs, manufacturing plans, and NC programs. Depending on its cardinality, a component type is represented by a double-shaped or single-shaped box (upper bound n or 1, respectively) with solid or dashed lines (lower bound 1 or 0, respectively). Note that the concrete model applies to both document groups and configurations.

The database described above serves as an *integration database* above the databases of application systems. The integration database maintains references into the databases of application systems where the contents of document revisions are stored physically. Provided that an application system supports external access to its database, this database is split into a private part and a public part the latter of which is controlled by the CIM Manager. Documents are exchanged between public and private part by Checkout and Checkin operations which control change propagation and provide for a well-organized cooperation between multiple engineers. Engineers release their work to the public by Checkin operations; others do not see inconsistent intermediate results produced in their private workspaces. For further details, the reader is referred to [9, 10].

3. Process integration

We have developed an approach to management of engineering design processes which is closely integrated with product management. The term *product-centered process management* emphasizes this point. By interpreting a configuration of interdependent components as a process net, we avoid the conceptual overhead which is implied by separating product and process structure and keeping the latter consistent with the former. Basically, each component of a configuration corresponds to a process which has to produce this component (i.e. a *component process*). Analogously, dependencies between components correspond to *data flows* between component processes (if A depends B, then data flows from B to A). Configurations are enriched by all information which is required for process management. In particular, this implies attachment of process-related attributes to elements of configurations (e.g. execution state of a component process).

Process management must cope with *dynamic evolution* of process nets. Only rarely may a process net be built up completely before execution is started. In many cases, decisions are made during execution which determine how to proceed. Either these decisions cannot be anticipated in their totality, or taking all possible execution paths into account yields a huge and complex process net which is hard to understand and to follow. Therefore, it is essential to support modifications of process nets during execution.

Evolution is constrained by a *schema* which captures a priori knowledge about configurations and process nets. For example, let us interpret the schema given in fig. 4 in terms of process management. The process of developing a single part involves preparation of one design and one manufacturing plan, respectively. However, the number of NC programming processes is not determined a priori. Furthermore, it depends on the lot size whether it pays off to produce a raw part in order to minimize cut-offs.

Process nets are *extended incrementally* as execution proceeds. A simple example is presented in fig. 5. The left-hand side of the figure shows an initial net which contains a design and a planning process which are connected by a data flow. The processes are identified by the names of the components which have to be developed (CAD and CAP, respectively). Note that the names are not qualified by indices. This means that the corresponding components are not yet bound to a specific version. The product management model supports partially bound configurations where some components are not bound to a specific version. The right-hand side of fig. 5 shows the state of the net after design and planning have been completed. The net has been extended with two NC programming processes which have been defined in the manufacturing plan. Terminated processes are displayed in grey; their indices correspond to the numbers of the versions which they have produced.

initial net   extended net

Fig. 5 Dynamic net extension

The "classical", conservative rule for defining the execution order states that a process can be started only after all master processes have been finished. However, enforcing this rule significantly impedes parallelism. In order to shorten development time, the conservative rule needs to be relaxed such that processes may be executed concurrently even if they are connected by data flows. Our approach to process management supports *concurrent engineering* [8] inasmuch as intermediate results may be pre-released to dependent processes as soon as possible.

To this end, each process maintains a *release set* indicating which dependent processes may access the result produced so far. An "all or nothing" approach to releasing results is too coarse-grained because it depends on the kind of a dependent process when data should be propagated from master to dependent. Therefore, the release set contains the types of component processes to which the result of the corresponding process has been (pre-) released.

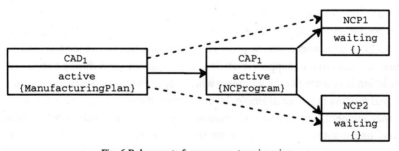

Fig. 6 Release sets for concurrent engineering

Fig. 6 illustrates the release strategy by means of a simple example. After a draft design has been prepared, this intermediate result has been passed to the planning process such that design and planning process may proceed concurrently. The planning process has in turn released the preliminary manufacturing plan to NC programming. However, NC programming cannot be started yet because geometric data produced in design are still too imprecise (i.e. data are not yet propagated along dashed flows).

The execution semantics of processes is partially defined by a *state transition diagram* (fig. 7). Normally, a process steps through the states in the middle column (Created, Waiting, Ready, Active, and Done). The left column consists of states denoting interruption (Blocked, Suspended) and abortion (Failed), respectively. Iterate transitions start the process life cycle again; Reuse skips activation if an existing component can be reused without modification. Undecided serves as an intermediate state in which reactivation is considered. For further details, the reader is referred to [10].

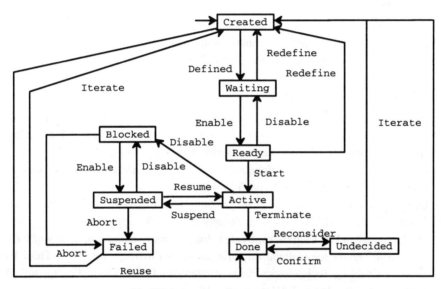

Fig. 7 State transition diagram for processes

4. Conclusion

Within the SUKITS project, we built a prototype of the CIM Manager which was completed in fall 1993 and is currently being redesigned and extended. In order to evaluate the prototype, two *sample scenarios* were investigated, namely development of single metal parts and of plastic parts produced by injection moulding, respectively. Both scenarios are rather complex; the corresponding schemas both comprise about 15 document types and 30 dependency types. Documents to be managed include e.g. CAD designs for different purposes (e.g. geometries of the single part to be produced, of the raw part to start with, of the positions of the working piece, etc.), manufacturing plans (CAP), NC programs (CAM), material requirements definitions, results of FEM simulations (CAE), etc. The CIM Manager was tailored to both scenarios by defining corresponding schemas and by integrating heterogeneous application systems.

The *SUKITS prototype* consists of a schema editor for adapting the system to a specific scenario, tools for editing, browsing, and analyzing version and configuration management data on the instance level, interfaces to CIM application systems to be integrated, and an OSI-based communication system [3, 4] gluing all components together in a heterogeneous environment (multiple types of machines, operating systems, and data management systems). The implementation of the SUKITS prototype comprises more than 60,000 loc written in Modula-2 and C; about 50 % of the code are dedicated to the implementation of operations on the integration database. The implementation of the prototype was made considerably easier by reuse of components developed within the IPSEN project [7], which is concerned with the construction of integrated software development environments. In particular, we have heavily used the GRAS database system [5] for the implementation of the CIM Manager's integration database.

5. Acknowledgements

This work was partially supported by the German Research Council (DFG) under the project title SUKITS (software and communication structures in technical systems). The author is indebted to all members of the SUKITS project.

6. References

1. W. Eversheim, M. Weck, W. Michaeli, M. Nagl and O. Spaniol, *The SUKITS Project: An approach to a posteriori Integration of CIM Components*, Proc. GI–Jahrestagung (Informatik aktuell, Springer Verlag, Berlin, 1992), pp. 494–503.
2. D. Harrison, A. Newton, R. Spickelmeir and T. Barnes, *Electronic CAD Frameworks*, Proc. of the IEEE **78–2** (1990), pp. 393–419.
3. O. Hermanns, *Data Access Protocols for Integrated Engineering Environments*, Proc. COMPEURO '93 (IEEE Computer Society Press, 1993), pp. 350–357.
4. O. Hermanns and A. Engbrocks, *Design, Implementation and Evaluation of a Distributed File Service for Collaborative Engineering Environments*, Proc. Third Workshop on Enabling Technologies: Infrastructure for Collaborative Enterprises (IEEE Computer Society Press, 1994), pp. 170–175.
5. N. Kiesel, A. Schürr and B. Westfechtel, *GRAS, a Graph-Oriented (Software) Engineering Data Base System*, Information Systems **20–1** (1995), pp. 21–51.
6. K. Kosanke, *The European approach for an Open System Architecture for CIM (CIM–OSA) – ESPRIT project 5288 AMICE*, Computing & Control Engineering Journal (May 1991), pp. 103–109.
7. M. Nagl, *Characterization of the IPSEN Project*, Proc. 1st International Conference on Systems Development Environments & Factories 1989 (Pitman Press, London, 1990), pp. 141–150.
8. R. Reddy, K. Srinivas, V. Jagannathan et al., *Computer Support for Concurrent Engineering*, IEEE Computer (January 1993), pp. 12–16.
9. J. Schwartz and B. Westfechtel, *Integrated Data Management in a Heterogenous CIM Environment*, Proc. COMPEURO '93 (IEEE Computer Society Press, 1993), pp. 248–257.
10. B. Westfechtel, *Integrated Product and Process Management for Engineering Design Applications*, Integrated Computer–Aided Engineering (1995), to appear.

Systems Integration

A Case Analysis of the Limits to Integration in Organizations Adopting Computer Integrated Manufacturing
    P. L. Forrester and J. S. Hassard

The Integration of Document Archives in a Manufacturing Automation Protocol (MAP) Environment
    A. Prosser

Communication Networks for Computer Integrated Manufacturing
    W. P. Lu and W. F. Lu

Manufacturing Information Enhancement through Data Integration
    A. E. James

Systems Integration — Coping with Legacy Systems
    V. Singh

A CASE ANALYSIS OF THE LIMITS TO INTEGRATION IN ORGANIZATIONS ADOPTING COMPUTER INTEGRATED MANUFACTURING

PAUL L FORRESTER

Department of Management, Keele University, Keele, Staffs, ST5 5BG, U.K.
E-mail: mna03@keele.ac.uk

and

JOHN S HASSARD

Department of Management, Keele University, Keele, Staffs, ST5 5BG, U.K.
E-mail: mna00@keele.ac.uk

ABSTRACT

The majority of empirical research into the design and operation of CIM has tended to be conducted from a technical perspective. Consequently approaches adopted are often limited in the extent to which they address the capital, human and technical resource constraints acting upon manufacturing companies. This paper describes a programme of empirical case-based research into CIM systems design. It concludes that not only are the organizational difficulties involved in CIM underestimated, but that CIM is perceived as a technological issue which runs into organizational difficulties rather than as an organizational change in itself.

Key Words: Integration, Organization, Design Process

1. Introduction

CIM can be viewed as an holistic approach to computerised systems and facilities development. From the management perspective, however, the question still remains as to whether, in the long term, such an approach to manufacturing systems design and operation can be achieved. It is most unlikely that such an holistic system can ever be realised when the manufacturing environment is always subject to change as a result of market demands, social and organizational change and technological innovations. In addition, interpretations of CIM vary considerably with the result that applications tend to have a high degree of specificity to the business and organizational circumstances of adopting businesses. For this reason the authors have used analytical methods used by organizational theorists to structure empirical evidence and explore the design process for CIM in a number of manufacturing companies. This paper describes the methods and frameworks used, and some of the conclusions from this analysis.

2. CIM and Business Organization

The relationship between technological innovation and company structure is one that has been widely explored in the literature on the management of organizational change. This suggests that rigid modes of communication exhibited by functionally differentiated firms can inhibit the innovation process. The realisation on the part of adopting companies, that CIM involves more than the technical integration of various computer systems is an issue that resonates with the wider literature on organizational change. Among the structural problems which arise are those related to functional differentiation, status inconsistencies, co-ordination flows and task interdependence (Child, 1984; Perrow, 1967; Thompson, 1967). In order to explore, therefore, the design process for technological developments a number of so-called "periodizations" of design have been developed. A model directed specifically towards process and product design was proposed by Utterbeck and Abernathy (1975), which demonstrates the benefit of distinguishing between different types of innovations. They advocated a framework of analysis for production units which distinguished between product, process and work organization design. This was used by Abernathy (1978) in his investigation of design and productivity improvement in the US automobile industry. Later the model was employed by Whipp and Clark (1986) in their investigation of the Rover SDI project and, to this, they added the periodization of conception, translation, commissioning and operation. In a similar vein a member of KAMG used this framework to examine innovations within Cadburys Limited, emphasising the relationship between new product and process innovations and work design (Smith *et al*, 1990).

A key element in the Keele project was the need to identify at which stage organizational issues were addressed. In some cases this might be before the decision to adopt the technology has been made. It is for this reason that use of the Whipp and Clark (1986) framework is being proposed. This model enables design to be set in a historical context and does not presuppose that the technology is a major determinant of organizational change. For the purposes of the Keele investigations into CIM development therefore, the four stages of design were identified as:

(a) conception, the envisioning of CIM;
(b) translation of the intent to adopt CIM into specifics through analysis and planning (the unravelling and rebundling of the elements of CIM);
(c) implementation, which involves finalising definitive specifications, tendering, procurement, construction and commissioning of CIM elements; and
(d) operation of CIM in practice, including the monitoring and analysis of performance and the achievement of operational refinements.

3. Research Framework

The research conducted had three main objectives: firstly, to ascertain the reasons for the introduction of CIM and the extent to which these are organizational; secondly, to use a temporal framework to analyse the CIM design process in a number of UK companies; and, thirdly, to contribute to debates on organizational change and its relationship with technological innovation.

The research objectives are directed towards examining the generic hypothesis that "the role of Computer Integrated Manufacturing (CIM) in production organizations is shaped by strategic choice rather than by technological imperative". This hypothesis locates the research within the broader debate between determinism and voluntarism. For empirical purposes, two sub-hypotheses were examined within the context of the main hypothesis: (a) that "CIM is introduced by management as a strategic technological innovation"; and (b) "CIM is a vehicle for organizational changes which are socially constructed as a necessary condition for technological change".

Although predominantly a process innovation, CIM should not be seen in isolation from product or work organization design. From the management and organizational perspective the question still remains as to whether, in the long term, such an holistic approach to manufacturing systems design and operation can be achieved. The testing of this hypothesis, therefore, entails an examination of the extent to which managers see the adoption of a concept seemingly as esoteric as CIM as a means to effect organizational change.

4. The Case of the Automotive Products Manufacturer

The company examined in this paper is that of a medium sized UK manufacturer of electrical components for the automotive industry. The company grew from a family business, but has for many years been part of a larger group. The company, however, still maintains a high degree of independence in policy and actions. It employs a total of around 4000 people spread across four manufacturing sites. The company was not slow to adopt computerised technology, having been pushed into this by the demands of its customers, the large vehicle assembly companies. This development, however, had been piecemeal, with a lack of central direction and very little view to compatibility and communication between different modules. The company now sees the integration of these 'islands of computerised technology' as the key to ensuring continuing manufacturing edge and responsiveness to customer needs. The integration of systems and parts of the business was essential in maintaining competitiveness and responsiveness to customer requirements, as the need to respond more efficiently and consistently to customers' demands for design flexibility and schedule changes increases.

The basic design of the company's product is very much within the domain of the customers, the car manufacturers, although more recently the increasing use of guest engineers is bringing more design responsibility to the company. The main concern in product design is the translation of customers' specifications, written using the customers' own terminologies and conventions, into a form that employees can understand and that can be directly entered into the computerised manufacturing control systems. The key to success is to do so with as little delay as possible. Also, argue the systems designers, 100 per cent accuracy is required in order that error-free prototypes and pre-production samples can be sent to the customer whilst the overall design leadtimes, from conception to manufacture, can be minimised. In order to analyze the relationship between the technology and the organization of the company, we shall trace the development of the CIM project in terms of the four stages identified by Whipp and Clark.

4.1 Conception and Translation

The company's 'CIM Project' was officially conceived in 1989. However, development of many of the components that make up CIM were initiated long before this: 1984 in the case of Computer Aided Design (CAD); 1987 in the case of the initiation of Computer Aided Production Management (CAPM) in the form of Manufacturing Resources Planning (MRPII); and 1988 in the establishment of an advanced engineering group. The responsibilities of the CIM Manager were largely confined to overseeing CAE and CAPM developments and to effect their integration. In 1989 the company conducted an assessment of its engineering knowledge in product design and development. It concluded that the current engineering and production systems were not adequately coordinated and suggested the key to maintaining the core business was the development of a fully integrated information system. It was here that the initial interest in CIM within the company was nurtured. Furthermore, it was recognized that the developments under the banners of CAD/EDI and CAPM constituted an underswell towards computerised information systems within the company. Having seen that the company had decided that the concept of CIM was to be 'translated' into merely bringing together existing and emergent systems through an internal project team, we can now move to an examination of how the project was 'commissioned'.

4.2 Commissioning of CAE

Turning firstly to computer-aided engineering (CAE), those components of the proposed system that are in existence at present, for example CAD and customer EDI, required further development and refinement before they were capable of functioning efficiently within an integrated system. For many of the components their development was starting from scratch. In the light of the requirements outlined above, the company invested in the use of 2D CAD. This was initially introduced in 1984 to assist in what was (and, to a large extent, still is) an almost entirely manual process. A considerable amount of the CAD programs comprise software written specifically to perform the functions required by the company. Expenditure on CAD since 1984 has been around £2 million. The first year saw the installation of CAD screens and supporting hardware and software, plus the training of staff. The emphasis at first was in the use of CAD for individual component design and tool design. As a result the first CAD designed component was released in 1985 and was closely following by the design of a machine tool on CAD. Component design was fully dependent on CAD by 1987 and 1988 saw the transfer of designs on paper to electronic design and drawings. The design of tools was also progressed, with press tools designed using CAD in 1986 and NC programming for a number of processes in 1987. By 1990 tool design, like that of components, was fully dependent on CAD for all new designs and most modifications. Only those modifications on tools with very old design drawings were carried out by conventional methods.

Given that the development of CAD has entailed substantial time, money and human resources, there is currently a debate within the company as to whether the benefits accruing from CAD have justified the expenditure. Design of the product is initially very much within the domain of the vehicle manufacturers, and so the CAD system acts more as a translator and representor of customer design details than an initial design tool. As a consequence of this, there has been questions as to whether CAD represents a huge investment for little direct and immediate benefit. The converse argument, whilst accepting that the measurable benefits deriving from CAD are not great at present, suggests that the main prize from CAD will come in downstream CAE developments as one of the key foundations of fully integrated CIM. This argument has still to be resolved. However the very presence of this debate is moving investment decisions from being made using an 'act of faith' to the position where future developments have to be justified and their measurable benefits stated in order for further expenditure to be appraised in quantitative terms.

EDI developments have been very closely linked to the evolution of CAD over the last few years. The main objective in developing EDI at present is to enable electronic design drawings to be directly sourced from the customer and then loaded automatically into CAD. The first milestone in the development of EDI was in 1988 when the company ran a number of trials with a major customer. By 1990 product design engineering communication through EDI existed with two customers. The company can now take electronic data in from customers, albeit on a tentative basis, process these through the various translator packages and then transfer directly into the CAD system. Though this requires further refinement, the company has earned considerable prestige from its customers for its pioneering work and application of EDI in the areas outlined above. There is also a high regard for the company's adoption within the automotive supply industry as a whole.

4.3 Commissioning of Computer Aided Production Management (CAPM)

Engineering systems, centring largely around the base of CAD and central mainframe databases, form one strand in the development of computerised systems. Another major part is the evolution of a computer aided production management system, particular emphasis being placed on the development of an MRPII system. Work on the introduction of MRPII began in 1988. Prior to this the company used an MRP (Materials Requirements Planning) system, although this was not operating satisfactorily and was difficult to upgrade because it used mostly in-house, bespoke software developed between 1978 and 1982. It also operated in batch mode with long time buckets and in no way could be viewed as a real-time CAPM system. Its level of control in a day to day dynamic sense was, as a consequence, very restricted. The company decided to move towards an MRPII system and chose to base the new planning and control system around a vendor's package, tailoring this to suit the needs of the company wherever possible.

Initial estimates of the times to develop the new CAPM systems were highly optimistic. The general impression was that it would take around eighteen months, and so the system was expected to be fully operational by the end of 1989. The current status is that only one of the company's four plants has gone 'live' and even this was restricted to MRP replacement rather than full MRPII implementation. Moreover the original justification and terms of reference for the introduction of MRPII were very loose, there being few explicit or quantified statements on the level or timing of these improvements and, therefore, little to measure MRPII implementation against in an evaluative sense in terms of adherence to plan and achieved benefits.

4.4 Commissioning of Other Developments

To support CAE and CAPM systems developments the company have recognized the need to develop appropriate supporting modules in administrative areas. Two recently conceived module developments warrant mention here. Firstly, as one of the goals of the CIM system is seen to be rapid response to customer needs, an efficient and effective computerised sales order processing system is required to front the interaction with customers. This module will not only provide input to the CAPM systems for scheduling purposes, but will also interact with the CAE systems, providing input in the case of orders for new or modified product designs and drawing data from the various parts of the engineering systems for use in MRP scheduling and the ordering of components. The second development is that of computerised financial modules, the current systems being perceived as a weakness in the company's operations and not providing an adequate supply of information for systems users and managers. At the time of writing a multi-disciplinary and user-led project team was being set up under the chairmanship of the Financial Director to address the design of this module.

Although the effort towards CIM within the company has been mainly directed towards engineering and operations control rather than using advanced production machinery technologies, many developments are currently underway in the machine automation area. The question of their integration is an important one, particularly regarding the linking of process technologies with CAE and CAPM modules. However, it is recognized that if CIM is to incorporate the concept of Computer-Aided Manufacturing (CAM) and its links to CAD (CADCAM), then the activities currently being conducted within manufacturing engineering departments must be integrated with the present CIM work at some time in the future.

4.5 Operation

The unforeseen complexity of the tasks involved when developing scheduling routines, a dynamic bill of materials and an inventory status file to operate with the requisite degrees of integrity has resulted in a severe overrun of this project. In the integration of CAPM systems with the CAE system there exists a gulf between the separate development teams on these projects, both in a technical and organizational sense. As such integration between CAE and CAPM within CIM is still some way from being realized. Despite the company's original intentions, therefore, CIM is still far from being in operation.

Finally, it needs to be stressed that he need to integrate systems has only truly been recognized in the last eighteen months and there currently exists a gulf between the engineering driven CAE systems developments on the one hand and the operations and management service driven CAPM control systems on the other. Organizational barriers and political tensions exist and have hindered the fusing of parallel systems developments, akin to what Abernathy (1978) has termed 'roadblocks to innovation'.

5. Conclusions

All the CIM projects examined by the authors (including the one presented here) were seen as being most likely successes. But, given that there were no criteria established at the outset against which to measure performance of the new system, these claims are hardly a surprise. In all cases CIM was conceived as a matter of faith. As the engineering director of the automotive products company put it: "I think we are doing it very much on a faith basis. I don't think we have sat down and set out the detailed cost benefits."

Our research indicates that managers and engineers alike frequently underestimate the organizational difficulties involved. Our view is that CIM should be seen not only as a technological entity, but also as an organizational approach because without the attention to detail to the organizational development implied by "integration" the full benefits of linking together CAE, CAD, CAPM, MRP, etc. will not be achieved. Therefore the implementation of CIM has to be preceded and accompanied by a high level appraisal of the organization, its objectives, structure and processes.

Despite the widespread availability of integrating technology that makes possible the computerised interface between CAE and CAPM, there exist a number of difficulties for CIM design and operation. CAE modules encompass those activities traditionally performed by designers and engineers, whilst CAPM replaces the manual activities of production scheduling, control and operation, historically within the domain of production management. Organizational integration and the elimination of barriers has therefore proven to be more difficult to achieve in practice than systems integration. For truly effective CIM to work in practice there needs to be a detailed appraisal of organizational structure and processes. The appropriate development and reorientation of the organization is needed to support and enable the technology to work for the maximum benefit of the business. The cases in this paper illustrate severe organizational difficulties at the CAE-CAPM interface. Modules have been developed in isolation, within different areas and only to be integrated at a later date. Attempting, then, to compel these isolated systems development teams to work together in harmony will, more often then not, pose organizational conflicts and power wrangling amongst the development personnel.

But, more than this, much of the research into the organizational issues of CIM to date has concerned itself with how to ensure the CIM technology is effective through the analysis of organizational strategies, structures and processes. This tends to miss the major issue. The point is not that managers and engineers underestimate the organizational difficulties involved in the implementation of CIM. Rather, CIM is perceived as a technological issue which runs into organizational difficulties, instead of being seen as an organizational change in itself.

6. References

W J Abernathy *The Productivity Dilemma*, Johns Hopkins University Press, Baltimore, 1978.
J Child *Organization: a Guide to Problems and Practice*, Harper Row, London, 1984.
C Perrow "A Framework for the Comparative Analysis of Organizations", *Amer. Soc. Rev.*, April, 1967, pp.194-208.
C Smith, J Child and M C Rowlinson *Reshaping Work: The Cadbury Experience*, Cambridge Univ. Press, Cambridge, 1990.
J D Thompson *Organizations in Action*, McGraw-Hill, New York, 1967.
J M Utterbeck and W J Abernathy "A Dynamic Model of Process and Product Innovation", *OMEGA*, 3(6), 1975, pp.639-656.
R Whipp and P A Clark *Innovation in the Auto Industry*. Frances Pinter, London, 1986.

THE INTEGRATION OF DOCUMENT ARCHIVES IN A MAP ENVIRONMENT

ALEXANDER PROSSER

Department of Industrial Information Processing, University of Economics and Business Administration
Augasse 2–6, Vienna, 1090, Austria
E-mail: alexander.prosser@wu-wien.ac.at

ABSTRACT

Business process reengineering is creating the need for increased information system integration. However, existing document archive systems are mostly proprietary, which makes it difficult to integrate document processing in distributed corporate information systems. On the basis of the ISO Document Filing and Retrieval (DFR) standard (ISO 1991), this paper discusses the technical integration of archives in an open MAP environment. Following the technical discussion, an enhancement to the corporate repository covering document archives is proposed.

Keywords: Document Archives, Distributed Databases

1. Introduction

In the past, application development mainly oriented by the isolated needs of the various departments of the business enterprise. In the meantime, users have come to understand a corporation not as the sum of its departments, but as the sum of its business processes and that each business process is accompanied by a flow of information which does not stop at organizational or application boundaries (Davenport and Short 1990, Harrington 1991). Integration so far has focused on records-oriented databases, but there is still a large and mainly unused potential for economies in document processing, for instance in the field of technical documentation. In many cases, it is the document and its layout and the underlying logical structure which shall be stored as such, not only some data records extracted from it. In the following, "database" will refer to data stores for records-oriented data (in most cases relational), in contrast to document archives. Today, archives are mainly stand-alone systems intended for a limited number of users. Usually, paper documents are simply scanned into the archive to be retrieved at need. But even powerful and highly reliable scanners cannot alter the fact that this is an extremely inefficient way of maintaining an archive, because the logical structure of the document (which was *known* when the document was produced) becomes unreproducible once the document is printed. Only an image — probably an optically recognized stream of characters — can be reproduced, but the underlying logical structure is lost for ever. Consider the example of a business process depicted in Fig. 1 as an event-driven process chain (Scheer 1994):

A company produces small series of machinery products with the layouts taken from a set of standard variants, probably with some customer-specific modifications. Therefore, the sales department does not only give a price quotation based on the cost accounting data taken from the commercial application *((1)* in Fig. 1), but may also include a technical documentation of the variant in question *(2)*. The documentation is taken from a document archive, but there is no link to the commercial data in the offer *(3)*. After the

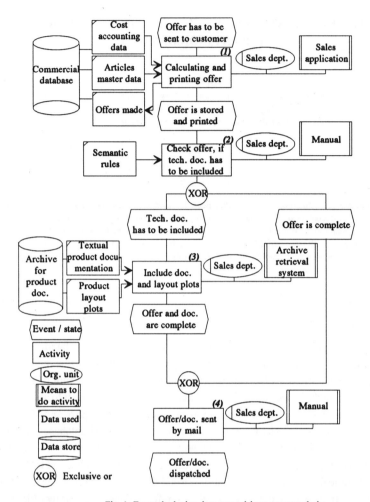

Fig. 1: Example depicted as event driven process chain

offer/documentation has been produced, the papers are sent to the customer by mail *(4)*. Again, it would be more efficient to send the documents electronically, but unless the customer has the same proprietary archive system as the supplier, this is pointless. As shown in this small example, business process redesign would often indicate integrated solutions, which cannot be implemented for lack of compatible products. The integration of document archives in open and distributed corporate information systems may be done in five steps:

1) A business process analysis to find out which document classes have to be integrated, where there are interfaces to databases, and which processes are concerned. The process chains are improved — thus establishing the requirements for the technical integration.

2) The planning of the technical integration of archives into distributed data processing; this is the stage the paper concentrates on.

3) As a first implementation step, standardized document description languages, such as SGML (ISO 1986), may be used; at this stage, the proprietary archive systems still remain in use, but an implementation-independent document description is already used.

4) Introduction of the DFR interface (also by enhancing existing archives).
5) Creation of a joint system environment for databases and document archives.

Sections 2 and 3 discuss the technical integration of document archive management systems (DAMS) in MAP/TOP, section 4 proposes the integration in the corporate repository.

2. Introducing ISO Document Filing and Retrieval, DFR

A DFR archive consists of groups hierarchically ordered in a tree structure whose top is a root group. Each group may contain documents and/or further groups. Each document consists of its content and a set of profile attributes, such as *author*, *subject*, or *document_ status*. If a document shall be accessible in several groups, it is stored redundancy-free in one group with only a reference being inserted in the other groups. The result of a search for documents may be stored permanently like a document. Therefore, possible entries in the archive are groups, documents, references, and search results. Each document may be stored in several versions connected by chained references.

The user services provided include *copy* and *move* operations for single entries or whole branches of the archive tree, creation, deletion, retrieval and modification of entries, the entries of a group may be listed. The *search* operation enables the user to search the archive or parts of its tree structure (i.e. a search domain) using search filters against the attributes of entries. (Un)Equality, ordering ($<$, $>$, $\leq$, $\geq$), or search for substrings may be used in filters. The user's access rights are checked for each single entry, the possible access levels range from *simple read* (without reading access right attributes), to *owner* (i.e. the user who created the entry and who controls the access rights to it).

DFR does not restrict the internal implementation, there will be no "DFR archives". Rather, existing archive systems will offer standardized DFR interfaces. A DFR interface can handle documents in any kind of content description. To some extent, the standard favours documents in ODA description (Office Document Architecture, see below), as the ODA document profile can be directly accessed by DFR services. But also data in other description conventions or any plain ASCII file may be stored.

ODA documents are defined by two hierarchical structures (ISO 1989): The layout structure defines a page-oriented document organization. Pages may be subdivided into frames which may contain more frames or "blocks"; blocks are the lowest-level elements in the layout hierarchy and may be text, images or line graphics. The definition of the logical structure is entirely up to the application designer (for it is completely application-specific); ODA only specifies that it must have a hierarchical tree structure. Each portion of the document content (paragraph, header, plot, etc.) is attributed to a block and a lowest-level element of the logical structure. Both structures are usually generalized as predefined generic structures.

3. Technical Integration

3.1. Applications Accessing Archives and Databases

The International Organization for Standardization's (ISO) Remote Database Access (RDA) specification (ISO 1993) provides a model for standardized communication between client applications and database servers. Since RDA is based on ISO 8822 presentation services and the corresponding set of lower layer services, it can easily be integrated into a MAP/TOP (Manufacturing Automation Protocol, Technical and Office Protocol) environment (Valenzano et al. 1992). DFR uses the same system environment, but it needs ISO's Remote Operations Service. For an introduction to ISO's upper-layer standards, see Hebrawi 1993, Prosser 1993)

As long as human users directly work with the archive (i.e. using the viewers/editors provided by most archive system), a syntactically exact definition of DFR groups and document classes stored in them is not needed. Once applications without direct human interaction access the archive, its internal organization must be unequivocal, but it should still support user-friendly direct human access. "Without direct human interaction" means that users work with applications in a formatted dialogue entering references to data stored in databases and text or plots from archives, with the application generating and manipulating documents, or the batch manipulation of documents.

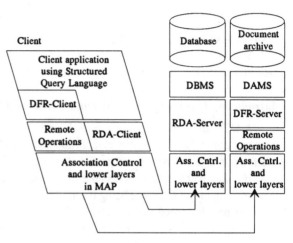

Fig. 2: Integrated application environment

This leads to the necessity of defining an archive schema which enables applications to identify document classes (section 3.2). DFR only defines a set of service primitives and their parameters, it does not specify any definite data manipulation language as RDA does define its use by SQL. Yet, the Structured Query Language SQL (ISO 1992, Date and Darwen 1993) has become the standardized means for querying databases. Most pre-relational databases offer SQL access, so do many existing archive systems. It is therefore reasonable to offer SQL access for the DFR interface as well (section 3.3). The result will be the integrated development and run-time system environment shown in Fig. 2.

3.2. Data Description in DFR Archives

Since direct human access to archives has to remain possible, the logical organization of the archive has to be enforced by the DAMS, not by the application programs. A database table is defined by its attributes — this automatically determines the exact structure of all records entered into the table. But from an archive's point of view, all documents are equal and may be stored in any group. Therefore, document classes have to be defined which are attributed to DFR groups. Applications accessing the archive supply the information necessary to identify the document class. For the implementor of a DFR/ODA archive, there are two fundamental options for document class definition:
- the fact that a document belongs to a certain generic logical structure or
- the domain of the attributes of the document profile.

Considering the seemingly obvious analogies (record – document) and (attribute – element of the logical structure), one would prefer the first method, but this variant has some serious disadvantages:

- Above-mentioned analogy does not take into account that documents may contain logical elements in repeating groups with varying repeating factors. Relational databases solve that problem by normalization (1 NF) (Date 1991), i.e. by defining a new table for the repeated attribute(s). But ODA documents cannot be "normalized", ODA does not know any mechanism for defining repeating groups in its content which are stored in an extra subdocument. The document is stored as a whole.
- Such a class definition may be used for ODA documents only.
- The requirements for class definition may be much more detailed than reference to the existing number of generic logical structures may allow.

In practice, the last point would lead to unnecessary duplication of identical generic logical structures just to be able to define document classes. Therefore, class definition based on the document profile attributes seems advisable. Also in this case, all ODA documents of the same class will contain reference to the same generic logical structure, but only as information for processing the document; several classes may share the same generic structure. The DFR document profile consists of two attribute sets, the basic DFR profile and the extended set which is identical to the ODA document profile. Since also non-ODA documents shall be handled by the DFR interface, only the basic profile with its 29 attributes will be used for class definition. Each of the profile attributes (a_1, .., a_{29}) has a given value domain, for a_1 being $D_{a_1} = \{v^1_{a_1}, \ldots, v^n_{a_1}\}$. The domain may only restrict the attribute to a certain data type, such as *integer* or *string*, or it may impose more explicit restrictions, even enumerate all allowable values. The Cartesian product of the sets of permissible values of the attributes gives the set P of n-tuples (n=29), where each tuple is a combination of allowable representations of these attributes: $P = D_{a_1} \times \ldots \times D_{a_{29}}$. The archive schema defines a class c by identifying a subset $P_c \subseteq P$; for the root class of the hierarchy, $P_c = P$.

A DAMS implementation should give the archive administration the opportunity to choose only some profile attributes for class definition. This distinguishes between attributes used for class definition and for information only. Each organization has a number of defined document types which can be grouped into a document hierarchy. The root class in each document hierarchy will be *document*. Descending from the root class, subclasses are defined which inherit properties (Coad and Yourdon 1991), from their superclass. Inheritance, in this case, means that all attributes used for definition will be present in *document* and that subclasses will redefine some of the properties inherited from their superclass by imposing further restrictions on the domain of some profile attributes. For the role of redefinition in inheritance, see (Hughes 1991).

A class c will have a set of subclasses S_c (its direct successors in the hierarchy), where each $c(i) \in S_c$ is defined by a set $P_{c(i)} \subseteq P_c$. It can be distinguished between leaf classes ($c / S_c = \{\}$) and intermediate classes ($c / S_c \neq \{\}$). The subclass definition of an intermediate class c (i.e. definition of all $P_{c(i)} / c(i) \in S_c$) will be called complete and disjoint, respectively, if the conditions stated in Eq. (1) and Eq. (2) hold:

$$\text{for i subclasses in } S_c, P_{c(1)} \cup, \ldots, \cup P_{c(i)} = P_c, \qquad (1)$$

$$\forall c(i), c(j) \in S_c \text{ there is no pair } (c(i), c(j)) / P_{c(i)} \cap P_{c(j)} \neq \{\}. \qquad (2)$$

A subclass definition of c which is complete and disjoint, ensures that a document which can be attributed to c can also be attributed to exactly one of its subclasses. If all

subclass definitions in the archive are disjoint and complete, all documents with valid profile attributes can be attributed to leaf classes. But completeness may be difficult to achieve, because if in a class, an attribute is defined rather loosely, e.g. as *string*, but its subclasses explicitly enumerate possible values, the subclass definition will hardly ever be complete. Three strategies may be chosen to solve this problem:
- The completeness criterion is abandoned, but the DAMS refuses an entry if it cannot be attributed to a leaf class. But this leaves the burden of guiding the user until he enters a valid attribute set which matches his intentions to the application.
- Completeness is only maintained for the subclass definition of *document*. Thereby, a document can at least be attributed to one of the highest-level subclasses. The document is stored in the lowest-level class it can be attributed to.
- An attribute whose domain in any subclass is defined by an explicit enumeration of allowable values has to be defined by explicit enumeration in all other classes, too; the final attribute definition in *document* will contain union of all permissible values explicitly enumerated in the leaf classes.

The second solution may be rather confusing to users because documents may be found on different levels of the hierarchy. A main goal of the class definition is to allow automated archive access and to enable users directly accessing the archive to find their way quickly to the documents they need. A hierarchy where only leaf classes actually contain the documents seems to be better suited to easy human access. The last of the suggested solutions seems to satisfy that requirement; it works particularly well with a bottom-up class definition.

3.3. Data Manipulation

Entry, deletion, retrieval, and update operations on single entries and operations on a selection of entries can be matched to SQL statements: a document entry is considered a tuple, its profile attributes and the content (as a whole) are regarded as attributes. A *search* can be expressed in SQL by

SELECT <dfr_attributes>
FROM <search_domain>
WHERE <dfr_search_filter>

The *move* and *copy* service may apply to single entries or groups (group structures). These operations on single entries may be expressed as the appropriate combination of SQL statements. But since group membership depends on the values of the profile attributes used for class definition, a *move* operation is in fact requested by making the appropriate changes to the profile. If, for instance, *document_status* is used for class definition and two subclasses are defined by the values *"draft"* and *"final_version"*, changing *document_status* in the document profile moves the document to another class and thereby also to another DFR group. On each modify, the DAMS must therefore dynamically check the set of profile attributes used for class definition and determine the DFR group the document belongs to.

4. Including DFR and ODA in the Corporate Repository

Fig. 3 provides a data model for such a repository (only the relevant part of the repository's underlying extended ER-model (Chen 1977, Scheer 1992) is shown, entity and relationship types are written in italics in the text):

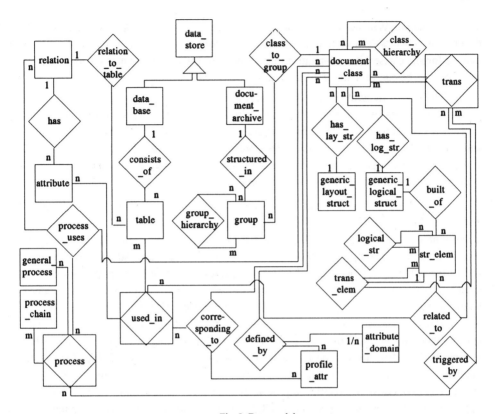

Fig. 3: Data model

Data stores may be classified into databases and document archives; they are organized in tables and a hierarchy of groups, respectively (*consists_of, structured_in*). Relational databases are used to store relations, archives store document classes. Relations may enter into several database tables as a result of replication or fragmentation decisions (*relation_to_table*). A table may not take up all the attributes (*has*) of a relation, depicted by *used_in* which is reinterpreted as an entity type in the model. Also a document class may enter into several groups, if it is replicated (*class_to_group*). An attribute in this relationship type may indicate whether the class is fully replicated or whether only a reference is entered. At this point, the similarities between databases and archives end.

The attributes of a relation used in a database table automatically define the structure of each record stored in the table. Provided ODA document description is used, each document class contains reference to the generic logical and layout structure used (*has_lay_str* and *has_log_str*). The generic structures impose some restrictions on documents, but documents sharing the same generic structures may vary considerably. They may contain the same kind of elements, but in different repeating factors or in different combinations. Each of these generic structures is *built_of* structure elements (only shown for the logical structure in Fig. 3) which are organized in a hierarchical tree (*logical_str*). The elements of a generic logical structure may be *related_to* attributes *used_in* database tables. This relationship may vary depending on the document class. Document classes are grouped in a *class_hierarchy*. A class is *defined_by* the fact that certain DFR profile attributes have to belong to a defined *attribute_domain*. Whether *attribute_domain* enters into *defined_by* in a 1- or n-relationship, depends on subclass definition. If all subclass definitions are disjoint and *defined_by* is only entered for leaf classes, it is a 1-relationship. Depending on the document class also a profile attribute may be *corresponding_to* attributes *used_in* database tables.

The goal of integrated archive and database applications is to improve the support of business processes by information systems. Corporate process chains consist of several processes, each started and ended by an event or state (see also Fig. 1). Since one and the same process may enter into several chains, processes are generalized to *general_process*. If a general process enters into a *process_chain*, the relationship *process* is generated. A *process_uses* relations and/or document classes.

In the course of some process chains, documents which accompany them may be transformed from one class into another (*trans*) by making the appropriate changes to the document profile (*triggered_by*). Each entry *trans*[c(i), c(j)] indicates that class transformation c(i) => c(j) is possible. But these changes cannot be made freely, they must reflect the company's processes and its business rules. In addition, the generic structures used by the classes involved have to be compatible, the generic structures used by c(i) must be unequivocally mapped to the elements of the structures used by c(j). That is also why c(i) => c(j) does not necessarily mean that the inverse transformation is possible as well. The rules by which the elements of one generic logical structure are mapped onto those of another generic logical structure can be found in *trans_elem*. For each pair of (transformation, element_of_the_source_generic_logical_structure) exactly one element of the target logical structure is defined (n:m:1-relationship between *trans*, *str_elem*, and *str_elem)*. The same applies to the elements of the generic layout structure (not shown in Fig. 3). These mapping rules enable application programs to effect the changes in the document structure automatically when the user requests a change in the document's class by changing its profile.

5. References

P. Chen, *The Entity–Relationship Approach to Logical Database Design* (Q.E.D., Wellesley, 1977)
P. Coad and E. Yourdon, *Object-Oriented Design* (Prentice-Hall, Englewood Cliffs, 1991), pp. 126–135
C.J. Date, *An Introduction to Database Systems I, 5th Ed.* (Addison-Wesley, Reading, 1991), pp. 532–540
C.J. Date and H. Darwen, *A Guide to The SQL Standard, 3rd Ed.* (Addison-Wesley, Reading, 1993)
T.H. Davenport and J.E. Short, *Sloan Management Review* Vol. 31 (1990), pp. 11–27
J. Harrington, *Organizational Structure and Information Technology* (Prentice-Hall, New York, 1991)
B. Hebrawi, *OSI Upper Layer Standards and Practices* (McGraw-Hill, New York, 1993)
J.G. Hughes, *Object-Oriented Databases* (Prentice-Hall, New York, 1991), pp. 96–98
International Organization for Standardization, *ISO 8879 Standard Generalized Markup Language* (ISO, Geneva, 1986)
International Organization for Standardization, *ISO 8613 Office Document Architecture (ODA) and Interchange Format* (ISO, Geneva, 1989)
International Organization for Standardization, *ISO 10166 Document Filing and Retrieval* (ISO, Geneva, 1991)
International Organization for Standardization, *ISO 9075 Database Language SQL* (ISO, Geneva, 1992)
International Organization for Standardization, *ISO 9579 Remote Database Access, Part 1, Generic Model, Service, and Protocol* (ISO, Geneva, 1993)
A. Prosser, *Standards in Rechnernetzen* (Springer, Vienna, 1993), pp. 236–281
A.-W. Scheer, *Architecture of Integrated Information Systems* (Springer, Berlin, 1992), pp. 52–55
A.-W. Scheer, *Business Process Engineering 2nd Ed.* (Springer, Berlin, 1994), pp. 10–84
A. Valenzano, C. Demartini, and L. Ciminiera, *MAP and TOP Communications* (Addison-Wesley, Wokingham, 1992)

Communication Networks for Computer-Integrated Manufacturing

Wen-Pai Lu

Manager
GTE PCS
245 Perimeter Center Parkway
Atlanta GA 30345

Wen F. Lu

Department of Mechanical Engineering/Intelligent System Center
University of Missouri-RollaRolla, MO 65401, USA
Email: wflu@umr.edu

Abstract

With the advent of communication technology especially in wide area networking, the Computer Integrated Manufacturing (CIM) has moved into a new era. This new era introduces a new concept in CIM, and is called Tele-Manufacturing. Tele-Manufacturing utilizes the advantages of networking both in local and wide areas to provide better communications and information sharing and control among equipment, controllers and human being. In this paper, the needs for and the benefit of Tele-Manufacturing will be discussed, followed by the generic architecture of Tele-Manufacturing,

1. Introduction

The advent of many different information technologies have made possible a new cycle of industrial revolution. These technologies allow us not only to retrieve information faster but also more efficient. One of the key components in the information age society is the advances in communication technology. Business entities use it to improve their productivity and communications between associates. The communication networks build upon these technologies bring people closer together without traveling to meet each other. They allow associates to meet at the same time from different places across the country or even in different part of the world.

The advances in communications not only help people to communicate better, the technology also provides capabilities that allow resources sharing and improve productivity. For example, the optical storage device or high speed laser printer is very expensive for a single person to use alone and with the help of networking, all associates in the same department can now enjoy using it. Furthermore, important records or documents may have to mail to each other before any discussion can start and now with the help of communication network, multiple associates can share the same files to do collaboration work together.

Communication technologies can also help to improve the productivity in manufacturing. The coordination of each process on the shop floor, the sharing of information among processes, the distribution of marketing material in real time to distributors and the exchange of design requirements and specifications among R & D group and other departments require a high speed, high efficiency and flexible communication network to support these high demands [1-5]. The manufacturing industry is now facing a new wave of revolution with communication networks. All equipment and device on the factory floor not only require computerization, but they also need the high bandwidth networking to provide distributed control and processing capability. This revolution will change the way the factory produce their products.

Many articles in the past has described how to provide communications in the manufacturing environment [6-9]. They discussed the planning, the design, the protocol and its network architecture. Most of them focus on the networking on the factory floor, i.e. the local area network. They have ignored the discussion on the interconnection of these network, especially with the metropolitan area network and wide area network. In this paper, we will focus on the wider area network technology such as Metropolitan Area Network (MAN) or the Wide Area Network (WAN) for CIM. We will also focus on how manufacturing will change in the information age society with the networking technologies available today and in the future, and the interconnection among these networks both in local and wide area arena, and introducing a new concept of Tele-manufacturing.

2. Overview

2.1 Computer Integrated Manufacturing (CIM)

According to Groover [23], computer integrated manufacturing is used to denote the pervasive use of computers to design the products, plan the production, control the operations, and perform the various business related functions needed in a manufacturing firm. It is mainly concerning with the information processing tasks at all levels and functions of factory and management.
- Computer-Aided Design:- CAD can be simply described as using a computer to convert a designer's idea into a formal representation. It includes both geometry representation and its analysis of the product.
- Computer-Aided Production Planning:- This function provides the link from CAD to manufacturing. Included with the production planning are process planning, master scheduling, requirements planning and capacity planning.
- Computer-Aided Production Control:- Production control is concerned with managing and controlling the physical operations in the factory to implement the manufacturing plans. The activities include the use of computers in shop floor control, inventory control, quality control, process control and other control activities.
- Computer-Assisted Business Function:- Examples of these business functions are sales and marketing, sale forecasting, order entry, cost accounting, and custom billing.

From the above descriptions, it is obvious that effective information processing and communication at all levels and functions are critical to the success of the production.

2.2 Communications Technology

The purpose of this section is to provide an overview of the communication technology that can be used in the Computer Integrated Manufacturing (CIM). A typical communication network architecture for CIM consists of a number of computing devices which range from mainframes for processing transaction documents such as files, records, etc. to low level control devices such as controllers. The architecture is divided into hierarchical level, which can be simply classified into three levels. Each level can further divide into many sublevels depends on the granularity of the partitions. These three levels are specified as enterprise level, facility level and shop floor level [4]. For devices and computers at each levels, there consists of computer networks, such as Local Area Networks (LANs) to link these devices. Most papers in the past [10-12] specify the use of the IEEE 802.4 token passing [13] network for the facility and shop floor levels, and the IEEE 802.3 Carrier Sense Multiple Access/Collision Detect (CSMA/CD) [14] for enterprise level. The IEEE 802.3 network is commonly known as

Ethernet [15].

Among these network, a device called gateway is used to interconnect these network. Depending on the types of network and the types of communications protocol used, a device called bridge or router will be used. In general, a bridge is used to interconnect two similar type of networks in different functions of domain, and a router is used to interconnect two different types of networks. For example, a bridge can be used for connecting networks at shop floor and facility level since both levels use the IEEE 802.4 token passing network whereas a router will be used between networks at facility and enterprise network.

Since devices at the facility and shop level are generally located in the same locations, a Local Area Network (LAN) will provide adequate communications functions for these devices. For communication at Enterprise level, or between Facility and Enterprise level or even at the Facility level, due to the distance between computing devices, LAN will be too short for such communications. Then, a Metropolitan Area Network (MAN) or a Wide Area Network (WAN) will be used. Before we discuss how the MAN and WAN are implemented in the CIM architecture, let us provide some introductory materials on these two technologies.

The MAN or WAN described in this paper will be focus on the public high speed network since high bandwidth network is desirable due to the volume transactions between components, and the public network is more flexible than the private. Two of the commonly used high speed public network services will be described in the following section are the frame relay and the Switched Multi-megabit Data Services (SMDS).

2.2.1 Frame Relay

Frame Relay is a connection-oriented, high throughput, low delay packet switching technology. [16-19]. The communications between two end-stations on the Frame Relay network require to set up a connection prior to the beginning of the conversation. Figure 1 shows the interconnection of LANs and workstations with a FrameRelay network. Frame Relay is a new type of Packet Network Service. It evolves from the X.25 network without the layer 3 functions, i.e., the network will not provide error correction, retransmission, or flow control. It only checks the validity of the frame received. If an error is found, the frame is discarded and no retransmission is requested. In addition, the packets transported within the Frame Relay network are not acknowledged. The less function performed by the Frame Relay Network is due to fact that the transmission technology is much better now. This allows a Frame Relay network to achieve approximately 10 times the packet throughput of the existing X.25 network.

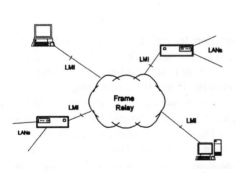

Figure 1. A Frame Relay Network Interconnects LANs and Workstations

2.2.2 Switched Multi-megabit Data Service (SMDS)

SMDS is a high-speed, connectionless, public, packet switching service [20-22]. Unlike Frame Relay, which is the connection-oriented network, the end-station does not require to establish a connection prior to the sending of data. The routing of this packet is based on its destination and source address. SMDS is targeted to interconnect LANs [20,21], computer systems and intelligent workstations across a metropolitan or wide area. Figure 2 illustrates typical SMDS network configurations, and how it is used to interconnect LANs and workstations. The interface protocol, SMDS Interface Protocol (SIP), is based on the IEEE 802.6 [22] standard which is a Metropolitan Area Network (MAN) standard developed by the standard committee who developed the IEEE 802.3 and 802.4 standard. SMDS is a cell switching network ,i.e., a packet from the higher layer protocol is divided into many small fixed size packets called cells, and the switch will relay the cell based on the information in the cell header. SMDS supports both T1 at 1.544 Mbps and T3 at 44.73 Mbps. The SMDS address is consistent with the international standard addressing format which is pretty similar to the current telephone number format. Additionally, for security purposed, SMDS provides source address validation, and both source and destination address screening.

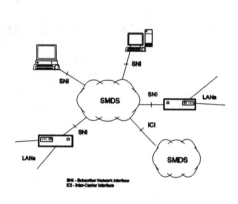

Figure 2. A SMDS network interconnecting with LANs and Workstations

3. Tele-Manufacturing

After the introduction of the communication technologies especially the wide area networking technology, the next obvious question is how these technologies will be used in the manufacturing, why they are importance, and how they fit into the future of manufacturing.

Manufacturing has gone through many revolutions. Products are becoming so complex that it is getting quite impossible that a single factory or vendor can make all components in their productions. For example, a computer company will have to depend on the other vendors to provide them the memory, disk drive, adapter board, main processor, and the monitor. A automobile factory will depend on vendors to supply windows glass, seat cushion, electrical components, and many others. The coordination between them is becoming a critical component in the whole process. It requires a high speed, reliable network to transfer information between them so that time critical mission can be accomplished.

Due to the cost cutting strategy many corporations are implementing, they start to consolidating the same departments located in different premises to a centralize place. For example, finance and marketing, or even personnel departments that may exist in each facility are now located in the same place preferably with the corporate headquarter. With such move, the coordination between finance and marketing and the factory floor become very critical since they are now located in different locations. A high speed wide area network become necessary not just preferable. This kind of operation starts to change the traditional communications link between enterprise level, such as the mainframe or minicomputer in the finance department, with the controllers in the facility level. Instead of using high speed local network, which is generally high speed and efficient, the wide area network with the same functionality is becoming essential. This also includes the communications between the Design or R&D center and the factory production.

In addition, in some cases that not all the departments or components in the enterprise level are located in the same facility. One of the example is that the R&D organization may locate separately from the corporate data processing center. In order to get the coordination between them, they do require a WAN or MAN to interconnect all these locations. Similarly, they do require the same type of speed and functions they have enjoyed in the past with the Local Area Network.

The timing of introducing new products into the market is becoming very critical. Generally, high glossy brochures are produced, and send to the dealers for distribution to the consumer. However, most of these updates or upgrade of current version may take place every three to six month. In order to have these colorful, fancy brochure ready when the products roll out, they need to be prepared at the same time the product was developed. However, many things may change during the development of the products, and it is very difficult to change the contents in the brochure. In order to convey this information to the consumers in real time, it becomes not just desirable but necessary to have a high speed communication network that links the vendors and the dealers so that the color image of the products or better of the video display of the products can be sent to the dealers' showroom during the new product introduction. This will get quicker attention from the consumers, and any modification to the product information can be changed in real time.

It is obvious that modernization of the manufacturing factory requires not only the upgrade of those computer processing system, and the controllers, but also many high speed, reliable networks, both the local and wide area networks. Computers and networking become an integral part of the manufacturing. This lead us to introduce a new concept in manufacturing. It is called Tele-Manufacturing.

Tele-Manufacturing is different from the communications in a factory using Local Area

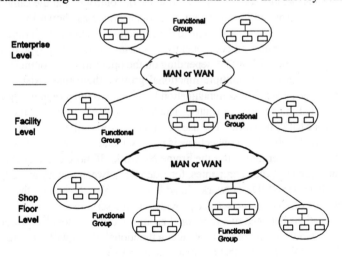

Figure 3. Tele-Manufacturing Architecture

Network. Tele-Manufacturing extend the limit of communications that all components in the CIM architecture [4] can be located in any place. The whereabouts of the components is not constrained by the limitation of the communication networks and it is assigned due to the policy, control, operations and management. The use of MAN or WAN should be performed as well as the LAN used. In practice, the components that belong to the same functional group may all be together that they will be connected with a LAN. The communication between these functional groups may use either LAN or MAN or even WAN. The architecture for the Tele-Manufacturing is depicted in Figure 3.

The main idea of the Tele-Manufacturing is to improve coordination in the manufacturing process. The Tele-Manufacturing is the manufacturing process that requires communications and networking, not only for process to process but also for humans interactions where each entity may locate in the same or different places. One of the usage of Tele-Manufacturing is that an expert in the R&D center may not have to travel to the manufacturing site to solve a problem. He or she can sit in the office to remotely diagnostics the problem occurred in the manufacturing process. It requires high speed network to allow the information about the manufacturing process, which may be graphical or image representation, can be sent back in real time for the expert to perform the work. Since the diagnostics work may only take an hours, however, the travel time may involve more than a day. In another scenario, a R&D person may be working with the engineers in the factory to finalize the detail of the product by using high performance workstations and high speed network. They may see the same design figure at the same time and they can make changes right at the spot. This will eliminate a lot of errors and iterations between them, thus save a lot of time in the product design stage. Therefore, with the help of advance networking, Tele-Manufacturing can increase productivity, and improve the quality of the product. Tele-Manufacturing can help to solve all problems mentioned above.

4. Network Interconnection and Integration

With the introduction of Tele-Manufacturing, we will briefly describe how MAN or WAN is used, and how these networks are interconnected. In the Tele-Manufacturing architecture, all components, which includes computers, mainframes, or controllers, at the same level within the same location are grouped into different functional groups as shown in Figure 4. Each functional group can also be divided according to their functions. Within each functional group, a LAN will be used to connect all components in the same group. The type of LAN can be either the IEEE 802.3 or the IEEE 802.4 depending on the operating environment. Since these components are combined into multiple functional groups, there may exist more than one functional groups at each level. The communication between two functional groups at the same level will be connected via a bridge or router. The MAN or WAN in the architecture is either a Frame Relay or a SMDS network. The use of either network will depend on the amount of connections required. If only a few connections is required, and all communications are within the same company, it is preferable to use the Frame Network. If more connections are needed, and the communications between companies, for example between factory and suppliers, then it is more desirable to use SMDS network. In addition, it is also depend on the availability of the network services offered by the service providers in the serving area.

For communications between two functional groups at different level, a bridge or a router may also be used. However, there are cases that a functional group will serve as the gateway between two MANs or WANs such that communications between the higher level and the lower level can be carried out. As shown in Figure 4, the functional group on the right hand side at the facility level is served as the gateway. It takes two routers, one to connect to the upper MAN/WAN, and one to the lower network. The connection may also be accomplished by using a router between two MAN or WAN. This configuration may require the cooperation from the service providers.

The function of the bridge or router become an important gateway between two functional group. Since the router may require to perform a lot of routing decision functions, it may require a very high performance device. A lower-end router may become a bottleneck even though the network in between is a high speed network. In this paper, we are focus on the network that runs at the very high speed, such as the Asynchronous Transfer Mode (ATM) network, which may operate at 155 Mbps and beyond. Furthermore, the detail of the interconnection between each functional group and the WAN using router will not cover in this paper. Such detail will be described in the subsequent paper.

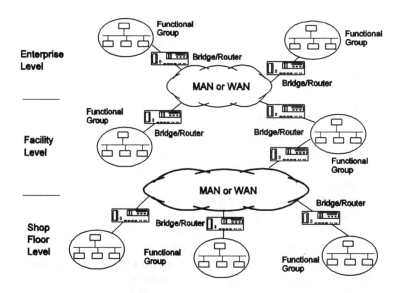

Figure 4. Network Interconnection in Tele-Manufacturing

5. Conclusion

In this paper, a new concept of computer automation manufacturing is introduced. This concept, which is called Tele-Manufacturing, extends the CIM architecture to include the interconnection with the MAN and WAN. This is due to the need that not all components in the manufacturing automation are located in the same place. The separation of these components are administrated by their functions, the policy of the corporation, and the operation and management of the manufacturing process. This paper provides a detail analysis of the advantages and benefits of Tele-Manufacturing in the automated manufacturing environment. Finally, an architecture of the Tele-Manufacturing is given followed by the ways these networks, both LANs and WANs, and the functional groups are interconnected.

References

[1] Kusiak, A., and Heragu, S.S., "Computer Integrated Manufacturing: A Structural Perspective," *IEEE Network*, vol. 2, no. 3, May 1988, pp. 14-22.

[2]   Daigle, J.N., Seidmann, A., and Pimental, J.R., "Communications for Manufacturing: An Overview," *IEEE Network*, vol. 2, no. 3, May 1988, pp. 6-13.
[3]   Chintameneni, P.R., et. al., "On Fault Tolerance in Manufacturing Systems," *IEEE Network*, vol. 2, no. 3, May 1988, pp. 32-39.
[4]   Messina, G., and Tricomi, G., "Manufacturing Communication Architecture," *Computers in Industry*, vol. 13, 1990, pp. 285-293.
[5]   George, L.J., and Mital, A., "Components of Computer Communication and the Role of Computers in the Automatic Factory," *Computers in Industry*, vol. 12, 1989, pp. 215-225.
[6]   Ray, A., et. al., "Communication Networks for Autonomous Manufacturing and Process Control," *Computer in Engineering*, vol. 2, 1987, pp. 39-45.
[7]   Ray, A., "Networking for Computer-Integrated Manufacturing," *IEEE Network*, vol. 2, no. 3, May 1988, pp. 40- 47.
[8]   Holland, J.R., "Factory Area Networks - The Key to Successful Factory Automation Strategies," *SME Autofact Europe Conference, 1983*, pp. 35-51
[9]   Jayasumana, A. P., Swunboriruska, K., and Herath, J., "Local-area Networks for Manufacturing," *ACC 1989*, pp. 523-527.
[10]   Hong, S., and Ray, A., "Analysis of the Priority Scheme in Token Bus Protocols," *ACC 1989*, pp. 511-516
[11]   Genter, W.L., and Vastola, K.S., "Performance of the Token Bus for Time Critical Messages in a Manufacturing Environment," *ACC 89*, pp. 534-539
[12]   Lee, S., and Ray, A., "Perturbation Analysis of a Token Bus Protocol for Network Performance Management," *ACC 89*, pp. 517-522.
[13]   ISO 8802-4: 1990 (ANSI/IEEE Std 802.4-1990), Information processing systems — Local area networks — Part 4: A bus utilizing token passing as the access method.
[14]   ISO 8802-3: 1990 (ANSI/IEEE Std 802.3-1990), Information processing systems — Local area networks — Part 3: A bus utilizing CSMA/CD as the access method.
[15]   The Ethernet - A Local Area Network, Version 1.0, Digital Equipment Corporation, Intel Corporation, Xerox Corporation, September, 1980.
[16]   Digital Equipment Corporation, et. al., Frame Relay Specification with Extensions Based on Proposed T1S1 Standards, Rev. 1.0, Sept. 1990.
[17]   Bhushan, Brij, "A User's Guide to Frame Relay," *Telecommunications*, July 1990, pp. 39-43.
[18]   Hemrick, C., Klessig, R., and McRoberts, J., "Switched Multi-megabit Data Services and Early Availability via MAN Technology," *IEEE Communications*, vol. 26, no. 4, April 1988.
[19]   Dix, Frances R., Kelly, M., and Klessig, R.W., "Access to a Public Switched Multi-megabit Data
[20]   Kramer, Michael and Piscitello, D.M., "Internetworking Using Switched Multi-megabit Data Services in TCP/IP Environments," *Computer Communication Review*, vol. 20, no. 3, July 1990, pp. 62-71.
[21]   Lang, L., and Watson, J., "Connecting Remote FDDI Installations with Single-mode Fiber, Dedicated Lines, or SMDS," *Computer Communication Review*, vol. 20, no. 3, July 1990, pp. 72-82.
[22]   ISO DIS 8802-6: 1991 (ANSI/IEEE Std 802.6-1991), Information processing systems — Local area networks — Part 6: A ring utilizing slotted ring as the access method.
[23]   Groover, M. "Automation, Production systems, and Computer-Integrated Manufacturing," Prentice-Hall, Inc. 1987.

Manufacturing Information Enhancement through Data Integration

Anne E James
*School of Mathematical and Information Sciences,
Coventry University,
Priory Street,
Coventry, UK*
E.mail: csx188@uk.ac.coventry

Over the last decade or so, the accessibility of computing power has led to the rise of multifarious information and engineering computer systems throughout enterprises. Often separate systems cover overlapping information domains and this gives rise to problems of information integration. One of the greatest challenges now facing the information engineering industry is the achievement of interoperability and data integration. This paper describes a dynamic approach to data integration which is based on a canonical binary model type and the use of predicate logic.

interoperability; binary conceptual model; predicate logic

1. Introduction

In any typical enterprise, one now finds many different types of information system. For example, the drawing and design departments might have one or more types of CAD system. The production department may have a system covering the manufacturing processes required for each product as well as CIM systems. The sales department will have an order processing system and the accounts department will have various financial systems. Some systems will have nothing in common but others will cover, at least in part, the same information domain. In order to maximise the advantages to be gained from the information technology explosion, such information systems should be integratable so that, security permitting, users of one system can have seamless acess to information from another system.

The idea of integration is that the users of one system will be able to view data from another system in the same way as they would see information from their own system. This can be difficult because each system will have its own way of modelling information. To overcome this problem, information integration requires the use of an canonical conceptual data model type which is capable of capturing the modelling concepts inherent in all the application models to be integrated. Each application model should then be translatable into and out of the canonical model type.

The problem of interoperability and integration is now justifiably being given much attention (Fulten 1992, Godwin 1994, Hsiao 1993, Kambayashi 1991, Schek 1993). Usually integration will be at the level of view only with mappings onto the various databases. It is generally accepted that the process cannot be fully automated and that there will be a PRI (person responsible for integration). The standard model must be rich or flexible enough to express all of the concepts found in the application models to be integrated. Since different types of models with new concepts are constantly being developed, it is not feasible to create a new standard model which we can be sure will be rich enough to capture all possible semantic concepts. This is where our approach differs from others. Most other work utilises a rich conceptual model type as the underlying canonical model (Fulten 1992, Kambayashi 1991). We propose that the canonical model be based on the binary data model (Senko 1976) and include just two modelling primitives: type and relationship. Types classify real-world objects and relationships are associations between types. Everything is expressed in terms of the two primitives. Predicate logic is used in the approach to describe the various models involved and also to formally express the translation and integration rules used. The advantage of such a reductionist approach is that it provides the independent base needed for the translation to richer models as may be required by varying applications. There are two basic approaches to integration. The first involves creating a federated data model which combines the various application models. Cross-application interaction is then always through the integrated model. The second approach involves dynamically building integrated subsets of the inter-application domain of discourse as they are required. The approach described in this paper falls into the latter category although similar techniques could be applied also to the former (James 1995).

The paper is organised as follows. Section two presents an example. Section three describes the translation process from application model to canonical model as well as the process of identifying conflicts and defining mappings between the models. Section four explains and illustrates how the canonical models can be combined and integrated into varying forms as required by different applications. Section five offers some conclusions and identifies further work.

2. The Example

The example to be used to illustrate the approach is that of a manufacturing enterprise. The enterprise has a production department, a quality department, a drawing department and a product design department that have independent systems which cover the overlapping information domains. The enterprise wishes to integrate these separate systems. The data held by each of these departments is given below.

2.1. The Production Department

The production department uses a relational database to hold information about products. For each product it holds the cutting-time, the drilling-time and the assembly-time. It also holds the size co-ordinates of height, width and depth for each product. It holds the information in two tables as shown below.

```
product ( product-id, cutting-time,drilling-time,assembly-time)
co-ordinates( product-id, height,width,depth)
```

2.2. The Quality Department

The quality department uses a relational database as shown below to record average monthly rejects for each stage in the production process.

    cutting-rejects(prod-id, ave-rejects)
    drill-rejects (prod-id,ave-rejects)
    assem-rejects (prod-id,ave-rejects)

2.3. The Drawing department

The drawing department uses a wire-frame CAD system to hold drawings of products. The user interface is object-oriented in that the user can issue a fixed set of commands to interact with a particular drawing. The user does not know anything about the way the data is organised internally but knows what commands can be used. For each stored drawing, a user can issue the command "draw" to produce the drawing, the command "co-ordinates" to have displayed a list of edge co-ordinates and the command "give-total-edges" to produce the total number of edges in a drawing. The data is defined as shown below.

```
object class drawing
{    ....
     operation draw {};
     operation co-ordinates{};
     operation give-total-edges{};          }
```

2.4. The Design Department

The design department uses a surface modelling CAD system to hold designs of products. Like the system of the drawing department, this system is object-oriented and defined as shown below. For each existing design, users can issue the commands "produce-drawing", "co-ordinates" and "total-surfaces". The command "produce-drawing" will produce the design, the command "co-ordinates" will produce a list of surface co-ordinates for the design and the command "total-surfaces" will give the total number of surfaces in the design.

```
object class design
{    ....
     operation produce-drawing {};
     operation co-ordinates{};
     operation total-surfaces{};            }
```

2.5. The Intersection of the Information Domains

Each application is concerned with products. The production and quality departments have overlaps with regard to product identification. The production department uses the term "product-id" and the quality department the term "prod-id". Each drawing in the drawing department and each design in the design department refers to a product. For these departments, each CAD system has a different operation to produce a drawing. The drawing system uses the term "draw" and the design system uses the term "produce-drawing". The operation "co-ordinates" is common to both systems but produces different data in each case. In one case it produces a list of edge co-ordinates and in the other it produces a list of surface co-ordinates. There is another difficulty for integration with regard to the term "co-ordinates" in that it is also used in the production department to refer to something quite different.

3. The Translation Process and the Mapping Definition

The integration process consists of three steps. The first step is to translate each of the application models into the canonical form. The second step is to define the mappings between the translated application models. The third step is the integration activity itself which works on the translated application models and the mappings.

3.1. The Translation Process

Translation rules have been developed for translating from models of a certain group to the canonical form as follows.

3.1.1. Relational to Canonical

(1) Each relation become a type
(2) Each attribute becomes a type
(3) Each relation-attribute relationship becomes a relationship called "has" whose first term is the type of the relation and second term the type of the attribute
(4) Each relation-relation relationship becomes a relationship "links-to" whose first term is the type of the main relation and second term the sub-ordinate
(5) Duplicate types are removed

3.1.2. Object-oriented to Canonical

(1) Each object class becomes a type
(2) Each operation becomes a type
(3) For each operation there will be a relationship "has" whose first term is the type of the object class and second term the type of the operation
(4) For each operation which is not retrieval or which has side-effects, there will be a relationship "is-update" whose first term is the type of the object class and second term the type of the operation
(5) For each operation which is retrieval but which does not yield a atomic value, there will be a relationship "is-multivalued" whose first term is the type of the object class and second term the type of the operation
(6) Duplicate types are removed

3.1.3. The Application Translations

The translations of each of the application models are shown below.

Production Department

type(product)   has(product,product-id)
type(product-id)   has(product,cutting-time)
type(cutting-time)   has(product,drilling-time)
type(drilling-time)   has(product,assembly-time)
type(assembly-time)   has(co-ordinates,product-id)
type(co-ordinates)   has(co-ordinates,height)
type(height)   has(co-ordinates,width)
type(width)   has(co-ordinates,depth)
type(depth)   links-to(product,co-ordinates)

Quality Department

type(cutting-rejects)
type(drill-rejects)
type(assem-rejects)
type(prod-id)
type(ave-rejects)
links-to(cutting-rejects,drill-rejects)
links-to(cutting-rejects,assem-rejects)

has(cutting-rejects,prod-id)
has(cutting-rejects,ave-rejects)
has(drill-rejects,prod-id)
has(drill-rejects,ave-rejects)
has(assem-rejects,prod-id)
has(assem-rejects,ave-rejects)

Drawing Department

type(drawing)
type(draw)
type(co-ordinates)
type(give-total-edges)

has(drawing,draw)
has(drawing,co-ordinates)
has(drawing, give-total-edges)
is-multivalued(drawing,draw)
is-multivalued(drawing,co-ordinates)

Design Department

type(design)
type(produce-drawing)
type(co-ordinates)
type(total-surfaces)

has(design,produce-drawing)
has(design,co-ordinates)
has(design,total-surfaces)
is-multivalued(design,produce-drawing)a
is-multivalued(design, co-ordinates)

3.2. The Mapping Definition

The PRI has to analyse each application and must identify areas of intersection. There are two areas of importance: synonyms and homonyms. The PRI must be able to identify different terms that refer to the same thing (synonyms) and same terms that refer to different things (homonyms). In the example, "product-id" and "prod-id" are examples of synonyms. "Co-ordinates" is an example of a homonym. The PRI has to define a mapping which will handle homonyms and synonyms appropriately in the production of the integrated conceptual model. A model for defining the mapping is expressed below in canonical form.

type (application)	.. to identify the different applications
type(synonym-group-id)	.. to identify a group of synonyms
type(synonym-id)	.. to identify synonyms
type(new-name)	.. to hold a new name
relationship(synonym)	.. to link a synonym id with the synonym
relationship(synonym-app)	.. to link a synonym with the application
relationship(synonym-group)	.. to group synonyms
synonym(synonym-id, type)	
synonym-app(synonym-id, application)	
synonym-group(synonym-group-id,synonym-id)	
type(homonym-id)	.. to identify a homonym
relationship(homonym)	.. to link a homonym id with the homonym
relationship(homonym-app)	.. to link a homonym with an application
relationship(homonym-rename)	.. to give a new name to a homonym
homonym(homonym-id, type)	
homonym-app(homonym-id,application)	
homonym-rename(homonym-id, new-name)	
relationship(type-rename)	.. to rename a type
type-rename(type,new-name)	

Homonyms need to be renamed in an integrated model so that their referents are distinguishable. Other types or relationships might need to be renamed to be more meaningful across applications. For the application considered the mapping may be expressed as follows using this model. The PRI has an important role in deciding appropriate names, operation functionality and the sub-ordinancy of "links-to" relationships although suggestions could be made automatically.

application(production)
application(quality)
synonym-group-id(sg1)
synonym-group-id(sg2)
synonym-id(syn-pd1)
synonym-id(syn-ql1)
synonym-id(syn-pd2)
synonym-id(syn-dr1)
synonym-id(syn-de1)
synonym(syn-pd1,product-id)
synonym(syn-ql1,prod-id)
synonym(syn-pd2,product)
synonym(syn-dr1,drawing)
synonym(syn-de1,design)
synonym-app(syn-pd1,production)
synonym-app(syn-ql1,quality)
synonym-app(syn-pd2,production)
synonym-app(syn-dr1, drawing)
synonym-app(syn-de1,design)
synonym-group(sg1,syn-pd1)
synonym-group(sg1,syn-ql1)
synonym-group(sg2, syn-pd2)
synonym-group(sg2, syn-dr1)
synonym-group(sg2, syn-de1)

application(drawing)
application(design)
new-name(measurements)
new-name(state-edges)
new-name(state-surfaces)
new-name(produce-wire-model)
new-name(produce-surface-model)
new-name(total-edges)
homonym-id(hom-pd1)
homonym-id(hom-dr1)
homonym-id(hom-de1)
homonym(hom-pd1,co-ordinates)
homonym(hom-dr1,co-ordinates)
homonym(hom-de1,co-ordinates)
homonym-app(hom-pd1,production)
homonym-app(hom-dr1,drawing)
homonym-app(hom-de1,design)
homonym-rename(hom-pd1,measurements)
homonym-rename(hom-dr1,state-edges)
homonym-rename(hom-de1,state-surfaces)
type-rename(draw, produce-wire-model)
type-rename(produce-drawing,
            produce-surface-model)
type-rename(give-total-edges,total-edges)

4. Integrating the Application Models

Given the canonical models and mapping model discussed in section 3, rules have been developed to integrate applications into either paradigm. The application canonical models must first go through a pre-integration stage where the mapping conflicts are resolved and then integration follows.

4.1. Pre-integration Stage

A copy is made of the canonical model of each foreign application to be integrated into some target application. For each copy the following rules are applied to resolve synonym and homonym conflicts and to carry out any other renaming required.

(1) Any synonym is replaced by the associated synonym in the target model
(2) Any homonym is renamed as specified in the mapping
(3) Any type involved in a type-rename mapping is renamed accordingly

4.2. Integration Stage

The integration strategy involving applying the following sets of rules, as appropriate, firstly to the target application canonical model and then in turn to the canonical models of each of the foreign applications to be integrated with it after they have been through the pre-integration stage.

4.2.1. Integrating to Object-oriented Model

The rules are as follows.

(1) Any type which occurs as the first term in a "has" relationship becomes an object class provided that object class has not already been included

(2) Any type which occurs as the second term in a "has" relationship becomes an operation for the object class with which it is related provided that operation has not already been included for that class

(3) Any type which occurs as the second term in a "links-to" relationship becomes an operation for the object class associated with the first term provided that operation has not already been included for that class

Thus if the production department's application were to be integrated into that of the drawing department, we would obtain the following.

```
object class drawing                    object class measurements;
{   ....                                {   ....
    operation product-id{};                 operation product-id{};
    operation cutting-time{};               operation height{};
    operation drilling-time{};              operation width{};
    operation assembly-time{}               operation depth{};}
    operation measurements{}
    operation draw{};
    operation co-ordinates{};
    operation give-total-edges{};}
```

Since the integration is at view level only, behind non-local operations there will be calls to the home application to retrieve the relevant data.

4.2.2. Integrating to Relational Model

The relational model can only handle atomic data items and since operations of the object-oriented model may yield complex data structures, special arrangements have to be made. Operations that yield a single value can be accommodated as attributes. Other operations cannot be represented directly in the relational model but our solution is to use a special relation to hold the names of such non-single-value-retrieval operations and their associated objects. The user can then invoke the operations in question through a special interface. The following rules cover integration to the relational model.

(1) Any type which occurs as the first term in a "has" relationship becomes a relation provided that relation has not already been included

(2) Any type which occurs as the second term in a "has" relationship and which does not occur as a second term in an "is-update" or "is-multivalued" relationship becomes an attribute for the relevant relation provided that attribute has not already been included for that relation

(3) Any type which occurs as the second term in a "has" relationship and which occurs in an "is-multivalued" relationship becomes an attribute value for "operation" in the special "operations" relation and the first term of the relationship becomes an attribute value for "relation" provided that pair of attribute values have not already been included

Thus if the drawing and design departments' applications were to be integrated into that of the quality department, we would obtain the following.

cutting-rejects (prod-id,ave-rejects)
drill-rejects (prod-id,ave-rejects)
assem-rejects (prod-id,ave-rejects)
product(product-id, total-surfaces,total-edges)
operations(relation, operation)

operations

relation	operation
product	state-edges
product	state-surfaces
product	produce-wire-model
product	produce-surface-model

5. Conclusion

This paper has given an overview of an integration process based on the binary data model and predicate calculus. Breaking models down to binary form provides for flexible canonical models from which varying application views of diverse model types can be built. Future work will involve developing further sets of rules specifically to handle more application model types and extending the mapping model to cover different types of conflict and offer alternative means of resolving them.

Acknowledgement

The author thanks her colleagues and fellow members of the Data Modelling research group at Coventry University, Richard Gatward, Frank Giannasi, Nick Godwin and Shirin Tahzib, for their helpful comments on an earlier draft of this paper.

References

1. J Fulten et al, *Technical Report on the Semantic Unification Meta-Model*, ISO TC184 SC4 WG3 N175
2. A N Godwin, S Tahzib and F Giannasi,"An Example using the SUMM with EXPRESS and Relational Models", *Proc. EXPRESS User Group Conference*, 1994
3. D K Hsiao, E J Neuhold and R Sacks-Davis, *Interoperable Database Systems*, IFIP Transactions, North Holland, 1993
4. A E James, "A Database Integration System and an Example of its Application", *Proc. Intl. Workshop on Database Re-engineering and Interoperability*, Hong Kong.1995
5. Y Kambayashi , M Rusinkiewicz and A Sheth , *First International Workshop on Interoperability in Multidatabase Systems*, Kyoto, 1991
6. H J Schek , A P Sheth and B D Czejdo, *Third International Workshop on Research Issues in Data Engineering: Interoperability in Multidatabase Systems*, Austria, 1993
7. M Senko, "Diam II, Semantic Binaries and Ansi Sparc", *Proc. Conf. Database Technology*, 1976

Systems Integration - Coping with Legacy Systems

VALDEW SINGH

CIM Centre
German-Singapore Institute
Nanyang Polytechnic
10 Science Centre Road, Singapore 2260
Tel : (65) 5685316
Fax : (65) 5621189
E-mail: valdew@technet.sg

ABSTRACT

Today systems integration has become a strategic issue of major concern. The aim is to facilitate sharing and dissemination of information (of common interest) across functional boundaries within the manufacturing enterprise. However, apart from the complexities involved in integrating closely related mission-critical "islands of computerisation", which are predominantly domain specific, many organisations are finding it increasingly difficult to maintain, modify and augment them in this rapidly changing world. *Without the provision for a higher degree of evolution and migration capability, these systems are in constant danger of being obsolete.* This paper reports on the methodology conceived and the software toolset which has been developed by the author to facilitate the adaptability of integrated manufacturing systems in response to changing needs. It addresses issues relating to flexibly structuring interactions among enterprise functions and information sharing across conventional functional boundaries.

KEY WORDS : *CIM, Integration, Modelling*

1.0 INTRODUCTION

Most contemporary software applications are conceived with Tayloristic principles in mind, this to address some focused aspect of the manufacturing domain [Scheer 1988]. They are implemented in a variety of forms and serve different purposes, many of which are typically classified under the general headings of PPC, CAD, CAM, CAE, CAPP, CAQ, DNC, SFC, etc. [Rembold *et al.* 1993, Scheer 1991]. They tend to exist as stand-alone software packages, each focused on enabling localised efficiency and productivity improvements with respect to different aspects of a manufacturing enterprise. Thus, there is little or no interworking among constituent software applications, particularly those associated with different functional areas of the enterprise, and do not promote synergy on a enterprise-wide basis [Moerman 1991, ITAP 1990, Weston *et al.* 1988]. This results in specialised departmentally-determined data organisation and contributes to data hoarding by individual departments and application, each with its own restricted and incomplete view of the enterprise goals [DTI 1993]. Companies are often in a dilemma when they need to :-
- accrue and consolidate fragments of information of common concern which are distributed and also duplicated among the various software applications; and
- facilitate information sharing and transfer between "islands of computerisation"

In today's highly competitive consumer-oriented market, there is increasing pressure on companies to be cost effective, responsiveness and to achieve better ("more informed") decision-making through effective dissemination and sharing of information and knowledge [Pheasey 1992, Weinberg 1989]. Thus, it is no longer reasonable to expect a single software application to fulfil its purpose without support or reference to data and events which are handled by other closely related application systems. This demands software interoperability, which requires the various islands of computerisation to be linked to facilitate functional interaction and sharing of resources of common interest. Software interoperability is a major sub-goal of systems integration and it is widely conceived as requiring data integration as well as functional integration [Singh and Weston 1994a, Hars 1990].

2.0 SOFTWARE INTEROPERABILITY AND LEGACY SYSTEMS

Contemporary software application are normally designed and implemented to address a set of problems prevalent in current company situations. In order to overcome the risk of becoming obsolete (i.e. to cope with situations other than those for which it has been designed), they are required to (i) adapt to changes in business needs; and (ii) conform to methods and standards adopted in current generation solutions.

However, with a previously installed base of a software, which is often referred to as legacy (or "as is") software, the possibility of modification or enhancement in functionality so as to upgrade and make it functionally effective again may be remote. Often this is due to a lack of proprietary knowledge, support and expertise to carry out necessary changes (such as amendments to the required source programs), thereby making the task arduous indeed [Singh and Weston 1993]. Furthermore, it is uncommon for software manufacturers to reveal and release information pertaining to the source programs of their developed applications. Understandably, this is to protect their vested commercial interest and rights to intellectual property embodied in their products.

Hence in seeking to facilitate information sharing, generally speaking to retrieve existing information (which can be a vital resource in support of the operation of the manufacturing enterprise), seems to be a more workable and pragmatic means of dealing with legacy software. However, in order to achieve an acceptable level of interoperability among such software applications, often the following prerequisites are essential :

- *that the information source (which forms part of the legacy element) can be independently accessed;*
- *that the information architecture and schema used by the legacy components are clearly understood in terms of their structure and composition.*

3.0 OVERVIEW OF METHODOLOGY DERIVED

A meta-level overview of the methodology adopted and developed by the author [Singh 1994] to support legacy systems and to enable interoperation of software applications is depicted in Figure 1. One of the underlying concepts adopted in the methodology is (as far as possible) to decouple functions from their information repositories so as to enable the information to be treated independently from the functional capabilities realised by software applications. This not only enables easier access to information but also decouples changes to application processes from those associated information systems. The following are central to the overall methodology :

(I) Use of an Integrating infrastructure (IIS)

The purpose of the IIS is to structure, service and simplify interconnection among component elements of software systems by (i) separating integration and application issues; (ii) providing inter-process communication services; and (iii) mapping of distributed processes on the physical resources contained within a target manufacturing system. *It can assume responsibility for maintaining a knowledge of integration details* (such as networks used, the hardware and operating systems that software components are run on, the location of an information fragment, etc.) so that software components themselves need only have knowledge of how to use the IIS (i.e. NOT OF EACH OTHER).

Although there are an increasing number of integrating infrastructures appearing on the market [Gould 1992,

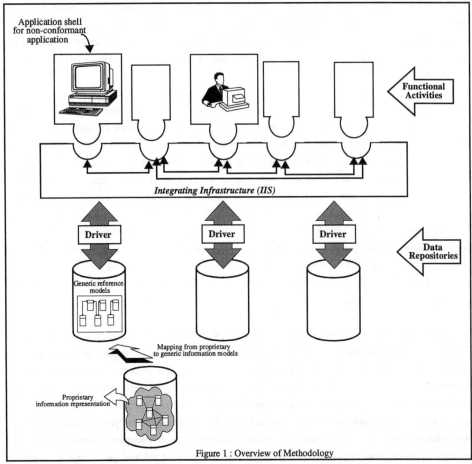

Figure 1 : Overview of Methodology

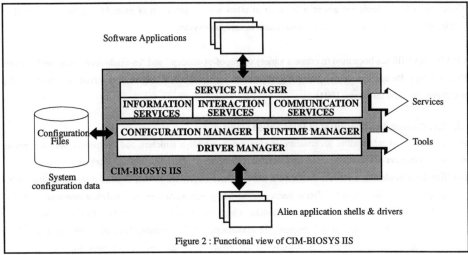

Figure 2 : Functional view of CIM-BIOSYS IIS

AMR 1991], they offer a restricted set of integration services and tools and are proprietary in nature. *Generally they provide a limited 'de facto standard' interface where so called conformant software applications (which are strictly compatible to the IIS) will be supported.* Currently this limits the potential advantage gained from proprietary forms of IIS and is particularly limiting with respect to the inclusion of legacy systems [Singh and Weston 1993].

Since 1986, the general requirements of IIS for manufacturing systems integration have been studied by researchers of the MSI Research Institute at Loughborough University of Technology. The author's research has also contributed to this study. In 1990, MSI research led to the development of the CIM-BIOSYS (CIM-Building Integrated Open SYStems) IIS which, as depicted in Figure 2, achieves a unification of general purpose computational integration mechanisms and tools [Weston 1993]. CIM-BIOSYS is charged with resolving differences in a physical system relating to heterogeneity, distribution and data fragmentation. Through the effort of MSI researchers, methods and software tools for dealing with certain classes of no-conformant (or alien[1]) applications have also been identified and produced. *'Drivers' and 'alien application shells' represent the particular software tools referred to here which respectively provide resources (such as databases and datafile systems) and legacy application components with sufficient capability that they can use the integration services of the CIM-BIOSYS IIS* (to facilitate data management, access, manipulation and presentation, and support inter-process communication).

The 'driver' flexibly maps software applications onto system resource, i.e. achieves a mapping between physically distributed databases and the logically integrated system-wide data repository. To-date a number of database 'drivers' have been developed to enable access from a number of commercial relational database management systems. These include ORACLE, Progress and Ingress. The 'application shell' serves as an front-end interface to the legacy component in order to bring it to a level of conformance to use the services of IIS. Through the 'application shell' operations such as start, stop, status update, queries are performed in a standardised manner. In addition, common integration services such as inter-process communication, and information management offered by CIM-BIOSYS IIS are also accessed and managed through this interface. These methods and tools are essential in order to allow for the inclusion of embedded legacy systems thus helping to safeguard the user's existing investment in computer systems.

CIM-BIOSYS IIS has been used to create a variety of 'proof-of-concept' and 'live industrial' integrated systems [Singh 1994]. Its use has demonstrated significant savings in the cost and time involved in manufacturing integration projects [SI Group 1994].

(II) Specification and Enactment of Information model
A relational-based information architecture is required to establish structure and uniformity whilst enabling sharing and transfer of information among functional components. The following are necessary :
(a) *High level modelling methods, based on a set of generic reference model, is used to facilitate integrated system design. The output consists of particular models of information entities and their interrelationships.*
   * A set of reference information models, which incorporate essential data that will very commonly be used in any discrete part manufacturing environment, have been identified as part of the author's research study [Singh 1994]. They are open to changes and can be

---

1. Alien - The term is used to imply that the application component is not inherently compatible with the CIM-BIOSYS IIS architecture.

modified and expanded when necessary. Together, these information models can provide the enterprise with a single coherent view of engineering, production and management information. Data will need to be stored with reference to these generic models and in a practical system will be stored in distributed data repositories to enable common access and usage by the various software components via 'drivers' over the IIS. Hence the data repository comprises a directory of shared elements where common data definitions are recognised throughout the enterprise as global data elements; they have globally understood inter-relationships. *The system-wide data repository would serve as the focal point for access of shared data* and provides users and software applications with a consolidated view of information - one that is independent of physical media or data location.

* In this research study, the $IDEF_{1X}$ entity-attribute relationship modelling tool [ICAM 1985] was used to represent the global schema of common information used. It is a composite view of a set of information models which corresponds to the generic reference models. The structure, content and entity relationships among the attributes of the information models are clearly described. The semantics are, therefore, made explicit so that there is common interpretation of the relationships among data items.

(b) *A second set of build tools are used to create and populate information models in a form which structures and enables control of information shared between software components during system operation* [Singh and Weston 1994b]. *The output descriptions are consistent with the information models created in (a).*

* The EXPRESS data modelling language [ISO 1993], which is an emerging standard used within the STEP (STandard for the Exchange of Product data), is used in this research to facilitate detailed information modelling. It was chosen because the author believed that it would enable implementation of a physical system in a manner which maintains the structure of the formal models. Through parallel research effort at MSI the following tools have been developed to exploit EXPRESS as a means of representing aspects of the information architecture of CIM systems and providing a means of managing change more effectively [Singh and Weston 1994b] :

- *EXPRESS to SQL Complier*

Using an EXPRESS to SQL Complier software tool, an EXPRESS information model can be directly compiled to generate SQL statements which will later be responsible for creating SQL compliant tables within the database. This approach is database independent (EXPRESS is not biased towards implementation) and can produce as output SQL statements which function for different relational database implementations, such as Ingress and ORACLE.

- *STEP Parser*

This tool enables the population of the tables (within different relational databases) with real data in the format specified by the EXPRESS model. This process is structured by information contained in the output files generated by the EXPRESS to SQL complier.

In addition an $IDEF_{1X}$ to EXPRESS transformation tool was developed by the author to enable the automatic creation of an EXPRESS based information model from $IDEF_{1X}$ entity-attribute relationship descriptions. It operates by directly mapping the entities of the

IDEF$_{1X}$ model into entities of the EXPRESS model. The establishment of this IDEF$_{1X}$ to EXPRESS computerised link not only offers a means of formally structuring information requirements but also supports design, build and change processes associated with them.

The overall approach conceived by the author offers the following advantages :
   (i) It helps simplify entity-attribute relationship modelling.
   (ii) Allows easy extension and modification of information models.
   (iii) Supports and encourages the reusability of information objects, thereby reducing the development and maintenance phases of projects.
   (iv) Implementation is database independent.

4.0 PROOF-OF-CONCEPT IMPLEMENTATION

A proof-of-concept system was built to demonstrate the interoperation of a number of typical software applications which are representative of the manufacturing control domain being considered in this research study [Singh 1994]. The main generic class of problem tackled in the system involves the transformation of a commercially available stand-alone proprietary CAPM package, namely MCC, into a more open system so as to enable interoperation with other software components. The following functional components are involved in the demonstration system :

- *Production planning*
- *Finite capacity scheduling*
- *Cell controller*
- *Decision support system for cost analysis*

The following inter-related meta-steps were supported by the author's methodology in a way which structured the development of the proof-of-concept demonstration system :

(I) Implementation of a system-wide information repository

Here the following activities are facilitated which are representative example activities typically involved in establishing and managing manufacturing information system :

(a) Identification and formal representation of information requirements

A global schema corresponding to system-wide shared information; i.e. shared between the functional components, was represented with reference to the generic information models. It represents a unification of information entities of common concern which will require to be accessed and updated.

(b) Mapping between local databases to the system data repository

Proprietary information, stored in the MCC relational database, which are of common interest to other functional components, were mapped onto the system data repository via reference to the global schema.

(c) Creation of relational tables and populating the system data repository

A set of build tools (developed together with other researchers in MSI) is used to developed the system data repository. This includes :
   - IDEF1X to EXPRESS transformation tool
   - EXPRESS to SQL Complier
   - EXPRESS STEP Parser

(II) Flexible interconnection of functional components via CIM-BIOSYS IIS

The task here is to interconnect the various functional components, which will normally be heterogeneous in nature and reside at different computer nodes of a distributed system, through database 'drivers' and 'alien application shells'.

(III) Establishment and configuration of an appropriate functional interaction capability
In this meta-step, a set of build tools is used to formally establish and configure associations between functional activities and their shared information requirements. This will be referenced during run-time to govern the dynamic behaviour of the system in a controlled and co-ordinated manner based on predefined sequences of activities and information needs established between functional components.

(IV) Establishment and configuration of user interface capabilities
A configurable 'application shell' was created to act as a front-end user interface. In the proof-of-concept demonstration system this 'application shell' was configured to
- provide a consistent and simple user interface to the various functional components, thereby providing an effective working environment for human interaction during run-time.
- provide human access to CIM-BIOSYS IIS, functional components and the set of tools, in a way which facilitates system construction, management and change.

5.0 SUMMARY
The methodology devised supports the formal design and development of interoperable systems where mechanisms and software toolset developed support legacy systems and enable them and other software components to "functionally interact" in a coherent manner by sharing information of common interest; this through accessing distributed data repositories in an efficient, highly flexible and standardised manner. It facilitates "soft" integration of functional activities over an integrating infrastructure in order to :
- help simplify typical integration problems found when using contemporary interconnection methods; and
- enable their reconfiguration and incremental development.

In conclusion, the author's work offers a realistic approach which can be considered to be part-way between the extremes of "open" and "closed" systems. *The aim is to provide a migration path towards "open" systems which are readily adaptable in the face of changing functional and operational requirements so as to avoid obsolescence.*

The key challenge for the next step of development is to offer a more formal and structured "model-driven" approach to facilitate design, implementation and change management of "as is" and "to be" systems.

6.0 ACKNOWLEDGEMENT
The author wish to fully recognise the contributions, in terms of conceptual thinking and availability of CIM-BIOSYS toolset, of other members of the MSI Research Institute at Loughborough University of Technology. Particular thanks in this area are due to Richard Weston, Paul Clements, Mike Leech and Ian Coutts.

REFERENCES

1. AMR (Advanced Manufacturing Research), March 1991, *Application Enabler*, Report, USA.
2. DTI, 1993, *Computer Integrated Manufacturing - A survey of Worldwide R&D*
3. Gould, L., August 1992, *CIM Interface Modules : A route to Open Systems*, Managing Automation, Vol. 26, No. 3, p. 443-455
4. Hars, A., 1990, *CIDAM - modules for the creation of CIM*, Procs. Sixth CIM-Europe Annual Conference, pp286-295

   ICAM, December 1985, *Information Modelling Manual IDEF1 - Extended*, ICAM Project Report (Priority 6201), D Appleton company, Inc, Manhattan Beach, California
5. ISO 1993, IOS DIS 10303-1, *Product Data Representation and Exchange Part 1 : Overview and Fundamental Principles*, International Organization for Standardization, Geneva
6. ITAP Technology Seminar, 1990, *Advances in Computer Integrated Manufacturing*, ITAP Technology Report No. 5/90, National Computer Board (Singapore)
7. Moerman, P. A., 1991, *The evaluation of technology in relation to products and markets : observations, considerations, experience, and solutions*, International Journal of CIM, Vol. 4, No. 1, p. 2-15
8. Pheasey, D., November 1992, *Competitive Manufacturing - 'A Vision of the year 2001'*, Procs Twenty-seventh Annual BPICS Conference, Birmingham, UK, p. 23-31
9. Rembold, U., Nnanji, B. O., Storr, A., 1993, *CIM*, Addison-Wesley, UK
10. Scheer, A., -W, 1991, *CIM - Towards the Factory of the Future*, Second Edition (Springer-Verlag)
11. Scheer, A., -W, 1988, *Computer Integrated Manufacturing - Computer Steered Industry*, First Edition (Springer-Verlag)
12. SI (Systems Integration) Group (LUT), February 1994, *Model Driven CIM : The design, implementation and management of Open CIM systems*, Loughborough University of Technology, SERC/ACME Review Report No. 2, Grant No. GR/H/22798
13. Singh, V., 1994, *Applications Integration for Manufacturing Control Systems with particular reference to Software Interoperability Issues*, PhD Thesis, Department of Manufacturing Engineering, Loughborough University of Technology
14. Singh, V., Weston, R. H., 1994a, *Functional interaction management : A requirement for software interoperablity*, Procs. of the Institution of Mechanical Engineers, Part B, Journal of Engineering Manufacture, Vol. 208, p. 289-305
15. Singh, V., Weston, R. H., 1994b, *Software Interoperability for Integrated Manufacturing, A Reference Model Driven Approach*, International Conference on Data and Knowledge Systems for Manufacturing and Engineering (DKSME' 94), Hong Kong
16. Singh, V., Weston, R. H., *New Generation of "Open" Manufacturing Control Systems for "Seamless" Integration in CIM*, Procs. International Conference on Computer Integrated Manufacturing (ICCIM' 93), Singapore, p. 309-321
17. Weinberg, J. C., 1989, *Linking the CIM Plan with Operations Strategy*, Procs. Conference Autofact' 89, Detroit, Michigan
18. Weston, R. H., 1993, *Steps Towards Enterprise-Wide Integration : a Definition of Need and First Generation Open Solutions*, International Journal of Production Research, Vol. 31, No., 9, p. 2235-2254
19. Weston, R. H., Gascoigne, J. D., Rui, A., Hodgson, A., Sumpter, C. M. and Coutts, I., 1988, *Steps towards information integration in manufacturing*, International Journal of CIM, Vol. 1, No. 3, p. 140

Implementation

Avoiding Human Error in CIM Implementations
   R. G. Hannam

Integrating CAPM and CAMM: A Bottom-up Approach
   K. F. Pun and J. A. Bull

A Step Towards CIM Using Intelligent Documents
   V. Raja

CIMS in Shenyang Blower Works
   R. L. Mu, K. M. Lin, J. S. Xue and H. B. Yu

AVOIDING HUMAN ERROR IN CIM IMPLEMENTATIONS

Roger G Hannam
Department of Mechanical Engineering
University of Manchester Institute of Science and Technology,
PO Box 88, Manchester M60 1QD, U.K.
E-Mail: Roger.Hannam@umist.ac.uk

ABSTRACT

The implementation of CIM involves very significant changes in companies' operations which need careful management. Studies of the introduction of new technology in the UK have shown that too frequently, not all the gains sought have been obtained. This paper identifies the planning phases of such implementations as frequently deficient and compares them with the Japanese approach to introducing new technology. Key differences are shown to lie in the human side of the implementations. Lessons are drawn from this comparison relating to the implementation of CIM. These are likely to be as applicable to other countries as to the UK.

(CIM, Human Factors)

1. Introduction

CIM has been part of the industrial vocabulary in the UK and elsewhere now for over 10 years. While some companies have implemented CIM comprehensively in some of their plants, these companies are few and are mainly aerospace companies or the major manufacturers of computers and computer accessories. These latter companies have implemented CIM both for the competitive advantages it offers but also as a means of promoting sales of their CIM related products, both hardware and software.

Most other companies are moving rather slowly towards greater integration and surveys in the UK suggest they are having difficulties doing so. The difficulties with CIM seem to continue difficulties with earlier technologies which resulted in computer-aided systems. While these computer assisted technologies are up and running in many companies in the UK, management consultants have surveyed their success and produced reports which state that there are many examples of investments which have not produced the benefits expected, (Anon 1985, 1987, 1989). Thus, in 1985, Brian Small of Ingersoll Engineers described the level of success of the applications of CIM, FMS, MRPII and similar sophisticated technologies as "often abysmal" (Challis 1985). In 1987 this type of comment was extended to cover companies across Europe and the USA (Anon 1987).

The difficulties reported with new technology take many forms but they particularly focus on managements' ability to implement new technology and manage the change processes involved, (Anon 1985, 1987). In the case of increasing the effective use of CAD, the difficulties have been identified to be managements' reluctance to educate themselves in CAD and then to listen to those in the company who understand the technology (Simmonds and Senker 1990). This is explained by the relatively autocratic and non-participative style of British management. Consultants have similarly studied implementations of Japanese methods such as quality circles and Just-in-time (Milsome 1993, Oliver and Hunt 1994) where they quote examples of some implementations that have not worked to companies' expectations and others that are judged failures. So what is wrong with these implementations and the companies that made them and what are the lessons for CIM?

This paper will answer these questions, drawing on the author's experiences of working in and with British industry over many years and contrasting this with his experiences of Japanese working practises which have been a subject of study since his first visit to Japan in 1973. Conclusions will then be drawn for CIM implementations which are relevant to all implementations and all countries .

2. Planning and Decision Making in British Companies

The planning of strategy in British companies is seen primarily as a Board level or senior management activity. The experience of the senior management is seen as making them eminently qualified to undertake this task. It is not uncommon to have a small sub-Committee or a Chairman's think-tank to think up and develop strategic initiatives. This may involve two or three key personnel such as the heads of the marketing, engineering and systems departments. The functional specialists need to be represented because the specialisms involved are fundamental to companies operations in most British companies. It should be pointed out that British industry is not unlike industry in other countries. It has its control hierarchy of Chief Executive (President), directors (Vice-Presidents) and managers of specialist functional departments.

At the next level down, possible new investments are planned. This will again generally be carried out by a small team who will be charged with preparing a feasibility study. If this is accepted by a Senior Executive Board, the team will subsequently prepare a more detailed report which will include a financial justification for the particular investment. The report will be in terms of specifications, details of what is proposed and a cost benefit analysis. A feature of these procedures is the reports will invariably be kept confidential to the team and to senior management. They are rarely disclosed to those likely to be affected by the investment or to those who might use it. Planning changes and decision making in relation to possible changes in systems or in operating procedures or new investments are seen by management to be their business as managers; they do not see that this activity should be shared.

Once the think-tank's or the internal team's report for a new investment has been accepted by the Board, a project manager or 'project champion' may be appointed and only then will any announcement may be made to the staff. One of the first jobs of the project champion is to start to explain and sell the proposed new technology or system to those who are to use it or who are affected by it.

In this paper, it will be argued that this sequence of events is the main cause for investments not reaching their full potential. It will be shown why, if those who are to use new investments are excluded from any role in their selection, they are often antagonistic or at best neutral to their success, and why, when no attempt is made to involve the staff concerned, those staff will use the new technology or system with little enthusiasm. To see why involvement is important, one only has to compare this approach with that of the Japanese.

3. The Japanese and Change

Many facets of Japanese working practises and Japanese company operation are now well-known and documented (Hannam 1993). These include the Japanese working long hours, for being company-men before being family men, for taking part in karaoke singing and drinking with colleagues after work, for many having life-time employment, for being paternalistically looked-after by their company, for only taking part of their holiday entitlement... and so the list could be continued. The aspects of company operation which will be discussed here are those which relate to the generation and implementation of new ideas and policies for change and new investments. These are those relevant to implementing CIM.

3.1 The Group Ethos

The Japanese have a long-established cultural tradition of operating in groups. These developed from household groups and self-help groups which existed in agricultural-based villages of the last century. The villages were governed by a forum of all the heads of the households. The group tradition may be argued to have developed from growing rice, in which mountain-sides were levelled and tiered for paddyfields. This needed collaboration. Further collaboration was then required to manage the water needed for growing the rice and for planting and harvesting. Groups structures thus readily transferred into industry and found application in the 1960s with the Toyota Production System (Ohno 1988) and Quality Circles (Ingle 1982). Groups have now existed throughout companies for many years with each employee being a member of a working group. These groups have a strong identity. (Employees will also belong to a year group comprising all those who joined the company from school or university in a particular year. Their year of joining the company determines their company seniority). The working groups will be based on the functional or departmental structures within a company.

The group or team has many functions but one of the main ones is to provide a channel of communication between the company and the employee and the employee and the company,- the two way element is emphasised deliberately. This puts a significant responsibility on those who lead groups. The group leader must also ensure that all members of the group participate and feel part of the group. Thus, the group leader must establish and maintain a good and personal relationship between himself and each group member,- (providing a key element in company paternalism). The terms 'boss' and 'subordinate' have not been used in this description because they are not appropriate words. The Japanese approach is based on a team approach with a team leader. This difference is important and its importance can be explained by using a sports team analogy. A captain of a baseball or a football team is both a player and one with responsibility for the rest of the team. Other team members may be more skilled, even better players but it is the captain who provides leadership and should ensure each of his players so collaborates that the team is successful. You cannot have a successful captain of an unsuccessful team. You can equally have a team of star players who are not successful.

The skills at the top of the list of those required of a Japanese manager/team leader are different to those of a Western manager. A Western manager will typically concentrate on company productivity targets, cost savings, head-count reductions. The most important skill of a Japanese manager is human relations; an ability to relate to others, to ensure each group member feels they are contributing, that their contribution is recognised and that the group functions effectively. The key fact about Japanese groups is that they interact very regularly (often once a day) and they involve two way communication. The team leader will certainly pass company information on to his group but on all other matters, he will expect and encourage participation from all members of the group. He will ensure that each member develops within the group and feels part of it, ensuring each feels significant and is recognised for their contribution.

3.2 The Company Think-tank

Groups have been used for many purposes in Japan but, apart from their use for communication, their best-known use is for quality circles and Kaizen (continuous improvement) activities (Suzaki 1987). Groups operate in all departments and provide a forum for technical discussions. For example, in a design department, they are used for colleagues to discuss their progress (and thus educate their colleagues) and to invite suggestions (and thus shorten design times). In a group, everyone is effectively consulted on everything and individual specialisms are continuously broadened. Groups can generate ideas and they can review and refine ideas which have been generated through the widespread use of suggestion schemes.

Ideas for change and improvements and investments can emanate from and are encouraged from all levels in an organisation, except the top. Top management expects to review and approve proposals referred to it but only after these have been widely discussed at lower levels. Thus, when a proposal is approved, it is approved on the basis that it has already been discussed widely and approved by other parts of the company. This is just one element of a system of consensus management in Japanese companies,. A significant part of consensus management is good and effective communication and the time taken to discuss matters fully and widely before they are agreed. The group structures are thus the key to the Japanese change mechanisms. The operation of groups and suggestion schemes can be seen to allow all members of a company to act as the company think-tank—not just a few executives or a project team. But how can this widely based think-tank acquire the competence to generate worthwhile proposals?

The Japanese achieve this in three ways, all based on developing an educated and informed workforce. This is achievable first by thorough on-the-job education which is achieved as has just been described. The second is achieved through the breadth of jobs. Few narrow specialist jobs occur in Japan. A CNC machine operator, for example, is responsible not only for setting and operating his machine, but for programming the machine, the initial maintenance of the machine and also cleaning the machine and the work area around it. (This approach explains why Japan has a much higher ratio of direct to indirect workers than in the West.) The third way is through job rotation; particularly in the first 10 years or so after joining a company. During this period, new employees, whether school leavers and university graduates, will hold a number of different jobs and gain knowledge and experience of different disciplines and different parts of the company. A career is not pursued in a single specialism. This facilitates employees having the experience and knowledge to make wide ranging and informed suggestions to the benefit of the company.

4. The British:Japanese Contrast

This brief description of the Japanese mechanisms of involving their staff and encouraging discussion of new developments and investments allows the main differences with British companies now to be summarised.

1. British companies are far less based on consensus and participation at all levels than Japanese companies.
2. Western managers see it as part of their job to take decisions. Japanese managers will more readily delegate decisions.
3. British companies are staffed by specialists. Staff and managers in Japanese companies have a breadth of experience.
4. British companies may communicate with their staff. Japanese companies communicate and listen to their staff.
5. Japanese company groups often meet daily. British companies would not consider this financially justified. Meetings would only be held to meet a perceived need.
6. British companies have small groups charged with making proposals for new investments or systems. Japanese companies have a company wide think-tank.
7. British employees are often wary of change. Change is the norm in Japanese companies.
8. The Japanese approach to managing change is part of their approach to management. The British see the management of change as a separate activity.

These observations are relevant to CIM because the implementation of CIM will involve a significant change in working practices for many companies. So what models of change are there that can help implementations of CIM to be successful?

5. Models of the Change Mechanism

The Japanese approach to management and hence the management of change has a ready model in the theses and findings of Western behavioural scientists, particularly Elton Mayo (George 1968), Maslow (Maslow 1970) and Herzberg (Herzberg, Mansner and Synderman 1959) who studied the motivational needs of individuals and the behaviour of groups. If the operation of the Japanese working group is studied in detail, it will be found that every motivational need (as determined for Western workers) is satisfied. These motivational needs are for achievement, recognition, responsibility, advancement and growth. It is the involvement of staff in an effectively run group that improves and contributes to a company's operation that gives satisfaction of all these needs. Remarkably, although the use of groups leads to the suppression of individuality, the operation of groups actually satisfies the needs for individuality through each member's active involvement. It is each's contribution to the group, rather than the job done, that satisfies motivational needs. Similarly, job rotation, the breadth of the jobs and on-the-job education contribute to each individual's growth. The result is a highly motivated workforce which will achieve success for the company. This is believed to be the key ingredient of Japanese industrial success. It should be noted that no study of the motivational needs of Japanese workers comparable with the Western studies have been found. This does not, however, invalidate the thesis—which has been accepted by Japanese directors of British companies. The current difficulties of Japanese companies can be attributed to the very high price of the Yen and the fall in the Japanese stock market. They are not related to the well-established group-based structures which have contributed to the economic success Japan has enjoyed.

The model of the British approach described is not so easy to specify but it has many facets of the Taylor model, captured in McGregor's Theory X (McGregor 1960). Theory X characterises employees as limited in their abilities to contribute, lacking creativity, preferring to be controlled and supervised, their prime motivator is money and they do not like working. This is now an extreme model but many aspects of this model are still prevalent in those companies who deliberately exclude most of their workers from significant involvement in discussions of any type. Very few have good two-way communications or see a need for it; (many are still establishing good one-way communications). Very few recognise their employees' broader job knowledge let alone exploit it. Only slowly is the benefit of a broader training and a broader job being appreciated. Few decisions are delegated, managers are expected to manage, rather than provide leadership. So how is this relevant to CIM?

6. The Implementation of CIM

In the introduction it was pointed out that CIM is both a system and a technology. The technology is concerned with networks, protocols, cabling and what has been termed the CIM architecture. This is the design template for interconnecting the software so that a company's data is managed in a coordinated way. The system partly relates to the software organisation , partly to the users and the user interfaces. Currently the technology appears to be in control in implementations because of the significant challenges to be solved in implementing the technology of CIM. But this cannot be an excuse for not thinking through the human dimension of CIM because the earlier quoted studies show that implementations of new technology will only be successful if they satisfy the motivational needs of those who are to use it.

So, is CIM to be so user friendly that it will offer no challenge to those who use it or provide any satisfaction for them? Will users gain the satisfactions they need from a computer-based workgroup rather than a round-the-table based group? Alternatively, will individuals be isolated from each other and just relate to their computer, software prompts and tables full of data? Will the technicalities of a CIM implementation be considered the reserve of specialists who will exclude the users?

A number of papers on CIM implementation have been published—many concentrate on the strategic aspects of CIM, others contain cost-benefit analyses and suggestions of means to satisfy accountants. This author would suggest these approaches are not enough because the benefits will not be fully realised without a CIM strategy which includes a Japanese-type human approach to planning, implementation and a continuous improvement of the system once implemented. Those who plan without the human dimension will be making a significant error.

7. Discussion and Concluding Remarks

This paper has shown how the Japanese approach to introducing new technology and to a company's management has much to commend it because it results in both a highly motivated workforce, a workforce who are informed and a workforce who will make any new technology work, rather than be fearful of it. The few British and American companies who have tried using the Japanese approach (or even part of the approach) have realised its benefits. Most of these companies have used the approach without appreciating that they were imitating the Japanese, they have just felt that good communications and employee involvement should be part of their management approach. This must be an integral element of the implementation of CIM.

A paper of this length is limited in the arguments which can be developed and in the explanations given. Full justification for the statements and arguments put forward in this paper can be found in Hannam (1993). The paper did not set out to provide a guide to CIM implementation, but rather to add to the requirements needed to be satisfied by an implementation. They are important requirements because a company's investment in CIM may be one of the largest investments it makes.

8. References

Anon, *Integrated Engineering*, (Ingersoll Engineers, Bourton on Dunsmore, UK, 1985).
Anon, *Technology In Manufacturing, How Much is Hype? How Much is Profitable? Why?* (Ingersoll Engineers, Bourton on Dunsmore, UK, 1987).
Anon, *Computer Integrated Manufacturing: Competitive Advantage or Technological Dead End?* (A T Kearney, London, 1989).
H. Challis, *Abysmal success rate of CIM, FMS*, Engineering News, 5 July, (1985) p.1.
C. S. George, *The History of Management Thought* (Prentice Hall Inc., Englewood Cliffs, 1968).
R. G. Hannam, *Kaizen for Europe, customising Japanese strategies for success* (IFS, Bedford, 1993).
F. Herzberg, B. Mansner and B. B. Synderman, *The Motivation to Work* (John Wiley & Sons, Inc., New York, 1959).
S. Ingle, *Quality Circles Master Guide* (Prentice Hall Inc., Englewood Cliffs, 1982).
A. H. Maslow, *Motivation and Personality* (2nd edition, Harper Row, New York, 1970).
D. McGregor, *The Human Side of the Enterprise* (McGraw-Hill, New York, 1960).
S. Milsome, *The impact of Japanese firms on working and employment practices in British manufacturing industry* (Industrial Relations Services, London, 1993).
T. Ohno, *Toyota Production Systems, Beyond largescale production*, (Productivity

Press, Cambridge, MA, 1988).

N. Oliver and G. Hunter, *The financial impact of Japanese production methods in UK production companies,* Research Papers in Management Studies, **24**, (The Judge Institute, University of Cambridge, 1994).

P. Simmonds and P. Senker, *Making more of CAD* (EITB Publications, Stockport, 1990).

K. Suzaki, *The New Manufacturing Challenge, Techniques for Continuous Improvement* (The Free Press, New York, 1987).

INTEGRATING CAPM AND CAMM:
A BOTTOM-UP APPROACH

K. F. PUN & J. A. BULL
Department of Manufacturing Engineering
City University of Hong Kong
Tat Chee Avenue, Kowloon, Hong Kong
E-mail: mefrankp@cityu.edu.hk

ABSTRACT

The traditional top-down approach of implementing Computer Integrated Manufacturing (CIM) has showed its drawbacks in designing system parameters which may not take full accounts of the dynamic operations and business requirements. This paper is intended to investigate the viability of a "bottom-up" approach of integrating both computer-aided production management (CAPM) and computer-aided maintenance management (CAMM) systems in small and medium-sized manufacturing enterprises. It focuses on exploring a feasible framework for the successful integration. The skeleton of object-oriented database management system (OODMS) is adopted, and the system parameters are defined using Entity Relationship Models (ERM) and Data Flow Diagrams (DFD). The applicability of the integration model and its potentials to link to higher level of CIM database are briefly presented in this paper.

Keywords: CIM, CAPM, CAMM

1. INTRODUCTION

Advanced manufacturing technologies have evolved over the last decades that hold the promise of revolutionising manufacturing. Manufacturing companies are increasingly attempting to integrate these technologies into computer integrated manufacturing (CIM) systems. Using the right means to achieve the implementation ends is essential. This paper attempts to explore the applicability of a "bottom-up" approach into implementing CIM, namely that of integrating systems from the bottom up and gradually expanding to the highest level of integration, after each stage has been successfully implemented. The integration of production and maintenance information management systems is specifically looked at, and a system model proposed. The model also accounts for preventive maintenance, and can be further expanded to include the costing system, vendor system, customer order system, shop-floor feedback system and so on.

2. TOP-DOWN VERSUS BOTTOM-UP APPROACHES

Implementing CIM drains on company's resources (particularly of money, and personnel in the early stages), and only a limited number of even the large organisations are able to justify and afford the massive on-going implementation costs. Consequently, there have been only a handful of successful CIM implementations. Most traditional CIM strategies employed have merely been a top-down approach, dictating the need for new compatible equipment to be purchased which can be controlled by the top end of the system. These have also meant designing the whole environment and system parameters in one go from a massive task which does not take the changing environment of the business into account (Noaker, 1993 and Rasmus, 1994). In the ever increasing competitiveness of global markets, the majority of small- and medium-sized companies cannot afford this top-down 'Computer-integrated' (CI) approach. A more realistic approach would be to build 'Integrated-manufacture' (IM) from the bottom up, and look to ways of integrating individual system, one at a time (Kawase, 1992). Figure 1 depicts the fundamental differences between top-down and bottom-up approaches.

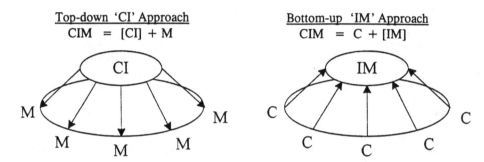

Figure 1: Comparison Between Top-down and Bottom-up Approaches

3. AN OODMS MEANS FOR INTEGRATION

Implementing CIM has been bombarded with technology. Centre to this has been the need for data and information and how they be managed in an effective and efficient way. Object-oriented database management systems (OODMS), which are characterised of a distributed relational architecture, have been developed following the stages of file systems, hierarchic systems and relational systems (Loomis, 1990). An OODMS stores data as objects, just as in the real world. The system can be completely defined by the end user (depending on application) and allow all data to be stored in objects as they really exist. This would be a definite advantage in the manufacturing environment where objects are often extremely complex and require a lot of data to describe all attributes (Odell, 1992).

An OODMS uses the Client/Server network architecture that each client computer database can update its information to a much larger database on a server computer. The server database would contain records listed against each cell number. This simple framework lends itself to the function of a manufacturing cell within the factory environment, where each manufacturing cell can have its own "client" database system which links to the much larger system of the whole factory. The very concept of this distributed network brings much more information handling control to the manufacturing cell level. Similar to the cellular manufacturing philosophy, this would make similar the whole process of material and resources scheduling across the factory. Each cell would be completely in control of its own scheduling and planning, allowing indirect cost departments (such as, personnel and finance) access to the information through the database. The OODMS serves as a feasible means to design an integrated system capable of managing both the production and maintenance functions of the organisation.

4. DESIGN AN INTEGRATED CAPM/CAMM INFORMATION SYSTEM

4.1 System Analysis and Design

In order to effectively model the procedures and notations of distributed relational data for an integrated CAPM/CAMM system, the OODMS together with structured systems analysis and design methodologies are used. The two principle techniques employed are firstly, Data Flow Diagrams (DFD) which are dynamic and show what happens to data; and secondly, Entity Relationship Models (ERM) which are static and show the relationships between different stores of data (Bull, 1994). Figures 2 and 3 show the notation used for constructing a DFD and ERM, respectively.

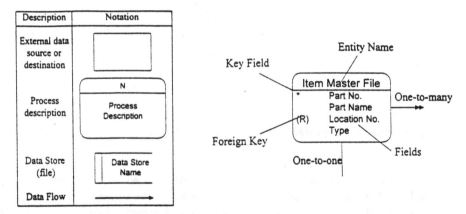

Figure 2: Data Flow Diagram Notation   Figure 3: Entity Relationship Model Notation

4.1.1 Data Flow Diagram (DFD)

The proposed model has three levels of DFD. The top level (i.e., level 0) defines the entire system as viewed from the outside, and shows all the external inputs and outputs to the system. The next level (i.e., level 1) breaks down the system as defined in the context diagram into the principle processes, data stores and flows of data. Sources and destinations of the data can also be listed. The bottom level (i.e., level 2) breaks down each of the processes into each operation. This can often be fairly complex, although it strictly defines how a databases programmer should then construct the required database management system (Burch, 1992). To further define the exact operation(s), a process description can be written in structured English using such statements as "IF","THEN","WHILE","ELSE","DO","AND","OR" and so on.

4.1.2 Entity Relationship Model (ERM)

An entity is something which data needs to be stored about. These individual data stores or entities are represented on a DFD in isolation, but an ERM shows the relationships between them. Each entity in the relational database is a table of data stored in a number of fields. Each line of the table is for a new record which must have an attribute in each field. Each entity has a key field, this should be unique to the entity and allows data to be sorted by that field and retrieved by other entities. In addition to this, if any other table features this same key field, it is called a Foreign Key or Related field, as the information is mirrored exactly from the original entity. Within the distributed relational model, there are three types of relationship including :

1. A 'one-to-one' relationship, which means where one entity relates to just one record in another entity.
2. A 'one-to-many' relationship, which means one entity is can be made up of information from many records in another entity.
3. A 'many-to-many' relationship, which means the related entities each consist of lots of records from the other. In this case it is necessary to normalise the data model, and create a link entity which is indexed by a key field (Gane, 1989).

The ERM can be further defined using a data dictionary (or table). The dictionary notation can take one of many forms, and show all field attributes being used. It indicates any optional inputs which should be used, and the coding systems be explained with the key field identification numbers.

4.2 Modelling an Integrated CAPM/CAMM System

Having regard the requirements of both CAPM and CAMM systems, the entities could be drawn up listing all the fields of data which need to be stored. The entity "Parts List" is a link entity, as a relational database system cannot handle a many to many relationship, which would normally exist between "Equipment" and "Part Inventory". The proposed integrated CAPM/CAMM system is depicted in figure 4 in abbreviated form, and the common entities as summarised as follows :

1. The Item Master File of the CAPM and the Parts Inventory File of the CAMM store similar data. Examples are part numbers, part descriptions, costs, stock locations, safety stock level features in both entities.
2. The Equipment Data File of the CAMM stores similar data to that of the Tool and Machine files of the CAMM.
3. The Works Order files of all types on both systems feature similar fields data which could easily be combined.
4. The Employee file of both systems is virtually identical.
5. The Schedule file of the CAMM and the Job Schedule file of the CAPM store similar information, all of which again could be combined.

With the use of coding in identification numbers in some of the fields, it could be ease to distinguish the difference between maintenance-related data and production-related data. The coding table represents how maintenance numbers be distinguished from production numbers when applied to the Works Order Number. Further validation could be added by only permitting the corresponding values as shown in Table 1. This is possible in many relational database systems.

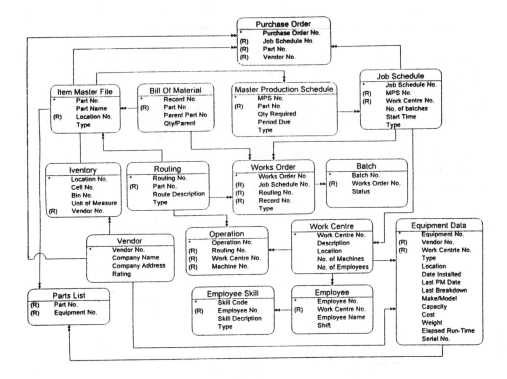

Figure 4: ERM of the Proposed Integrated CAPM/CAMM System

1st Char	2nd Char	3rd	4th	5th	6th	7th	8th
M (Maintenance)	E (Electrical)	n	n	n	n	n	n
	M (Mechanical)	n	n	n	n	n	n
	C (Chemical)	n	n	n	n	n	n
C (prev. Main.)	E (Electrical)	n	n	n	n	n	n
	M (Mechanical)	n	n	n	n	n	n
	C (Chemical)	n	n	n	n	n	n
P (production)	A (Level-0 BOM)	n	n	n	n	n	n
	B (Level-1 BOM)	n	n	n	n	n	n
	C (Level-2 BOM)	n	n	n	n	n	n
	D (Level-3 BOM)	n	n	n	n	n	n
	E (Level-4 BOM)	n	n	n	n	n	n

n = sequential chronological integer

Table 1: Coding Example for Work Order Number Identification

4.3 Developing the Integrated CAPM/CAMM Model

A series of Data Flow Diagrams are created to show all the inputs and outputs to the system, the processes of the system, and the flow of data, etc.. These diagrams allow the database programmer to develop a programme to control and manage all the data stored in the relational database. Figure 5 depicts a context diagram for the integrated system, which the external sources and inputs to the system (such as, suppliers, customers, and the shop-floor) are directly controlled by each system within the cell environment, except that the management function would operate through the higher level combined server level. All the processes which the system must carry out in order to extract the desired information from the necessary data store (or tables) are determined in the next level diagram. The system operations would then be further specified down to the lowest level with respect to the integrated requirements and various parameters.

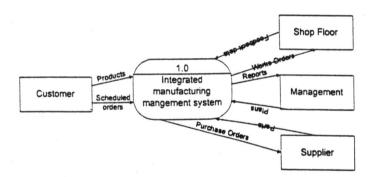

Figure 5: <u>Level-0 Data Flow Diagram of the Integrated System</u>

The proposed system requires one desktop PC for each manufacturing cell and a "host" server computer, probably a work station or equivalent. Desktop PCs in the offices of managers in the indirect departments of the organisation could be connected to the host computer to allow them to extract relevant data for subsequent analysis. Borland's Paradox for Windows 4.5 was used to build the database system for the PC environment. The software has an object-oriented programming language for developing scripts for the database operation, and making the best use of user-friendly graphical interfaces. The system can also be developed for use on networks, and as the additional capability of generating Structured Query Language (SQL) code, allowing it to access and update any information stored in the Unix environment which a host server computer running Oracle or Sybase would do.

5. CONCLUSION

It is believed that integrating the maintenance information systems of a business, whilst implementing a productive maintenance strategy across the business is the most solid form of integrated foundation on which to build. The proposed information system combines both maintenance and production functions, and the bottom-up integration tends to be more suitable for the majority of the small and medium-sized companies. There are always of structuring a company-wide database but one of distributed forms using each manufacturing cell as a remote database. The ERM model for the system has been successfully created, using the Paradox software. Once a fully working database had been modeled at the cell level, the database could be integrated with the larger CIM database, using the SQL tools feature of Paradox to link up with Oracle, allowing the smaller database to update and access data on the CIM database. There are still many holes left to fill, but the proposed model offers a potential solution to the integration of both CAPM and CAMM systems, and form the basis of any more complex models.

Acknowledgements

The authors would like to gratefully acknowledge the City University of Hong Kong for supporting the work described in this paper through the provision of a Small-scale Research Grant (Project no.903204).

References

1. Noaker, P., "Meet the CIM innovators", *Journal of Manufacturing Engineering*, November 1993, pp.47-50.
2. Rasmus, D., "The once and future CIM", *Journal of Manufacturing Systems*, March 1994, pp.62-68.
3. Kawase, T., "IE Problem Solving", *Journal of Ebina Plant Management*, Fuji Xerox, Vol. II, 1992, pp.109-117 (in Japanese).
4. Loomis, M., "ODBMS vs. Relational", *Journal of Object-Oriented Programming*, July/August 1990, pp.79-82.
5. Odell, J., "Managing object complexity, Part 1: Abstraction and generalisation", *Journal of Object-Oriented Programming*, September 1992, pp.79-82.
6. Bull, J.A., *Feasibility Study on the Integration of CAPM and CAMM for the Successful Implementation of CIM*, Project Report (No.903204), Department of Manufacturing Engineering, City University of Hong Kong, July 1994.
7. Burch, J., *Systems Analysis, Design and Implementation*, Boyd & Fraser, Boston, 1992.
8. Gane, C., *Rapid System Development Using Structured Techniques and Relational Technology*, Prentice-Hall, New Jersey, 1989.

A STEP TOWARDS CIM USING INTELLIGENT DOCUMENTS

Vinesh H. Raja.
Tel: (01203) 523924
Fax: (01203) 524307
E-mail: esrgr@csv.warwick.ac.uk

Warwick Manufacturing Group
International Manufacturing Centre
University of Warwick
Coventry CV4 7AL
United Kingdom

Abstract.

In many engineering companies, hard copy documents are still being used to distribute engineering information. There are notable advantages along with considerable shortcomings in using hard copy engineering documents. The review of recent literature has shown that hard copy engineering documents can influence the efficiency of a company. In order to maintain the competitive edge of the company, it is necessary to cope with the existing "hard copy" problems.

To address these problems, a pilot study on *Intelligent documents for Computer Integrated Manufacturing* (IDCIM) has been carried out. The basic idea is that all the information shown within the intelligent documents is retrieved from up-to-date data sources. All the intelligent documents can be maintained automatically and are instantly available throughout the company over the computer network. This paper describes the investigation of Intelligent Documents for CIM concept in the context of manufacturing information systems integration.

Key Words:- CIM, CADCAM, ERP

1. Introduction.

The substantial improvements in computer and manufacturing technologies facilitates the manufacturers to maintain competitive edge in this global dynamic manufacturing environment. The advances in computer systems technology result in cheaper PCs, workstations, as well as the application software becomes widely available. All these emerging technologies encourage the use of computers and related technologies in manufacturing industry. As a result, more computerised manufacturing applications have been developed in order to automate many manufacturing activities . However, the design of these systems are emphasised in some local solutions such as systems that deal with production planning and control, product and process design, etc. Such systems are using closed architecture, incompatible platform and heterogeneous database that results in the formation of 'islands of automation'. The "islands of automation" have the problems of lacking the control of interactions and data inconsistencies between subsystems.

The integration of information flow between the various manufacturing functions becomes a popular domain in current research and development. In the recent years, research efforts have focused on the integration of information systems for manufacturing. The objective of such research is to provide an integrated manufacturing system that its subsystems can "talk" to each other efficiently although their operating platform and data format are incompatible [Harhalakis, 1991; Weston, 1991; Kang, 1992; Lim, 1992]. On the other hand, commercially available software packages (such as, ORACLE RDBMS, ORACLE CASE TOOLS and CIMLINC LINKAGE) are becoming more powerful in the context of information integration. These types of new technologies facilitate the integration of "Islands of automation" and the idea of Computer integrated manufacturing (CIM) becomes more realistic.

However, in real life the success of an integrated manufacturing system does not only depend on the technical aspect of systems integration but also the incorporation of human factor is considerable in such a dynamic manufacturing environment.

The objective of this research is to introduce the concept of "Intelligent Engineering Documents" within the context of CIM. The concept of Intelligent documents allows the engineers to access information from a variety of data stores. All they need to do is just use the engineering workstation to prepare or retrieve whatever engineering documents they want, including the operation sheet, tool set-up instructions, machine manuals etc. In fact, the main feature of intelligent engineering documents within the context of CIM is that it provides the function of engineering document management in CIM environment. The term "engineering document management" can be described as the process of getting the right document to the right person at the right time [Griffiths. 1993], it can improve the flow of information between "islands of automation" when it is implemented effectively. The intelligent engineering document concept is expected to provide updated engineering documents to the engineers incorporating CAD drawings, process plans, tooling details, etc.

In this pilot study, the idea of Intelligent engineering documents is proposed. The architecture and prototype of Intelligent document for CIM (IDCIM) has been developed and implemented using CIMLINC's LINKAGE version 3.1.

2. CIM Definition.

CIM can be considered as the logical organisation of individual engineering, production and marketing and support functions into a computer integrated system [Bunce, 1985].

The aim of CIM is to achieve effective integration of various computer-based manufacturing processes that were previously viewed as islands of automation [Forrester, 1991].

3. Intelligent Documents.

Engineering Documents contain data formatted to suit the needs of engineers' and presented on a suitable medium for a particular purpose. The functions of *Intelligent Documents* are basically the same as the traditional documents, but the key difference is that all the information regarding the required intelligent documents is retrieved from a up-to-date data sources.

Intelligent Documents can be created within the intelligent documents environment. Users are able to alter or view the intelligent documents according to their login identification. The data sources and the intelligent documents are bi-directionally linked so that data can be retrieved from or stored into the data sources through the use of intelligent document software. The data includes text, relational or non-relational and graphical data. All types of information from multiple data sources, for instance, CAD software, ERP software, process planning package, etc. can be accessed and merged into the intelligent documents. Within CIM, it is assumed that all the information regarding the manufacturing activities are prepared using the computers and all the information is available through the computer network. Therefore, it is more realistic to implement intelligent documents within CIM.

Figure 1 illustrates the concept of intelligent documents within the context of CIM.

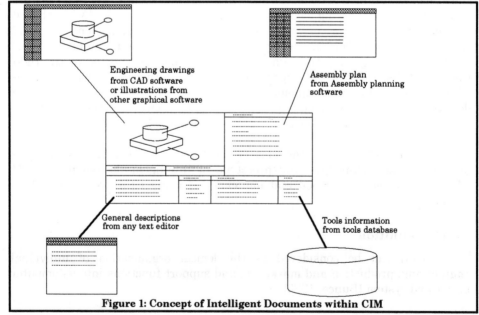

Figure 1: Concept of Intelligent Documents within CIM

4. Architecture of Intelligent Documents for CIM. (IDCIM)

Currently, it is quite difficult to describe the intelligent documents within the context of CIM since there is no common framework and vocabulary. Moreover, it is a new research interest and a review of literature led the author to conclude that very little work has been done on this subject. In order to describe the IDCIM architecture within the context of CIM, a layered architecture, which is proposed by the author, is introduced. Basically, the IDCIM architecture is a vertical system architecture, which is similar to the ISO OSI network architecture. The IDCIM architecture is a means to describe the architecture of intelligent documents for CIM. The IDCIM architecture's five layers are sufficient to describe IDCIM, they are namely:-

(1) Hardware layer
(2) Kernel layer
(3) Intelligent Documents layer
(4) Domain layer
(5) Application layer

The five layers IDCIM architecture is shown in Figure 2. Each layer is dependent on the functionality provided by the layers below it. Higher layers are more subject to change due to either a new application being added to the system, or due to the changes in the lower layers. Lower layers are therefore more stable and generic.

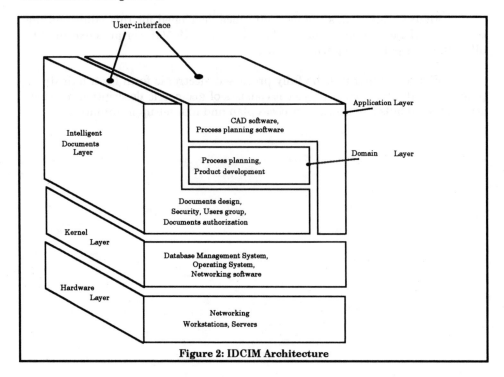

Figure 2: IDCIM Architecture

5. Linkage.

CIMLINC's Linkage is a new generation software which helps engineering and manufacturing staff to build automated systems for work instructions, simplifying engineering changes and increase the speed of product development. LINKAGE is actually an integrated set of software product which includes multimedia functionality, a raster editor, scripting tools, database interfaces, electronic mail, mark-up and redlining. The original idea for the design of LINKAGE was to :-

- Provide a common format for information.
- Allow users to re-use all existing computer based data.
- Be implementable by user.
- Capture and preserve procedures.
- Operate in a heterogeneous computer environment.
- Be scalable from small to large implementations.

6. The Pilot Study.

The pilot study looked at the use of intelligent documents for CADCAM, Process Plan, Bills Of Materials, and tool set-ups. Obviously, the intelligent document in this study provides the integration between CADCAM system, Process Planning system, ERP system and the tool set-up system. Hair dryer was chosen as the product for this study.

Figure 3 illustrates the user interface developed for this study. As the user moves the cursor into a particular a box, a LINKAGE script is executed to allow the user perform a particular task.

Figure 4 illustrates the fully processed electronic form for a hair dryer project. For the hair dryer project it consists of geometric description, product structure and bills of materials, process plan and tool set-up information.

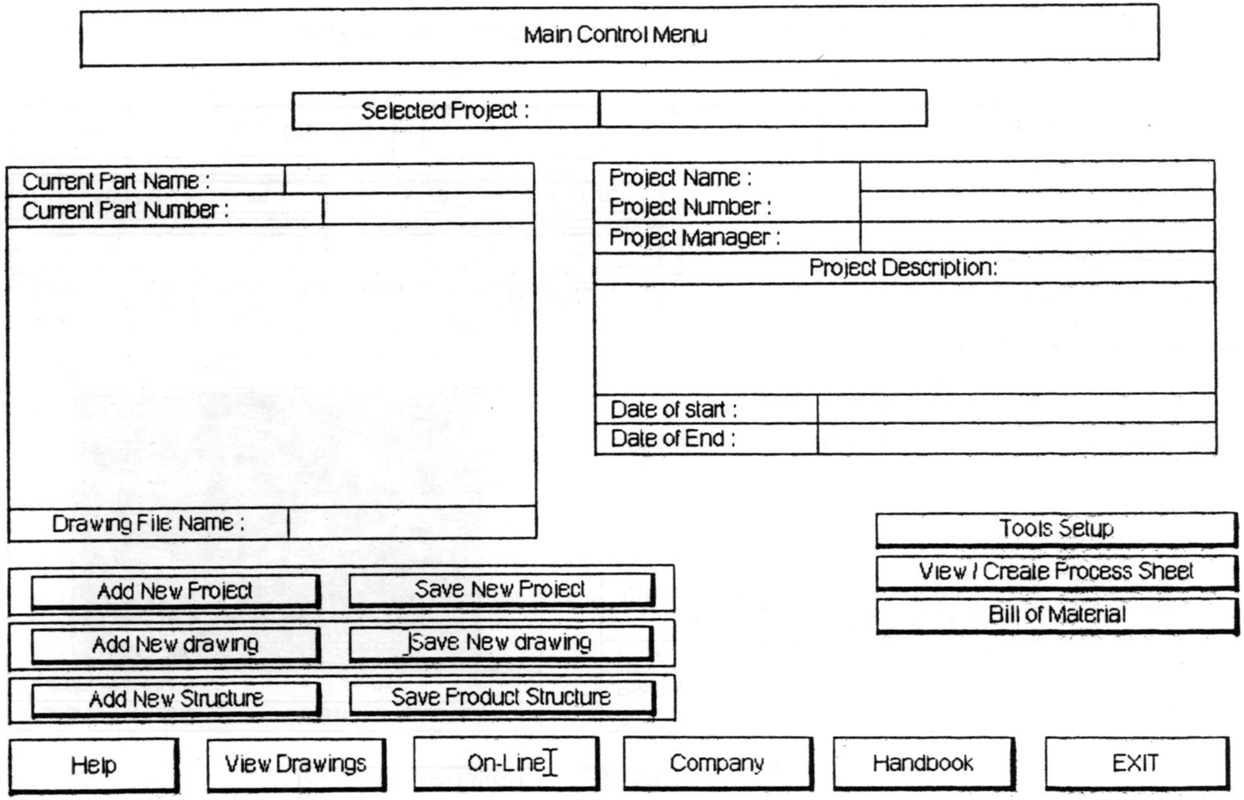

Figure 3 User Interface

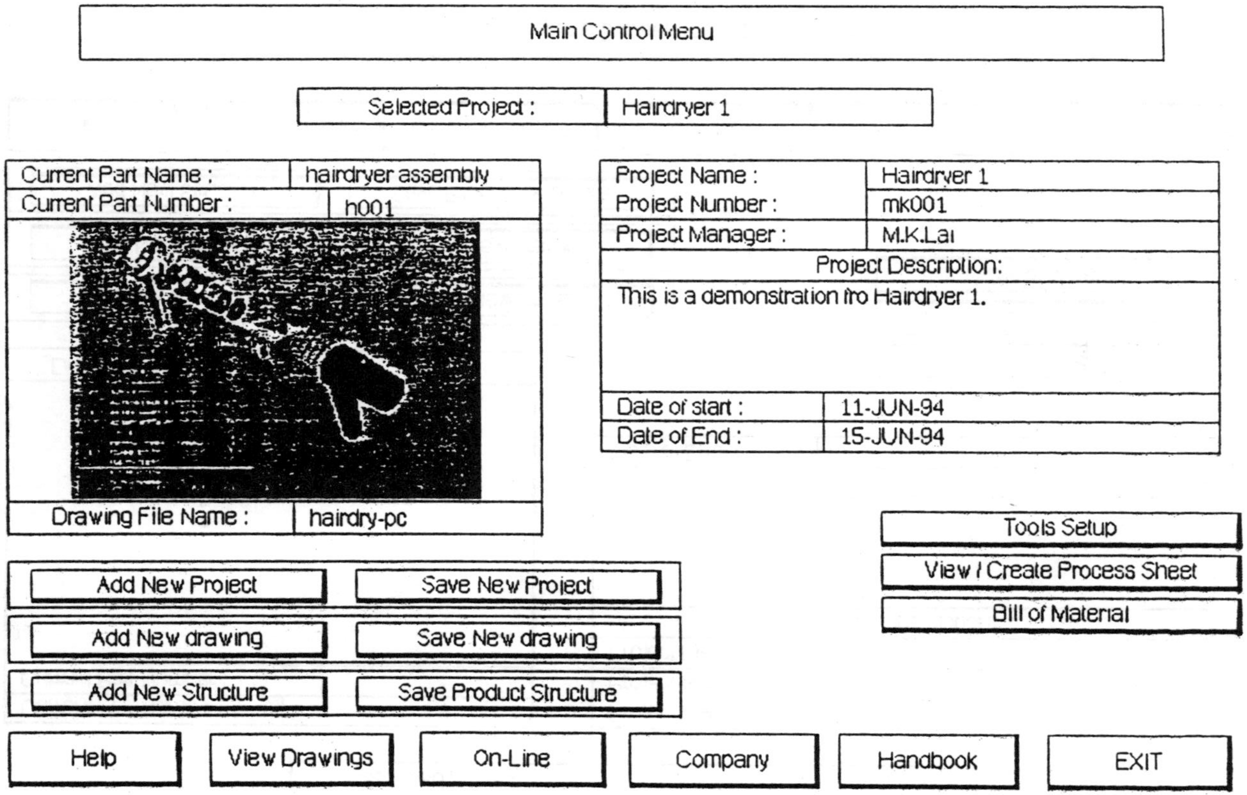

190

Main Control Menu

Selected Project : | Hairdryer 1

Project Name :	Hairdryer 1
Project Number :	mk001
Project Manager :	M.K.Lai
Project Description:	
This is a demonstratio from Hairdryer 1.	
Date of start :	11-JUN-94
Date of End :	15-JUN-94

Current Part Name :	hairdryer assembly
Current Part Number :	h001

Drawing File Name : | hairdry-pc

Tools Setup
View / Create Process Sheet
Bill of Material

Add New Project — Save New Project
Add New drawing — Save New drawing
Add New Structure — Save Product Structure

Help | View Drawings | On-Line | Company | Handbook | EXIT

Figure 4 Fully Processed Electronic Form for the Hair Dryer Project.

7. Conclusions.

The integration of CADCAM, Process Planning, ERP and tool set-up has been demonstrated by the use of intelligent documents. Successful implementation of intelligent documents will provide a step towards Computer Integrated Manufacturing.

Acknowledgments.

The author would like to acknowledge the Warwick Manufacturing Group for providing the facilities.

References.

Bunce, P. "Planning for CIM", The Production Engineer, vol. 64, No.2 February, 1985.

Forrester, P. and Hassard, J. "The development of a CIM strategy within the context of manufacturing flexibility in Electrical Engineering", Computer Integrated Manufacturing ICCIM '91, p.14, World Scientific Publishing Co. Ltd. 1991.

Griffiths, A., "Who needs engineering document management?", Information Management and Technology, 1993, vol. 26, No. 4, 176-179.

Harhalakis, G., Lin, C.P., Mark, L. and Muro-Medrano, P.R., "INformation Systems for Integrated Manufacturing (INSIM): a design methodology", Int. J. of Computer Integrated Manufacturing, 1991, vol.4, No 6, 351-363.

Lim, B.S., "CIMIDES - A computer integrated manufacturing information and data exchange system", Int. J. of Computer Integrated Manufacturing, 1992, vol.5, No 4 & 5, 240-254.

Weston, R.H., Gascoine, J.D., Rui, A., Hodgson, A., Sumpter, C.M. and Coutts, I., "Steps towards information integration in manufacturing", Int. J. of Computer Integrated Manufacturing, 1991, vol.1, No 3, 140-153.

CIMS IN SHENYANG BLOWER WORKS

Jinsong XUE   Haibin YU
Shenyang Institute of automation, Chinese Academy of Sciences, No. 114 Nanta St.
Shenyang, Liaoning 110015, P. R. China

and

Kemin LIN   Ruilin MU
Office of CIMS, Shenyang Blower Works, No. 36 North Yunfeng St. Tiexi
Shenyang, Liaoning 110021, P. R. China

Abstract

In the face of the keen and varied market competition, Computer Integrated Manufacturing (CIM) has been paid serious attention in most of the industrialized countries. Shenyang Blower Works (SBW) is the largest enterprise engaged in developing, designing and fabricating turbine compressors, blowers and large fans in China. SBW plans its production according to orders due to its singal/small batch production mode. Since the 1990's, SBW has begun to adopt CIM going through the steps of feasibility study, primary design, detailed design and partial implementation-- breakthrough project. A relative perfect computer support environment for computer integrated manufacturing system in Shenyang Blower Works (SB-CIMS) and primarily integrated application software system have been built, and played important roles in gaining remarkable economic benefits and SBW's entering the rank of technologically advanced enterprises. The implementation of CIMS is a process by stages. With the completion of the next stage of CIMS project in SBW, the SB-CIMS, we believe, will play more important roles in product design, production and business managements.

Key Words: SB-CIMS, breakthrough project.

1. Introduction

Shenyang Blower Works is the largest enterprise engaged in developing, designing and fabricating turbine compressors, blowers, and large fans in China. It is also a mainstay of the State Council and a technologically advanced enterprise in the fan industry of China. SBW's products are used as key equipments in large installations in the petroleum industry, chemical industry, metallurgical industry, coal industry, mining industry, light-textile industry, power stations, scientific research, and national defenses among others. The products of SBW are in operation in many large enterprises, such as Baoshan Iron and Steel Corp., Anshan Iron and Steel Corp., Panzhihua Iron and Steel Corp., Daqing Petrochemical Complex, Shanhai Petrochemical Complex, Zhenhai Petrochemical Complex, and Qilu Petrochemical Complex.

Turbine compressors are main products of SBW and are made-to-order in a single/small batch mode. Owing to their great variety, complex structure and thousands of components, raising product outputs and labor productivity to meet the needs of the market is difficult.

Since the 1970's, SBW has begun to apply computers to product design, engineering drawing, and material management. A CAD/CAM system and a production management information system for turbine compressors have been developed, and numerical control machine tools have been used in the production processes. The applications of advanced technology have not only brought hope and remarkable economic benefit but also brought up a group of qualified technicians in every specified field, especially in computer applications, for SBW.

In the face of the keen and varied competition in the domestic and foreign market, it is necessary to find new management modes and technical methods within the enterprise in order to enhance the production capacity as a whole, improve the technical level, shorten the period of delivery, improve the quality, reduce the cost, and increase the manufacturing flexibility, which encouraged SBW to adopt CIMS, the best and most available resoluation, we believe[1][2].

In 1990, the Chinese National Science and Technology Commission authorized SBW to become a major CIMS application factory[2], with the completion of the feasibility study[2], primary design[3], detailed design[4] and partial implementation[5] with the help of the outside experts, the prologue to the CIMS project in SBW has since been launched.

In 1993, SBW was selected as a breakthrough application factory of 10 key application factories, according to the guiding principle "benefit driven and making key points stand out" put forward by Chinese National Science and Technology Commission and the "863"* Expert Committee on the field of automation. So far, A relative perfect computer support environment for CIMS and primarily integrated application software system have been built, and played an important role in SBW's gaining remarkable economic benefit and entering the rank of technologically advanced enterprises.

2. The Way of Implementing the SB-CIMS

First, CIMS project headquarter was set up to ensure the SB-CIMS to be implemented successfully. The whole organization mechanism is depicted as following chart.

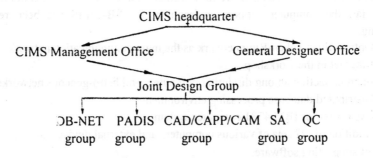

Fig.1 SB-CIMS organizational mechanism

The CIMS headquarter is led directly by the director of SBW. The CIMS management office is a branch department of SBW responsible for SB-CIMS project. The general designer office is composed of the famous experts from several institutes

*863---Nationl High Technology Program of China.

and universities, which direct technologically Joint Design Group to accomplish the tasks about the Network and Database (DB-NET) System, Production Adminastration and Decision Information System (PADIS), CAD/CAPP/CAM (3C) System, Shop Automation (SA) System, and Quality Control (QC) System.

The joint design group is composed of 60 more experts from 6 universities or institutes and 150 staff members from SBW's dozen divisions.

The principle is put into effect that the factory plays a dominate role in developing and implementing CIMS. Totally SBW and the "863" Expert Committee have respectively invested RMB 31,100,000 and RMB 2,190,000 in SB-CIMS; It is being gradually accomplished the transition from the technical depedency on Shenyang Institute of Automation, Chinese Academy of Sciences and 6 other universities to reliance to SBW itself. These fully reflected the above principle.

It is also emphasized that the SB-CIMS project is carried out by stages because of its complex nature. After the feasibility study, primary and detailed design was finished, the breakthrough project was set according to the principle "benefit driven and making key points stand out". The remarkable economic benefit from the breakthrough project make SBW adopt CIMS more firmly. The second stage of breakthrough project is now being carried out and will have be finished by the end of 1995.

3. Design Scheme of SB-CIMS

3.1 Computer Support Environment

Computers has been widely used in SBW and many individual application softwares have been applied well in production information management, product design. It is necessary to build a computer support environment suitable for integrating PADIS, CAD/CAPP/CAM and SA, shown in Fig. 2.

The whole information of SB-CIMS is integrated based on Command Data Facility (CDF). The system structure for information integration is shown in Fig. 3.

So far, the computer support environment for SB-CIMS has been realized, including:
- Optical fiber Token-Ring network as the main communication network
- Ethernet in the workshop
- Interconnection among dissimilar computers and hetro-geneous networks
- Distributed database processing environment
- Expansion for IBM 4381 computer system
- Addition of 97 sets of various computers and external devices, and 75 kinds of supporting software.

Altogether, there are more than 320 sets of computers and auxiliary equipment in the computer network of SBW.

DB-NET group is reponsible for building the computer support environment and the whole data integration.

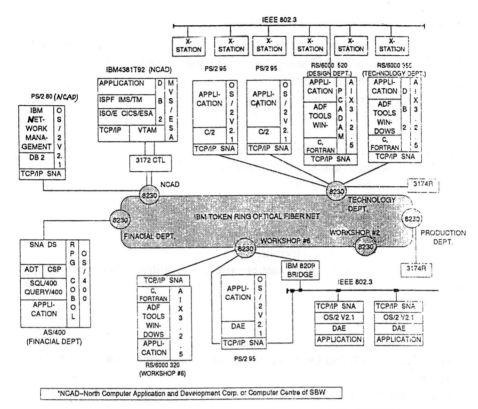

Fig. 2 Computer support environment for SB-CIMS

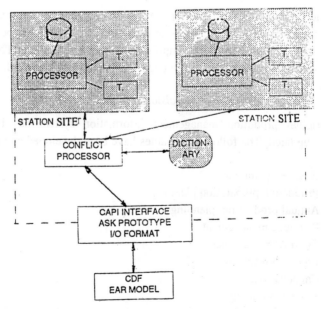

Fig. 3 System Structure for whole data integration

3.2 System Decomposition Based on Functions

SB-CIMS is composed of PADIS, 3C and SA three main subsystems, covering the main functions of SB-CIMS except for the automated material handling system.

```
                business planning
PADIS-----production management
                resources management


            product design information
3C     -----process planning information
            NC programming


            shop controller
SA     -----cell controller
            workstation controller
```

In addition, a overall quality control system is distributed over these three subsystems.

4. Breakthrough Project

The objectives of the SB-CIMS breakthrough project are:

a) To develop a Product Quotation System for turbine compressors in the CIMS environment;

b) To extend the CAD applications and integrate CAD/CAPP/CAM primarily;

c) To improve and extend the PADIS.

In general, we hope to gain remarkable benefits through the realization of these three targets.

With the completion of the development, implementation and application of the SB-CIMS breakthrough project, a hierarchically structured CIMS has been built with two dimensions in the direction of application and information. The following application subsystems have been developed or expanded and improved.

4.1 Production Adminastration and Decision Information Subsystem

The original prodution management information system has been greatly expaned and updated. The following modules have been developed and added to the subsystem:

- Business planning
- Production preparation planning
- Annual production planning
- Resource management
- Contract management
- Production statistics
- Capacity balance
- Key parts tracing

For more than one year's opertion, the realization of the whole production process tracing management from preparation to manufacturing by computer has greatly improved resource utilization.

4.2 Engineering Design (CAD/CAPP/CAM) Subsystem

The primary integration of CAD, CAPP and CAM has been realized, while the CAD/CAM system for DH and MCL serials of products has been greatly expanded. The following modules have been developed and added to this subsystem:
- BOM generation
- Impeller process drawing
- Impeller CAPP

Thus, the engineering design of the major products (turbine compressors), whose output value is over 60% of the general output value of SBW, can be completed by computers. The coverage of CAD application now has reached to:

DH type compressor	90%
auxiliary	60%
instrument system	90%
MCL type compressor	70%
auxiliary	60%
instrument system	80%

The realization of product design and process planning automation can shorten the product design period from 6 months to 3 months and increase product design capacity by 100% compared to that in 1992.

4.3 Product Quotation System

This system contains:
- Technical proposal based on user's inquiry form
- Financial quotation based on product assemblages
- Business quotation based on the market conditions

The system can gengrate quotation files in accordance with the American Petroleum Institute (API) standard. The system can also shorten the quotation period from 6 weeks to 2 weeks.

4.4 Shop Automation Subsystem

In SA system, the shop controller and cell controller as well as newly developed simulation software at workstation level have laid sound technological foundations for further shop automation.

In addition, the following systems have been designed or developed according to the overall plan of SB-CIMS:
- Comprehensive information code system
- GT code software for parts
- Framework of a quality control system in accordance with ISO 9000
- Workshop quality control software
- Central monitoring system

5. Benefits From the CIMS

With the implementation of the breakthrough project of SB-CIMS, the business and production processes in SBW from product quotation, ordering, designing, process planning, material purchasing, production preparation, product manufacturing to packing have been integrated closely. A new management mode, which is based on the part planning and market requirements, has been built and the organization mechanism in SBW has been optimized.

There is also remarkable economic benefit produced from the CIMS project.

In comparison with 1992, the output value and output in 1993 were increased by 57% and 54% respectively;

The output value and output in 1994 are expected to increase by 134% and 90% respectively.

The period of product delivery can be shortened from 18 months to 11--12 months.

The capital occupied by inventory has been reduced by 29%.

The equipment utilization has been increased by 13.9% and the labor productivity has been nearly doubled.

6. SB-CIMS's Prospects

The application of SB-CIMS has brought remarkable economic benefits to SBW, which has inspired the personnel of SBW to continue their CIMS project firmly. Looking ahead, the application of CIMS in SBW will bear fruits and a modern SBW is coming rapidly.

7. Reference

1. J. Xue, C. Wang and R. Mu, Progress in the Journey Developing SB-CIMS, International Journal of CIM, Vol. 7, No.4, 1994, p.242--248.
2. SB-CIMS Joint Design Group, Feasibility Study Report for SB-CIMS, 1991
3. SB-CIMS Joint Design Group, Primary Design Report for SB-CIMS, 1992
4. SB-CIMS Joint Design Group, Detailed Design Report for SB-CIMS, 1993
5. SB-CIMS Joint Design Group, Technical Document on SB-CIMS Project, 1994

Computer Aided Logistics

The Effects of Transportation Interval in International Global
Complementary Production Systems
   S. Hiraki, K. Ishii and H. Katayama

A Decision Support Tool for Designing Software Interfaces Between
Logistics Centres and Associated Systems
   E. J. Fletcher and R. Brunner

THE EFFECTS OF TRANSPORTATION INTERVAL IN GLOBAL COMPLEMENTARY PRODUCTION SYSTEMS

Shusaku HIRAKI
Faculty of Economics, Hiroshima University, Higashisenda-machi, 1-1-89, Naka-ku, Hiroshima 730 Japan.
E-mail: hiraki@ue.ipc.hiroshima-u.ac.jp

Kazuyoshi ISHII
Department of Industrial Engineering, Kanazawa Institute of Technology, Ohgigaoka, 7-1, Nonoichi-cho, Ishikawa-ken 921, Japan.

and

Hiroshi KATAYAMA
School of Science and Engineering, Waseda University, Okubo, 3-4-1, Shinjuku-ku, Tokyo 169 Japan.

Abstract: This paper aims to analyze the effects of transportation lead time and interval in the international co-operative global complementary production systems.

Keywords: global complementary production system, logistics, pull type ordering system

1. Introduction

In recent years, "international co-operative global complementary production systems" (ICGCPS) for mutual development have been widely developed in order to improve the international coordination and divisions of labor in the automobile industry. It is a global production, transportation and inventory system with several production bases located in several countries. Each production base produces a final product by means of assembling components such as the engine, transmission, accelerator and so on. In order to get the advantages of scale merit, each production base produces only special kinds of components with the total demand required all the participating countries, and supplies them to the other production bases. In an ICGCPS, as the components are transported among the participating countries each other, much longer transportation lead time is required to procure components from the other production bases, because they are generally supplied by marine transport over a long distance. With this in mind, this paper investigates the effects of transportation lead time and interval on the variation of ordering and inventory quantities by the following procedures: (1) formulate a pull type ordering system based on the concept of the just-in-time production system, (2) clarify the properties of distribution of ordering and inventory quantities at each stage and stock point, and (3) analyze the effects of transportation lead time and interval.

2. Modeling an ICGCPS

2.1 Schematic diagram of the ICGCPS

We focus on one of the production bases participating in the ICGCPS and call it the "home production base" (HPB). The other production bases are called the "local production bases" (LPB). Let us consider that the ICGCPS consists of four stages: the final assembly process (stage 1), the transportation process of component (stage 2), the subassembly process of component (stage 3) and the machining process of part (stage 4). Figure 1 shows the schematic diagram of the ICGCPS considered in this paper.

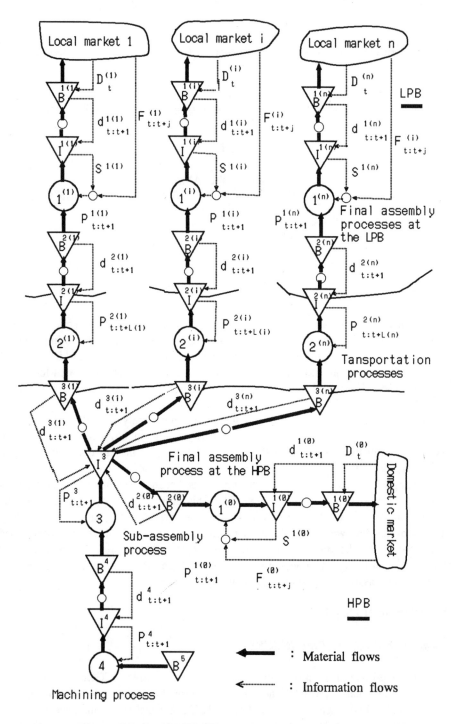

Fig.1 Schematic diagram of the ICGCPS.

In this figure, we define notation as follows:
(1) Final assembly process at the HPB:

$D_t^{(0)}$: demand for the final product in period t at the home production base.

$F_{t:t+j}^{(0)}$: the forecasted demand for the final product for the (t+j)-th period forecasted at the end of the t-th period.

$B_t^{1(0)}$: on-hand inventory quantity of the final product for the market available at the end of the t-th period.

$d_{t:t+1}^{1(0)}$: withdrawal ordering quantity of the final product, which is required at the stock point $I^{1(0)}$ at the end of the t-th period and delivered to the stock point $B^{1(0)}$ at the beginning of the (t+1)-st period.

$I_t^{1(0)}$: inventory quantity of the final product fabricated by stage $1^{(0)}$ at the end of the t-th period.

$P_{t:t+1}^{1(0)}$: production ordering quantity of the final product at stage $1^{(0)}$, which is calculated at the end of the t-th period, placed into production during the (t+1)-st period and completed at the end of the (t+1)-st period.

$S^{1(0)}$: safety stock of the final product at the stock point $I^{1(0)}$.

$B_t^{2(0)}$: on-hand inventory quantity of the component for the final assembly process available at the end of the t-th period.

$d_{t:t+1}^{2(0)}$: withdrawal ordering quantity of the component, which is required at the stock point $I^{2(0)}$ at the end of the t-th period and delivered to the stock point $B^{2(0)}$ at the beginning of the (t+1)-st period.

(2) Final assembly and transportation processes at the LPB:

In the ICGCPS, as the demand for the final product at each LPB is not so enough, the component is not necessary transported to the LPB every period. Let us denote K(i) the transportation interval from the HPB to the LPB 'i', and n the number of the LPB. The other notation is defined as follows.

L(i) : the transportation lead time from the HPB to the LPB 'i'.

$D_t^{(i)}$: demand for the final product in period t at the LPB 'i'.

$F_{t:t+j}^{(i)}$: the forecasted demand for the final product at the LPB 'i' for the (t+j)-th period forecasted at the end of the t-th period.

$B_t^{1(i)}$: on-hand inventory quantity of the final product at the stock point $B^{1(i)}$ at the end of the t-th period.

$d_{t:t+1}^{1(i)}$: withdrawal ordering quantity of the final product, which is required at the stock point $I^{1(i)}$ at the end of the t-th period and delivered to the stock point $B^{1(i)}$ at the beginning of the (t+1)-st period.

$I_t^{1(i)}$: inventory quantity of the final product fabricated by stage $1^{(i)}$ at the end of the t-th period.

$P_{t:t+1}^{1(i)}$: production ordering quantity of the final product at stage $1^{(i)}$, which is calculated at the end of the t-th period, placed into production during the (t+1)-st period and completed at the end of the (t+1)-st period.

$S^{1(i)}$: safety stock of the final product at the stock point $I^{1(i)}$.

$B_t^{2(i)}$: on-hand inventory quantity of the component for the final assembly process at the stock point $B^{2(i)}$ available at the end of the t-th period.

$d_{t:t+1}^{2(i)}$: withdrawal ordering quantity of the component, which is required at the stock point $I^{2(i)}$ at the end of the t-th period and delivered to the stock point $B^{2(i)}$ at the beginning of the (t+1)-st period.

$I_t^{2(i)}$: inventory quantity of the component transported by stage $2^{(i)}$ at the end of the t-th period.

$P_{t:t+L(i)}^{2(i)}$: transportation ordering quantity of the component at stage $2^{(i)}$, which is calculated at the end of the t-th period, placed into production during the (t+1)-st period and completed at the end of the (t+L(i))-th period.

$B_t^{3(i)}$: on-hand inventory quantity of the component for the transportation process available at the end of the t-th period.

$d_{t:t+1}^{3(i)}$: withdrawal ordering quantity of the component, which is required at the stock point I^3 at the end of the t-th period and delivered to the stock point $B^{3(i)}$ at the beginning of the (t+1)-st period.

(3) Sub-assembly and machining processes at the HPB:

I_t^3 : inventory quantity of the component fabricated by stage 3 at the end of the t-th period.

$P_{t:t+1}^3$: production ordering quantity of the component at stage 3, which is calculated at the end of the t-th period, placed into production during the (t+1)-st period and completed at the end of the (t+1)-st period.

B_t^4 : on-hand inventory quantity of the part for the sub-assembly process available at the end of the t-th period.

$d_{t:t+1}^4$: withdrawal ordering quantity of the part, which is required at the stock point I^3 at the end of the t-th period and delivered to the stock point B^4 at the beginning of the (t+1)-st period.

I_t^4 : inventory quantity of the part fabricated by stage 4 at the end of t-th period.

$P_{t:t+1}^4$: production ordering quantity of the part at stage 4, which is calculated at the end of the t-th period, placed into production during the (t+1)-st period and completed at the end of the (t+1)-st period.

2.2 Formulation

Using the notation defined above, we formulate an ordering model based on the concept of the pull-type ordering system [1],[2],[3]:

2.2.1 Modeling for the domestic final assembly process

(1) On-hand inventory quantity of the final product at the stock point $B^{1(0)}$:

$$B_t^{1(0)} = B_{t-1}^{1(0)} + d_{t-1:t}^{1(0)} - D_t^{(0)} \tag{1}$$

(2) Withdrawal ordering quantity of the final product from the stock point $B^{1(0)}$ to $I^{1(0)}$:

$$d_{t:t+1}^{1(0)} = D_t^{(0)} \tag{2}$$

(3) Fabricated inventory quantity of the final product at the stock point $I^{1(0)}$:

$$I_t^{1(0)} = I_{t-1}^{1(0)} + P_{t-2:t-1}^{1(0)} - d_t^{1(0)} \tag{3}$$

(4) Production ordering quantity of the final product:

$$P_{t:t+1}^{1(0)} = F_{t:t+2}^{(0)} + (F_{t:t+1}^{(0)} - P_{t-1:t}^{1(0)}) - I_t^{1(0)} + S^{1(0)} \tag{4}$$

(5) On-hand inventory quantity of the component at the stock point $B^{2(0)}$:

$$B_t^{2(0)} = B_{t-1}^{2(0)} + d_{t-1:t}^{2(0)} - P_{t-1:t}^{1(0)} \tag{5}$$

(6) Withdrawal ordering quantity of the component from the stock point $B^{2(0)}$ to $I^{2(0)}$:

$$d_{t:t+1}^{2(0)} = P_{t-1:t}^{1(0)} \tag{6}$$

2.2.2 Modeling for the final assembly and transportation processes at the LPB

(1) On-hand inventory quantity of the final product at the stock point $B^{1(i)}$:

$$B_t^{1(i)} = B_{t-1}^{1(i)} + d_{t-1:t}^{1(i)} - D_t^{(i)} \qquad (i=1,2,\ldots,n) \qquad (7)$$

(2) Withdrawal ordering quantity of the final product from the stock point $B^{1(i)}$ to $I^{1(i)}$:

$$d_{t:t+1}^{1(i)} = D_t^{(i)} \qquad (i=1,2,\ldots,n) \qquad (8)$$

(3) Fabricated inventory quantity of the final product at the stock point $I^{1(i)}$:

$$I_t^{1(i)} = I_{t-1}^{1(i)} + P_{t-2:t-1}^{1(i)} - d_{t-1:t}^{1(i)} \qquad (i=1,2,\ldots,n) \qquad (9)$$

(4) Production ordering quantity of the final product:

$$P_{t:t+1}^{1(i)} = F_{t:t+2}^{(i)} + (F_{t:t+1}^{(i)} - P_{t-1:t}^{1(i)}) - I_t^{1(i)} + S^{1(i)} \qquad (i=1,2,\ldots,n) \qquad (10)$$

(5) On-hand inventory quantity of the component at the stock point $B^{2(i)}$:

$$B_t^{2(i)} = B_{t-1}^{2(i)} + d_{t-1:t}^{2(i)} - P_{t-1:t}^{1(i)} \qquad (i=1,2,\ldots,n) \qquad (11)$$

(6) Withdrawal ordering quantity of the component from the stock point $B^{2(i)}$ to $I^{2(i)}$:

$$d_{t:t+1}^{2(i)} = P_{t-1:t}^{1(i)} \qquad (i=1,2,\ldots,n) \qquad (12)$$

(7) Transported inventory quantity of the component at the stock point $I^{2(i)}$:

$$I_t^{2(i)} = I_{t-1}^{2(i)} + P_{t-L(i):t}^{2(i)} - d_{t-1:t}^{2(i)} \qquad (i=1,2,\ldots,n) \qquad (13)$$

(8) Transportation ordering quantity of the component at the transportation process:

$$P_{t:t+L(i)}^{2(i)} = \begin{cases} \sum_{k=0}^{K(i)-1} d_{t-k:t-k+1}^{2(i)} & (t \in T_1) \\ 0 & (t \in \bigcup_{k=2}^{K(i)} T_k) \end{cases} \qquad (i=1,2,\ldots,n) \qquad (14)$$

where, $T_k = \{k, K(i)+k, 2K(i)+k, \ldots\}$ $(k=1,2,\ldots,K(i))$

(9) On-hand inventory quantity of the component at the stock point $B^{3(i)}$:

$$B_t^{3(i)} = B_{t-1}^{3(i)} + d_{t-1:t}^{3(i)} - P_{t-1:t+L(i)-1}^{2(i)} \qquad (i=1,2,\ldots,n) \qquad (15)$$

(10) Withdrawal ordering quantity of the component from the stock point $B^{3(i)}$ to I^3:

$$d_{t:t+1}^{3(i)} = d_{t:t+1}^{2(i)} \qquad (i=1,2,\ldots,n) \qquad (16)$$

2.2.3 *Modeling for the sub-assembly and machining processes at the HPB*

(1) Fabricated inventory quantity of the component at the stock point I^3:

$$I_t^3 = I_{t-1}^3 + P_{t-1:t}^3 - d_{t-1:t}^{2(0)} - \sum_{i=1}^n d_{t-1:t}^{3(i)} \qquad (17)$$

(2) Production ordering quantity of the component at the sub-assembly process:

$$P_{t:t+1}^3 = d_{t-1:t}^{2(0)} + \sum_{i=1}^n d_{t-1:t}^{3(i)} \qquad (18)$$

(3) On-hand inventory quantity of the part at the stock point B^4:

$$B_t^4 = B_{t-1}^4 + d_{t-1:t}^4 - P_{t-1:t}^3 \qquad (19)$$

(4) Withdrawal ordering quantity of the part from the stock point B^4 to I^4:

$$d_{t:t+1}^4 = P_{t-1:t}^3 \qquad (20)$$

(5) Fabricated inventory quantity of the part at the stock point I^4:

$$I_t^4 = I_{t-1}^4 + P_{t-1:t}^4 - d_{t-1:t}^4 \qquad (21)$$

(6) Production ordering quantity of the part:

$$P_{t:t+1}^4 = d_{t:t+1}^4 \qquad (22)$$

Without loss of generality, we assume that the transportation ordering quantity of the component is calculated at the end of the periods $t \in T_1$ in Eq.(14). We call the mathematical model given by Eqns.(1)–(22) ordering model for the ICGCPS.

3. Effects of Transportation Lead Time and Interval

3.1 Distributions of ordering and inventory quantities

In order to clarify the effects of transportation lead time and interval on the variance of ordering, withdrawal and inventory quantities of the ordering model for the ICGCPS formulated in Section 2, we consider the following conditions:
(1) Demand for the final product at the HPB (i=0) and the LPB (i=1,2,...,n) is stationary time series with exponential autocorrelation as follows:

$$E(D_t^{(i)}) = \mu_i, \quad Var(D_t^{(i)}) = \sigma_i^2, \quad Cov(D_t^{(i)}, D_{t+j}^{(i)}) = \lambda_i^j \sigma_i^2 \qquad (i=0,1,2,...,n)$$

where, μ_i and σ_i^2 are the mean and variance of demand, $Cov(D_t^{(i)}, D_{t+j}^{(i)})$ is the covariance between demand $D_t^{(i)}$ and $D_{t+j}^{(i)}$, and λ_i is the autocorrelation coefficient. We assume that demands $D_t^{(i)}$ and $D_t^{(k)}$ ($i \neq k$) are mutually independent.
(2) Forecasted demand for the final product at the HPB (i=0) and the LPB (i=1,2,...,n) is calculated by the autocorrelation method as follows:

$$F_{t:t+j}^{(i)} = \lambda_i^j D_t^{(i)} + (1 - \lambda_i^j)\mu_i \qquad (i=0,1,2,...,n)$$

Table 1 shows the variance of ordering, withdrawal and inventory quantities at each stage and the stock point.

3.2 Effects of transportation lead time and interval

From Table 1, we can observe the following:
(1) The transportation interval affects the variance of the inventory quantity at the stock points $I_t^{2(i)}$ and $B_t^{3(i)}$ and the transportation ordering quantity $P_{t:t+L(i)}^{2(i)}$.
(2) Only the variance of the transported inventory quantity is affected by changes in transportation lead time. This fact is the same result pointed out in [4].

When the transportation interval from the HPB to the LPB 'i' is K(i), the transported inventory quantity changes every K(i) period. It takes the maximum value at the end of such the periods that t−L(i) belongs to the set T_1, and takes the minimum value at t−L(i) $\in T_{K(i)}$. Hence, it is important to investigate the variation of transported inventory quantity at the end of the periods t−L(i) $\in T_{K(i)}$ in order to assure the reliability of the final assembly process at the LPB 'i'.

Let us define the amplification of the transported inventory quantity as follows:

$$Amp(I^{2(i)}) = Var(I_t^{2(i)})/Var(D_t^{(i)}) \qquad (i=0,1,2,...,n)$$

Figure 2 shows the relationship between the amplification of the transported inventory quantity at the end of the periods t−L(i) $\in T_{K(i)}$ and the transportation lead time and interval with respect to the autocorrelation coefficient. As is evident from this figure, we can see that the variation of transported inventory quantity ($Amp(I^{2(i)})$) increases according to the increase of the transportation lead time and interval. It also increases according to the increase of the autocorrelation coefficient. This means that we must include a large buffer inventory at the transported stock point if the transportation lead time and interval

Table 1. Variations of ordering, withdrawal and inventory quantities

	Stage	Home production base (HPB)				Local production base (LPB)		
		1	2	3	4	1	2	3
$B_t^{k(i)}$	Variance	σ_0^2	$b_0\sigma_0^2$	—	$\sum_{i=0}^{n} b_i\sigma_i^2$	σ_i^2	$b_i\sigma_i^2$	$e_i\sigma_i^2$
$d_{t:t+1}^{k(i)}$	Variance	σ_0^2	$b_0\sigma_0^2$	—	$\sum_{i=0}^{n} b_i\sigma_i^2$	σ_i^2	$b_i\sigma_i^2$	$b_i\sigma_i^2$
$I_t^{k(i)}$	Variance	$a_0\sigma_0^2$	—	0	0	$a_i\sigma_i^2$	$c_i\sigma_i^2$	—
$P_{t:t+L(i)}^{k(i)}$	Variance	$b_0\sigma_0^2$	—	$\sum_{i=0}^{n} b_i\sigma_i^2$	$\sum_{i=0}^{n} b_i\sigma_i^2$	$b_i\sigma_i^2$	$d_i\sigma_i^2$	—

Where, $L(i)=1$ $(k=1,3)$, $L(0)=1$ and

a_i	$2-3\lambda_i^2+\lambda_i^4$	$(i=0,1,2,\ldots,n)$	b_i	$1-2\lambda_i+2\lambda_i^2+4\lambda_i^3-2\lambda_i^4-2\lambda_i^5$	$(i=0,1,2,\ldots,n)$

c_i : $(L(i)+m-2)-2\lambda_i+4\lambda_i^3+2\lambda_i^4+2\{(\lambda_i+\lambda_i^2)(1-\lambda_i-\lambda_i^2)\lambda_i^{L(i)+m-2}+\sum_{k=1}^{L(i)+m-3}\sum_{h=1}^{k}\lambda_i^h\}$ $(t-L(i)\varepsilon T_m; i=1,2,\ldots,n;$ $m=1,2,\ldots,K(i))$

$K(i)-2\lambda_i+4\lambda_i^3+2\lambda_i^4+2\{(\lambda_i+\lambda_i^2)(1-\lambda_i-\lambda_i^2)\lambda_i^{K(i)}+\sum_{k=1}^{K(i)-1}\sum_{h=1}^{k}\lambda_i^h\}$ $(t\varepsilon T_1; i=1,2,\ldots,n)$

d_i : 0 $(t\varepsilon\bigcup_{h=2}^{K(i)} T_h; i=1,2,\ldots,n)$

$(K(i)+1)-2\lambda_i+4\lambda_i^3+2\lambda_i^4+2\{(\lambda_i+\lambda_i^2)(1-\lambda_i-\lambda_i^2)\lambda_i^{K(i)+1}+\sum_{k=1}^{K(i)}\sum_{h=1}^{h}\lambda_i^h\}$ $(t\varepsilon T_1; i=1,2,\ldots,n)$

e_i : 0 $(t\varepsilon T_2; i=1,2,\ldots,n)$

$(m-2)-2\lambda_i+4\lambda_i^3+2\lambda_i^4+2\{(\lambda_i+\lambda_i^2)(1-\lambda_i-\lambda_i^2)\lambda_i^{m-2}+\sum_{k=1}^{m-3}\sum_{h=1}^{h}\lambda_i^h\}$ $(t\varepsilon T_m; i=1,2,\ldots,n; m=3,4,\ldots,K(i))$

are large, especially when the autocorrelation coefficient is large. As the inventory holding cost is a function of the buffer inventory and the transportation cost is a function of the transportation interval, it is necessary to trade off these two costs in order to improve the system productivity. This is the topic for the further discussion.

4. Conclusion

In this paper, we investigated:
(1) Considering not only the transportation lead time but also interval, a production, inventory and transportation model based on the concept of the "pull-type" ordering system was formulated for the ICGCPS.
(2) The effects of the transportation lead time and interval on the variation of ordering, withdrawal and inventory quantities, especially at the transported stock point was clarified using the time series forecasting.

References

[1] S. Hiraki, "Comparative analysis of ordering models for an international co-operative global complementary production system", *International Journal of Production Economics*, (in preparation)

[2] O. Kimura and H. Terada, "Design and analysis of pull system, a method of multistage production control", *Int. J. of Prod. Res.*, **19(3)**, pp.241-253 (1981).

[3] S. Hiraki and H. Katayama, "Designing of an international co-operative global complementary production ordering system", *New Directions in Simulation for Manufacturing and Communications*, Eds. S. Morito et., ORSJ, pp.519-525 (1994).

[4] S. Hiraki, K. Ishii, K. Takahashi and R. Muramatsu, "The Effects of transportation lead-time and safety stock in international co-operative knockdown production systems", *Prod. Planning & Control*, **1(4)** pp. 209-221 (1990).

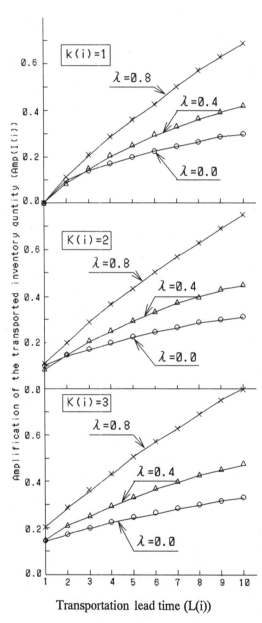

Fig.2 Relationship between $Amp(I^{2(i)})$ at the end of the period $t-L(i) \in T_{K(i)}$ and the transportation lead time $(L(i))$ and interval $(K(i))$.

A DECISION SUPPORT TOOL FOR DESIGNING SOFTWARE INTERFACES BETWEEN LOGISTICS CENTRES AND ASSOCIATED SYSTEMS

E.J. Fletcher

School of Computing, University of Sunderland, Green Terrace,

Sunderland, SR1 3SD, United Kingdom.

E-mail: E.Fletcher@Sunderland.ac.uk

and

R. Brunner

Hewlett - Packard GmbH, Postfach 1430, 71004,
Boblingen, Germany.

E-mail: rbrunner@hpbbn.bbn.hp.com

ABSTRACT

Recent years has seen the development of several very large logistics centres in which fully automated warehoueses are incorporated into the logistics chain of company operations. Such logistics centres can transform the warehouse from a cost contributor to a value added role within the company. However designing the software interfaces for such systems has proved a difficult and expensive task and there is evidence that a complexity limit has been reached with traditional design methods. This paper reports on an analysis of five such projects in Europe which has led to the identification of the elements which are common to all such projects and the specification of a full 'Query Model' for the interface design. A prototype decision support tool has been developed based on the query model which has been tested on a group of design engineers.

Keywords : Logistics-Centres, Software-Design, Expert-Systems.

2. Introduction

The conceptual ideal of an optimised warehouse [Duncan 1985] which requires no labour for operation, which is in complete control and which takes little space can only be partially realised. In the first generation of warehouses companies were focused on the space factor with the aim of storing the maximum quantity of goods in the minimum space. This was succeeded by the second phase in which attention moved to achieving fast and random access to goods at any time. The latest generation of warehouses focuses on integrating the warehouse into the company logistics chain [Grossmann 1994]. The success of logistics, material handling, and warehousing is determined by the integration and synchronisation of material flow, information flow and capital flow. Logistics and material handling systems can be defined in terms of using the right methods to provide the right amount of the right material, at the right place, at the right time, in the right sequence, in the right position, in the right conditions and at the right cost [White 1987]. In this logistics context the warehouse plays a strategic role in a companies success.

For many companies warehousing trends are towards centralised distribution systems with decentralised production buffer warehouses. Besides the classic warehousing functions the modern warehouse now includes enhanced functions.

Examples are automated order accumulation, packaging of goods and carrier schedule management. These additional tasks support the bridging functionality of the warehouse between the receiving and shipping of goods. It is now almost impossible to define the 'standard' functionality of a warehouse, however one can outline the characteristics required to achieve warehousing excellence. Briefly these are :

- **Customer service**; Customer service and distribution efficiency are a key factor in success.
- **Flexibility**; The warehouse needs to respond to changing product volumes and mixes, marketing strategies, throughput and service requirements.
- **Quality**; Identification of all sources of error and damage through the operations of receiving, shipping, storing, packing, conveying and transporting.
- **Technology**; New technologies open up new opportunities for rationalisation, automation and improvements in efficiency.
- **Inventory**; The maintenance of accurate inventory identification and quantities and of optimum stock levels.
- **Integration**; The warehouse is fully integrated into the logistics chain.
- **Cost**; Ideally the warehouse should have a 'value added' role rather than as a contributor to costs.

Thus the modern warehouse linked to the logistics train of the company is a complex entity whose design is likely to differ for each implementation. The success of the implementation can have a profound impact on companies performance. The linking of a centralised warehouse to the associated companies material and information flow systems effectively makes it a new entity usually referred to as *a logistics and distribution centre*.

3.   The Logistics & Distribution Centre

Today's conventional warehouses in the sense of automated warehouse systems (AWS) are characterised by a lower degree of system complexity, functionality, and autonomy when compared to logistics and distribution centres (LDC). The LDC is characterised through a higher degree of complexity, functionality and autonomy and is almost fully automated in its warehousing functions. Besides various automated storage systems, LDC automation can include additional functions such as automated picking, sorting, packaging or palletising. In its conception the LDC is focused on distribution functions to improve delivery service. Besides elementary warehousing functions the LDC can include additional functional features, for example order accumulation or carrier schedule management. Considering organisational and data processing technical aspects, the LDC is integrated as an element in the logistic chain, but it is not integrated into associated systems. The LDC warehouse management system operates as an autonomous system to improve system availability. This system autonomy necessitates a warehouse order buffer to guarantee smooth operation and to bridge interruptions in communication with the associated systems. In the LDC the WMS has the all-embracing control and management of all processes and functions, which establishes a very high degree of transparency and reliability of material and information flow. All functions and services to and from the LDC are exchanged through the external interfaces (Fig. 1). So LDCs can be considered as autonomous service modules, which offers its warehousing and distribution functions to associated systems.

Interface design complexity results from the divergent objectives between the system autonomy on the one hand and the data access and modification possibilities on the other hand. In general system autonomy with an order buffer functionality restricts data access and leads to more complex requirements concerning communication between systems. Additional interface problems can arise out of the requirement to handle exceptional situations. This aspect can increase the degree of complexity of the external DP-interfaces to associated systems. Caused by the delayed information processing, communication interruptions, or data lost, "out-of-sync situations" must be handled between the communication partners. In these cases the master systems must be able to generate reconciliation to correct data. Additionally, error handling processes must be established in the communication between LDC systems and CEDP-environments to handle failures and errors and to reject these specific interface transactions.

A further consideration is the fact that the more complex the required functionality in the LDC is, the higher the effort to link to the CEDP-systems. This means that also less complex associated systems must process the whole range of all possible interface transactions, even when it is not required for their internal system processing. This comment is based on the generic functional consideration, which do not consider system specific aspects of associated systems.

Further effects on the interface design are often based on system process related aspects. Examples are process restrictions, which are based on the "point of no return" time in processing of orders or internal system processes, which have an impact on the stock management which is not known to the warehouse external system environment.

A detailed examination has been undertaken of the interface design of five of the largest LDC projects in Europe. The projects were from widely different industrial sectors (e.g. Computer peripherals & instrumentation, plumbing components, Aircraft spares) This has demonstrated that it is possible to codify the design principles for the software interface into a set of rules all be it a large set ! These rules form the basis for the Decision Support System.

4. Interface Rules and the query model

The application of the interface rules will produce different results for each project depending on the input parameters that describe the requirements and restrictions of the LDC project. The output of applying a set of interface rules and input parameters produces a design for the EDP-interfaces. The interface rule parameters that specify requirements and restrictions can be classified into two types:

1. CEDP environment related parameters, 2. LDC system related parameters.

Figure 1

The CEDP environment related parameters describe all requirements and restrictions that result from the existing company environment mainly determined by the CEDP-systems. The following list outlines some examples of this type of parameter:

CEDP system functions, Product structure, Data volume, Data exchange requirements.

On the other hand the LDC system related parameters are focused on all requirements and restrictions based on the required and offered LDC services and functionality. Examples for this type of parameter are:

Degree of system autonomy, Required warehouse elements, Warehousing processes, WMS functions.

The interface rule parameters can lead to different consequences for the design of the EDP-interfaces. Possible consequences can influence the type of required interfaces, on the data structure of the interfaces, and finally on the data element assignments to specific data levels. Detailed consideration by the authors of the transaction interface rules shows the system to be complex and only an overview and an example are presented here. The following major interface types can be specified :

Receiving (Rcpt), Shipment (Shpt), Part Status (Part_stat), Part Master (Part_mast).

Besides these central interface types, in the EDP-interface design of LDC systems two further interface types might be necessary in some circumstances. In comparison to the central interface types these interface types are optional in the LDC design. These optional interface types are:

Code Table (Cd_tbl) & Report (Rprt).

Each of these six interface types are determined by different function and communication requirements, which are mainly characterised by the related major transactions. Further design aspects can result from additional interface type related requirements (e.g. status information, rejection handling, ...).

The major transactions can be classified into functional and operational transactions. The major transactions of the receiving and shipment interface types and the part status major transaction are functional transactions, because these major transactions request or are triggered through warehousing functions. All remaining major transactions are operational transactions required for operation of the system, but which do not have a direct impact on warehousing functions. Besides this classification, major transactions are determined by their communication direction in the communication system between the WMS of the LDC system and the CEDP-environment. The communication partner that generates the transaction is the sender system and the other communication partner that receives the transaction is the receiver system.

Besides these interface type related characteristics there are also transaction related characteristics. Every transaction is characterised through the type of data defined by the period of validity of the data. Based on this data type definition, data can be distinguished into "period data" and "moment data". Period data is data valid for a longer time span before processing. This data can be modified by the sender system after the data transfer to the receiver system. Moment data is processing time exact data that cannot be modified by the sender system after the transaction is generated because moment data represents data of a specific time. A further characteristics of transactions is whether or not they are inventory changing transactions. This means additional data exchange requirements concerning the data synchronisation because of the inventory booking processing in the WMS and the CEDP-environment.

Process functional dependencies between transactions can exist. The functional dependencies between transactions is based on the transaction relation to the same warehousing or WMS process function. For examples there is a functional dependency between the receipt-expectation and receipt-acknowledgement major transactions that are both related to the receiving process in warehouse systems. Based on a full analysis of the interface requirements a detailed query model for interface design was established. This model runs to over 200 rules for the more complex interface types. All rules of the different interface types have been expressed in an IF-THEN-ELSE format. Figure 2 shows an example of a single rule as expressed in the query model :

```
Do the associated systems always know, which data elements of the transferred receipt
expectation are effected by changes that require an update in the WMS system ? [Q]
IF "yes" THEN
    1. rcpt-expt-(OT)-data-mod basic txn required [R]
    2. For the data element modifications, will additionally an interface function be
provided which updates all data elements of the transferred data ?[Q]
    IF "yes" THEN
            data level modify function required [R]
        ELSE
            data level modify function not required [R]
        ENDIF
ELSE
    1. Data level modify function required [R]
    2. In the case of data element modifications in which the associated systems
know exactly which elements to modify do you wish to implement an interface
function which updates these elements only ? [Q]
    IF "yes" THEN
            rcpt-expt-(OT)-data-mod basic txn required [R]
        ELSE
            rcpt-expt-(OT)-data-mod basic txn not required [R]
        ENDIF
ENDIF
```

Figure 2

All the rules have been codified into a design document *'Query model of the interface rules'*. A prototype Decision Support System (DSS) based on the query model was then implemented in PDC Prolog.

5. The Decision Support System

The major aim of the prototype DSS was to demonstrate the functionality of a fully implemented DSS, it was also an objective on the one hand to prove that the full system could be built and on the other hand that this could be done on the basis of the design document *'Query Model of the Interface Rules')*. Based on these criteria the prototype implementation was structured into a vertical and horizontal implementation part.

The objective of vertical prototype implementation is to show all important system functions for one interface type module. The major focus for this part of the implementation is on the interface design process with the determination of the basic transactions, data structures and elements, and communication transactions. Additionally the user-interface to add new data element interface rules is also implemented. Finally vertical prototype implementation contains system functions to display and print the output result files generated during the interface design process. For the implementation of this part of the prototype the shipment interface module was selected because of its module complexity.

The objective of the horizontal part of the prototype is to show the implementation of the basic transaction process over all modules to get an idea of the functionality of the fully implemented DSS. Besides the system functions to display and print the basic transaction related output result files, further system functions are implemented to explain the basic transaction decisions. The implementation of the explanation functions was possible because of a different implementation method used for the horizontal part of the prototype.

The prototype was implemented in DOS version 3.31a of PDC Prolog. The particular implementation language was chosen because of the close cooperation between the University of Sunderland and the Prolog Development Centre in Copenhagen. It is also intended to develop a full DSS for commercial use and this may be developed in PDC Visual Prolog.

6. Prototype Evaluation

Having completed the DSS prototype it was necessary to test its effectiveness with potential users of the system. A further objective was to judge the market potential for such a system and to obtain the requirements for a fully developed system.

Based on these objectives the tester profile can be defined. On the one hand it is necessary to have experienced testers, who are experts in this field because they have managed one or more of the WMS interface problems or because they have considerable experience in the software environment of warehouse systems. On the other hand it is also recommended to get evaluations from inexperienced people, who have no knowledge about designing WMS interfaces, because this allows a comparison of their opinion ratings with these of the experts and it gives an alternative view on the effectiveness of the DSS prototype.

Two types of evaluation session were developed :

Case study evaluations are evaluations on the basis of real interface design problems. In these project related evaluations sessions the implemented DSS prototype is tested on real interface designs of WMS projects. With the help of these evaluation sessions the interface rules are tested and they are checked, if the generated DSS output is in accordance with the real interface design. Case study evaluations are time intensive because of the project related DSS prototype sessions and the required comparisons of the design results.

Non project related evaluations in which the participants do not test the DSS prototype on the basis of interface design problems of specific WMS projects. Instead the sessions are based on a fictitious WMS interface design example, which is shown during a DSS prototype demonstration. It was intended that only half of the time that is usually necessary for case study evaluation sessions is required for the non project related evaluation sessions.

Nineteen case study evaluations and eight non project related evaluations were undertaken. The results of the interface designs developed and the answers to an associated detailed questionnaire were analysed in terms of the experience profile of the testers.

The case study evaluations which allow a direct comparison to be made between the output of the DSS and results obtained from the five warehouse projects which formed part of the study showed a high level of agreement with the design results. The experts concerned also reported a significant reduction in the time required to develop the design and an increased reliability in the design. In the responses to the questionnaire it was also clearly revealed that the expert users saw the major benefits of the system in supporting design work for full LDC systems and as having far less value for the simpler type of warehouse system. Whilst the system was seen to have benefits in simulating alternative designs and adapting existing designs its major benefits were seen to be in the area of new project design. Less experienced users commented unfavourably on the lack of help function facilities in the system and there was some criticism that the user had to have both warehouse design and software design experience to use the tool effectively or for two people with the relevant experience profile to co-operate.

In order to develop a commercial version of the tool as well as help facilities, reporting facilities, rule editing and rule addition functions will need to be added to the existing prototype in addition to a full GUI.

7. Conclusion

A detailed investigation of the software interfaces between automated warehouses and the associated CEDP systems for five large logistic centre projects has been undertaken. This has shown that there is a high degree of commonality in the requirements for the software interface design for such systems. A full query model for the design of the interfaces has been developed in terms of IF-THEN-ELSE rules. This query model has been used to design and implement a prototype expert system based decision support tool to aid the design of such interfaces. The tool has been tested on a group of design engineers drawn from the five projects and others from unrelated project areas. The results of case study tests have shown a high degree of correlation between the results of using the tool and the designs developed by traditional methods but the tool produced a significant reduction in design effort and an increase in 'first pass accuracy'. Considerations are now being given to developing a commercial version of the tool.

References

DUNCAN, L.S. [1985]. "System Design Trends in Automated Warehousing". *In: Proceedings of the 6th International Conference on Automation in Warehousing*, 15-17 October 1985, Stockholm, Sweden, Cotswold Press Ltd, Oxford, England, 1985, p.23-34.

GROSSMANN, G. [1994]. "Braucht Logistik Lager?". *Zeitschrift für Logistik, Zürich*, 1994, **vol. 63**, no. 1, p.5-10.

WHITE, J.A. [1987], "World-class Warehousing: The Competitive Edge". *In: Proceedings of the 8th International Conference on Automation in Warehousing*, 6-8 October 1987, Tokyo, Japan, Cotswold Press Ltd, Oxford, England, 1987, p.37-59.

BRUNNER R. "Externe Schnittstellen von Lagerverwaltungssystemen". *Deutsche Hebe- und Fördertechnik - dhf, Ludwigsburg*, 1993, **vol. 39**, No 9. pp 58 - 61.

PRODUCT DESIGN

PRODUCT DESIGN

Geometrical Design

A Prototype System For Early Geometric Configuration Design
 X. H. Guan, D. A. Stevenson and K. J. MacCallum

Feature Based Design of Free Form Products
 S. R. Mitchell and R. Jones

Geometry-Oriented Information Modeling for Mechanical Products
 W. Z. Zhang, F. Y. Wang, H. Ding and S. Z. Yang

A PROTOTYPE SYSTEM FOR EARLY GEOMETRIC CONFIGURATION DESIGN

X. GUAN, D. A. STEVENSON, K. J. MACCALLUM
CAD Centre
Department of Design, Manufacture and Engineering Management
University of Strathclyde, 75 Montrose Street
Glasgow G1 1XJ, Scotland, U. K.
E-mail: x_guan@cad.strath.ac.uk

ABSTRACT

In this paper, we present a prototype system that has been developed to support geometric configuration of objects at the early stages of design. Guided by the general principle of minimum commitment, this system assists in the iterative development of alternative geometric configurations based on approximately or precisely defined information. The system has been evaluated in the context of computer enclosure design.

Keywords: geometric configuration, early geometric design, computer aided design.

1. Introduction

A computer aided design system suitable for early stages of design should be an integrated environment that offers a wide range of capabilities in assisting the development and evaluation of early design concepts, as well as their detailing in subsequent stages. This requires the system to support not only functional and other relevant design activities, but also geometric information processing. Our research goal has been to investigate a system that assists in the development of early geometric concepts.

At the early stages of geometric design, a designer's attention and interest is mainly on exploring a variety of possible geometric configurations of a product, where geometric configuration refers to the total geometric structure of the product consisting of the approximate or precise geometry of components of the product and their overall spatial arrangement in forming the structure (Guan 1993). These alternative configurations are developed at an abstract and approximate level, and are evaluated with respect to certain criteria. As a result, the most suitable ones are selected for further detailed design and evaluation. This development–evaluation–refinement process may iterate many times until the best concept is derived and ready for full design. Geometric information available during the early stages is usually a mixture of both vague and precise information, where vague can be characterised as approximate, abstract, or as incomplete. Free-hand sketches and diagrams are used frequently during the design process for effective expression, communication, and recording of both geometric and non-geometric aspects of or information about the product being designed (Tovey 1989, Ullman 1992). These sketches, and their use in the design process, highlight a desire to explore and investigate design options or concepts without commitment to exactness or detail.

Although various Computer Aided Design (CAD) systems, such as the traditional and now widely used geometric modelling systems and more advanced parametric, variational and feature based systems, have been developed to support the modelling of product geometry, they usually require complete, concrete and precise definitions on the geometry which are only available at the detail stages of the design process. Thus, by requiring from designers greater commitment than they can make or are willing to make at the early concept design stages, these systems are not well suited to deal with early geometric configuration problems.

In contrast, we have adopted a principle of minimum commitment modelling which does not force a designer to make commitments prematurely, i.e. earlier than necessary, desired or appropriate. Here, a commitment refers to a decision regarding components, arrangements, sizes or positions. Any decision which uses vague information is regarded as less committing than precise information.

This paper presents a prototype system for supporting the modelling of geometric configurations. Adopting the minimum commitment principle imposes important requirements on the modelling capabilities of the system. These requirements are discussed in Section 2, followed by an overview of the structure of the system in Section 3. We illustrate, in Section 4, the use of the system through examples, and in Section 5 raise issues for further investigation.

2. Modelling Requirements

To support the minimum commitment modelling of geometric configurations, the following basic requirements were established for the system:
- Permit the use of various types of geometric information which may be vague or precise;
- Support incremental refinement of approximate size and location of components;
- Support simultaneous incremental development of multiple approximate or precise geometric configuration models;
- Support the handling of conflicting or inconsistent geometric information.

A major requirement for the user interface to the system was that it should facilitate, or at least not hinder, the user's access to and interaction with the modelling utilities of the system. A graphical user interface (GUI) was deemed appropriate, but with the clear understanding that the user-accessible functionality of the system should not become embedded in, and so dependent on, any one particular user interface front end.

Through an examination of the early stages of design, we identified the various types of geometric information involved during the process of geometric configuration and the possible forms in which such information is given by designers (Guan 1993). During the initial development phase, we decided to concentrate on providing the system with the capabilities of handling primitive shapes, approximate and precise size information given in the form of inequalities (e.g. width ≤ 15.2, depth ≈ 13.4), ranges (e.g. height = [22.5, 23.4]), and equalities (e.g. radius = 9.4), as well as abstract and precise location information given in the form of spatial relationships (e.g. above, right and behind) and point position (e.g. (12.0, 22.4, 21.7)).

3. Structure of the System

Figure 1 illustrates the various modules of the prototype system. They have been implemented on a SunSparc platform running Sun Common Lisp with the Common

Lisp Object System (CLOS). The prototype system also accesses, directly or indirectly, three non-Lisp packages: a geometric modeller ACIS (Spatial Technology Inc. 1992), a constraint solver CLP(R) (Heintze 1991) and a graph editor EDGE (Paulisch and Walter 1990). Our approach has been to define an interaction shell around the modelling core (which is also referred to as the application system), consisting in the complete suite of modelling operations by which any outside agent can manipulate and access the models under construction, and to develop a GUI which is linked to this interaction shell. This allows the application system to be accessed in two ways: directly through the interaction shell commands, and, more conveniently, through the GUI. A most significant advantage of this is that it allows the GUI and the application system to develop in parallel.

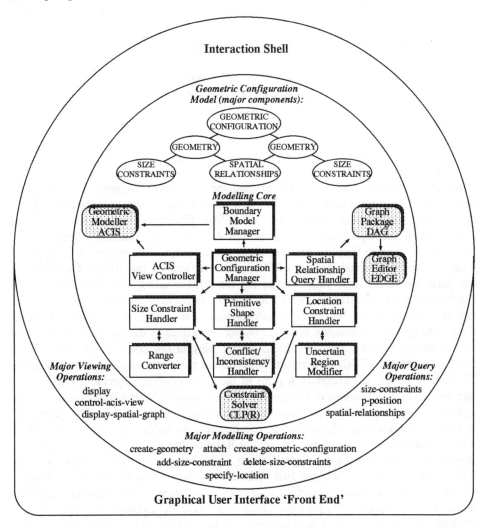

Figure 1. Modules of the Geometric Configuration System.

A Geometric Configuration Model is a cluster of CLOS objects that are linked to one another as defined by a developed representation framework (Guan 1993). These objects are instantiated with the known geometric information about a product. The model describes the total geometric structure of a product as a geometric configuration, confined to a geometric configuration space, which consists of geometry objects that represent the components of the product and are spatially related to one another. Based on a comparison of a number of approximation handling techniques, we have chosen the interval based method (Moore 1966) to represent uniformly the approximate and precise size and location definition of components. Using this method, size information of types e.g. depth ≈ 30 and height = 40 is captured by ranges as [28.7, 31.3] (assuming the degree of approximation is 2.6) and [40, 40]. Here, a range is an interval of real numbers defined by a lower and an upper bound. The location of a component is characterised in the model by a datum point on the component, which is situated in a 3D uncertain region defined by three ranges capturing the allowable x, y, and z co-ordinates of the datum point.

To handle such information, a reasoning mechanism has been developed based on constraint management technique combined with interval calculation (Guan 1993). The various modules in Figure 1 have been designed to provide the necessary handlings based on the mechanism. When using the system, the user establishes the geometric model of a component by specifying[a] its shape type, size constraints, and possibly the location of the component in the configuration through spatial relationships or point position. The Geometric Configuration Manager is responsible for dispatching such information to other modules and managing the established models. The Primitive Shape Handler accesses basic definitions of a set of primitive shapes, currently cuboid, cylinder, sphere, frustum and prism. The Size Constraint Handler manages the processing of the set of size constraints sent down by the Geometric Configuration Manager. It tries to simplify some of them and dispatches the others, in appropriate forms, to the Constraint Solver CLP(R) to solve and to the Range Converter to transform the results into value ranges for the corresponding size parameters. The Conflict/Inconsistency Handler deals with possible syntactical and some semantic errors that may exist in size or location information. The Location Constraint Handler transforms spatial relationships or point positions specified by designers into constraints on the bounds of the uncertain regions of the relevant components and solves these constraints using CLP(R). The results obtained are passed on to the Uncertain Region Modifier to update the uncertain regions of the corresponding objects. Based on the size and location definitions of components, the Boundary Model Manager generates or updates the corresponding boundary geometric models by use of the ACIS geometric modeller. The user's graphical display requirements on ACIS are set up by the ACIS View Controller. Supported by the Spatial Relationship Query Handler, a user may inspect the spatial relationships specified for a given geometric configuration, such as finding the relationships between two geometry objects in a configuration and finding all the spatial relationships specified for a configuration. This handler has been implemented using the DAG package (Donaldson 1993), developed based on the EDGE program, which consists of Lisp based utilities for making and manipulating directed acyclic graphs.

---

[a] Default values are used for unspecified parameters.

A widget-based graphical user interface has been developed using Lispview, a SunCL binding to OpenLook windows, and the MINDER system (Stevenson 1990). It provides active displays of application objects (i.e. changes in values of displayed objects invoke automatic updates of the displays), form-fill panels for composing and executing modelling operations defined in the interaction shell, and view control panels for the ACIS wire-frame and EDGE graph displays.

4. Examples of Using the System

In this section, we briefly illustrate the use of the system through examples taken from computer enclosure design: the process of spatially interconnecting together various components of a computer system via a mechanical structure to satisfy various requirements—functional, service, ergonomic, safety, environmental, etc.

Suppose we have some information about the geometry of a Power Supply Unit and would like to construct a rough model for it (we either are not interested in a precise model or are not yet sure of the exact size). The shape of the component can be approximated by a cuboid with the size of width $\approx$ 21.08 (expressed as width $^\wedge$= 21.08 in the system), depth in range [14.02, 15.50] (expressed as depth = 14.02 -> 15.50) and height $\approx$ 3.05. We do not yet want to consider its location in the whole enclosure design. Figure 2 shows the use of the system in making this model by issuing the corresponding commands through the GUI. Note that the modelling operations can also be performed by typing directly into the Lisp Listener as Lisp commands.

To present the associated size approximation, the system, through the ACIS window, displays two boundary models of the component which can be interpreted in two ways: (a) the actual boundary of the component is currently known to lie in the space bounded by the two displayed boundary models (inclusive), whose ultimate position in the space is determined by the precise sizes which would become known or certain at a later stage; (b) all cuboids whose boundaries fall between the two boundary models (inclusive) satisfy the size requirements of the component and can therefore be selected as a geometric model of the component at some later stage when necessary or desired. Support for working with such approximate information in the system does not force a user to make commitment to unnecessary or unavailable details, and thus leaves the solution space as open as possible. It is assumed that later incremental refinement or configuration constraints will reduce the uncertain ranges until a precise model is defined.

Figure 2 also shows the main panel of the system which provides menus of commands that allow the user to construct, modify and inspect various geometric configuration models. It also presents active lists (initially empty) of the modelling entities such as geometries, configurations and relations, as they are created by the user during the session.

Having established the rough model, suppose we now wish to place it on the right of an Electric Fan whose geometric model, geometry2, has also been constructed. This can be achieved by invoking the operation specify-location. As a result, a geometric configuration, geometric-configuration1, will be constructed by the system in which the Power Supply Unit is located on the right of the Electric Fan.

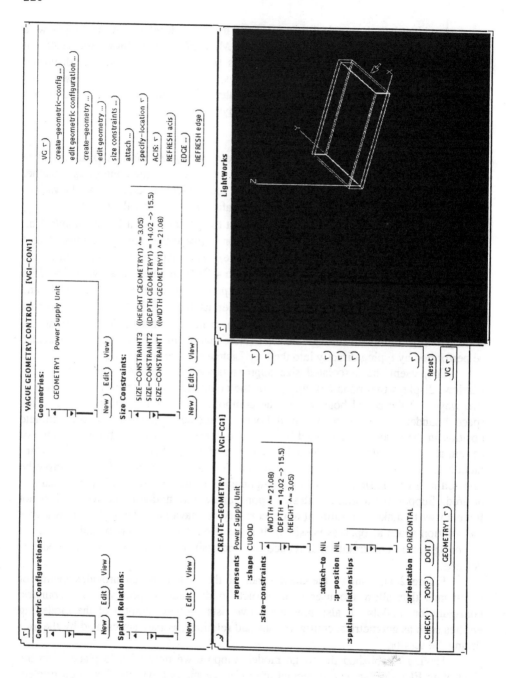

Figure 2. Constructing the Geometric Model of the Power Supply Unit.

The width of the Power Supply Unit has been between 14.02 and 15.50 (Figure 2). If we at this stage discover or decide that this depth should be ≥14.65, this new information can then be used to refine the value range of the depth from the initial specification of [14.05, 15.50] to [14.65, 15.50]. If, however, a new value is supplied that is outside the previous range, say depth ≥ 16, the system detects and notifies the user of the inconsistency, and disallows the change. If the user does require the change to be made, the existing conflicting constraint must first be deleted.

The system is also able to detect certain obvious spatial conflicts. For example, if a Mother Board was specified to be behind a Hard Disk Drive and on the right of the Electric Fan in the configuration in a previous step, then specifying the Mother Board to be in front of the Hard Disk Drive in the same configuration will be detected as a conflict, and the user is prompted to either withdraw from the operation, or enforce the new position.

Figure 3 shows a rough computer enclosure design—geometric-configuration1—with all the defined spatial relationships, incrementally developed using the system. Note that it displays only one of the possible configurations allowed by the location constraints. In this default display, components are shown in the left-low-front corners of the corresponding location uncertain regions. By changing the current configuration being worked on during a modelling session, one can also investigate simultaneously a number of different geometric configurations for the same product.

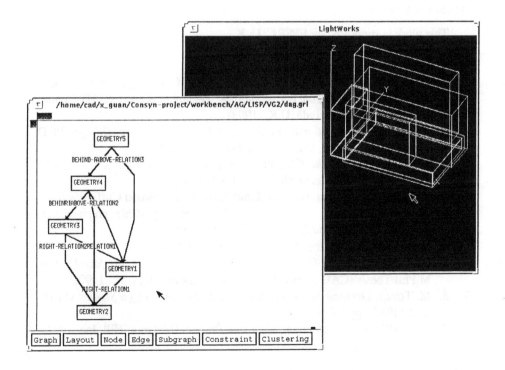

Figure 3. geometric-configuration1 - A Preliminary Enclosure Design

Throughout this approach, the uniform definition of approximate and precise geometry underpins the minimum commitment principle. In particular, it allows commitments to complete and exact size and location specifications to be deferred as desired or necessary.

5. Conclusions

In this paper, we have described a prototype system for supporting the development of early geometric configurations following a minimum commitment modelling principle. The system currently supports a user's modelling, incrementally and non-sequentially, of multiple approximate or precise geometric configurations using primitive shapes, simple inequality, range and equality types of size constraints, and abstract primitive spatial relationships.

To adapt to various practical configuration problems, further research and development is required to enhance the capacity of the system, in particular for modelling more complex size constraints, spatial relationships, shapes, and different orientation. A sketch-based GUI would further facilitate the use of the system. To fully support the development of early geometric configuration concepts, research should also be carried out to provide utilities for evaluating and conserving geometric configurations. Finally, investigation is proposed to integrate with those systems that deal with other aspects of concept design, such as functional modelling, towards providing an integrated conceptual design support environment.

6. Acknowledgement

This work is supported by EPSRC, U. K.

7. References

1. I. Donaldson, *The CONCEPT FRAME SYSTEM: an Object-oriented Data Representation for Concept Modelling*, Technical Report (CAD Centre, University of Strathclyde, U.K., 1993).
2. X. Guan, *Computational Support for Early Geometric Design*, Ph.D Thesis (CAD Centre, University of Strathclyde, U.K., 1993).
3. N. Heintze, et al, *The CLP(R) Programmer's Manual version 1.1* (IBM Thomas J. Watson Research Centre, USA, 1991).
4. R. T. Moore, *Interval Analysis* (Prentice-Hall, Englewood Cliffs, 1966).
5. F. N. Paulisch and F. T. Walter, EDGE: an extendible graph editor, *Software Practice and Experience* **20**, S1 (1990).
6. Spatial Technology Inc. *ACIS: Interface Guide* (1992).
7. D. A. Stevenson, *An Output-Oriented Approach to User Interface Design*, M.Phil Thesis (CAD Centre, University of Strathclyde, U.K., 1990).
8. M. Tovey, Drawing and CAD in industrial design, *Design Studies* **vol.10**, no.1 (1989), pp.24-39.
9. D. G. Ullman, *The Mechanical Design Process* (McGraw-Hill, Inc. 1992).

FEATURE BASED DESIGN OF FREE FORM PRODUCTS

S R Mitchell and R Jones
Department of Manufacturing Engineering, Loughborough University of Technology
Loughborough, Leicestershire LE11 3TU, UK
E-mail: S.R.Mitchell@LUT.AC.UK

Abstract

Many parts with free form or sculptured surfaces are frequently designed by crafting prototypes. This requires considerable skill and expertise. Few formal methods are available for the design of such products, but they may be effectively modelled using computer based surface modelling systems. Analysis of such products reveals that their design if often based on a number of dominant features. This paper describes the construction of a feature hierarchy for sculptured parts that can be generated based on primary extended form, secondary blend and tertiary cosmetic features. The use of extended form feature methods enables feature substitution, modification and parametric variation throughout a set. The methods have been verified using common examples such as golf clubs, shoe lasts and crockery.

Keywords: Features, free form, sculptured, CAD

1. Introduction

A considerable amount of research effort has been expended over recent years on feature based design and manufacture systems (Refs. 1,2 and 3). Much of the work is based on the fact that design features also possess manufacturing or process capability enabling process planning to take place concurrently with the design process. In almost every instance these systems have been developed or based on prismatic parts where the feature library consists of planes, holes, pockets, steps, slots, bosses etc. The features are parametrically defined allowing feature instances to be combined to form components or products. Parametric feature based computer aided design systems such as Pro-Engineer (Ref. 4) and CADDS5 (Ref. 5) are now commercially available.

A considerable amount of design is concerned with modification of existing products and a significant advantage of a feature based design system is the ability to create new products by feature extraction, modification or substitution. Prismatic parts lend themselves conveniently to this design approach. Although opportunities exist with such feature based systems for the use or application of features having sculptured surfaces, the sculptured feature is seen as a particular part instance and would not be parametrically defined. Furthermore, the ability to combine such features into a total sculptured product has not been addressed. This is unfortunate since there are many instances of sculptured parts where a feature substitution approach to design would be beneficial.

It is apparent that there has been little work concerned with feature based design of sculptured parts and there are a number of significant reasons why this has not taken place. Sculptured parts have fewer scientific or engineering applications and are generally considered to be the domain of the artist. This is exacerbated by the fact that a considerable amount of design takes place by handcrafting prototypes with the finished article often becoming the design specification. In addition, engineering computer based modelling systems for sculptured parts are concerned with surface manipulation and these systems are inevitably complex and require considerable expertise to use compared with solid modellers.

2. Design by Features

Commonly used products with sculptured surfaces will often have been remodelled many times during their lifetime. For sculptured parts this modification process is time consuming with often the whole part requiring remodelling since part features are generally joined together in a continuous manner. Analysis of sculptured parts (Refs. 6 and 7) reveals that they have several dominant or primary features which strongly affect the design and these will be joined together by blend features to give a smooth part. The primary features will have become established over a long period and will be particular to a trade or product family. These features may be emphasised by protrusions, depressions or significant changes in curvature and the terminology used to describe them will be product specific.

It may be worth considering this terminology with respect to a common example. The human face can be considered to be a sculptured object which has obvious dominant features such as the nose, eyes, lips, chin etc. It is however difficult to specify exactly where the nose finishes and the cheek begins since there is a blend or secondary feature joining them together. Analysis has revealed that it is possible to develop a feature anatomy for a particular product based on three feature types:

I. Primary (dominant) features.
II. Secondary (blend) features.
III. Tertiary (cosmetic) features.

Figure 1 shows a simple anatomy for a human face. It is evident that a feature hierarchy is required for a design system since each feature will have specific association with its near neighbours. Clearly changing or substituting a particular feature will affect the adjacent features which may also require modification.

3. Extended Form Feature Methods

In 1991 work at Loughborough University commenced in an attempt to develop a formalised approach to feature based design of sculptured parts. It quickly became apparent that a generic feature anatomy for all parts could not be achieved. However, analysis revealed that a product group or family could have its own feature anatomy which would enable acceptable substitution and manipulation. For example, it would be possible to change a nose on a face and although the character would change the face would still be acceptable. Unfortunately direct feature removal and substitution can either result in surface gaps or unacceptable creases, edges or lines. A method is therefore required that will enable acceptable feature substitution.

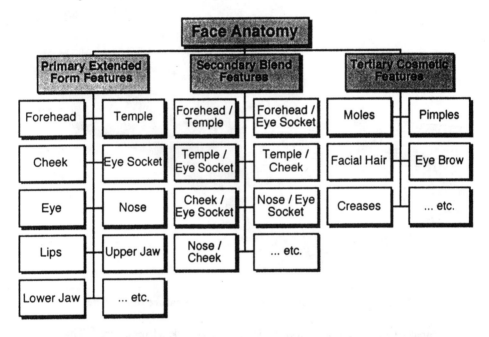

Figure 1. Sculptured Feature Face Anatomy (Symmetric Half)

In this work to date no attempt has been made to develop an elegant solution for feature identification. The features used have been those commonly accepted within a particular product industry. For a sculptured part the significant problem is to decide the extent of the feature. The extended form feature method (EFFM) (Ref. 6) was developed to overcome this problem. Primary features are identified and those parts of the feature which are confidently known are digitised and developed as a surface. This surface is then extended far beyond its nominal boundaries to produce the extended form feature. Design proceeds by positioning primary features with reference to a global product datum such that the surfaces intersect. The excess surfaces are then trimmed to their boundaries with subsequent secondary blends to give aesthetic acceptability, Figure 2. shows an example of an extended form feature for a shoe last. The contribution of the EFF surface can be seen as the unshaded area.

4. Design System

The research has used the Delcam International DUCT 3-D surface modelling software as the foundation for developing a prototype model design system. This CAD system is particularly useful because it has free form surface manipulation and visualisation through to CNC machine path generation for artefact manufacture (a particularly important aspect of the design iteration process). However the techniques and methodology developed could be implemented using other advanced CAD systems.

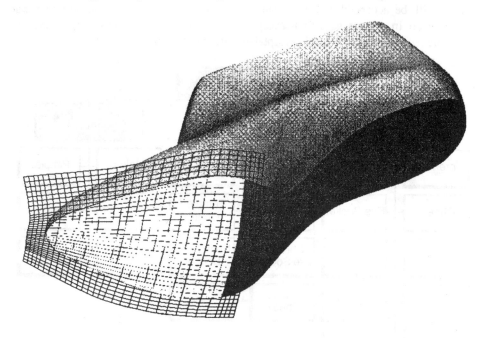

Figure 2. Shoe last Extended Form Feature (Sole Toe)

Rapid on-screen visualisation of designs is considered necessary. This requires workstations with adequate processing capability and advanced graphics presentation. During this research programme a number of workstations have been used with the software currently residing on Silicon Graphics Indigo2 Extreme hardware.

The design system model was developed using iron golf clubs as the first example and has been verified using shoe lasts as a further example (other products such as crockery and sanitary-ware have also been considered). Golf clubs and shoe lasts have proved to be particularly suitable since they both come from traditional craft based industries. Furthermore they are normally produced in sets with some parametric variation throughout. In both instances there are general design guidelines or rules, however, many of the styling or fashion features, although identifiable from size to size (club to club) will not have a linear relationship. It is also interesting to note that much of the feature terminology is the same for these examples (e.g. heel, toe, sole) even though the shape and functionality are different.

The feature anatomy upon which to develop a design system requires a full set of features to describe an object. The data structure is such that these features can be added to or removed as fashion changes and each feature could embody a sub-anatomy if required (Ref. 8). In the case of the iron golf club system 13 primary features and 20 secondary blend features were used. A ladies shoe last design system would require 13 primary features and 13 blend features. An example of a feature anatomy for a ladies shoe last is given in Figure 3.

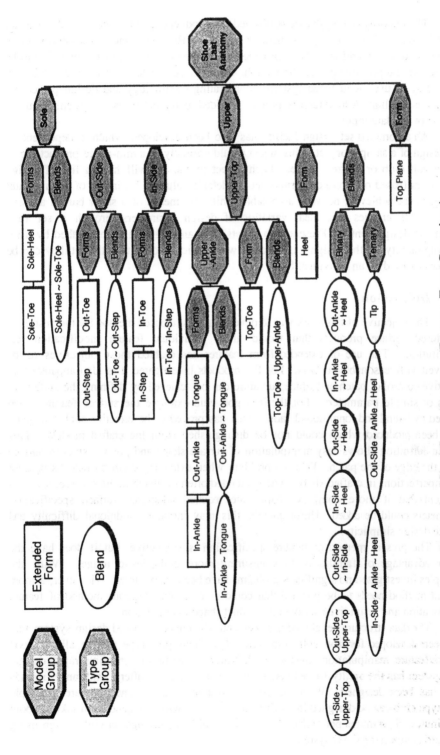

Figure 3. Shoe Last Anatomy (Feature Type Groupings)

The majority of designs are performed by "product experts", who in fashion or art related industries are seldom scientists or engineers (Ref. 9). One of the objectives of the system design therefore was to provide an environment or interface which would be easy to use and promote design innovation. This has been achieved by the development of an easy to use menu based system incorporating terminology and design procedures which are familiar. A help facility is also provided to remind users of system status or to prompt for data input.

An automated set design facility has also been developed, which is dependent on the design of a sample object. This object would normally be a mid-range product e.g. a 5-iron golf club or a size 7 shoe. Each object in the set will have a list of design parameters (based on product norms) set to default values. Variation of each default setting can be achieved, between acceptable limits, by means of a slider bar on a menu form. This enables non-linear variation of design parameters through a set thus enhancing design control. Presentation of sets for examination is accomplished by a set viewing facility. If the designer is unhappy with particular objects in the set they can be further modified on an individual basis.

5. Discussion

The objective of this work was to develop a feature based method for the design of sculptured surface products that would allow individual feature modification and substitution. The use of extended form features has enabled these objectives to be achieved with reasonable success. For the products investigated it is also apparent that effective control of patch or surface curvature can be achieved by defining the surface in terms of simple parameters. For the iron golf club example the surface features were defined by lateral and longitudinal radii. This has proved to be effective such that parts have been produced which could not be distinguished from the crafted models. This simple definition allows easy manipulation of surface shape and gives a smooth surface up to the edge of the patch. This is a problem when extending digitised patches because the imperfections in crafted shapes when extended often results in numerous creases. It is apparent that if more complex shapes are required additional surface specification parameters could be added. Unfortunately, this may introduce additional difficulty and reduce design simplicity.

The present method of feature specification is subjective, but it does have the major advantage of familiarity thus ensuring more enthusiastic usage. With the examples investigated no significant problems have been encountered. However, a more elegant method needs to be pursued that could give an unambiguous method of feature specification and result in easier overall product shape manipulation.

The data structure developed has been embodied in an advanced design system that has been developed for iron golf clubs (Ref. 10). This prototype system allows good surface/feature manipulation together with feature substitution enabling rapid design. The system has the ability to generate complete sets of clubs after a mid-range iron (5-Iron) has been designed. With the present system hardware and software a simple prototype club can be designed in as little as one hour with set generation taking about 30 minutes (3_ minutes per club). This gives significant savings on club design using standard engineering CAD systems.

6. Conclusions

The research has shown that it is possible to decompose many products having sculptured surfaces into feature anatomies. However, the features chosen will normally be product family specific and not have generic applicability. The work has also shown that it is possible to simplify complex surfaces to enable parametric control and still achieve shapes comparable to crafted articles. This formalised approach to sculptured part design can yield significant savings in design effort and time, enabling fast new product introduction into the market.

7. Acknowledgements

The authors' would like to thank:
- The UK Engineering and Physical Sciences Research Council (EPSRC) for providing grant funding.
- Dunlop-Slazenger and Delcam for their support and funding.
- The Manufacturing Engineering Department at Loughborough University of Technology and its engineering support staff.

8. References

1. J. J. Shah, 1989, *Philosophical Development of Form Features Concept*, Dept. of Mech. and Aerospace Eng., Arizona State Uni., Tempe, AZ85287.
2. J. J. Shah , 1972, *Features in Design and Manufacturing* (A. Kusiak ed.) J. Wiley & Sons 1992.
3. O. W. Solomans, F. J. A. M. Van Honten and H. J. J. Kals, 1993, *Review of Research in Feature Based Design*, J. of Manuf. Systems, Vol. 12, No. 2, pp 113-32.
4. M. James, *Prototyping CAD*, CADCAM, July 1991.
5. *Computervision, Introduction to CADDS5 Parametric Design*, Prime Computervision, DOC38800-1LA, 1991.
6. R. Jones, S. R. Mitchell, and S. T. Newman, 1993, *Feature Based Systems for the Design and Manufacture of Scupltured Products*, Int.J.Prod.Res., Vol. 31, No 6, pp 1441-1452.
7. S. R. Mitchell, R. Jones and S. T. Newman, 1994, *A Structured Approach to the Design of Shoe Lasts*, Journal of Engineering Design, submitted Aug. 1994.
8. S Mitchell, R. Jones and C. J. Hinde, 1994, *An Initial Data Model Using the Object Oriented Paradigm for Sculptured Feature Based Design*, accepted for publication Sept. 1994, Journal of Research in Design.
9. *High-Tech comes to Golf Clubs*, 1989, Tooling and Production, July.
10. S. R. Mitchell, S. T. Newman, C. J. Hinde and R. Jones, *A Design System for Iron Golf Clubs*, 1994, Science and Golf 2, (Proc. 2nd World Scientific Congress of Golf) pp 390-395.S.

Geometry-Oriented Information Modeling for Mechanical Products

Zhang Wenzu
School of Mechanical Engineering, Huazhong University of Science & Technology,
Wuhan, P.R.China 430074

Wang Fengyin
Gintic Institute of Manufacturing Technology, Nanyang Technological University,
Singapore 2263

Ding Hong and Yang Shuzi
School of Mechanical Engineering, Huazhong University of Science & Technology,
Wuhan, P.R.China 430074

ABSTRACT

Traditional solid models of a component do not meet the requirement for a number of engineering activities in the product cycle such as manufacturability analysis, process planning, design synthesis, analysis, and optimization. In this paper, a modeling method for mechanical products is presented based on a new scheme to represent mechanical components and assemblies. In the geometric model, not only geometrical, but also semantic information of a component is included, providing more data for subsequent design applications. A prototype system based on this method is introduced.

Keywords: Product Modeling, Design with Features, Constraint Based Representation.

1. Introduction

There are two types of information flow in a manufacturing system for mechanical products. One is the geometry-oriented and the other is administration-oriented [1]. In traditional manufacturing processes, the geometry oriented product information is represented by drawings and shared by all departments involved. The administrative information is transferred among departments orally or by written reports. The product information related to certain departments is understood and extracted by people. In computer-based manufacturing systems, product models represented by computers are needed and shared by all the departments through sharing a common database. Accordingly, the product information related to a certain department should be understood and able to be extracted by computers. In order to do so, the CAD/CAM system must be made more intelligent; the product information is required to be expressed by the form of knowledge, understandable and able to be extracted by computers in the way of knowledge processing. Presently, however, most product information is still processed under people's control; the computer is only used as a tool. This paper focuses on the geometry-oriented product information modeling. The aim is to enrich the product information model to sustain computer integrated manufacturing. A preliminary research on product modeling and knowledge representation is to be presented.

2. Related Work

Most of the previous work in product modeling can be broadly separated into component modeling and assembly representation. For component modeling, early work was focused on the modeling of components' nominal shapes, i.e., geometrical (solid) modeling [2]. The geometrical information provided by it is widely used in geometry design, finite element analysis, mass-property calculation, static and dynamic interference checking, NC code generation and verification, etc.. At the same time, its applications are limited greatly because it lacks the information attached to the components' nominal shapes, such as dimensions, tolerances, and surface roughness, etc., which are involved in most engineering tasks in reality. Some work has been done on geometrical modeling to add the above mentioned information to it and gain the complete description of components. Meanwhile, in order to support decision-making in engineering activities, such as process planning, the feature-based model has been proposed. Its basic motivation is the observation to use elements with engineering meanings directly instead of low level elements—vertices, edges and faces—in the geometrical model. The feature-based modeling methods have two major groups: feature extraction [3] and design with features [4]. In the former features are extracted from a geometrical model. Because there is intertwinement among features, it is difficult to extract all the features successfully. In the latter a designer is carried out with features, as the terms suggest. Because features are uniquely defined, it often limits the designers' thoughts. Both of the two groups face the classification of features, and it is difficult to do so.

Previous work in representation of assembly can be broadly classified into semi-manual and automatic assembly generation. For the semi-manual approach [5], the user first defines the geometrical model of every component, then specifies the location and orientation of every component in the assembly structure. For the automatic approach [6], the user must set mating conditions among components, then the locations and orientations of them are automatically determined. In general, the complete assembly information is not obtainable, such as the information for assembly tolerance analysis.

In this paper, we present a modeling method according to a new scheme, which divides the components' information into semantic layer and geometrical layer. The semantic layer information describes the shape elements with complete meanings, such as the surface of a hole which is perpendicular to a coordinate plane, etc., and all the information attached to the nominal shape of the components. The geometrical layer information is related to topology and geometry information. The information on the two layers is maintained to be consistent. For assembly modeling, the present method gives a complete description to the product's specification, which includes the components of the product, the number of components in the product, the materials of the components, the mating relations and fitting properties among components, dimensions and so on.

3. Modeling in the Component Level

The basic information of a component includes the nominal shape information, which includes the topological and geometrical information and a set of shape elements, and some technological information attached to the nominal shape, such as dimensions and tolerances, etc.. The topological and geometrical information is a solid model of the component and the split edge structure [7] is adopted in the present prototype system. A set of shape elements, which is defined according to the relationship in Euclidean space, represents the primary features of mechanical parts and composes of the semantic layer information with technological information, such as dimensions, datums, tolerances and surface roughness, etc., attached to them. By such shape elements, the shape of components can be constructed and the topological and geometrical information can be generated. The link between the two layers is also generated. During defining shape elements and constructing the shape of a component with such elements, every shape element must be ensured so that its corresponding primitive elements in the geometrical layer should have the same semantic information attached to the shape of the component. This way, the semantic information can be treated as attributes of the shape elements and represented in the semantic layer.

In modeling a component, shape elements describing the component are generated first. These elements can be planes or lines, which are parallel or perpendicular to a coordinate plane, general planes or lines, spatial points, cylindrical surfaces, conical surfaces, or spherical surfaces which are defined by these planes, lines or points. At the same time, the constraints among these elements, such as dimensional constraints, are generated. Then, the shape of the component is defined with these elements, and the semantic layer and geometrical layer are generated. When the constraints among the shape elements are changed, the size, location and orientation of the shape elements, and the geometrical information in the geometrical layer are also changed accordingly. Finally, the semantic information is attached to the shape of the component on the semantic layer.

As an example, a method of layered sweep to model prismatic components is presented. The process of layered sweep is opposite to the projective process. The details of modeling prismatic components using layered sweep method are given in the following section.

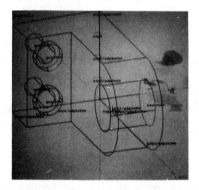
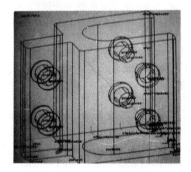

Fig. 1 Shape Construction by layered sweep operations

3.1. Constructing a Two-Dimensional Shape on a Plane

To define shape elements on a plane means to generate and select datum first, followed by defining new shape elements relative to the current datum. Every shape element defined can be selected as a datum to define a new one. During defining a new shape element, a set of constraints is generated to link the element with its datum. At the same time, these constraints are satisfied by DeltaBlue algorithm[8]. A 2D shape can be represented by a number of shape elements. During defining a 2D shape, some new shape elements may be generated, too.

3.2. Layered Sweeping of 2D Shapes

A datum face is selected among faces of a 2D shape to sweep other faces along their own normal directions and a 3D shape is generated. As variables of the dimensional constraints, the sweeping distances are set by the user. In the process of sweeping, the semantic and geometrical information, as well as their links are generated.

3.3. Operations on 3D Shapes

When the generated shape of a component does not satisfy certain requirements, some structures need either to be added to or removed from the shape until these requirements are satisfied (see Fig.1).

3.4. Renewing Layout and Modifying Dimensional Constraints

During constructing a shape, the selection of dimensional constraints depends on whether the shape elements can be defined conveniently. The dimensional constraints only describe the sizes of the shape. Dimensions not only reflect the sizes of the shape, but also influence the machining process when the components are manufactured. After a shape is generated, therefore, the layout of the dimensions must be renewed and some dimensions are modified to gain proper size of a component's shape. This is implemented by changing the privileged strengths and the values of variables of dimensional constraints and satisfying these constraints with DeltaBlue algorithm [8]. During renewing layout and modifying dimensional constraints, the semantic and geometrical layer information are modified accordingly.

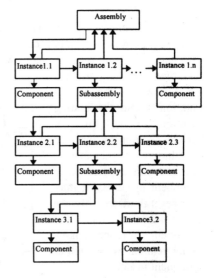

Fig. 2. Assembly structure of a component

3.5. Representation of Semantic Information

For every primitive element or every set of primitive elements in the geometrical layer, a corresponding shape element and its relative shape elements are recorded in the semantic layer. For a surface of a hole feature, for example, several corresponding faces are recorded in the geometrical layer, and its center line in the semantic layer. Every primitive element or every set of primitive elements with semantic information, such as tolerances, etc. in the geometrical layer, has corresponding shape elements in the semantic layer, and the latter is treated as attributes of the shape elements.

4. Modeling in the Product Level

Before the assembly of a product is defined and represented, its components must be defined, and the mating features of a component with other components, which are attached to the semantic layer of the components, need to be specified. Then, the assembly is modeled. During these procedures, all the mating relations among components are generated.

4.1. Component and Subassembly Chain

Currently, the description in the component level is first generated interactively or obtained from a file. The component chain of the product is then generated. Since a component or a subassembly may appear more than once at different locations with different orientations in an assembly, the concept of instance is used. The location and the orientation of every component in the component chain are related to its own coordinate system.

Every subassembly is represented by a subassembly chain. It may have several instances. These instances can be some components, or some subassemblies as well. After the component chain and the subassembly chain are generated, the assembly can be described with components and subassemblies in both chains. An assembly may consist of several instances of components and subassemblies (see Fig. 2).

4.2. Mating Relations Among Components

All mating relations among components are represented by the component instance chain and the mating relation chain. A mating relation between the component instance 1 and the component instance 2 is formed by the mating condition 1 and the mating condition 2 (see Fig. 3). Since a component instance may have several mating relations with several other component instances, it may have several mates. Every mate indicates that the component instance has a mating relation with another component instance.

In generating these subassemblies and the assembly, mating relations and mating conditions are generated for every new component instance between it and other component instances.

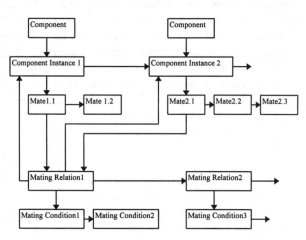

Fig. 3 Representation of mating relations among components

5. A Prototype System for Product Modeling

According to the concepts and methods presented above, a prototype system was implemented at Huazhong University of Science and Technology. It includes five parts, i.e., component shape modeling; dimensional constraint management; semantic information management; assembly definition; and file management. For modeling the shapes of components, both the semantic and geometrical information are generated. The former has only the shape elements and their constraints. According to the characteristics of different components, the modeling method may be different.

Generally, mechanical parts are classified into the prismatic components, turning components, special components (such as gear type, spring type etc.), and standard components. The basic method of modeling a prismatic component has been described above. For a turning component, its rotational section is first defined on a plane, the section is then rotated and a turning shape is generated. Finally, some standard structures, such as a keyway, groove, etc., are added and the expected shape of a component is obtained. For special components, such as a gear, the shape of its teeth can be defined by several parameters, and the other parts of the gear can be defined like the turning components. The standard components can be determined by several parameters.

The types of shape elements in the semantic layer of the components determine the coverage of components which can be modeled. Different components may be modeled with different methods, but they must be defined by the instances of the existing types of shape elements in the semantic layer.

6. Summary

In this paper, a modeling method is proposed for geometry-oriented product information representation. Its component information is described by two layer information: semantic and geometrical layer information. For the information in a product level, a scheme for representing assemblies is given. Finally, a prototype system is introduced.

Acknowledgments

The research work, supported by the National Natural Science Foundation of China (NSFC) and National Science and Technology Board of Singapore (NSTB), is a part of the earlier stage work of a joint research project (# Gintic-C-C93-113) between Huazhong University of Science & Technology, China, and Gintic Institute of Manufacturing Technology, Singapore.

References

1. G. Spur, F. L. Krause and P. Armbrust, *Int. J. Mech.Tools Des. Res.*, **26** (1986) 171~178.
2. A. G. Requicha, *Computer Surveys*, **12** (1980) 437~464.
3. S. Joshi and T. C. Chang, *CAD*, **20** (1988) 58~66.
4. J. R. Dixon, J. J. Cuningham and M. K. Simmons, in *Intelligent CAD I*, ed. H. Yoshikawa (North-Holland, 1989).
5. L. Z. Liebenman and M. A. Wesley, *IBM J. Res. Develop.*, **21** (1977) 321~333.
6. K. Lee and D. C. Gossard, *CAD*, **17** (1985) 15~19.
7. Y. E. Kalay, *CAD*, **21** (1989) 130~140.
8. B. N. Freeman-Benson, J. Maloney and A. Borning, *Comm. of ACM*, **33** (1990) 54~63.

Product Development

Solid Freeform Manufacturing as Part of the Integrated Product Development Process
    K. H. Grote, J. L. Miller and T. K. Pflug

Interactive Multiple Criteria Decision Making for Product Development
    A. Jeang and D. R. Falkenburg

An Integrated Computer-Aided Gear Design System
    J. Y. Lai and W. F. Lu

Development of a Framework System for Tool Intergration in a Product Information Archive
    S. Sum, D. Cheng, D. Koch, D. Domazet and S. S. Lim

Design Support for Configuration Management in Product Development
    B. Yu and K. J. MacCallum

Reverse Engineering — Modern Tool for Product-Planning
    B. Schumacher and P. Pscheid

Solid Freeform Manufacturing as Part of the Integrated Product Development Process

Professor Dr.-Ing. Karl-H. Grote
Institut für Maschinenkonstruktion ,Konstruktionstechnik-,Otto-von-Guericke-Universität
Universitätsplatz 2, 39106 Magdeburg, Germany
E-mail: kgrote@engr.csulb.edu

Jeffrey L. Miller, MsEng.
McDonnell Douglas Aerospace, 1510 Hughes Way
Long Beach, CA. 90810-1870, USA

Dipl.-Ing. Thomas K. Pflug
NC-Gesellschaft, Helmholtzstr. 22
89021 Ulm, Germany

ABSTRACT

Rapid Prototyping has emerged as a valuable tool. Its advantages over conventional manufacturing processes for quick product visualization enable companies to spend less money and time in bringing products from computer to the table top. To effectively utilize this manufacturing technique to its full potential, its process capabilities must be understood. The advances in accuracies, resins, and casting techniques necessitate the requirement to have this data available for the Integrated Product Team. This paper will discuss the importance of relating process capabilities to product features and present a method for automated producibility analysis utilizing component features and process knowledge capture.

1. Introduction

Solid Freeform Manufacturing (SFM), or the technologies by which complex shapes may be built using layer manufacturing techniques, is one of many applications of Rapid Prototyping (RP) [1]. RP is a general term which refers to methods of quick validation of engineering concepts or designs. Other examples of this methodology include a development approach whereby traditional barriers were torn down as was done by McDonnell Douglas Corporation to design, build, and successfully flight test the Delta Clipper Single Stage Rocket Technology demonstrator in only 24 months. Another area where RP is being applied is in software development. Software developers are using techniques to quickly generate code to perform complex tasks without the traditional cycle of manual code generation and extensive debugging. All RP applications share the common goal of bringing quality products to market in the shortest period of time at a competitive price.

SFM is becoming an increasingly valuable tool in the Integrated Product and Process Development (IPPD) environment. IPPD is the process by which Integrated Product Teams (IPTs) representing diverse disciplines convert customer requirements, through process capability data, into first time quality products. The IPPD concept is not new, however, it has gained considerable momentum in the last 10 years and is being adopted by many US companies [2]. The ultimate objective of the product development process is to develop quality products which meet all performance requirements, can be manufactured cost effectively, and on schedule.

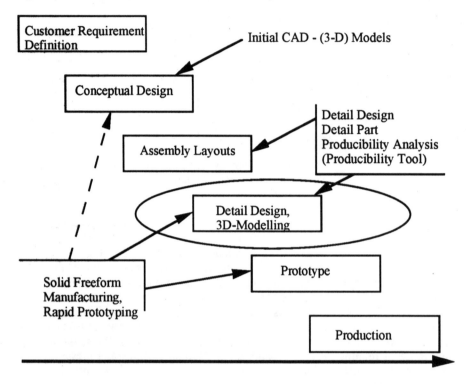

Figure 1: Product Development Cycle

SFM is an enabling technology which provides several tools by which the IPT members can quickly convert engineering data from a Computer Aided Design (CAD) environment into a physical product which can be used for fit checks, visual aids, or as molds for cast parts. The producibility function is the IPPD discipline which focuses on the total product life-cycle cost along with product evolution and is most interested in the SFM capabilities. Producibility is a process accomplishment that enables a product satisfying both functional and physical objectives to be repeatable fabricated, assembled, and tested in the total quantities required, on schedule, and at optimum cost. Producibility is critical in the early stages of the product development cycle as it is commonly stated that approximately 80% of the total life cycle cost of a product is determined in the first 5% of the conceptual design effort [3].

A major element of producibility is Design for Manufacture and Assembly (DFMA) which focuses on the mechanics of product manufacture. For SFM, DFMA focuses on the best way to design parts to take full advantage of the SFM processes. This may include hard supports on the model, draft angles, or wall thickness. The simplest application for SFM is for visual aids which have loose tolerances, surface finish, and material property requirements. For this case, producibility is not an issue. The parts are being built to support product visualization and do not have a mechanical function. Producibility becomes an issue where parts directly produced from the SFM technology are going to be used for a fit check or working prototype. For this case, tolerances, surface finish, and material properties are of great interest because the SFM parts may support loads or perform a mechanical function. The final case where SFM is used, is to generate molds to make parts produced by casting processes. For this case, the characteristics of the SFM part are critical where they contribute to the properties of the final part. The final cast parts will be used on a working prototype or production product. It is for the final two instances where producibility analysis using process capability data is required.

2. Process Capabilities

Process Capability is a tool which can be used to keep a manufacturing process under statistical control or provide data to the engineering community about the attainable parameters of a process. Process capability analysis originated out of Statistical Process Control (SPC) theory. SPC is a philosophy that uses statistical techniques and applications in all areas of production, design, maintenance, and service to make continuous improvements in quality and productivity by reducing variation in all processes. A process capability study is an analysis of data from control charts to determine if the statistical variation in the process is well within the upper and lower specifications of a part characteristic. A control chart is a graph showing the performance of a characteristic for a critical aspect of a process over a period of time using measured observations or sample data. Prior to performing the capability study, the process must be brought under statistical control by eliminating the special causes of variation that are indicated on control charts. The process capability is then determined using various capability indices that compare expected statistical variation to specifications for the part characteristic [4].

In recent years, Motorola, Inc. has been using six sigma (6σ) analysis as the statistical methodology to achieve quality products to meet their objective of customer satisfaction. 6σ is a statistical term which can be simplified to mean 3.4 defects per million opportunities. The term opportunities for error is defined as any action performed or neglected during the creation of a unit of work, where it is possible to make a mistake which may ultimately result in customer dissatisfaction. An example of customer dissatisfaction is an out-of-tolerance condition which may occur on a part. This is a typical type of error which may occur on a part fabricated using conventional machining or an SFM technique. The practice of determining the probability of successful manufacture using the concept of defects per unit is being adopted by many companies. Extensive information on the details of 6σ analysis may be found in reference [5].

Process capability is critical to the design community for developing products which can be manufactured with the highest probability of achieving first time quality. For this process to succeed, the producibility and manufacturing engineers must provide process capability information to the design engineers. However, the process capability data must be in a form which can be efficiently used in the product development cycle. A form that has proven to be effective utilizes feature based design. Feature based design is dependent on an engineering model consisting of a collection of geometric features. The process capability information relative to the manufacturing process selected for the product feature can be used for risk assessment. The feature based risk assessment will use a combination of statistical techniques such as 6σ analysis to produce a relative value in terms of defects per unit or other method that enables a probability to be associated with the successful manufacture of a feature.

3. Product Features

A product feature is a measurable or definable parameter or geometry that describes an element or aspect of a component. Examples of product features include holes, fillets, chamfers, and pockets. Examples of feature parameters include length, width, height, diameter, and radius. An engineering CAD model will need to be broken down into features prior to producibility analysis. This can be accomplished using pattern recognition algorithms which work directly with solid model information from a CAD system. The features and all parameters that describe those features will be identified and stored in a data structure which facilitates retrieval of the feature parameters. The data structure will provide all information about the features for producibility analysis. Additional work done on feature recognition may be found in reference [6].

A producibility risk assessment can be performed using a combination of the product features and the process capabilities associated with manufacturing each feature. The results of this analysis will provide information to the IPT demonstrating that the part can be manufactured by the process chosen with no defects or with acceptable risk. Because of the tremendous amount of data required to perform a thorough producibility analysis to ensure a product can be manufactured cost effectively with minimal risk, a method for automating the process is required.

4. A Method for Automating Producibility Analysis

A method for automating producibility analysis is currently under development using the concepts of component features, process capabilities, and rule based analysis to function as a tool to assist the IPTs. The primary objective of this effort is to provide a design team with an engineering model producibility checker directly linked to the CAD system. The system will use a combination of rule based analysis and actual process capability data to perform the producibility function.

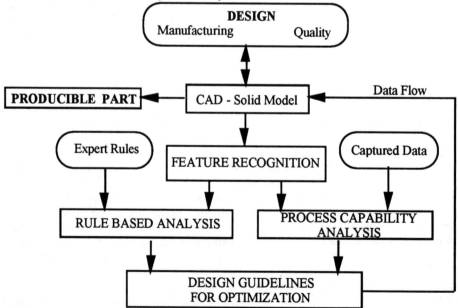

Figure 2: Computer Automated Producibility Analysis

Additional guidelines include a user friendly interface to the CAD system and a robust (expandable) database. Figure 2 presents a graphical illustration of the automated producibility analysis methodology.

Initially, an engineering solid model is created using a CAD system. This model should contain the desired geometry of the end product. This implies that this tool will be used in the detail design stage of the product development process where an adequate description of the product is available for analysis. However, this tool may be applied for individual feature analysis if the data is available earlier in the product development process. Next, the engineering data is converted to usable format for analysis. The .stl (stereolithography format) works well because of the nature of the faceted description of the component [7]. This is also advantageous for parts that will be built using a SFM technology that accepts this data format. Next, the part is broken down into a collection of features using a pattern recognition technique. The features are then assembled into a data structure which contains all parameters of the features detected. This data structure will then be used to support expert rule and process capability based analysis.

Rule based analysis, a method for comparing features of a part to established parameters defined by experts and reporting the results, is performed on the detected features (Figure 3).

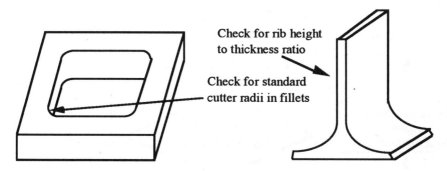

Other checks for Rapid Prototyping:
Check for tools, as drills and reamer sizes
Membrane thickness
Web thickness
Support conditions
Part orientation

Figure 3: Rule based feature analysis

In the case of stereolithography, the most popular SFM process, this may be a minimum wall thickness determined from experience with the post curing cycle [8]. For a conventional manufacturing process, this may be a hole diameter or corner radius determined by the use of standard cutting tools. Rule based analysis can be performed on any feature that can be mathematically defined. Once the feature is defined, it can be detected and compared to a set of rules developed by domain experts. Using this methodology, a wide variety of features may be analyzed.

Feature based process capability analysis is performed using captured SPC data. For this analysis, the engineering model feature parameters are compared to collected (actual) data using the statistical techniques previously discussed. Since exact matches of detected features to actual data will not be common, the software is designed to extrapolate using rules defined for feature similarities.

The process capability database is updated through the use of feature based inspection. Selected features are inspected with the results being entered into the database. This cycle continuously updates the data for accurate process capability analysis.

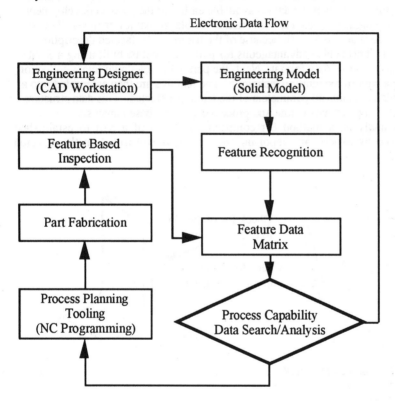

Figure 4: Feature based process capability analysis

The software then provides output to the designer. The engineer will be provided with recommended tolerances or parameters for the features detected, access to the statistical or design rules that make up the decision, and both graphical and text information for easy comparison to the CAD model. The combination of this data will enable the engineer to analyze the part to determine if the achievable tolerances, determined from statistical assessment and expert defined rules, will meet the performance requirements of the design. The same results may be reviewed by the entire IPT for recommendations to enhance the producibility of the product. This iterative process can be repeated until a design which meets all criteria is defined.

Extensions to this system include feature based parametric cost estimating and generative feature based planning. Feature based parametric cost estimating is accomplished through establishing a set of rules, or using the actual data, to provide a cost estimate for each of the product features. By summing the cost to produce each feature, with the cost of the raw material, provides an estimate of the total cost of the part. Generative process planning can be accomplished using a similar method. After the features have been identified, the manufacturing process steps required to achieve those features can be determined and a manufacturing process plan generated. Generative process planning will benefit parts built using conventional manufacturing methods more than those built by SFM techniques because of the multiple process steps required. Development work done in the area of generative process planning may be found in reference [9].

5. Application to Solid Freeform Manufacturing

SFM technologies are continuously improving. Resins, accuracies, and casting techniques are maturing on a consistent basis. For IPTs to effectively utilize these new technologies, their process capabilities must be captured. The IPT members must have this data to effectively choose between available processes. Extensive literature exists which compares tolerances of several SFM techniques. Numerous case studies also exist that discuss the tolerances achieved for various parts. This data must be made available to the IPTs in a form that can be used for decision making early in the product development process. Other areas for decision making include choosing between SFM and a conventional manufacturing method. Often, this decision is product application dependent. However, if either method will achieve a suitable part, the decision is determined by cost and schedule. In order to get to this point, the engineer first needs to know if the SFM technique is capable of achieving the same results as an established conventional manufacturing process. The producibility tool described can be used to distinguish between conventional manufacturing processes and SFM techniques using the same feature based design parameters.

6. Conclusion

As new methods of production are developed, process capabilities must be captured and analyzed to support the product development process. This can be accomplished by using digital data available at the earliest stages of the design process. The companies that adopt the principle of linking process capability data with producibility analysis will produce products with the least risk. Those that use a combination of computer automation and expert knowledge will be the leaders in this field.

7. References

1. K.-H. Grote and D. Kochan, *Neue Wege zur Produkt- und Prozessinnovation durch Solid Freeform Manufacturing*. 1. Internationaler Anwenderkongress - Proceedings, Technical University, Dresden, Germany, October 28-30, 1993.

2. Jeffrey L. Miller, *Software Needs for Aerospace Manufacturing*. McDonnell Douglas Corporation, MDC 93H1406, October 1993, also presented at Weltraum Institut Berlin, Germany, November 2, 1993.

3. Eugene S. Rosenbaum and Frank D. Postula, *Computer-Aided Engineering Integrates Product Development*. 1991 AACE Transactions.

4. Lawrence A. Hamilton, *Statistical Process Control Handbook*. Developed for McDonnell Douglas Corporation, May, 1989.

5. Mikel J. Harry and J. Ronald Lawson, *Six Sigma Producibility Analysis and Process Characterization.* Motorola University Press, Addison-Wesley Publishing Company, Reading Massachusetts, 1992.

6. Jan H. Vandenbrande and Aristides A. G. Requicha, *Spatial Reasoning for the Automatic Recognition of Machineable Features in Solid Models.* IEEE Transactions on Pattern Analysis and Machine Intelligence, **Vol. 15, No. 12,** December 1993.

7. 3D Systems, Inc., *Stereolithography Interface Specification.* Valencia, California, October 1989.

8. Paul F. Jacobs, *Rapid Prototyping and Manufacturing, Fundamentals of Stereolithography.* Society of Manufacturing Engineers, Dearborn, Michigan, 1992, pp. 334.

9. Joo Y. Park and Behrokh Khoshnevis, *Real-Time Computer-Aided Process Planning System as a Support Tool for Economic Product Design.* Journal of Manufacturing Systems, **Volume 1/No. 2.**

INTERACTIVE MULTIPLE CRITERIA DECISION MAKING FOR PRODUCT DEVELOPMENT

Angus Jeang

Department of Industrial Engineering
Feng Chia University, Taichung
Taiwan, Republic of China

and

Donald R. Falkenburg

Department of Industrial and Manufacturing Engineering
Wayne State University, Detroit
Michigan, 48202, USA

ABSTRACT

The objective of this paper is to build a multiple criteria decision making methodology for product development. Procedures to form the set of indifference feasible alternatives to provide the designer flexibility in design are developed. However, ways of establishing suitable criteria for the methodology will not be addressed. The advantages of this approach are that the designer can narrow down the list of candidates for consideration at an early stage, as well as provide a possible direction for design improvement. Criteria values can be measurable or non-measurable. The methodology is in a form which is easily adapted to computer programming and can be used to create screening-level alternatives for assistance in an interactive design process. As a result, an effective evaluation for design activities becomes possible in a multiple criteria design process.

Key Words: Design, Methodology, Comparison

1. Introduction

Previous works in evaluation of product design are mainly dependent on a single common criteria such as cost, so that the best design among various alternatives is the one which costs the least (Dewhurst and Boothroyd 1988, Zenger and Dewhurst 1988). However, these single factor criteria models are only rough estimates of design considerations. This is especially true because the engineering estimates may not include other useful and even necessary factors. Determinations based on consideration of a single criteria may well result in a wrong design decision. In addition, the estimated costs from some models do not reveal the actual reasons for the cost differences among alternative designs, since the cost functions are found from a few variables which are not necessarily directly related to design criteria(Jansson, Shankar, and Polisetty 1990). Therefore, we are not able to locate potential areas for design improvement by observation of these given variables. Because of this, an alternative approach for product design evaluation is needed.

In general, designers prefer to have as many feasible designs as possible in the early stages in order to allow for change as the design team encounters complexities (Andreasen and Hein 1987, Gross 1989). Design changes occur as a result of criteria trade-off among the criteria in consideration. Since the decisions made during the design process are not permanent, this flexibility in making design changes facilitates an evolutionary decision making process(Pahl and Beitz 1988, Rabin 1986). The product design process is therefore a dynamic process which involves adding or deleting design criteria, changing the associated criteria value, and adjusting preference priority for relevant criteria. In order to maximize the effectiveness of the design process, it is necessary to use the high speed analysis capabilities of a computer. In consideration of the above discussion, this paper provides an effective multiple criteria evaluation methodology which enables designers to narrow down the alternatives to a limited number of candidates for further consideration.

This paper is divided into four sections, with the first being this introductory section. Section two presents the evaluation methodology for multiple criteria decision making. A discussion of the methodology follow in Section three. Finally, a conclusion is presented in Section four.

2. Development of Evaluation Methodology

Most real-world problems involve multiple criteria which, particularly for engineering design, can be used to establish an evaluation base for feasible alternatives(Zeleny 1982, Radford and Gero 1988). In order to effectively identify the superior designs among them, we suggest the application of the Pareto optimal approach.

A feasible solution to a multi-criteria problem is a Pareto optimal if no other feasible solution exists that will yield an improvement in one criterion without causing a degradation in at least one other criterion. In other words, some alternatives are eliminated from further consideration because they are definitely inferior when compared to the remaining alternatives. The Pareto optimal set was originally presented by Hart et al.(Pareto 1906). In this research, the set of superior alternatives is determined by a pairwise comparison of the feasible alternatives for each criterion. The alternatives within the superior set (or Pareto set) are indifferent, because no single alternative dominates the remaining ones, which is the definition of a Pareto optimal.

For the purpose of explanation, we should define the set and elements as follows. Let P be a set of Pareto alternatives where P_i is a member, or alternative, of the set. That is, $P = \{ P_1, P_2, P_3, ..., P_n \}$. Let A be a set of feasible alternatives where A_i is a member of the set. That is, $A = \{ A_1, A_2, A_3, ..., A_m \}$. Let C be a set of criteria, or element, associated with each alternative A_i, where C_k is a single criterion. That is, $C = \{ C_1, C_2, C_3, ..., C_r \}$. For example, the set C shown in Table 1 is the partial criteria for mechanical component design. The criteria from Table 1 collectively represent the geometrical complexity of a part and are purely dependent on geometrical parameters(Jeang 1992). They are defined in such a way as to be applied in the general case instead of only specific cases. However, we are not attempting to show all possible design criteria for effective manufacturing. Instead, we will apply these set of criteria as an example for explaining our methodology.

Table 1. Examples of Effective Manufacturing Criteria for Mechanical Component Design

Crt.#	Criteria Description	Reasons for These Criteria
1	Number of features in a part	The more features in a part, the more direct and non-direct activities will be involved. Because of this, it is desired to minimize the number of features in the design of a part.
2	Available contact space for holding activities	Workpiece holding problems usually occur whenever the holding devices block the tool path in machining operations or the workpiece has to be in a position to provide a suitable tool feeding direction. Caution in selecting the contact spot between fixture and workpiece is necessitated. Hence, it is best to design a part which has open space for convenient holding.
3	Number of workpiece turning operations	Turning the workpiece to different faces provides a suitable position or orientation for tool engagement. Associated activities for workpiece holding, such as clamping, may also required. Such these operations are considered nonproductive, it is best to place the maximum number of features on the least number of faces to satisfy the design requirements.
4	Number of tool changing operations	Machining of various features may require of different tools. For this reason, simple and similar features in a part design are preferred.
5	Total distance among features	The distance is defined as the distance between features on the same face. The total distance is a collective measurement of tool travel distance associated with each feature and also indicates the degree of convenience in holding operations.
6	Average distance per feature	The average distance is a measure of the feature distribution.
7	Average feature per turning	The ideal design is to place as many needed features on as few faces as possible. In other words, it is best to avoid spreading the features onto different faces if the design requirements permit it.
8	Average volume per feature	This is a measure of material removal for each feature.

| 9 | Total volume removed from the part | The larger the volume removed from the part, the longer the time for machining operations. |
| 10 | Direct distance among the faces which contain features | This criterion reflects the additional time and costs of turning through longer distances. |

Definition 1 - Let A be a set of feasible alternatives which contain elements A_1, A_2, A_3, ..., A_m. If and only if A_i is superior to A_j, then all criteria in A_i must not be inferior to the criteria in A_j and at least one criteria in A_i must be superior to the corresponding criteria in A_j. If and only if A_i is inferior to A_j, then all criteria in A_i must not be superior to the criteria in A_j and at least one criteria in A_i must be inferior to the corresponding criteria in A_j. If and only if A_i is indifferent to A_j, then at least one criterion in A_i must be inferior to the corresponding criterion in A_j. Also, at least one criterion in A_i must be superior to the corresponding criterion in A_j. Superior criteria are ones closer to a desired value (or target), and inferior criteria are ones further from the desired value (or target). Hence, it is not always such that the superior criteria value must be greater than the other one. For example, the less weight for a design is a good design in some cases. Conversely, greater electrical resistance is better in some designs etc.. Namely, they are dependent upon the expected target in the particular design. Also, it is not necessary that the associated value for each criteria be measurable; it could be represented by " Yes ", " No ", " Satisfactory ", or " Not Satisfactory ". In this case, the ones with " Yes ", or Satisfactory " will be superior than the ones with " No ", or " Not Satisfactory ". This very phenomena of immeasurability necessitates the application of pairwise comparison to form the Pareto set.

Definition 2 - If no single alternative is superior or inferior to any other alternative within the feasible set, then we consider this feasible set as a Pareto optimal set. In other words, none of the alternatives within the set are globally superior.

The following is an example illustrating the use of the Pareto's optimal approach.

Example: Given four feasible alternatives $A = \{ A_1, A_2, A_3, A_4 \}$. There are three criteria, or elements, associated with each alternative: C_1 = volume, C_2 = weight, C_3 = cost. That is $C = \{ C_1, C_2, C_3 \}$. For this example, assume that smaller criterion values are preferable. The values of each criteria within the individual alternatives are:

$C_{A1} = (20, 5, 1.5), C_1 = 20, C_2 = 5, C_3 = 1.5$
$C_{A2} = (15, 7, 1.2), C_1 = 15, C_2 = 7, C_3 = 1.2$
$C_{A3} = (12, 3, 0.8), C_1 = 12, C_2 = 3, C_3 = 0.8$
$C_{A4} = (8, 4, 1.0), C_1 = 8, C_2 = 4, C_3 = 1.0$

Alternatives A_3 and A_4 are superior to alternatives A_1 and A_2, because each criterion in the former ones is always less than in the latter. However, we cannot discriminate between Alternatives 3 and 4, since neither one is superior in all criteria. Therefore, we form a Pareto optimal set P which will be moved to the next stage for further analysis, where $P = \{ A_3, A_4 \}$. In this way we can limit the number of alternatives for detailed comparison. Now, we have the set, where, $P = \{ A_3, A_4 \} = \{(12, 3, 0.8), (8, 4, 1.0)\}$. Clearly, A_3 and A_4 are indifferent.

In the above example, each alternative has only four criteria. It is not difficult to discriminate alternatives by observation. In cases where the number of alternatives is large and the associated criteria for each alternative is also large, it may not be possible to form the Pareto set by casual observation. It then becomes necessary to write the developed methodology as a computer-based algorithm. To achieve this goal, the methodology must be expressed in a logical way which can be codified and implemented on a computer. The following theorems and rules are developed as bases to convert the process of discriminating between alternatives into computer readable algorithms.

Theorem 1 - If one alternative, A_i, from set A is superior to one member, P_j, in set P, then, A_i is either indifferent or superior to the remaining members in P.
Consider the following example.

Example: Given $P = \{ P_1, P_2, P_3 \}$, $C_{P1} = (2, 1, 3)$, $C_{P2} = (4, 4, 1)$, $C_{P3} = (3, 5, 1)$, $C_{A2} = (2, 2, 3)$, $C_{A3} = (5, 5, 4)$. Considering a greater criteria value as superior, it follows that A_2 is superior to P_1, but it is indifferent to P_2 and P_3. Also, A_3 is superior to P_1, P_2, and P_3.

Theorem 2 - If one alternative, A_i, from set A is inferior to one member, P_j, in set P, then, A_i is either indifferent or inferior to the remaining members in P.

Theorem 3 - If one alternative, A_i, from set A is indifferent to one member, P_j, in set P, then, A_i could be superior, indifferent, or inferior to the remaining members in set P.

From the above theorems and the definition of a Pareto optimal we have the following rules for forming the set P.

Rule 1 - Let A_i be one member in set A and P_j also be one member in set P. If A_i is inferior to P_j, then by the definition of a Pareto optimal and Theorem 2, A_i is definitely not considered for addition into set P.

Rule 2 - Let A_i be one member in set A and P_j also be one member in set P. If A_i is superior to P_j, then P_j must be discarded. Also, by Theorem 1 and the definition of a Pareto optimal, we have to continue comparing A_i with the remaining members in P and discard all the members which are inferior to A_i.

Rule 3 - Let A_i be one member in set A and P_j also be one member in set P. If A_i is indifferent to P_j, then by Theorem 3 we have to continue comparing A_i with the remaining members in P. Unless a member in P is found superior to A_i, then A_i becomes a member of P. Otherwise, A_i is dropped from further consideration.

Based on the above theorems and rules, it becomes possible to form a pairwise comparison algorithm for the purpose of establishing the set of Pareto optimal alternatives. This algorithm can be implemented with standard computer languages.

Initial Step - Pick any one member, A_i, from the feasible set,
   F, to be a member of the Pareto set P. The
   remaining members of F form the set A.

Step 1 - Pick any one member, A_i, from the set A and
   compare with any member P_j from the set P.

Step 2 - If A_i is inferior to P_j, go to Step 3
   If A_i is superior to P_j, go to Step 4
   If A_i is indifferent to P_j, go to Step 5

Step 3 - Discard A_i from the set A. Go to Step 6

Step 4 - Discard P_j from the set P. Compare the
   A_i with the other members in P.
   If A_i is either superior or indifferent to the
   other members in P, discard all inferior
   members from P and retain the indifferent

members. Then, form A_i and the indifferent members as a new Pareto set P. Go to Step 6.

If A_i is superior to all members in P, then A_i itself forms a new Pareto set. Go to Step 6

Step 5 - Compare A_i with the other members P_j in P. Let $P_i = P_j$. Go to Step 2.

Step 6 - If A becomes an empty set, go to Step 7. Otherwise, go to Step 1.

Step 7 - Comparison process is terminated.

Lacking an algorithm such as is described above, we would have difficulty forming the Pareto set. Hence, we can first let the design be described by a set of criteria from which a Pareto set can be formed by the methodology introduced in this section. Use of this method enables avoidance of unnecessary considerations which are beyond the conceptual stage in the design process.

3. Discussion

Single criterion models for product design may, in some cases, be of assistance when comparing different alternatives. However, beyond the comparison of the single criteria they do not provide the reasons why one alternative is a better design, nor do they give any indication of potential directions for design improvement. This is because the criteria in the model may not accurately express design considerations and reasoning. Additionally, engineering estimates for this criteria may not be accurate, which could easily result in a wrong design decision. Therefore, it would be better to have a multiple criteria evaluation system through which the product designer can more accurately evaluate alternatives. By identifying the necessary multiple set criteria for product design and evaluate them with the methodology developed in section 2, the most prominent candidates for further consideration will result.

The decisions made during a design process are not permanent, but rather are part of an evolutionary decision making process. Designers may frequently add oreliminate some criteria, or change the criteria values, or even their priority, in the design process. It is also preferable to have as many feasible design options as possible in the process, in order to facilitate change as the design team encounters various problems and difficulties. In addition, complex interactions between various criteria may exist and may be such that excessive efforts or time are required when executing the design process. Hence, a methodology to evaluate and form potential candidates for consideration is needed. One way to consider the above discussion is to provide an evaluation methodology which is based on a set of predefined criteria which can accurately reflect the design aspects under consideration. With such a method the problem is made more manageable and dynamic, and is more effective in analyzing alternatives than a single criteria system.

Because the use of multiple criteria enables design reasoning, it is possible to perform a more thorough comparison among alternatives. Additionally, using multiple criteria also allows the location of potential areas for design improvement, since differentiating between alternatives becomes more complex as the number of criteria and alternatives is increased. Also, because designing is an evolutionary decision making process, it call for the ability to make real time changes in design parameters, criteria re-prioritization, and a continual comparison of alternatives. Hence, development of an effective comparison methodology is necessary.

Design criteria involve a range of uncertain measures. With the example given in Table 1, there is some design uncertainty as to the size and the location of geometrical features. But some criteria are simply guidelines for the designer to follow. With such preferences, the decision is clear cut whether to accept or reject recommendations based on design guidelines. For example, the centerline of a hole to be drilled must always be normal to the surface of the part to avoid tool bending and breakage; in the endmilling slots, the depth should not exceed the diameter of the cutter, through holes are preferred to blind holes, because through holes make the cleaning of chips easier(Botz 1986). Hence, designs which follow the guidelines and their use in the evaluation methodology are superior than ones which do not. Other issues such as quality, functionality, assemblage, and flexibility in a mechanical component design, are not considered in our example. However, the methodology introduced in section 2 can still be applied if these additional criteria are included. Criteria from Table 1 and guidelines from producibility handbooks can be combined to create a new set of criteria for use in this evaluation system. As information about design criteria is evaluated through the developed methodology, a set of designs is formed from which the best combinations are determined.

In our evaluation methodology, the determination of " superior " or " inferior " in a comparison can be decided by the designer as needed. This can be done interactively throughout the design process. Criteria or other considerations for product design can be added and/or deleted as necessary in a real time manner as well. The criteria value can be countable or non-countable as determined by the necessary product design. This dynamic property is particularly important because designs are not created through an individual decision but in fact are achieved through an on-going decision making process(Banares-Alcantara 1991). In this way, the designer can use the developed methodology as presented in this paper to narrow the field of candidates for consideration and then modify them through design improvement until satisfaction is reached.

The design is not done in isolation; it involves interactions between humans, and design aides such as computers. The effectiveness of this communication is influenced by the medium, how the medium is being used, and how the information is provided and structured(Sanders and McCormick 1987). The developed system in this paper also provides feedback for existing design. The manner and structure of information provided through this methodology can assist the designer in executing design improvements through the enumeration method. This is an effective and time saving task through which decision time and trial and error periods can be reduced. An effective, flexible, and intelligent design thereby becomes possible.

4. Conclusion

Product design is a process of converting a set of requirements into the detailed geometry of a part. The process of converting all problems into a uniform cost is difficult. Hence, there is a degree of uncertainty in cost estimation. Also, such a cost-based approach makes reasoning about design improvement difficult. Therefore, we introduce a multi-criteria evaluation methodology. The criteria presented can be used not only in the evaluation and comparison of alternative designs, but also as an aid in design reasoning and design improvement.

In order to reduce the decision uncertainties, a methodology is developed to form a Pareto set which contains the indifferent alternatives. The criteria used in the example represent the partial consideration in relevant machining and manual motions to remove material for given geometrical features. Hence, based on the criteria defined in Table 1, the best alternative from the current evaluation system also represents the best design in terms of effective machining operations. The methodology is not limited to use with the criteria defined in section 2; it remains valid for other sets of criteria as well.The methodology is developed with programmability and computer assistance throughout the design process in mind. Furthermore, the methodology developed in section 2 can be applied to other tasks beyond the boundaries of product design as well.

5. References

1. P.Dewhurst and G. Boothroyd, Early Cost Estimating in Product Design, *Journal of Manufacturing Systems*, **Vol.7, No.** 3(1988).
2. D.Zenger and P.Dewhurst, Early Assessment of Tooling Costs in the Design of Sheet Metal Parts, *Report# 29 of Product Design for Manufacture*, (Department of Industrial and Manufacturing Engineering, University of Rhode Island, Kingston, Rhode Island 02881, 1988).
3. D.G.Jansson, S.R.Shankar, and S.K. Polisetty, Generalized Measures of Manufacturability, *Design Theory and Methodology - DTM'90, The 1990 ASME Design Technical Conferences - 2nd International Conference on Design Theory and Methodology*, **DE-Vol. 27**, (Chicago, Illinois, 1990, September 16-19), PP.85-96
4. M.M.Andreasen and L.Hein, *Integrated Product Development*,(IFS Ltd, UK, 1987).
5. N.Gross, *Engineering Design Methods*, (John Wiley & Sons, 1989).
6. G.Pahl and W.Beitz, *Engineering Design - A Systematic Approach*, (The Design Council, London, 1988).
7. M.J.Rabin, *Goals and Priorities for Research in Engineering Design*, (A Report to the Design Research Community, ASME, 1986, July).
8. M.Zeleny, *Multiple CriteriaDecision Making*, (McGraw-Hill Book Company, 1982).
9. A.D.Radford and J.S.Gero, *Design by Optimization*, (Van Nostrand Reinhold, 1988).
10. V.Pareto: Manuale di economia politica, *con una introduzione ulla scienza sociale*, (Societa Editrice Libraria, Milan, Italy, 1906).
11. A.Jeang, An Evaluation Methodology for the Mechanical Parts Designed Using Form Features, *Ph.D Dissertation*, (Department of Industrial and Manufacturing Engineering, Wayne State University, Detroit, Michigan, 48202, 1992).
12. R.W.Bolz, *Production Processes - The Productivity Handbook*, (Conquest Publications, 1986).
13. R.Banares-Alcantara, Representing the Engineering Design Process: Two Hypotheses, *Computer-Aided Design*, **Vol. 23, No.9**(1991).
14. M.S.Sanders and E.J.McCormicK, *Human Factors in Engineering and Design*, (McGRAW-Hill Book Company, 1987).

An Integrated Computer-Aided Gear Design System

Jau-yeu Lai

Department of Mechanical Engineering / Intelligent System Center
University of Missouri-Rolla, Rolla, MO 65401, U.S.A.

Wen F. Lu

Department of Mechanical Engineering / Intelligent System Center
University of Missouri-Rolla, Rolla, MO 65401, U.S.A.
Email: wflu@umr.edu

Abstract

The objective of this research is to develop an integrated computer-aided gear design system with solid model representation. This system integrates gear design formulas with database, solid model and dialog box. Six design modules including spur gear design, bevel gear design and helical gear design, etc. were developed. AutoCAD with AME was adopted as the tool to implement such an integrated system. The system provides a simplified and systematic way for gear design with friendly human-computer interaction. One example of designing the gears for automobile transmission was given to illustrate the practical usage of this CAD system.

Introduction

In the design stage of a product, engineers are always facing a tremendous amount of new information and variables which should be manipulated and compromised each other to a feasible degree; such as assembly dimension, mechanism function, available materials, tolerances, manufacturability, cost, etc. All the information and variables involved can be integrated as a design goal, and various factors will influence its completeness. The complexity of the design stage can be greatly reduced if computers are used to integrate various functions and activities in the design. With the Computer-aided design, human errors can be eliminated, tedious draftings can be simplified, the computational time and information manipulation can be reduced in the design process.

Just like the function of wheels, gears are indispensable and it is difficult to find its exact replacements especially in the conditions that higher power transmission are required. Therefore, in the study of power transmitting, gears have long been playing an important and supreme role. In the design of gear, because of its complex geometry, engineers are frequently represent the drafting in simplified symbolic notation. One significant defect in doing so is that engineers might lose the sense of realistic geometry in conjunction with design equations. Furthermore, many information and knowledge are associated with the design of gears. Design engineers not only have to figure out the requirements and dimensions of each gear, but also have to consider the mechanics associated with available materials under specific loading condition. In addition, they have to communicate with manufacturing engineers regarding the producibility of the gears designed. With the assistant of computer in gear design, all these factors that affect the design of the gear can be integrated in a concurrent engineering environment. The complex geometry of the gear can be also represented precisely for designers to preview its spatial relationship and for manufacturing engineers to determine its manufacturability.

Gear design have long been studied by many engineers and scientists [TOWN 92]. Many standard textbooks provide design formulas for the gear design [DEUT 75, SHIG 86, SPOT 85]. Andrews and Argent [ANDR 92] recently had developed a computer-aided method for obtaining optimum strength of spur gears through iterative strength calculations. Osward et al. [OSWA 93] used CAD and a numerical procedure for minimize dynamic effects on high-contact ratio gears. For the representation of the gear teeth profile, Huston et al. [HUST 94] presented differential geometry for the involute teeth profile of various types of gears.

Although gear design are continuously studied, not many works have focused on the development of an integrated gear design system with solid model representation in general purpose applications. The objective of this research is to develop a methodology to integrate the design processes of gears into a computerized environment. The design procedures which previously performed on paper from equation calculation to part drawings will now primarily be fulfilled with precision by computer aided gear design system. AutoCAD Release 12 from Autodesk® with AME™ (Advanced Modeling Extension) Release 2.1 was chosen as the tool to develop such system. Solid models are used to represent various gears. The developed computer-aided design processes are user friendly through the improved Graphical User Interface, i.e., the interactive dialog box. Database, such as material properties, machining data and tolerances were used to build "Look-up tables" and were integrated into the system.

The Framework

An overview of the common framework of the integrated computer-aided gear design system is shown in Fig. 1. With the design requirements, feature is selected in the component module. Twelve features commonly used in gear design were created in the component module as shown in Fig. 2. Once the initial design parameters are input, the designer modules

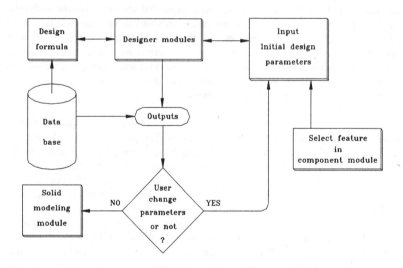

Fig. 1. Common Framework of Gear Design.

will activate with the design formula and database support. Users will be able to change the design parameters if the output results are not satisfied. The final output of the gear design will be displayed graphically through solid modeling module. Sample of solid models of gear design can be seen in Fig. 3(a) and Fig. 3(b).

Fig. 2. Features in Component Module.

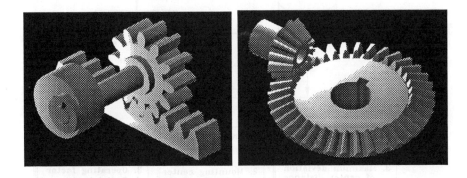

Fig. 3(a). Combination of Rack, Spur Gear, Shaft Rectangular Keyway and Cam.

Fig. 3(b). Two Bevel Gears For Transmit Power Between Perpendicular Shafts.

1. Design Modules

The purpose of the design module is to integrate the design formulas and the database in the computer-aided system. It also links its output to the solid model module. Algorithms have been developed [LAIJ 94] to determine the dimensions, material, operating conditions and manufacturing information of the gear component. Presently, there are six design modules in the system.

(a) Spur gear design module

The structure of the spur gear designer module is shown in Fig. 4. In the procedure of designing gears, trial and error are inevitable because the size, tooth form, and dimensions of the gear must be known before the actual loads and stresses can be determined. There are many possible design procedures that yield satisfactory results, such as the design formulas provided by AGMA (American Gear Manufacturers Association). This paper uses only a portion of the AGMA standards as the design basis. Note that one common character of all those gear design procedures is that they were mostly derived from empirical formulas. In this design program, those procedures have been rearranged into design steps as algorithms [LAIJ 94]. Furthermore, all teeth profile in this system will be generated by genuine involute function which precisely represent the tooth profile. This will further enhance the function and utility of solid modeling.

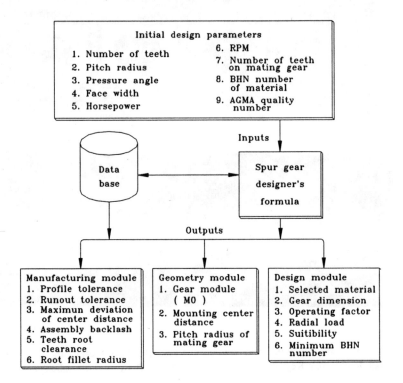

Fig. 4. Spur Gear Designer Module.

Before the design, users should aware of the gear geometric factors for deciding design parameters. Factors may include gear ratio, gear size, pitch diameter, face width and mounting center distance, etc. After all the necessary initial design parameters are decided, outputs are just a click away. Information provided in the design module will pop-up as sub-menu which can be roughly divided into manufacturing, geometry and design modules. Information in the design module will assist engineers in knowing what other design choices they have in order to achieve the design goal. Users will have the alternatives to select either increase or decrease the pitch diameter and the face width, or to check the operating factor, or to change the material with different BHN number. After calculation, appropriate BHN number will be retrieved from database for the design parameters to meet the design goal. The information in geometry module will reveal some important geometric characters of the gears under design. And the manufacturing module will provide some typical numbers in machining and assembly of the designed gears. Information such as profile tolerance, runout tolerance, maximum deviation of center distance, assembly backlash and finally, root fillet radius can be retrieved from the database. Detailed screen of spur gear design module are shown in Fig. 5(a) and Fig.5(b).

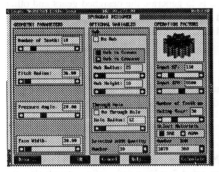
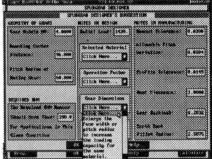

Fig. 5(a). Dialog Box of Spur Gear Design Module.

Fig. 5(b). Sub-Dialog Box of Spur Gear Design Module.

(b) Bevel gear designer module

The bevel gear design module is similar to the spur gear design module except for design formula and database used. The geometry of bevel gears are more complex that spur gears.

(c) Helical gear design module

Helical gears are widely used in high speed power transmitting. Again, the helical gear design module is similar to the spur gear design module.

(d) Shaft design module

Shafts are widely used in conjunction with gears in many industrial applications, and therefore, a shaft design module is also included. When designing a power transmitting shaft, engineers need to know the energy the shaft is carrying, materials that are appropriate for manufacturing and, perhaps optional, factor of safety. ASME Code for Design of Transmission Shafting provides specific information for shaft design. The maximum allowable shearing stress will be retrieved according to the selected material properties from database. The corresponding manufacturing tolerance range will also be given.

(e) Rectangular key design module

Keys are other components that commonly used in gear design. In the key design, some initial design parameters like shaft radius and torque must be cited from the shaft design module to combine the design of shaft and key together. It is capable of generating both external and internal keyways on cylindrical objects but only the external type has design formulas and database support in the present system.

(f) Spline design module

Similar to the constrain function of a key, splines are capable of providing axial movement for the assembled parts. This module includes internal spline design and external spline design, where the details of spline design can be found in [LAIJ 94]. All the design modules have dialog box similar to the spur gear design module shown in Fig. 5(a) and Fig. 5(b).

2. Database

There are nine database tables used in this design system, namely, material properties, K versus BHN value, profile tolerance of spur and helical gear, runout tolerance of spur and helical gear, allowable center deviation of spur and helical gear, typical backlash of spur and helical gear, root fillet radius of gear teeth, typical backlash of bevel gear and finally, Lewis form factor. These databases are used to support each data demanded in the design system and are available in [LAIJ 94].

3. Solid models module

The function of the solid modeling module is to generate the designed gears and other accessory parts. By the representation of solid model, users can easily figure out the spatial relations of each designed part and their assemblies.

System Implementation

In this paper, AutoCAD Release 12 was adopted to implement the computer-aided gear design system. It was used to develop solid modeling, database and dialog box in the system.

1. Solid modeling

AME in AutoCAD Development System™ (ADS) was used to develop solid models. In the gear design, all gear teeth profiles were generated first by a series of points which were derived from involute function to construct the teeth contour. Then the Boolean operation was employed to union the teeth cross section as a region to be extruded and/or rotated by specific type of gear. The teeth thickness of each gear under design has been calculated prior to the construction of teeth contour, i.e., teeth thickness was complied with the specific number of teeth. Finally, all teeth were made to union with a gear blank which was derived from the gearing fundamentals. The solid modeling came from Boolean operation sometimes is known as Constructive Solid Geometry (CSG). Where as, the procedures about how a solid object was formed will be recorded as a hierarchy called CSG tree. In the system, most solid objects will create through the combination of AME commands, with AutoLISP as the command organizer to buildup this design system. In addition to using the CSG modeling technique to define a composite model, AME also maintains a Boundary Representation (B-rep) to display the solid objects. In this research, CSG was used for Boolean operation while B-rep was was used to display the solid models.

2. Database

The AutoCAD SQL Extension™ (ASE) allows users to access non-graphic data in various external databases. To support the development of external applications incorporating non-graphic data, ASE provides the AutoCAD SQL Interface™ (ASI). ASI enables the users to develop AutoCAD-specific applications that work with various Data Base Management System (DBMS) products. In the present design system, ASI was adopted to retrieve data values from databases built in dBASE III environment. For instance, the material database were retrieved one at a time in their row order by the initial selection from user. Then the properties of designated material such as modulus of elasticity, yield stress, etc, will be retrieved in their column order by the request of the specific design formula. Each SQL retrieving function was built into design algorithm written in AutoLISP. When executing the design module, the combination of SQL commands will be compiled by AutoCAD through the interface provided by ASI. Database tables used in this research were also designed to be modular and resource-sharing, allowing the greatest flexibility and most efficient use of resources.

3. Dialog box

For increasing the design efficiency and productivity, AutoCAD's programmable interactive dialog system, known as Proteus, was adopted by the design system. The appearance of Proteus dialog boxes depends on the GUI (Graphical User Interface) in which they are used. They are portable between different platforms such as Microsoft Windows™, DOS or UNIX® without the need to be modified. In the design system, two programming languages were required to create a dialog box. The first was the Dialog Control Language (DCL). Its function was simply controls the statement of the design procedure layout in the dialog box. The second language was the one needed to compile the DCL, which might be AutoLISP or an ADS application written in C language. Benefits of using dialog box are that it provides an overview of all the design parameters and those design variables can be re-modified without the need to execute the entire program again. Examples of dialog boxes can be found in Fig. 5(a) and Fig. 5(b).

Example

To modify the performance of an engine, engineers of X motor company found that the 2nd gear and its coupler gears on the countershaft and secondary shaft of the transmission system used to serve the engine need to be redesigned to meet the new specifications. The design requirements are given as follows:
1. The maximum output of the engine is 110 horsepower at 5500 rpm.
2. Using spur gears for the coupler and helical gears for the 2nd gear.
3. Gear ratio between the two shafts of the coupler gear should be 7/5.
4. Gear ratio between the two shafts of the 2nd gear should be 3/5.
5. Face width of each gear should not exceed 40 mm.
6. Pitch diameter of each gear should not exceed 80 mm.
7. 180 mm of spline should be provided for the 2nd gear on countershaft.
8. All gears must have convex hubs with 10 mm in height.
9. Length of the first segment of shafts should be 60 mm.
10. The two pair of gears must be separated by at least 80 mm.
11. Speed of countershaft will be the same as engine output.

Steps for solving this problem by using the gear design system are listed as follows:
Step 1: Use shaft design module to construct the raw shapes of countershaft and secondary shaft.
Step 2: Resume the design data from the shaft design module and use the rectangular key designer to construct keyways on the countershaft and secondary shaft.
Step 3: Use the external spline design module to construct the spline portion of the countershaft and then union the original shaft and the external spline segment together.
Step 4: Use the spur gear design module to construct the two coupler gears.
Step 5: Use the helical gear design module to construct the two 2nd gears.
Step 6: Use internal spline design module to create internal spline on the designated helical gear.
Step 7: Check all geometric factors to see if they are all within constraints. Items including face width, pitch diameter. Also check if the mounting center distance of the two pair of gears are the same.

The final drawing of all designed parts is shown in Fig. 6 and the rendered view of their assembly is shown in Fig. 7. Note that the assembly of each designed part must be performed by users, just like dimensioning. In this example, practical operating conditions of a transmission gear box on vehicles have been implemented as an engineering design problem. In the process of solving the problem, some iterations might be required to obtain satisfactory results. However, once the user is becoming familiar with the design system and the overall design procedures, the time spent in design can be reduced.

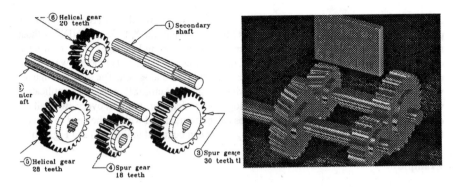

Fig. 6. Drawings of Designed Parts from Example.

Fig. 7. Rendered View of Parts Assembly.

Conclusions

An integrated computer-aided gear design system was presented in this paper. This computer-aided design system integrated design formulas with databases, solid models, and dialog box. It provides a systematic approach to design gear geometry with friendly human-computer interaction and generates design output with realistic gear geometry in the form of solid model. AutoCAD with AME was adopted as the tool to develop such a computer-aided design system. Example was given to illustrate the practical usage of the computer-aided gear design system in the transmission of automobiles. The present design system could be easily extended into a more general and powerful CAD system by including more functions and informations.

Reference

[ANDR 92]   Andrews, G. C. and Argent, J. D., "Computer-aided optimal gear design," ASME International power transmission and Gearing Conference, DE-Vol. 43-1, pp. 391-396, 1992.

[AUTO 92]   AutoCAD Release12 reference manual, Autodesk, Inc, 1992.

[DEUT 75]   Deutschman, Aaron D., Michels, Walter J. and Wilson, Charles E., Machine Design-Theory and Practice, Macmillan Publishing Inc., 1975.

[HUST 94]   Huston, Ronald L., Dimitrios Mavriplis, Oswald, Fred B. and Liu, Yung Sheng, "A basis for solid modeling of gear teeth with application in design and manufacture" Mechanism and Machine Theory Vol. 29, No. 5, pp. 713-723, 1994.

[LAIJ 94]   Lai, Jau-yeu, "A Parametric and Feature-Based Design System for Mechanical Transmission Parts," Master of Science Thesis, Department of Mechanical Engineering, University of Missouri-Rolla, 1994.

[OSWA 93]   Oswald, F. B., Townsend, D. P., Lin, Hsiang Hsi, and Lee, Chinwai, "Computer-Aided Design of High-Contact-Ratio Gears for Minimum Dynamic Load and Stress," ASME Journal of Mechanical Design, March, Vol. 115, pp. 171-178, 1993.

[SHIG 86]   Shigley, Joseph E. and Mischke, Charles R., Standard Handbook of Machine Design, McGraw-Hill, Inc., 1986.

[SPOT 85]   Spotts, M. F., Design of Machine Elements, 6th Edition, Prentice-Hall, Inc., 1985.

[TOWN 92]   Townsend, Dennis P., Douley's Gear Handbook, McGraw-Hill, Inc., 1992.

DEVELOPMENT OF A FRAMEWORK SYSTEM FOR TOOL INTEGRATION IN A PRODUCT INFORMATION ARCHIVE

Stephen Sum, David Cheng
Concurrent Engineering Unit, SIEMENS Pte Ltd
2 Kallang Sector Singapore 1334

Dorothee Koch
Fraunhofer-Institut für Arbeitwirtschaft und Organisation
Nobelstr 12c D-70569 Stuttgart
Federal Republic of Germany

Dragan Domazet, Lim Seng San
GINTIC Institute of Manufacturing Technology
Nanyang Technological University, Singapore 2263

Abstract

The described framework system has the goal of providing an integration platform for engineering tools to interact. Engineering tools exchange information via the data repository of the framework system. In the European research project *ESPRIT EP6896 Concurrent/Simultaneous Engineering System (CONSENS)*, a Product Information Archive (PIA) is being developed based on the object-oriented database system of the framework, the Object Management System. The Product model is based on a STEP compliant schema. This has been achieved by developing an Application Resource Model (ARM) for the required product information according to a user requirements analysis. The ARM then was mapped to the Integrated Resource Models of STEP which resulted in an object-oriented STEP compliant model. The main objective of PIA is the integration of the product information flows between parallel teams using the framework for product development. This is provided by an interface consisting of a library of functions that enable tools within and external to the framework to access PIA and exchange up-to-date product information. Additionally an X Motif based interface provides human users with direct access possibilities. The framework has been tested by the integration of various tools which support product development.

1. Introduction

Framework systems provide the software environment which supports the interaction of software applications from different vendors. The framework adopted in the CONSENS project is based on CAD Framework Initiative (CFI) specifications and has been implemented in the ESPRIT project JESSI Common Frame (JCF). The Product Information Archive (PIA) is a realisation of a central repository which allows different tools to exchange data on an integrated platform. The framework has been shown to be the ideal platform for heterogeneous tools integration[13].

Advantages of object-oriented engineering databases have been researched in [8]. In [11], a framework for the integration of the product data from different CAD system is described. The approach to adopt an international standard such as the Standard for the Exchange of Product Model Data (STEP)[6] for archiving of product data has been described in [3].

The importance of a Product Information Archive (PIA) is underlined by the fact that in the Deutsche Aerospace, the product information system of the TORNADO aircraft manages approximately 6 million configurations for 800 aircrafts in service. The development of PIA is an effort to define a universal object-oriented model for those product data that are essential in the information flows between tools in Concurrent/Simultaneous Engineering, and the model can be easily enhanced by company specific features. It is currently being tested with data concerning the wing of an aircraft and an electronic ABS module. By adopting STEP in its modelling approach, PIA aims to be the archive for a wide range of engineering tools, not exclusively for CAD systems. The archive is suitable for VLSI, aerospace, automotive industry and other domains.

The specifications of PIA [16],[17] focus on the conception, engineering and design phase, as the highest industrial benefit is gained in the early engineering stages. Taking into consideration that data exchange takes place across a heterogeneous and distributed environment, it is the objective of PIA to generate different views of the product for the different applications, ensure consistency of the representations and handle distributed management of the product information.

The development of PIA evolved through the following phases:

- *Analysis of user requirements*:

The *Deutsche Aerospace* and *AEG* who are members of the *CONSENS* project consortium have identified specific products such as the X31-A fighter aircraft and the electronic ABS controller for the analysis. Requirements for a generic product model for Concurrent/Simultaneous Engineering were analysed.

- *Specification of information model*:

Selection of a modelling approach such as STEP. An Application Reference Model (ARM) was specified based on the requirement analysis. It is independent of any physical implementation (e.g. the database storage technology). Subsequently the Integrated Resources of STEP(particularly part 44) were used to develop the ARM into a STEP compliant model. This model has been described in EXPRESS and EXPRESS-G and is the conceptual model that was implemented in PIA..

- *Implementation of a prototype*

To show the feasibility of the concept, a first prototype of PIA was demonstrated in September 1993. The second version was finished in August 1994. It is implemented on the framework object-oriented database, the Object Management System (OMS). It supports interactive use and provides a Scheme language[5] procedural interface for data access. In the third prototype which is to be presented in 1995, a STEP compliant schema will be supported together with a C and SDAI [7] based interface.

2. The Purpose of PIA

In order to have an integrated platform for supporting concurrent and simultaneous engineering, it is necessary that the information flow between the product development group working in parallel is integrated by providing all components of the system concurrent access to the same product information. This is achieved by having an integrated Product Information Archive (PIA) which stores the development of the products in all its different versions and variants.

The development process from the point of view of PIA works in the following way: The product development process starts with the concept of an idea of a product. A version ID and the time frame for this version of the product are defined. Then several teams start working in parallel, and each creates a first version of the product structure. These are written into PIA as different variants of the product version defined in the beginning. They are stored as different variants, so that they can be referred to later in the product development process. After a certain period of working, the teams must harmonise their work and decide which variant will be the one to be worked on in the next development iteration. This variant then is the input for the further work of the teams to improve the product structure in order to achieve an optimum with respect to quality, costs, manufacturability, layout of production facilities, etc. Finally, after several such steps, one (or several) variants are decided to be the official release of the product. Any development of the product after this is stored in PIA as new (temporal) versions or (customer) variants thereof.

3. Description of the PIA Product Model

PIA aims at managing the product information that is essential for the integration of information flows in concurrent engineering. At the present state it includes the product structure as well as master part data and process plans. The product structure and attributes are represented as structured objects in the object-oriented database.

Different stages of development in the product model encountered during the engineering design process are captured by corresponding representations within the product model. For each component it is possible to maintain different versions, reflecting the evolution of the product development. The knowledge on how engineering designs are modified and under what constraints is an important piece of information for concurrent work. The product model also includes different variants for components on each level of the product structure as well as the correct time dependencies. This integrated structure model allows the generation of different application specific views on the products.

The data model consists of a generic part which is suitable for any type of product. It can be customised to an even closer fit for a specific company by enhancing it with a company-specific sub model.

3.1 Reason for selection of an Object-Oriented Archive

Alternatives for the implementation of PIA have been investigated, such as:
- Using a relational database

Recent research results have shown that object-oriented database systems are more suitable for complex engineering data.
- Maintaining separate databases for each tool

A large overhead would be necessary for writing interfaces for each tool to access the external databases, and for ensuring data consistency in this heterogeneous system. It was decided to use the framework repository, the Object Management System (OMS), as the base for the implementation of PIA. OMS provides the advantages that object-oriented databases have over engineering applications[4].

3.2 Functionality

PIA offers the functionality for the creation and development of product data:

- the product structure of nested components, i.e. products, assemblies and single parts
- master part data for each component
- process plan
- version history of components
- customers variants of components

The tool interface and user interface of PIA include for instance the following functions:
- creation of new components
- input of master part data
- creation of new versions and variants for existing components
- creation and modification of the product structure
- creation and modification of the process plans

The concept of a bill of material in PIA is illustrated in Figure 1.

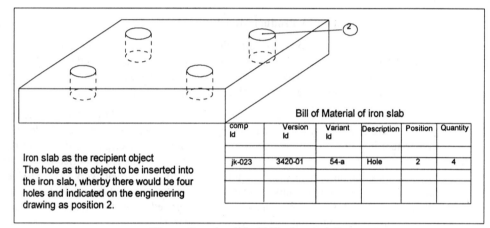

Figure 1: a simplified bill of material.

3.3 STEP modelling of PIA

In order to provide an open system platform, the exchange of product information should be based on an international standard. ISO 10303 Standard for Exchange of Product Model Data (STEP) is adopted due to its wide spread acceptance in the industry and completeness.

The PIA data model has been made partial STEP compliant based on ISO 10303 part 41, part 44 and part 49. ISO 10303-41 specifies the structure for the representation of products and their properties. It covers the identification, grouping, versioning, shape representation and the application context of a product. ISO 10303-44 covers the product structure, product concept and configuration management. ISO 10303-49 concerns process plan data.

The Application Resource Model of PIA was mapped onto an Application Interpreted Model which was modelled in EXPRESS-G and EXPRESS (see Table 1).

The PIA data model has not been completely mapped to STEP Integrated Resources since not all features necessary for PIA are represented in STEP. the connection between generic and the company specific product model is out of the scope of STEP. STEP does not differentiate between temporal development versions and customer variants as PIA does. PIA provides the possibility to categorise components according to their features in an object-oriented class tree. The instances of the lowest classes will by inheritance possess all features of the generic and the company specific product model, and the classification can be done according to specific user needs. This is not supported by STEP.

Table 1: an extract of the PIA to AIM mapping

PIA ARM	AIM	Source
Component	product	41
Name	product.name	41
Identification	product.id	41
Versions (Component to Version)	product.of_pia_version	
Version	pia_version	
Version_ID	pia_version.version_id	
From_To	pia_version.from_to	
Component (Version to Component)	pia_version.of_product	
Variants (Version to Variant)	pia_version.of_product_version	
Variant	product_version	41
Variant ID	product_version.id	41
Entry	quantified_assembly_component_usage	44
Quantity	quantified_assembly_component_usage	44

4. The Internal Data Model

The framework repository: the Object Management System (OMS) supports an Internal Data Model (IDM). The IDM is based on a set of objects of different granularity. The properties of the objects can be described using typed attributes. The IDM provides object classes of different granularity which constitute a three level hierarchy consisting of Primitive Objects (PO), Complex Objects (CO), and Partitions. POs are objects without contents, COs are object with contents, each partition is contained in a Top Level Container (TLC) which contains the attributes and relations of the COs.

Container Objects are further subdivided into the categories:
- Structured Complex Object (SCO) which contains a cluster of POs.
- Unstructured Complex Object (UCO) which contains a byte stream and behaves like a UNIX file.
- External Complex Object (XCO) which contains a reference to a UNIX path name.
- Empty Container Object (ECO) behaves like a complex object without content. It is used to model relations between COs.

The storage system of OMS is based on the Wisconsin Storage System (WiSS). Since OMS is object-oriented while the WiSS is page-oriented, an object cache is realised on top of the page-oriented buffer pool of WiSS. The data are distributed into Partitions which are stored in special directories called Background Databases (BDB). The BDB is distributed over the network via NFS.

The data definition language of the IDM, the Tool Integration Description Language (TIDL) is used to specify the database schema which consists of object types and relations. TIDL adopts an object-oriented approach and adheres to the syntactic conventions of C++. As TIDL is not a programming language, it is restricted to the descriptive element of C++.

5. EXPRESS to TIDL

For the creation of the data model in the OMS, a TIDL script with the schema class definitions are required. The EXPRESS/STEP environment (EXE) [9] of framework system provides an EXPRESS-to-TIDL compiler to process data models textually represented in EXPRESS according to ISO 10303-11 [7]. The front-end of the compiler parses and checks the EXPRESS model, while the back-end maps the data definition part of the analysed model to the TIDL code. The TIDL script then initialises the database according to the given EXPRESS model. The resulting TIDL script contains the class definitions with their data members and the specifications of the abstract relationships that are required to implement the data types and relationships of the EXPRESS model.

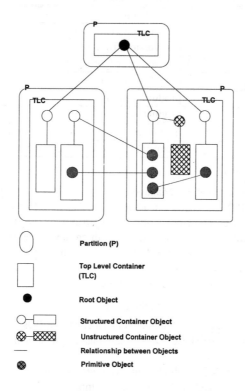

Figure 2: The Internal Data Model

The present prototype of the compiler contains some restrictions. The rules, functions and procedures declared in the EXPRESS model are not mapped to TIDL. Additionally the back-end of the compiler does not create instances of the data types nor checks the constraints on instances. These constraints will be resolved when the EXE incorporates a SDAI-generator in future work.

The following shows an example of the EXPRESS inheritance mapping with an abstract supertype *super* with two subtypes *sub1* and *sub2*. Abstract supertypes are indicated in TIDL by a private constructor.

EXPRESS
SCHEMA tree
ENTITY super
ABSTRACT SUPERTYPE OF
(ONEOF (sub1, sub2));
a: NUMBER;
b: STRING(26);
END_ENTITY;
ENTITY sub1 SUBTYPE OF (super);
SELF\super.a: INTEGER;
x: LOGICAL;
END_ENTITY
ENTITY sub2 SUBTYPE OF (super);
attr: STRING(129);
END_ENTITY;
END_SCHEMA

TIDL

class TREE_x_SUPER {
private:
TREE_x_SUPER() { };
NUMBER A;
char B[26];
};
class TREE_x_SUB1: TREE_x_SUPER {
INTEGER A;
LOGICAL X;
};
class TREE_x_SUB2: TREE_x_SUPER {
char ATTR[129];
};

6. Tools Integration

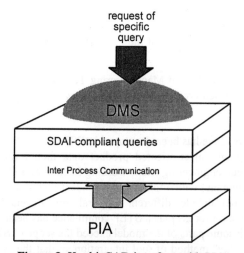

Figure 3: KnobieCAD interface with PIA

The integration of tools into the framework whereby tools can access the OMS directly, demonstrates the possibility of shared concurrent access to product information for different, parallel engineering processes. Different levels of integration are supported by the framework. The framework supports a C and SDAI procedural interface to OMS and Inter Process Communication (IPC) with the PIA message server.

A Testplan Generator for optimising electronics design testing strategies using Failure Mode and Effects Analysis (FMEA) access the PIA via the C procedural interface[2]. The data extracted are passed on to embedded SQL queries which update the local relational database of the Testplan Generator.

The Knowledge-based Intelligent CAD (KnobieCAD) system developed by DELOS, Italy [1],[14] has demonstrated the possibility of exchanging product data via IPC. A client/server architecture is used in which a data driver implemented as a client available to KnobieCAD modules communicates with the server residing in PIA. First the data driver filters any geometrical data not required by PIA; followed by a request message to PIA. When PIA acknowledges, the product structure is transferred to PIA in a predefined format such as a STEP physical file. PIA would de-serialise the message and uses the OMS Object Manipulation Language to update the object graph.

7. Future Work

The on-going research investigates how to integrate heterogeneous database systems on a logical level through a software layer called the Database Federation Services (DBF). The framework system will support Database Federation Services in the future [10]. Each existing database system will be a component database (CDB). The DBF will support each CDB schema with a powerful shared data model. The shared data model will be able to capture the concept as well as the semantics of the component data model [12]. Further investigation will be carried out to adapt PIA to the Database Federation Services to support tools residing on different CDBs.

In CONSENS, an Engineering Data Management System (EDMS) - the product Sherpa - is coupled with the framework. In order to integrate the product data, project management and process management functionality in the framework system, it will be possible to extend the current implementation of the coupling of the Product Data Management system to include the PIA [15].

In Singapore, the GINTIC Institute of Manufacturing Technology (GIMT) is exploiting the results of PIA in a project titled Product Information Knowledge Server (PIKS).

8. Conclusion

A Product Information Archive has been developed in ESPRIT EP6896 CONSENS with the objective of supporting an object-oriented product data model for integrating product information flows between engineering teams working in parallel. PIA provides its different users (tools as well as humans) with application-dependent views on the structured product model. The product model includes different temporal versions and customer variants, as well as master part data and process plans. STEP compliance has been achieved for parts of the PIA model, although some parts of the model exceed the scope of STEP. The framework supports the "plug and play" method of tool integration so that the system can be flexibly used also in existing environments. The third prototype of the system will be installed at different user sites (DASA, TEMIC and ASF in Germany, Hidrosorefame in Portugal, Alcatel in Belgium) in January 1995.

References

[1]   Pugliese, L., Architectural Specification of the Intelligent CAD, ESPRIT III 6896 CONSENS Deliverable D7.1, DELOS, Italy, February 1994

[2]   Jeschke, K., Architectural specification for Design for Quality, ESPRIT III 6896 CONSENS Deliverable D13.1, IWF, Germany, February 1994

[3]   Grabowski, H., Anderl, R., Malle, B., "Long term archiving of produt data model", *IFIP Workshop on Interfaces in Industrial Systems for Production and Engineering*, 1993.

[4] Hurson, A.R., Pakzad, S.H., Cheng, J.B., "Object-oriented database management systems, evaluation and performance issue", COMPUTER, IEEE CS Press, June 1993.

[5] IEEE 1178-1990 Standard for Scheme Programming Language.

[6] ISO/DIS 10303-1 Industrial automation system - product data representation and exchange, part 1: overview and fundamental principles, 1993

[7] ISO/WD 10303 Part 22 Standard Data Access Interface, May 1993

[8] Koch, D., Warschat, J., "An integrated vehicle documentation system based on an object-oriented database system", Proceedings of the *25th ISATA*, Florence, Italy, 1992

[9] Löschner, Elke, Detailed Functional Specification of the EXPRESS/STEP environment, CADLAB, Paderborn, Germany, June 1994.

[10] Radeke, E. "Federated database systems - an Introduction". Cadlab Report 5/93.

[11] Shah, J. J., Urban, S. D., Raghupathy, S. P., Rogers, M.T., "Product data Integration framework for synergistic CAD Systems", *IFIP Workshop on Towards World Class Manufacturing*, 1993.

[12] Sheth,A., and Larson, J. "Federated database systems". ACM Computing Surveys, Vol 22, No.3, September 1990.

[13] Sum, S., Ibold, C., "Information technology support for concurrent and simultaneous engineering - tool integration in a meta-framework", Proceedings of the *IEEE Region Ten Ninth International Conference, TENCON '94*, Singapore, August 1994.

[14] Sum, S., Singh, A.K., Ibold, C., "Development of an intelligent framework system for Business Process Reengineering", Proceedings of the *Second Singapore International Conference on Intelligent Systems, SPICIS '94* Singapore, November 1994.

[15] Sum, S., Singh, A.K., "Coupling of a information management system in a framework for concurrent product data handling", Proceedings of the *1994 International Computer Symposium, ICS '94*, Hsinchu, Taiwan, December 1994.

[16] Koch, D., Functional Specification of the Product Information Archive, ESPRIT III 6896 CONSENS Deliverable B3.2, IAO, Stuttgart, Jan 1994

[17] Koch, D., Diepold, T., Schumacher, B., Architectural Specification of the Product Information Archive, ESPRIT III 6896 CONSENS, Deliverable D3.1, IAO, Stuttgart, Feb 1994.

Decision Support for Configuration Management in Product Development

Bei Yu†, Ken J. MacCallum†
CAD Centre
University of Strathclyde
75 Montrose Street
Glasgow G1 1XJ, UK
Email:bei@cad.strath.ac.uk, ken@cad.strath.ac.uk

March 9, 1995

Abstract

In product development, configuration is the task of determining a complete set of components and their relationships such that the resulting product structure satisfies all requirements and constraints. It has two aspects: configuration design which is the process of creating configurations, and configuration management which is the process of maintaining a consistent configuration under change. For configuration management in particular it is necessary to understand the complex interdependencies between decisions and relationships between selected components.

This paper presents an approach to configuration management based on the use of AI techniques to manage consistency. The approach uses an architecture which requires background knowledge sources of product families and generic product breakdown structures, and configuration constraints. As decisions are made the system will manage consistency by making other decisions, or removing available decisions from the decision set. The main achievement of the approach is to maintain logical consistency of configurations for a designer as a design develops.

1 Introduction

In engineering product development, configuration is viewed as the task of determining a complete set of components, design decisions or options and their relationships such that the resulting product structure satisfies all requirements and constraints. For whole life cycle of product development, configuration can be considered as two aspects: *Configuration Design* and *Configuration Management*. Both configuration design and configuration management are complex processes for many products, particularly when the product structure is complex in terms of a large number of

⁰†Member of the Strathclyde Intelligent Design Systems Forum.

components with different relationships, the configuration problem will be significant. Configuration design is the process of creating configurations, in which it concerns with the components selection and the ways of configuring components. In contrast, configuration management is the process of maintaining a consistent configuration under change, in which it concerns with the configuration consistency. Especially when the decision of selecting components is changed, the configuration management should trace all the decisions which related to changed decision and revise them if necessary for maintaining consistency among components and decisions.

In practice many products reuse past designs or components. Since most products are changed depending either on their functionality or on particular requirements, the products are renewed incrementally rather than being changed totally to a new one. Reuse and adaptation of previous products is very important in the design process. Adapting established configuration to new requitements, functionalities, or technologies, requires an approach to configuration management rather than design; that is maintaining the consistency of configuration under change, rather than simply selecting.

The concept of configuration management pervades the whole life cycle of a product. Throughout the design and manufacture cycle, configuration changes may take place. After initial launch of a product, product versions and variants may be introduced. For some products, particularly large one-off capital goods, the product while in-service will undergo configuration changes. For many organisations, configuration management is a complex and time-consuming process.

Most work on the application of computers to configuration have concentrated on configuration design, producing systems which will generate a valid configuration using design rules from a set of design requirements. However, there is little work on configuration management support. In this paper, we will emphasise supporting configuration management in the product development life cycle.

This paper presents an approach to configuration design management, based on a concept of interactive, flexible and effective decision support. The approach presented in this paper is to apply a Reason Maintenance System technique which uses a nonmonotonic reasoning mechanism. It has been developed and tested in the design life cycle through an experimental world of a domestic heater configuration design. Work is now proceeding with integration of this mechanism within a practical context of a product development life cycle of sootblowers.

2 Configuration Management in Product Development

In engineering product development, the process of whole life cycle of a product starts from the market or customer requirements, into design specification stage, then though conceptual design, into detail design, and on to manufacturing, eventually to sales phase[Pugh 1991]. As a generic design activity, configuration design is viewed as the tasks concerning different relationships, and interdependencies among product components, design decision and options, so as to form a consistent product structure or model that satisfies all requirements and constraints. Configuration design takes place in each design stage. For different development stages, configuration design carries out individual tasks, i.e., it is concerned with different kinds of relationships within different decisions and components. In the design stage, for example, the configuration process begins by examining the product family which includes all parts and components, makes decisions on selecting components from it, and combines these components into a consistent artifact. In the manufacturing stage, configuration is more about assembly based on a sequence of orders or manufacture decisions.

Configuration can be a bottleneck in the product development process. The market demands are for short lead times and improved quality of the product If the configuration space is large, it can take a long time to search, choose and make correct decisions. Complexity of the product family and configuration information also leads to difficulty in the configuration process, particularly in maintaining consistency and dependency with change i.e. configuration management task. Under time pressure it is easy for designers to change a design feature and overlook a "knock-on" effect of the change.

Most new products are obtained by incremental product development; therefore reusing previous design concepts and design knowledge is an important aspect in the design process. To do this successfully, the designer needs to know which previous concepts and knowledge can be reused, and how it can be applied. The interdependencies of decisions, however, is not always explicit from past designs, making a further source of error[MacCallum 1992].

Information changes which cannot be known in the beginning also delay the whole product development time by changing components in the late production stage. If the components have to be changed for some reasons in the manufacture stage, the configuration process in the design stage would need to be done again. These types of changes are unacceptable in a "right first time" design process.

There is, therefore, a need for configuration management to manage these changes to ensure that a consistent configuration(s) is maintained. The task of configuration management is to maintain a consistent configuration under change. However configuration management cannot be considered separate by from configuration design. They have interrelationships with each other. Configuration design takes place through configuration management to get a consistent configuration(s) as a design solution.

The key aspect of this problem for which a designer needs help is not in designing itself, but in maintaining consistency across design decisions. Although this is not a new observation, providing decision support for consistency maintenance during configuration design is still an important problem.

3 Decision Support Approach to Configuration Management

The purpose of the configuration support system is to aid a designer manage the process of defining the configuration of new products. As an assistant, the approach of interactive decision support rather than automatic generation is adopted. This is in contrast with many systems for configuration design in which solutions are generated automatically based on a set of requirements. In the interactive approach, a designer or design team drives the process of identifying systems or components which form part of a new design, but rely on the knowledge within the system to guide the process and to maintain consistency across a set of decisions.

To support the designer, the system is able to make inferences based on product and company knowledge together with knowledge of the requirements. It uses a reasoning mechanism to provide a solution within the given constraints. Because the design process is iterative, further information may be added as the product is defined. This requires the system to operate nonmonotonically so it can modify its assumptions as more information is provided. These changes must be monitored so that decisions can be justified. The output from the system will be a design configuration model which can be used both as a basis for change and variant control, and for design analysis and detailing.

To fulfill its role, the system needs to contain various types of knowledge together with a means for reasoning with this knowledge. Therefore a knowledge representation for product and configuration knowledge, and an inference engine for the reasoning within the knowledge are required.

3.1 Knowledge Source Structures

The overall goal of using the system is to produce a product breakdown structure for a new product which is a legal combination of the selected components, from a series of design decisions. In order to formalise knowledge about the configuration support system, the system specification is presented in terms of its input/output aspects:

The inputs of the configuration management system are considered to consist of:

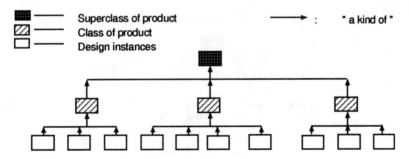

Figure 1: Product Family Classification Tree

- **Product Family Classification Trees (PFCT)** each of which is a tree structure that presents a product class from an abstract level to product instances. Each node in the tree represents a product class with its parts breakdown. The ancestor-descendant relationship of two classes is presented as "a kind of", i.e., a class of product is a kind of the superclass of product. Fig. 1 shows the structure of a general product family classification tree.

  There will be several product classification trees which are related to one another. In other words, all existing parts and components which might be configured can be found in the given knowledge sources in terms of their own Product Family Classification Trees.

- **Design Constraints Knowledge** which is a set of decision constraints related to subsystems, parts and components to be selected. These constraints can be separated into two types, one of which presents the dependencies among the Product Family Classification Trees, and the other presents a set of limitations on the possible combinations of subsystems, parts and components that are feasible in a single design. Such restrictions can be represented as AND, OR and NOT relationships, and reduce the possible choices at design time.

- **Decisions** form the interactive input during a design session. Each decision represents either a choice of a particular system, module or component from the PFCTs, or a propositional logic combination in terms of assertions in logic terminology. It can be changed at a later stage in terms of the configuration process. After each decision the system propagates the effect of the decision using its PFCTs and constraints knowledge to maintain consistency.

The outputs of the system are considered to consist of:

- **Product Breakdown Structure (PBS)** which includes *a components list* and *a hierarchic structure*. All the attributes, features and properties of selected components are recorded in the component list. The PBS is represented as an "AND" hierarchical tree, in which the overall relationships among parts or components are indicated in this structure. Links in the structure are viewed as "a part of" (see Fig. 2).

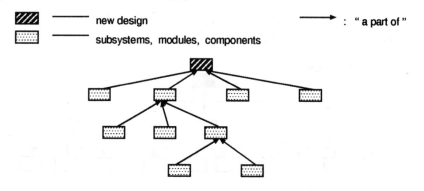

Figure 2: Product Breakdown Structure

- **Explanation of the Structure** which involves all the reasons for each selected part or component and decision history.

3.2 Configuration Design Process

In this formalisation of the system, configuration design is the process of deriving an instance of the PBS through a series of decisions on choice sets to best meet a set of requirements. The configuration design process therefore is shown to be the generation of an object which is composed of smaller sub-objects that together meet the required functionalities. The process starts from a set of requirements that need to be satisfied by the product to be configured, then obtains a suitable decision network for the case which allows a designer to make decisions on component choices selection. Eventually a product breakdown structure along with its explanations is produced based on the decisions.

During configuration design, each decision can be shown as an individual consistent statement that relates to the previous decision. Whenever a decision is made, the previous dynamic constraints will be tested to check either if they need to be relaxed or if new constraints are propagated. In addition, fixed, given constraints might be added depending on inference from previous decisions. On the other hand, the consistency among the decisions, and among the selected components is maintained to ensure the validity of the decision sequence based on the constraints knowledge and design requirements. The process continues until the decision sequence satisfies all the requirements and constraints given by the designer.

As the solution of configuration, the product structures is being created while the configuration process occurs. This product structure is an "AND" logical hierarchy, and combines all the selected components with "a part of".

3.3 System Architecture

Fig. 3 shows the system architecture which supports configuration management. In Fig. 3, *PFCTS Constructor* organises the domain knowledge about components and their relationships into the PFCTs. The design constraints and requirements can be represented based on the generic constraints representation. During the configuration design process, a decision network at system level is created by *Decision Network Constructor* based on the components selection in the PFCTs domain that refers to the constraints. The selected components are configured into an instance of PBS structure as the configuration solution in the *PBS Builder*.

As the core of system, the *Inference Engine* controls the configuration process, makes inferences on the decisions and maintains the consistency among the components and decisions. It uses the Reason Maintenance System (RMS) which is a non-monotonic reasoning mechanism to assist the designer to do configuration. The purpose of the RMS is to assist the problem solver, which operates on a body of domain knowledge to make inferences according to some problem-solving procedure, making and maintaining these inferences[Doyle 1979][McAllester 1990][Kelleher 1988].

As a problem solver, the task of the decision processor is concerned with the manipulation of the configuration knowledge sources, as well as the maintenance of consistency and dependency either once a decision is made or a decision is changed. In addition, it supports the configuration task to obtain a solution, by checking if the selected components meet the requirements and constraints.

The LTMS maintains a table of decisions and justifications for the decisions. It deduces the new justification for a made decision in order to constrain the further unmade decision. Meanwhile, it checks for consistency of a decision, that is, revises the beliefs that cause contradiction in the made decisions. The Dependency Directed Backtracking algorithm is used to identify and retract the decisions that underlie a contradiction.

As a constraint-based reasoning technique, the Forward Checking algorithm is used to increase the power of the inference engine by constraint management. The algorithm eliminates incompatible branches early on in the search so that the efficiency is improved compared with chronological backtracking.

More details about the inference engine design has been described in [Yu 1994].

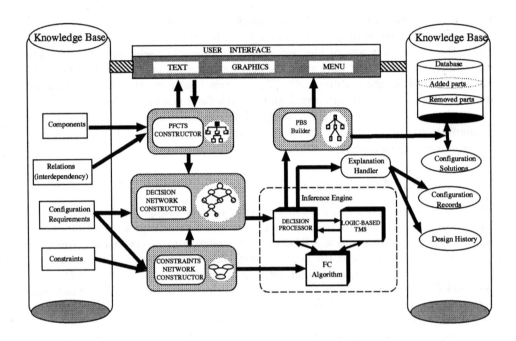

Figure 3: Configuration Management System Architecture

4 Discussion

A goal of the approach described in this paper is to develop a knowledge-based system prototype which will assist in the configuration management task. The configuration knowledge modelling enables the product and process information to be structured as domain-independent formalisms. The system inference engine is able to make inference relying on the configuration knowledge structure either to identify parts or components which are selected to form a product structure, or to indicate the decisions which should be made. The idea in the design of the inference engine is to use a RMS technique which has a nonmonotonic reasoning mechanism to manage constraints and maintain consistency among selected components or decisions.

The decision processor can call LTMS to maintain consistency among the decisions if necessary. In order to cope with the nonmonotonic configuration process, a nonmonotonic reasoning mechanism is used for maintaining consistency effectively. Specifically this mechanism is used in the case of either the conflicts which appear when a decision is made, or when there is a contradiction in the updated situation in which case a decision needs to be changed. It maintains consistency dynamically i.e. maintains the updated consistent situation for each step of the configuration process.

The constraints based reasoning technique enables the decision processor to be able to handle different logical constraints. The Forward Checking algorithm is able to reduce the boundary of decision selection to decrease the complexity of consistency maintenance and avoid unnecessary mistakes[Shanahan 1989]. In addition, the decision processor handles the local constraints propagation for constraining the next decision. The task of the Forward Checking algorithm is concerned with handling constraints before any decision is made. The benefit from this mechanism is to decrease the complexity of reasoning systems by problem decomposition, so that the constraint management ability in the inference engine is increased.

The approach presented in this paper can provide a full explanation facility during configuration process which records the decision sequence and reasons for each decision. A Truth Maintenance System is shown to be the most appropriate mechanism for an explanation facility because it records not only belief but also its inference. When the explanation is requested, it reasons backward to find all the reasons in each level until the initial premises.

5 Summary

This approach has been tested on several examples of configuration domains, such as domestic heaters, telephone sets and sootblowers. In these examples, the product information about domestic heaters and telephone sets has been structured into the configuration knowledge tree i.e. the *Product Family Classification Trees*. The inference engine accesses these trees along with a set of well-defined constraints to support the designer to make decisions on selecting components for these product domains. It maintains consistency among components and decisions during configuration, so it ensures that the breakdown structures of heater, telephone set and blower meet the design requirements and satisfy all the constraints. In practice, it has been shown that this approach is an appropriate mechanism for a configuration management system[Yu 1992(2)]. The whole configuration management system is being built by using this approach. The architecture which incorporates the reason maintenance approach and integrates with existing product structures is being developed for this purpose.

References

[Doyle 1979] J. Doyle, **A Truth Maintenance System.** *Artificial Intelligence*, No. 12, pp231-272, 1979.

[Kelleher 1988] G. Kelleher, and B.M. Smith, **A Brief Introduction to Reason Maintenance Systems.** *Reason Maintenance Systems and Their Applications,* Ellis Horwood Limited, ISBN 0-7458-0482-9, pp4-20, 1988.

[MacCallum 1992] K.J. MacCallum, B. Yu, A. Frederiksen, and D. McGregor, **A System For Supporting Design Configuration.** *Proceedings of Artificial Intelligence in Design'92,* Kluwer Academic Publishers, ISBN 0-7923-1799-8, 1992.

[MacCallum 1993] Ken J. MacCallum, Bei Yu and Fiona MacDanald, **A System For Configuration Management in Design.** *Proceedings of Expert System'93 Conference,* Leeds University Press, pp109-118, December, 1993.

[McAllester 1980] D.A. McAllester, **An Outlook on Truth Maintenance.** *AI Memo 551, MIT AI Laboratory,* 1980.

[McAllester 1990] D.A. McAllester, **Truth Maintenance** *Proceedings of AAAI-90,* pp1109-1116, ISBN 0-262-51057-x, 1990.

[Pugh 1991] Stuart Pugh, **Total Design**, Addison-Wesley Publishing Company, ISBN 0-201-41639-5, 1991

[Shanahan 1989] M. Shanahan, R. Southwick, **Search, Inference and Dependencies in Artificial Intelligence.** Ellis Horwood Limited, ISBN 0-7458-0488-8, 1989.

[Yu 1992(a)] Bei Yu, **Evaluation and Critique for Existing Configuration System.** *Internal research report, CAD Centre, University of Strathclyde,* CADC/R/92-05, CONFIG/R/92-02, July, 1992.

[Yu 1992(2)] Bei Yu, **The Use of Artificial Intelligence Techniques for Configuration in Engineering Design.** *Internal research report, CAD Centre, University of Strathclyde,* CADC/R/92-23, CM/R/92-03, October, 1992

[Yu 1994] Bei Yu and Ken J. MacCallum, **Reason Maintenance for a Configuration Management System**, *Expert System'94 Conference Proceedings: Research and Development in Expert Systems XI*, SGES Publications, pp173-185, ISBN 1-899621-01-6, 12-14th December, 1994.

REVERSE ENGINEERING - MODERN TOOL FOR PRODUCT-PLANNING

Authors:
Dr. Bernd Schumacher
Prof. Dr. Peter Pscheid
Interstate Institute of Technology St. Gallen (ISG), Tellstr. 2, CH-9000 St. Gallen, Switzerland
E-Mail : pscheid@ dial-switch.ch

ABSTRACT

"Reverse Engineering" can among many other applications be a versatile tool to systemize product-innovation in congruence with market requirements and in respect to competition. The method originally developed in software-redesign while the authors analyse applicability to mechanical/electrical engineering tasks.

Keywords: Reverse-Engineering, Product-Planning, Mechatronics, Bench-Marking

The method "Reverse Engineering", very different from "RE-Engineering" or "Business RE-Engineering" activities, is applied to product planning or project-management, intending rationalisation and cost-reduction to products or production-organisations. As shown in Fig. 1, it changes the traditional approach and begins with an intensified investigation of market requirements and existing competition (best practice world wide) The main goal is, to quickly result in an comprehensive new product-view, allowing a world wide marketing. The planning however should stay away from conceptual considerations.

When the authors wanted to find out about the "state of the art", the advantages and limitations of 'Reverse Engineering" during development and launching of mechatronic products, they realised soon, that many institutions intuitively use this method but a systematic description is totally missing. Best practice was found in Taiwan's ITRI (Industrial Technology Research Institute) and in Singapore. Derived from discussions with several scientists and applying experts, this method may become defined as:

Scientific analysis of existing products in order to achieve a new product which is better, cheaper and more marked-oriented.

- Systematic market inquiry on best practice and weaknesses as well as of all important design features of an existing competition to plan a new, quick, cost-attractive, reliable and innovative product.

- Goal must be, to develop enough market focus and customer satisfaction to result in the own companies success, to overcome organisational friction, delays and prejudgements and to de-block mental fixation from internal controlling to a world-view.

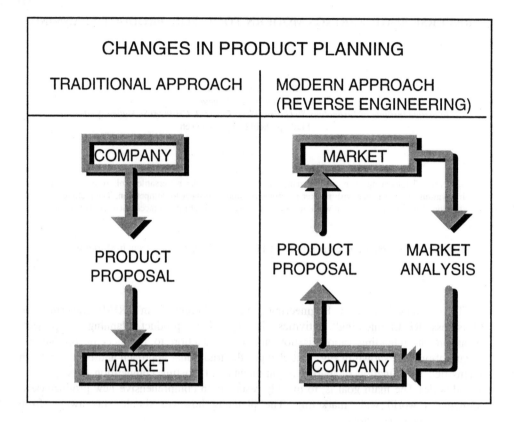

Fig. 1: Changes in Product Planning

The research in different Asian countries cleared, that this definition stays different from any identity with the "KAI ZEN" philosophy of "continuous improvement" and also wants to stay away from bad minded straight copying of competitors design or structures. Applicability is possible, where existing products serve already a market. Bottom-up research from new ideas and subsequent product-generation is not a field for „Reverse Engineering". The method forms a clear alternative approach to traditional product-development, which starts in full secrecy, is organised through several phases of development and testing until it becomes presented to public with its official new product launch. This procedure, standardised in strict rules (German VDI-recommendations 2220 - 2225), is a helpful tool for the very few 100 % bottom-up product-inventions.

"Reverse Engineering", and therefore also its name, starts in the opposite direction, it starts from existing products in the world market and is therefore only a tool in areas, where a certain market level was already gained. This, however, is the case for closely 95 % of the products. The different procedure in steps clearly shows Fig. 2.

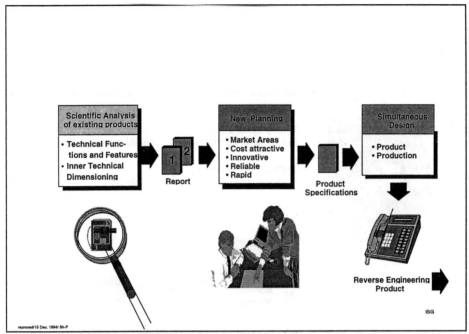

Fig. 2: Reverse Engineering Procedure

The detailed review of products in the market, the research of market potentials and the analysis of inner design as well as sizing has to respect, of course, existing intellectual property rights. A separate study for those has to be executed and is carefully done if a company wants a serious reputation.

The method of "Reverse Engineering" should reasonably speed up the product planning. The neutral description, giving functional and feature descriptions as well as demands on pricing, quality, delays and mandatory prescription, should stay away from deep conceptual rules and considerations. The result will enable the internal departments to parallelize product, process, quality and market-planning (Simultaneous Engineering) and focuses all activities to quality and to marketing success. Often not clearly envisaged defects of the method, are problems of incorporating innovative steps and time-synchronisation of parallel project branches. It is further important, to limit the volume of innovation inside such a project to a time-frame, which can be managed and surveyed well enough to exclude trouble to the schedules and failures in product design. Any larger "crash" within a running project will cause a reset. This means heavy time and money losses.

Fig. 3 presents a good example how time delays for a new product-launch, however, may vary. The digital printed circuit board can become "Reverse-Engineered" impressively short within three weeks. The telephone housing on the other side needs the manufacture of injection-moulds and is a skilled hand worker's job for months. Generally, mechanical parts will need excessively more time than others. Software can offer extremely short-hand new releases and would be the leader in development cycle time. Too many new software-releases, however, risk to confuse more than helping the customer.

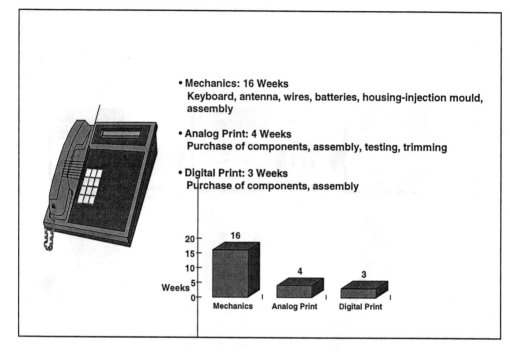

Fig. 3: Variable time-frame of development-elements

So: "Reverse Engineering" has to be accompanied by an intelligent management.

At ISG (Interstate Institute of Technology St. Gallen) we are convinced, that "Reverse Engineering" bears still a wide potential for further development. As mechanical projects prove to be most time consuming elements, when studying mechatronic developments, we would very much like to research for the definition, the practical refinements and the introduction in teaching of "Reverse Engineering". It needs to find organisational tools and to speed up the mechanical part-development. We are interested to insert this research in an international co-operation, also to overcome local mentality and to focus on the "RE"-method to global thinking.

Fig. 4 once again shows the limits of the "Reverse Engineering" method as a project-definition instrument. The method uses in the first step "Product Benchmarking" may be also some "Process-Benchmarking" but needs to end in an innovative, quick, economic and reliable new product definition.

"Reverse-Engineering" is undoubtedly an excellent tool for small and medium-sized companies, which normally lack of potentials to develop everything in house. When touring ASIA the authors found out that "Reverse-Engineering" besides its already described characters grew up and got important in this world region because of the typical industrial structure with very many small industrial units. Their time frames are tight and small and project-volumes are of minor size.

```
┌─────────────────────────────────────────────┐
│             REVERSE ENGINEERING             │
│                                             │
│           PRODUCT SPECIFICATION             │
│           PRODUCTION LIMITS                 │
│       LEGAL PRESCRIPTIONS & STANDARDS       │
│           QUALITY REQUIREMENTS              │
├──────────────────────┬──────────────────────┤
│  PRODUCT CONCEPTION  │      ECONOMICS       │
│                      │                      │
│     MODULARITY       │     PRODUCTION       │
│         &            │         &            │
│    STRUCTURING       │   QUALITY PLANNING   │
└──────────────────────┴──────────────────────┘
```

Fig. 4: "Reverse Engineering" as initial part of a product-renewal

Governments could most effectively help such industries not only by the organisation of good infrastructures (Science/Industry-Parks), but also in centralising the first and second step of the "Reverse Engineering" procedure under governmental control (also payment) in the hand of qualified experts. Complex mechanical structures that need heavy installations may furthermore be analysed annually with a standardised program and could be done quick, as the procedure, the installations and the staff are already prepared. It should, however, also be recognised, that such governmental support is interfering with free market forces and builds up strategies of competition between nations.

The more we engage to clear understanding of "Reverse Engineering" methods for mechanical and electrical product developments, they will become more commonly used and accepted world-wide. This will at the same time eliminate misuse and its advantages may be gained everywhere.

Moulded Parts Design

A HyperCAD Expert System for Plastic Product Design
   J. Borg and K. J. MacCallum

A Framework to Develop an Expert Injection Mold Planning System for Early Product Design Decisions
   K. S. Chin and T. N. Wong

A HYPERCAD EXPERT SYSTEM FOR PLASTIC PRODUCT DESIGN

JONATHAN BORG
Department of Manufacturing Engineering, University of Malta
Msida, MSD 06, Malta
E-mail: jjborg@unimt.mt

Prof. K.J. MacCALLUM*
CAD Centre, DMEM, University of Strathclyde, 75 Montrose Street
Glasgow, G1 1XJ, Scotland
E-mail: ken@cad.strath.ac.uk

ABSTRACT

The plastics' industry is realizing that the traditional trial and error approach to product design is an inefficient process which can lead to high reject rates. This paper presents an approach which supports the design process of thermoplastic products. An analysis of the process involved revealed that assistance could be applied during what is being termed the transition design phase. The outcome of this project is a tool nicknamed IMDA, based on the fusion of C.A.D., Hypermedia and Expert Systems, which resulted in providing suitable assistance during the transition design phase.

Key words: Design for X, Transition Design Phase, HyperCAD Expert System.

1. Introduction

In injection moulding[1], the way and ease with which the plastic flows into the mould is of paramount importance in determining the quality of a part. Since the poor design of features[2] such as ribs and bosses, making up an article (see Fig.1) can give rise to a number of defects such as sink marks[3], the quality of a moulded part basically derives from its design. A consequence is that for a successful design solution, one cannot design a product without considering the related mould tool design. This means that there is a need to utilize the past experience and guidelines which link the product and mould tool design. Different experts are hence involved in this complex design process, namely:

- Product design engineers, who primarily look at functions, such as bosses to reinforce holes, snap fits to assemble parts etc.;
- Mould tool engineers, who are concerned mainly with the design and construction of the mould tool;
- Process engineers, who are mostly concerned with the flow and behaviour of the injected material in the mould;
- Others, such as electrical engineers for designing the electrical circuitry of products (eg. in business communication systems).

---

* Member of Strathclyde Intelligent Design Systems Forum

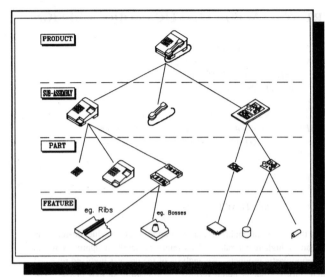

FIG.(1) - Typical Product Structure Tree

These experts take different perspectives at the candidate design using different variables in design evaluation. It can be appreciated that difficulties arise when part designers do not have the right expertise, which helps them generate a design solution which is also acceptable to the tooling and processing engineers. These issues give rise to a number of basic questions such as:

- How can a plastic product designer be assisted in taking better decisions?
- To what extent should a CAD system provide plastic product design guidance?
- What system architecture should a CAD system have to assist such decisions?
- How can a CAD system identify relations between the product design and mould tool design?

2. Product Design Process Analysis

Through the evaluation of design process models,[4][5][6][7] the model illustrated in Fig.(2), was developed, which correlates well with the actual process involved in the design and manufacture of plastic parts.

In order to satisfy a need, product specifications are identified and defined (stage 1). These result in the generation of primitive conceptual designs (stage 2) to try and satisfy the functional need. These are initially mental models held by the designer in the form of ideas. These models are then input to the next phase, during which there is a transition process, from the conceptual design to the detailed design. During this transition, the

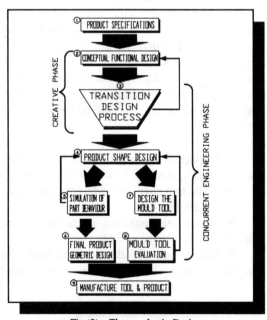

Fig.(2) - Thermoplastic Design Process Model

aim is to convert the product's functional design into a geometric design. One can say that stages 2 and 3 are basically the creative phase in this design process. The output of the "transition design process" will be a detailed part model, normally for injection moulding in the form of a detailed drawing, acceptable to the different perspectives of experts involved. The design process then proceeds with two concurrent processes: simulation of part behaviour (stage 5) and mould tool design process (stages 7 & 8). Once these stages (5, 6, 7 & 8) converge to an acceptable solution, the mould tool and hence part can be manufactured (stage 9).

2.1 The Role of Transition Design

As illustrated in Fig.(3), during this process, the aim is to convert the product's functional design into a geometric design, this being achieved by a number of sub-design processes running concurrently, namely:

- Design for Manufacture: to ensure that the part's geometric design can be easily manufactured by injection moulding;
- Design for Assembly: because thermoplastics parts can be joined using a variety of assembly techniques.
- Design for Least Cost: to minimize tool design and production costs;
- Design for The Environment: so as to consider several issues, including plastic recycling.

It should be noted that all these sub-processes take the form "Design for X" and thus the transition design phase could include other DFX processes, suitable for other product domains. With this

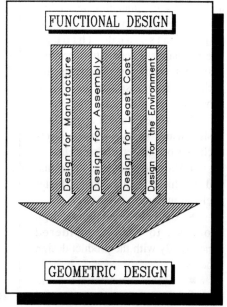

Fig.(3) - DFx Processes In The Transition Design Phase

background, then one can say that standards, design guidelines, constraints and the designer's past experience all help the designer in taking fundamental decisions[8] to converge towards an acceptable detailed product shape design.

2.2 Existing Computer Based Design Tools

Commercial tools currently available in the injection moulding domain include software for material selection[9], CAD for detailed design, Finite element analysis[10], mould tool design CAD[11], CADCAM for tool manufacture[12] and expert systems[13] to assist the injection moulding machine setting. It is evident that the designer does not have access to a tool which provides assistance during the early design stages (stages 1, 2 & 3), which can help generate a part design ideally right first time. This and previous arguments therefore lead to the conclusion that assistance would be beneficial during the transition design phase, where designers have to make fundamental decisions. In this phase, the designer and the computer-based tool required should work in concert, with basically each doing what they are best capable of doing.

3. An Alternative Design Tool

The tool required would then be a useful design assistant if it can participate actively during the transition design phase by reasoning, checking facts related to the design and hence recommending alternative ways of designing a part. With these arguments and through questionnaires sent to product designers, main specifications for such a design assistant are that it should:

■ Provide a natural approach to design using terms of part objects such as ribs, gussets, bosses etc.;

■ Provide guidance in textual and graphical form to enable better decision making;

■ Allow the mould tool design process to be considered concurrently with the product design process;

■ Take constraints into consideration eg. manufacturing and costs;

■ provide both 2D and 3D product visualisation.

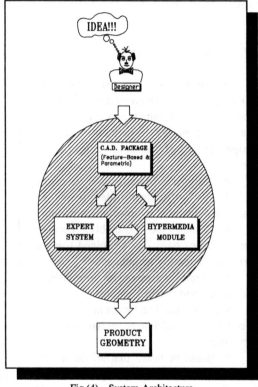

Fig.(4) - System Architecture

Through these specifications, the project identified that the "Injection Moulding Design Assistant (IMDA)" tool should therefore be based on the fusion of (see Fig.(4):

(i) a feature-based modelling approach, so that the designer can freely select features such as ribs from a CAD library. Also, the features should be parametrically defined, to minimise user input during their generation.

(ii) an Expert System for capturing and reasoning about expertise associated with injection moulding product design. In particular, the expert system should check for design conflicts when product features are being added;

(iii) a hypermedia[14] module, which employs textual and graphical displays to connect related chunks of design information, so as to effectively assist the complex decision-making process.

With such an architecture, the designer will not only be able to generate part geometry, but will also have access to design guidelines, reference to standards and constraints, this leading to less iterations during the stages following in the life-cycle process. The expert system component is responsible for firing the necessary rules and providing the information necessary for decision making in a hypermedia format. The expert system, will also automatically pass on the recommended dimensions to the CAD package in order to parametrically generate the feature geometry. The IMDA architecture will therefore enable the designer to lead the design process.

4 System Implementation

It was decided to use AutoCAD^{tm}, a commercial CAD package and employ its internal programming language AutoLISP^{tm}, to develop an expert system (ES) and a hypermedia module (HM). This meant that problems of communication between three commercial applications (CAD, ES & HM) would not be encountered during the implementation phase. Furthermore, AutoLISP code can be readily transported from one AutoCAD system running on a PC to another system running on a different hardware platform, this making it more attractive during the implementation phase.

4.1 CAD Component

A number of algorithms were developed for the parametric generation of profiles for features such as ribs, based on guidelines provided by plastic material suppliers[15][16]. Taking a simple algorithm, depending on the nominal wall thickness (NWT) beneath a rib, then parameters making up the rib are related as illustrated in Fig.(5). In this case, the CAD system acquires the values of NWT, halfbase, 'c' and 'e' from guidelines held in the expert system's facts list with the user inputting the desired draft angle (drft) and the starting position (sp) of the rib. These values are then used to determine the other parameters as illustrated in Eq.(1).

Fig.(5) - Parametric Rib Feature

$$PT1_x = SP_x + halfbase + e; \quad PT1_y = SP_y;$$
$$PT2_x = SP_x + halfbase; \quad PT2_y = SP_y;$$
$$OPP = Tan(drft) \times c;$$
$$PT3_x = PT2_x - OPP; \quad PT3_y = SP_y + c;$$
$$PT4_x = SP_x; \quad PT4_y = SP_y + c \quad (1)$$

4.2 Expert System Component

Developing the expert system component within AutoCAD, required firstly, the capturing of related expertise and secondly, building an inference engine. For the injection moulding transition design phase, a production system was considered a suitable approach, with rules taking the form:

- Design Guidelines: These can be said to be independent from the product domain, and have the form of:

IF distance of hole from edge < Hole diameter
THEN do not accept hole definition.

- Product Attributes: These can be said to be product domain dependent, as they vary from one product to another. Such knowledge is required to check the product's functionality. Rules in this category have the form of:

IF telephone handle's length < 200mm and > 100mm
AND telephone handle's width < 60mm and > 20mm
THEN handle is comfortable.

- Manufacturing Constraints: These can be said to be product independent, as they are more related to the product's form features, rather than to its functionality. For instance, a rule in this category might be:

IF hole orientation is at an angle to the parting line
THEN An expensive split cavity mould is required.

In the case of the IMDA system, product guidelines had to be mapped in geometrical form onto the part design. Consider a simple feature definition rule of a hole, used by the system to determine the optimal position of holes having a radius 'R', in a rectangular plate (with parameters, width, breadth and height). The following algorithm, based on Fig.(6) was employed:

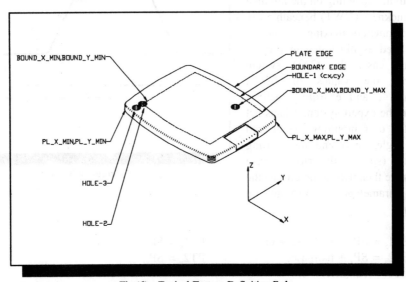

Fig.(6) - Typical Feature Definition Rule

1. Determine plate coordinates (pl_x_min, pl_y_min, pl_x_max and pl_y_max).
2. Determine 'acceptable boundary' based on the rule saying that 'the minimum distance from the plate edge of a hole should be at least equal to the hole's diameter'.
   2.1 Determine boundary coordinates:
   eg. bound_x_min = plate_x_min + (3 x hole radius)
   2.2 Draw boundary
3. User inputs centre point position of hole (eg. hole-1 with centre point cx,cy).
4. Check whether centre point is within boundary
(If cx < bound_x_min Then reject hole) or (If cx > bound_x_max Then reject hole) or
(If cy < bound_y_min Then reject hole) or (If cy > bound_y_max Then reject hole)
5. If hole is acceptable, then draw a hole with the inputted radius and depth.

The other rules in the expert system were structured and implemented in a format similar to the one explained, this making it possible to map rules in a geometric form (see Fig.6) with respect to facts related to the part being designed.

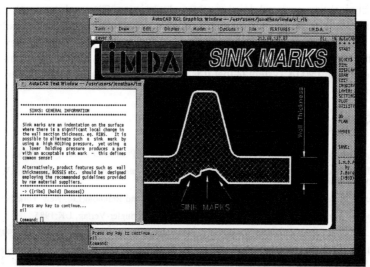

Fig.(7) - Hypermedia Module

4.3 Hypermedia Component

The implemented hypermedia module displays textual information in AutoCAD's text window and graphical images in AutoCAD's drawing editor window (see Fig.(7)). This module contains 'related chunks' of captured expertise, in textual and graphical form, but unlike in the expert system module, reasoning on this expertise is performed by the designer, rather than through an inference engine. The captured domain knowledge was utilized to establish hot-spot links allowing the designer to browse through related chunks of information, thereby making the decision making process more effective. Animation was used were necessary to effectively explain certain problems/terminology (eg. the formation of weld lines), this making hypermedia a very effective component in the IMDA system architecture for decision making purposes.

5. Discussion And Conclusions

As implemented, the Injection Moulding Design Assistant (IMDA) system:
- makes use of goals/constraints to keep track of the design process;
- automatically provides design recommendations generated by the Expert System to the feature-based CAD package;
- assists the designers' decision making process by linking the product and mould tool designs through the hypermedia module ; this also makes the IMDA design tool useful for teaching purposes;
- maps design rules to the product geometry;
- provides recommendations regarding the processing phase, in the form of a concurrency list;
- runs on different hardware platforms (PCs and Sun).

Evaluation of the system revealed that when using IMDA, more time was being spent by designers on thinking and taking decisions, rather than wasting time browsing through technical literature and doing routine calculations. The evaluation also revealed that information not thought of by the designers in the first place, was easily made available for other experts involved in the process through the "concurrency list", a feature which proves beneficial during teamwork design. All in all, the key to the overall contribution in the IMDA system is that it is based on the fusion of CAD, Expert System and Hypermedia components in a single architecture.

6.0 References

1. C. Weir   *Introduction to Injection Molding*, U.S.A.: Society of Plastics Engineers, Inc.1975.

2. G.E.M. Jared   *Recognizing and using geometric features* in *Geometric Reasoning*, edited by J.Woodwark. USA:Oxford University Press, 1989, p.169.

3. R.D. Beck, *Plastic Product Design*, Second Edition, Van Nostrand Reinhold Ltd. New York:1980.

4. M.J.French,   *Conceptual Design for Engineers*, published by Springler-Verlag, UK:1985.

5. A.Ertas, J.C.Jones,   *The Engineering Design Process*.   John Wiley & Sons, Inc. New York:1993.

6. E.Lim, B.Barnes, *A case in integrating CAE tools in injection molding design* in *Proc. of the International Conference on Computer Integrated Manufacturing ICCIM '91*, Singapore, edited by B.S.Lim. p. 235 - 238. Singapore: World Publishing Co. Ltd., 1991.

7. ICI. *Table 1 - Summary of Selected CAD Techniques* in *Injection Moulding* (Technical Information N102, Thirteenth Edition), p.13. Published by Imperial Chemical Industries, 1991.

8. C.V. Starkey, *Engineering Design Decisions*. Published by Edward Arnold, UK:1992.

9. Monsanto Campus II - *Plastics Database software* distributed by Monsanto Europe S.A., Belgium.

10. R.V.Wilder, *Computer Simulation works - and works well - in the real world*. In *Modern Plastics International*, **Vol.19**, No.7, Jul. 1989, p.26 - 29.

11. O.Heuel, *CAD/CAM Aspects In Die and Mold Making*, in *TRANSFER*, **Vol.4**, 1985.

12. *Injection Moulding CAD/CAM System* in *American Machinist*, April 1989, **Vol.133**, p.13-14.

13. A.Shenoy, U.Shenoy, *Expert Systems in Plastics Processing* in *Materials Engineering*, November 1988, **Vol.105**, p.33-36.

14. H.Maurer, *An Overview of Hypermedia and Multimedia Systems*, in *Virtual Worlds and Multimedia*. UK:John Wiley & Sons Ltd., 1993, p.1 - 12.

15. Du Pont, *Design Handbook - Module I, General Design Principles for Du Pont Engineering Polymers*, Du Pont, Switzerland:1992.

16. Dow, *Designing With Thermoplastics*, Dow Chemical Company. USA:1992.

A FRAMEWORK TO DEVELOP AN EXPERT INJECTION MOLD PLANNING SYSTEM FOR EARLY PRODUCT DESIGN DECISIONS

KWAI-SANG CHIN
Department of Manufacturing Engineering, City University of Hong Kong
Tat Chee Avenue, Kowloon, Hong Kong. Fax : (852) 2788 8423

and

T N WONG
Department of Industrial & Manufacturing Systems Engineering, University of Hong Kong
Pokfulam, Hong Kong. Fax : (852) 2858 6535

ABSTRACT

This paper aims to explore the applicability of expert systems technology to today's competitive environment in injection molding product development with respect to the increasing trend of concurrent engineering practice. The proposed framework ESPIMP-1 is the phase one of a research work focusing in developing expert system for injection mold design and development with emphasis in the early design decisions. The ESPIMP-1 covers the areas of plastic material selection and injection mold design. With the inputs of rough part design features and requirements, the system will automatically select the appropriate plastic material and generate the major injection mold design features.

1. Introduction

The plastic products manufacturing industry has been growing very rapidly in recent years. The growth will be accelerated by the tendency of substituting plastics for metal which is appearing throughout the world. The injection molding process is one of the most effective and advance molding process for plastic products. In traditional practice, the mold design relies on the experience of engineers and designers. It is always the case that many changes of the product are required during the mold design stage due to the manufacturability requirements of the mold as well as the molding production which undoubtedly leads to longer lead time and higher cost. The key problem is that there is little consideration for mold design/making at the early product development stage.

The injection mold design has to deal with the issues of product requirements, material selection, and mold design simultaneously as they interact with each others. Changing one aspect for better results, for instance product design feature, may have a negative effect on the other influencing factors, for instance the mold design. The design process involves a substantial practical knowledge component (heuristic knowledge). Knowledge and expertise of more than one specific area are required to have an optimum solution. It thus relies heavily on the human experts, the product designers and mold designers, who are required to have a high standard of specific knowledge, experience and judgement. Unfortunately, the growing demand in industry for such experienced designers and engineers far exceeds the supply.

Expert systems (ES) can be defined as computer programs that capture the expertise of human experts in particular application domains. They are designed to manipulate information, including knowledge, facts and reasoning techniques, in a high level way, and emulate or assist users to solve problems that normally require the abilities of human experts. There is obviously a potential to utilize ES technology to solve the product development problems at the early design stage. The authors have developed an implementation framework ESPIMP-1 for using ES to help making the decisions of material selection and injection mold design at the early stage of plastic parts development in a concurrent approach, i.e. simultaneous consideration of the feasibility of mold design is made prior to the confirmation of product design details.

2. Brief Review of Plastic Material Selection and Injection Mold Design

2.1 Plastic Material Selection

Whenever a new product development project begins, the design engineers starts with a conceptual design of the part and then decides which plastic material should be used. Designers must select the material not only with the customer requirements, such as physical functions, aesthetics, durability, etc., but also that will produce a high quality part in an efficient manner. The selection requires extensive knowledge and experience. The basic steps of the plastic material selection process are i) determine the needs and conditions to be fulfilled, ii) search alternative materials, and iii) evaluate the alternatives in terms of needs and conditions.

2.2 Injection Mold Design

Once the material is determined, the mold designers will start the mold design by determining the number of cavity and the location of parting line. A molding machine will then be chosen based on some simple calculation. This is followed by the selection of dimensions and material of the mold base; design of the feed system which consists of a sprue, runner and gate; design of cooling system and ejection system.

3. Application of Expert Systems in Injection Molding

Researchers have started to adopt ES in solving the injection molding problems in recent years. For instance, several systems for problem diagnosis of injection molding, such as those of Shanghai Jiao Tong University [Qiang et.al. (1991)] and New Jersey Institute of Technology [Jan & O'Brien (1992)], etc., are being developed. Capturing injection molding part design features from CAD models, advising plastic material selection, automating the mold design process, developing design for manufacturability in mold design, etc. are popular research topics. GERES [Nielson (1986)], CIMP [Jong & Wang (1989)], HyperQ/Plastic [Beiter et.al. (1991)], the ES of Shanghai Jiao Tong University [Ying & Ruan (1991)], PLASSEX [Agrawal & Vasudevan (1993)] etc. were developed for selecting plastic materials based on part requirements. Most systems possess searching mechanisms and heuristic rules to assist designers in selecting a candidate material by both quantitative and qualitative evaluations. They were, however, developed in a standalone manner, not integrated into the part design, mold design or process planning, that unable to address issues of part design for moldability and mold design. Systems like IMPARD [Vaghul (1985)], ICAD [Clinquegrana (1990)], the ES of Drexel University [Tseng et.al. (1990)], the ES of University of Massachusetts at Lowell [Meckley et.al. (1992)], etc. were developed for injection mold design. They are, however, limited to simple parts and not mature enough to cover general mold design issues. More importantly, part design details, such as three-dimensional geometrical profile and dimensions, are compulsory inputs to these systems so they are not appropriate for the early product planning purpose. There is a potential need to develop an expert planning system for the injection mold in an integrative manner for early design decisions.

4. The Proposed ESPIMP-1 Framework

In real life, the mold designers are always required to determine the mold designs according to a rough part design or merely a mock up, i.e. in the absence of part design specification and other molding details, at the early design stage for evaluation of different design alternatives or for replying a customer quotation in a very short time interval. A framework to develop an expert system called ESPIMP-1 (Expert Injection Mold Planning System) has been proposed to help the design engineers in selecting the plastic materials and determining the major injection mold design features in such an environment. The ESPIMP-1 mainly consists of two modules; namely, the Expert Plastic Material Selection Module (ESMATL) and the Expert Mold Design Module (ESMOLD). The overall structure of the ESPIMP-1 is shown in figure 1. The inputs to the system are the product requirements initiated by the external customers or internal development. If the part material is not yet specified in the product conception stage, the ESMATL will help making the material selection decision based on the product requirements. Once the material is selected, the material properties together with the product requirements become the inputs to the ESMOLD to determine the design of injection mold. The outputs of the ESMOLD are the features and characteristics of the injection mold that on one hand are important for proceeding to detailed mold design work and on the other hand can be extended to determine the mold cost estimation, mold making process planning, production molding planning, etc..

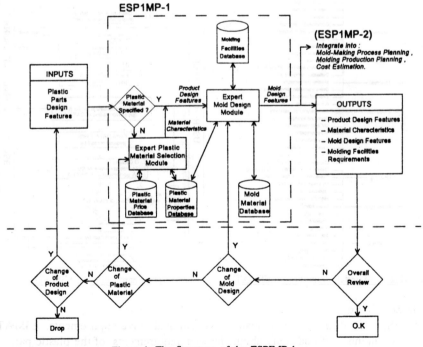

Figure 1 The Structure of the ESPIMP-1

5. The Plastic Material Selection Module (ESMATL)

5.1 General Structure

The selection module aims to assist the designers in selecting a set of thermoplastic materials those comply with the product requirements, including both quantitative and qualitative requirements. It is designed for product designers with little experience in plastics. The decision making logic follows the usual process used by the designers in material search and selection. Designers usually prioritize the material properties according to product requirements and then choose the most important property for the first iteration of search. In the searching process, those materials whose properties do not conform to the requirement will be eliminated. Another property will be used for next iteration until all properties match or there are no more material for search.

The functions of the ESMATL include (a) searching individual plastic material property when a specific plastic material is selected by users, (b) selecting the appropriate materials by inputs of quantitative property value or inputs of qualitative property requirements through interactive questions, and (c) ranking the alternative materials. The existing properties included in this system are divided into eight categories, namely; physical, mechanical, electrical, thermal, environmental, molding, assembly and finishing. The last three categories are concerned with the manufacturability of the molding part. Each category consists of several properties. There are totally 42 properties, as listed in figure 2.

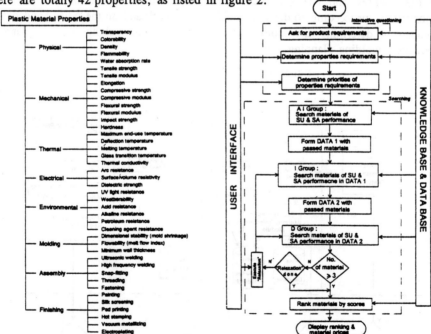

Figure 2 Plastic Material Properties in ESMATL   Figure 3 The Search Flow in ESMATL

5.2 The Search Logic

The ESMATL deals with both quantitative and qualitative requirements. ESMATL firstly asks a number of questions about the user requirements of the plastic part. If the input is a quantitative value, it is simple to search alternative materials by comparing the required value with the property databases of individual material in the system. In addition to storing up all the quantitative value of individual property of each material, the system also builds in a comparative qualitative measure of each property by classifying their performance in four grades, namely; SU (superior), SA (satisfactory), M (marginal) and US (unsatisfactory). The users are also required to prioritize the property requirements of target material into three grades, namely; AI (absolutely important), I (important) and D (desirable). The system will then begin to search appropriate materials according to the information gathered. The searching sequence follows the logic shown in figure 3.

In case there is no sufficient material selected in the first searching process or the selected materials are not in the favour of the users, the system can do "relaxation" to seek alternative materials. For quantitative requirements, the "relaxation" will not be done automatically. For qualitative requirements, if the number of materials selected is less than three, the system will do "relaxation" automatically until sufficient materials are found or the "relaxation" process is completed.

5.3 The Ranking Mechanism

The objective of the ranking is to prioritize alternative materials relative to the order of importance of their attributes to the designers. It combines multiple attributes into a single measure and ranks the candidate materials by this measure. Weighting scales, AI, I and D, which reflect the designer requirements are used to evaluate the relative importance among different material property requirements. The following quantitative scoring system is proposed for the ranking.
Priorities : AI = 3 points, I = 2 and D = 1
Performance ratings : SU = 3 points, SA = 2, M = 1, US = 0
Score = Summation of the products of the priorities and the performance ratings

6. The Expert Mold Design Module (ESMOLD)

6.1 The System Logic

The main objective of the ESMOLD is to enable the design engineers to create the major injection mold design features quickly according to the inputs of product requirements and plastic material characteristics. Figure 4 illustrates the flow chart of the system logic of ESMOLD. A series of questions are set to ask for the part design features. Based on the part features such as part size, major part shape and undercut requirements, etc., the system will automatically determine the mold structure, number of cavity, mold insert material, designs of gating, ejection and cooling systems. The logic basically follows the similar decision making approach being used by the professional mold designers in the industry. The user has the options to integrate the execution of ESMOLD with ESMATL or just to run the ESMOLD alone. In the integration of ESMATL, the properties information of the selected plastic material will be passed from ESMATL to ESMOLD through the heritance property of expert system in making the mold design decisions.

6.2 The Knowledge Base

The ESMOLD uses the feature approach to bridge the information flow between product design and mold design. There are four feature groups in ESMOLD to describe the product requirements of plastic part, namely; geometric features, outer shape, appearance and plastic materials.

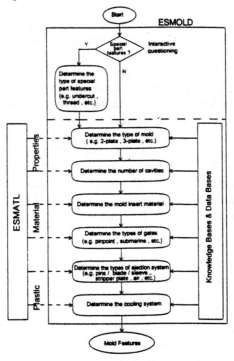

Figure 4 The System Flow of ESMOLD

Table 1 shows the items in the geometric features and outer shape groups. The features of injection mold are categorized into six major groups including mold construction type, special mechanism, mold material, feed system, cooling system and ejection system. Table 2 show part of the standard mold design features in ESMOLD. Having regard to different product features and mold designs, decision rules are developed for determining the mold construction, special functional mechanism, choice of mold material, runner and gate design, ejection and cooling systems design, etc.. Table 3 shows examples of these rules.

Geometric Features :
G1 = Part without special features
G2 = Light non-cylindrical internal undercuts
G3 = Light cylindrical internal undercuts
G4 = Permissible depth of non-cylindrical internal undercuts
G5 = Permissible depth of cylindrical internal undercuts
G6 = Light external undercuts
G7 = Permissible external undercuts
G8 = Light internal threads
G9 = Permissible internal threads
G10 = Light external threads
G11 = Permissible external threads
G12 = Part size (length / width / height)

Outer Shape :
F1 = Cylindrical shape
F2 = Semi-circle
F3 = Rectangular block / cubic
F4 = Ellipse
F5 = Pyramid
F6 = Cone
F7 = Irregular shape

Table 1 Plastic Part Features

Mold Type :
X1 = Two-plate mold
X2 = Three-plate mold
X3 = Runnerless mold

Special Mechanism :
Y1 = No special function
Y2 = Sliding split action by angle pin / cam
Y3 = Split core for cylindrical internal undercut
Y4 = Split core for non-cylindrical internal undercut
Y5 = Unscrewing (rack and pinion)
Y6 = Unscrewing (electrical or hydraulic motor)

Feeding Systems :
Z1 = Side gate/ sprue gate/ disk gate/ ring gate
Z2 = Submarine gate
Z3 = Pin-point gate
Z4 = Insulated runner
Z5 = Hot runner (heater)

Mold Material :
M1 = Tool steel
M2 = Alloy steel
M3 = Stainless steel
M4 = Medium carbon steel

Ejection Features :
E1 = Ejector pin / blade
E2 = Ejecting sleeve
E3 = Stripper plate
E4 = Air-assisted ejection

Cooling Systems :
C1 = Cooling the shallow core / cavity inserts
C2 = Cooling the deep core / cavity inserts

Table 2 Standard Injection Mold Design Features

IF (Plastic Part Features & Other Conditions)	THEN (Mold Structures)
1. Common plastic part without any specific feature; OR 2. (Light cylindrical internal undercut OR Light non-cylindrical internal undercut OR External undercut OR Internal thread's depth OR External thread's depth < 2% of Diameter / Width / Length) AND the plastic material is HDPE OR PP; OR 3. (2% < Light cylindrical internal undercut OR Light non-cylindrical internal undercut OR External undercut OR Internal thread's depth OR External thread depth < 3% of Diameter / Width / Length) AND the plastic material is HDPE.	Y1 = No Special Mold Features Required
1. Light external undercut < 2% of length/width, but the plastic material is not HDPE or PP; or 2. 2% < Light external undercut < 3% of length/width, but the plastic material is not HDPE; or 3. External undercut > 3% of length/width	Y2 = Sliding splits action by angle pin/cam

Table 3 Excerpted Rules for Part Features & Mold Types

7. Conclusion and future work

This paper described a proposed framework to develop an expert system for injection molding part development at early design stage. The system helps design engineers selecting thermoplastic materials according to the product requirements and determining the mold design features. During the material selection and mold design processes, feedbacks can be conveniently given to the original conceptual product design for modification. It thus realizes the concurrent engineering philosophy at the early design stage and can minimize the early design errors so as to reduce subsequent costly reworks. The authors' current research is the development of a prototype expert system based on the proposed ESPIMP-1 framework. The future direction of this research is of two folds, namely; further enhancement of the ESPIMP-1 and extending the ESPIMP-1 to the mold making process planning, mold cost estimate and production molding planning.

8. References

Agrawal D. & Vasudevan P.T. (1993), PLASSEX : An Expert System for Plastic Selection, Advances in Polymer Technology, Vol. 12, No. 4, 1993, pp 419-428.

Beitet K., Krizan S. & Ishii K. (1991), HyperQ/Plastics : An Expert System for Plastic Material and Process Selection, Computers in Engineering, ASME, Vol. 1, 1991, pp 71-76.

Clinquegrana, D.A. (1990), Knowledge-based Injection Mold Design Automation, DEng thesis, University of Lowell.

Jan T.C. & O'Brien K.T. (1992), Reducing Surface Defects in the Injection Molding of Thermoplastics Using a Hybrid Expert System, Proceedings of ANTEC'92, SPE, 1992.

Jong W.R. & Wang K.K. (1989), An Intelligent System for Resin Selection, SPE, Proceedings of ANTEC'89, pp 367-370.

Meckley J.A., McCarthy S.P. & Cinquegrana (1992), A Knowledge-Based Computer Optimization for Three Plate and Hot Runner Mold Designs, SPE, Proceedings of ANTEC'92, pp 2342-2343.

Nielson E.H., Dixon J.R. & Simmons M.K. (1986), GERES : A Knowledge Based Material Selection Program for Injection Molded Resins, Computers in Engineering, ASME, Vol.1, 1986, pp.255-261.

Qiang H., Ying S. & Ruan X. (1991), Knowledge-Based System for Problem Diagnosis of Injection Moulding, Proceedings of the 11th International Conference on Production Research (ICPR), 1991.

Tseng A.A., Kaplan J.D., Arinze O.B. & Zhao T.J. (1990), Knowledge-Based Mold Design for injection Molding Processing, Proceedings of the 5th IEEE International Symposium on Intelligent Control, 1990, pp 1199-1204.

Vaghul M.D., Zinsmeister G.E. & Simmons M.K. (1985), Expert Systems in CAD Environment, Computers in Engineering, ASME, Vol.2, 1985, pp.77-82.

Ying S. & Ruan X. (1991), Application of Expert System into Injection Polymer Selection, Proceedings of the 11th International Conference on Production Research (ICPR), 1991, pp 1855-1858.

CAD/CAM & CAE

CAD/CAM & CAE

CAD/CAM Technology

A Computational Tolerancing Scheme for CAD/CAM
   K. M. Yu

Application of an Environment for International Collaborative CAD/CAM
   G. C. I. Lin, Y. C. Kao, H. C. Liaw, R. S. Lee, L. S. Chen
   and D. Y. Zhang

3D Graphic Collision Control as a Function of CAD/CAM Systems
   Z. J. Bao

A Computational Tolerancing Scheme for CAD/CAM

YU, Kai-Ming
Department of Manufacturing Engineering, Hong Kong Polytechnic University, Hong Kong

ABSTRACT

The paper proposed a mathematical scheme to represent various tolerance types being used in traditional engineering practice. The scheme is suitable for computer implementation in CAD/CAM system. The paper first describes the state-of-the-art of contemporary tolerancing methodology. A mathematically precise definition for both dimensional and geometrical tolerances is then given. Discussion in relation to the tolerancing principles given in Engineering Standards are finally made.

1. Introduction

Currently, tolerancing and metrology is said to stand at a watershed towards a mathematical based know-how. On one hand, the contemporary state-of-the-art geometric modellers are lagging in providing a complete facility for dimensioning and tolerancing. On the other hand, Engineering Standards on dimensioning and tolerancing specify largely the syntax of the graphical annotation symbology. The semantics of the various types of dimensions and tolerances are scattered in the Standards in an inconsistent manner. In order to make the full benefit of the current geometric modelling and Coordinate Measuring Machines capabilities, tolerancing theories up to the point of mathematical rigour is necessary and essential to future technological advancement.

In Engineering Standards [1], **dimension** is defined as a numerical value expressed in appropriate units of measure and indicated on a drawing and in other documents along with lines, symbols, and notes to define the size or geometric characteristic, or both, of a part or part feature. **Dimensional tolerance** (also called limit or parametric tolerance) is defined to be the total amount by which specific dimension is permitted to vary. **Geometrical tolerance** is the general term applied to the category of tolerances used to control form, profile, orientation, location, and run-out. This is introduced to overcome the two main drawbacks of dimensional tolerance: the lack of means for dealing with imperfect form, and ambiguous references.

In addition, three principles in Engineering Standards are used to generalize some characteristic properties. They are the Principle of Independency, the Principle of Envelope [2] and the Maximum Material Principle [3]. The **Principle of Independency** states that dimensional tolerances and geometrical tolerances are independent of each other. However, the Standards do not say whether the independency principle can also be applied to the different types of geometrical tolerance. For example, should tolerances in form and orientation of an axis be interpreted separately? A positive answer would enable any proposed theory on tolerances to differentiate the two types of geometrical tolerances succinctly. The remaining two principles are applicable when mutual dependency of size and other geometrical properties (i.e. form and location) would not cause serious discrepancy (to save efforts in too detail a tolerance specification) or should be considered together (as in fitting of a hole and a shaft). The **Envelope Principle** requires that "the envelope of perfect form at maximum material size of the feature shall not be violated." The **Maximum Material Principle** is employed in the assembly of features of size (e.g.

holes and shafts) such that the geometrical tolerance may be increased (lower manufacturing cost) if the size has not reached its maximum. This is argued from the fact that the clearance between mating parts are due to deviations in both sizes and geometries. The difference between the envelope and independence principles is that the former imposes the constraint of perfect form at maximum material condition.

Whether or not the four dimensional tolerance types and the fourteen geometrical tolerance types [1] are necessary and sufficient to model all kinds of manufacturing tolerances is unknown. The Engineering Standards only list out the practicing guidelines, which is empirical in nature, without elaborations on the theoretical foundation. Though this is acceptable to human interpretation, deeper understanding and more precise definitions are required for computer implementation.

A number of researchers have also proposed various methods to implement dimensions and tolerances into CAD/CAM systems. For tolerances, Requicha [4,5,6] proposed normal offset for modelling the tolerance zone. The offset zones are constructed by Minkowski operators. The implementation was done by introducing new nodes for features and datums into a boundary representation called variational graph. The theory did not support dimensional tolerance nor all types of geometrical tolerance. Srinivasan [7] uses general sweeps to define the tolerance semantics. Virtual boundaries are introduced to bound three types of tolerance zones. However, the scheme is limited to size tolerance and roundness. Turner and Wozny [8] have proposed a mathematical interpretation of tolerance. Instances of in-tolerance parts are associated with points in a normed vector space over the real numbers. Distance metric can then be used to establish a correspondence between the tolerance specification and the in-tolerance vector space region.

In this paper, the approach developed by Yu, Tan and Yuen [9,10] is explained with emphasis on the tolerance aspect.

2. Geometrical Meaning of Tolerance

Unlike dimension, tolerance is not a piece of geometric information. It is, in fact, closely related to engineering applications, say in tolerance value allocation and assembly analysis.

Tolerance in engineering usage is also not to represent non-ideal geometry. Rather, it is a stipulation on the acceptance of non-ideality. Non-ideal geometry is difficult, if not impossible, to model because of its unlimited possibilities and time dependent variations in the working environment. On the other hand, the acceptance of non-ideality is measurable and may be modelled. For instance, the acceptance may be modelled as a tolerance zone with boundary made up of ideal geometric entities.

Thus, tolerance is considered as the allowable (or acceptable) variation on a nominal geometry such that the class of toleranced geometry is functionally equivalent. Variational class is represented by a tolerance specification. The allowable variation of tolerance is also expressed quantitatively. Instead of investigating the criteria for functionally equivalent variation, which is a tolerance analysis problem (e.g. assembly fittings and statistical quality control), the discussion here will concentrate on how variations are being applied to nominal geometry.

2.1. Variation in Geometrical Construction

It can be shown that a solid may be represented as a geometrical construction graph with nodes consisting of geometric entities and scalars [10]. Thus, it is natural to attach variations (tolerances) directly to the nodes (i.e. geometrical entities and scalars). Since geometrical entities are inter-linked in a geometrically meaningful way, their variations would also inherit the same kind of inter-links.

It was also shown that the nominal geometries of geometrical entities may be represented by vector equations. To formulate deviations of the geometric entities from ideality, one may apply differential calculus (a tool to examine the effect on the function due to changes in its parameters) on the parameters of the equations. For example, for a variable point in space, which is a function of a set of characteristic point(s), direction(s) and scalar(s), i.e. $\mathbf{P} = \mathbf{P}(\mathbf{P}_0, \mathbf{w}_0, s)$, differentiating the equation gives

$$d\mathbf{P} = \frac{\partial \mathbf{P}}{\partial \mathbf{P}_0} d\mathbf{P}_0 + \frac{\partial \mathbf{P}}{\partial \mathbf{w}_0} d\mathbf{w}_0 + \frac{\partial \mathbf{P}}{\partial s} ds = d\mathbf{P}(\mathbf{P}, d\mathbf{P}_0, d\mathbf{w}_0, ds).$$

Using the difference operator, Δ, to denote variation of the variables, the equation becomes $\Delta \mathbf{P} = \Delta \mathbf{P}(\mathbf{P}, \Delta \mathbf{P}_0, \Delta \mathbf{w}_0, \Delta s)$. Similarly, $\Delta \mathbf{w} = \Delta \mathbf{w}(\mathbf{w}, \Delta \mathbf{P}_0, \Delta \mathbf{w}_0, \Delta s)$ for $\mathbf{w} = \mathbf{w}(\mathbf{P}_0, \mathbf{w}_0, s)$.

In the $\Delta \mathbf{P}$ function, the $\mathbf{P}$ term is used to group those $\mathbf{P}_0$, $\mathbf{w}_0$ and s terms remain unvaried. This is similarly interpreted for $\mathbf{w}$ in $\Delta \mathbf{w}$.

For instance, the variational vector equations of the toroidal surface is:
$[(\mathbf{P}-\mathbf{P}_0)\cdot(\mathbf{P}-\mathbf{P}_0)-R^2-r^2][(\Delta\mathbf{P}-\Delta\mathbf{P}_0)\cdot(\mathbf{P}-\mathbf{P}_0)-R\Delta R-r\Delta r]$
$= 2R\Delta R\{r^2-[(\mathbf{P}-\mathbf{P}_0)\cdot\mathbf{w}_0]^2\}+2R^2\{r\Delta r-(\mathbf{P}-\mathbf{P}_0)\cdot\mathbf{w}_0[(\Delta\mathbf{P}-\Delta\mathbf{P}_0)\cdot\mathbf{w}_0+(\mathbf{P}-\mathbf{P}_0)\cdot\Delta\mathbf{w}_0]\}$

Note that it is $\mathbf{P}+\Delta\mathbf{P}$ rather than $\Delta\mathbf{P}$ that defines the position of the variational point of the geometrical entity with respect to the global coordinate system. In other words, $\mathbf{P}+\Delta\mathbf{P}$ defines the boundary of variational surface.

The representation of tolerance as variation of geometrical construction graph would more truly reflect the nature of tolerance than a simple mechanical analogy of non-rigid mechanism by Hillyard [11]. The method is also superior than applying offsets to boundary entities (mainly surfaces) since variations (tolerances) may also be applied to construction entities, such as centres and axes. Unfortunately, applying variations to nodes of the geometric construction graph (e.g. surfaces) in addition to the leaves (e.g. intersection curves), may result in non-incidence of the geometries involved. Nevertheless, a geometry with small deviations would be functionally equivalent to the nominal one if all variations (tolerances) are within their acceptable limits.

2.2. Tolerance as a Special Case of General Variation

Provided that they are differentiable, $\Delta\mathbf{P}_0$, $\Delta\mathbf{w}_0$ and Δs may be any complicated functions. However, for design purposes a complicated function is difficult to analyze.

Theoretically, a tolerance can be formulated from any combination of the following variations, i.e. $\Delta\mathbf{P}_0 = \mathbf{0}$ (zero vector) or $\Delta\mathbf{P}_0 \neq \mathbf{0}$, $\Delta\mathbf{w}_0 = \mathbf{0}$ or $\Delta\mathbf{w}_0 \neq \mathbf{0}$ and $\Delta s = 0$ or $\Delta s \neq 0$. Nevertheless, in Engineering Standards, the following two **simplifications** are common:

(1) only the scalar terms are allowed to vary,
(2) the variations are restricted to some analytical simple shapes.

Tolerance is normally used to model small deviations (typically around 0.1 to 0.01 times the order of magnitude of the dimension value) from nominal values. (The tolerance values may come from $\|\Delta\mathbf{P}_0\|$, $\|\Delta\mathbf{w}_0\|$ or $|\Delta s|$.) Tolerance variation is also normally restricted to a zone for inspection purposes. Mathematically, the tolerance zone may be interpreted as a point set. Strictly speaking, $\Delta \mathbf{P}_0$, $\Delta \mathbf{w}_0$ and Δs are not point sets. To circumvent this, one needs to define $\Delta \mathbf{P}_0$, $\Delta \mathbf{w}_0$ and Δs in accompany with certain characteristic points and directions (e.g. $\mathbf{P}_0 + \Delta \mathbf{P}_0$). This enables $\Delta \mathbf{P}_0$, $\Delta \mathbf{w}_0$ and Δs to have physical locations in the Euclidean space.

In addition, the three variational terms have different metric measures. $\Delta \mathbf{P}$ which is defined as $\mathbf{P'} - \mathbf{P}$ ($\mathbf{P'}$ is the variational counterpart of $\mathbf{P}$ and $\mathbf{P} + \Delta \mathbf{P} = \mathbf{P'}$) has the unit of length while Δs has the unit of length or angle measure depending on the dimension types. On the other hand, $\Delta \mathbf{w}$ which is defined as $\mathbf{w'} - \mathbf{w}$ ($\mathbf{w'}$ is the variational counterpart of $\mathbf{w}$) is a direction and does not have a well-defined unit of measure.

2.2.1. Geometrical Meaning of Dimensional Tolerance

Simplification (1) stated in the last section implies that $\Delta \mathbf{P}_0 = 0$, $\Delta \mathbf{w}_0 = 0$ $\Delta s \neq 0$, resulting in only dimensional tolerance. Since dimension is defined as a vector, this also implies that only its magnitude would be varied. Thus, the direction of the dimensional tolerance would be the same as the dimension. In general, a dimensional tolerance zone may be one, two or three dimensional depending on the geometric entity under consideration and the number of dimensions involved (i.e. scalar-direction pair) in the defining vector equation.

In general, three types of dimensional tolerances are common and may be represented as:
1. Linear dimensional tolerance = $\Delta d \mathbf{w}$,
2. Angular dimensional tolerance = $\Delta \theta \mathbf{w}$,
3. Radial dimensional tolerance = $\Delta R \mathbf{n}$.

The value of Δd, $\Delta \theta$ and ΔR may be positive, negative or zero. Also, dimensional tolerances are usually stated in pair to indicate the maximum and minimum limits for the dimensions concerned. The different combinations of the tolerance values would result in the familiar expressions of tolerance as unilateral, symmetric bilateral, asymmetric bilateral or limits of size.

2.2.2. Geometrical Meaning of Geometrical Tolerance

Geometrical tolerances may be represented as any deviation in the characteristic $\mathbf{P}_0$ and $\mathbf{w}_0$ of the geometrical entities, i.e. $\Delta \mathbf{P}_0$ and $\Delta \mathbf{w}_0$. In practice, simplification (2) is involved.

In order to distinguish between the different types of geometrical tolerances, a simple method would be to control the values of the $\Delta \mathbf{P}_0$ and $\Delta \mathbf{w}_0$ terms. Table 1 shows the different types of geometrical tolerances: form tolerance allows variation in form, for example, a (real) straight line needs not be straight. In orientation or location tolerance, the form should not vary, that is, a straight line remains straight but $\mathbf{w}_0$ and $\mathbf{P}_0$ may vary, thus, $\Delta \mathbf{P}$ = constant (i.e. vary by the same amount) for all points of the geometrical entity. No such restriction exists for $\Delta \mathbf{P}$ in form tolerance. Similar tolerance zone formulations apply to run-out tolerance. The difference is, however, in the inspection stage where different inspection instrument and steps are used. For instance, a dial gauge reading (when the component is turned in one revolution) may be taken to be the run-out tolerance. Strictly speaking, run-out tolerances do not control any single geometric characteristic, they are in fact a composite of other geometric tolerance types.

The tolerance zone formulations depend on the type of geometrical tolerances and the geometrical entities being varied. In addition, a geometrical entity may have limited types of geometrical tolerance (for example, flatness may only be associated with a plane).

Table 1. Relationship between geometrical tolerances, ΔP_0 and Δw_0.

geometrical tolerance	ΔP_0	Δw_0	notes
form	$\Delta P_0 \neq 0$	$\Delta w_0 \neq 0$	form will vary, $\Delta P_0, \Delta w_0 \neq$ constant when non-null $\forall P \in$ geometric entity
orientation	$\Delta P_0 = 0$ preferred	$\Delta w_0 \neq 0$	form will not vary, $\Delta P_0, \Delta w_0 =$ constant $\forall P \in$ geometric entity
location	$\Delta P_0 \neq 0$	$\Delta w_0 = 0$	
run-out	$\Delta P_0 \neq 0$	$\Delta w_0 \neq 0$	form will vary (composite), $\Delta P_0, \Delta w_0 \neq$ constant $\forall P \in$ geometric entity

NB: 1. $\Delta P_0 \neq 0 \not\Rightarrow \Delta w_0 \neq 0$
2. $\Delta w_0 \neq 0 \Rightarrow \Delta P_0 \neq 0$ not preferred
$\Delta w_0 \neq 0 \not\Rightarrow \Delta P_0 \neq 0$ preferred

2.2.3. Single and Related Features in Geometrical Tolerances

Engineering Standards have classified geometrical tolerances into two groups: those that apply to single (or individual) toleranced feature and those that relate the toleranced feature to some datums (Table 2). In other words, the geometrical entities of the former group (e.g. form tolerance) do not need any reference datum in specifying the geometrical tolerance (e.g. roundness). The vectorial formulation discussed so far may be applied to related feature geometrical tolerance: orientation and location tolerances are applicable to those geometrical entities constructed by copying or sharing other entities' characteristic P_0 and w_0 while run-out tolerances relate the toleranced feature to the rotational axis during inspection. Clarification is, however, required for geometrical tolerances of form since the toleranced entity may not be related to other entities.

Form is an **intrinsic** property, which may be described without reference to the surrounding space (i.e. datum). The geometrical construction method is able to determine the appropriateness of form as an intrinsic property. For example, if a plane is constructed by the three points form, the flatness of the plane is an intrinsic property and the geometrical tolerance for flatness may be considered as a single-feature geometrical tolerance. For other construction methods, like point-normal form and translational copy (where external datum is referenced), it is better to consider flatness of the plane as related-feature geometrical tolerance. Similar consideration may be applied to other single-feature geometrical tolerances, for example cylindricity and roundness. In fact, in defining geometrical entities, their characteristic P_0 and w_0 which are directly involved in the construction, are considered as the local datum references. In such cases, the form tolerances should be referenced to these characteristic P_0 and w_0.

Whether or not a tolerance zone may be specified without external reference to a coordinate system or other geometric entities depends on the tolerance types. Obviously, an external reference is required for orientation, location and run-out geometrical tolerances. For form tolerance, the only possible condition for having no external reference is when the tolerance zone is expressible by a normal offset. For other cases (e.g. parallelpiped tolerance zone for straightness of axis), external datum references are required.

Another consideration which has important practical consideration is whether or not form tolerance may be inspected as an intrinsic property, that is, without the need for an inspection datum. No rules on this aspect were stipulated in Engineering Standards. For instance, some verification methods given in ISO5460 [12] require external inspection datum references while other do not. It all depends on the inspection methods used. For example, a plane is flat in all directions. However, when the plane is part of a solid (boundary or construction entity) and if for some manufacturing errors, it is made to deviate in orientation only, then it may no longer be considered as "flat" relative to the solid.

Table 2. Datum features in geometrical tolerances

features	geometrical tolerances	characteristics
single features	form	straightness, flatness, circularity, cylindricity
single or related		profile of any line, profile of any surface
related features	orientation	parallelism, perpendicularity, angularity
	location	position, concentricity, symmetry
	run-out	circular run-out, total run-out

2.2.4. Sufficiency and Dependency of Tolerance Types

This section will discuss whether the tolerance types listed in Engineering Standards are sufficient and inter-dependent to each other based on the variational geometry approach.

The **sufficiency** aspect may be easily answered. Geometrical constraints can be classified into metric constraints, geometric properties, and shapes. Thus, the possible geometrical characteristics to be controlled were already discussed. Metric constraints or sizes were controlled by dimensional tolerances. Shapes or forms were controlled by form geometrical tolerances. The geometric properties in Euclidean space are location and altitude which are controlled as geometrical tolerances with notations ΔP and Δw respectively.

The independency aspect, however, has no definite answer. It mainly depends on the particular situation under consideration. The **Independent Principle** states that dimensional and geometrical tolerances are to be interpreted independently (say in tolerance analysis). In practice, it is difficult for dimensional toleranced geometric entity to have variation in size only but perfect in form, location and orientation. The envelope principle and the maximum material principle may be used to allow for dependency between dimensional and geometrical tolerances. These principles are, however, originally designed to reduce the gauging cost and manufacturing cost respectively.

In Table 1, form, orientation and location geometrical tolerances are defined to be independent of one another. This is made possible because of the mathematical formulations. However, recommendations in Engineering Standards [2] state that in certain cases, a geometrical tolerance characteristic may be used to control other types (Table 3). Nevertheless, no clear guidelines are given on the inter-dependency of geometrical tolerance types. In fact, it may not be appropriate to treat positional tolerance as a composite tolerance to control straightness, roundness, flatness, parallelism, perpendicularity, angularity and coaxiality. Indeed, for computer implementation purposes, it is more straightforward to restrict each geometrical tolerance type to only one geometrical characteristic.

Table 3. Inter-relationships between different tolerance types.

tolerance		relationship with others
types	characteristic	
dimensional	linear, angular, radial	-may control geometrical tolerances
form	straightness, flatness, circularity, cylindricity, profile of a line, profile of a surface	-cylindricity is combined of straightness, circularity & parallelism (or diameter variation along the axis)
attitude	parallelism, perpendicularity, angularity	-may control form -parallelism may control straightness -squareness is a particular case of angularity
location	position, concentricity, symmetry	-can control form & angularity -concentricity & symmetry are particular cases of position tolerances
composite	circular run-out, total run-out	-circularity, concentricity, flatness or perpendicularity can contribute to run-out tolerances

NB: Inter-relationship between different geometrical tolerance characteristics are extracted from [13].

Conceptually, the different dimensional and geometrical tolerance types are functionally unique and are distinguishable (though they may not be inspectable) using the theory discussed. It is recommended that they should be explicitly and separately specified. For untoleranced size, form, location or orientation, less stringent default values are assumed (since no manufacturing process is perfect). If an explicit tolerance control is used, size, form, location or orientation will be explicitly specified. The unspecified tolerance types can then be assumed to have the default tolerance.

There are some views which expressed that geometrical tolerance is more preferred than dimensional tolerance. From the variational geometry theory, geometrical and dimensional tolerances are different and they have different origins: one from Δs while the other is from ΔP_0 and Δw_0. Though sometimes their effective tolerance zones are indistinguishable, there are also situations in which the difference is obvious. For instance, tolerance zones for angular tolerance and angularity are different. In fact, to omit dimensional tolerance would make the geometric characteristics to be incompletely controlled.

3. Conclusion

This paper explains a variational vector geometry approach to interpret tolerance. The semantics so defined is completed in the sense that it covers all types of dimensional and geometrical tolerances given in Engineering Standards. The scheme is geometrically natural as it utilizes the Euclidean concept of position, orientation and scalar sizing. Besides, it allows clear differentiation of the various geometrical types as needed in using the Independency Principle. The differentiation can also be relaxed if necessary in employing the Envelope Principle and the Maximum Material Principle as the effects of the variational vectors on position, orientation and scalar are "combinable" - a property of vectors. All these make the scheme superior to the offset and sweep approaches. In addition, the approach can also be used to represent dimension based parametric design. Thus, a unified scheme to represent large and small variation is possible in geometric modelling system.

Acknowledgements

The author wishes to thank the Mechanical Engineering Department, University of Hong Kong and the Manufacturing Engineering Department, Hong Kong Polytechnic University for providing the support in preparing this work.

References

1. The American Society of Mechanical Engineers, ANSI Y14.5M-1982 *Dimensioning and tolerancing*.
2. International Organization for Standardization, ISO 8015-1985(E) *Technical drawings - Fundamental tolerancing principle*.
3. International Organization for Standardization, ISO 2692-1988(E) *Technical drawings - Geometrical tolerancing - Maximum material principle*.
4. A. A. G. Requicha and S. C. Chan, "Representation of geometric features, tolerances, and attributes in solid modelers based on constructive geometry", *IEEE Journal of Robotics & Automation*, Vol.RA-2, N3, Sep. 1986, pp.156-166.
5. A. A. G. Requicha, "Toward a theory of geometric tolerancing", *International Journal of Robotics Research*, v2, n4, pp.45-60, Winter 1983.
6. A. A. G. Requicha, "Representation of tolerances in solid modelling: issues and alternative approaches" in *Solid Modeling by Computers: from theory to applications* edited by J. W. Boyse and M. S. Pickett, Plenum, 1984, pp.3-22.
7. V. Srinivasan, "The role of sweeps in tolerancing semantics", *Manufacturing Review*, v6 n4 Dec 1993 pp.275-281.
8. J. U. Turner and M. J. Wozny, "A mathematical theory of tolerances" in *Geometric Modeling for CAD Applications* edited by M. J. Wozny, H. W. McLaughlin and J. L. Encarnacao, Elsevier 1988 pp.163-187.
9. K. M. Yu, S. T. Tan and M. M. F. Yuen, "An alternative approach to interpreting and representing dimensions and tolerances" in *CAPE'90* edited by J. A. McGeough, London November 1990 pp.40-58.
10. K. M. Yu, *Dimensioning and tolerancing in geometric modelling*, PhD thesis, Dec. 1990, University of Hong Kong.
11. R. C. Hillyard, *Dimensions and Tolerances in Shape Design*, Ph.D. Thesis, May 1978, University of Cambridge.
12. International Organization for Standardization, ISO 5460-1985(E) *Technical drawings -Geometrical tolerancing - Tolerancing of form, orientation, location and run-out - Verification principles and methods -Guidelines*.
13. British Standards Institution, BS308 Part 2 : *Recommendations for dimensioning and tolerancing of size*, 1992.

Application of an Environment for International Collaborative CAD/CAM

Grier C.I. Lin
Yung-Chou Kao
Hon-Chy Liaw
Centre for Advanced Manufacturing Research, School of Manufacturing and Mechanical Engineering University of South Australia, Pooraka SA5095, Australia

R.S. Lee
L.S. Chen
National Cheng Kung University, Tainan, Taiwan

Dayong Zhang
Qikdraw Systems Pty. Ltd., Adelaide, South Australia, Australia

ABSTRACT

The concept of computer supported collaborative work (CSCW) is achievable. Since the 1980s, the need in developing the CSCW environment has been recognised by various researchers, such as behavioural scientists, computer scientists, application developers and users. However, most of them focused on the technical writing, text editing, and electronic mail message transmission, ie two dimensional graphics and text. To enable the CSCW to accommodate 3D CAD/CAM, the authors have explored an international collaborative environment that links geographically distributed locations: Taiwan and Australia. Two examples have been used to test the proposed environment internationally and domestically. UNIX, LAN and Internet are adopted in this international collaborative integration.

Keywords: International Collaboration, CSCW

1. Introduction

In the past two decades, CAD/CAM technology has been successfully diffused into the manufacturing industry resulting in significant improvement in productivity and competitiveness. However, the existing CAD/CAM application is based on the computer environment developed decades ago. The development of CAD/CAM is based on CAD. CAD can be traced back to the Automatic Programmed Tools (APT) project started in the 1950s at the Massachusetts Institute of Technology (MIT). (Rembold, U., et al 1993) APT is a fixed text format created to represent the workpiece geometric shape for numerically controlled machines in high-precision operations. APT is not designed for interactive operation in nature. Although the early 1970s saw the commercialised CAD/CAM package, nearly at the same time (1969) that the United States Department of Defence (DoD) launched the experimental ARPANET (Davison 1988), the CAD/CAM technology was still designed as a single user application tool. The CAD/CAM user can only communicate with the computer's CPU, as indicated in Figure 1. Therefore, the CAD/CAM technology is restricted to single location application! Even though the CPU can be a timeshared mainframe system, the users can not communicate among themselves. With LAN becoming more popular, CAD/CAM systems started to be implemented on network servers, which allow the user to run from any station linked to the server. However, the

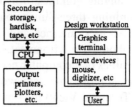

Figure 1. Typical CAD/CAM system

CAD/CAM user (designer) can only 'talk' with the computer. Users can not 'talk' with one another. The interaction between users is still the same as before.

The advancement in the computer and communications technologies have made computer assisted work potentially the most convenient and the most powerful tool for geographically distributed group of people who requires communication (Hiltz and Turoff 1978). The prevailing trends of increasing business internationalisation and international competition have been well recognised. The role of product development is becoming more important than ever before. Not only do they have to meet the consumers' needs at a lower cost and price-performance, higher quality and better after sale services, but also be able to cope with competition on a global basis. Therefore, design methodologies, market information, consumers' requirements, current technologies, and design facilities, etc, all need to be thoroughly considered.

Groover (1987) points out that the trends in the future automated manufacturing factory includes shorter product life cycles, increased emphasis on quality and reliability, more customised products, new materials, growing use of electronics, pressure to reduce inventories, outsourcing, just-in-time production, point-of-use manufacture, and greater use of computer in manufacturing. Pawar (1994) also mentions that 80 per cent of a product's cost is committed during the design phase. With companies having multi-location operations, a need has arisen for a collaborative CAD/CAM computer system. Such a system must provide a team work environment. The environment should enable remote designers to work together and to communicate among themselves on a common design activity. (Lin 1993) This implies a design environment has to be created to make full use of all the related domain expertise, technologies and resources available in the world to ensure that the design meets market requirements. To achieve such a challenging task, many issues have to be considered, eg extending CAD/CAM functionality, featuring the function for ease of instantly exchanging the design and manufacture ideas among geographically dispersed designers. Therefore, at least five key disciplines: distributed systems, communications, human-computer interactions, artificial intelligence (AI), and social theory must be considered (Ellis, Gibbs, and Rein, 1991), to implement the current CAD/CAM technology to be a groupware CAD/CAM for collaborative design.

Unfortunately, such a groupware CAD/CAM environment is not available yet. Although in the mid 1980s saw a series of researches on the Computer Supported Cooperative Work (ACM 1986, 1988, 1990, 1992), most of these concentrated themselves primarily on creating the environment for co-authoring, documentation, and message passing. For the computer-aided collaboration, only some of them supported collaborative design. For example, VideoDraw (Tang and Minneman, 1991) for sharing two dimensional sketch drawings, TOPES (Pferd, Peralta, and Prendergast, 1979) for exchanging graphics, drawings and text, and Teledesign (Shu and Flowers, 1992) for examining three-dimensional CAD groupware interface issues. Abdel-Wahab, et al, (1988) uses internet and UNIX™ interprocess communications in implementing a collaborative environment. This environment enables a team of remote users on different machines of a computer network to collaborate synchronously using familiar single-user tools. A prototypic framework (Abdel-Wahab and Feit, 1991) of distributed system XTV (X Terminal View) for synchronously sharing X window applications among a team of remotely dispersed users at workstations running X window and interconnected by the internet was proposed enabling the single-user application to be shared. Gay, et al

(1993) proposed a concurrent CAD environment by using PC-NFS™ based communication between local area networked PC and Sun workstation. This system uses only message-based notification instead of real time communication.

To make better use of the modern computer and communications technology in achieving CAD/CAM collaboration, UNIX™, Local Area Network (LAN), and Internet are used. An international collaborative CAD/CAM environment for collaborative product development and real time modification has been implemented in this paper. A new product designed by collaborative CAD/CAM designers at one side of the world can be tested, analysed and manufactured at the other side of the world almost immediately. The potential benefits will be the elimination of unnecessary international travel time and associated expenditure, the availability of world wide expertise, the reduced lead time to market, the production of the right product at the right place, the elimination of product shipment time, the increased competitiveness and profit, etc.

In the following sections, firstly, the authors briefly present a configuration for the international CAD/CAM collaborative environment. Secondly, the procedures in implementing the collaboration are explained. Thirdly, two examples in demonstrating the collaborative strategies are described. Finally, the authors summarize the explored collaborative environment.

2. The system configuration

To establish a cooperated environment for the international CAD/CAM collaboration, a team project was launched in 1992. A system configuration has been established, as indicated in Figure 2 (Lin, kao, and Lee 1994). This system consists of the UNIX™, LAN, and Internet. Each side, Australia and Taiwan, can be considered as one group separately. Each group has three workstations in LAN. Both groups are linked by internet, and the communication can be text message passing and oral conversation. The 'talk' command in Sun is used as the tool for real time text conversation under the normal network load. Telephone is used for oral communication (Resnick 1993). The CAD software Advanced Solids Modeller (ASM), developed by Qikdraw Systems Pty Ltd, is installed in each workstation for the daily design tasks.

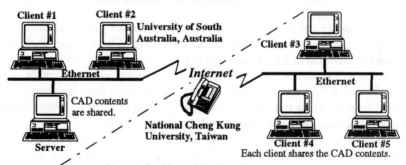

Figure 2. The Network Configuration

3. Realisation of the Network System and Collaborative Outlines

Computer, networking, and communication technologies are the enabling tools in establishing the environment for the international collaborative CAD/CAM. The data transmission media used in the system, as indicated in Figure 2, is based on the TCP/IP infrastructure. (Santifaller 1991) The TCP/IP is developed by the United States Department of Defence (DoD) for use in resource sharing computer networks. TCP stands for Transmission Control Protocol. IP refers to Internet Protocol. The two acronyms together describe not only the protocols, but they also refer to the set of services that collectively allow computer systems to exchange files, electronic mail and interactive character streams. The networking within this international collaborative CAD design system is to support the data communication (Davison, 1988).

3.1. Example 1: Engine block design

To verify the idea for the international collaborative CAD technology, several workshops have been conducted. These workshops link Australia and Taiwan. The procedures of such an international link are explained as the following.

A typical engine block, as shown in step 5 of Figure 3, that consists of four components, cylinder lining, piston, connecting rod, and crank shaft, is used for demonstrating the international CAD collaboration. The role of client #2 in example 1, as indicated in Figure 2, is assigned to assist client #1 forming a connecting rod design subgroup. Similarly, Client #3, Client #4, and Client #5 are subgroup for designing crank shaft. There are five steps for the collaboration after every participant is ready for the collaboration. The server who is also the coordinator rings client #3 to start the international CAD/CAM collaboration.

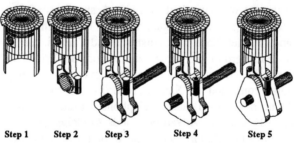

Figure 3. Collaborative design procedures of engine block

The server shows the design of assembled *cylinder* and *piston* on his own screen (Step 1 in Figure 3), as the collaboration starts and asks for the contribution of connecting rod from client #1. A message is passed to client #1 from the server either by 'talk' command in Sun workstation, telephone, or direct conversation (if the server and client #1 are at the same location). Responding the need from the server, Client #1 contributes his connecting rod and sends a message back to the server immediately after the connecting rod has been contributed to the common data base.

The server accesses the connecting rod from the common data base and attaches it onto the assembly, as indicated in Step 2 of Figure 3. Then the server asks client #3 overseas over the international telephone call to contribute the *crank* component. Client #3 contributes the *crank* into the common data base afterwards and informs the server. The server assembles the crank, as indicated in step 3 of Figure 3.

The first version of the engine block assembly consisting of the cylinder, piston, connecting rod and the crank, as indicated in Step 3 of Figure 3, can be seen by the visitors attending the workshop. However, client #1 apparently did not consider the bolt holes. Moreover, the crank shaft component that is contributed by client #3 is seemingly an old design. Therefore, the server needs to ask client #1 to modify the connecting rod, and also ask client #3 to contribute newer design alternative of crank shaft. It means that the engine block design has come to the stage of dynamic collaboration between the server and client #1, and between the server and client #3, separately.

The server, after sets the connecting rod to be silently Hot-Linked, notifies the client #1 to modify the design of connecting rod. Client #1 revises the connecting rod immediately on his own local ASM and contributes the modified connecting rod into the common data base again. With the Hot-Linked function, the server does not need to access the modified connecting rod, the system will automatically renew the connecting rod, as indicated in Step 4 of Figure 3.

Similarly, the server Hot-Linked the crank before he asks the client #3 to contribute alternative crank. Client #3 contributes a new crank into the common data base. The revision of the engine block is hence accomplished.

The on-line collaborative engine block design, the real-time connecting rod modification under negotiated agreement, and design alternatives of crank have been shown during the demonstration emulating the simplified collaborative design activities.

In this example, the authors have successfully networked geographically dispersed designers, but the data transmission speed through internet could not be controlled. Although the file sizes of the old and new cranks are 45155 bytes and 43121 bytes, respectively, the transmission time takes from 5 minutes to 16 minutes approximately. The file size of connecting rod is 125589 bytes that takes only a few seconds for the transmission. Therefore, more reliable transmission line is needed. Moreover, there are difficulties as the server asks the client #1 to make the bolt holes in the connecting rod, because the authors use only audio medium for the communication. Therefore, the media that allow the geographically dispersed designer to see the images, graphics, or CAD geometry are required. We solve some of these problems in example 2.

3.2. Example 2: Collaborative CAD

To avoid the unreliable internet, LAN is explored for the CAD collaboration. ShowMe™ and SharedApp™ are added to enrich the environment. ShowMe™ is a multimedia tool and has three functions, Audio, Whiteboard, and Video. To save the bandwidth of the network, as suggested by Chen, et al, (1992) that audio should be given higher priority over video, the authors use only Audio for voice communication. Whiteboard is used as graphics discussion panel among the conferees. SharedApp™ is a tool to share single user applications under X windows among networked computers, eg COMIX (Babadi, 1993). SharedApp™ is used to allow all the conferees to watch the collaborative progress.

The system configuration, as shown in Figure 4, consists of three designers: camr1, camr2, and camr3. Each designer accesses their own interactive workstation and runs ASM separately as daily design tool.

The collaborated procedures, as shown in Figure 5, show the proposed strategies for the collaboration. The directory DB1-1 is shared by camr1 and camr3 as the common database. Every designer, including the initiator (chairperson), runs their local ASM, namely LASM1, LASM2, and LASM3 separately. Camr2 and camr3 work on their own design until they are notified to join the CAD collaboration. The appointment that the initiator asks camr2 and camr3 for joining the collaboration is made by e-mail firstly, and 'talk' command is used to reach both of them. During the 'talk' session, the initiator asks camr2 and the camr3 to run ShowMe™ and SharedApp™ preparing for the collaboration. Later on, the initiator invites camr2 and camr3 into the ShowMe™ and SharedApp™ conference. The 'Audio' is used for the oral communication. The 'Whiteboard' is used as the graphics discussion panel. Whiteboard can snap the image on the display screen and attach the image into the Whiteboard window itself. The conferees can use their mouse pointer to 'mark' the region of snapped image for visual discussion. The colours of the mouse pointers are arranged to be red, for camr1, green for camr2, and blue for camr3 respectively for easy recognition.

All the collaboration concerned affairs are activated and initiated by the initiator, that is to say, camr1 is the chairperson in the CAD collaboration. SharedApp™ is used for 3D CAD discussion in compensating the insufficiency of 2D Whiteboard. To access the shared ASM (SASM), the designer must 'Take control' of the SASM. However, only the initiator can keep the control of SASM at any time. The screen layout of the devised system environment is shown in Figure 6. Only part of the engine block image is snapped onto the Whiteboard™ for graphics-rich discussion. The assembly window is available for further discussion and allows the real access for all the collaborators via SharedApp™.

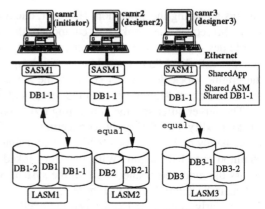

Figure 4. The System Configuration for CAD Conference

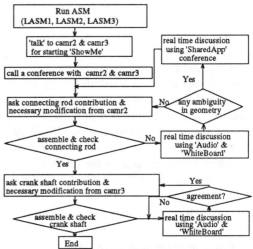

Figure 5. The collaborative procedures

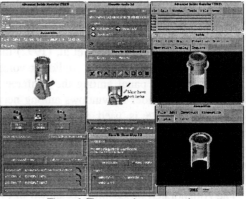

Figure 6. The screen layout example

4. Summary

Network-based communication is now a global phenomenon. The data transmission speed depends heavily on the traffic load of the internet that is being shared by millions of users all over the world (Ishida, et al, 1993). From the example 1, a more reliable network, is needed in the international collaboration, especially when the real time interaction is required. Oral communication medium is insufficient for the CAD geometry discussion.

LAN environment is explored to temporarily avoid the unreliable internet. Multimedia tool ShowMe™ Whiteboard is used for the graphic discussion panel, as shown in example 2. ShowMe™ Audio can be replaced by the conference telephone. SharedApp™ can be used for the three-dimensional computer aided design collaboration. This environment can also be applied to other software-based collaboration and served as the communication tools for education and training.

Most of the manufacturing industry type of developing country is Original Equipment Manufacture (OEM). OEM type manufacturing industry consists of small to medium sized work forces that can work semi-independently but geographically separated. These companies can become the precious outsources and the major supplier of special components for the multi-national enterprises. The computer-based network and communication system, eg the explored system in this paper, is the bridge. More efficient communication link, eg ISDN, is needed. ISDN implementation is the next step.

In this paper, the authors have constructed a collaborative design environment for CAD/CAM collaboration, and extended the existing single-location CAD/CAM to multi-location application. The CAD software ASM has been customized according to the requirements for international collaboration. The integration of a remote machining system and the constructed collaborative CAD environment is under development.

5. Acknowledgements

The authors appreciate the Australian Federal Department of Employment, Education and Training for supporting this project under its Targeted Institutional Link Program. Thank also Camtech Pty Ltd, Australia, for supporting the ShowMe™ and SharedApp™.

6. References

Abdel-Wahab, Hussien M., Guan, Sheng-Uei, and Nievergelt, Jay (1988) 'Shared Workspaces for Group Collaboration: An Experiment Using Internet and UNIX Interprocess Communications', IEEE Comm Magazine,Vol.26,No.11, pp10-16, Nov 1988

Abdel-Wahab, Hussien M., Feit, Mark A. (1991) 'XTV: A framework for sharing X window clients in remote synchronous collaboration', Proc. of TRICOMM '91, IEEE Conf. on Comm. Software: Communications for Distributed Applications and Systems, 18-19 April 1991, Chapel Hill, NC, USA

ACM, (1986), Proc. CSCW'86, Austin, Texas, USA.

ACM, (1988), Proc. CSCW'88, Portland, Oregon, USA.

ACM, (1990), Proc. CSCW'90, LosAngeles, California, USA.
ACM, (1992), Proc. CSCW'92, Toronto, Canada, USA.
Babadi, Aliasghar (1993) 'COMIX: A Tool to Share X Applications', Proc. of Second Workshop on Enabling Technologies Infrastructure for Collaborative Enterprises, pp. 192-6, April 1993
Chen, Mon-Song, Shae, Zon-Yin, Kandlur, Dilip D., Barzilai, Tsipora P., and Vin, Harrick M. (1992) 'A Multimedia Desktop Collaboration System', GLOBECOM '92, Comm. for Global Users, IEEE Global Telecommunications Conference, pp. 739-746
Davison, John M. (1988) 'An Introduction to TCP/IP', Spring-Verlag New York Inc., ISBN 0-387-96651-X, pp. 1-6.
Ellis, C.A., Gibbs, S.J., and Rein, G.L. (1991) 'Groupware - Some Issues and Experiences', Communications of the ACM, January 1991, Vol. 34, No. 1, pp. 38-58
Gay, Robert K.L., Seet, P.K., and Lee, B.S. (1993) 'Network-based Concurrent Design Environment for Distributed-Based CAD', Computing & Control Engineering Journal, December, 1993, pp. 253-260
Groover, Mikell P. (1987) 'Automation, Production Systems, and Computer Integrated Manufacturing', Chapter 27, Prentice-Hall Inc., ISBN 0-13-054610-0
Hiltz, Starr Roxanne, and Turoff, Murray (1978) 'The Network Nation - Human Communication via Computer', Addison-Wesley Pub. Co., Advanced Book Program.
Ishida, Haruhisa and Landweber, Lawrence H. (1993) 'Internetworking', Communications of the ACM, August 1993, Vol. 36, No. 8, pp. 28-30
Lin, Grier C.I. (1993) 'Management of Advanced Manufacturing Technology', Manufacturing Management Technology Seminar Lecture Notes, National Cheng Kung University, Tainan, Taiwan, July.
Lin, Grier, C.I., Kao, Y.C., and Lee R.S. (1994) 'On the Real Time International Collaboration of CAD/CAM', The Third International Conference on Automation Technology, July 1994, pp.281-284
Pawar, Kuwant S., Menon, Unny, and Riedel, Johann C.K.H. (1994) 'Time to Market', Integrated Manufacturing Systems, Vol. 5, No. 1, pp. 14-22
Pferd, W., Peralta, L.A., and Prendergast, F.X. (1979) 'Interactive Graphics Teleconferencing', IEEE Computer, Vol. 12, No. 11, November 1979, pp. 62-72
Qikdraw Systems Pty. Ltd. (1993) 'ASM Reference Manual'
Rembold, U., Nnaji, B.O., Storr, A. (1993) 'Computer Integrated Manufacturing and Engineering', Addison-Wesley Publishing Company Inc., Chapter 7.
Resnick, Paul (1993) 'Phone-Based CSCW: Tools and Trials', ACM Transactions on Information Systems, Vol. 11, No. 4, October 1993, Pages 401-424
Santifaller, M. (1991) 'TCP/IP and NFS Internetworking in a UNIX Environment', Addison-Wesley.
Shu, L. and Flowers, W. (1992) 'Groupware Experiences in Three-Dimensional Computer-Aided Design', in Proceeding of ACM 1992 Conference on Computer Supported Cooperative Work, October 31 to November 4 1992, Toronto, Canada, pp. 179-186
Tang, John C., and Minneman, Scott L. (1991) 'VideoDraw: A Video Interface for Collaborative Drawing', ACM Transactions on Information Systems, **Vol. 9**, No. 2, April 1991, Pages 170-184

3D Graphic Collision Control as a Function of CAD/CAM Systems

Zhuojun Bao

Institute for Computer Application in Planning and Design (RPK)
University of Karlsruhe
Kaiserstr. 12, D-76131 Karlsruhe, Germany
E-mail: bao@rpk.mach.uni-karlsruhe.de

ABSTRACT

A new method for the computer aided collision control between moving bodies is integrated as a new function in a CAD/CAM system. It starts from the computer interior representation model as a source of information. In solid modeling, the B-Rep or CSG trees of components can be converted into the recursive hierarchical approximation of octrees respectively. Thus the collision control of moved bodies can be reduced into determination of the intersection between motion cubes from Octrees. A discrete collision control is based on static min-max-tests by subdividing a given time interval into sufficiently small pieces. A dynamic and continuous collision control is realized by introduction of an effective space of a moving component for each time interval however. The effective space is determined by a novel method for modeling swept volume along the moving trajectory. Using the octrees leads to a quick elimination of non-relevant nodes. In case of a possible collision, the relevant no-black (grey) cubes can be tested recursively, until a given resolution or minimal cube is reached. Consequently, the collision zone can be recognized quick by the relevant nodes.

The proposed method is useful for the collision control in engineering design and planning of robotics, mechanics, manufacturing and assembling process.

Keywords: CAD/CAM, Solid Modelling and Simulation, Collision Recognition.

1. Introduction

Because of the demand on efficient integrated CAD/CAM systems such a system requires amongst the basic functions of solid modeling, kinematic modeling, NC programming and FEM an additional function for universally applicable collision control of kinematical mechanisms (Figure 1). All of CAx-systems or functions can be integrated on the basis of a product model, which is composed of different partial models. The product modeling system generates and describes produc model data, that means all relevant information of the different phases of the product life cycle, their interrelations and product dependent views. The formulation of an integrated product model offers many advantages. These are in particular the avoidance of data redundancy and the consistency after modification of the product [10]. A collision free motion is an important aim for the design and the planning of manufacturing processes and of robot motion. The temporal and spatial intersection or coincidence of two objects is understood by the term "collision".

In many industrial environments it is necessary to ascertain whether or not there is collision between components. In products made up of assemblies of components, and in product manufacturing and testing facilities, there are many potential collision problems

[4]. In current practice an attempt is made to control such unwanted collisions before the fact by using drafting methods. Engineering drawings of the product assembly or test facility are prepared to show the various components in a number of views and possible positions. If collisions are detected on these drawings, modifications are made and new drawings are prepared.

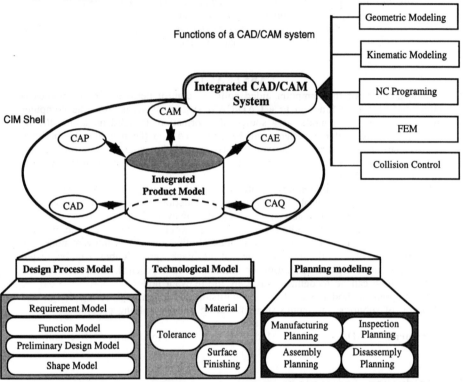

Figure 1:   Functions of an integrated CAD/CAM system

Unfortunately, the two-dimensional drafting medium does not always show collisions among three-dimensional objects, especially when these objects can move relative to one another. When drawings fail, the collision problems appear in the prototype stage or when facilities are set up for production. Correcting these problems at this point is expensive and time consuming. In order to avoid these unwanted collision problems, first, a true three-dimensional representation of the objects to be checked for collision is needed. Second, a method of using this representation to tell when and where such collisions occur has to be developed.

Most mechanical components are characterized by the relationships among the fairly simple surface types. But today products become more and more complex so that the complexity of their surfaces has to be regarded. For computer aided collision control among solids, a representation is needed that takes advantage of the simple nature of the analytic surfaces and, in addition, can represent the complex components with the sculptured (free form) surfaces. Modeling a swept volume is important in simulating the collision between a moving solid and its environment. In this article an effective space is introduced and determined by a novel method for modeling swept volume. Based on this approach, a tool has been developed for collision control of moving solids using computer graphics.

2. Concept of a new graphic collision control

The main idea of the method is that a complex collision problem of solid bodies will be reduced to more simple and solvable problems. The overall solution is represented by adding partial results. There are hierarchical approximation models for that: quadtrees, octrees. As a 3d general process the octrees are chosen to represent the geometric approximation of bodies. Every both simple and complex body can be modeled approximately using the octree structure. Using the recursive hierarchical approximation of octree representation the collision problem of moving bodies can be subdivided into collision problems between cubes. Consequently, the intersection of a body with the domain (effected space) of another moving body is reduced to the distance determination between cubes of body octree and cubes of effective space octree, which can be realized easily. Thus, a report about occurrence of the collision can be fast attained. Figure 2 shows the conception of the graphic collision control.

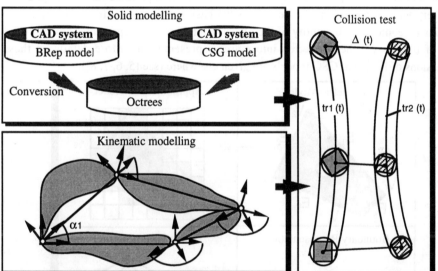

Figure 2: Collision control process on the basis of a distance comparison between cubes using octrees

2.1 Solid modeling using octree representations

The description of moved bodies and static obstacles is the key to a collision control method. A contented collision control is only possible, if solid models are used for geometric modeling. A octree is a hierarchic binary cell model with a tree structure. It contains three node types (b, w, g): black, white and grey nodes for leaves of the approximation tree (Figure 3).
- **White node:** is defined as a node that is completely outside the object.
- **Black node:** is completely inside the object.
- **Grey node:** is partially interior and exterior to the object.

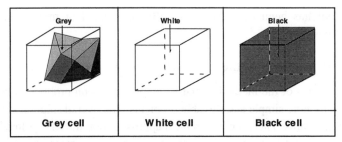

Figure 3: Node types of the octree primitiva

Octree representations are solid models which are generated by a recursive subdivision of a finite cubic universe and represent the tree of the subdivision process. The root of the octree represents the universe, which is subdivided in eight leaf nodes so that they can be described in a simple form. Leaf nodes can be one of the three defined node types. Grey nodes are nonterminal nodes and are recursive bisected then until terminal nodes or a resolution (a minimal node) are reached. Freeform surfaces of an object can be described approximatively by black and grey nodes. Figure 4 shows the octree representation of an object. Octrees suit the approximate representation of complex objects very well, because black node type as the basic form of the octree representations contains a geometric cubic entity of the object. Both boundary representations and CSG trees of objects can be converted into octree representations. More details on these conversions of Brep and CSG models are given elsewhere (see [5, 6, 7, 8]).

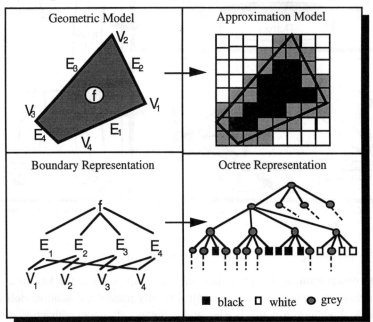

Figure 4: Converting boundary representations of objects into extended octrees

2.2 Simulation of body motion and collision control

Kinematic connections between bodies are defined by a kinematic chain using joint types. A library of time functions that corresponds to drive functions of the kinematic chain can be built for independent degrees of freedom. Every effect of drives on every body can be calculated using matrix methods of transformation [2].

For the kinematic modeling the movement of a body point relative to another body can be represented with a homogeneous transformation matrix in a coordinate system (frame). The trajectory $\vec{tr}(\vec{x}, t)$ of a body point $\vec{x} = (x_1, x_2, x_3)^T$ is described by a matrix vector product:

$$\vec{tr}(\vec{x}, t) = A(t) \cdot \vec{x}$$

$$\vec{tr}(\vec{x}, t) = \begin{pmatrix} tr_1(\vec{x}, t) \\ tr_2(\vec{x}, t) \\ tr_3(\vec{x}, t) \\ 1 \end{pmatrix}, \quad \vec{x} = \begin{pmatrix} x_1 \\ x_2 \\ x_3 \\ 1 \end{pmatrix},$$

$$A(t) = \begin{pmatrix} r_{11}(t) & r_{12}(t) & r_{13}(t) & v_1 \\ r_{21}(t) & r_{22}(t) & r_{23}(t) & v_2 \\ r_{31}(t) & r_{32}(t) & r_{33}(t) & v_3 \\ 0 & 0 & 0 & 1 \end{pmatrix}$$

$$= \underbrace{\begin{pmatrix} 1 & 0 & 0 & v_1 \\ 0 & 1 & 0 & v_2 \\ 0 & 0 & 1 & v_3 \\ 0 & 0 & 0 & 1 \end{pmatrix}}_{\text{Translation matrix V}} \cdot \underbrace{\begin{pmatrix} r_{11}(t) & r_{12}(t) & r_{13}(t) & 0 \\ r_{21}(t) & r_{22}(t) & r_{23}(t) & 0 \\ r_{31}(t) & r_{32}(t) & r_{33}(t) & 0 \\ 0 & 0 & 0 & 1 \end{pmatrix}}_{\text{Rotation matrix D}}$$

$$= V \cdot D$$

$r_{ij}(t)$ (i, j=1, 2, 3) is the rotation part, $v_i(t)$ (i=1, 2, 3) is the translation part. The dynamic transformation matrix A(t) represents the rotation and translation on a coordinate frame originally aligned with the reference coordinate frame.

Given an body described by a reference coordinate frame as in Figure 5, and a transformation representing the position and orientation of the body's axes, the body can be simply reconstructed, without the necessity of transforming all the points, by noting the direction and orientation of key features with respect to the describing frame's coordinate axes. By drawing the transformed coordinate frame, the body can be related to the new axis directions. The movement of a body can be simulated by computer visualization using transformation matrixes.

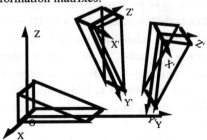

Figure 5: Transformed body

The inversion some such transformation matrixes is reduced to multiplication of inversion of rotation and translation matrix. Because of a orthogonal matrix, the Inversion of the rotation matrix D is obtained by transposition of lines with columns. The translation matrix V is inverted by negation of the column vector which represents the translations.

$$A \cdot A^{-1} = V \cdot D \cdot D^{-1} \cdot V \cdot V^{-1} = V \cdot V^{-1} = I$$
$$A^{-1} = D^{-1} \cdot V^{-1}$$

$$= \begin{pmatrix} r_{11}(t) & r_{21}(t) & r_{31}(t) & 0 \\ r_{12}(t) & r_{22}(t) & r_{32}(t) & 0 \\ r_{13}(t) & r_{23}(t) & r_{33}(t) & 0 \\ 0 & 0 & 0 & 1 \end{pmatrix} \cdot \begin{pmatrix} 1 & 0 & 0 & -V_1 \\ 0 & 1 & 0 & -V_2 \\ 0 & 0 & 1 & -V_3 \\ 0 & 0 & 0 & 1 \end{pmatrix}$$

Often a dynamic collision control is reduced to static tests by subdividing a given time interval into sufficingly small pieces. Using octree representations the collision control of moved bodies is reduced to the collision control of motion cubes.

Discrete collision control

For every small time interval the collision control is made once. The fundamental elementary operation for the collision control process can be put down to collision test of two along the centre trajectories carried bodies for a time interval. For that, using octree representations both bodies can be modeled. So the collision control can be reduced to a distance determination between cubes from octrees.

A cube or cuboid is generally represented by

Cuboid = $(x_{min} \leq x \leq x_{max}) \cap (y_{min} \leq y \leq y_{max}) \cap (z_{min} \leq z \leq z_{max})$

This results in the min-max-test of two cubes or wrapping cuboides: An intersection between two cubes (terminal nodes) takes place, if the intervals which are determined by their minimum and maximum show an intersection in all coordinate axes. A mathematical representation of this condition is described by (Figure 6):

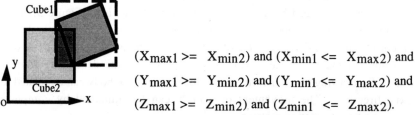

$(X_{max1} >= X_{min2})$ and $(X_{min1} <= X_{max2})$ and
$(Y_{max1} >= Y_{min2})$ and $(Y_{min1} <= Y_{max2})$ and
$(Z_{max1} >= Z_{min2})$ and $(Z_{min1} <= Z_{max2})$.

Figure 6: Representation of the intersection between two terminal nodes

The distance between body trajectories $\Delta(t) = |tr1(t) - tr2(t)|$ is a function of the time. A given time interval can be subdivided into sufficiently small pieces. Bodies are rekusive divided hierarchically into cubes. Trajectories of cubes refer only to centres or main diagonals of wrapping cuboides which are placed around the cubes by minima and maxima. For each small piece a collision is had to test. The method of the interval calculation [3] is used for the numeric evaluation. From that the following cases of collision can be distinguished:

- $W \cap x; x \in \{W, B, G\}$. No collision exists if one of two cubes is white.

- B ∩ B. One collision happens if the condition of an intersection between two black cubes is met for a time interval.
- x ∩ y; x, y ∈ {W, B, G}. If the intersection condition between two cubes is not satisfied, then no collision takes place. In another case a collision is possible. The relevant no-black cubes (grey nodes) can be divided until a given resolution or minimal cuboid is reached. So the relevant no-black cubes can be tested forward, if a result about the exacter collision recognition is necessary.

Octree's node type *Grey* are often filled only partly by objects. Because of the coarse octree structure it will result in an inexactness of the collision control. For that there are possible collision situations (Figure 7). Therefore, the defining a minimal length of cubes for octree representations has to be connected with the exactitude for the collision recognition. Thus, octree is useful for a graphic collision control, particularly in case of no collision or of a collision. Consequently, a collision zone can be determined by the two relevant cubes for the collision. The total collision zone can be built by adding of results.

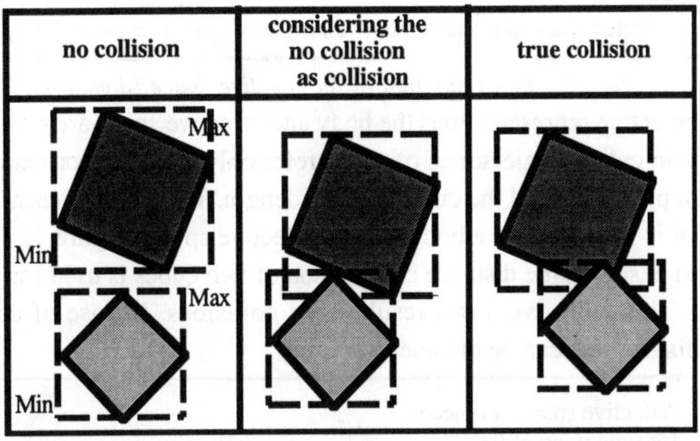

Figure 7: Possible collision situations for collision control

Continuous collision control

Only using the static distance determination of bodies the wanted result can not be obtained, because the motion continuity is not in all cases recorded. Because of the motion continuity a continuous contemplation is required urgently to give a statement about collision as reliable and as accurate as possible.

The minimum and maximum of a moving cube or terminal node describe a function of the time. The continuous control of collision is based on the computation and comparison of distances between two moved bodies or cubes where the time as a highly important factor has to be taken into consideration [1, 2]. If the time dimension is projected into the volumetric track, the collision control for a small time interval is reduced to a problem of the distance determination between a three-dimensional body and a volumetric track which is determined by the motion of another body.

The determination of a body position over a time interval leades to a volume described as an effective space (domain) which is presented with function SKIN of the CAD system DICAD (Figure 8). Inside the effected space the body or cube is moved so that the continuity of the motion is included. The timewise continuous contemplation and test allow a reliable discovery and delimitation of all possible collision cases.

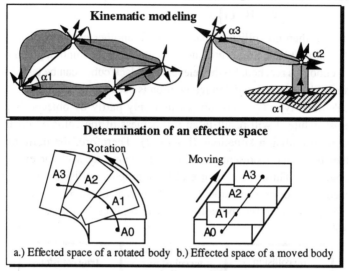

Figure 8: *Determination of an effective space (domain)*

Using the octree representations the body and effective space are recusively divided hierarchically into cubes. Trajectories of cubes refer only to centres or main diagonals of cubes which are placed around the cubes by edge length. The above presented method for collision control is used just for a body and an effective space (Figure 9). Consequently, the collision test based on the distance comparison of two cubes is used for a body and an effected space. This will give a fast result about collisions. In case of collision a corresponding collision zone can be obtained as result.

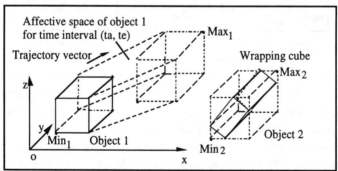

Figure 9: *Collision control between an effective space (domain) and an object*

3. Summarization

In this paper a new method for graphic collision control of moving objects was presented (Figure 10). It was implemented in C using PHIGS for graphic representation. The discrete or continuous collision control are based on the intersection between cubes resulting from the octree approximations of objects. The use of the octree representations leads to a quick elimination of non-relevant cubes. In case of a collision a collision zone is represented only by the relevant cubes. The generated collision information can be useful to change the geometric or kinematic properties of objects to avoid the collision.

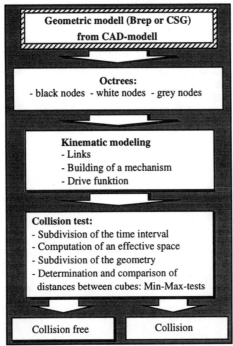

Figure 10: Representation of the collision control process

References

1. H. Grabowski; Z. Bao: DFG Research Report: *Algorithmische Kollisionsprüfung für die Simulation von Bewegungsabläufen.*. RPK, Universität Karlsruhe, (Dezember 1992).
2. R. Paul: Roboter Manipulators: *Mathematics, Programming, and Control.* (MIT Press, Cambridge, Massachusetts and London, England, 1981).
3. S.P. Mudur; P.A. Koparkar: *Interval Methods for Processing Geometric Objects.* IEEE Computer Graphics and Applicationc, **Vol. 4**, Nr. 2, (1984).
4. J.W. Boyse: *Interference Detection Among Solids and Surfaces.* Comm. of the ACM, **Vol. 22**, No. 1, (1979).
5. D. Meagher: *Geometric Modelling Using Octree Encoding.* Computer Grapgics and Image Processing, **19** (1982), PP. 129-147.
6. K. Yamaguchi; T.L. Kunii; K. Fujimura: *Octree Related Data Structures and Algorithms.* IEEE Computer Graphics & Applications, January (1984), pp. 53-59.
7. H. Samet; M. Tamminen: *Bintrees, CSG Trees, and Time.* Computer Graphics of the ACM, **19** (1985), pp. 121-130.
8. I. Gargantini: *Linear octrees for fast processing of three-dimensional objects.* Computer Graphics and Image Processing, **Vol. 20**, (1982), pp. 365-374.
9. H. Grabowski; Z. Bao: *Graphic and Analytic Collision Control Using Polytree Representations.* Proceeding of ICARCV'94, Singapore, November (1994).
10. H. Grabowski; R. Anderl; V. Holland-Letz, B. Pätzold; A. Suhm: *An Integrated CAD/CAM-System for Product and Process Modelling.* International Symposium on Advanced Geometric Modelling for Engineering Applications, (North Holland 1990), pp. 403-420.

CAE

Computer Aided Geometric Design of Ship Hull form Using
Bi-Cubic B-Spline Surface
   M. R. Bin Mainal and Y. Bin Samian

Virtual Tryout of Die and Molds Using Numerical Simulation
with PAM-STAM™
   L. T. Kisielewicz

The Best Casting Plane of Simple Polyherdron
   W. Wang, J. Y. Wang, W. P. Wang and X. X. Meng

Simulation of Conjugate Profiles in Gear Shaping Using
Computer Graphics
   S. V. R. Surya Narayana, V. Jayaprakash and M. S. Shunmugam

COMPUTER AIDED GEOMETRIC DESIGN OF SHIP HULL FORM USING BI-CUBIC B-SPLINE SURFACE

MOHD. RAMZAN BIN MAINAL

and

YAHYA BIN SAMIAN

Faculty of Mechanical Engineering
Universiti Teknologi Malaysia
Locked Bag 791, 80990 Johor Bahru, Malaysia

ABSTRACT

This paper deals with the implementation of bi-cubic B-spline surface technique to ship hull form development. An interactive computer program is developed to assist the designer to create fair hull forms. The ship hull surface is divided into three sections, that is, forebody section, middle section and after body section. Each section will then be divided into patches that can be represented by a single bi-cubic B-spline surface patch and later connected smoothly. Several examples that are either single patch or combination of a few patches of the ship hull surface are given to illustrate the function and effectiveness of the computer program in generating the hull surface of the ship.

Keywords: CAD, Surface Modelling

1. Introduction

At the preliminary design stage, the ship hull form is designed to satisfy the required performances such as speed, displacement, powering, stability, etc. Its offset values are important draft in ship design and becomes the basis in ship production. In practice, the hull form can be generated by either from the standard series, the parent form transformation method or the form parameter or sketches method.

From the mathematical point of view, ship hull forms can be described by two approaches. One is to describe a hull form with a set of curves and the other is to describe it with surfaces. The curve approach has long been used and proved to be adequate for preliminary design stage. However, owing to the requirement of 'cross fairing' procedure to smooth the surface and inability to describe the free-form surface perfectly with mathematical meaning, the curve method is no longer suitable for production purpose.

In order to overcome the defects resulting from the curves approach, surfaces approach is regarded as a promising method. In this approach a ship hull form is defined by means of surfaces directly, not by means of curves. Since the Coons patch was proposed (Coons 1967), several mathematical theories for free-form surface description have been developed (Bezier 1972, Riesenfield 1973, Barnhill 1977). Such approach normally divides the hull surface into several patches, then combined them to construct a fine and practical ship hull surface by the continuities of position and tangent plane.

2. Surface Generation Technique

There are various theories for free-form representation that have been actively researched in the recent years. As far as ship design is concerned, the following requirements should be satisfied so that surface generation technique can be utilised effectively:
- to express the surface by easy mathematical expressions,
- to have sufficient continuity,
- to have flexibility for representation for various shapes, and
- to be able to manipulate shapes easily.

To fulfill these requirements, the bi-cubic B-spline surface technique is used in this study to develop the ship hull form.

2.1 B-spline Surface

B-spline surface is the natural extension of the Bezier surface. The surface is generated as a weighted sum of control points forming a network usually outside the surface. A Cartesian product parametric B-spline surface is given by:

$$Q(u,w) = \sum_{i=1}^{n} \sum_{j=1}^{m} B_{i,j} N_{i,k}(u) M_{j,l}(w) \quad (1)$$

where $N_{i,k}(u)$ and $M_{j,l}(w)$ are the B-spline basis functions in the biparametric u and w directions respectively. The $B_{i,j}$ is the defining m x n polygon net. The definition for the basis functions is given by:

$$N_{i,}(u) = \begin{cases} 1 & \text{if } x_1 \leq u < x_{i+1} \\ 0 & \text{otherwise} \end{cases}$$

$$N_{i,k}(u) = \frac{(u-x_1)N_{i,k-1}(u)}{x_{i+k-1} - x_i} + \frac{(x_{i+k} - u)N_{i,k-1}(u)}{x_{i+k} - x_{i+1}}$$

$$M_{i,k}(w) = \begin{cases} 1 & \text{if } y_j \leq w < y_{j+1} \\ 0 & \text{otherwise} \end{cases}$$

$$M_{i,k}(w) = \frac{(w-y_j)M_{j,l-1}(w)}{y_{j+1-1} - y_j} + \frac{(y_{j+1} - u)M_{j+1,l-1}(w)}{y_{j+1} - y_{j+1}} \quad (2)$$

where,

x_i, y_j — the elements of a uniform knot vector,
$x_i = \{0\}$ for $1 \leq i \leq k$
$x_i = \{i - k\}$ for $k + 1 \leq i \leq n+1$
$x_i = \{n - k + 1\}$ for $n + 2 \leq i \leq n + k + 1$
$y_i = \{0\}$ for $1 \leq i \leq 1$
$y_i = \{i - 1\}$ for $1 + 1 \leq i \leq 1 +1$
$y_i = \{n - 1 + 1\}$ for $n + 2 \leq i \leq n + 1 + 1$

k, l — the order of the B-spline surface in the u and w direction,

n, m — are one less than the number of polygon net points in the u and w directions respectively.

B-spline surface have several advantages for representation of surfaces such as:
a. Since a B-spline basis has local supports, the B-spline surface has the better facility to make alteration.
b. The degree of the basis function is independent of the number of vertices, so the low degree B-spline surface can represent a complex form.
c. There are no specific geometric restrictions to the surface and only a relatively small number of points are required to describe even complicated geometries.

2.2 Bi-Cubic B-Spline Surface

Different types of control 'handles' are use to influence the shape of a B-spline surface. Such control can be achieved be changing the type of knot vector, changing the order k and l of the basis function, changing the number and position of the defining polygon vertices, using multiple polygon vertices and using multiple knot values in the knot vector. In order of the surface will be utilised. Hence, the surface of fixed order is controlled by the location of the polygon net points and the number of multiple points at a particular net point.

Referring to Eqs. (1) and (2) for the B-spline surface definition, the bi-cubic B-spline surface mean that the surfaces degree in u and w direction are both third degree. In other ward, the order of the surface in u and w direction are both fixed as fourth order. Thus a Cartesian product parametric of a 4 x 4 bi-cubic B-spline can be represented by:

$$Q(u,w) = \sum_{i=1}^{4} \sum_{j=1}^{4} B_{i,j} \, N_{i,4}(u) \, M_{j,4}(w) \qquad (3)$$

where the basis functions and knot vector are similar to Eq. (2). Several properties of such equation are as follows:
a. The order of the surface is each parametric direction is equal to the number of defining polygon vertices in that direction,
b. The continuity of the surface in each parametric direction is two degree in both the u and w directions,
c. The surface is invariant with respect to an affine transformation,

d. The influence of a single polygon net vertex is limited to 2 spans in each parametric equation, and
e. There are no interior knot values, that is, the knot vector in both u and w directions are equal to [0 0 0 0 1 1 1 1].

Thus, the continuity of the surface is good enough for defining most shape of the ship hull. The 4 x 4 control vertices will then fixed the open knot vector with no interior knot values. This reduces the B-spline surface control 'handles' to only controlling of the multiple defining the polygon net. These sixteen control vertices are adequate to define a bigger patch of the ship hull such as the midship section.

3. The Computer Module

The ship hull surface beneath and above the design waterline in longitudinal direction will be divided into three sections, that is, forebody section, middle section and afterbody section as shown in Figure 1. Each section will then be divided into patches that can be represented by a single bi-cubic B-spline surface patch and later connected smoothly. Since the conditions for combining several patches should have continuities of position and same tangent plane, the joining edges should have showed the same tangent vector as shown in Figure 2.

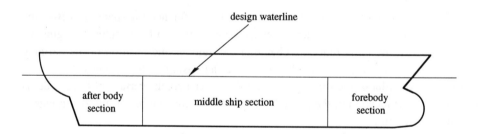

Figure 1 Patches Division of a Ship Hull Surface

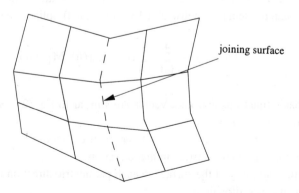

Figure 2 Condition of Patches Combination

The computer module developed in this study for the hull form definition consist of six modules such as input, interpolation, transformation, blending, plotting and output. Figure 3 shows part of the computer module flowchart. The input data for this program is the control vertices in three dimensional consisting of coordinates x, y and z. As for preliminary ship design, station offsets are used as input data.

The interpolation module contains all the calculations and routines of B-spline surface algorithms based on Eqs. (1) and (2). The knot vector which is stored in an array will then be used in the basis function calculations. Using such functions, surface data point on the patches is then generated utilising the bi-cubic B-spline surface algorithms.

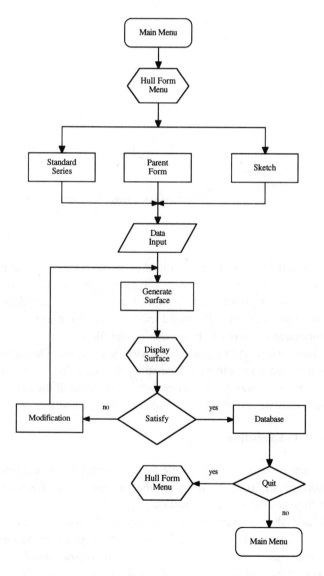

Figure 3   The Computer Module Flowchart

The results from the interpolation module is a list of three dimensional coordinates that may not be suitable for presentation purposes on the display device. Thus the transformation module converts the coordinates expressed in a coordinate system centred at the viewpoint. Figure 4 shows the correlation between the world coordinate system and the eye coordinate system. Having applied the viewpoint transformation, it is still left with a list of three dimensional coordinates, thus needing a perspective transformation to produce a list of two dimensional or screen coordinates.

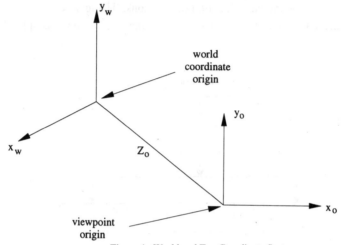

Figure 4 World and Eye Coordinate System

In the blend module, the surface data file of every patch have to be prepared before such process can be carried out. The system will prompt the user to key in the 'master file' which is then use for plotting, while the rest are the 'secondary files' which are required to be blended. Basically, the blend module will merge and rearrange all the connectivity information of every patch into one 'master file'.

The ship hull surface will be plotted as a wireframe model. Plotting a wireframe model requires only the connectivity information, that is, a list of two dimensional coordinates generated from transformation module. The output of the system consist of either the disk storage (data output) and the screen (graphical output).

4. Examples of Execution

Several examples that are either single patch or combination of a few patches of the ship hull surface are given to illustrate the functions and effectiveness of the computer program in generating the hull surface.

Figure 5 and 6 show the input and output surface of a semi-cone. The sharp edge at the bottom of the cone is the result of defining 4 multiple points at the same position. An even more complicated geometry shapes are shown in Figures 7 and 8.

A typical example for a ship hull surface is given in Figures 9 and 10. This example is generated by combining the three sections which are the stern section, midship section and the bow section. Stern and bow are constructed from combining two patches and the midship section consist of three patches. All these patches are then blended into one whole ship as displayed in Figure 10.

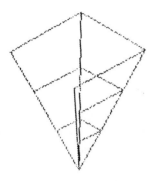

Figure 5 Control Vertices for a Semi-Cone

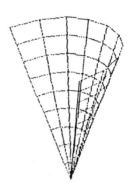

Figure 6 Output Display for the Semi-Cone

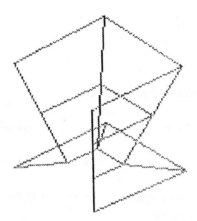

Figure 7 Control Vertices for a Complex Shape

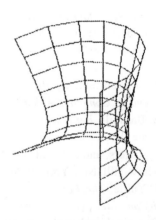

Figure 8 Output Display for the Complex Shape

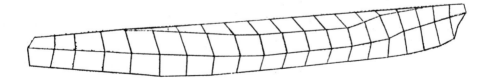

Figure 9 Control Vertices Polygon Net for Hull Surface

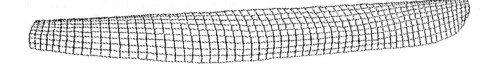

Figure 10 Hull Surface Constructed from 7 Patches after Interpolation

5. Conclusion

This paper has described the ongoing research and development effort concerning with ship hull form representation with bi-cubic B-spline in Universiti Teknologi Malaysia. Based on the examples given, bi-cubic B-spline have been found to be adequate to represent fair ship hull forms and to design new forms from scratch. Though the computer program developed in this study is considered extensive, several interactive and user friendliness functions are planned to be installed to improve its effectiveness.

References

1. S.A. Coons, *Surfaces for Computer-Aided Design of Space Forms*, (Project MAC, Massachusett Institute of Technology, 1967).
2. P. Bezier, *Numerical Control - Mathematics and Applications* (John Wiley London, 1972).
3. R. F. Riesenfield, *Applications of B-Spline Approximation to Geometric Problems of Computer-Aided Design*, (Ph.D. Thesis, Syracuse University, 1973).
4. R. E. Barnhill, *Representation and Approximation of Surfaces*, (Mathematical Software III, Academic Press, New York, 1977).
5. D. F. Rogers, and J. F. Adams, *Mathematical Elements for Computer Graphics*, (McGraw-Hill, New York, 1976).
6. I. M. Yullie, *The Forward Design System for Computer Aided Design Using a Mini-Computer* , (RINA, Spring Meetings, London, 1978).
7. M. R. Mainal, and Y. Samian, *On the Application of Numerical Modelling for the Hull Form Development*, (First Regional Conference on the Engineering Application of Numerical Modelling, Malaysia, 1994).
8. Y. Samian, *Mathematical Modelling for Ship Lines*, (Numerical Analysis Seminar, Malaysia, 1991).

VIRTUAL TRYOUT OF DIE AND MOLDS
USING NUMERICAL SIMULATION WITH PAM-STAMP™

L.T. KISIELEWICZ
Nihon ESI
Well Uehara Bldg 5F, Uehara 2-47-18, Shibuya-ku, Tokyo 151, Japan

ABSTRACT

The paper summarizes the situation of die and mold design in the perspective of recent technological trends. It introduces a new tool tryout technique based on numerical simulation of the manufacturing process and illustrates it with several industrial examples.

1.0 Introduction

The mass production of steel parts using sheet metal forming technology is a long established practice relying on the experience of the manufacturing engineers. Nevertheless such technology is not troublefree and the complexity of adjusting tools and manufacturing process can lead to months of physical tryouts at the workshop.

Requirements for new materials and more complex part shapes are extending the boundaries of this technology's field beyond its traditional experience limits. Designers face the challenge of tuning manufacturing processes for configuration in which no one has a lot of experience in time frames which are being reduced continuously.

CAE can support the engineers and offer an effective alternate approach to the traditional physical tryout at the workshop. A proper simulation tool and methodology which addresses the issues encountered by the designers are needed for it. In addition, the industrial usage of such methodology can be successful only if some criteria are fulfilled.

2.0 Industrial needs

2.1 Stamping process

Stamping or sheet metal forming with presses is widely used in many fields of manufacturing industries with parts sizes ranging from a few millimeters to several tens centimeters. It has been used for decades but remains far from being a straightforward and troublefree method except for some very simple parts. For some complex parts, adjusting the manufacturing process including tryout of dies and molds can take several months.

The outcome of physical sheet metal forming depends on a complete sequence of operations including holding, binding and forming the blank with a single or a multiple stroke process. Then after withdrawal of the formed blank some springback can develop as well as after trimming the blank.

2.2 Technology trends

In the recent years, efforts have been made by the automotive industry to reduce the weight of passenger cars to limit the gas consumption. This led to using materials not previously used in automobile body parts (high strength steels, aluminums, plastic composite) and to limit the number joints thus increasing the complexity of parts [Fig.1].

These trends have required intensive research in the material forming technology and in significant changes with respect to the classical approach for which some engineers and technicians developed decade long experiences. The new solutions coupled with the general objective of reducing the overall project duration (typically from 5-6 years to 3-4 years for a passenger car) strains the tool design engineers by requiring more complex tool designs in shorter periods of time.

The field of non metallic materials is also expanding even though the production cost of such new materials is still braking its expansion. Press forming these materials is a common way for manufacturing although it raises some different difficulties than with metals [11][12].

These technology trends go beyond the accumulated experience of manufacturing engineers and without other means require a large number of tryout tests before an acceptable (and more than often not optimized) solution can be found. A new method for designing tools is needed to cope with these technological trends.

Figure 1 - Complex body part made of HS steel

Figure 2 - Cracked part during forming

Figure 3 - Forming Limit Diagram example

2.3 Potential troubles

Troubles than can arise during a sheet metal forming are numerous and complex.

Metal tearing. Metal rupture during the forming process results from an excessive microdamaging of the material. In practice this situation is checked out by comparing the main strain locus versus the metal specific Forming Limit Diagram [Fig.3].

Metal instabilities. When too much material is fed into the die during the forming, bulges and wrinkles can develop either transiently or permanently.

Shape deviations. Differences between the design shape and the final shape after forming result mostly from springback effects.

Dimension tolerances. Final thickness distribution may lead to rejection of the part during the assembly part is they do not match the design specifications.

Surface appearances. Whereas not critical for structural reasons, bad surface appearance can result in a blunt like appearance.

Edge appearances. Rugged edge, lip or ear formations at the edges during the manufacturing process can be unacceptable if they cannot be removed by trimming.

Residual strains. If residual strains are too high (safety margin against rupture too little), while the forming process appears adequate, the part may suffer a rupture either during a further assembly step or during the initial operation.

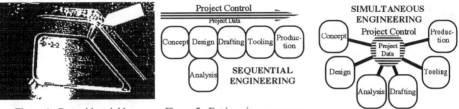

Figure 4 - Part with wrinkles   Figure 5 - Engineering processes

2.4 Practical issues

In the more general scope of simultaneous engineering which tends to have all technical teams involved in the design of an artifact working simultaneously [Fig.5], the main needs of the tool and manufacturing departments are listed as follows [13].

Cost savings in tryouts. By reducing reliance on trial-and-error tests at the workshop in the process of tuning tools and forming processes, direct financial gains are obtained by reducing the amount of material to be used during the tests and indirect gains by relieving the schedule of the press to be used during the tests.

Time savings in tryouts. A reduction of the tryout phase will have a direct impact on the overall project duration as it is one of the major time consuming tasks (sometimes over 75% of time).

Reliable technical decisions earlier. Technical decisions on the tools and forming process, if taken early in the overall design cycle, can be taken into account by the other design teams before bearing too much cost impact.

Optimization of the stamped parts. Accounting for manufacturability of the parts, an optimization would consider production cost which could have a significant influence on the cost of final artifact (market share) and on the finances of the company.

Accurate data for selecting process parameters. Somewhat related to the optimization of parts, a proper selection of process parameters would be eased by a capability for fast assessment of parametric variations.

Detailed insight in the physical phenomena. A detailed understanding of the physics of the physics would enable the engineers to take faster decision based on rationale rather than on empirical know-how.

3.0 Specifications for simulation

3.1 Physical phenomena

For a numerical simulation to be considered as an acceptable means for the tool designers a number of criteria must be fulfilled. All the main physical phenomena must be accounted for [Fig.6] and all the potential troubles that can develop must be predicted.

Material rheology modeling is one the major conditions for it. It may be necessary to take into account phenomena that are usually neglected such as the anisotropic plasticity, the strain rate effects and the strain softening effect of material damaging.

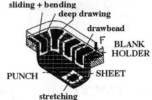

Figure 6 - Main phenomena in stamping

3.2 Results expected

The results expected by engineers from simulation of forming processes are related to the technical decisions he has to take. They must be supported by numerical results.

Manufacturability. The manufacturability relates to the range of process parameters like the binding force under which no wrinkles nor material tearing develop [Fig.7]

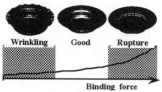
Figure 7 - Manufacturability range for blankholding pressure

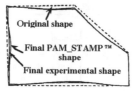
Figure 8 - Blank contour

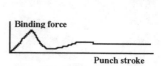
Figure 9 - Typical binding force time history

Cutting pattern of blank. The cutting pattern can be critical for the manufacturability and must be validated by the blank contour after forming as predicted by the simulation [Fig.8].

Process parameters (drawbeads, lubrication). The material flux during the process can be controlled using a number of parameters including the lubrication condition, the detailing of drawbeads and in some cases the orientation of blank material orthotropy axes with respect to the tools.

Bulging of tools. The tuning of tools such as local bulging or smoothing is usually done for controlling springback and local material behavior.

Selection of press. The selection of the press to be used in production is based on the range of forces to apply on the blankholder and on the punch. It is to underline that, contrary to the feeling of many, the binding and punch forces are usually not constant and can vary widely during the forming process [Fig.9].

Assessment of tools wear. The assessment of the tools wear which determines their lifetime in terms of number of parts that can be manufactured depends on the pressure at the contact areas [Fig.10].

Surface quality. The surface quality which is essential for the visual aspect of the formed part depends on the development of transient folds during the forming process [Fig.11], on the residual plastic strains [Fig.12] and on the lines of first contact between the blank and the tools [Fig.13].

Dimension tolerances. They depend on the final thickness of the part and need to be predicted as accurately as possible [Fig.14].

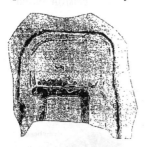
Figure 10 - Contact pressure in an oil pan

Figure 11 - Transient folds in a front fender

Figure 12 - Final residual strains in a rear fender

4.0 PAM_STAMP ™ solution

4.1 Solver

PAM_STAMP™ is a special purpose CAE solution using finite element space discretization and an explicit time integration scheme. It has been thoroughly validated in the scope of large size R&D projects in Europe and Japan with some major automakers.

Figure 13 - First contact between punch and front fender

Figure 14 - Final thickness distribution in oil pan

Figure 15 - Adaptive meshing (h-adaptivity) example

Modeling the geometry of the tools and blank is done using the finite element technique. The density of mesh must be defined in order to account for the phenomena to be simulated. Especially small size wrinkles can only be predicted is enough elements are present in the wrinkle wave. In PAM_STAMP™ adaptive meshing capabilities have been introduced helping the user to obtain accurate results even if some meshing errors were introduced [Fig.15].

The material properties (for steel) in PAM_STAMP™ has been developed, after extensive research and validation, using some of the latest development in material rheology modeling. The physical model selected are those providing the most robust and wider range of validity in sheet metal forming simulations accounting for anisotropic plasticity, strain hardening and strain rate effects.

Friction between the tools surfaces and the blank is accounted for by using a Coulomb model with friction coefficient dependent on the contact pressure and on the sliding velocity. These dependencies can be non linear and provide very accurate results provided these dependency laws can be properly calibrated.

Drawbeads in PAM_STAMP™ can be modeled in two ways. Either a direct geometrical modeling [Fig.16] in which the real geometry and deformation of the blank under the drawbeads are simulated. Or using an equivalent model in which a special restraining and material straining condition is set at the location of beads [Fig.17].

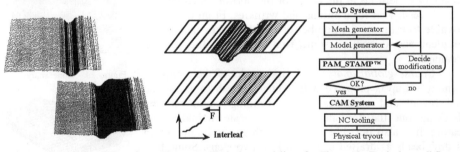

Figure 16 - Direct modeling of drawbeads

Figure 17 - Equivalent modeling of drawbeads

Figure 18 - Typical overall flow chart of simulation

4.2 Simulation process

A typical overall flow chart of the simulation process is depicted on figure 18. This chart shows the sequence and iteration loops that are usually applied when numerical simulation is used to reduce the number of physical tryouts of die and molds.

The modeling of tools is usually done using one among many CAD systems. These models originally made to generate NC machining can also be used for the numerical simulation of the stamping process.

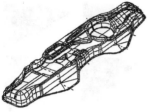  

Figure 19 - CAD model of tools   Figure 20 - Regular mesh   Figure 21 - Stereolithographic mesh

Meshes must be provided for all parts to be included in the model (all tools and the blank). Traditionally, finite element meshes are prepared by special software modules which use a geometry defined in a CAD format [Fig.20]. Mesh generation can be a time consuming task. Some CAD systems have released a capability to prepare automatically meshes regulated by the angle of contiguous facets [Fig.21]. PAM_STAMP™ has been enabled to use such meshes thus reducing the duration from a few days to a few hours.

Once the mesh are prepared, some additional data need to be provided before the simulation of the forming can start. These include the input of material parameters (which can be retrieved from a data bank), the contact conditions and the sequence of operations (kinematics of tools).

When assessing the manufacturability of a part, a four step approach with gradually increasing detailing is recommended in PAM_STAMP™.

Line analyses. Two dimensional models of stripes of the blank are selected in critical areas [Fig.22]. These models includes portions of the tools and provide information to assess the local blank cutting pattern, the binding and forming forces and especially a first estimate of the range of acceptable binding forces.

Figure 22 - Line analysis

Zone analyses. Local three dimensional models are prepared for areas where potential problem can develop (wrinkling, folding, tearing) [Fig.23]. By defining appropriate boundary conditions at the edges of the portion, such models provide results on local strains and wrinkles. From these it is possible to estimate the number of punch strokes that would be ideal and determine some first values of the process parameters.

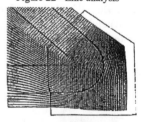

Figure 23 - Zone analysis

Coarse global analyses. A first model of the complete forming set-up is prepared with a coarse mesh to assess the tool layout and estimates of the forming parameters by checking out the formation of large scale wrinkles and tearing. It is also possible to refine at this step the definition of drawbeads as drafted using the first two steps. Some first estimate of the springback can also be obtained from these analyses.

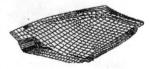

Figure 24 - Coarse analysis

Fine global analyses. For the final verification of the parameters selected, and some limited adjustment through parametric variation, a three dimensional fine mesh model of the complete forming set-up is prepared. This will confirm the values of strains, thicknesses and small size wrinkles if any. Detailed springback results can be obtained if the first results are not deemed sufficient.

Figure 25 - Fine analysis

Post-processing is the phase during which the results of the simulation are analyzed before a technical decision is made. In the case of sheet metal forming simulation, specific output format are desirable to ease the tasks of the engineers. such as automatic display of main strains in FLD format and stress or thickness distributions along sections.

4.3 Success criteria

Technical criteria. The main technical criterion for a successful usage of CAE is the predictiveness of the approach. This requires a validation of the method by comparison between test results and numerical predictions and a proper modeling of the phenomena including all significant physical effects.

Economical criteria. The essential economical criterion in implementing a CAE solution is to obtain an attractive Return on Investment as compared to alternate solutions. Accounting for direct investments and indirect expenses, and for direct benefits and indirect gains the RoI must be convincing for executive financial officers.

Human criteria. A key criterion for the successful introduction of CAE solution is the acceptance by the people designated to use it. This issue is not, as in any issue related to human motivation, a proved science but rather an art. Acceptance is based on confidence and on a personal advantage in terms of self-esteem.

5.0 Examples of application

5.1 Oil pan [8]

The oil pan [Fig.26] presented two trouble during its first physical tryouts, wrinkling in one of the lateral faces and over thinning (45%) in the deepest drawn zone. The simulation with PAM_STAMP™ showed that wrinkles developed in an area where the contact between the punch and the blank occurred very late in the process. The thickness changes proved important in their influence on the binding force. The correction of the troubles required tool changes to alter the sequence of contacts with the blank.

Figure 26 - Oil pan model

5.2 Wheel house [10]

During the firsts physical tryouts of the aluminum wheel house [Fig.27] material tearing developed. Lack of extensive experience in drawing aluminum would have led to carrying out such a very large number of tests that changing the material from aluminum to steel was considered. The key parameter that was adjusted with numerical simulations was the binding pressure that eventually was dropped to 10% of the original value.

Figure 27 - Wheel house model

5.3 Exhaust manifold [9]

The exhaust manifold [Fig.28] was a new design based on deep drawing rather than casting to obtain a lighter part. The PAM_STAMP™ study was carried out before any physical tryout to eliminate unrealistic forming solutions. Three blank cutting shapes, two sheet materials and two lubrication conditions were investigated numerically before a physical tryout was organized. Most configurations consi-

Figure 28 - Model of manifold

dered led to material rupture and only a forming with two halves blanks proved to work. The only alternate solution was to use a milder stainless steel material.

6.0 PERSPECTIVES

6.1 New forming processes

Following the simulation of stretching, deep drawing and stretch drawing situations with steel materials, first simulation of new forming processes have been carried out successfully. This includes blow forming where the punch is replaced by highly pressurized gas, hydroforming where the die cavity is filled with a fluid allowing to distribute the stretching on larger portions of material, or forming of floor mats where the blank is made of an elastomeric support with fibers.

6.2 Coupling with structural analyses

Structural analyses are being carried out for many years for passenger car crash. These analysis are based on ideal modeling, i.e. not accounting for the real physical state the parts are after manufacturing. These idealizations resulted from the impossibility to determine realistic values of the strains and thicknesses to be used as initial conditions. Numerical simulation for metal forming provides as a by-product these initial conditions. Some first assessment of these possibilities are being investigated [15] and should in a short term provide a procedure to improve the accuracy of the structural analyses.

7.0 Conclusions

While sheet metal forming is the most frequently used metal forming process among mechanical industries and has been in use for many decades, it is not trouble free. The technological trends complexifies further this process. The paper has listed some of the major troubles and requirements of die and mold design engineers and the criteria for using numerical simulation as a tool design means.

A numerical simulation solution based on finite element modeling has been introduced with example showing the reliability of the solution and some ways of using it for trouble shooting in the process of die and mold design.

Perspectives in using this new numerical tool have been briefly outlined based on some preliminary R&D applications.

8.0 References

1. E.Haug, *Innovative Methodology for Sheet Metal Stamping*, UKD'94
2. F. ElKhaldi, *Numerical Simulation of Plastic Forming*, JCMR (1993)
3. W.C. Hsieh, *Die Design for Stamping of Rear Floor Panel*, 2nd ICDMT (1992)
4. L.T.Kisielewicz, *Critical Issues for CAE in Sheet Metal Forming*, ICES'92
5. L.T.Kisielewicz, *CAE Solution for Sheet Forming Simulation*, ICCME (1992)
6. S.Aïta, *Formability Analysis using Forming Simulation*, NUMIFORM'92
7. L.Penazzi, *Material Characterization for the Simulation of Forming*, IDDRG '92
8. S.Aïta, *Industrial Sheet Metal Forming Simulations Issues*, ComPlas3 (1992)
9. F.Lehringer, *Formability Analysis for an Automotive Exhaust Manifold*, PAM'93
10. C.Westerling, *Simulation of Forming Car Body Parts*, PAM'93
11. A.K.Pickett, *Industrial Press Forming of Fibre Reinforced Thermoplastic*, UKD'94
12. T. Hayashi et, *Thermoforming of Fiber Reinforced Thermoplastics*, PUCA'93
13. L.Recke, *Simulation of Forming in Industry*, Sem. Virtual Manuf. (1993)
14. L.T.Kisielewicz, *Advantages of CAE and Success Criteria*, Crash Symp. (1993)
15. V.Richet, *Numerical Analysis of Folding Collapse of Hexagonal Profiles*, PAM'93

THE BEST CASTING PLANE OF SIMPLE POLYHEDRON

Wang Wei
Dept. of Computer Science, Fudan University, Shanghai, P.R.China.

Wang Jiaye
GINTIC Nanyang Technological University, Singapore, 2263.
E-mail: gjywang@ntuvax.ntu.ac.sg
Dept. of Computer Science, Shandong University, Jinan, P.R.China.

Wang Wenping
Dept. of Computer science, Hongkong University, Hongkong.
E-mail: wenping@csd.hku.hk

Meng Xiangxu
Dept. of Computer Science, Shandong University, Jinan, P.R.China.

ABSTRACT

The castability of a polyhedron can be defined as that the boundary of the polyhedron can be divided by a plane into two polyhedrons, and every facet of the two polyhedrons is compatible with one direction each. The plane is called casting plane. Usually, if the polyhedron has n vertices, there are $O(n^2)$ casting planes, and they can be found in $O(n^2 \log n)$ time and linear space [1]. The $O(n^2)$ casting planes are not always suitable for casting or injecting. Some of the casting plane are more suitable for casting than the others. In this paper the criteria of the best casting plane is studied. An algorithm for finding the best casting plane in $O(n^2 \log n)$ time is presented.

1. Introduction

Computational geometry has wide application in computer graphics, CAD, VLSI, etc. The application in the area of automated manufacturing, such as gravity casting, NC machining, automated welding and stereolithograph, is getting interesting[1,2,3,4,5,6].

In [1] the castability of a simple polyhedron is studied. If the boundary of a polyhedron can be divided by a plane into two polyhedrons, and every facet of the two polyhedrons is compatible with one direction each, the polyhedron is castable. The plane is called casting plane. In order to cast a polyhedron, the prototype of the polyhedron is separated into two parts by any casting plane, and they can be used to construct upper and lower parts of a cast mold (ref. figure 1. Figure 1 is from [1]). Liquid metal is poured into the opening until it fills the cavity of the cast mold. For a castable polyhedron, after the metal solidifies, the object can be removed from the cast mold without breaking the cast mold and the object.

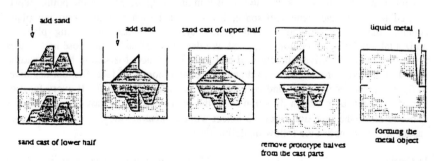

Figure 1

There are three versions of the castability problem [1]. They are

1. The two cast parts must be removed from the object by one translation each, in opposite
directions, and normal to the casting plane (orthogonal cast removal).

2. The two cast parts must be removed from the object by one translation each, and in opposite
direction (opposite cast removal).

3. The two cast parts must be removed from the object by one translation each, in arbitrary direction (arbitrary cast removal).

For arbitrary cast removal, Bose (ref. [1]) proves that there are $O(n^2)$ casting planes and they can be found in $O(n^2 \log n)$ time and linear space. For both cast removal in orthogonal and in opposite direction, the number of the casting plane and the time complicity is $O(n)$ and $O(n^2)$ respectively (ref. [1]).

However among the casting planes found in [1], some of them are more suitable for casting. For example, the cast in figure 2c is better than the one in figure 2b for casting process of the object in figure 2a. If the cast in figure 2c is used, the resistance will be smaller, the distance of removal of object will be shorter, and there are also several other advantages. In this paper criteria representing the suitability of the casting plane for casting is studied. The algorithm for finding the best casting planes is presented. The time complicity of finding the best casting planes for orthogonal, opposite and arbitrary is $O(n^2), O(n^2)$ and $O(n^2 \log n)$ respectively. The result of this paper can also be applied in injection mold.

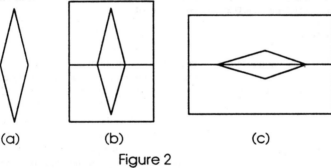

(a)   (b)   (c)

Figure 2

2. The criterion of the best casting plane

In this paper the same notations as [1] are used. Let P be a polyhedron with n vertices and let h be a plane. ∂P denotes the boundary of P. For any facet f of P, the closed half-space bounded by a plane supporting f is denoted by $\Psi(f)$, and such that for any point in f, $\Psi(f)$ intersects the interior of P in an ε-neighborhood of the point. $\Psi_0(f)$ is used to denote the same half-space, but translated such that the bounding plane contains the origin. Without loosing generality, assume that the casting plane is horizontal, otherwise, a rotation is applied. We denote by h^+ and h^- the close half-spaces above and below h. Given direction **d** and facet f, the facet f is said to be compatible with **d**, if the inner product between **d** and the outward normal of facet f is non-negative. In figure 3, the facets ab, bc and cd compatible with the direction oe. If every facet of $h^+ \cap \partial P$ is compatible with a direction, then the upper part of the casting mold can be removed along the direction from the object without breaking the object and the upper casting mold. It implies the following definition.

Definition 1 A simple polyhedron P is castable if there exists a plane h such that the polyhedron can be divided by a plane into two polyhedrons, and every facet of $h^+ \cap \partial P$ is compatible with a direction, and every facet of $h^-\cap\partial P$ is compatible with another direction.

Let

$$\xi^+(h)=h^+\cap\{\bigcap_{f\in h^+\cap\partial P}\psi_0(f)\} \qquad \xi^-(h)=h^-\cap\{\bigcap_{f\in h^-\cap\partial P}\psi_0(f)\} \qquad (1)$$

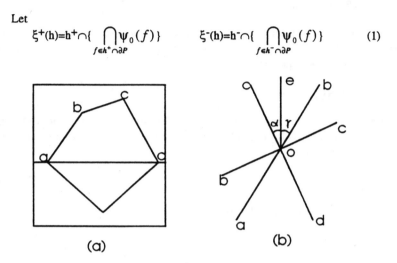

Figure 3

In figure 3b the wedge ado is $\xi^+(h)$ of the polyhedron in figure 3a. If $\xi^+(h)$ is not empty, then the facts of $h^+\cap\partial P$ will be compatible with any direction from any point Q belonging to $\xi^+(h)$ to the origin O. It is therefore that if both $\xi^+(h)$ and $\xi^-(h)$ are not empty, the object is castable, and h is a casting plane. If $|\alpha| = |\gamma|$, the minimum angle between direction oe and the facets of $h^+\cap\partial P$ is maximum. The resistance of removing upper part of casting mold depends on the minimum angle between the removing direction and the facets of $h^+\cap\partial P$. The bigger the minimum angle is, the more easily the part of mold removes. The condition $|\alpha| = |\gamma|$ is equivalent to that the minimum distance between any point Q on oe and the support plane of the facets of $h^+\cap\partial P$ is maximum. The direction oe satisfying the above condition and pointing to the outward of $\xi^+(h)$ is called optimum direction of h^+. In order to simplify the problem the point Q is assumed to be on the casting plane. Now we describe the method to find the point Q (ref. Figure 4).

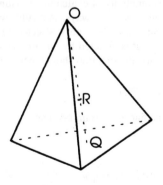

Figure 4

For given casting plane h

$$Ax+By+Cz+D=0,$$

The facets of $h^+ \cap \partial P$ is denoted by f_i (i=1,2,...k), k<=O(n), $\omega(f_i)$ is used to denote the support plane of f_i, and translated such that the plane contains the origin O (figure 4). Assume that the normal equation of $\omega(f_i)$ is

$$A_i x + B_i y + C_i z + D_i = 0,$$

and for any point in $\xi^+(h)$, $A_i x + B_i y + C_i z + D_i > 0$. The distance from a point (x,y) to the plane $\omega(f)$ is equal to $A_i x + B_i y + C_i z + D_i$. The coordinates (x,y) of the point Q satisfy the following linear program

max m (2)

$m - (A_i x + B_i y + C_i z + D_i) \geq 0$, i=1,2,...k, (3)

$Ax + By + Cz + D = 0$. (4)

It is equivalent to the linear program

max m (5)

$m - (A_i x + B_i y + C_i (-(D+Ax+By)/C) + D_i) \geq 0$, i=1,2,...k (6)

This problem can be solved in $O(k) \leq O(n)$ time [7]. The direction QO is the best direction for the casting plane h to remove the upper part of the casting mold. The same method is applied to find the best direction and the minimum angle A_{lh} for the polyhedron $h^- \cap \partial P$. let

$$A_h = \min(A_{uh}, A_{lh}). \quad (7)$$

Since angle A_h reflects the resistance of removing both upper and low parts of the casting mold, it is used as a criterion to evaluate the casting planes. The bigger the angle A_h is, the more easily the casting mold removes from the object. To evaluate a casting plane h, firstly the best casting directions QO of h^+ and h^- must be calculated by (5) and (6), then calculate (7). The time complexity of evaluating a casting plane is O(n).

The criterion can be defined in several ways. Actually, when the object removes from the casting mold, the resistance is not only affected by the angle between the moving direction and the facets, but is affected by the area of the facets. If the area of the facet is considered, the best moving direction should make

$$\min\{(A_i x + B_i y + C_i z + D_i)/area(f_i)\} \quad (8)$$

maximum for all $f_i \in h^+ \cap \partial P$. The linear program become

max m, (9)

$m - (A_i x + B_i y + C_i (-(D+Ax+By)/C) + D_i)/area(f_i) \geq 0$, i=1,2,...k (10)

where $f_i \in h^+ \cap \partial P$. The solution of problem (9) and (10) is a point $Q(x_q, y_q)$ on the casting plane, and it can be solved in O(n) time. If the upper casting mold is removed along the direction QO, the resistance will be smaller. Let R(xr,yr) be on line OQ and such that |OR|=1 (ref. figure 4). As we described above,

$$R^+ = \min_{f_i \in h^+ \cap \partial P} (A_i x_r + B_i y_r + C_i (-(D+A x_r + B y_r)/C) + D_i)/area(f_i) \qquad (11)$$

and

$$R^- = \min_{f_i \in h^- \cap \partial P} (A_i x_r + B_i y_r + C_i (-(D+A x_r + B y_r)/C) + D_i)/area(f_i) \qquad (12)$$

reflect the resistance of removing the upper and low parts of the casting mold respectively, the criterion for evaluating the casting plane can be

$$\min\{R^+, R^-\} \qquad (13)$$

The casting plane with maximum value of $\min\{R^+, R^-\}$ is called the best casting plane.

3. The algorithm for calculating the best casting plane.

Firstly, the version of non-opposite cast removal is considered. For a polyhedron with n vertices, there exists $O(n^2)$ casting planes, and they can be found in $O(n^2 \log n)$ time. For every casting plane the criterion (13) must be calculated, then we can choose the best casting plane with the largest criterion (13). To calculate the criterion (13), point Q is necessary to be found by solving (9)(10). The time complexity of solving (9) and (10) is $O(n)$. The total time to calculate the criteria of the $O(n^2)$ casting planes is $O(n^3)$ time. It is therefore that the time complexity of for finding the best casting plane is $O(n^3)$. Now an algorithm for finding the best casting plane in $O(n^2 \log n)$ time is studied. The following lemma is proved in [1].

Lemma 1 If a simple polyhedron P is castable with casting plane h and in non-opposite directions then h contains an edge of P.

Since castability with respect to a plane h is only determined by the facets of P that intersect h^+ and the ones that intersect h^-. If h is a casting plane for P, then h can be perturbed if this does not involve new facets intersecting h. In case of orthogonal cast removal, the only perturbation allowed is translation. It follows that (ref. [1])

Observation 1 For castability with orthogonal cast removal, we may assume that the casting plane contains at least one vertex of P. For opposite and arbitrary cast removal, we may assume that the casting plane contains at least three vertices of P.

Partially best casting plane h_s is defined as that it is a casting plane with greatest value of (13) among the casting planes containing a side s of P. From lemma 1, the best casting plane can be found among the partially best casting planes h_s (s∈ the set of sides of P). From observation 1 the partially best casting plane h_s can be found among the planes containing a side s of P and a vertices of P. Now the method for finding the partially best casting plane h_s is described. Project all the planes containing edge s of P and a vertex V_i of P to the plane being vertical to edge s (ref. figure 5). If plane SV1 rotates along edge S in anti-clockwise to SVn-2, it sweeps all the vertices of P, the plane SV1 is chosen as the initial plane for sorting. Sort the planes SVi (i=1,2,....n-2) by the angle between the planes and the plane SV1.
Let

$$R_j^+ = \max_{f_i \in SV_j^* \cap \partial P} (A_i x_r + B_i y_r + C_i (-(D+A x_r + B y_r)/C) + D_i)/area(f_i). \qquad (14)$$

and

$$R_j^- = \max_{f_i \in SV_j^- \cap \partial P} (A_i x_r + B_i y_r + C_i(-(D + A x_r + B y_r)/C) + D_i)/area(f_i). \quad (15)$$

where $SV_i^+ = \{SV_j: j \geq i\}$ and $SV_i^- = \{SV_j: j \leq i\}$

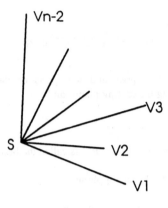

Figure 5

Lemma 2 When j increases from 1 to n-2, if SV_j is casting plane, then R_j^+ decreases, and R_j^- increases.

proof.
  Since $SV_j^+ \cap \partial P \supset SV_{j+1}^+ \cap \partial P$, from (11), it is obvious that R_j^+ decreases, when j increases. Similarly, the conclusion of that R_j^- increases, when j increases can be proved. It completes the proof.
  From lemma 2, when j increase from 1 to n-2, either $\min_{0<j<n-1}\{R_j^+, R_j^-\}$ is monotonous (ref. Figure 6(a)), or $\min_{0<j<n-1}\{R_j^+, R_j^-\}$ increases firstly, then decreases (ref. Figure 6(b)). Assume that $\min\{R_j^+, R_j^-\}$ is maximum at $j=j_0$, j_0 can be found by using a method similar to the bisection method. It needs $O(\log n)$ steps. For each step time $O(n)$ is used to calculate $\min\{R_j^+, R_j^-\}$, the total time for finding a partially best casting plane is $O(n \log n)$.

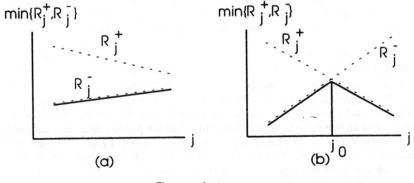

Figure 6

Since there are $O(n)$ edges of P, $O(n^2 \log n)$ time is necessary to calculate all the partially best casting planes. From $O(n)$ partially best casting planes the best casting plane can be found in $O(n)$ time. The total time for finding the best casting plane is $O(n^2 \log n)$.

For opposite cast removal there are only $O(n)$ casting planes, they can be found in $O(n^2)$ time. Since to calculate the criterion (13) can be completed in $O(n)$ time. The best casting plane of the $O(n)$ casting planes can be found in $O(n^2)$ time. For orthogonal removal the moving direction of a casting plane must be vertical to the casting plane, the best moving directions of h^+ and h^- need not to be calculated, the time complicity for calculating the criterion (13) of a casting plane is $O(1)$. If $O(n)$ casting planes are already found. The best casting plane can be found in $O(n)$ time. If the time of finding the $O(n)$ casting planes is considered, the time complexity of finding the best casting plane is still $O(n^2)$. From the above discussion, it follows that

Theorem 1. The best casting plane for arbitrary cast removal can be found in $O(n^2 \log n)$ time. For opposite and orthogonal removal the best casting plane can be found in $O(n^2)$ time.

4. Conclusion

Bose (ref. [1]) studied algorithms for finding the casting planes. If a casting mold is constructed based on the casting plane, the casting object can be removed from the cast mold without breaking the cast mold and the object. Some of the casting planes found by the algorithm provided in [1] is not suitable for casting. In this paper considering the casting plane with smaller removing resistance and removing distance, we define two criteria of evaluating casting planes, and study an efficient algorithm for finding the best casting planes among the casting planes found in [1]. They can be found in $O(n^2 \log n)$, $O(n^2)$ and $O(n^2)$ for arbitrary, opposite and orthogonal removal respectively. The time complexity is exactly the same as that for finding the casting planes by the algorithm in [1]. The method described in this paper can be extended to define new criteria considering other practical requirements, and can be used to design algorithms to find desirable casting planes.

6. Reference

1. Prosenjit Bose, David Bremner, Mare van Kreveld, Determining the Castability of Simple Polyhedra, *ACM Symposium on Computational Geometry* 1994.
2. Bose,P.,van Kreveld, and G,Toussaint, Filling Polyhedral molds. *Proc.3rd WADS* (1993),Lect. Notes in Comp. Science 709, Springer-Verlag,pp.210-221.
3. Bose,P., and G, Toussaint, Geometric and Computational aspects of injection molding. Proc.3rd Int. Conf. on CAD and Computer Graphics (1993),Beijing, China, pp.237-242.
4. Fekete,S.P.,and J.S.B. Mitchell, Geometric aspects of injection molding, manuscript,1993
5. McAllister,M.,and J.Snoeyink, Two dimensional Computation of the three dimensional reachable egion for a welding head. *Proc. 5th Canadian Conf. on Comp. Geom* (1993),pp.437-422.
6. Asberg, B., G. Blanco, P.Bose,J.Garcia, M. Overmars, G. Toussaint, G. Wilfong, and B. Zhu, Feasibility of design in Sterolithography. To appear in: *Proc. 13th Symp. on FST TCS*,1993.
7. Preparata, F.P., and M.I. Shamos, *Computational Geometry- an introduction.* Springer-Verlag, New York,1985.

SIMULATION OF CONJUGATE PROFILES IN GEAR SHAPING USING COMPUTER GRAPHICS

S.V.R.Surya Narayana
Research Scholar

Dr.V.Jayaprakash
Asst. Professor

Dr.M.S.Shunmugam
Professor

Department of Mechanical Engineering
Indian Institute of Technology
Madras - 600 036
INDIA
E-mail: mech7@iitm.ernet.in

ABSTRACT

Conjugate profiles are profiles which transmit motion by a prescribed function. Both analytical and graphical methods are used to arrive at the conjugate profile for a given profile. In this paper use of computer graphics in determination of conjugate profiles has been explained with examples related to gear tooth profiles which are generated practically using gear hobbing or shaping processes. Modifications on the conjugate profile corresponding to a change in the given profile are also explained. Undercutting of gear tooth profile using simulation has been modelled.

Keywords : Conjugate profile, Graphical Simulation, Gear shaping

1. Introduction

Two profiles are said to be conjugate if they transform motion by a prescribed function. The function may be a constant angular velocity ratio as in the case of gears or in general it may be non-linear as in the case of cams and followers. The use of conjugate profiles has wide applications in the areas of gears, cams, mechanisms and linkages. Analytical and graphical methods are used to obtain the profile conjugate to a given profile [1,2,3].

In this paper use of computer graphics in determining the conjugate profile has been explained. Gears can be manufactured accurately in generation cutting as compared to form cutting processes [2,7]. In generation cutting the cutter and gear will be of conjugate shapes. Given the cutter profile and the basic kinematic motions involved in the generation process the gear profile can be determined. The methodology to calculate the conjugate profiles has been outlined and use of simulation in the determination of conjugate profiles has been dealt with examples in the area of gear shaping.

2. Determination of Conjugate Profile

Conjugate profiles can be calculated using coordinate transformation techniques and theory of envelopes. Given the profile $[r_1]$ in vector form, it is possible to calculate the conjugate shape $[r_2]$. Analytical and graphical methods used to compute the conjugate gear tooth profile have been explained with reference to rack type shaping.

Fig 1 shows the rack type shaping process. The straight sided rack is conjugate form to the involute gear tooth profile. The generating motions involved are the rotation of gear and the translation of the rack. The cutter is radially fed into the work for the required depth of cut.

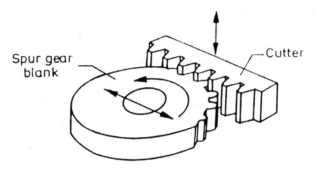

Fig.1 Spur gear and it's conjugate rack in rack type shaping

The coordinate systems $S_1(x_1,y_1,z_1)$, $S_2(x_2,y_2,z_2)$, $S_3(x_3,y_3,z_3)$ are connected to the tool, gear blank and the fixed frame respectively as shown in Fig 2. The figure shows the pitch circle of the gear in mesh with pitch line of the rack. The steps (a) to (e) explain the methodology to calculate the conjugate profile.

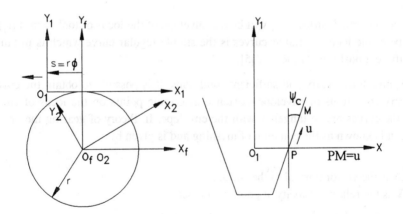

Fig 2. Kinematic motions involved while generating gear tooth profile with rack

Step a :

The profile of the tool $[r_1]$ has to be represented in system S_1 in parametric form as follows.

$$r_1(\theta) \in C^1 \qquad (1)$$

where θ is the parameter and C^1 denotes that the curve is continuous and first derivative exists. In rack type shaping, the rack with pressure angle ψ_c can be represented with parameter u as shown in figure.

$$x_1 = \frac{\pi m}{4} + u \sin\psi_c \, ; y_1 = u \cos\psi_c \qquad (2)$$

where m is the module of the gear.

Step b :

The motions of the tool and blank have to be represented by rotation or translation of systems S_1 and S_2 with the using coordinate transformation.

The kinematic motions of gear and rack cutter are represented as translation 's' of rack system S_1 and rotation ϕ of blank system S_2. by the relation

$$s = r\phi \qquad (3)$$

Step c :

As a result of the kinematic motions, a locus $[r_\phi]$ of the tool shapes $[r_1]$ is obtained in gear blank coordinate system S_2, ϕ being the parameter involving the kinematic motions.

Using the kinematic motions, the locus of rack tooth form $[r_1]$ in system S_2 is obtained as

$$[r_2] = [M_{21}][r_1] = [M_{2f}][M_{f1}][r_1] \qquad (4)$$

Step d :

The generated surface $[r_2]$ will be the envelope of the locus of tool shapes $[r_1]$. The envelope of the locus of planar curves is the simple regular curve which is in tangency with different positions of the loci [5].

Applying necessary and sufficient conditions, it is possible to obtain the envelope. Necessary conditions of envelope existence determine points on the locus of curves at which the curves are in tangency with the envelope. In theory of gearing the necessary condition is known as the equation of meshing and is given by

$$\mathbf{N.V} = 0 \qquad (5)$$

where **N** is the vector normal to the surface,
and **V** is the relative velocity at point of contact.

In rack type shaping, equation of meshing $N_1 . V_1 = 0$ gives

$$u - r\phi \sin\psi_c + \frac{\pi m}{4}\sin\psi_c = 0 \qquad (6)$$

As parameter u is varied ϕ can be calculated from equation (6) and gear profile can be calculated from equation (4).

Step e :

Sufficient conditions of envelope existence determine the singular points where envelope does not exist as a regular curve. The sufficient condition is given by $N_2=0$, where vector N_2 is the normal to the generated surface. This condition gives the singular points beyond which undercut occurs.

Applying the sufficient condition in the case of rack type shaping i.e vector $N_2 = 0$ results in a limiting value of ϕ, thus limiting the tool parameter u beyond which undercut occurs.

$$\phi_1 = \frac{\pi}{2z} - \tan\psi_c$$

$$u = -\frac{\pi m}{4}\sin\psi_c + r\phi_1 \sin\psi_c \qquad (7)$$

Using this methodology, given the tool surface, the generated gear surface can be determined. Similarly, in the reverse transformation, given the gear tooth surface $[r_2]$, it is possible to determine the tool surface $[r_2]$ which would be a useful information in reverse engineering for the tool designers.

In the simulation procedure, the kinematic motion of the rack and gear resembles the rolling of rack on a gear. As the pitch line of the rack rolls on the gear pitch circle, the rack profile can be graphically plotted at different rotation angles, thus obtaining the locus of tool shapes in gear coordinate system. Since the gear tooth profile is the envelope of loci of rack profile, the conjugate gear profile can be seen as a tangential curve to the locus of tool shapes. Fig 3 gives the rack tool and the simulated conjugate tooth profile.

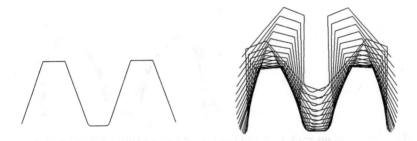

Fig 3. Rack tool and conjugate tooth profile

A rack undercuts the gear tooth profile if the number of teeth on gear are less than a certain number. For 20° pressure angle, undercut occurs when the number of teeth is less than 22. Fig 4 shows three different gear profiles having number of teeth 10, 22 and 60 as generated by same rack. Undercut is clearly seen on the gear profile having number of teeth 10 while undercut is not there on gear profile with number of teeth 60. The gear profile having number of teeth 22 is the limiting case for undercut.

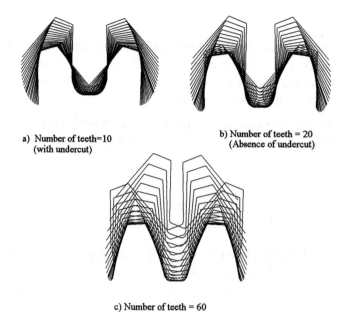

a) Number of teeth=10 (with undercut)

b) Number of teeth = 20 (Absence of undercut)

c) Number of teeth = 60

Fig 4. Gear tooth profiles

The tip of gear profiles is often chamfered in order to reduce impact loads while meshing. In order to generate the chamfer on the gear profile, a semitop tool profile is used. Fig 5 shows the rack tool with semitop feature and the generated gear tooth profile with chamfer.

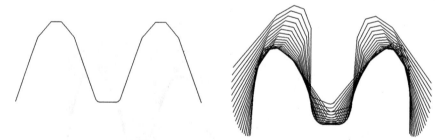

Fig 5. Semitop on rack tool and conjugate tooth profile with chamfer

3. Reverse Transformation for Calculation of Cutter Profile

Using reverse transformation techniques, given the gear tooth form it is possible to determine the tool profile. The gear profile [r_2] is taken as an involute curve and the locus of the curve in tool coordinate system S_1 is determined and the envelope to that locus is obtained as a straight sided rack.

The involute gear tooth form in coordinate system S_2 is expressed as

$$x_2 = r_b(\sin(\theta+\xi) - \theta\cos(\theta+\xi))$$
$$y_2 = r_b(\cos(\theta+\xi) + \theta\sin(\theta+\xi)) \tag{8}$$

where r_b is the base circle radius of gear, θ is the parameter of gear profile, and angle

$$\xi = \frac{\pi}{2z} - \tan\psi_c + \psi_c \tag{9}$$

The kinematic motions remain same but the locus of gear tooth form $[r_2]$ has to be calculated in rack coordinate system S_1 as given below.

$$x_1 = x_2\cos\phi + y_2\sin\phi + r\phi$$
$$y_1 = -x_2\sin\phi + y_2\cos\phi - r \tag{10}$$

The equation of meshing is given by

$$\theta + \xi - \phi = \psi_c \tag{11}$$

Using equations (10) and (11) the conjugate rack profile can be calculated analytically. Using the methodology given above, the locus of gear profile at different rotation angles has been plotted and the envelope obtained is seen as a straight line. Fig 6 shows the conjugate rack profile as obtained in simulation and the rack tool profile (offset) as obtained by analytical method.

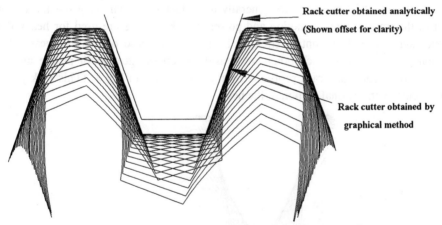

**Fig 6. Gear tooth space profile and conjugate rack
(Comparison of graphical and analytical methods)**

4. Case Studies

Case 1:

The conjugate profile depends on the given shape as well as the kinematic motions of the two profiles in engagement. Pinion type shaping is a generating process in which the cutter and gear rotate as if two gears are in mesh. The kinematic motions involved are rotation of gear and cutter in a timed relationship given by $\phi_g / \phi_c = r_c / r_g$, where r_c and r_g are the pitch circle radii of cutter and gear respectively. Using the simulation procedure, the cutter profile for involute gear profile is obtained as another involute as shown in Fig 7.

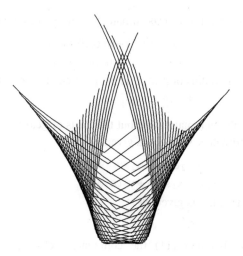

Fig 7. Simulation of pinion type shaping

Case 2 :

Sprockets and Novikov gears generally have double circular arc geometry and are used in the transmission of motion and power. Novikov gears are used for heavy duty applications since their contact strength is more than those of involute gears. Two circular arcs tangent to each other are considered on the gear tooth profile and the corresponding conjugate rack profile is obtained as shown in Fig 8. (This rack profile also resembles the normal section of the hob).

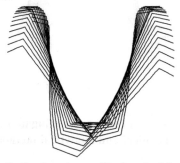

Fig 8. Conjugate profile for double circular arc profile

Case 3:

Machine elements with serrated features are used for transmission of motion and torque in different applications. Splines with straight sides are used for rear axle shafts, universal joints and machine tool spindles. The profile on the spline is taken as straight line and the conjugate rack profile is obtained by the simulation procedure. Fig 9 shows the straight line on the spline and the conjugate rack form as the envelope.

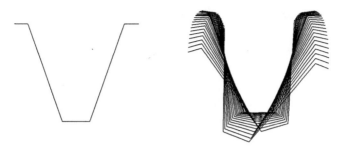

Fig 9. Conjugate rack for straight sided spline

5. Conclusions

In this work, considering the kinematic motions involved in rack and pinion type shaping, conjugate profiles are calculated for the given cutter/gear profiles using graphical simulation. A generalised method to determine the conjugate profiles graphically by using computer simulation techniques is presented. This method can be used to calculate the tool profile for various gear tooth profiles in generation cutting. This method can be extended to other fields like cams, mechanisms and linkages also. This simulation procedure can be extended to meshing and tooth contact analysis of gears. The modifications produced on gears due to different features on the tool profile can also be simulated.

6. References

1. F.L.Litvin, "*Theory of Gearing* ", **NASA** Publications, 1989
2. P.Rodin, "*Design and Production of Metal Cutting Tools*", **MIR** Publishers, Moscow.
3. R.L.Huston., "*Computer Simulation of Gear Manufacturing Processes*", **NASA,** CR185200.
4. F.L.Litvin, "*Methods for Generation of Gear Tooth Surfaces and Basic Principles of Computer Aided Tooth Contact Analysis*", Proc. of Computers in Engg., **Vol 1**, 1985, pp 556-564.
5. A.Goetz, "*Principles of Differential Geometry*", **Addison Wesley** Publishing Company, 1970.
6. Chung-Biau Tsay, "*Helical Gears with Involute Shaped Teeth: Geometry,Computer Simulation,Tooth contact Analysis,and Stress Analysis*", Trans of ASME Jr.of Mech. , Trans. and Auto. in Des., Dec 1988, **Vol 110**, pp 482-491.
7. V.Arshinov,G.Alekseev, "*Metal Cutting Theory and Tool Design*",**MIR** Publishers, 1970

CAD Tools Development

An Algorithm For Nesting Patterns In Apparel
    J. Y. Wang, D. Y. Liu, E. W. Lee, T. H. Koh and M. B. Maswan

A 3D-Clipping Algorithm for Form Feature Volume Extraction
    S. R. P. Rao Nalluri, V. Vani and B. Gurumoorthy

An Experiment in Integrating CAD Tools
    M. Bounab and C. Godart

Parametric CAD Based on Relation Model
    Z. Y. Ou, J. Liu and B. Yuan

Evaluation of Four CAD Systems Using Analytic Hierarchy Process
    S. Agarwal, Y. T. Lee and S. B. Tor

Joint Application of B-Spline and Bezier Methods for Surface Modeling
    X. G. Ye, D. Q. Li and F. Y. Wang

Tool Paths for Face Milling Considering Cutter Tooth Exit/Entry Conditions
    Y. S. Ma

AN ALGORITHM FOR NESTING PATTERN IN APPAREL

JIA YE, WANG *, DING YUAN, LIU, ENG WAH, LEE, THONG HWEE, KOH
GINTIC, Nanyang Technical University, Singapore.2263
E-mail: gjywang@ntuvax.ntu.ac.sg
(* Dept. of computer science, Shandong University,P.R.China 250100)

MASJURI BIN, MASWAN
Institute of Technical Education, Singapore.

ABSTRACT

An algorithm for finding the first contact points between two polygons is presented. Under certain condition the algorithm can be completed in linear time, and is optimal. An approach reducing the number of potentially intersecting laid out polygons is discussed. These two aspects are essential for reducing the response time in the process of interactive nesting of polygons.

1. Introduction

Nesting of two dimensional shapes has a wide range of applications in the real world. For example in shipbuilding, sheet metal stamping, apparel industry and so forth. But the requirements of layout in different industry is quite different. For example in the apparel industry the number of different patterns to be laid out is considerably large, and the expected efficiency of the material is very high. Conversely, in sheet metal stamping usually one or two kinds of patterns are needed to be laid out. There is a great amount of literature discussing the optimization of layout.

Some of the algorithms presented can only be applied to the case where the number of different patterns is one or two (ref. [1],[2]). It is practical for some applications, such as sheet metal stamping. In apparel industry to lay out as many as forty different patterns on a fabric is common. Furthermore the expected utilization ratio is usually between 86% to 91%, and the shape of the patterns are irregular. Since it is NP-hard for layout in 2D plane, to find a practical algorithm for optimization of layout patterns in the apparel industry is impossible, and it is also difficult to find a heuristic algorithm to obtain such high utilization ratio (ref. [4],[5]). Interactive layout is therefore the practical approach in the apparel CAD/CAM system these days.

When a chosen pattern is to be laid out, the user drags the pattern to a certain position on the screen by mouse, a direction is then specified. The pattern is going to move in the direction until it touches the other polygons laid out, then the pattern settles this position. Usually the apparel patterns are composed of lines and curves, therefore the polygons approximating the patterns may have many sides. At a later stage where a great deal of polygons have been laid out, all of them can have the potential first contact with the moving polygon. If a naive method is applied to calculate the first contact point, we will certainly not have a fast enough response which is one of the most important requirements for an interactive process. Thus in order to shorten the computing time two problems must be solved. One is to minimize the time complexity of finding the first contact points between a moving polygon in a direction specified by the user and another laid out polygon. The second problem is to reduce the number of laid out polygons potentially touching the moving polygon.

In [3] an optimal algorithm for finding the first contact points between two polygons under a condition that the bounding box of one polygon must not intersect with another polygon is presented. However this condition is too strict to apply it in apparel manufacturing.

In this paper a practical optimal algorithm for finding first contact points of a moving polygon towards another polygon is presented. An approach reducing the number of potentially intersecting laid out polygons is also discussed.

2. Algorithm finding first contact point

In this paper polygons are represented by a sequence of vertices (x_i, y_i), and they are assumed tobe directioned. When a travel is taken on the boundary of polygon P, if the inside of P is on the right-hand side of a traveler, the direction of the traveler is called positive. That is, if the polygon and a circle is topological homeomorphism, the clockwise direction is positive.

Let R be a polygon laid out, and P be the moving polygon. The direction of movement of polygon P is assumed to be left horizontal and towards polygon R (Figure 1). Now we consider a relatively simple case, i.e. if polygon P moves any distance in the opposite direction, it will not intersect polygon R (ref figure 1). The more complicated cases (eg. Figure 10) will be discussed later. It is clear that if the first contact points are calculated by naive method, the time complexity is $O(n^2)$, where n is the number of sides of P and R.

There are two steps in this algorithm for finding the first contact points of P and R. The first step is to create monotonous polylines of P and R (dotted lines in figure 1). These polylines are composed of some horizontal line segments and certain sides or their part of P and R where the first contact points may be on them. The second step is to find the first contact points of the two monotonous polylines.

In order to create the monotonous polyline of polygon P, assume that Pt and Pb are the top-most and bottom-most points of P respectively, and R is on the left side of P. The sides of the monotonous polyline of P are required to satisfy the following conditions.

1. For any point (x,y) on the sides, $y(Pb) <= y <= y(Pt)$. where y(Pi) is the y coordinate of point Pi.

2. Rays directed from any point on any side of the monotonous polyline of P going towards left do not intersect any sides of P.

3. Horizontal sides are used to link above line segments to form a polyline.

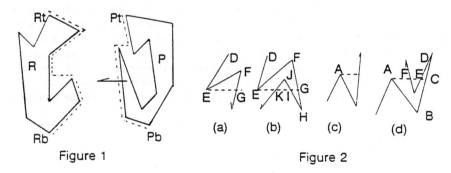

Figure 1   Figure 2

A double linked list LP is used to record the temporary and final results of the monotonous polyline. The component or individual element of LP is made up of

```
struct point
{float x,y;
  point pre,next;
},
```
where (x,y) are the coordinates of a vertex of the polyline, pre and next are two pointers pointing to the previous and next components of LP respectively. Linking two vertices stored in adjacent components of LP forms a side of the polyline.

To construct the monotonous polyline of P, the sides of left part of the boundary of P between vertices P_b and P_t are studied. A travel is taken on the boundary of P from P_b to P_t in positive direction. As one travels, some new components storing vertices which the traveler has just passed are inserted, and some components storing those vertices which fall on the right side of those just created sides of the monotonous polyline are deleted. Hence LP will keep all the left most sides which the traveler passed. The first component of LP stores the vertex Pb, and the y-coordinate of the component in LP is always equal to or less than that of the next component. A pointer M points to the last created component. If the component pointed to by M is going to be deleted, M points to its previous or next adjacent component before deleting. Whether M points the previous or the next components dependents on if the traveler goes up or down. If the traveler goes up, let M point to the previous component, otherwise M points to- the next component. For example, In figure 2(b) when one travels on IJ and JK, M points to I

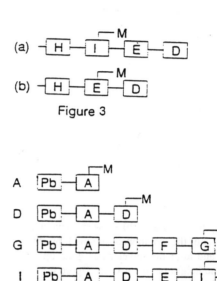

Figure 3

Figure 4

before the traveler reaches F from point E. Since E must be deleted, when he gets to point F, M points to M->pre, that is point A. As the traveler travels, LP is created step by step. Figure 4 shows LP, when the traveler reaches to the vertices A, D, G, I and Pt on the path showed in Figure 5. When the traveler goes down to a new vertex, the previous components of M (M->pre, M->pre->pre,..) must be checked in sequence, if the vertices stored in them fall in the right side of the last travel (ref. figure 3(a)), then the traveler goes down to K, since I is in the right side of K, I should be deleted, M points to M->next, that is point E. In figure 2(d) M points to E,
eling side. If it is true, the components are deleted (e.g. F in figure 5). This process will continue, until a previous component of M being not on the right side of the last traveling side is found. If the traveler goes up the next component of M (M->next) and its following should be checked in sequence. When the monotonous polyline of P is created, some vertices passed by the traveler will be storedin LP, and some vertices will

not. A criterion must be given to determine if the vertices of **P** passed by the traveler should be stored in LP or not. A flag is used to indicate whether in the next several steps the vertices of **P** the traveler passed will (flag=1) or will not (flag=0) be stored in LP. The characteristics of making flag changing from 1 to 0 are listed below.

1. The traveler goes up and turns down to the right side of the previous side in LP (figure 6 (a)).
2. The traveler goes down and turns up to the right side of the next side in LP (figure 6 (b)).
3. The traveler goes up and passes by the right side of the vertex pointed by M->next (figure 6 (c))

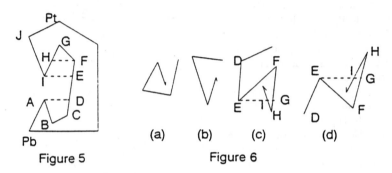

Figure 5   Figure 6

Figures 7 (a) and (b) show the contents of LP and the component pointed by M, before and after the traveler has gone up to the same height of point G in figure 6 (c). After G is deleted, the traveler passes by the right side of vertex E pointed by M->next in LP.

4. The traveler goes down and passes by the right side of the vertex pointed by M->pre (figure 6(d)).

It is clear from figures 7 (c) and (d) which show the contents of LP and the component pointed by M, before and after the traveler has gone down to the same height as vertex G of figure 6 (d).

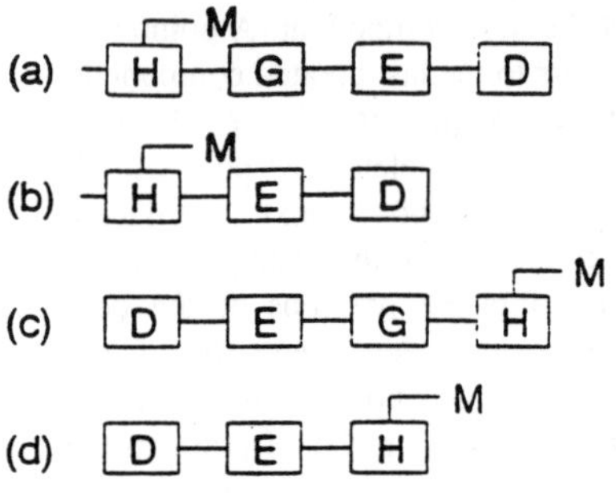

Figure 7

The characteristics of making flag changing from 0 to 1 are listed below.

1. The traveler goes down and passes by the right side of the vertex pointed by M (figures 2 (a),(b)).

Figures 3 (a) and (b) show LP and the component M pointed to, before and after the component storing I (figure 2 (b)) is deleted. According to the rule described above for changing the pointer M, before the component storing I is deleted, M changes to point to E. After the component stored I is deleted, the traveler passes by the right side of E which pointed by M.

2. The traveler goes up and passes by the right side of the vertex pointed by M (figures 2(c),(d)).

The time complexity of creating the monotonous polyline of **P** is liner. The number of sides in the polyline is equal to or less than O(n), where n is the number of sides of **P**. Similarly, a monotonous polyline of polygon **R** can be created.

Once the monotonous polylines of **P** and **R** are created, we travel up from Pb and Rb around the monotonous polylines of **P** and **R** simultaneously, and calculate the horizontal distances between vertices and sides of the monotonous polylines of **P** and **R**. The points on **R** corresponding to the minimum horizontal distance are the first contact points (figure 8). This process can be completed in linear time.

From the above description, it follows theorem 1.

Theorem 1. If a moving polygon **P** intersects a polygon **R** when it moves in direction **d**, and it does not intersect **R**, when it moves in the opposite direction, then the first contact points can be found in $O(m+n)$ time, where m and n are the numbers of sides of the polygon **P** and **R** respectively.

The algorithm described above is also optimal in time complexity.

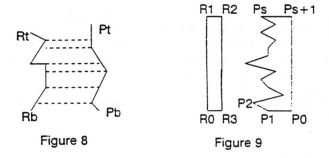

Figure 8   Figure 9

Theorem 2. The lower bound of time complexity of finding the first contact points for two polygons **P** and **R** is $O(m+n)$.

Proof.
Assume that

$$x_1, x_2, x_3, \ldots, x_s,$$

are s real number in the interval (a,b). construct a polygon P with vertices $\{P_0, P_1, P_2, P_3, \ldots P_{s+1}\}$ (ref. figure 9), where

$$P_0=(b+1,0),\ P_{s+1}=(b+1,s)$$

and

$$P_i=(x_i,i),\ i=1,2,3,\ldots s.$$

Let R be a rectangle with vertices $\{R_0, R_1, R_2, R_3\}$, where

$$R_0=(a-2,0),\ R_1=(a-2,s),\ R_2=(a-1,s),\ R_3=(a-1,0).$$

Assume the first contact points are P_{ik} $(k=1,2,\ldots k_0)$, when polygon P moves towards the right. The minimum number of $x_1, x_2, x_3, \ldots, x_s$ must be x_{ik} $(k=1,2,\ldots k_0)$. If an algorithm can find the first contact points in less than $O(s)$ time, then the minimum of s numbers can be found in less then $O(s)$ time. it contradicts that the lower bound of time complexity of finding the minimum of k numbers is $O(k)$. It is therefore that the lower bound of the algorithm for finding the first contact points of the polygons P and R is

$O(\max(m,n)) = O(m+n)$.

This completes the proof.

In order to solve more complicated cases (ref. figure 10), firstly we identify local monotonous boundary segment (LMBS) of polygon P. LMBS of P is a monotonous part of the boundary of P, when one travels on the segment in positive direction, the y-coordinate always increases. In figure 10 BA, DC, FE, HG and JI are the LMBS of P. LMBS is different to monotonous polyline of P described above. The letter is composed of some left most sides of P and horizontal lines linking the sides to be a monotonous polyline. LMBS is a monotonous part of the boundary of P. From every end vertex of LMBS draw a horizontal line to the left (ref. the doted lines in figure 11). These horizontal lines are called constructive lines. The polygons bounded by the constructive line segments and the LMBS of P are called local polygons (figure 11 Q_0, Q_1, Q_2, Q_3, Q_4). In every local polygon Q_i a local monotonous polyline of R can be created. To do so, a traveling on R in positive direction carries on. When traveler enter to Q_i , a similar approach as creating the monotonous polyline of polygon R described above is applied to create the local monotonous polyline of R in Q_i. The difference is that only the sides of R in Q_i are eligible to be the sides of the local monotonous polyline of R in Q_i . After the local monotonous polylines of R in Q_i (i=1,2,3,...) are created, in every polygon Q_i local first contact points of the LMBS, a part of the boundary of Q_i, and the local monotonous polyline of R in Q_i can be found by the same approach described above. Among the local first contact points, those having minimum horizontal distance to P will be the first contact points of P and R.

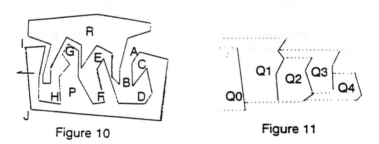

Figure 10   Figure 11

To create the local monotonous polylines of R in local polygons Q_i, one travels around the boundary of R in positive direction. Enter or leave polygon Q_i, only when one crosses certain horizontal constructive line segments. If there are k local polygons Q_i, in order to reduce the time of determination of which constructive line is crossed, the constructive horizontal line segments are sorted by the y coordinate. It takes $O(1)$ time to determine if the side of R the traveler being on crosses a horizontal constructive line, except the first crossing. For the first crossing it takes $O(\log(k))$ time to find two adjacent constructive lines that the traveler being between them. To create the local polygons of P takes $O(k*n)$ time, where n is the number of the sides of polygons P and R. The time complexities of creating LMBS of P, local monotonous polylines of R, finding local first contact points and global first contact points are $O(n)$. Sorting the horizontal line segments needs $O(k*\log(k))$ time. The total time complexity is therefore equal to $O(k*n)$. It follows theorem 3.

Theorem 3. The time complexity of finding the first contact points of two polygons is $O(k*n)$, where n is the number of the sides of the polygons, k is the number of the local polygons of the moving polygon.

In applications, particularly in apparel industry, usually k=O(1). In this case the time of finding first contact points of two polygon is still O(n).

3. Reducing the number of polygons calculating first contact points.

In previous paragraphs, an efficient algorithm finding first contact points is described. When a great amount of polygons have been laid out on a fabric, and one more polygon is going to be laid out. If the first contact points of the moving polygon with all laid polygons must be found, it is time consuming. We must reduce the number of polygons for which the first contact points with the moving polygon must be calculated.

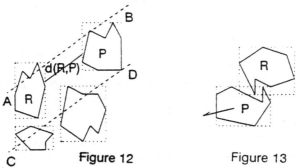

Figure 12   Figure 13

Assume that the moving polygon is P. The area between AB and CD, two support lines being parallel to the moving direction of P, is called potentially intersecting zone (PIZ in short) of P (Figure 12). Those polygons except P itself the bounding boxes of which intersect PIZ of P are called potentially intersecting polygons. Calculate the distances d(R,P) (ref. figure 12) in the moving direction between the bounding boxes of P and the potentially intersecting polygons R. For the case of figure 12, the moving direction is point to down and left, d(R,P) can be calculated by the following formula.

$$d(R,P)=\max((x_{min}(P)-x_{max}(R))/|\cos(\alpha)|, (y_{min}(P)-y_{max}(R))/|\sin(\alpha)|).$$

where α is the angle of the moving direction and a horizontal line, $x_{min}(T)$ and $x_{max}(T)$ are the minimum and maximum x-coordinates of the vertices of polygon T, $y_{min}(T)$ and $y_{max}(T)$ are the minimum and maximum y-coordinates of the vertices of polygon T. d(R,P) can be negative. Even if it is negative, P may touch R, when P moves in the moving direction (e.g., In figure 13). From figure 12 and figure 13, it follows that if and only if

$$d(R,P)*|\cos(\alpha)|>-(x_{max}(R)-x_{min}(R))-(x_{max}(P)-x_{min}(P)) \text{ and}$$

$$d(R,P)*|\sin(\alpha)|>-(y_{max}(R)-y_{min}(R))-(y_{max}(P)-x_{min}(P)), \qquad (1)$$

polygon P may touch polygon R. Sorting the potentially intersecting polygons T that satisfies (1) by the distance in the moving direction d(T,P). Beginning from the polygon that has smallest distance with P, find the first contact points with P in turn, and record the minimum distance d in moving direction between the first contact points and P. This process will be ceased until all the polygons in the sequence are calculated or a polygon T with d(T,P)>d is met. After that, if polygon P moves a distance of d, the minimum moving distance recorded, in the moving direction, P just touches the first contact points. Since only those laid polygon the bounding boxes of which intersect PIZ and (1) is satisfied can be in the sequence of potentially intersecting polygons, and actually after calculating the first contact points for first two or three polygons in the sequence is implemented, the condition d(T,P)>d already becomes true. It is therefore that only for two or three laid out polygons the first contact points should be calculated.

4. Conclusion.

In this paper two main aspects of reducing time complexity of interactive nesting are discussed. The methods presented are successfully applied to implement nesting in apparel industry (ref. figure 14). Rapid response is performed for a polygon to find a closest position to the laid out polygons in the specified direction. The methods can also be applied in other industries such as shipbuilding.

5. Reference

1. A.Y.C.Nee etc., Computer Aided Layout of Metal Stamping Blanks, Proc. Instn Mech Eng. Vol 198B, NO.10,1984.
2. Yu Huagang, Xiao Xiangzhi, Xiao Jingrong, "Optimization of Blanking Layout." Journal of Huazhong University of science and Technology, vol.15 No1 1987.
3. J.Y.Wang, "An Optimal Algorithm of Finding First contact between translating polygons." Chin. J. of Computer. No2 1992.
4. Alain Mangen and Nadine Lasudry, "Search for the Intersection Polygon of any Two Polygons: Application to the Garment Industry." Computer Graphics Forum 10 ,PP195-208 1991.
5. C. Sechen, A. Sangiovanni-Vincentelli, "The TimberWolf Placement and Routing Package." IEEE Journal of Solid-State Circuits 20(2), p.510 (April 1985)

A 3D-CLIPPING ALGORITHM FOR FORM FEATURE VOLUME EXTRACTION

S. R. P. RAO NALLURI

Department of Mechanical Engineering, Indian Institute of Science
Bangalore, 560 012, India
E-mail: srp@mecheng.iisc.ernet.in

VEGESNA VANI and B. GURUMOORTHY

Department of Mechanical Engineering Indian Institute of Science
Bangalore, India.
E-mail: bgm@mecheng.iisc.ernet.in

ABSTRACT

Extraction of volume corresponding to a recognised form feature is an important step in feature extraction. Face extension (Sakurai and Gossard 1988, Dong and Wozny 1988) and edge extension (Falcidieno and Ginnini 1987) techniques that have been used earlier have limited scope. Only, features with convex volumes and those, that do not create new faces on updating the part solid have been addressed by the above techniques. In this paper a general algorithm based on the concept of 3D clipping is presented. This algorithm has been implemented and tested as part of a domain independent form feature recognition system.

Keywords

Form-feature, volume extraction, 3D-clipping

1. Introduction

1.1. Feature concept

The concept of feature evolved from the desire to devise methods for integration of part geometric model with applications such as process planning, group technology (GT) classification and NC programming (Shah 1991). Features are high level abstractions of part geometry such as holes, slots and ribs around which engineering knowledge and expertise are structured. Hence automation of reasoning involved in applications such as process planning requires part description in terms of these features rather than basic topological entities (face, edge and vertex) that are available in geometric models. These features are referred to as form features (Shah 1991) as they describe the form and shape of a part. Automatic feature extraction is one of the approaches to realise feature models.

1.2. Automatic Feature Extraction

In this approach, form features of interest to an application are automatically extracted from the geometric model of a part. Automatic feature extraction has fundamental significance from the perspective of integrating product development cycle. Reasoning involved in downstream applications cannot be automated without correct interpretation of part design at the appropriate level of abstraction. Through feature extraction, it is possible to generate interpretation of part geometry at the correct level of abstraction for complete automation. The main attraction of feature extraction is that existing CAD models can be used. The feature extraction process consists of two steps: recognition of the form feature and extraction of the volume associated with the feature.

2. Feature Volume

The volume associated with the feature is useful in two ways. In automatic NC path generation, feature volumes are required for automatic generation of the cutter location (CL) file from the extracted features (Chang and Bala 1991, Balaji and Gurumoorthy 1992). Volume associated with a feature is also useful in the feature extraction process itself, as it can be used to handle interacting features (Sakurai and Gossard 1990). For example, in the object shown in figure 1, recognition of the two blind slots is straightforward in all approaches reported in literature. However most

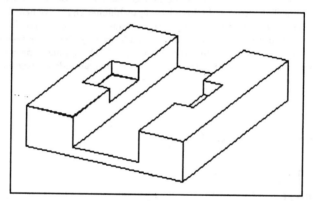

Figure 1 : Interacting features

approaches based on graph matching (Joshi and Chang 1988, Marefat and Kashyap 1990) or rules (Henderson 1984) would be unable to recognise the through slot (in the absence of specific templates or rules for this). If however the recognised feature is deleted from the part model, as is done in some approaches (Sakurai and Gossard 1990, Dong and Wozny 1988), recognition of through slot becomes straightforward. Therefore, by deleting the recognised feature and using the updated part model for further extraction, problems due to interacting features can be avoided. Face extension and edge extension have been used in the earlier approaches to update the part solid during extraction. Sakurai and Gossard (1990) have implemented the wireframe fleshing algorithm, proposed by Markowsky and Wesley to extract the volume associated with general features.

A more general approach to updating the part solid is to find the volume corresponding to the recognised feature and adding/subtracting the volume to/from the part model. If the recognised feature is negative, the feature volume is added to the part and if the recognised feature is positive, the feature volume is subtracted from the part.

Approaches based on convex decomposition and those based on CSG models obtain the feature volumes directly. However in the convex decomposition approach the volumes may have to be processed further as they may not be meaningful. The use of CSG models for feature extraction has not been popular because of the non-uniqueness inherent in CSG and because, to obtain the features in terms of the evaluated entities on the part, boundary evaluation is required. Volume decomposition work done by General Dynamics 1985 is also based on face extension approach. A given part solid is decomposed into machinable volumes. In this work face extension approach has been used only for simple parametrizable features.

3. Feature definition and classification

We define a form feature as a set of faces with distinct topological and geometric characteristics. We have modelled creation of a form feature as addition/subtraction of a solid (feature-solid) to/from another solid (base-solid). Here the feature-solid is the solid used for creating the feature and the base-solid is the solid on which the feature is created. We define the feature-solid as the exact minimum volume required to create a negative feature.

A feature is created by subtraction/addition of a feature-solid from the base-solid. In this process, some faces are newly created in the base-solid and some existing faces are modified. The newly created faces are referred to as created-faces and modified faces are referred to as shared-faces with respect to the created feature. In figure 2, faces 7, 8, 9, and 10 are created-faces and faces 1, 2, and 5 are shared-faces with respect to the feature created.

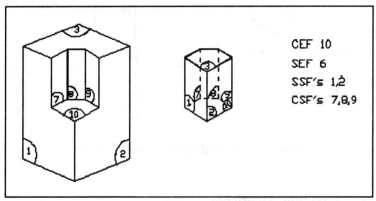

Figure 2 : Face classification

A face in the part model can be the created face of one and only one feature. The whole process of feature extraction is about correct association of faces in the part model with their respective features. A recognised feature therefore is represented by lists of created faces and shared faces, respectively.

4. Definitions and Concepts

Supporting Face: A face in a solid is classified as 'supporting' face if and only if the entire solid lies on one side of the plane containing the face.

Face-Face Classifications: A face 'f1' of a polyhedral solid can be classified with respect to the plane of another face 'f2' as above, below, intersect or on depending on whether vertices of f1 are all above, all below, some above and some below, or all on, respectively with respect to the plane of face 'f2' (figure 3). Each face of the solid is classified with respect to all the remaining faces and stored in Face Classification Table (FCT) which is a nf*nf (nf=number of faces in solid) matrix. FCT gives the spatial distribution of faces of the solid which is useful in geometric reasoning.

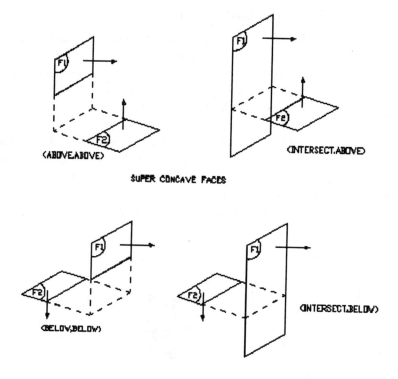

Figure 3 : Spatial classification of faces

5. Volume Extraction Algorithm

In this section the algorithm to extract the volume of the recognised feature is presented. The algorithm extracts the feature volume by clipping the bounding box (Bbox_S), that encloses the part-solid, with respect to the created and shared faces of the feature recognised.

Algorithm Extract_Feature_Volume(created and shared faces, Bbox_S)

STEP1: Form a set Fc, consisting of the created and shared faces of the feature. For each shared-face in Fc
    {

	IF the adjacent created-faces are CONVEXLY connected
	Reverse the orientation of that shared-face
	IF feature is positive
	Reverse the orientation of all created-faces
	}
STEP2:	Construct face classification table (FCT) and mark all supporting faces in Fc
STEP3:	For each supporting-face fi, in Fc
	{
	split Bbox_S about the supporting-face fi.
	Discard the below portion.
	Assign Above portion to Bbox_S.
	Remove the supporting-face fi from the face set Fc.
	}
STEP4:	**IF** there is a non-supporting face fp, in face set Fc
	Partition(Fc, Bbox_S,fp, Volume_list)
	ELSE
	Add Bbox_S to Volume_list
STEP5:	Feature_volume = Merge_Volume(Volume_list)
END Algorithm Extract_Feature_Volume	

Procedure Partition(Fc, Bbox_S, Partitioning face fp, Volume_list)

STEP1:	Split Bbox_S about the plane of the face fp into two parts (S_above and S_below).
STEP2:	Split faces in Fc about the plane of the face fp into two sets, Fc_above and Fc_below.
STEP3:	construct Face Classification Tables fct_above and fct_below.
STEP4:	For each supporting face in Fc_above
	{
	Split the solid S_above about the supporting face (fp)
	Discard the below portion.
	Assign above portion to solid S_above.
	Remove the supporting face from the face set Fc_above.
	}
	IF there is a non-supporting face, fp_above, in face set Fc_above
	Partition(Fc_above,S_above,fp_above,Volume_list);
	ELSE
	Add solid S_above to the Volume_list;
STEP5:	For each supporting-face in Fc_below
	{
	Split the solid S_below about the supporting face.
	Discard the below portion.
	Assign above portion to solid S_below.

           Remove the supporting face from the face set Fc_below.
       }
       IF there is a non-supporting face, fp_below, in face set Fc_below
              Partition(Fc_below, S_below, fp_below, Volume_list);
       ELSE
              Add solid S_below to the Volume_list;
END Procedure Partition

Procedure Merge_Volume merges all the volumes in the list, Volume_list. It merges two volumes if they have a face lying on the same plane. If there are more than one volume in the list, then each volume will have face/faces that lie on the same plane as face/faces in some other volume.

Splitting a solid about a face involves determining the two parts of the solid which are above and below the plane containing the face. The computations involved are line-plane intersections and one point classification. The complexity of this operation depends on the number of faces in the solid. It may be noted that in the clipping process the number of faces in the solid will be six to start with (bounding box) and the maximum number of faces of the solid can only be equal to the total number of shared and created faces of the feature.

5.1 Implementation details

The volume extraction algorithm has been implemented as part of a feature extraction program. The part model is represented by the winged-edge data structure (Baumgart 1982). Features are represented as a list of created faces and a list of shared faces.

The program has been developed in C under UNIX/X environment and runs on a Sparc platform. The volumes extracted for, the features in the CAM-I benchmark solid and another part are shown in figures 4 and 5. In figure 4, the first box shows the part model and the subsequent boxes show the extracted feature volumes. In figure 5, the feature volumes extracted from the CAM-I benchmark solid are shown in the same figure in their corresponding locations in the part.

6. Discussions and conclusions

Techniques based on either edge extension (Falcidieno and Ginnini 1987) or face extension (Sakurai and Gossard 1988) work for simple (single end-face) features with convex volumes. Redundant computations, involved in generating faces from edges in the algorithm of Markowsky and Wesley, are eliminated in the current algorithm. The algorithm presented in this paper, can handle complex features with multiple end-faces and features with non- convex volumes. In order to increase the scope of the present algorithm, features with free-form surfaces have to be handled. The difficulty arises in the splitting of the solid about a face with free-form geometry. One approach is to polygonise the free- form faces and use the above algorithm. However if the free-form geometry has to be maintained in the feature volume, further research is necessary.

7. Acknowledgements

Support for this work by ADA, Bangalore, through grant number ADA/ME/BG/18 is gratefully acknowledged.

8. References

1. Dong, X. and Wozny, M., *FRAMES, a frame-based feature extraction system*, Proc. Int. Conf. Computer Integrated Manufacturing, (Rensselaer Polytechnic Institute, USA, 1988) p. 296.
2. Falcidieno, B., and Giannini, F., *Automatic Recognition and Representation of Shape-Based Features in a Geometric Modelling System*, Computer Vision, Graphics and Image Processing, **48** (1989) p. 93.
3. General Dynamics Corporation, *Volume decomposition algorithm*, CAM-I Report, r-82-ANC-01 (1985).
4. Henderson, M. R., *Extraction of feature information from three dimensional CAD data*, PhD dissertation, (Purdue University, USA, 1984).
5. Joshi, S. and Chang, T. C., *Graph-based heuristics for recognition of machined features from 3D solid model*, Computer Aided Design, **20** (1988) p. 59.
6. Markowsky, G. and Wesley, M.A, *Fleshing out wire frames*, IBM J. Research and Development, **24** (1985) p. 582.
7. Sakurai, H. and Gossard, D. C., *Shape feature recognition from 3D solid models*, ASME Intl. Computers in Engineering Conf. (San Francisco, USA, 1988).
8. Sakurai, H. and Gossard, D. C., *Recognizing shape features in solid models*, IEEE Computer Graphics and Applications, (1990) p. 22.
9. Shah, J. J., *Assessment of features technology*, Computer Aided Design, **23** (1991) p. 331.

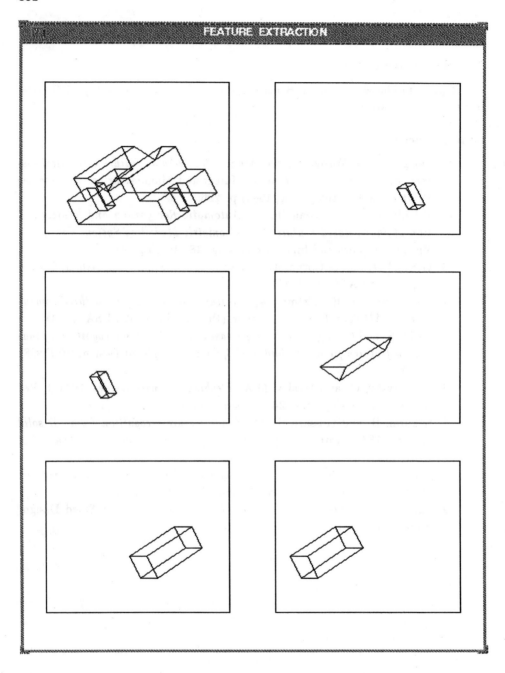

Figure 4 : Feature volumes extracted from test-solid 1

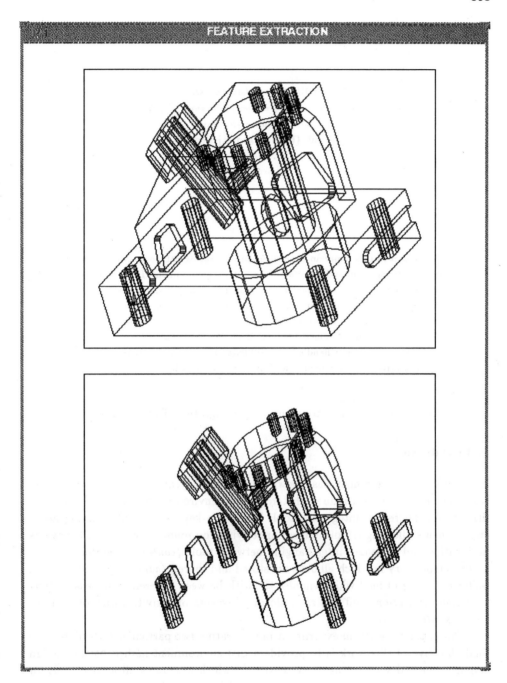

Figure 5 : Feature volumes extracted from test-solid 2

AN EXPERIMENT IN INTEGRATING CAD TOOLS

Malek BOUNAB and Claude GODART
Centre de Recherche en Informatique de Nancy (CRIN/CNRS)
BP 239, 54506 Vandœuvre-lès-Nancy, France
E-mail: {bounab, godart}@loria.fr

ABSTRACT

We present in this paper an experiment in integrating CAD tools in the DMMS (Design Management and Manufacturing System) environment. Data exchange and sharing between tools, leading to their interoperability, are backed by a common PCTE (portable common tool environment) based repository.

The final objective of this work is to build a CIM environment which integrates all the tools handling products life cycle steps. We have limited our experiment to design process step handled by PROPEL and SPEX CAD tools respectively representative of the mechanical and the automation skills.

Keywords: CIM Environment, Tool Integration, Data Sharing.

1. Introduction

CAD tools development has known, these recent years, a tremendous expansion. Their proliferation induced heterogeneous environments composed of monolithic tools. Therefore, integrating tools becomes a key issue to offer homogeneous environments allowing (i) an uniform underlying communication system for events notification and messages interchanging between tools *(control integration)* (ii) an uniform repository in which all data are stored and shared *(data integration)* (iii) an uniform "look and fell" which enables to swith between different tools easily *(presentation integration)* and (iv) an uniform processing of activities with other tools *(process integration)*.

This paper presents an experiment in integrating two particular CAD tools. The final objective of this work is to provide a CIM environment named DMMS (Design Management and Manufacturing System)[1,2,3,4] which integrates all needed tools for products management during their life cycle. We focus in this paper particularly on data integration by integrating PROPEL[a] and SPEX[b] CAD tools. Section 2 describes the integration architecture. We present in section 3 PROPEL and SPEX tools separately and model data used by these tools. The integration of these tools on top of PCTE[c] is developed in section 4 and a production scenario showing tools interoperability is given in section 5. Section 6 presents some conclusions and future work.

2. The Integration Architecture

DMMS environment aims to link different concurrent product design functions. From a product life cycle point of view, all these functions cooperates widely during the design step.

The design function deals with the relationship between the mechanical and automation skills which cooperate to design a product.

The manufacturing function aims to execute the manufacturing of the designed part. From a life-cycle point of view, many shop assistance operations must be realized such as reconfiguring, reprogramming and maintaining products.

The management function deals with the management of data exchanged between the previous functions and to control the manufacturing process by invoking and chaining the right tools at the right time.

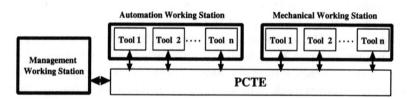

Figure 1: The Integration Architecture

We focus here on the design function which is mainly achieved by the mechanical and the automation working stations (figure 1). Each working station is composed of a set of coherent and cooperative CAD/CAM tools integrated on top of PCTE[5].

3. The Separate Tools: PROPEL and SPEX

The tool integration experiment involves the integration of PROPEL and SPEX CAD tools. These tools are respectively representative of the mechanical and the automation skills. PROPEL covers the manufacturing process plan (generated from the part geometrical description and the set of tools and machines composing the workshop) while SPEX realizes automation equipment functional specification, behavioral

---

[a] PROPEL is a product of ITMI
[b] SPEX is a product of TNI, CRAN and SPIE-TRINDEL
[c] PCTE is an ECMA standard providing an open repository for software development. We use in our experiment an implementation of PCTE 1.5 named Emeraude V12.5, product of GIE EMERAUDE

description, executable code generation and operative system emulation.

3.1. The Mechanical Design Tool: PROPEL

PROPEL[6] is an expert system providing process plan generation. PROPEL generates a manufacturing process plan from the geometrical part description (faces, slots, profiles...) and the workshop definition. A workshop is described by a set of machines (lathes, milling machines ...) and the manufacturing tools present in the workshop (drills, face-cutters..). Process plan, part, machine and tool descriptions are textual files which are either created by the tool (for the process plan) or by the user (part, machine and tool descriptions). We have extracted the data contained in these files, by reverse engineering, in order to model data handled by PROPEL. We have considered only input and output data of PROPEL because it is a closed tool and is therefore considered as a "black box". We must define input and output interfaces providing PROPEL interaction with PCTE integration repository. After extracting these information, we build the corresponding PCTE schemas.

3.1.1. Part Description and Modeling

The description of a part in PROPEL contains the definition of basic features composing this part and their physical and geometrical characteristics. An example of a part description file, representing only "up" and "down" faces belonging to part "totem", is given in figure 2.a. The part schema include all the features which can compose a part. To each feature, we have to define all its parameters. We have followed a top-down approach to break down a part into a set of features managed by PROPEL. Therefore, the schema describing a part contains all the features which may occur in its description. We have defined, from the objet type *feature*, all its sub types. The part break down continues until the leaf features representing part physical components. The PCTE schema of a part is given in figure 2.b.

3.1.2. Manufacturing Process Plan Description and Modeling

The manufacturing process plan represents the operations to be processed to achieve the manufacturing of a part. These operations are structured in phases corresponding to machines involved in the part manufacturing.

From the process plan (figure 2.c), we build, by reverse engineering, the process plan schema depicted by figure 2.d. Each part (represented by the entity *part*) has one associated process plan (*process-plan* entity) linked to a *phase* entity. Each *phase* is linked to a set of *sub-phases* which are, in their turn, linked to *operations*. Each *operation* is composed of *toolings* processed on *features*. We associate, for each *phase*,

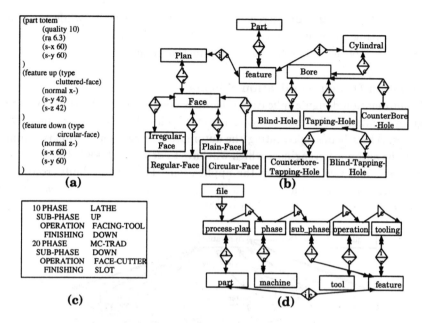

Figure 2: Part Description and Modeling

a *machine* and, for each *sub-phase*, a *feature*.

3.2. The Automation Design Tool: SPEX

SPEX[7] is an environment allowing the specification, the design and the prototyping of automation production systems. Software engineering concepts such as structuring, modularity and reusability necessary to automation component definition are also provided by this environment. SPEX manages two kind of automation components which are the *functional box* (FB) and the *functional diagram* (FD). A functional box is a "behavioural" unity[d] producing one or many output values from a set of input values and parameters.

Building a SPEX Automation Design Application

We use a top down approach to define a SPEX application. That means that all FDs are initially empty and are completed by the automation engineer while needed information are received from the mechanical engineer. SPEX automation design application considers four basic activities occuring in a product design. These activities are the TRANSPORT activity which handles transport of products between machines during the manufacturing step, the TRANSFORM activity which considers the effective manufacturing of a product, the STORE/UNSTORE activities which ensure product storage and unstorage before its use and finally, the CONTROL activity which controls data flow between the above activities. The information given by the

---

[d] A "behaviour" may be a graphcet, a ladder or a C program.

mechanical engineer are considered in the transforming activity while the controling, transporting and storing/unstoring activities are defined by the the automation engineer, the robotics engineer and the production manager.

Automation Design Application Schema

The automation design application schema is deduced from the SPEX automation design application. The meta schema instantiation defines the set of used FD object types. The schema representing the automation design application is completed later by the automation engineer which will have to define each FB behaviour connect different FBs and FDs making up the application. The schema built from the automation design application comprises *Tool, Station, Control, Transform, Transport* and *Store/Unstore* object types (figure 3).

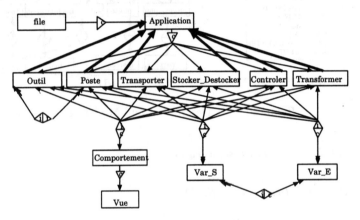

Figure 3: PCTE Schema of the Automation Design Application

4. Tool Integration on top of PCTE

SPEX and PROPEL tool integration, from data point of view, consists in defining object types through which these tools can cooperate and interoperate. These object types are then used to integrate tool respective schemas. Schema integration, as defined in [8], consists of three main steps. Firstly, we have to define all common object types of different schemas we want to integrate. Secondly, we have to solve name (homonymy and synonymy) and structure conflicts. Thirdly, we process the effective schema integration by juxtaposing the different schemas. In our case, we call the resulting schema *"exchange schema"*. Object types common to PROPEL and SPEX schemas are *machine, tool* and *sub-phase*. In addition to these schemas, we have integrated to the exchange schema a set of management information modeled by the *management schema* (figure 4). This schema is made up object types related to a project (object type *project*) for which we define, on the one hand, its specifications (object type *specifications* containing a graphical description of the part being designed), and on the other hand, persons allowed to work in this project (object type

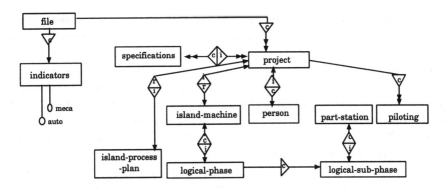

Figure 4: Management Schema

person). We define, in the management schema, a logical project representation by introducing the notion of part-station (object type *part-station*), logical phases and sub-phases (object types *logical-phase* and *logical sub-phase*). These informations are intended to link logical and physical aspects of a project.

Object types *part* (called *project* in the management schema), *process-plan* (called *island-process-plan* in the management schema) and *application* (called *piloting* in the management schema) are common object types allowing the integration of the process plan, the automation design application and the management schemas.

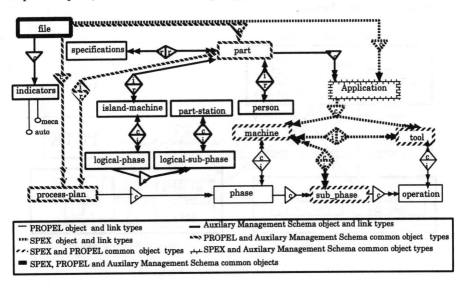

Figure 5: Exchange Schema

5. The Production Scenario

The objective of the production scenario depicted by figure 6 is to achieve parts design. The production scenario consists of two main steps which are the mechanical and the automation design steps managed by the management working station. This one allows the mechanical working station to begin its processing (1). From the information given by the part description file (2), we instantiate the part PCTE schema (3). PROPEL generates then, from the part description file and workshop machine and tool descriptions, the manufacturing process plan file (4). We extract from this file information allowing the process plan schema instantiation (5). The end of the mechanical design step is notified to the management station (6). This one is read by the management working station (7) which can now allow the automation station to start the automation design step (8).

Automation design step takes results generated by the mechanical design step (Process plan objects) (9) to build, on the one hand, a SPEX automation design application (10) and, on the other hand, a PCTE automation design application schema. We have extended SPEX in such a way that it can create, access and manipulate PCTE object type instances. This extension, which we have named SPEX-PCTE interface, generates a SPEX automation design application by counting *sub-phase* object type instances belonging to the PCTE process-plan schema SPEX-PCTE module creates as much FDs as sub-phases which are linked to each other using their input/output variables. Finally, a schema representing a PCTE automation design application is generated. This schema is compiled to be instantiated (11). The end of this step is notified to the management working station (12). All notifications are done thanks to *meca* and *auto* notification attributes belonging to *indicators* object type in the exchange schema.

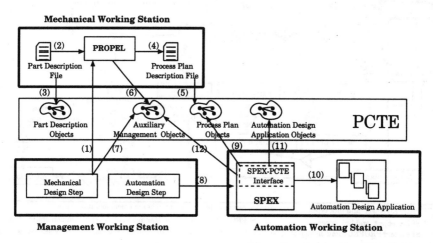

Figure 6: The Production Scenario

6. Conclusions and Future Work

This paper describes an experiment in tool integration, involving the integration of two different, but complementary tools whose integration was never anticipated. From a data integration point of view, PCTE provides basic mechanisms to manage object versions, to manipulate composite objects and, in a limited way, to manage schema evolution and multiple object points of view. The management of instances after schema restructuring, complex type and semantic constraint definitions are yet to be supported by PCTE.

Object types through which the different tools can cooperate was a critical task because tool interaction depends on object types granularity represented in the exchange schema. A coarse granularity limits the number of managed object types but penalise tools interactions due to a non enough detailed view of other tools managed objects. On the other side, a fine granularity allows an efficient interaction between tools by enhancing schema semantics but increase the set of object types we have to deal with and decrease the PCTE global access performances to objects.

We plan to extend PCTE data model, initially designed for software engineering requirements, to cover CIM requirements such as fine granularity object management and type instance management while evolving schemas. PCTE manages concurrent accesses using locking mechanisms and transactions. These mechanisms are efficient for short time transactions but are particularly unsuitable for long term design processes which can take several hours and even several days. The integration to DMMS of long term transactions, such as those implemented in COO[9], is actually studied.

7. References

1. G. Morel and P. Lhoste. *Prototyping a Concurrent Engineering Architecture*, volume 1 of *TSI Press Series*, pages 163–167, 1994.
2. M. Bounab, J. C. Derniame, C. Godart, and G. Morel. DMMS: A PCTE Based Manufacturing Environment. In *Proceedings of PCTE'93 International Conference*, November 1993.
3. M. Bounab. Tool Integration in Heterogeneous Environments: Experimentation in a Manufacturing Framework. Phd Thesis, National Polytechnical Institute of Lorraine, October 1994. (in french).
4. M. Lombard. Contribution to discrete part manufacturing engineering : prototyping a concurrent engineering architecture for manufacturing integrated systems. Phd Thesis, University of Nancy I, February 1994. (in french).
5. L. Wakeman and J. Jowett, editors. *PCTE: The Standard for Open Repositories*. Prentice Hall, 1993.
6. J.P. Tsang. Planning by plan combinaison: application to process plan generation. PhD Thesis, INPG, 1987. (in french).
7. H. Panetto, P. Lhoste, G. Morel, and M. Roesch. SPEX : Du Génie Logiciel pour le Génie Automatique. In *4th International Workshop: Software Engineering & its Applications*, pages 211–221, Toulouse, Dec. 1991.
8. C. Batini, M. Lenzerini, and S.B Navathe. A Comparative Analysis of Methodologies for Databases Schema Integration. *ACM CS*, 18(4), 1986.
9. C. Godart. COO : A Transaction Model to Support COOperating Software Developers COOrdination. In *Proc. 4th ESEC*, pages 361–379, Garmisch (Austria), September 1993. Springer Verlag. LNCS, Nr 717.

PARAMETRIC CAD BASED ON RELATION MODEL

ZONGYING OU,
JUN LIU
and
BO YUAN

*Department of Mechanical Engineering, Dalian University of Technology,
Dalian, 116024, P.R.China*

ABSTRACT

This paper deals with relation model processing technique, which is the base for generating parametric CAD drawings. A special graph data structure has been developed for representing the entities and constraint relations. The intelligent parametric processing is also introduced in this paper.

Keywords: Parametric CAD, Variation Geometry, Relation Model

1. Introduction

In engineering practice, more than 70% of the new designs are so called adapting designs and series designs, i.e. the constructions of the new designs are similar to a prototype product design, only some dimensional parameters of the new designs are different and, probably, part of the construction has to be made little modification.

Relation model CAD technique is a very efficient technique for these similar designs. A common relation model can be built with a prototype design and shared between the similar designs. A new similar design drawing then can be generated from the common relation model after input new data and a following intelligent processing, which is often called parametric or variant processing.

2. Design Drawing Data Base

Design drawings are used for defining and representing design products completely and are very important documents in production. In convention CAD systems, CAD database stores a design drawing as a collection of primitive entities and predefined compound entity groups, and the entities and entity groups are treated independently of each other.

To be parameterized, the database of a drawing could be reorganized into two categories: topology (skeleton structure) database and geometry database (fig. 1). Topology database keeps the records of composed entity types and their connection relations, which are intrinsic for a specific prototype drawing; geometry database stores the coordinates of key points of entities and other geometrical data, which are changeable during the parameterized processing.

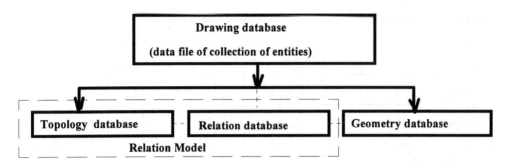

Fig 1 Drawing data base

3. Relation Model and the Data Structure

From the considerations of possessing similar performance functions and technology of manufacturing, entities of paramertric(variant) drawings come from same prototype drawing are related each other with specific relations (constraints). Some of these relations are explicit and some are implicit. Typical relations in a drawing are: a line tangent to a circle or an arc; a line keeps horizontal; a line keeps vertical; a line parallels to other line; a line is perpendicular to other line; two arcs or circles are concentric; the distance of two entities keeps a defined value, etc. .

The collection of relations of a drawing can be named relation database of a drawing. Relation database and topology compose a relation model of a drawing. All the similar designs share a common relation model but different dimension parameters. Building a relation model is the first step for a parametric design.

The basic data for a parametric design could be classified into three classes: entities; key point coordinates and dimension parameters; relations. Entities are segments geometrically composed a drawing; key point coordinates and dimension data are used for defining positions or sizes of entities; relations are constraints between entities. These three class data themselves can be recorded dynamically by using three double lists. Entities will be treated as objects with references to key points. A special graph data structure is used for representing relation of entities and constraints for a relation model, entities are vertexes in the graph and constraints are edges connecting the entities. The data structure should be compact and convenient for searching from entities to their subjected constraints or vice versa. An adjacency multilist data structure is used for representing this graph, fig. 2 shows data structure for the drawing shown in Fig. 5.

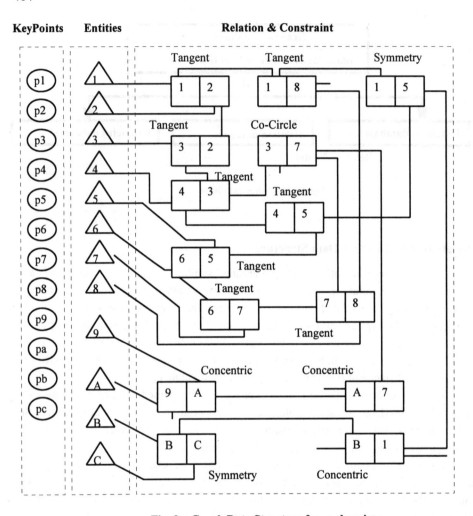

Fig 2 Graph Data Structure for a drawing

4. Geometry Reasoning and Equations Solving Processing

To generate a new variant design drawing from a prototype drawing is a procedure in which the entities of the prototype drawing are moved and adjusted to fit newly given data (parameters and relations). This is a complicated intelligence processing procedure. There might be many techniques that could be applied to this processing; however, geometry reasoning technique and equation solving technique are very important among them (Light and Gossard 1982; Aldefeld 1988; Verroust, Schonek and Roller 1992). Combining these two techniques with other expert system technique will form a powerful technique for relation geometry processing. We use the following techniques and strategies:

(1) Constraint-driven is important reasoning mechanism besides data-driven and object-driven. If all the variables related to a constraint are known and comply with this

constraint, then this constraint relation should be canceled from the reasoning relation list.

(2) The constraint relations should be tested whether a constraint relation can be solved separately. The singly solvable constraints should be processed first before processing procedure with a group of constraints.

(3) After singly processing, the intelligent processing should implement with a group of geometry entities and related constraint relations. The number of unknown geometry variables (geometry freedom) should be equal to the number of equations of related constraint relations within this group. This group could be called solvable group. There are two basic approaches to search entities for a solvable group: entities follow a connected contour and entities follow a defined relation chain. The searching procedure is a dynamically procedure of evaluating geometry freedoms and constraint equations. Whenever an entity is added to a group, the geometry freedom increases; however, the number of constraint relations also increases.

(4) A line has 4 freedoms, a circle has 3 freedoms and an arc has 5 freedoms. To simplify the processing and building the constraint equations' procedure, entity circle and entity arc could be replaced by special line segment or its combination. These line segments have the same freedoms and key points as the replaced circles or arcs, and are called equivalent line(s). The equivalent line for a circle is a line segment with one end point fixed in circle center and the length of the line segment is equal to the radius of the circle. The equivalent lines for an arc are two line segments that are from the center point of the arc to two end points of the arc respectively. We can use equivalent line(s) to represent the geometry element and constraint relation (e.g.,tangent) as shown in Fig. 3. The dashed lines are equivalent lines for representing the primitive element and constraint. Equivalent line(s) can equivalently describe freedoms, key points and other information of elements.

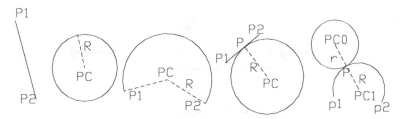

Fig 3. Representation of equivalent entity

We can represent all element in a drawing with their equivalent lines, then deal with these lines. When equivalent lines are all known, the original drawing would be known totally.

(5) The equation of a constraint relation depends on the constraint feature of the relation. A rule base should be built to formulate the knowledge of reasoning or transferring the constraint relations into equations. For example, two rules about tangent relation are shown below:

i). If two circles (arcs) are tangent, the equation is:

$$\sqrt{(C1x - C2x)^2 + (C1y - C2y)^2} = |R1 \pm R2| \qquad (1)$$

C1, C2 are the center point of circle 1 and circle 2 respectively; R1, R2 are the radius; C1x, C1y, C2x and C2y are x, y coordinate of C1 and C2.

ii). If a line and a circle (arc) are tangent at the point P1, and an end point of the line is Pm, the equation is:

$$\sqrt{(Cx - Pmx)^2 + (Cy - Pmy)^2} = \sqrt{R^2 + (P1x - Pmx)^2 + (P1y - Pmy)^2} \qquad (2)$$

C is the center point of circle; R is the radius; Cx, Cy, Pmx and Pmy are x, Y's coordinate of point C and Pm.

(6) The constraint equations in the general case are nonlinear. To solve a group of simultaneous equations, numerical method (i.e. Newton method) has to be used.

5. Implementations

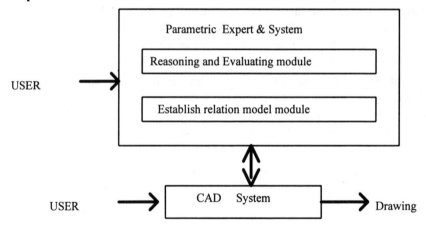

Fig 4 System architecture

A prototype expert system based on relation model technique has been developed in Mechanical Design Division, Dalian University of Technology. There are two modules in this expert system: establishing relation model module and evaluation (reasoning and solving) module. The expert system works combined with a CAD system(fig. 4). Establishing relation model module is also an interface module between the expert system and a CAD system: getting data from CAD system, then recognizing, confirming and modifying (defining) constraint relations; transferring evaluated data to CAD system then generating a new drawing. Fig. 5 shows the expert system working display.

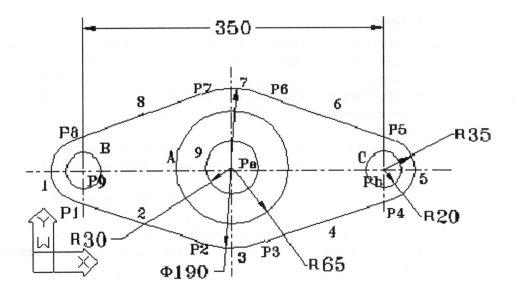

Fig 5 Parametric processing working display

When we input some new dimension values, and do parametric processing, the drawing will be changed as shown in Fig. 6.

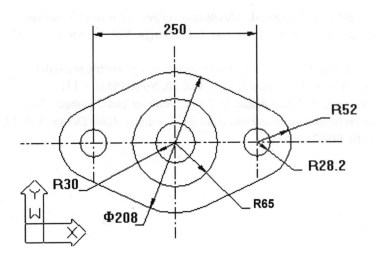

Fig 6 After parameterized processing

6. Conclusion

(1). The database of a drawing in a CAD system could be divided into two categories: topology database and geometry database. The entities of a parametric (variant) design drawing will keep some relations with each other, the collection of these relations can be named relation database for a drawing. A relation model can be established based on topology database and relation database.

(2). The parametric procedure is an intelligent processing procedure, in which the entities of a prototype drawing are moved and adjusted to fit newly given data (parameters and relations).

(3). A special graph data structure has been developed for representing relation of entities and constraints within a relation model.

(4). Combining geometry reasoning and equation solving techniques with other expert system technique will form a powerful relation model processing technique.

7. Acknowledgments

The work reported in this paper is supported by the CHINA Natural National Science Foundation under grant number 69174034. The authors would like to acknowledge the researchers and graduate students in CAD & CG Lab, DUT, for studying and implementing this system.

8. References

1. R. Light and D. Gossard, *Modification of geometric models through variational geometry*, Computer Aided Design, **Vol. 14, No 4**,(1982), P.209
2. B. Aldefeld, Variation of geometry based on a geometric reasoning method, Computer Aided Design, **Vol. 20, No 3**,(1988), P.117
3. A.Verroust, F. Schonek and D, Roller, *Rule-oriented method for parameterized computer aided design*, Computer Aided Design, **Vol. 24, No 10**, (1992), P.531

EVALUATION OF FOUR CAD SYSTEMS USING ANALYTIC HIERARCHY PROCESS

S. Agarwal, Lee Y. T. and Tor S. B.
School of Mechanical and Production Engineering
Nanyang Technological University, Singapore
E-mail : m9308e53@ntuvax.ntu.ac.sg

ABSTRACT

In this paper we study the capabilities of four pieces of software to use it as a tool for 'design for manufacturing.' The four pieces of software selected are Pro/Engineer, Unigraphics, ICAD and Catia. Until now no quantifiable data is published on comparison of CAD software. In this paper we will provide quantifiable data for comparing these four software using Analytic Hierarchy Process (AHP).

These four pieces of software are selected for comparison based on their suitability to be used as a feature-based modeller and their ability to allow more knowledge to be added to it.

Keywords : CAD Software, Analytic Hierarchy Process, Design for Manufacturing.

1. Introduction

It has been pointed out that the present trend is towards design for manufacturing. Design for manufacturing can only be achieved if the manufacturing constraints are considered during the design stage. One way to check the manufacturability of a product is to generate the manufacturing features while the geometric features are generated. Thus the CAD software that already can check the geometric integrity of the product, will also check the manufacturability of the product.

In this paper we study the capabilities of four pieces of software to use it as a tool for fulfilling the above objective.

In selecting the software a multicriterion decision making problem was encountered. It involves consideration of many subjective factors to best achieve the goal. We use a multicriterion decision making technique, Analytic Hierarchy Process (AHP), to handle the problem. AHP is discussed in section 2. In section 3 the salient features of CATIA, Unigraphics, ICAD and Pro/Engineer are studied. Here we give some subjective comments about the four pieces of software. In section 4 the objective and the criteria are identified in order to come to a conclusion from subjective comments. The problem of software selection is broken down in hierarchy and solved using AHP. Finally section 5 gives the conclusion.

2. Analytic Hierarchy Process

Analytic Hierarchy Process (AHP) is a systematic procedure for hierarchically representing the elements of any problem. It organizes the basic rationality by breaking down a problem into its smaller and smaller constituent parts and then guides decision makers through a series of pairwise comparison judgements. These judgements are documented and can be re-examined. They are then translated into numbers to express the relative strength or intensity of impact of the element in the hierarchy. The AHP includes the procedures and principles used to synthesize the many judgements to derive priorities among criteria and subsequently for alternative solutions [1]. Validation studies for AHP have been performed [2].

2.1 Model formulation and pairwise comparison

The first step in the analytic hierarchy process is to decompose the problem into a hierarchy consisting, minimally, of a goal, criteria, and alternatives. The example of hierarchical decomposition of the CAD system selection problem is shown in Figure 1. Once a hierarchic or network representation of the problem is complete then the elements of a problem are compared in pairs with respect to their relative impact ("weight" or "intensity") on a property they share in common.

The second step is to establish the priorities among the elements at each level of the hierarchy. In order to compare two elements at a time a matrix is made. The elements are compared using the scale of relative importance [1]. For example in Table 1, criteria 2, i.e., *the amount of geometric knowledge presently associated with the model once created*, is given the value of 5, i.e. *strongly more important*, when compared with criteria 4, i.e., *the openness or the accessibility of the data inside the software*.

Once the values of pairwise comparisons are entered in the matrices, mathematical computations are done to calculate the priority vectors, consistency ratio and the final decision, as described in the next section.

2.2 Computations

In order to get the priority vector for any matrix, eigen vectors of the matrices are calculated and then normalized to unity. The software 'Expert Choice' [2], developed by Saaty does this, thus saving the time consuming job of calculating the eigen vectors.

Next, to make sure that the decisions made are consistent, an index of consistency which provides information on the extent of seriousness of violations of numerical and transitive consistency is used. This is then compared with the average consistencies of the random matrix of the same order to get the consistency ratio (CR). The value of a CR less than 10% is acceptable. Otherwise the judgements need to be taken again.

The principle of synthesis is now applied. Priorities are synthesized from the second level down by multiplying local priorities by the priority of their corresponding criterion in the level above. These are then added for each element in a level according to the criteria it affects. This gives the composite or global priority of that element which is then used to weigh the local priorities of elements in the level below and so on upto the bottom level.

The next section explains the CAD software selection problem with subjective comments about the four pieces of software involved.

3. Salient features of four CAD systems

In this section first we identify the criteria for comparison. Then we present general comments on the solid modeling capabilities of the four CAD systems followed by the specific comments related to feature based capabilities of the four pieces of software.

Some of the CAD systems in the market which are available in NTU are CATIA, Pro/Engineer, Unigraphics and ICAD. These software are reviewed on their suitability for design for manufacturing. It is worth noting at this point that not all criteria but only the ones important for feature based design and manufacturing are considered in evaluating the software. Criteria like how good a software is for a particular application domain are not considered.

3.1 Common features

The systems can create solids via Boolean operations using primitives, barring Pro/Engineer which has other better ways to do it. Surfaces can be sewn to give a solid. A solid can be split into two solids. Catia, Unigraphics and ICAD support primitives like cube, sphere, cone, prisms and torus. Also solids can be created by sweeping a profile.

Specific comments about the four software are given in the following sections. Besides solid modeling capabilities, surface modeling capabilities are also discussed which may be of importance for solids made up of complex free form surfaces.

3.2 CATIA

Catia is developed by Dassault Systems, a subsidiary of a major French aerospace company and supported by IBM. The 'CATIA Solids Geometry' supports solid modeling using a hybrid boundary representation (B-rep) and constructive solid geometry (CSG) solids modeller [4].

The history of operations and definition of primitives is stored in CSG representation of the model. The CSG tree may be reconstructed or edited to make a new solid and the primitives can also be modified to modify the shape of a final solid.

Surface types modeled include surfaces of revolution, ruled surfaces, constant and variable radius fillets, blended surfaces, drafted surfaces, wrapped and unwrapped surfaces, net surfaces, and swept curve driven surfaces. Surfaces are modeled with Bezier curves or non-uniform rational B-Spline (NURBS). The boundary representations of solid models and curved surfaces are approximated by planer facets. The accuracy of the facets is user-controllable. When Boolean operations are performed, the faceted representation provides faster response time. Exact geometry for underlying surfaces is stored for analysis and to obtain cross sections. It supports the importing and exporting of files in the IGES format.

3.3 Unigraphics

Unigraphics (UG) is from Electronic Data Systems Corporation, USA, originally released in January 1985, it is developed on the solid modelling kernel Parasolids, underlying which is the boundary representation [6, 7]. In Unigraphics solids can be created via Boolean operations, but neither the CSG tree nor the object hierarchy (of the features) are maintained. Once the solid is created it loses its original identity. After the solid has been created, features such as holes, pockets, slots and chamfers can be added. But the identity of a feature is lost once it has been added. Since it is based on boundary representation, access to specific loops, edges, faces and solids is easily available and thus it can support local modifications such as:

- translating, rotating, or subdividing faces
- replacing a face with a different type
- deleting a face or a set of faces
- tapering groups of faces

Although this software rates highest among the four compared in integrating design with manufacturing (Table 8, section 4.2.2.), once the design is completed, its internal data format is not structured enough to identify the features individually. Thus an external program cannot read the data, reason about it and do automated process planning.

3.4 ICAD and Concept Modeller

ICAD and Concept Modeller is supported by ICAD, Inc. ICAD is a tool that allows knowledge to be embedded. It connects externally to a solid modeller for geometric computations and display tasks. The solid modeller can be changed, linked and de-linked at any time in ICAD while the knowledge of the model stays in ICAD.

As with UG, ICAD Solids uses Parasolids as the solid modelling kernel. It requires network access to a UNIX workstation running Parasolid. Whenever a solid part is instantiated in the "ICAD Browser" (user interface), the "ICAD Solids Designer" automatically connects to the solid modeller. ICAD is like a shell, into which more knowledge can be added. This software allows knowledge to be captured while the designer is working, in the form of words and symbols. Also, external programs can be linked to the product model. In this software there is the disadvantage that the designer should know the full details of manufacturing, and the programmer who creates the application in it should know Lisp. Only then can a design be made, or the full capabilities of the software can be utilized. It does not provide the user geometric capability and user friendliness that other systems already have. It is knowledge-based but does not have the geometric intelligence of its own to the extent that Pro/Engineer has.

ICAD is a knowledge based engineering system that allows for the symbolic description of product model information (physical attributes such as geometry, material types, and functional constraints) and process information (such as analysis, manufacture, and test). [5]

3.5 Pro/Engineer

Pro/Engineer, from Parametric Technologies founded in May, 1985, provides a feature-based user interface based on engineering / manufacturing terminology.

Pro/Engineer does not support primitives but uses feature-based construction techniques. It supports the following surface types:
- cylinders, planes, tori, spheres, and conics;
- ruled surfaces and surfaces of revolution; and
- sculptured surfaces.

Sculptured surfaces are defined by bi-cubic polynomials and includes special surfaces such as fillets, corner fillets, special NURBS-based surfaces, and patches. All Pro/Engineer surfaces may be trimmed.

It uses a parametric, object-oriented database which models functional, topological, geometrical, and other relationships between disjointed geometry.

Features can be user defined and its structure allows incorporation of manufacturing data. Assembly layouts define the relations between parts and sub-assemblies and provide an automatic assembly capability. Pro/Engineer uses this information to assemble the components of an assembly. When changes are made to dimensions of parts, or to a part's location, it automatically propagates the effects of these changes. The salient features of this software which make it suitable for feature-based modeling are as follows:
- Parts are defined in terms of form-features selected from a library, or in terms of special features defined by users.
- Form-features are dimensioned as they are defined. These dimensions locate a form-feature relative to other form-features. If any of these dimensions are changed, then all other related form-features are repositioned or changed as required.
- The relationships between form-features are carried over between parts in assemblies.
- Any branch in the model's relationship tree of form-features may be changed and a new tree, resulting in the some evaluated model, is formed.

Thus, Pro/Engineer is a parametric, form feature-based solid modeling software system. It is dimension-driven and captures geometric relationships between assemblies, parts, features, and parameters [4]. In the next section AHP is used in order to analyze all the factors (criteria) as per their relative importance and hence come to a conclusion.

4. Analytic Hierarchy Process for CAD software selection

To come to a decision on the software selection from subjective comments (as given in section 3) AHP is chosen as the technique. A model is developed which can also be used to take the expert views of other people working in the same area.

4.1 Problem formulation

The problem is decomposed into a simple hierarchy by identifying the objective and the criteria. The objective will serve as the goal to be fulfilled. The criteria one level below in the hierarchy and finally the alternatives as the lowest level of hierarchy as shown in Figure 1.

4.1.1 Objective

To select the best software tool among Pro/Engineer, Unigraphics, ICAD and CATIA, for 'design for manufacturing', so as to enable concurrent considerations of manufacturing constraints while designing. The choice of these four software is based on their ability to be used as feature based modeling tool and availability in NTU.

4.1.2 Criteria

The following criteria were identified as the key factors in selecting a tool for feature based modelling:

1. The ease of creation of an unambiguous & 'intelligent' model.
2. The amount of geometric knowledge presently associated with the model once created.
3. The way of representation of data inside, Is it object oriented representation. or not.
4. The openness or the accessibility to the data inside the software.
5. The forms of external interfaces that the software can support.
6. The extent to which more knowledge can be added to the existing model.
7. The existing integration to the manufacturability of the part once the design has been completed.
8. Market life of the software.

All the criteria here have a specific meaning. For example the word 'intelligent' in criteria 1 refers to how easily the constraints on geometry and design can be encoded and ease of creating an unambiguous model, etc. Similarly the criteria 3 is more geared towards the style of programming and the comparison is not based on advantages and disadvantages of boundary representation, CSG or hybrid representation.

Since AHP requires that 7 ± 2 factors only be compared at a time (based on the study of human psychology that a human mind can only compare maximum 7 factors at a time), presently market life is not considered in the pairwise comparison.

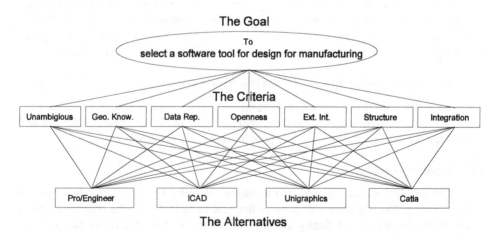

Figure 1 : The decomposition of the CAD software selection problem into a hierarchy

4.2 Pairwise comparison

Following the above hierarchical decomposition of the problem, we do the comparison of two elements at a time. The numbers entered in the table are necessarily subjective and objective results are then arrived from it. First, comparison is made of the criteria, two at a time, to identify the relative importance of each criteria with respect to the goal in hand. Then, the four pieces of software are compared with respect to each of the criteria. This second comparison (Table 2 - 8) yields an objective estimate of how good the software is for the specific criteria for which it is being compared, in the form of normalized priority vectors.

4.2.1 Pairwise comparison of the criteria

In the Table 1 below, the criteria are compared two at a time and the numbers indicate the importance of criteria number in the row over the criteria number in the column. The criteria numbers correspond to the serial number of the criteria identified above. For example, (Table 1) the number 3 in row 5 and column 6 shows that criteria number 5 - the form of external interface the software can have, is *moderately* more important then criteria number 6 - the extent to which knowledge can be added to the existing model.

The numbers in this table translate the subjective feelings gained by studying and experiencing the software into quantifiable data. The priority vectors calculated show the relative importance of each of the criteria which has been calculated using a combination of the principles of mathematics and human psychology. Two criteria are compared at a time and rating is done as per the 'Scale of relative importance' [3].

Table 1 : Judgements and priorities with respect to the goal - **to select a software tool for design for manufacturing. Which criteria is more important or has a greater impact?**

	1 Unambiguous	2 Geo. Know	3 Data rep.	4 Openness	5 Ext. Int.	6 Structure	7 Integration	Priority Vector
1 Unambiguous	1	5	7	8	7	7	9	.490
2 Geo. Know.	1/5	1	2	5	6	6	2	.186
3 Data Rep.	1/7	1/2	1	4	5	5	3	.140
4 Openness	1/8	1/5	1/4	1	1/2	2	2	.048
5 Ext. Int.	1/7	1/6	1/5	2	1	3	1	.056
6 Structure	1/7	1/6	1/5	1/2	1/3	1	2	.038
7 Integration	1/9	1/2	1/3	1/2	1	1/2	1	.043

Consistency Ratio = 0.101

where 1, 2, 3 .. in the column and row headings refer to the serial numbers of the criteria previously identified in section 4.1.2. This table compares the relative importance of the various criteria with respect to the given goal.
Matrix entry indicates that *row* element is
1. *Equally*   3. *Moderately*   5. *Strongly*   7. *Very strongly*   9. *Extremely*
more *important* than *column* element, unless the numbers are reciprocals.

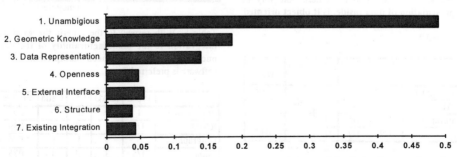

Figure 2 : Relative importance of the 7 criteria in selecting a software tool for design for manufacturing

The priority vector and the consistency ratio is calculated using the Software Expert Choice. As discussed in section 2.2, since the consistency ratio is 10.1%, the decisions made in this matrix are consistent. The priority vectors calculated in Table 1 are depicted in the Figure 2.

4.2.2 Pairwise comparison of the alternatives

The pairwise comparison of two alternatives at a time, with respect to each of the criteria is done as shown in Tables 2 - 8 below.

Table 2: With respect to the criteria: **the ease of creation of an unambiguous & 'intelligent' model**, which software is preferred ?

	Pro/E	ICAD	UG	Catia	Priority Vec.
Pro/Engineer	1	5	3	9	.566
ICAD	1/5	1	1/3	5	.127
Unigraphics	1/3	3	1	7	.267
Catia	1/9	1/5	1/7	1	.040

Consistency Ratio = 0.063

As a subjective comment on Table 2 it can be said that Pro/Engineer allows constraints to be easily added from the user interface of the solid modeller, while in ICAD the constraints have to be programmed inside LISP. Solid modeling is easily done from the user interface in UG and Pro/Engineer.

Table 3: With respect to the criteria: **the amount of geometric knowledge presently associated with the model once created**, which software is preferred ?

	Pro/E	ICAD	UG	Catia	Priority Vec.
Pro/Engineer	1	5	3	9	.573
ICAD	1/5	1	1/3	3	.110
Unigraphics	1/3	3	1	7	.271
Catia	1/9	1/3	1/7	1	.045

Consistency Ratio = 0.032

Table 4: With respect to the criteria: **the way of representation of data inside, is it object oriented representation or not, etc.**, which software is preferred ?

	Pro/E	ICAD	UG	Catia	Priority Vec.
Pro/Engineer	1	3	5	9	.558
ICAD	1/3	1	5	7	.303
Unigraphics	1/5	1/5	1	3	.095
Catia	1/9	1/7	1/3	1	.043

Consistency Ratio = 0.065

Table 5: With respect to the criteria: **the openness or the accessibility to the data inside the software**, which software is preferred ?

	Pro/E	ICAD	UG	Catia	Priority Vec.
Pro/Engineer	1	1/3	5	7	.290
ICAD	3	1	7	9	.583
Unigraphics	1/5	1/7	1	3	.085
Catia	1/7	1/9	1/3	1	.042

Consistency Ratio = 0.061

Table 6: With respect to the criteria: **the forms of external interfaces that the software can have**, which software is preferred ?

	Pro/E	ICAD	UG	Catia	Priority Vec.
Pro/Engineer	1	1/3	5	7	.324
ICAD	3	1	3	9	.510
Unigraphics	1/5	1/3	1	5	.128
Catia	1/7	1/9	1/5	1	.039

Consistency Ratio = 0.125

Table 7: With respect to the criteria: **the extent to which more knowledge can be added to the existing model**, which software is preferred ?

	Pro/E	ICAD	UG	Catia	Pri. Vec.
Pro/Engineer	1	1/5	5	7	.242
ICAD	5	1	7	9	.640
Unigraphics	1/5	1/7	1	3	.078
Catia	1/7	1/9	1/3	1	0.040

Consistency Ratio = 0.114

Table 8: With respect to the criteria: **the existing integration to the manufacturability of the part once the design has been completed**, which software is preferred ?

	Pro/E	ICAD	UG	Catia	Pri. Vec.
Pro/Engineer	1	9	1/3	5	.290
ICAD	1/9	1	1/9	1/3	.039
Unigraphics	3	9	1	9	.596
Catia	1/5	3	1/9	1	.075

Consistency Ratio = 0.069

4.3 Computations

After calculating the individual priority vector and making sure that all the consistency ratios are around 10% or better (in this case it varies from 3% to 12.5%) the priorities are synthesized for the two levels and the results obtained. From this the final decision is computed and the results are as shown in Figure 3.

4.4 Sensitivity analysis

It can be seen that the results obtained above are based on the judgements given during the pairwise comparisons. What if some of the figures do not correctly weigh the elements? Sensitivity analysis is done to see the effect of changing some of the judgements.

The bar chart "Alternatives" in Figure 4 depicts the choice of alternatives if the weights of each of the criteria are as shown in the "Criteria" bar chart, which corresponds to the judgements in Table 1.

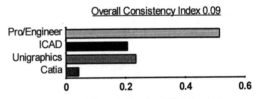

Figure 3 : Results of Final Synthesis

The graph in Figure 4 shows that Pro/Engineer is the best choice when the weight of the first criteria, the unambiguous creation of the model, is 0.49 (represented by the vertical line at 0.49). Now if 0.49 is changed to any value in the range 0 - 1, the first

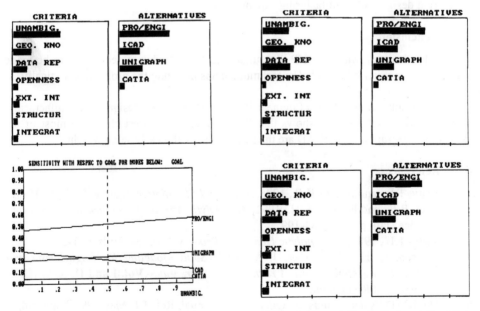

Figure 4 : The weights of criteria and alternative as per table 1 and the sensitivity analysis of criteria 1

Figure 5 : The weights of criteria and alternative after dynamically changing the weights

choice is still Pro/Engineer. While Unigraphics is the second choice at 0.49. If the weight of this criteria (the first criteria) is reduced below 0.37 then it can be seen from the graph that ICAD would be the second choice, and Unigraphics will go to third choice. Similar kind of graphs are also drawn for the other six criteria.

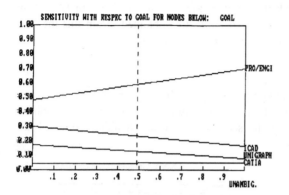

Figure 6 : The sensitivity analysis of criteria 1 as per second pair of bar chart in Figure 5

Figure 5 show how the choice of alternatives will be affected by changing the weights of the criteria on the left bar chart. The lengths of the bars on the "Criteria" bar chart can be dynamically changed (by an expert who has knowledge / idea of the range) to view the change in the resulting weights of the alternatives.

Graphs similar to the graph in Figure 4 can be drawn for any pair of bar charts developed dynamically. Graph in Figure 6 is an example based on the second pair of bar chart in Figure 5.

Looking at the graphs it can be concluded that Pro/Engineer is the best choice to be used as a tool for development of a integrated feature-based modeling tool, which has concurrent design and manufacturing capabilities.

5. Conclusion

It has been shown that AHP is a suitable technique when a decision is to be made with subjective criteria in mind. The technique helps in seeing the problem analytically.

The results show that Pro/Engineer is the best choice for design for manufacturing. The sensitivity analysis done shows that Pro/Engineer is the best choice in most of the situations. Unigraphics and ICAD are the next choice with nearly same weight.

6. References

1. Saaty, T.L., and Kearns K.P., 1985, *Analytical Planning*, Pergamon Press, p. 19-62.
2. Dyer, R.F., and Forman, E.H., 1991, *An Analytic Approach to Marketing Decisions*, Prentice Hall, p. 75 & 184-199.
3. Saaty, T.L., 1980, *The Analytic Hierarchy Process*, McGraw Hills, p. 1-34.
4. Johnson, R. H., and Turner, J. A., 1989, *Solid Modeling for Engineering and Manufacturing Applications, A review and buyers guide*, **Vol. I and II**, CAD/CIM Management Roundtable.
5. The ICAD System Solids Designer User's Manual, Rel. 4.1, May 1994, Cambridge, Massachusetts.
6. Modeling User Manual, EDS Unigraphics, Ver. 10.2, December 1993.
7. Unigraphics Essentials User Manual, EDS Unigraphics, Ver. 10.2, December 1993.

Joint Application of B-spline and Bezier Methods for Surface Modelling

Ye Xiangao

*School of Mechanical Engineering, Huazhong University of Science and Technology,
Wuhan, P.R. of China, 430074*

Li Dequn

*School of Mechanical Engineering, Huazhong University of Science and Technology,
Wuhan, P.R. of China, 430074*

Wang Fengyin

*Gintic Institute of Manufacturing Technology, Nanyang Technological University,
Nanyang Avenue, Singapore 2263*

ABSTRACT

This paper introduces a specific non-uniform bi-cubic B-spline surface that has the advantage of representing complex surfaces with sharply varying cross-sections. B-spline and Bezier methods are applied jointly in the free-form surface modeling. The transformation from B-spline representation to Bezier representation is emphasized.

1. Introduction

Uniform B-spline and non-uniform B-spline surfaces have been widely used in free-form surface modeling systems to represent complex surfaces. Uniform B-spline surfaces have uniform parameters in both U-direction and W-direction. The non-uniform B-spline surfaces that are used in many surface modeling systems have unified sequence of knots. It is often difficult to represent the surfaces encountered in some engineering practices, such as aircraft, automobile and ship manufacturing engineering by using these two types of surfaces. This is because the traditional methods applied in these engineering practices are to give points of surface according to parallel cross-sections whose shapes are described by different sequences of points. The uniform B-spline basic function or unified non-uniform B-spline function is impossible to fit to the varying distributions of points in all sections.

A specific non-uniform B-spline surface is used in the present free-form surface modeling system to describe complex surfaces. This kind of non-uniform B-spline surface has the same sequence of knots in one direction, i.e. a set of unified basic functions is taken in one direction. In the other direction, the non-uniform B-spline surface can have different sequences of knots, i.e., different sets of basic functions are taken. The direction in which the surface has the same sequence of knots is called longitudinal direction or W-direction, while the other direction transverse direction or U-direction. This kind of specific surface can satisfy the requirement of engineering practices mentioned above, because in each section different non-uniform sequences of knots can fit to the distribution of surface points.

The subdivision algorithms for Bezier curves and surfaces are easier and faster. As a result the rendering and finding intersections of Bezier surfaces are easier to be implemented. Because of this reason, two methods are jointly used in the surface modeling. When describing a free-form surface, B-spline method is used and the vertices of B-spline curves and surfaces are stored. For rendering of surface and finding intersection lines, Bezier method is employed. Discrete B-spline method is used to transform B-spline surface into Bezier surfaces.

2. Non-uniform B-spline surfaces

In the free-form surface modeling, cubic non-uniform B-spline curves, and bi-cubic non-uniform B-spline surfaces are used to represent complex free-form surfaces. The method is described below.

2.1. Cubic non-uniform B-spline curves.

Assume that we have a set of knots $\{t_i\}$ (i=-2, -1, 0, ..., n+1, n+2), that corresponding B-spline basic functions are $N_{i,4}(t)$ (i=-3, -2, -1, 0,......, n-2, n-1), and that there is a set of vectors $\{V_i\}$ (i=-1, 0,......, n, n+1). The cubic non-uniform B-spline curve can then be represented as:

$$r(t) = \sum_{i=-3}^{n-1} N_{i,4}(t) \times V_{i+2} \qquad t_0 \le t \le t_n \qquad (1)$$

Where $N_{i,4}(t)$ can be calculated by using de-Boor-cox's recursive equation, and r (t) and V_i are vectors. The notations for vectors are omitted for simplicity.

2.2. Bi-cubic non-uniform B-spline surface

As described above, a specific bi-cubic non-uniform B-spline surface is used in the free-form surface modeling. The way of constructing this kind of surfaces is described below.

In the U-direction (transverse direction), there are different sequences of knots $\{U_{i,j}\}$ (i=-2, -1,......, n+2; j=-1, 0, ..., m, m+1), while in W-direction (longitudinal direction), the sequence of knots remains the same $\{W_j\}$ (j=-2, -1, ... , m+1, m+2). $\{V_{i,j}\}$ (i=-1, 0, ..., n+1; j=-1, 0, ..., m+1) is an array of vectors (vertices), The cubic B-spline curves in transverse direction can be represented to be:

$$Q_j = \sum_{i=-3}^{n-1} N^j_{i,4}(U) \times V_{i+2,j} \qquad U_{0,j} \le U \le U_{n,j} \qquad (2)$$

The basis functions $N^j_{i,4}$ of the j-th transverse curves are in correspondance with the j-th sequence of knots.

Taking the corresponding points on the U-curves as vertices and applying a set of unified basis functions in the W-direction, a B-spline curve in longitudinal direction can be created,

$$P_u(U_c, W) = \sum_{j=-3}^{m-1} N_{j,4}(W) \times Q_{j+2} \qquad W_0 \leq W \leq W_m \qquad (3)$$

As U varies along the U-direction, The B-spline curve $P_u(U_c, W)$ will generate a surface,

$$P(U, W) = \sum_{j=-3}^{m-1} N_{j,4}(W) \times Q_{j+2} \qquad W_0 \leq W \leq W_m \qquad (4)$$

From the construction process, it can be seen that this kind of specific B-spline surface is very useful to describe the complex free-form surfaces which are designed according to the parallel sections. In each section, different sequences of knots are very useful to describe different shapes of sections.

3. Transformation from B-spline curve to Bezier curve

Because the subdividing algorithm of Bezier curve is easy to implemented, cubic non-uniform B-spline curves are transformed to Bezier curves so that the rendering and finding intersections of B-spline curves and surfaces can be more efficient. According to the discrete B-spline algorithm [2], the equations for transformation from cubic non-uniform B-spline curve to cubic Bezier curves can be derived.

Cubic B-spline curves are represented by Eq. 1. At knot t_l (l= 0, 1,......, n), new knots are inserted till t_l becomes triple-knot. The new sequence of vertices $\{V_i\}$ of the identical curves must satisfy the following conditions:

(1) If $t_{l-1} < t_l < t_{l+1}$

$$V'_l = \frac{t_{l+1} - t_l}{t_{l+1} - t_{l-2}} \times V_{l-1} + \frac{t_l - t_{l-2}}{t_{l+1} - t_{l-2}} \times V_l$$

$$V'_{l+2} = \frac{t_{l+2} - t_l}{t_{l+2} - t_{l-1}} \times V_l + \frac{t_l - t_{l-1}}{t_{l+2} - t_{l-1}} \times V_{l+1} \qquad (6)$$

$$V'_{l+1} = \frac{t_{l+1} - t_l}{t_{l+1} - t_{l-1}} \times V'_l + \frac{t_l - t_{l-1}}{t_{l+1} - t_{l-1}} \times V'_{l+2}$$

(2) If $t_{l-1} < t_l = t_{l+1} < t_{l+2}$

$$V'_{i+1} = \frac{t_{l+2}-t_l}{t_{l+2}-t_{l-1}} \times V_l + \frac{t_l-t_{l-1}}{t_{l+2}-t_{l-1}} \times V_{l+1}$$

$$V'_i = V_i \qquad (j < l+1)$$
$$V'_i = V_{i-1} \qquad (j > l+1) \qquad (7)$$

(3) If $t_{l-1} < t_l = t_{l+1} = t_{l+2} < t_{l+3}$, no knot is inserted, and $\{V'_i\}$ is the same as $\{V_i\}$.

For the sequence of knots $\{t_i\}$ $(i = -2, -1, \ldots, n+2)$, Let $t_{-2} = t_{-1} = t_0 < t_1 \le t_2 \le \ldots \le t_{n-1} < t_n = t_{n+1} = t_{n+2}$. From t_1 to t_{n-1}, new knots are inserted till t_l ($l = 1, 2, \ldots, n-1$) becomes a triple-knot. Thus, the cubic non-uniform B-spline curve can be identically redefined by n cubic Bezier curves.

The vertices of Bezier curves $\{Q^j_i\}$ $(i = 1, 2, \ldots, n; j = 1, 2, 3, 4)$ corresponding to the vertices of B-spline curve are shown in Fig. 1 and given to be:

$$Q_2^{i+1} = \frac{t_{i+2}-t_i}{t_{i+2}-t_{i-1}} \times V_i + \frac{t_i-t_{i-1}}{t_{i+2}-t_{i-1}} \times V_{i+1}$$

$$Q_3^i = \frac{t_{i+1}-t_i}{t_{i+1}-t_{i-1}} \times Q_2^i + \frac{t_i-t_{i-1}}{t_{i+1}-t_{i-1}} \times V_i \qquad (8)$$

$$Q_4^i = \frac{t_{i+1}-t_i}{t_{i+1}-t_{i-1}} \times Q_3^i + \frac{t_i-t_{i-1}}{t_{i+1}-t_{i-1}} \times Q_2^{i+1}$$

$$Q_1^{i+1} = Q_4^i$$

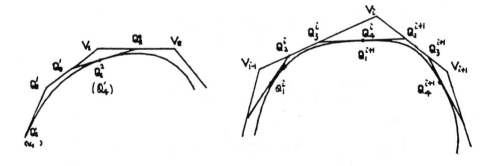

Fig. 1 Transformation between B-spline and Bezier curves

These n cubic Bezier curves are identical to the previous cubic non-uniform B-spline curve. If the sequence of knots of B-spline curve has no multiple-knot, the two adjacent Bezier curves must have the second order continuity (curvature continuity) at the joining point. This conclusion can be proved as follows.

According to [3], the conditions for the second order of continuity at the joining point between two cubic Bezier curves are:

$$a'_1 = \alpha \times a_3 \quad (\alpha > 0) \tag{9}$$

$$a'_2 = -\beta \times a_2 + \eta \times a_3 \tag{10}$$

$$\beta = \alpha^2 \tag{11}$$

Where a'_1, a'_2, a_2, a_3 are vectors as showed in Fig. 2. Again, the notations for vectors are omitted for simplicity.

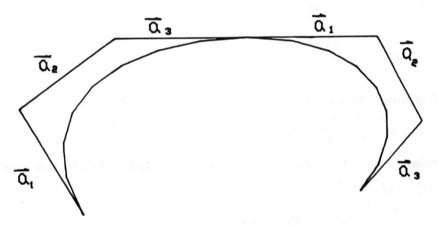

Fig. 2 Continuity of two adjoining Bezier curves

Let us prove that two adjoining Bezier curves, which are transformed from cubic B-spline curve, will meet the conditions given in (9), (10) and (11), if their knots have no multiple-knot.

From Fig.2, it is obvious that the following relations exist:

$$a'_1 = Q_2^{i+1} - Q_1^{i+1}$$

$$= Q_2^{i+1} - (\frac{t_{i+1} - t_i}{t_{i+1} - t_{i-1}} \times Q_3^i + \frac{t_i - t_{i-1}}{t_{i+1} - t_{i-1}} \times Q_2^{i+1})$$

$$= \frac{t_{i+1} - t_i}{t_{i+1} - t_{i-1}} \times (Q^{i+1}_2 - Q_3^i)$$

$$a'_3 = Q_4^i - Q_3^i$$

$$= (\frac{t_{i+1} - t_i}{t_{i+1} - t_{i-1}} \times Q_3^i + \frac{t_i - t_{i-1}}{t_{i+1} - t_{i-1}} \times Q_2^{i+1}) - Q_3^i$$

$$= \frac{t_i - t_{i-1}}{t_{i+1} - t_{i-1}} \times (Q^{i+1}_2 - Q_3^i)$$

So that $a'_1 = \alpha \times a_3$ and $\dfrac{t_{i+1} - t_i}{t_i - t_{i-1}} > 0$. Equation (9) is thus satisfied.

Replacing Q_3^{i+1} and Q_2^{i+1} according to equation (8) in the expression $a'_2 = Q_3^{i+1} - Q_2^{i+1}$, we have

$$a'_2 = \frac{t_{i+1} - t_i}{t_{i+2} - t_{i-1}} \times (V_{i+1} - V_i).$$

It is apparent to have a relation,

$$a_2 = \frac{t_i - t_{i-1}}{t_{i+1} - t_{i-2}} \times (V_i - V_{i-1}).$$

Similarly, we can also replace $Q_4^i - Q_3^i$ according to equation (8) in the expression $a_3 = Q_4^i - Q_3^i$, resulting

$$a_3 = \frac{(t_i - t_{i-1})(t_{i+1} - t_i)}{(t_{i+1} - t_{i-1})(t_{i+1} - t_{i-2})} \times V_{i-1} +$$

$$\frac{t_i - t_{i-1}}{t_{i+1} - t_{i-1}} \times (\frac{t_{i+2} - t_i}{t_{i+2} - t_{i-1}} - \frac{t_i - t_{i-2}}{t_{i+1} - t_{i-2}}) V_i +$$

$$\frac{(t_i - t_{i-1})^2}{(t_{i+1} - t_{i-1})(t_{i+2} - t_{i-1})} \times V_{i+1}$$

Let $\beta = (\dfrac{t_{i+1} - t_i}{t_i - t_{i-1}})^2$, and $\eta = \dfrac{(t_{i+1} - t_i)(t_{i+1} - t_{i-1})}{(t_i - t_{i-1})^2}$, equation (10) is thus satisfied.

4 Transformation from non-uniform B-spline surface to cubic Bezier surfaces

The specific bi-cubic non-uniform B-spline surface mentioned previously can be turned into bi-cubic Bezier surfaces in the following steps. New knots are first inserted in the U-direction till each knots becomes triple-knot, and the new array of vertices are obtained correspondingly. Then, new knots are inserted in the W-direction to obtained the vertices of cubic Bezier surfaces (patches). In this step, the non-uniform B-spline surface is identically redefined by an array of Bezier surfaces (patches).

The bi-cubic non-uniform B-spline surface can finally be represented as $n \times m$ bi-cubic Bezier patches:

$$P(U,W) = \{P^{(k,l)}\} \qquad (k=1, 2, ..., n;\ l=1, 2, ...,m) \qquad (12)$$

and

$$P^{(k,l)} = \sum_{i=1}^{4}\sum_{j=1}^{4} V_{i,j}^{(k,l)} \times B_{i,4}(u)B_{j,4}(v) \qquad (0 \le u \le 1; 0 \le v \le 1)$$

The rendering of free-form surface can thus be complemented by using division algorithms of Bezier surface efficiently. To find the intersection between two non-uniform B-spline surfaces, we can turn the two B-spline surfaces into Bezier patches, then find the intersection between Bezier patches by division-conquer algorithms. It is obvious that there are more vertices of Bezier patch than those of B-spline surface to represent one free-form surface. Therefore, we use the Bezier patch temporarily and store vertices of B-spline surface.

5 Conclusions and examples

Applications of this free-from surface modeling system show that the specific non-uniform B-spline surface discussed above has the advantage of describing complex free-form surfaces, especially for the surfaces which have sharply varying sections. This is because the specific non-uniform B-spline surfaces allow different sequences of knots for each section and it is convenient to describe each section in its own sequence of knots. Fig. 3. and Fig. 4. show two examples of free-form surface.

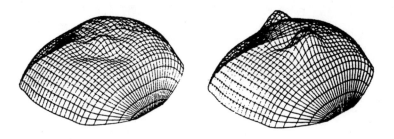

Fig. 3 Two surfaces with local bumps

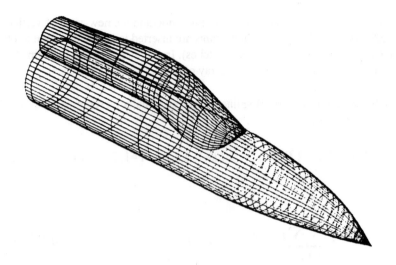

Fig. 4 A surface of fuselage

6. References

1. ID Fax and M.J.Pratt, *Computational Geometry for Design and Manufacture*. (Ellis Horwood, Chichester, 1979).

2. Xu P.J., *Journal of Zhejiang University,* (Zhejiang University Press, Hangzhou, 1984) (In Chinese), p.98

3. Su B.Q. and Liu D.Y., *Computational geometry*, (Shanghai Science and Technology Press, Shanghai, 1986). (In Chinese).

TOOL PATHS FOR FACE MILLING CONSIDERING CUTTER TOOTH ENTRY/EXIT CONDITIONS

YONG-SHENG MA

Dept. of Mechanical Engineering, Ngee Ann Polytechnic, 535 Clementi Rd.
Singapore 2159, Singapore
E-mail: may@np.ac.sg

ABSTRACT

Cutter tooth entry/exit conditions, in interrupted cutting, affect cutting performance considerably. In this work, tool path generation for face milling, using window-frame based approach, is investigated. Two objectives are pursued individually, the minimum total travel distance and keeping favourable cutter tooth entry/exit conditions during cutting.

Key words: Milling, Tool Path, CAD/CAM.

1. Introduction

In milling, brittle tool failure is closely related to cutter tooth entry/exit conditions due to the interrupted nature. The roles played by cutter tooth entry/exit conditions depend upon different work materials (Barrow, Ghani and Ma 1993).

Entry problem. When a cutter cuts into the workpiece, depending on the tool geometry and orientation of the tool and workpiece, the initial contact between the tool and the workpiece can be at a point, along a line, or over an area. Experimental results suggest (Opitz and Beckhaus 1970) that the

α_{en} —— entry angle
α_{ex} —— exit angle

Fig.1 Entry angle and exit angle

tool life is influenced by the form of initial contact which is in turn determined by the entry angle (Fig.1) of the cutter. It is recognised that the initial impact between tool and workpiece plays an essential role in tool failure when cutting very hard materials, such as chilled cast-iron and ultra high strength steels (Barrow, Ghani and Ma 1993).

Exit problem. The cutter tooth exit conditions are sensitive to tool failure when machining carbon and low alloy steels, stainless steels and the Co, Ni, Ti alloys etc (Barrow, Ghani and Ma 1993), because that negative shearing, or 'foot forming', occurs immediately prior to the cutter exiting the workpiece (Yellowley 1974, Pekelharing 1984). At low cutting velocities, negative shearing contributes to chip adhesion, which in turn leads to adhesive pitting on the rake face, at high velocities, it contributes to tool failure because of sudden unloading (Yellowley 1974).

The magnitude of the reversed tensile stress is determined by the effective chip thickness. Kamaruddin (Kamaruddin 1984) predicted the relationship between the negative shearing angle β and the exit angle θ by applying classical mechanics (Fig.2) as: $\beta=(\theta+\alpha-\pi/2)/2$. This equation predicts that the greater the exit angle θ is, the more severe 'foot

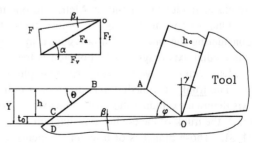

Fig.2 Geometry at exit in interrupted cutting

forming' is. Regarding the cutting geometry, the cutter tooth exit angle directly determines the effective chip thickness and whether 'foot forming' occurs.

When cutting light alloys, brasses and many forms of cast iron, negative shearing does not cause tool failure. This is because the chip adhesion at low velocities is insufficient to cause pitting on the rake face while at high velocities the compressive stress, even when reversed, is too low to cause cutting edge chipping.

Cutting Modes. Up-milling and down-milling are commonly encountered (Fig.3). In down-milling, the exit process is smooth, i.e. the effective chip thickness reduces gradually. In up milling the entry process is smooth. Hence, different cutting modes are preferred for different materials.

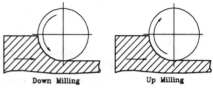

Fig.3

Types of Tool Paths. To generate tool paths, two approaches are normally used, i.e. staircase (or 'zigzag') milling, and window-frame based (or 'spiral') milling. However, when using zig-zag tool path, the cutting mode changes frequently. Zig milling is derived from zigzag milling by replacing the sequence of alternate movements by a sequence of cutting movements in one direction. Due to tool lifting and machine's re-positioning error, the surface finish is affected. The staircase method also involves a lot of transient cutting geometry when entering and leaving the workpiece, so the cutter tooth entry/exit conditions are neither under control nor predictable.

In the second approach (Fig.4), the surface is milled along a profile equidistant to the surface boundary, with the tool stepping inwards for the next pass. This approach is able to keep a constant width of cut and favourable entry/exit angles during most of the cutting time, and the cutter can machine surfaces without lifting the cutter as long as the cutter can access the area. However, there exists a covering problem in traditional CAM systems when the stepover is larger than the cutter radius, and the cutter tooth entry/exit angles are never considered (Ma 1994).

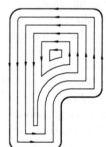

Fig.4 Spiral milling

2. Medial axis diagram

The concept of medial axis diagram is originated from Voronoi diagram. A simple definition of the Voronoi diagram is given in (Held 1991): "The Voronoi diagram of a set P of points in the plane is a partition of the plane such that each region of the partition represents (the sub-set of) points that are closer (under the Euclidean Metric) to one member of P than to any other member".

One extension of sets of points is to take line segments into consideration. Considering the interior of a polygon, a <u>medial axis diagram</u> is defined in (Lee 1982): "the medial axis diagram of a polygon is the set of Voronoi segments less the segments incident with concave vertices". An example is shown in Fig.5.

The <u>trunk</u> of a medial axis diagram is defined as the set of segments in the medial axis diagram excluding those segments which are linked to end points of the polygon. In Fig.5, the trunk of the medial axis diagram consists of $M_1 M_2 M_3 M_4 M_6 M_8 M_9$, $M_6 M_7$ and $M_4 M_5$. Those branches of bisector segments on the medial axis diagram that are linked with end points of the polygon, and without crossing any junction point of the diagram, such as M_1q_4, M_1q_5, and M_2q_6 etc, are referred to as the <u>end branches</u>. Each segment in the medial axis diagram is the bisector of two elements on the polygon. The area width of a polygon w is defined as a function of a point P on the medial axis diagram (M), w(P), (P∈M), it equals to the sum of the minimum distances from the given point P to its two corresponding boundary elements on the polygon. In Fig.5, Q_1 is a point on the medial axis segment M_2M_3, the area width at Q_1 equals to d_1+d_2, where d_1 and d_2 are the minimum distance from Q_1 to its two corresponding segments, i.e. q_6q_7 and q_4q_3.

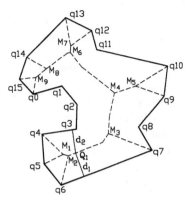

Fig.5 Medial axis diagram of a simple polygon

3. Problem analysis

In this work, tool path generation is started by offsetting the boundary profile. The drive profiles are first generated by offsetting the boundary profile, layer by layer, until the last offset profile can not be shrunk any more. The offset distance is the specified radial width of cut b. These drive profiles are then used to generate cutter Centre Location (CL) profiles by offsetting them by a distance b-R, where R is the cutter radius.

When the b/D (D, cutter diameter) ratio is less than 0.5, and the cutter tooth entry/exit conditions are not considered, similar to the traditional approach, the tool path generation is merely a process of offsetting. However, more often than not, a b/D ratio larger than 0.5 is preferred, and for some work/tool material combinations, the cutting performance will be improved considerably if the cutter tooth entry/exit conditions are favourable. In these cases, simple offsetting techniques are not satisfactory.

A simple L-shaped surface is shown in Fig.6. Drive profile DP_0 is the boundary profile, DP_1, DP_2, DP_3 are the inner layers of drive profiles. The work material between two drive profiles is supposed to be machined by one cutting pass. Please note that narrow part of the area enclosed by DP_2 where the area width (w) is less than 2b, and the area, enclosed by DP_3, need special treatment.

Assuming b/D ratio is greater than 0.5, all the cases where 0<w≤2b, can be classified into three geometry patterns, i.e. when 2b≥w>2R, 2(b-R)<w≤2R, and 0≤w≤2(b-R).

Pattern 1: When 2b≥w>2R. Since w≤2b, if the cutter follows the U shaped CL profile (Fig.7), then the area can be fully covered by the going (first) and the returning passes. However, the actual radial widths for the first cutting pass and the returning cutting pass are different. For the first pass, since w>2R, i.e. w>b, the radial width is always b (Fig.7(b)). Therefore the cutter tooth exit angle is

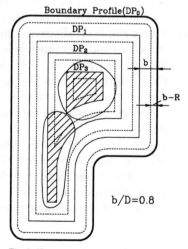

Fig.6 Narrow areas enclosed by drive profiles

kept favourable (0°). The returning pass clears the material left behind by the first pass. The actual radial width of cut equals w-b, and b≥w-b>2R-b. Its value changes according to the component geometry, i.e. the value of w. This means that the exit angle for the returning pass may change considerably. Suppose w_2 is the distance from the cutter centre to the exit edge (Fig.7(c)), then $w_2=w_1$-R. Since $2R≥w_1>2(2R-b)$, $R≥w_2>3R-2b$. If b/D=0.75, then $R≥w_2>0$, i.e. the cutter tooth exit angle may vary in a range from 0° to 90°. Since w>2R, it is impossible to cover the area in one pass.

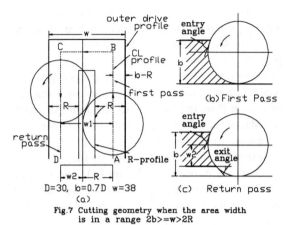

Fig.7 Cutting geometry when the area width is in a range 2b>=w>2R

Pattern 2. When $2(b-R)<w≤2R$. When using simple offsetting procedures (Fig.8), because the CL profile ABCD exists (w≥2(b-R)), the cutter will then traverse along the first and the returning passes (AB and CD). Similar to the previous pattern, the cutter tooth exit angle may change considerably even for the first pass, and it is possible that the first pass removes all the material in the area, leaving the second pass to cut air. Note that, since w≤2R, then cutter should be able to machine the whole area with one pass.

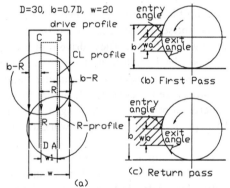

Fig.8 Cutting geometry when the area width is in a range 2R>=w>2(b-R)

Pattern 3. When $0<w≤2(b-R)$. This case is shown in Fig.9. The CL profile is twisted (negative) because of the narrow width of the outer drive profile. Traditionally, this CL profile is deleted, and therefore this narrow area will be left un-cut. This problem is caused by large b/D ratio (>0.5).

It can be predicted that keeping favourable entry/exit conditions increases cutter travel length, therefore the entry/exit conditions are considered only when they are necessary. In this work, two methods are developed:

(1) to keep favourable cutter tooth entry/exit conditions as much as possible while minimising cutter travel distance is considered as well;

(2) to minimise cutter travel distance as much as possible without considering the entry/exit conditions.

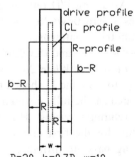

Fig.9 Cutting geometry when the area width is in range 2(b-R)>=w>0

4. The first method

Assume that, when considering cutting strategies with respect to the cutter tooth entry/exit conditions, down milling is preferred. Obviously, up milling can be achieved by simply reversing the cutter traversing direction on the CL profile.

In Fig.10(a), the basic idea of the first method is shown. ABCD is a drive profile, the area width of this profile is w, and 0<w≤2b. Suppose the included angles at B and C are not very sharp. XY, YC, and YB are medial axis segments of the drive profile. XY is the trunk of the medial axis

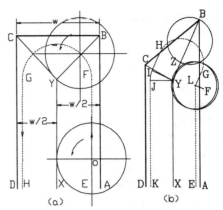

Fig.10 Tool paths around the trunk of the medial axis diagram and compensation edges

diagram of the drive profile. It should be noted that if the cutter cuts tangentially along XY for the first and the returning pass, i.e. the cutter traverses the profile EFGH, the cutter tooth exit angle is then kept as a constant (0°). If the cutter rotates counterclockwise, the cutting mode is then kept as down milling.

When there is any sharp corner, compensation is needed. In Fig.10(b), there is a sharp corner at vertex B. Because of this sharp corner, the trunk of the medial axis diagram has to be extended, so that the cutter covers the sharp corner completely when it moves tangentially along the modified trunk. The modification is straight forward. The distance from the junction point Y to each of the corresponding vertices on the drive profile (here B and C) is checked to see whether they are within a circle centred at Y, with a radius of D. If either of them is outside this circle (B here), then this vertex (B) cannot be covered if the cutter moves to Y only. Hence a further part of the branch (YZ out of YB, where ZB=D) is used as an extension of the trunk, hence the cutter paths is extended as well. In Fig.10(b), the CL profile becomes EFLFGHIJK. A minor compensation edge (dual directional) FL is inserted to ensure all the material at corner Y can be removed.

4.1. Covering of the inner-most areas

There is always at least one inner-most drive profile that cannot be offset by a distance b. A simple example is shown in Fig.11. ABCDEA is the inner-most drive profile, the medial axis diagram is shown as well. Here XYZ is the trunk, XA, XE, YB, ZC and ZD are end branches. For covering an inner-most area, the

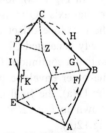

Fig.11 Tool Path for the inner-most drive profile

trunk of the medial axis diagram and if necessary, part of the end branches (to compensate sharp corners) are used to generate the CL profile. In Fig.11, no compensation is required. The R-offset profile (FGHIJKJF) for the trunk of the medial axis diagram (XYZ) is then generated as the cutter CL profile to cover the inner-most area. JK is the compensation segment for the corner at Y.

Basically, the maximum area width for the inner-most drive profile is 2b (otherwise a new layer of drive profile can be generated). The trunk of the medial axis diagram divides the area into two parts with equal area widths, so the maximum radial width for each cutting pass is b. Obviously, the cutter can cut a width up to D on each side, therefore the area can be covered fully.

4.2. Covering of narrow areas enclosed by an intermediate drive profile

Use the example shown in Fig.4. After removing the first two layers of material, to machine the next layer, more cautions should be exercised (Fig.12(a)). There is an area between two drive profiles where the area width varies between 0 and 2b. ABCDEFGA is the outer drive profile, $A_2 B_2 C_2 D_2 F_2 G_2 A_2$ is the inner-layer drive profile.

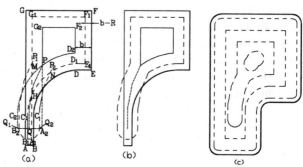

Fig.12 Tool path to machine a L-shaped surface

This inner drive profile intersects itself at point P, loop $A_2 B_2 C_2 P A_2$ is negative. The cutter is supposed to traverse along CL profile $A_1 B_1 C_1 D_1 E_1 F_1 G_1 A_1$. This CL profile is also twisted, a negative loop is generated, i.e. $A_1 B_1 C_1 H A_1$. If the cutter simply traverses the valid part of the CL profile, i.e. $H D_1 E_1 F_1 G_1 H$, then the cutter leaves some un-cut material, and the cutter tooth exit angle changes considerably as discussed previously.

The method proposed is to use the negative loop of the inner drive profile. The corresponding area enclosed by the <u>outer</u> drive profile can be identified as an area with the width 0<w<2b. Hence, the medial axis diagram of such a negative loop is exactly what is of importance.

In Fig.12(a), loop $A_2 B_2 C_2 P A_2$ is the negative loop of the inner drive profile, its medial axis PQ is used. Based on PQ, the CL profile to cover this narrow area which is the envelope offset of PQ by a distance of R can be obtained, i.e. $P_1 Q_1 Q_2 P_2$. These newly generated CL segments must be merged with the valid part of the outer CL profile so that a continuous CL profile can be generated. The merging method is simply selecting the outer envelope profile of these overlapped parts of the CL profile as a whole. In Fig.12(a), the final CL profile after merging is $Q_1 Q_2 N D_1 E_1 F_1 G_1 M Q_1$, which is also shown in Fig.12(b). The overall tool path layout for the given example is shown in Fig.12(c).

5. The second method

In this method, cutter tooth entry/exit conditions are not considered, and the cutter travel length should be saved wherever possible. The difference between the first and the second methods occurs only when the area enclosed by the drive profiles has a width (w) less than 2b. The basic idea of the second method is that, when w≥2R, the cutter traverses the b-R offset of the outer drive profile. When 0<w≤2R, the cutter's path is switched to the medial axis of the area, which can be obtained by analysing the negative loop of the R-offset of the drive profile. In Fig.13(a), ABCDE is a drive profile, FGHIJ is the offset of ABCDE by a distance of b-R (b-R offset), while KLMNS is the offset of ABCDE by

a distance of R (R-offset). KLMNS intersects itself at point T, and a negative loop is generated (TLMNT). Since the R offset for this area is negative, then 0<w≤2R. TW is the medial axis of loop TLMNT. In Fig.13(b), the cutter is navigated to traverse on the course of FUTWXWYWTVJ where U and V are the projection points of T on both sides of the b-R offset profile, while WX and WY are compensation branches for the corner at C and D (Fig.13(b)). TW is the shortest path to cover the narrow area.

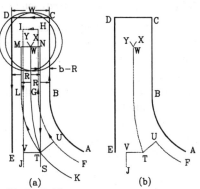

Fig.13 Cutting a narrow area by traversing along the medial axis

5.1 Covering of the inner-most areas

The narrow and wide areas in inner-most areas can be detected by generating the R-offset of the inner-most drive profile. There are three possible cases (Fig.14):

Case 1: The R-offset is valid and there are no negative loops. In this case, as shown in Fig.14(a), the drive profile is ABCDEA, its R-offset $A_2 B_2 C_2 D_2 E_2 A_2$ is positive, then the CL profile to clear the area enclosed by drive profile ABCDEA is the b-R offset, i.e. $A_1 B_1 C_1 D_1 E_1 A_1$.

Case 2: The R-offset is completely invalid. In this case (Fig.14(b)), since the R-offset is negative, the area width is less than 2R, therefore, if the cutter traverses along the medial axis of the drive profile, the cutter can cover the whole area in one go. In Fig.14(b), $A_1 B_1 C_1 D_1$ is the negative R-offset of the drive profile ABCD. The medial axis of this area is calculated first. Trunk PQ is used as the effective CL profile. The compensation for sharp corners has to be considered at the same time.

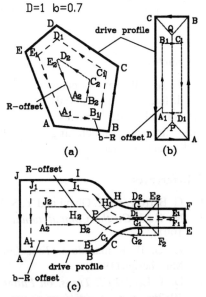

Fig.14 Strategies to deal with innermost drive profiles

Case 3: The R-offset has positive and negative loops. As illustrated in Fig.14(c), the R-offset of the drive profile, i.e. $A_2 B_2 D_2 E_2 F_2 G_2 H_2 J_2$, intersects itself at point P. This means that the area corresponding to the positive loop $PB_2 A_2 J_2 H_2 P$ can not be machined by one cutting pass as the area width is larger than 2R, while the area corresponding to the negative loop $PD_2 E_2 F_2 G_2 P$ is narrow enough to be covered by one cutting pass.

Based on the above observation, the b-R offset is used as the CL profile for the wider area; while for the narrower area, the medial axis of the area is used. These two parts of the CL profile are merged by a couple of switch edges, which link the constriction point P and its nearest points on the b-R offset on the both sides.

5.2 Covering of narrow areas enclosed by an intermediate drive profile

In Fig.15(a), a narrow area whose width is less than 2R, is identified by offsetting the drive profile by a distance of R, the negative loop is given by $PG_2 A_2 B_2 P$, where P is the self intersection point. This narrow area can be machined in a single pass, i.e. driving the cutter along the medial axis of the negative loop, i.e. PTU. For the compensation of the two sharp angles at G and A on the drive profile, two small compensation edges, UV and UW, are attached to the end point U of the medial axis PTU (Fig.15(b)). Two switch edges, PS and PQ, are used to change the cutter's track from the b-R offset to the medial axis PTU. They are the shortest edges from P to the b-R offset. The final CL profile of this layer is shown in Fig.15(c).

6. Conclusion

Using different tool paths for different work materials will smooth cutting process, and improve machining quality and productivity.

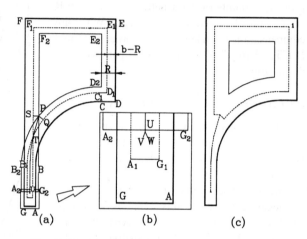

Fig.15 Tool path to machine a narrow area enclosed by a intermediate drive profile

7. References

1. G. Barrow, A. K. Ghani and Y. S. Ma, Proc. of 6th IMCC, Hong Kong, March, 1993.
2. M. Held, *On the Computational Geometry of Pocket Machining*, (Springer-Verlag, 1991).
3. A. G. B. Kamaruddin, *Tool Life in Interrupted Cutting*, PhD Thesis, Univ. of Manchester (1984).
4. D. T. Lee, *IEEE Trans. on Pattern Analysis and Machine Intelligence*, **No.4**, (July 1982).
5. Y. S. Ma, *The Generation of Tool Paths in 2½-D Milling*, PhD Thesis, Univ. of Manchester (1994).
6. H. Opitz and H. Beckhaus, *Ann. CIRP*, **Vol.18**, (1970) p.257.
7. A. J. Pekelharing, *Ann. CIRP*, **Vol.33(1)** (1984) p47.
8. I. Yellowley, *The Development of Machinability Testing Methods with Specific Reference to High Strength Thermal Resistant Work Materials*, PhD Thesis, Univ. of Manchester (1974).

CAD/CAM Applications

CAD-Based Intelligent Robot Workcell
   G. C. I. Lin and T. F. Lu

CAD-Based Robotic Welding System with Enhanced Intelligence
   T. Kangsanant and R. G. Wang

Efficient Toolpath Geometry for CNC Turning Operations
   R. N. Ibrahim, P. K. Kee and S. Cabarkapa

Determination of Analytic Features from Laser Scan Data
   S. E. Ebenstein

CAD-Based Intelligent Robot Workcell

Grier C. I. Lin
Professor of Manufacturing Technology & Management
Centre for Advanced Manufacturing Research, University of South Australia
Adelaide, South Australia, Australia
grier.lin@unisa.edu.au

Tien-Fu Lu
School of Manufacturing and Mechanical Engineering, University of South Australia
Adelaide, South Australia, Australia

ABSTRACT

This paper presents a CAD-based intelligent robot workcell that consists of a neural-network-based 3D force/torque sensor, image system, a six-axis industrial robot, CAD package, and database. The capability of the CAD package is extended to provide more useful infomation, such as images, robot path coordinates, etc. A neural-network-based 3D force/torque sensor is developed for solving the uncertainty in locating parts and assembly products - the difference between simulation models and the physical world. This proposed image system can be used to eliminate the needs of barcode, accurate part feeders and the tedious image training process for object recognition. It can also search objects randomly located on a workbench and derive their 3D coordinates. This system and its performances will be detailed in this paper.

Keywords: intelligent, 3D force/torque sensor, robot

1. Introduction

Over the past few years, the manufacturing industry has exploited the use of AI technology . This has been motivated by the competitive challenge of improving quality while at the same time decreasing costs and reducing design and production time.

Nowadays, product components which are to be processed have to be placed in fixed pre-determined positions to be picked up for machining or assembly in most industrial applications. Accurate feeders, fixtures or ancillary toolings are needed for such kind of automation systems. CAD/CAM, robot, partially manual off-line inspection, sensors, barcode, pallets, parts feeders, etc. are always adopted for factory automation. They have been integrated together for some time, but still without enough flexibility and intelligence to automatically adapt itself for changes, such as new products, the difference between CAD simulation models and the real physical world operation.

Efforts have been made to improve the flexibility and intelligence of existing manufacturing systems in order to meet the shorter lead time in production and the quick product model changes for diversified and individualised customer requirements. Many CAD-based systems and intelligent system prototypes have been proposed for fulfilling the requests. (Yi and Chung 1990) proposed a collision-free and optimal sequence planning scheme for a spot welding task using CAD data. (Kangsanant and Lin 1992) integrated a welding robot and AutoCAD for user to program the welding tasks from within a CAD environment. (Meijer and Jonker 1991) presented the philosophy and architecture for an intelligent assembly cell named DIAC. (Sekine, et al 1991) proposed an intelligent body assembly system for Nissan's new production system by integrating information network, production process and CAD/CAM/CAE. The path and orientation errors created by off-line programming, CAD simulation for robot assembly, welding simulation, for instance, existed. In these cases, accurate facility installation is needed. Time-consuming, tedious calibration has to be taken in order to obtain satisfactory system performance.

There is still a long way to go for a system to possess human-like intelligence. Much useful data is residented at product designing stage. Therefore, if this data can be used properly, a flexible and intelligent system can then be achieved easier. With these in mind, this research focuses on the development of a CAD-based intelligent robot workcell which can quickly accommodate highly dynamic and varying conditions. Basically, this system is developed by following the intelligent system scheme proposed by (Lin and Lu 1994).

2. System Configuration

This CAD-based intelligent robot workcell consists of the following major components:
1) Neural-network-based 3D force/torque sensor: this provides 3D force/torque information or robot fine motion movement instructions for robot control, while performing positioning and assembly tasks interacting with environment.
2) image processing and analysis system: this is developed under SUN Sparc 10 platform. This image system accesses and processes the images from both ASM (Advanced Solid Modeller) and real camera to find out the right component randomly placed on the workbench. The position and orientation of components can be derived from camera for robot to pick up on-line. It possesses the capability to search through the workbench to find out the demanding workpart.
3) Robot: The robot used here is IRB 2000, a six-axes industrial robot manufactured by ASEA BROWN BOVERI. C programs are developed for communicating with other programs or facilities via PC-NFS.
4) CAD package: The CAD package, used here, is ASM provided by (Qikdraw 1994). It can be performed under both SUN platform and PC. SUN version is adopted for this project. Camera function simulation for generating images has been developed. Robot and manufacturing systems simulation environment is under development.
5) Common Database: All the related data are kept here for accessing. It is placed under SUN at this moment.

3. Force/Torque Sensor

A bar type 3D force/torque sensor has been developed in the authors' previous work, (Lin and Lu 1994), is adopted in this proposed system. It has been proven that the neural-network-based calibration method is able to overcome the manufacturing errors, sensor's coupling effects, incomplete mathematical model, noise, environment effects integrated into, and so on, in computer simulation, (Lin and Lu 1994), and our later experiments.

One of the application examples that is under development is shown below in Figure 1. This sensor can render force/torque or raw strain gauge readings while force/torque applied on it. Then, the job classifier switches the connection among the corresponding neural networks to generate proper output for robot fine motion adjustment by referring to the job instructions.

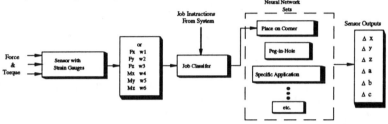

Figure 1: Neural-Network-Based 3D Force/Torque Sensor Application Example

4. CAD System

The main goal of CAD-based systems is to enable the system to obtain and use as much information from CAD as possible in order to increase the system planning efficiency and reduce human efforts.

ASM is used for designing and assembling components as well as simulating robot and manufacturing systems. The useful data can then be rendered after simulation is carried out, such as images, path coordinates, and so on.

4.1. Image Generation

To enable easy integration, communication between ASM and image system, the viewing parameter setting operation is implemented to reflect the way a real camera is operated, and the way the vision system processes images from ASM and pictures from the camera. The following needs are considered in order to set ASM for obtaining images similar to camera images: a) The projection plane (viewing plane): analogous to the film on the camera; b) Projection Type: perspective projection; c) Viewing Distance: the distance between the project plane and the viewing position; d) Viewing Angle: analogous to the lens angle of a camera; and e) Window Size: the width and height of the window. By setting these viewing parameters to be as simular as those in camera, the image captured from ASM should be highly similar to that from camera.

In Figure 2, they are from ASM. They are the same objects as those which shown in Figure 4. The CAD generated images do not have image distorsion problems but camera images do. How to accomplish an easy method to overcome such problems is taken by the image system for object recognition.

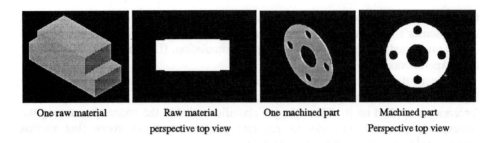

| One raw material | Raw material perspective top view | One machined part | Machined part Perspective top view |

Figure 2: Images from CAD (ASM)

4.2. Robot Simulation

The reasons for simulating the robot's movement and manufacturing systems are: a) To generate path coordinate data files as actual robot input; and b) can design system layout in advance. Part of the simplified ABB IRB 2000 robot workcell model is extracted and shown in Figure 3.

An internal model of a product assembly can be created in ASM. This model has all the information about the product final assembly position, and orientation of every component as well as the hierarchical information of the assembly. With this data rendered from CAD, the image system, and the assistance from neural-network-based 3D force/torque sensor for motion adjustment, robot can be coordinated to fulfill the assembly demands.

5. Image System

Image systems are becoming widely used in automatic manufacturing systems such as robotics and product inspections. For example, (Buurman 1992) presented an object recognition system for intelligent robot cell. He compared the wireframe images from real camera with a number of wireframe models from CAD to achieve objection recognition. To save more time and achieve an easier recognition method, 2D perspective vision for object recognition and 3D pose determination are developed at this moment. It processes the images from both ASM and real camera to find out the right component randomly located on the workbench. It can also derive on-line the component position and orientation from camera images for the robot to pick components up and send them to the processing machines. A searching function has been developed for the robot to go through the workbench and find the demanding object. The expensive part feeders, which used to be needed, are then not necessary.

The major functions of this image system can be divided into the followings :
1) Image Snapping or Loading: System loads the images generated by ASM from the database if it is invoked by ASM. If it is invoked by system integration programs, camera is activated to snap images;
2) Processing Criteria Setting: Some criteria needed to be set before processing is taken for proper results. They are: * area of interest for processing * size threshold to get rid of small objects (noise), etc;

3) Processing and Measuring: Based on the criteria setting, the system then processes and measures the images to derive image characteristics, such as object area, hole numbers, maximum length, etc;
4) Image Code Generating: Image codes can then be generated by manipulating those parameters derived at step 3;
5) Deriving Coordinates: According to those data obtained at step 3, workpart coordinates are derived for robot;
6) Searching: The next position or adjusted position will be reached until the demanding part is found.

One example of camera images is shown in Figure 4. The image code for the raw material is 1558 and for the machined part is 51042.

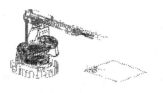

Figure 3: Simplified ABB IRB 2000 Robot Workcell Model in ASM

Raw material and Machined part camera top view
Figure 4: Camera Images

6. Robot Control

Robot control can be divided into three functional catalogues. The first is to read in CAD simulation results and enable the robot to follow the CAD generated path. The second is to adjust robot fine motion movement while the robot is interacting with the environment. Robot adjusts its fine motion by the force/torque information and the instructions from the neural-network-based 3D force/torque sensor to accommodate the difference between CAD path data and the physical world, and to accomplish the demanding task. The third part is a vision-based robot for guiding and picking up those components randomly located on the workbench. Robot picks up the demanding component after the image system searches through the workbench, finds out the right component and derives the coordinate of that component. The coordinate systems among the camera, gripper, and robot system is shown below in Figure 5. It is used for deriving coordinate data for robot control. The meanings of those terminologies are:

image zero point (zp): This point is the zero position in the camera image
Gripper (g): Here, it means the reference point on the top of the gripper
Camera (c): The original is located at the centre of the lens
$\vec{r}_{g \to zp}$: The vector from gripper reference point to image zero point
$\vec{r}_{zp \to o}$: The vector from image zero point to the object located on workbench
Δx_{im}: The x direction distance from zero point to object centre in camera image
Δy_{im}: The y direction distance from zero point to object centre in camera image

Therefore, the amount for gripper to move to grip the object randomly located on the workbench is:
$$\vec{r}_{g \to o} = \vec{r}_{g \to zp} + \vec{r}_{zp \to o}$$

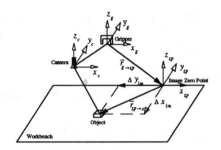

Figure 5: Coordinate System for Robot and Image system

7. System Integration and Communication

Integration:
Calibration procedures always have to be taken when a system is integrated. In this proposed system, some easy procedures are still needed for system integration in this proposed system. But the accuracy is not needed to be very precise, because this proposed system can accommodate such errors automatically. This calibration only needs to be taken once, if nothing is changed later.

a) Calibration for image distorsion: It is known that images are distorted in camera, especially the position far away from the centre line of camera. The centre of the camera image is probably not just below the central line of the camera lens. If this happens, the distortion on the 4 corners and edges will be different. Here, a simple procedure is taken to adjust the horizontal and vertical offset to align the image centre with the lens centre line for reducing camera distorsion effects. Anyway, this proposed image system can overcome such kind of image distorsion effects.

b) CAD environment setting: There is no image distorsion problem in CAD. Therefore, what has to be done is going in ASM to set window size, viewing angle, viewing distance, etc as discussed in section 6.1 for obtaining proper images.

c) Camera Position: Camera is attached on the robot. Therefore, robot is moved to different positions to find out the proper height for camera to obtain 1:1 image.

d) Coordinate Transfer: To complete the coordinate transfer between derieved image coordinate and actual robot coordinate by referring to section 6.

e) Modify data file: Data file that stores robot, image system initializing data has to be modifed. Then, system will adjust itself while it is initialized.

Communication :

a) Robot programs in PC with image programs in SUN: PC-NFS is adopted for PC and SUN communication.

b) PC and robot controller: RS232 is used to establish the serial communication link between robot controller and robot.

c) Image analysis program with image acquisition hardware: BIT 3 is adopted to link image acquisition hardware with SUN for anlayzing programs to access and process those images.

d) Robot program with 3D force/torque sensor: Lab-PC+ board is selected for acquiring signals from strain gauges.

8. System Working Principle

The system working sequence is off-line simulation, on-line running, and then assembly. The first is the product design, CAD image generation, robot job simulation for coordinates generation, etc that is classified as system off-line simulation. The second is robot on-line running. Recognizing, searching, loading, unloading and robot moving among machines are included in this part. The third is robot on-line assembly by following simulation results. These procedures are drawn in Figure 6.

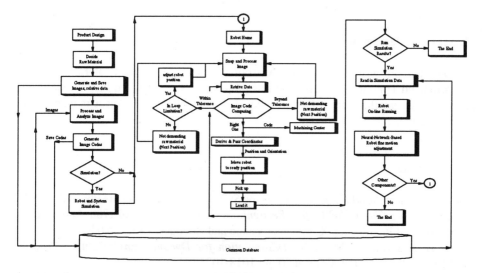

Figure 6: System Working Procedure

For the part one, the designer designs products in ASM. The image files of machined parts, their final assembly coordinates and the image files of the raw material are all rendered and saved into the common database when the product design is completed. Image system is then invoked to process those image files rendered from ASM to extract characteristics and generate image codes for object recognition. All this data will then be saved back to the common database again. CAD simulation for robot and manufacturing systems will then be performed. Numbers, setting of robots and machines can be easily modified. Manufacturing process can be simulated and the path coordinates, robot motion sequence including wait, machine serving sequence, and so on can be generated.

For part two, robot program in PC and image processing, analysis program in SUN will then be activated. Robot, attached to a camera, goes to the snap position above the workbench to search through workbench and snap images. These image files are then processed and analyzed as those from ASM. Comparision is then taken to recognize demanding workpiece. Object position and orientation is then derived to be picked up and sent to machining centre.

After the machining process, the robot performs assembly task by following the path simulated and planned in ASM. With the force/torque information, assembly task can be accomplished even though there are some coordinate data errors resulting from the difference between CAD models and physical world.

9. Summary and Future Tasks

A CAD-based intelligent robot workcell that has been partially achieved is detailed in this paper. This workcell integrates a neural-network-based 3D force/torque sensor, image processing and analysis system, a six-axis industrial robot, CAD package, and database. Flexibility and intelligence are demonstrated. Their integration performance is described. The integrated system will no longer need accurate feeders, fixtures or ancillary tooling normally associated with "automated" assembly.

10. Acknowledgements

The authors wish to thank the Australian Federal Department of Employment, Education and Training for supporting this project under its Targeted Institutional Link Program. The auxiliary from Qikdraw Pty. Ltd., especially the project manager Mr. Dayong Zhang.

11. References

1. *ASM-Advanced Solid Modelling, Qikdraw Systems,* Qikdraw systems Pty. Ltd., 1994
2. Buurman, J., "*The DIAC Object Recognition System,*" SPIE Vol.1708 Applicaitons of Artificial Intelligence X: Machine Vision and Robotics, 1992, pp.641-652
3. Fok H. and Kabuka M.R., "*CAD/Simulation for Dynamic Enviroments,*" Proc. 5 IEEE Int. Symp. Intell. Control 90, pp.1018-1025
4. Kangsanant T. and Lin F. S., "*CAD/CAM Technique for Robotic Welding,*" Proceedings 5th ASEE International Conference ECGDG 1992, Melbourne, Australia, pp.341-346
5. Lin G. C. I., and Lu T. F., "*Neural Network for Active Compliance Mechanism Calibration,*" International Conference on Data and Knowledge Systems for Manufacturing and Engineering, at The Chinese University of Hong Kong, pp.497-502, May 2-4, (1994)
6. Lin G. C. I., and Lu T. F., "*Construction for an Intelligent Product Assembly and Inspection System,*" The Third International Conference on Automation Technology, July, 1994, **Vol 3**, pp. 13-17
7. Meijer B. R. and Jonker P. P., "*The Architecture and Phylosophy of the DIAC,*" International Conference on Robotics and Automation Vol.3, 1991, pp.2218-2223
8. Sekine Y., Koyama S. and Imazu H., "*Nissan's New Production System: Intelligent Body Assembly System,*" International Congress and Exposition, Detroit, Michigan, February, 25-March 1, 1991, No.910816
9. Yi S. and Chung M. J., "*Optimal Teaching for Spot Welding Robot Using CAD Data,*" SICE'90, July 24-26, Tokyo, pp.837-840

CAD-BASED ROBOTIC WELDING SYSTEM WITH ENHANCED INTELLIGENCE

T KANGSANANT and R G WANG

Department of Electrical Engineering
Royal Melbourne Institute of Technology, GPO Box 2476V, Melbourne 3001, Australia

ABSTRACT

This paper concerns the development of an intelligent planning and programming system for an industrial welding robot within an AutoCAD environment. It is an extension to the initial work done earlier. The intelligence in workpiece analysis and path planning together with welding data expert system makes this system flexible and easy to use in robotic welding applications.

KEY WORDS: welding robot, AI, off-line programming, CAD/CAM

1. Introduction

Automation has become an undoubtable trend of modern industry. As a key component of the industrial automation, the robot, in terms of development and applications, is just in full swing. There are several major interests in the research area of robotics. One of them is to minimise the user interactions required in programming and planning of robot motion since the complexity in robot manipulation and programming has impeded robot's spread and utilisation. To achieve this, considerable work has been put into study of off-line interactive graphic programming, planning and simulation systems [Jacobs 1984].

An off-line robot planning and programming system is expected to be able to automatically handle most of work in the whole process which normally have to be performed by the robot operator, the robot programmer, and the process engineer. In other words, it should have some expert knowledge or intelligence. AI techniques, which have grown up rapidly both in its theory studies and practical applications, can be employed to facilitate advanced features of robot off-line planning and programming systems. However, most robot off-line programming systems developed so far are still limited to processing simple or specific cases and need a high level of human involvement in handling complicated problems such as recommending optimal process parameters and planning collision-free robot motion, etc. On the other hand, a few systems with some intelligent functions have shown their advantages and promising development prospects [Wloka 1986, Buchal 1989, Kangsanant 1993]. At least the following aspects in robot off-line planning and programming systems can benefit from the use of AI techniques:

- *Workpiece analysis.* In interactive graphic systems, it is always a problem that how to exactly inform the computer of users' intention or task requirements through simple graphic pointing or basic information data input. A workpiece analysis module can make intelligent decisions for this problem. Examining AI application areas, we can find that fruitful researches on knowledge-based generative process planning and workpiece recognition will stimulate many ideas for the development of the workpiece analysis [Joshi 1987].

- *Task planning and path planning.* This area itself is an AI research area combined with robot kinematics and some advanced mathematic problems [Paul 1981, Buchal 1989].
- *Recommendation of optimal process data.* Expert systems will be the best technique to realise this function since in practice the domain expert is deterministic to an optimal process [Kangsanant 1993].

This paper presents an intelligent planning and programming system for an ABB IRB2000 welding robot in AutoCAD environment. The above three functions have been included in this system. It is an extension to the initial framework done earlier [Kangsanant 1992(1)], in which a straight-line welding path can be specified graphically and corresponding robot program can be automatically generated. In addition, robot simulation, collision detection and welding data expert system, at a prototype stage, were also provided. The intelligence added to the initial framework makes the system more general and powerful. In the following sections, the integrated system will be briefly described first. Then the discussion will be focused on major intelligent components in this system.

2. Integrated Intelligent Robotic Welding System

The complete intelligent robotic welding system includes a 6-axis industrial welding robot ABB IRB2000 fitted with a laser seam tracker and its controller ABB S3, as well as a 486-type PC. There is a communication link between the robot and the computer, which is based on communication facilities of ABB Off-line Programming software (OLP3). The robot off-line planning and programming system has been developed under AutoCAD R11 and Advanced Modelling Extension (AME 2.0).

There are five modules in this off-line planning/programming system:

❏ *pre-processor: automatic workpiece interpretation*
   In the pre-processor, the welding task will be specified graphically by the user and the workpieces will be analysed by an expert-system based mechanism with the support of solid modelling techniques to establish all necessary information relating to welding process and robot motion. The information will be stored in a common database.

❏ *path planning*
   Depending on the information in the common database, the path planning module will plan a robot movement for given welding paths. The planned robot motion will be collision-free and will have a minimum cycle time.

❏ *welding data expert system*
   While the robot motion is planned, the welding process is also optimised by welding data expert system. A set of welding parameters will be recommended for each welding path to achieve a quality weld.

❏ *robot simulation*
   Simulation can be run to visually check the sequence of the motion before the associated robot program is generated.

❑ *post-processor*

The post-processor will translate the planned robot motion and the welding process parameters into an executable robot program, and download it to the robot controller via the communication link.

In this paper, only automatic workpiece interpretation and welding path planning are discussed. Another intelligent module, the expert system for welding data, which has been previously developed [Kangsanant 1993], will not be included.

3. Pre-processor: Automatic Interpretation of Workpiece

The interpretation of workpieces in this system includes two main parts - welding path detection and welding-relating feature recognition of workpieces.

When the user picks a point on the workpiece drawing to specify a welding path, the whole edge that the point lies on is selected. Actually, expected welding path may be only the part of this edge. And sometimes, the point picked is at an impossible place because of artificial mistakes, inaccurate drawings or inaccurate picking place. The welding path detection module is designed to recognise and describe a welding path that the user really wants, according to the user's picking point on the graphic model of workpieces.

There are several basic features of workpieces used in determining the welding parameters and welding seam track mode. Previously they used to be input manually, which requires the user's welding knowledge and some measurement work. It is sometimes inflexible, especially for 3D curved path, as its features vary on different segments and cause the changes of process data. In this system, the welding-related features, such as plate thickness, joint type and welding position, are acquired automatically from 3D workpiece drawings by the feature recognition module.

3.1 Welding Path Detection

When the user picks a point on the workpiece drawing to specify a welding path, the module will firstly find out all solid edges and faces passing the picking point. Normally two edges or one edge plus one face can be captured. The overlapped part is then measured. The overlapped part should not be zero. Further detection is done by checking the interference between all other workpiece solids and the overlapped part to exclude some segments of this overlapped part which are intercepted by some of those solids. Then a real welding path is obtained. Actually, this system defines three basic path types:

- path formed by the overlap of an edge and a face (eg T-joint)
- path formed by the overlap of two edges (eg butt joint)
- gapped path - In dealing with this case, the module will temporarily move a solid against its counterpart over the width of the gap, then process the seam as normal non-gapped type as above. Afterwards, the solid will be moved back to its original place.

In implementation, an IF-THEN production rule is set up for each type of path. The premise of a rule contains detailed geometric and topological facts to precisely represent the path definition. These facts are asserted by analysing the workpiece as described in section 3.3.

If the point picked is not on any edge, the module will search around that point to find possible solid edges that will satisfy some welding path definition. If the path selected is found not to coincide with any path definitions during the verification of rule conditions, corresponding error messages and possible causes will be displayed to help the user in making corrections.

3.2 Workpiece Feature Recognition

There are altogether 12 features, divided into 4 groups, which the workpiece feature recognition module is able to process:

- ❶ *Plate thickness (base material)* - Various shapes of plates can be detected including round or oblong type as well as plates with filleted or chamfered corner.
- ❷ *Joint type* - Four types can be detected including fillet, butt, lap+ and lap-.
- ❸ *Welding position* - Five types can be recognised including horizontal, flat, vertical up, overhead and 3D. The 3D case corresponds to the 3D-curved welding path, along which the welding position gradually varies.
- ❹ *Others* - For the 3D curved welding path, there are some features varying on different segments and leading to different welding process parameters, such as concave/convex face and face-up/face-down. Therefore, a 3D welding path will be divided into several segments by this module according to changes of their associated features.

The general geometric definitions of these features have been studied and corresponding definition rules have been built. When the module is going to recognise some kind of feature, corresponding group of rules is called. Verifying each geometric and topological condition in these rules, the process will not stop until all rules have been run or there is a rule which a conclusion is reached successfully. Then the feature is recognised by this rule.

3.3 Structure and Implementation of the Module

The automatic interpretation of workpieces has a reasoning mechanism and a supportive analysing mechanism. Fig. 1 illustrates an overview of automatic interpretation of workpieces and its peripheral functions.

The reasoning mechanism is based on expert system IF-THEN production rules. The rules are set up according to the geometric definitions of welding paths and workpiece features respectively. A separate rule to recognise each type of welding paths and features is needed. Since the reasoning procedure using backward chaining is time consuming, a data driven or forward chaining reasoning scheme is employed in this system to reduce the computational effort involved [Joshi 1987].

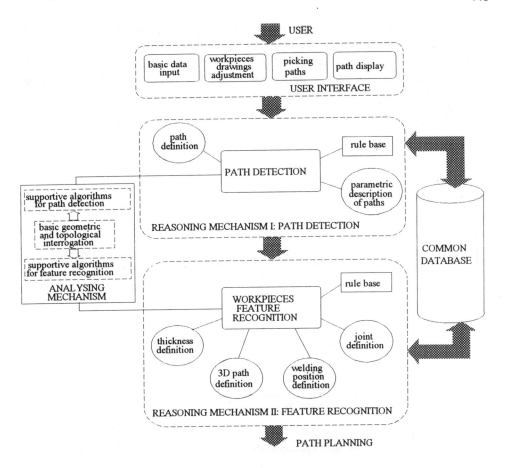

Fig. 1 Automatic Interpretation of Workpieces

To verify the conditions in these rules, various geometric and topological information of workpiece is required. The information, for examples, includes solid interference, parallel faces, face normal, overlapped part of edges, edges bounding a face, edge segment inside/outside a solid, point projection on a face, etc. A set of workpiece analytical techniques as well as geometric and topological interrogation functions have been developed in the analysing mechanism, on the basis of solid modelling technique provided by AutoCAD, to support this IF-THEN rule reasoning mechanism.

A temporary *fact database* is set up when the analysing mechanism is called in forward chaining module to verify the rules. Every fact asserted will be stored in the *fact database*. Since some facts are common in different rules, the fact that has been asserted by the analysing mechanism in previous rule will not be asserted again.

4. Welding path Planning

A general robot path planning problem can be simplified in welding robots because of its natural constraint: the motion path of robot end-effector is fixed along welding path designated. The path planning for welding paths is then only required to search a collision-free robot motion along a given spatial path. In addition, only two orientation variables of welding torch are allowed to be adjusted in a limited range in order for quality weld. One is welding direction. The other is travel angle, which typically can be changed from 5° to 30°. These limitations result in a smaller variable space to be searched for collision-free robot arm configuration, which simplifies the welding path planning as well.

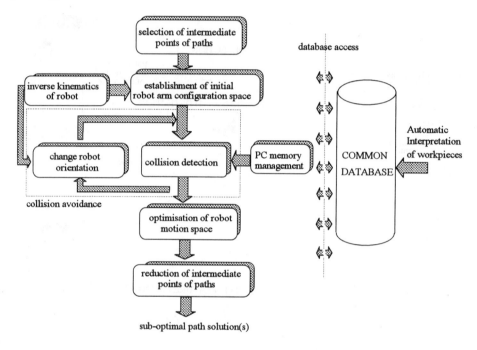

Fig. 2 Welding Path Planning

4.1   Collision Detection and Avoidance Strategy

A welding path is described by a set of carefully-selected intermediate points (start and end points are treated as special intermediate points). Collision detection is carried out graphically among the links of the robot solid model and between the robot graphic model and workpiece/table solid models on each intermediate point using AutoCAD AME solid interference function [Kangsanant 1992(2)]. Basically, the strategy of searching for collision-free welding robot motion is based on adjustment of the orientation of the robot end-effector (ie. welding torch) at each intermediate point, which is somewhat similar to Buchal's method [Buchal 1989]: Under a certain welding direction, for a chosen solution configuration and a minimal travel angle, a unique inverse solution can be found for the first point and then the robot solid model is translated and rotated accordingly. If the solution is kinematically feasible and does not cause interference, a solution with the same configuration and travel angle will be obtained for the next point on the path. This is repeated for each successive point until a joint limit is exceeded, or an interference occurs. The travel angle of the welding torch is then increased until a feasible non-interfering solution is found, or until the travel angle limit is exceeded. If no feasible solution is found, the above procedure will be repeated for the other welding direction. The process is finished when feasible solutions are found for the entire path under one welding direction, or the path is divided into two subpaths with different welding directions and feasible solutions are found for each subpath (The case in which a path is divided into more than two subpaths will not be considered since it is not common in practice). If no such solutions are obtained, the process will be repeated with a different configuration until feasible solutions are found or all configurations have been tried. Fig. 2 shows the diagram of this process.

4.2 Path Solution Optimisation

When a path has to be divided into two subpaths, there are probably a number of feasible solutions depending on the location of the intermediate point. An optimisation based on the minimum cycle time for a path, will be applied to select an optimal or sub-optimal solution from all possible path solutions. This optimisation is implemented by processing all feasible solutions obtained on each intermediate point. Basically, it is based on the following rules:
- For each path, the number of subpaths should be minimised
- In case of two subpaths, the robot idle running distance from the first subpath to the second subpath should be as short as possible. In a specific situation, for an example, supposing this idle distance is along the part of path between the end point of 1st subpath and the start point of 2nd subpath, the following solutions will be taken as the shortest idle distance can be attained:
    (1) If the start point of 1st subpath is at one of the ends of welding path and the 1st subpath is chosen as long as possible (because in this case the idle distance is along 2nd Subpath which is shortest)
    (2) If the start point of 1st subpath is *not* at one of path ends and the 1st subpath is chosen as short as possible (because in this case the idle distance is along 1st Subpath which is shortest)

5. Results

Each of modules as well as integrated system have been tested extensively using some specimens with typical plate shapes, path types and welding positions. It has been found that the system developed is capable of processing straight-line and arc paths, planar and cylindrical faces as well as 3D geometry both in the analysis and planning. In addition, the program generated is collision-free and the cycle time is minimised. In contrast to the initial work, this intelligent off-line programming and planning system is able to handle more complicated workpiece drawings, multiple straight/curved lines or their combination. Various path types, such as gapped path and the path formed by overlapped part of two edges, etc. can be recognised and processed. Most technical data relating to geometry and topology of workpiece, such as plate thickness, welding position and joint type can be extracted automatically from the 3D drawing. The test also shows that it is more efficient to use this system than manual path design and programming, especially for a complex assignment.

Since the system uses a great deal of solid interference and solid motion functions, some problems associated with the processing speed and PC memory can occur. Remedies have been studied and an improvement has been obtained (eg. by periodic memory purge, the times of running AME solid interference function without PC crash can be increased by about 700 times). However, these problems, which are inherently caused by the PC limitations, cannot be entirely eliminated. For more complicated applications, it may be desirable to transfer part of this system to a workstation or minicomputer. Also during the test, it has been found that workpiece location adjustment is an important strategy in collision avoidance. It is expected that the further research will pay attention to this aspect.

6. Conclusion

An intelligent off-line planning and programming system for a welding robot has been developed and tested successfully. The system operates in an AutoCAD environment and is based on a robot-independent approach. The intelligent functions in the system make the system more powerful, more user-friendly as well as easier to use.

It has been found that the employment of AI techniques in robot off-line programming and planning system can provide a more complete solution for a given task with a minimum intervention or special knowledge from the operator. The system is able to extract some welding information from CAD drawings and suggest collision-free and minimum-cycle-time solutions for robotic welding.

References

Buchal, R. O. et al. Simulated Off-line Programming of Welding Robots, *The International Journal of Robotics Research*, **Vol. 8**, No. 3, June 1989, pp. 31-43

Jacobs, M. P. Off-line Robot Programming: A Current Practical Approach, *Robots 8 Conference Proceedings*, 1984, pp. 4.1-11

Joshi, S. et al. *Intelligent Manufacturing System*, Chapter 6: Expert Process Planning System with a Solid Model Interface, 1987, pp. 111-136

Kangsanant, T. et al. CAD-based Expert System for Robotic Welding, *12th World Congress IFAC*, **Vol. 1.1**, 1993, pp. 241-244

Kangsanant, T. et al. CAD/CAM Technique for Robotic Welding, *Proceedings 5th International Conference ECGDG*, 1992 (1), pp. 341-346

Kangsanant, T. et al. CAD-based Software for Collision Detection of Robot Movement, *Proceedings 5th International Conference ECGDG*, 1992 (2), pp. 475-480

Paul, R. P. *Robot Manipulators: Mathematics, Programming, and Control, The Computer control of Robot Manipulators*, Cambridge, Mass, MIT Press, 1981

Wang, R. G. *CAD-based Intelligent Robotic Welding System*, Master Degree Thesis, RMIT, 1994

Wloka, D. W. ROBSIM - A Robot Simulation System, *IEEE International Conference in Robotics and Automation*, **Vol. 3**, 1986, pp. 1859-1864

EFFICIENT TOOLPATH GEOMETRY FOR CNC TURNING OPERATIONS

R. N. IBRAHIM, P. K. KEE and S. CABARKAPA
Department of Mechanical Engineering, Monash University
Caulfield East, Victoria 3145, Australia
E-mail:patrick@amtc01.eng.monash.edu.au

ABSTRACT

Optimisation of efficient toolpaths related to CNC-based turning operations for complex irregular geometry are proposed and discussed. Graphical and analytical models were used to study conventional parallel and biaxial contour toolpath geometries to predict the time efficient toolpath geometry. Given that the optimum machining conditions are available in CAM software packages or CNC machine controllers [1], the proposed consideration of the toolpath geometry will further improve the production rate by reducing the machining time following the reduced toolpath contour lengths for the same amount of material removal and similar machining conditions. A comparison of conventional parallel and biaxial contour toolpath geometries has shown considerable machining efficiency advantage of the latter for CNC turning operations.

Keywords: CAD/CAM, Toolpath, Turning

1. Introduction

It is possible to manufacture majority of parts on Computer-Numerically-Controlled (CNC) machines by using a Computer-Aided Design (CAD) generated wire-frame, surface or solid models. These CAD models are used to generate the toolpaths in conjunction with Computer-Aided Manufacturing (CAM) packages. For the generation of toolpaths for complex irregular shapes such as grooves of different depths, concave and convex turning operations, additional information related to the tool shape and the related modifications have to be provided by an experienced machining operator.

The graphical capabilities of the dedicated computers are limited and they can be enhanced by using external CAD and CAM systems. These CAD/CAM systems are based on a technique called computer graphics that allows the creation and display of an object in graphical form on a computer screen. The CAD model can be further upgraded by incorporating software that performs the numerical control function and permits the user to define on the screen a cutter path based on part profile and machine operations. A part machining program in APT or other higher-level NC programming languages defining the sequence of machine operations and the cutter motion can then be generated automatically and transferred to the CNC machine.

Using a CAM program, elementary toolpaths information is created by constructing a set of toolpaths in a specific sequence and adding tool definition and machining parameters. A CAM program uses the graphics data (eg. coordinate points describing the geometry) from a CAD system to generate the toolpaths and attaches information related to tools and machining parameters. This also groups the toolpaths pertaining to types of tools used and the shapes to be generated. This information is then pre-processed to a neutral file. At this stage, the set of defined toolpaths is not specific to any particular make of CNC machine, and it requires further post-processing to convert it into the format compatible to the controller of the CNC machine. The generated neutral file is linked to the postprocessor program of a particular machine to generate the NC code for that machine. This technology is being developed to make it software and hardware independent [2].

Despite their power and versatility, CAD/CAM systems cannot automatically carry out a complete job of design and manufacturing processes and analysis. Their role is to assist an engineer during a highly complex design and manufacturing process. The result of computer-aided design is to create a geometrical data base (or CAD model) that defines in numerical form, the part to be manufactured. The CAM part is then used to generate the information required to manufacture the part on the basis of the geometrical database. During this stage, a process planner provides information related to the sequence of machining operations, starting point, machining specifications, cutting tools and other paths. The CAD/CAM system then carries out the necessary calculations and translates the calculated results and the information compatible to a CNC machine controller.

In this paper, the CAD/CAM modifications are suggested for turning complex irregular shapes such as tapered, circular and elliptical envelopes, based on the toolpath geometry. Both conventional parallel and biaxial contour toolpath geometries will be used to generate these envelopes, and numerically compared on the basis of the total toolpath contour length to determine which toolpath geometry will result in minimum machining time for the same amount of material removed.

2. Toolpath Geometry

Toolpath geometry is dependent on the shape to be machined, the type and dimension of the tool used, the machining process and the surface finish. Most of the CAD/CAM systems use the geometric data from a CAD model to generate a source machining program comprising toolpath geometry and machining processes. A point to point motion of the tool is determined by the coordinates of the end point obtained from this CAD/CAM model. The toolpath can be generated using wire-frame, surface and solid models [3, 4 and 5].

Depending upon the CAD model, the toolpath geometry can comprise of linear segments, higher order curves, discontinuous irregular shapes or any combination of these. Linear geometry of a toolpath can be generated by one axis for parallel or biaxial CNC machine control for tapered surface generation in turning or machining operation. Generation of higher order toolpath geometry requires a minimum of biaxial control of a CNC machine. Discontinuous irregular toolpath geometry also requires a minimum of biaxial and interrupted control of a CNC machine.

In this paper, the two generic (ie. linear and higher order) toolpath geometries related to CNC turning operation have been studied with a view to minimise the total length of toolpath contours. This study assists in optimising toolpath contours thus generating savings in machining time.

3. Linear Toolpath Geometry for Turning Operations

Plain turning operation in a CNC turning centre is achieved by one axis control of the axis parallel to the axis of turning for a constant (or equal) depth of cut. Toolpath contours for optimised depths of cut and feed rates will be parallel and unique. The algorithms for determining the optimum cutting feed and speed for a given depth of cut in turning operations have been developed by Armarego and Kee [1] in the form of a user-friendly CAM software package. For tapered surface, the toolpaths can be considered as shown in Figs 1a and 1b.

3.1 Parallel Cuts

Fig 1a shows several tool cuts parallel to the axis of the job. The length of cuts are governed by the tapered envelope. These toolpaths are automatically generated by most of the CAM packages based on the depth of cuts and other relevant machining specifications. The finishing cuts have to be made along the tapered surface as biaxial cuts. The following analysis of the toolpath represents the total toolpath length as a function of depth of cut (d), number of cuts (n) and semi tapered angle (α). Assuming that all the cuts are of the same depth (d) and only one final cut is made along the tapered length, then the number of ($n - 1$) parallel cuts required is given by:

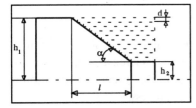

Fig 1a: Parallel Cuts of Linear Toolpath Geometry for Tapered Turning

$$(n-1) = \frac{(h_1) - (h_2 + d)}{d} = \frac{h_1 - h_2}{d} - 1 \quad (1)$$

$$\text{total number of cuts}: n = \frac{h_1 - h_2}{d} \quad (2)$$

$$L_a = l[(n-1) + \sec(a)] + d \tan(a) - d \cos(a) \Sigma (n-1) \quad (3)$$

where L_a is the *total length for parallel cutting*

3.2 Biaxial Cuts

The second option to create tapered surface is obtained by using biaxial cuts parallel to the tapered surface as shown in Fig 1b. Generation of these toolpaths is not supported by majority of CAM packages and the wire-frame CAD drawing can be used to specify the toolpath in a CAM package. For the same depth of cut (d), the number of cuts (n_1) required and the total length of the toolpath contours for biaxial cutting of tapered

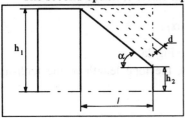

Fig 1b: Biaxial Cuts of Linear Toolpath Geometry for Tapered Turning

surface are given as follows:

$$\text{total number of cuts } n_1 = \frac{(h_1 - h_2)\cos(a)}{d} \quad (4)$$

$$L_b = n_1 l \sec(a) + n_1 d \tan(a) - d \cot(a) \Sigma (n_1 - 1) \quad (5)$$

where L_b is the *total length for biaxial cutting*.

From Eqs. (2) and (4), it is apparent that number of cuts required in biaxial cuts (parallel to the tapered surface) is less than the number of cuts required in parallel cuts. Depending upon the difference in (n), (n_1) and (α) values, Eqs. (4) and (5) can be calculated to determine the total toolpath contour length.

On the basis of Eqs. (2), (3), (4) and (5), optimum toolpath length for known optimum machining conditions [1] can be determined. For the same machining conditions, the machining time will basically depend upon the toolpath geometry and in a majority of the cases, can result in substantial machine time savings if toolpath contour lengths are predicted using Eqs. (2), (3), (4) and (5).

4. Higher Order Toolpath Geometry

For the sake of simplicity, the second order circular and elliptical curves will be considered as a shape of a part to be generated by turning process.

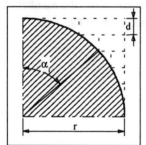

Fig 2a: Parallel Cuts of Circular Toolpath Geometry for Turning

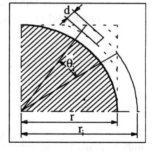

Fig 2b: Biaxial Cuts of Circular Tolpath Geometry for Turning

4.1 Circular Cuts

For turning a circular shape, the two types of cuts, ie. parallel and biaxial cuts can be made as shown in Figs 2a and 2b. It should be noted from Fig 2a that the total number of cuts required to generate the circular profile using conventional parallel toolpath geometry is the sum of the parallel cuts and one biaxial cut (for finishing) following the contour of the circular shape. Mathematically, the total length of the toolpath contour for conventional parallel cutting is given as:

$$L_a = \frac{\pi r}{2} + \sum_{i=1}^{n-1} r(1 - \sin \alpha_i) \quad (6)$$

where *number of cuts* $n = r/d$ and r, d, α_i as defined in Fig 2a.

When biaxial cuts are considered as in Fig 2b, the total length of the toolpath contour is given as:

$$L_b = \sum_{i=0}^{n} r_i \theta_i \quad (7)$$

where r_i and θ_i are as defined in Fig 2b.

4.2 Elliptical Cuts

Fig 3a shows another second-order elliptical surface created by a series of parallel cuts with the final finishing cut being biaxial. Fig 3b shows the same surface being created by parallel contour biaxial toolpaths. The toolpath modifications as shown in Figs 3a

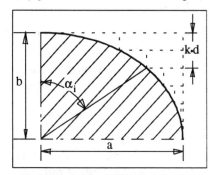 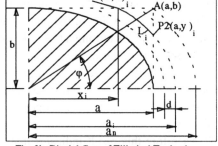

Fig 3a: Parallel Cuts of Elliptical Toolpath Geometry for Turning

Fig. 3b: Biaxial Cuts of Elliptical Toolpath Geometry for Turning

and 3b were made using a commercially available CAD package which were then imported in a CAM package to specify the toolpath contours. It can be shown mathematically from Figs 3a and 3b that the total toolpath contour lengths L_a and L_b for parallel and biaxial cuts are given as:

$$L_a = \sum_{k=1}^{n-1}\left[a - \frac{a\sqrt{b^2 - (b-k\cdot d)}}{b}\right] + \frac{Z}{2} \tag{8}$$

$$\text{and } L_b = \frac{Z}{4} + \sum_{i=1}^{n-1}\int_{x_i}^{q}\sqrt{1 + \frac{b_i^2 x_i^2}{a_i^2(a_i^2 - x_i^2)}}\, dx \tag{9}$$

$$\text{where } Z = p(a+b)\left(1 + \frac{l^2}{4} + \frac{l^4}{64} + \frac{l^6}{256} + \frac{25 l^8}{16384}\right) \tag{10}$$

$$\text{and } \lambda = \frac{a-b}{a+b} \tag{11}$$

and a and b are the semi-major and semi-minor axes of an elliptical profile respectively, and x_i is as defined in Fig 3b.

5. Brief Numerical Study of Parallel and Biaxial Cuts

For the purpose of this paper, a brief numerical comparison of the total toolpath contour lengths L_a and L_b (based on Eqs. 3, 5, 6, 7, 8 and 9) using parallel and biaxial cuts has been made for tapered, circular and elliptical profiles. The numerical comparisons have been based on the percentage increase (Π) in the value of L_a (using parallel cut) with respect to the corresponding L_b (using biaxial cut), where:

$$\Pi = \left[\frac{L_a - L_b}{L_b}\right] \times 100 \qquad (12)$$

The depth of cut per pass for each of the three profiles considered has been fixed to 0.3 mm. Fig 4 shows the value of the percentage increase of L_a wrt L_b as a function of the semi-tapered angle for turning a tapered profile. The tapered length in this case is 60 mm. Interestingly, the graph in Fig 4 shows a wide scatter of points across the range of semi-tapered angles (from 1° to 45°) considered. When these scattered points are curve fitted, a decreasing hyperbolic function resulted. It can be seen that at low semi-tapered angle of less than 6°, the value of Π is significantly higher (ie. between 3 and 8 %) when compared to large semi-tapered angle of greater than 6°, where the value of Π is minimal. Therefore this result indicates that machining time for tapered turning can efficiently reduced using biaxial cuts only for low semi-tapered angles.

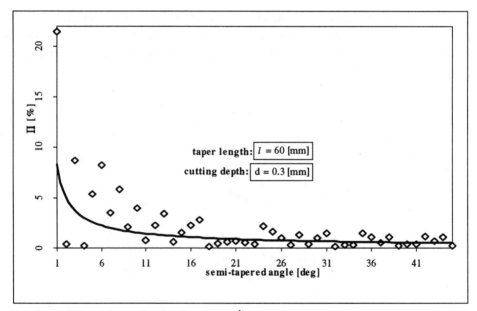

Fig 4: Graph of Π vs semi-tapered angle α for tapered turning

When turning a circular profile, the increase of L_a wrt L_b is plotted as a function of the radius of the circular component, as shown in Fig 5. The graph in Fig 5 indicates that as the radius of the circular profile increases, the value of Π decreases, hence suggesting that significant machining time savings can be realised using biaxial cuts for circular profile at lower radii. At higher radii (ie. radius greater than 100 mm), the value of Π becomes negligibly low (less than 0.4 %).

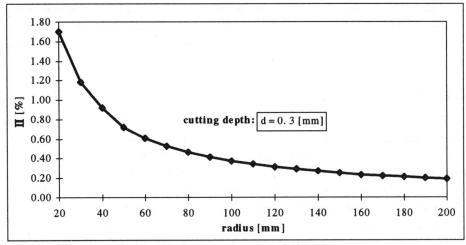

Fig 5: Graph of Π vs radius r for circular turning

Fig 6 shows the percentage increase of L_a wrt L_b as a function of the semi-major axis, for turning an elliptical component for three ratios of b/a, ie. 0.6, 0.7 and 0.8 respectively. It appears that the percentage increase of L_a wrt L_b is higher for lower value of b/a ratio when compared to higher value of b/a. Nevertheless, for each case of b/a ratio, the value of Π decreases when the value of the semi-major axis, a of the elliptical profile increases. By comparison with tapered and circular profiles, the value of Π for elliptical profiles is significantly higher, ie. between 5 % and 29 %. Further experimental studies using Hitachi-Seiki CNC turning centre and CADDSMAN (CAD/CAM software) for the toolpath contours shown in Figs 3a and 3b for the same geometry of the elliptical profile revealed that more than 10% reduction in production time for the modified toolpath using biaxial cuts shown in Fig 3b can be realised.

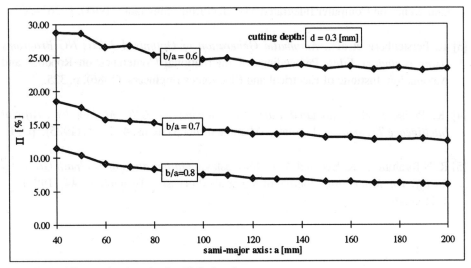

Fig 6: Graph of Π vs semi-major axis a for elliptical turning

From the above numerical comparisons, it can be concluded that for both linear toolpath and higher order curved toolpaths, it is more economical to generate toolpath contours using biaxial cuts rather than parallel cuts from machining time and cost points of view.

6. Conclusions

The optimisation of toolpaths for both linear and higher order geometries using parallel and biaxial cuts with turning operations has been presented. The optimum cutting conditions (ie. cutting speeds and feeds) for a given depth of cut, tool-workpiece material combinations and other relevant workpiece dimensions can be determined by using the developed algorithms [1] for turning operations. A substantial gain can be achieved by considering the minimisation of the toolpath contours.

The brief numerical study has shown that biaxial cuts gave lower total toolpath contour length when compared to parallel cuts for tapered, circular and elliptical profiles. Interestingly, the percentage increase of total toolpath contour length using biaxial cuts with respect to parallel cuts for generating elliptical profile was significantly higher when compared to tapered and circular turning. Careful study of toolpath geometry can result in considerably savings in lead time, machining and production time leading to a more cost effective production using CNC machine tools.

7. References

[1] E.J.A. Armarego, and P.K. Kee, *Development and Assessment of Alternative Multi-pass Turning Optimisation Strategies for CAM Applications*, 4th Int. Conf. on Metal Cutting and Automation, China, (1989), p. 1.

[2] J. Balthazar and A.K. Shrivastava, *Expert System for Optimal Drawing Data Exchange Between CAD/CAM Systems.*, Proceedings of the First International Conference on Computer Integrated Manufacturing (Singapore, 1991), p. 263–269.

[3] R. Ferstenberg et al ., *Automatic Generation of Optimised 3-Axis NC Programs Using Boundary Files*, Proceedings of International Conference on Robotics and Automation, Institute of Electrical and Electronics Engineers, (1986), p. 325.

[4] K. Preiss et al., *Automated Part Programming for CNC Milling by Artificial Intelligence Techniques*, Journal of Manufacturing Systems, **4** , No.1, (1985), p. 51.

[5] R. N Ibrahim, P. K. Kee and A. K. Shrivastava: *Optimisation of Toolpath Geometry for Efficient Machining* , Journal of Materials Processing Technology, **44**, (1994), p. 215–226.

Determination of Analytic Features From Laser Scan Data

Samuel E. Ebenstein
Materials Research Laboratory, Ford Motor Company
Ford Motor Company, Dearborn, Michigan 48121-2053, USA
E-mail: eben@sl0047.srl.ford.com

Abstract

A method is presented for determining the location and orientation of geometric features from high density scan data. The method can also be used with any high data density collection devices such as laser scanners or Moiré interferometry systems.

KEYWORDS   Analytic Features   Laser Scanning

Introduction

Laser scanning and Moiré interferometry systems are being used to collect high density data. This data can be used to for capturing the geometry of engine parts such as combustion chambers, intake and exhaust ports. The geometry consists of surfaces such as the combustion chamber surface which can not be easily described mathematically since they are complex free form shapes and require many measurements to accurately describe them. However other features such as the position and orientation of valve guides, valves, and valve openings can be represented in closed form as conic sections, basically circles and ellipses. We shall call these features geometric features. The precise location and orientation of these geometric features is necessary for successful reverse engineering. They can be used in building a feature based CAD model, since CAD modeling packages usually have basic primitives for constructing a cylinder. A method is presented to allow the accurate determination of the orientation and location of such features as valves and valve guides. They can be applied to any circular or elliptical feature which contains a planar (i.e. valve guide) or partially planar face (intake or exhaust valve).

General Description of the Method

1. The geometric feature is basically a circle or ellipse and as such is planar. The first step is to determine this plane or one parallel to it. We shall call this plane the reference plane. The feature will be contained in the reference plane, as for example in the case of a valve guide, or above/below the reference plane as in the case of a valve opening.

2. Pick a rectangular patch on the reference plane with an interactive graphics package.

3. Find the equation of a plane which best fits the data in a least squares sense. Take all points that are close (within a given tolerance) of this plane. Use this new set of data to recompute the equation of the plane.

4. Calculate the distance of all points in the scan data set from this reference plane. Display this derived distance data in a histogram format and allow the user to select a range from the histogram. In general peaks on the histogram will correspond in a obvious fashion to features.

5. Divide the scan data into two subsets. Those within the chosen range from the histogram and the rest. Display this binary data with two colors. This step enables the user to visually determine if he has accurately determined the plane. If the plane has not been correctly determined, the user may return to step 2 and repeat the procedure.

6. Divide the distance data into two subsets. Those points whose derived distances are within the selected range, and those outside it. Display the resulting binary image in two colors, say black and white. If the image correctly defines the desired feature, continue and allow the user to select either color. Otherwise return to the previous step and select a new range from the histogram.

7. Usually a color of the desired image will contain more than one feature. An algorithm called SCCA (Sequential Connected Components Algorithm) is used to descriminate between the features. The distinct features are then displayed with different colors. The user can then use the graphical interface to pick only the feature of interest.

8. An algorithm is then used to calculate the boundary points of the feature. If the feature is in a horizontal plane, the methods discussed in [3] can be used to calculate the precise dimensions of the analytic feature (circle or ellipse). If not, then map the data to a horizontal plane and the previous methods can be used.

Finding and Verifying the Reference Plane

The feature may be embedded in this reference plane for example the planar face of a valve as in figure 1 for example, or may be characterized by falling away very sharply from the reference plane. This second case occurs when finding a valve location from a valve hole. In this case either the top plane of the valve opening can be used or the plane determined by the valve guide. Either plane can be used since the planes are parallel. In either case a graphical user interface is used to pick a small rectangular patch on the reference plane. Then least-squares techniques are used to find a mathematical equation which best represents this reference plane. The user can select a tolerance and all points which are within this tolerance are then considered to be on this reference plane. The user is given graphical feedback so he can visually determine whether or not the reference plane is accurate enough to continue with the feature finding process. indicates all points that are within tolerance of this plane. As figure 2 shows the reference plane does a good job of identifying the planar portion of the valve. If the reference plane is not satisfactory, the user can pick a new reference patch.

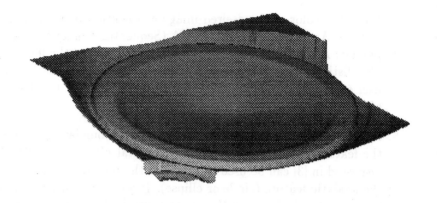

Figure 1: Valve with Planar Face

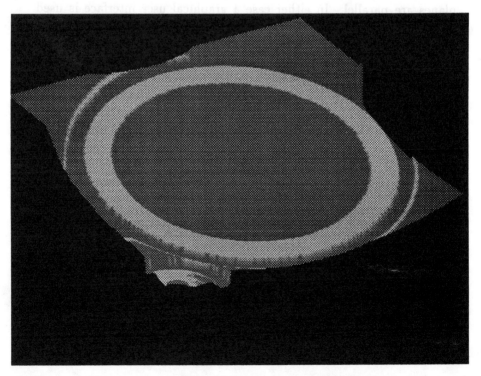

Figure 2: Area Within Tolerance of the Plane

Computing the Histogram and its Use

The signed distance of each point from the plane can be determined from the following formula

$$dist(z) = (z - (a*x + b*y + c))/\sqrt{a^2 + b^2 + 1} \qquad (1)$$

where the equation of the plane is

$$z = a*x + b*y + c \qquad (2)$$

In figure 3 the user can select a range of the histogram to determine the precise feature location. By choosing the range

$$-0.38 <= dist <= .15 \qquad (3)$$

(where dist is defined as in equation 1) the user can obtain a binary image as in figure 4. This image was created by setting all points whose distances are within the given range on the histogram are set to 0. All other points are set to 1. At this point the user can select either black or white from the binary image. If the user selects black subsequent processing will enable determination of the radius of the valve. If white is selected the user will be able to determine the radius of the valve portion which is interior to the planar portion. Figure 5 shows the components of the valve face as determined by the SCCA algorithm. Figure 6 shows the results of picking the annular region from the previous figure.

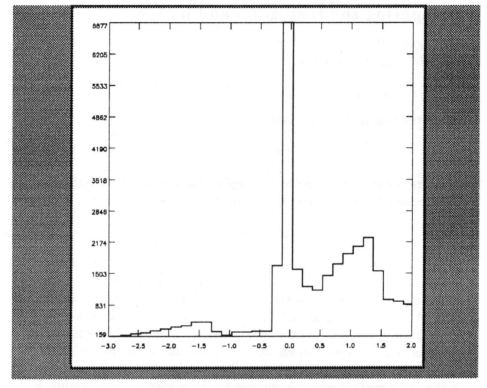

Figure 3: Histogram of Distances from the Plane

Figure 4: Area Within Tolerance of the Plane

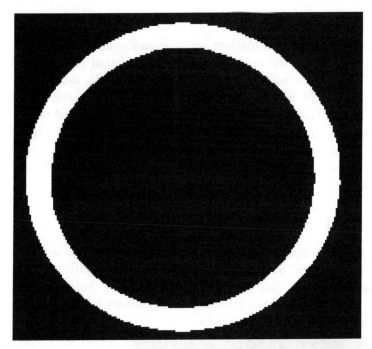

Figure 6: Picked Component (The Valve)

References

[1] *IMSL Library Reference Manual*. IMSL, Inc., 1982.

[2] K. M. Brown and J. E. Dennis. Derivative free analogues of the Levenberg-Marquardt and Gauss algorithms for nonlinear least squares approximations. *Numerische Mathematik*, 18:289–297, 1972.

[3] Samuel E. Ebenstein and Gregory H. Smith. Determination of Circular Features from Laser Scan Data. Technical report, Ford Motor Company, 1992. Technical Report No. SR-92-23.

[4] Ramesh Jain and Brian G. Schunck. *Machine Vison*. University of Michigan, 1990.

[5] K. Levenberg. A method for the solution of certain non-linear problems in least squares. *Quarterly of Applied Mathematics*, 2:164–168, 1944.

[6] D. W. Marquardt. An algorithm for least-squares estimation of nonlinear parameters. *SIAM Journal on Applied Mathematics*, 11:431–441, 1963.

SCCA

Whether the user has chosen black or white from Figure 4 a method is necessary to indicate to the computer which area of that color should be analyzed since the screen contains several areas of the same color. Fortunately the SCCA algorithm from binary image processing see [4] can be used to number the connected components in the derived binary image, and each component can be displayed in a different color. Figure 5 shows this components (here color has been replaced by gray-scale), and the components are clearly differentiated. The user can then select the large gray component to indicate the valve.

Finding the Feature Boundary

Once the geometric feature has been determined, it is necessary to determine the boundary points. At this point another algorithm from binary image processing can be used to calculate the border of the feature. See [4] page 54. Then standard least squares techniques can be used to calculate feature parameters.

Figure 5: Components Image (including noise)

Computer Aided Process Planning

Catch — A Practical Computer Aided Tolerance Charting System
    G. A. Britton and K. Whybrew

Model Based Planning and Calculation for the Manufacture of Dies and Moulds
    H. K. Toenshoff and J. Trampler

Computer Aided Tolerance Control in Process Planning
    Y. K. Chow and W. M. Chiu

DAVE Tolerance Charting

Society of Manufacturing Engineers
One SME Drive ● P. O. Box 930 ● Dearborn, Michigan 48121 ● 313/271-1500

CATCH - A PRACTICAL COMPUTER AIDED TOLERANCE CHARTING SYSTEM

Dr. G.A. Britton,
School of Mechanical and Production Engineering,
Nanyang Technological University, Nanyang Avenue,
Singapore 2263
E-mail: mgabritton@ntuvax.ac.sg

Dr. K. Whybrew,
Department of Mechanical Engineering,
University of Canterbury, Private Bag, Ilam, Christchurch,
New Zealand
E-mail: k.whybrew@mech.canterbury.ac.nz

ABSTRACT

The paper describes the implementation of a computer aided tolerance charting system developed in conjunction with Sundstrand Pacific Aerospace Pte. Ltd. Minimum conditions for computer aided tolerance charting in industry are proposed. It is argued that systems that do not satisfy these requirements have no practical usefulness in industrial applications.

Key Words: tolerance charting, rooted-tree graph, process planning.

1. Introduction

In previous publications we have described a tolerance charting algorithm using rooted tree di-graph techniques (Whybrew, et al. 1990; Britton et al. 1992). We have argued that tolerance analysis is essential for any practical computer-aided process planning system for metal cutting (Sermsuti-Anuwat et al. 1995). We believe that our algorithm offers considerable advantages over traditional methods based on chaining calculations, (Wade 1967; Drozda & Wick 1983). When used for manual charting, our algorithm offers an easily understood technique for visualisation of the sequence of machining operations and for calculating tolerance stacks on machining operations and stock removals. The algorithm has been taught to undergraduate mechanical engineers at the University of Canterbury for the last four years. Previous classes were instructed using chaining methods. Students found these methods notoriously difficult to use and charts were rarely constructed correctly. It is now unusual to encounter an incorrectly constructed chart. The advantages of our algorithm for CAPP accrue from the computational simplicity of the algorithm.

The success of our algorithm encouraged us to collaborate with Sundstrand Pacific Pte. Ltd. in Singapore to develop the algorithm into an interactive, computer-aided, tolerance charting system satisfying the quality assurance requirements of the high precision machining industry. The result of this collaboration is a program we have called 'CATCH', an acronym of Computer-Aided Tolerance Charting, that is evocative of 'catching' tolerance control problems before a process plan is committed to production. Part of a typical tolerance chart produced by CATCH is shown in Figure 1.

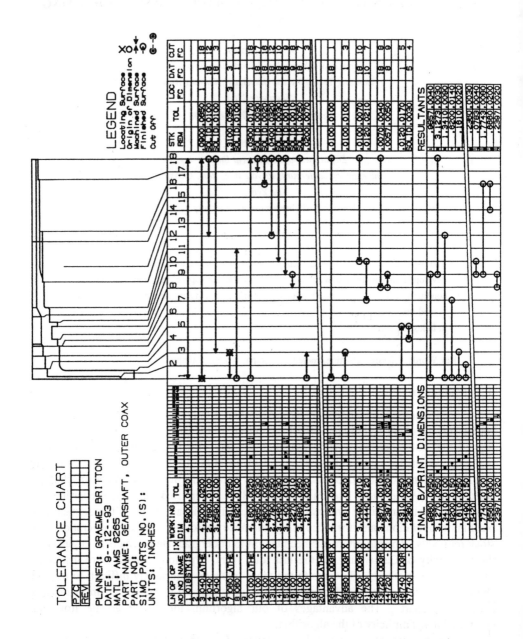

Figure 1: A Computer Generated Tolerance Chart

The full chart cannot be reproduced here because of its size. Charts for components produced by Sundstrand can be considerably more complex. Typically the number of faces ranges from 11-26, the number of blueprint dimensions ranges from 10-25, and the number of operations ranges from 30-50. A small proportion of the charts have 26-40 blueprint dimensions and 27-41 faces. On average, 4 charts are produced each day and manual charting takes 1 to 2 days per chart. The productivity gains from automation of tolerance charting are therefore considerable.

We have modified our original technique, described in Whybrew et al. (1990), to encompass the practical process planning experience of Sundstrand. CATCH is written in GRIP programming language and interfaces with UGII CAD software (GRIP and UGII are proprietary products of EDS, a commercial CAD vendor). It includes all practical manufacturing processes that affect part dimensions.

2. Tolerance Charting System Requirements

The basic requirements of a computer-aided tolerance charting system are:

(a) Location surfaces, reference surfaces, and cut surfaces for each operation must be clearly identifiable in the display.

(b) Calculation of tolerance stacks on the resultant dimensions and stock removal must be performed automatically.

(c) Working dimensions to achieve the design specification and stock removals must be calculated automatically.

(d) Resultant dimensions must be checked for conformance with design specifications.

Our original algorithm satisfies these basic requirements. Our experience with Sundstrand has enabled us to define the following additional requirements considered essential for an industrial, computer-aided, tolerance charting system.

(e) Tolerance charting must be interactive. The user must be able to quickly generate and evaluate alternative process plans. Consequently the tolerance charting software needs to be interfaced with the CAD software used to model the part.

(f) The tolerance charting technique must be capable of handling all manufacturing operations that have an effect on dimensions, and all design and manufacturing dimension specifications. This includes all conventional and unconventional machining operations, heat treatment operations, plating operations, angled cuts, radii breakout cuts, geometric tolerances, simultaneous operations, stock issue operations, and solid operations.

(g) The system must allow the user to set a stock removal allowance. Other researchers (Ngoi and Ong 1993) have adapted our technique and criticized it because we use the stock removal allowance as input. Ngoi and Ong make an erroneous assumption that the stock removal allowance can be calculated by setting it equal to the tolerance stackup on the stock removal. This is not the case!

Stock removal is a technological property of the manufacturing process. The stock removal allowance is a recommended amount of material for removal for optimum performance of the process. It is normally related to depth of cut. If it is too large then the cutting forces will be too high and multiple roughing cuts may be required. If depth of cut is too small the surface of the workpiece may deform locally without a chip being properly formed. In grinding this is known as 'ploughing'. In turning and milling the surface is burnished by the tool instead of being cut. With work hardening materials this will cause surface hardening and result in rapid tool wear, and may also cause micro cracks in the workpiece surface. Most modern cutting tools require a minimum depth of cut and feed for the chip breaking action to be effective, failure to satisfy these conditions will result in departure from the ideal surface finish.

The process planner must be able to confirm that the stock removal allowance is acceptable by comparing it with the tolerance stackup on the removal.

(h) The system must allow the user to set resultant dimensions that are different from the blueprint dimensions provided there is sufficient tolerance allowance to do this. The working dimension computation often produces working dimensions that are difficult to machine to and check. The process planner should be able to change a working dimension to a more convenient value. This in turn will affect the value of the resultant dimensions. An example of this is shown in Figure 1 on the second to last line of the blueprint dimensions: the blueprint dimension is 0.6560" but the resultant dimension is 0.6660".

(i) The system should be able to generate models of the workpiece at any stage in the operation sequence. This allows the process planner to check if the operations are correct. Our current technique calculates the positions of all surfaces independently and hence they can be used directly to modify the model of the workpiece.

(j) The system should be able to provide a hard copy output of the chart.

(k) The system must allow filing and retrieval of past charts.

(l) The system must remember the steps completed by the user both during program execution and when a user exits the program. This is important in a commercial context because process planners are often interrupted during planning.

It is our contention that systems that do not satisfy these requirements have no practical usefulness in industrial applications.

3. The Rooted-tree Technique

Modifications to the original rooted tree algorithm have been fully described in Britton et al. (1994): a summary is presented here.

The foundation of the rooted tree technique is a surface labelling system that uniquely identifies each cut surface in an operation sequence. In CATCH a four digit numbering system is used. The numbering convention is summarised in Figure 2. The process planner identifies the blueprint surfaces on a geometric model of the part using a cursor. CATCH automatically assigns the surfaces with the first two digits of the label in ascending order from left to right, e.g., 01 in the label 0102 indicates that the surface is the left hand end of the part. The second pair of digits in the label indicates the number of times the surface has been modified in the operation sequence. The first occurrence of the surface, (either first cut or preformed surface) is labelled 01, the next 02, and so on. The label 0102 identifies the surface as having been produced by the second operation on surface 01. The datum surface for the first operation in the sequence is uniquely identified by setting the last two digits of its label to 00. CATCH automatically labels surfaces using this convention as operations are input by the process planner.

This method of identification allows the process sequence to be represented by a rooted tree graph as shown in Figure 3. The graph is read from top to bottom. Each operation is indicated by a line between a datum surface and a cut surface. At branch nodes operations are placed in sequence order from left to right. This assists interpretation of the graph. Each operation can be numbered according to its position in the operation sequence (line number on the tolerance chart). These numbers are not included in Figure 3 for clarity of exposition.

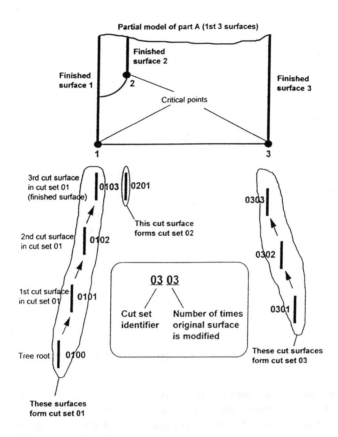

Figure 2: Surface Labelling Convention

There is a considerable amount of information contained in the rooted tree graph.

First, the whole graph is a succinct summary of the process sequence.

Second, important datum surfaces are easily identified: they are datums for two or more operations, e.g., surface 1802. These datum surfaces are extremely important for process planning and fixture design and are the basis of the automated process planning system developed by Sermsuti-Anuwat (Sermsuti-Anuwat 1992; Sermsuti-Anuwat et al. 1995).

Third, the cut surface numbering system clearly shows how many operations have been performed in producing each surface.

All of the above information is essential for process planning and can be easily read from the graph because of our labelling technique .

A limitation of our technique is that the tree must be rooted. This limitation is in fact an advantage and can be used to identify invalid operations. All operations must start either from the first datum surface or a surface resulting from an operation. Any gap in the tree indicates an attempt to locate on a non-existent surface. This rooted tree constraint is used to check operations for their feasibility when they are input into the system. An operation that violates the constraint is considered bad process planning and is rejected.

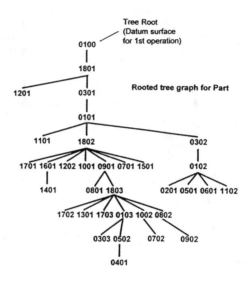

Figure 3: Rooted Tree Graph Representation Convention

The tolerance on a dimension between any two surfaces is calculated by tracing the path on the tree between the labelled nodes identifying the two surfaces. The tolerance on the dimension is the sum of the tolerances of the working dimensions in the path. Stock removal is the dimension between a cut surface and the surface with a label one unit smaller in the final digit. The tolerance on the stock removal is the sum of the tolerances of the working dimensions in the path between these two nodes. The current version of CATCH uses worst case tolerance calculations but the technique is similar if statistical tolerancing is required.

String variables and pointers are used to represent the rooted tree in our software. Each operation has a string variable, which is a serial list of the four digit labels of the surfaces contained in a continuous, sequential path from the tree root down to and including the surface resulting from that operation (this is the reversed path). For example, consider cut surface 0401 in Figure 3. Its reversed path is 0100-1801-0301-0101-1802-0901-1803-0103-0502-0401. The reversed paths can be visually displayed as part of the chart. They are used to identify branch nodes in the tree. The pointers are used to rapidly traverse the tree during the calculations.

We argue that this technique is an efficient and practical method because most commercial CAD systems use rooted tree graphs to represent CAD data. That is, our rooted tree representation technique is fully compatible with commercial, solid modelling, CAD systems.

4. The CATCH Interface

The overall procedure for tolerance charting using CATCH follows the sequence "Enter face data", "Enter blueprint data", "Enter operation data", "Check tolerance stackup", "Check stock removal allowance", and "Calculate working dimensions". The user interface is via a menu structure, shown in Figure 4, which constrains the user to input data in the correct order. The program is fully interactive and on screen editing allows the user to edit virtually all chart data. At each stage in the procedure the user is guided by the menu structure.

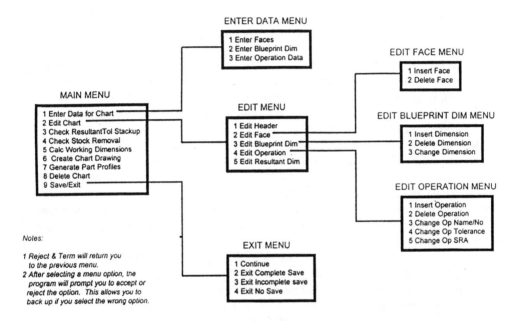

Figure 4: Menu Structure for CATCH

The chart is progressively constructed as data is entered. When data entry is complete, resultant tolerance stacks are calculated and compared with the blueprint dimensions and tolerances. If a blueprint tolerance is not achieved the relevant blueprint line on the chart is highlighted. It is not possible to progress to the next stage in the charting procedure until the process plan has been modified to satisfy the blueprint specification. Stock removal allowances are then checked in a similar manner. Working dimensions are not calculated until the process plan has been proved to satisfy blueprint tolerances and stock removal allowances.

The program includes procedures for producing chart documentation.

5. Conclusions

We believe that CATCH satisfies the requirements for a practical tolerance charting system. It offers considerable reductions in the time to produce a tolerance chart.

It is worth noting that the tree and computation algorithms are simple and are a relatively small part of the program code, less then 15% of the total code. Most of our effort has been directed at the user interface and error trapping functions which constitute approx. 85% of the code.

6. Acknowledgements

The authors gratefully acknowledge the support given by Sundstrand Pacific Aerospace Pte. Ltd. and the assistance of the technicians in the CAD/CAM Laboratory, MPE, Nanyang Technological University.

7. References

G.A. Britton, K. Whybrew, and Y. Sermsuti-Anuwat, "A manual graph theoretic method for teaching tolerance charting", *The International Journal of Mechanical Engineering Education* **20** 4 (1992), pp. 273-285

G.A. Britton, K. Whybrew and S.B. Tor, "An industrial implementation of computer-aided tolerance charting" to be published in *International Journal of Advanced Manufacturing Technology* (accepted for publication August 1994)

T.J. Drozda and C. Wick (eds), *Tool and Manufacturing Engineers Handbook,* Vol 1, (Society of Manufacturing Engineers: Michigan 1983).

B.K.A. Ngoi and C.T. Ong, "A complete tolerance charting system", *International Journal of Production Research* **31** 2 (1993), pp. 453-469.

Y. Sermsuti-Anuwat, *Computer-aided process planning and fixture design (CAPPFD)* (PhD. thesis, University of Canterbury 1992).

Y. Sermsuti-Anuwat, K. Whybrew and H. McCallion, "CAPPFD - A tolerance based feature sequencing CAPP system", to be published in the *Journal of Systems Engineering* **5** 1 (1995), pp. 2-15.

O.R. Wade, *Tolerance Control in Design and Manufacturing* (Industrial Press, New York, 1967).

K. Whybrew, G.A. Britton, D.F. Robinson, and Y. Sermsuti-Anuwat, "A graph-theoretic approach to tolerance charting", *International Journal of Advanced Manufacturing Technology* **5** (1990), pp. 175-183.

Model based Planning and Calculation for the Manufacture of Dies and Moulds

Prof. Dr.-Ing. Dr.-Ing. E.h. Hans Kurt Toenshoff
Institute of Production Engineering and Machine Tools, University of Hannover,
Schlosswender Str. 5, 30159 Hannover, Germany

and

Dipl.-Ing. Joerg Trampler
Institute of Production Engineering and Machine Tools, University of Hannover,
Division of Production Management
Email: trampler@mail.ifw.uni-hannover.de

Abstract

This paper aims at presenting a method which facilitates the planning of machining of mould cavities with respect to an economic point of view. Based on a geometric solid model, material volumes are determined which have to be removed from a blank. After each cavity altering process the model is updated according to the process progress. By accessing technical and economic enterprise databases, machining times and production costs are calculated from the model. A planning system supports the supply of necessary information and provides a planning desktop, which is used for the representation of Working Units. Working Units represent bundled operations and are chained to one another and used for calculation of total machining times and costs. Alternative machining operations can be planned, composed to one another and exchanged. The entire Working Units are stored in an order folder and are used for offer calculation, manufacture and process control.

Keywords: CAPP, Die Manufacturing

1. Introduction

The die and mould manufacturing mainly is a domain of small and medium sized enterprises. The order structure is characterised by single and small series manufacturing. Since each order actually represents a new product development for the company, production planning on the basis of existing part lists and process plans is not possible. Thus design and process planning do not provide any supporting data. But the increasing demand to shorten the throughput time as well as increasing the readiness of supply, flexibility and product quality requires an exact pre-order cost calculation and process planning for the product to be manufactured. A prerequisite for this is an enterprise wide computerised data processing alliance, which allows rapid access to all relevant enterprise data. Appropriate CAx-applications (e.g. CAD, CAPP, CAM-tools) must be modules within this information alliance. Unfortunately there are hardly any computer based applications available on the market up to now, which are sufficient for the special requirements of die and mould manufacturing. A survey, which was carried out by the IFW in German die and mould manufacturing companies [Tönshoff 1994b], confirms this image by showing a need to catch up in terms of CAx-software in the areas of process planning, NC-programming and machine control.

2. Offer Calculation and Process Planning

The pre-order cost calculation in general does not belong to a certain company department. For this reason there is no specific data stored in most enterprises, which allow the determination of production costs for dies and moulds. Practically most of the determined costs per unit are based on the estimation of staff members in the design and manufacturing departments. Furthermore, the offer calculation is more difficult in cases of special product inherent features. Therefore the planning depth is very low in the production planning due to a lack of standardised operating sequences in the manufacturing department. In addition to that, the great complexity of the mould cavity geometry requires a high percentage of manual work [Tönshoff 1994a]. A precise pre-order cost calculation is only possible with the support of detailed process planning.

Without suitable instruments, such as the pre-order cost calculation on the base of a reliable process planning, the business risk is difficult to limit. So the aim must be to carry out process planning activities in an efficient combination together with the pre-order cost calculation and the manufacturing. This is the only way by which a rapid offer in case of a request can be realised. Detailed process plans must be produced within a time span which normally is required for pre-order cost calculation. On the basis of the produced documents, reliable cost calculation can be made on the one hand. On the other hand, complete manufacturing documents would already be available, which in case of an order validation allow a rapid manufacturing start.

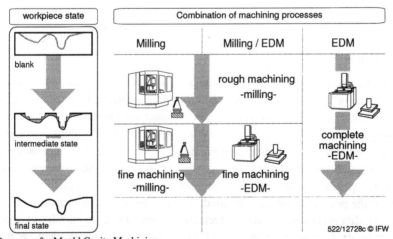

Fig. 1: Processes for Mould Cavity Machining

3. Milling or Electrical Discharge Machining

The scope of this paper is limited to the manufacturing of the mould cavity. Due to the geometric complexity the machining of the cavity and the selection of appropriate manufacturing sequences is rather difficult. For this reason a separation from other domains of workpiece manufacturing is necessary. Therefore this text does not consider those machining operations that are carried out outside the mould cavity.

Milling and EDM are regarded as the predominant processes in the manufacturing of dies and moulds. Because of the achievable high material removal rates, the milling process is of special significance, although it is limited in terms of technology and economy, if a filigrane contour or high surface qualities are to be produced. In the latter case a long machining time has to be accepted [Altan 1993].

The advantage of EDM is the fact that the production of various workpiece geometries is practically unrestricted since electrodes are available even for the most filigrane machining cases. Another advantage is the possibility to carry out EDM processes without constant supervision, for example on an unmanned shift or on the weekends. This way labor costs can be reduced and the throughput time can be decreased. On the other hand, compared with the milling process a lower material removal rate and thus the higher machining time have to be taken into account, as well as the additional effort of the electrode planning and manufacture [Buchholz 1993].

In order to achieve a short throughput time it is aspired to manufacture mould cavity by milling as wide as possible, ideally carrying out a complete milling machining. If this is not possible both processes have to be combined [Hernandez 1989]. Another possibility is a complete EDMing, if technological and economic factors make this seem efficient (fig. 1).The decision which machining processes and sequences really are suitable for the manufacture of a specific mould is yet made on an empirical basis. CAx-applications that support such a decision have not been realised up to now.

4. Determination of Processing Times and Production Costs

The mould cavities of casting tools often are of great geometric complexity. An evaluation of the machining process in terms of economic efficiency and throughput time can not take place without considering the workpiece geometry [Tönshoff 1990a].

In a new approach of computer based process planning it is now possible to use the CAD model of the mould cavity for the determination of machining operations. At the beginning of the planning activities this model represents the geometry of a blank and with increasing planning progress it approaches more and more the final workpiece state. The concept of the constantly updated workpiece model is the basis of the *Model based Process Planning* [Tönshoff 1992].

In order to be able to calculate the production times and costs, the volume to be removed must be determined first (fig. 2). Approaches for the determination of estimated values for the pre-order cost calculation based on solid models have already been realised earlier [Hernandez 1989, Lampkemeyer 1992]. Yet the influence of technological and machine specific data could hardly be considered since the calculation did not take place on the base of detailed process plans. Performing a milling process instead of EDMing, the depicted approach primarily considers strategies that aim to select appropriate cutting tools [Eversheim 1988].

Therefore access to a tool database exists, that contains geometric as well as technological information [Kuhn 1992]. The latter vary depending on the material of the workpieces. Using this data and an approach to calculate the removable chip volume per time during front milling [Paucksch 1993], machining times can be calculated. This requires optimal cutting conditions, where the tool's main axis must be perpendicular to the workpiece's surface. Even in 3-axis milling this ideal case is achieved rarely, because the machined surfaces are usually curved. Then the milling tool is no longer completely in contact with the surface. Such deviations are not considered so far.

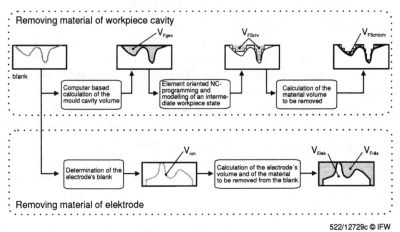

Fig. 2: Removing Material Volumes by Milling and EDM

For rough milling purposes the *Element oriented NC-Programming* has been developed at the IFW [Tönshoff 1990]. The volume to be removed is approximated by geometric primitives such as cuboids or half-cylinders. For these primitives machining time optimised milling cycles are generated automatically. Using the generated NC data, machining times can be calculated very easily and precisely. For the EDM process the determination of machining time is based on material removal rates which can be extracted from a technological EDM module depending on the used material of blank and electrode and the required precision [Bonte 1994].

For the determination of additional times, used for clamping, fixing or tool exchange, machine specific data from a machinery database is used. Along with technical information it contains further organisational data like machine hourly rates. Based on this information and associated operator costs, machining costs can be calculated from the times that were determined before. Adding the material costs for the blanks results in the production costs.

The costs that are caused in the pre-manufacturing departments are not considered in this calculation. Due to the use of computer support they are not higher than costs for ordinary offer calculation, although the plans are more detailed. The selection of the machining processes and operations only has small influence on the planning effort in this approach.

5. Working Units and the Costs Structure

The method presented above for determination of the to be removed material volumes with following time and cost calculation is carried out for single Working Units. Machining operations belonging together logically and being carried out on the same machine are bundled to Working Units for better handling and calculation. A Working

Unit comprises the set of machining operations that transform a workpiece from a certain cavity state *S[n]* to the following state *S[n+1]* without changing the used machine (fig. 3). In special cases a Working Unit can consist of only a single operation or can include the entire mould cavity machining. Consequently Working Units represent objects that are characterised by two workpiece states. This way they quantify the machining progress considering the removed material volume. They also contain characteristics that describe the used machinery, needed machining times and caused costs.

Since the manufacture of dies and moulds includes a high percentage of manual work, a determination of costs that is restricted to the use of machine hourly rates is not satisfactory. Furthermore during manufacturing progress additional blanks might be needed, e.g. for the manufacture of electrodes for a following EDM process. These costs influence the production costs of the actual workpiece. In order to be able to perform a costs-per-unit calculation according to the selected manufacturing process the costs of blanks, machines and labor have to be taken into account.

A separation of machine and labor costs seems necessary in order to be able to compare machining operations based on different grades of automation. The costs that are caused by a worker might be independent from the machining time. Therefore it must be calculated separately. That is necessary, for example, if a process like sinker EDM need not be supervised by a worker during the entire performing time. A separation is required also if machines can be set up or workpieces can be fixed without interrupting a running process: These operations result in labor but not machine costs.

Due to this way of structuring the costs it is possible to compare sequences of Working Units. Of course their start- and final-state have to be identical. The content of work and the way it is performed can be different. Even a make-or-buy decision is supported by this approach.

6. Plannings-Desktop for Calculation and Evaluation

All planning activities are based on a geometric solid model delivered by a CAD system describing the mould cavity. It is transfered by a geometric interface. Taking the spatial extension of the cavity into account, a suitable blank is determined first. Its data can be taken from a corresponding database. Using the blank together with the raw, a sequence of machining operations can be determined. Afterwards suitable machinery is selected from the specific database and assigned to the operations. The calculation of setting-period-, auxilliary-process- and labor-time is done by accessing machine specific tables from the same database.

For the determination of the machining time a distinction is necessary whether it is an EDM, a rough milling- or a fine-milling process. In order to carry out an EDM process the material removal rates have to be taken from a EDM database. The machining time is calculated using the volume of the material to be removed, which has been determined before.

For the rough milling process - there is no difference between machining the actual workpiece or an electrode blank - a NC programme is generated automatically. It describes the machining of a specific part of the workpiece, that has been determined by the user before. Using this NC data, process and auxiliary process time can be calculated relatively exact. Furthermore the NC programme is used to update the computer based workpiece model to the state of the real workpiece after having executed the NC programme. The generated model is a basis for the determination of the process time that is needed for the fine machining. An automatic generation of NC programmes, as for the rough milling, is not possible yet. Since mould cavities mostly are described by free form surfaces they do not allow that. Not to increase the expenditure of time for the user, the calculation of processing time is done as described in chapter 4.

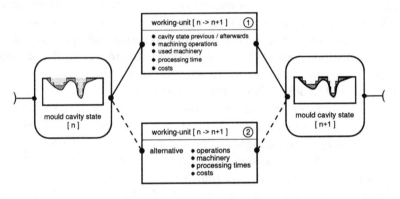

Fig. 3: Definition of a Working Unit

Operations executed in a sequence on the same machine and belonging together logically, are bundled as Working Units. The rough machining *wu-1* in fig. 4 is one example for a Working Unit. Its characteristics are: the entire processing time, the production costs as well as the workpiece state before the first and after the last machining operation.

After having executed the Working Unit *wu-1*, a workpiece state is reached that can be used to determine the geometry and volume of the electrode for the following EDM operation. All operations needed for manufacturing of the electrode are bundled to *wu-2*. The electrode in the example in fig. 4 is considered as being manufactured on a single machine. In the following *wu-3* the previously produced electrode is used for the finishing of the actual workpiece.

The specified Working Units can be put down on the desktop and connected in a row. Based on the process steps, the entire processing time and production costs for the workpiece can be calculated. The results can be compared with the values of alternative sequences. Alternative sequences can be built, generating and connecting further Working Units, which contain information that is redundant to parts of the first approach. Its length does not have to match the length of the compared sequence. It is also possible to define alternative Working Units that cover only parts of previously generated process steps.

An example of an alternative for the above mentioned Working Units *wu-1*...*wu-3* would be the entire machining of the mould cavity on a single machine by milling operations. A second example is the changing of machinery. A new Working Unit must be created if the work of *wu-2* is to be done on a machine other than *M05*. The result probably would be different processing time and production costs. All the Working Units, that share the same start- and end-state of the mould cavity, can be exchanged on the desktop. In this manner different combinations of machining operations, that lead from blank to finished workpiece, can be evaluated considering costs and times.

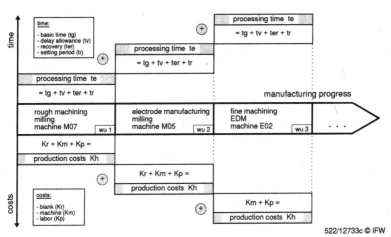

Fig. 4: Working Units carrying Cost and Time Information

The functions for generating, connecting and calculating of the Working Units are supported by a planning-desktop. Furthermore the module provides the information required for the calculations. This service is realised by connections to various databases all over the company. Also the data generated by the system, like process plans or NC data, can be supplied where they are needed.

All Working Units that are placed on the desktop as well as the machining operations, which are included, are stored in an order folder. In this way the generated information of an order request can be used for offer calculation and also during the manufacturing. Before starting the manufacturing, changes in the sequence of operations can be made without great effort. Having started the manufacturing process, the data can be used for the purpose of manufacturing control. In case of machine failures, appearing bottlenecks or changing order priorities, an alternative manufacturing path can be obtained by demand.

7. Summary

In die and mould making companies process planning is done only in a rough way because of the structure of the orders. The result is inaccurate offer planning because the needed precise information from detailed process plans is missing. Therefore a reasonable offer can not be made without spending much time on process planning and cost calculation. In the above approach of model based process planning and calculation the necessary machining operations are determined using a computer based solid modeler. The geometric model of the mould cavity is updated each time a process is finished. Besides the model of the workpiece, connections to technical and economic databases all over the company provide the necessary input of information. The approach facilitates the evaluation of Working Units, which consist of bundled machining operations, from an economic point of view. Alternative Working Units can be generated for the comparison with respect to processing time and production costs. The generated information can be used because of their wide range both for offer calculation and manufacturing control.

Acknowledgements

The research work described in this article is funded by the German Research Council (DFG).

References

1. Altan, T. et al.: Advanced Techniques for Die and Mold Manufacturing. Annals of CIRP (1993), vol 42/2, S. 707-716.
2. Bonte, A.; Emmer, T.: Technologiemodul für das Fräsen und Erodieren. Der Stahlformenbauer (1994), Nr. 2, S. 36-39.
3. Buchholz, B.; Pudig, C.; Stockter, R.; Trampler, J.: Rechnerunterstützte Konstruktion und Arbeitsplanung im Formenbau. Seminar des SFB 300, Univ. Hannover, 1993, S. 6/14-6/20.
4. Eversheim, W. et al.: Werkzeugauswahl im Griff. Industrie-Anzeiger 85 (1988), S. 38-41.
5. Hernandez-Camacho, J.; Beckendorff, U.; Gehring, V.: Fräsen oder Erodieren - eine Verfahrensabgrenzung. AV 26 (1989), Nr. 6, S. 230-233.
6. Kuhn, R.: Technologieplanungssystem Fräsen. Dissertation, Univ. Karlsruhe, 1992.
7. Lampkemeyer, U.: Objektschemata und Methoden für ein rechnerintegriertes Angebotsplanungssystem im Werkzeug- und Formenbau. Dissertation, Univ. Hannover, 1992.
8. Paucksch, E.: Zerspantechnik. 10. Auflage, Vieweg Verlag, 1992, S. 185ff.
9. Tönshoff, H. K.; Becker, M.; Hernandez-Camacho, J.: Planung der Werkzeugfertigung., HFF-Bericht Nr. 11 (1990), S. 20.1-20.12.
10. Tönshoff, H. K.; Becker, M.: Element Oriented NC-Programming for the Manufacture of Dies and Molds. CIRP (1990).
11. Tönshoff, H. K.; Lampkemeyer, U; Brunkhorst, U.: Technisches Informationssystem zur Vorkalkulation und Lieferterminbestimmung. Angebotsplanung im Werkzeug- und Formenbau. Produktion und Management - wt 84 (1994), Nr. 1/2, S. 42-47.
12. Tönshoff, H. K.; Trampler, J.; Pudig, C.: Technische EDV im Werkzeug- und Formenbau. wt-Produktion und Management (1994), Nr. 9, S. 405-408.
13. Tönshoff, H. K. et al.: Werkzeugeinsatz in der Fertigung - Planungsmodell fuer die Hohlformbearbeitung. VDI-Z 134 (1992) Nr. 5, S. 103-108.
14. Trampler, J.; Becker, M.: Modellgestützte Arbeitsplanung im Werkzeugbau. VDI-Z 136 (1994), Nr. 5, S. 48-50.

COMPUTER AIDED TOLERANCE CONTROL IN PROCESS PLANNING

Y.K. CHOW

*Hutchison Telephone Co.Ltd, 23/F, Citicorp Centre, 18, Whitfield Road,
North Point, Hong Kong.*

W. M. CHIU

*Department of Manufacturing Engineering, The Hong Kong Polytechnic University,
Hung Hom, Kowloon, Hong Kong.
Email: mfwmchiu@hkpucc.polyu.edu.hk*

ABSTRACT

This paper describes a methodology for the development of an automatic tolerance charting system. A path tracing algorithm is used to identify the dimension chains for the functional and stock removal equations which are then solved by the Gauss-Jordan method to obtain the working dimensions. A linear programming (LP) model is formulated for the optimization of working tolerances. The objective function seeks to maximize the cumulative tolerance of each working dimension. The LP model is then solved by the Revised Simplex method.

1. Introduction

It is not feasible to manufacture a dimension to an exact value. Tolerance must be placed on the blue-print (BP) to restrict the variations to permissible limits. The choice of design tolerances affects the manufacturability of the part. The selection of manufacturing processes and their sequences affect the accuracy of the finished part through process tolerance stacking. To insure manufacturability of parts being made to BP specifications, effective tolerance control is required. The task of transforming raw material into components satisfying the specified BP form and dimension is done by a tentative process plan. Feasibility of the tentative process sequence is checked via the tolerance chains. A tolerance chain shows how individual cuts can be combined to produce each BP dimension. It yields a set of linear algebraic equations showing the relationship between each BP dimension and the individual cuts that contribute to it. This accumulation, regardless of whether dimensions from individual cuts are added or subtracted, is referred to as "tolerance stackup". Whenever a series of successive operations contributes to the final dimensions and tolerances, the tolerance stackup problem can never be eliminated.

2. Graph-Theoretic Properties of Dimensions in a Tolerance Chart

There are three types of dimensions in a tolerance chart [1, 2]: BP dimensions, stock removals, and working dimensions. If the axial surfaces of a part are represented by vertices in Graph Theory, and the dimensions between the surfaces are edges, three different trees can be generated.

a. Blue-print Dimensions Tree (BP Tree) - In order to define a part with n axial surfaces completely, (n-1) distinct dimensions are required. These (n-1) dimensions should be chosen so that no existence of redundant or undetermined dimensions

should result. In Graph Theory, n vertices and (n-1) edges form a tree structure if these vertices are all connected.
b. Stock Removal and BP Dimensions Tree (SRBP Tree) - A BP dimensions tree can be extended to form another tree by adding stock removals on the appropriate vertices. Supposing there are m machining cuts in a tolerance chart with n axial surfaces, and m > n, then (n-1) machining cuts should correspond to solid cuts, and the rest [m-(n-1)] machining cuts should correspond to stock removal cuts.
c. Working Dimensions Tree (WD Tree) - The third tree is created from the working dimensions. It should have (m+1) vertices which are common with that of the SRBP tree, and m edges corresponding to the m machining cuts.

Any edge in the SRBP tree can form a circuit with edges in the WD tree, and the circuit is one and only one. As there are m edges in the SRBP tree, there will be m distinct circuits; [m-(n-1)] for stock removals, and (n-1) for BP dimensions. Each circuit represents a dimension chain, and all the circuits correspond to all the dimension chains in a tolerance chart. Figure 1 shows a simplified tolerance chart of an engine sleeve [2]. The combined SRBP tree with WD tree is shown in Figure 2.

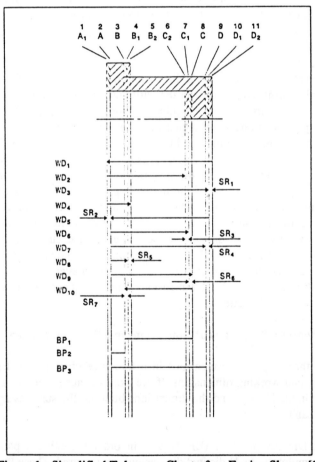

Figure 1 : Simplified Tolerance Chart of an Engine Sleeve [2]

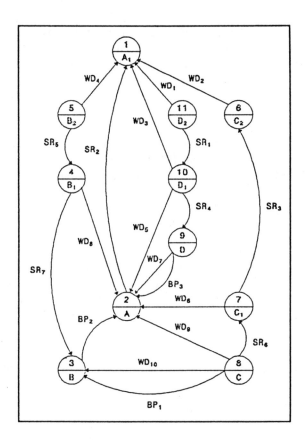

Figure 2 : The Combined SRBP Tree with WD Tree of the Engine Sleeve [2]

2.1. Fundamental Tolerance Equation

The difficulty of producing or maintaining dimensions increases as tolerances become smaller and as the magnitude of dimensions become larger. The ISO System of Limits and Fits classified the severity of tolerances into 20 standard tolerance grades (IT01, IT0, IT1, . . . , IT18 with decreasing severity of tolerances). The International Tolerance Grades provide a basis for comparing the precision attainable with the various manufacturing processes. Figures were given in BS 1916 : Part II which relates tolerance grades with common manufacturing processes operating under average conditions. For instance, drilling has IT12, broaching has IT7, and fine grinding has IT5.

In order to compute the expected achievable tolerance t of a design size D for each IT value, Farmer and Harris [3] have derived the following Fundamental Tolerance equation:

$$t = (0.450 \sqrt[3]{D} + 0.001 D) \, 10^{(IT-16)/5} \qquad (1)$$

Although this method of determining the expected achievable tolerance is based on the ISO System for calculation of tolerance fits between circular holes and shafts, it will be assumed that it can be applied to machining tolerance without causing serious inaccuracies.

2.2. Functional Equations

During the transfer from design dimensions to working dimensions for a process plan, the most important criterion is that the BP specifications must not be violated. The functional equations [3] that represent the relationships of design specifications with their relevant working dimensions and tolerances are written as follows:

$$[BP] = [C] [WD] \tag{2}$$
$$[bp] = [|C|] [wd] \tag{3}$$

where
[BP], [bp] = column vectors for BP dimensions and tolerances respectively, (n-1 x 1),
[WD], [wd] = column vectors for working dimensions and tolerances, (m x 1),
[C], [|C| = matrices of functional equations for dimensions and tolerances, (n-1 x m).

2.3. Stock Removal Equations

The stock removal equations for dimensions and tolerances [4] can be written as

$$[SR] = [A] ([M] - [G]) [WD]$$
$$= [D] [WD] \tag{4}$$
$$[sr] = ([|M|] + [G]) [wd]$$
$$= [|D|] [wd] \tag{5}$$

where
[SR], [sr] = column vectors of stock removal dimensions and tolerances (m x 1),
[A] = working dimension type matrix (m x m),
[M],[|M|] = matrices of machining equations for dimensions and tolerances, (m x m),
[G] = Unit matrix, when the i-th operation has no previous operation, i.e.,
  $(G_{i,i}) = 0$ (m x m),
[D], [|D|] = Matrices of stock removal equations for dimensions and tolerances (m x m).

3. Tracing Method for Functional and Stock Removal Equations

The procedure for establishing the functional equations is as follows [4]:

a   Trace upwards from both ends of a given BP dimension.
b. When one of the end traces of the BP dimension encounters an arrowhead of a working dimension, trace along this dimension to its locating surface, this dimension is a relevant dimension.
c. Then continue to move both traces upwards until either another arrowhead is encountered and step b. is repeated, or until the two traces converge and the tracing is then stopped.

The rule used to determine the sign of each working dimension is as follows:

Trace starts on BP dimension	Tracing long working dimension	
	From left to right	From right to left
Left side	+	-
Right side	-	+

The procedure for determining the stock removal equations is done by replacing the given BP dimension by the given working dimension in question and that the trace starts from the last operation. Given a process plan with m working dimensions and (n-1) BP dimensions, (n-1) of the working dimensions should correspond to solid stock removals. This is reflected in (n-1) non-converging traces of the working dimensions. An example trace for the stock removal of the engine sleeve is shown in the SR_6 trace for dimension WD_6 of Figure 3 [2].

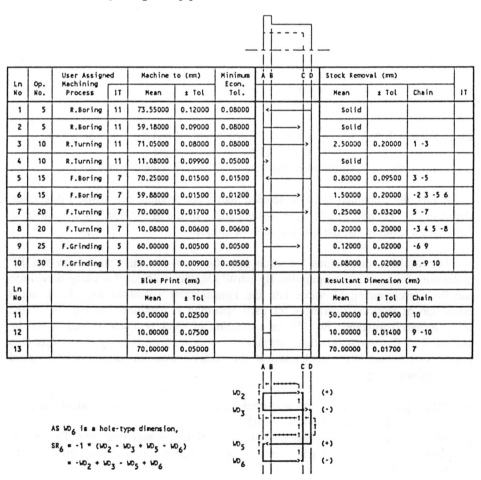

As WD_6 is a hole-type dimension,

$$SR_6 = -1 * (WD_2 - WD_3 + WD_5 - WD_6)$$
$$= -WD_2 + WD_3 - WD_5 + WD_6$$

Figure 3: Modified Tolerance Chart of the Engine Sleeve

3.1. Identification of Types of Working Dimension (TD)

A working dimension which makes its related dimension smaller after it is machined is defined as "shaft-type" working dimension (TD = +1). If it makes the related dimension larger after it is machined, it is defined as "hole-type working dimension (TD = -1). Taking this convention, the type of working dimensions can be identified by using the following rule:

Direction of machined surface	Locating surface < Machined surface	Locating surface > Machined surface
Leftward	-1	+1
Rightward	+1	-1

3.2. Solution for Working Dimensions

As the stock removal Eq. (4) and Eq. (5) have (n-1) rows of 0's corresponding to solid stock removals, so the stock removal equations and functional Eq. (2) and Eq. (3) can be combined to obtain m linear equations:

$$[SRBP] = [P] [WD] \tag{6}$$
$$[srbp] = [|P|] [wd] \tag{7}$$

where [SRBP], [srbp] = column vectors of combined SR and BP dimensions and tolerances respectively, (m x 1) and
[P], [|P|] = combined matrices of SR and BP equations (m x 1).

In fact, the combined SRBP dimension matrix corresponds to m distinct dimension chains in the WD tree by adding m distinct edges of the SRBP tree. So the equations must have a solution and the solution must be unique.

The working dimensions for each operation can be calculated from the BP dimensions for nominal size purpose. By substituting this calculated working dimension and the corresponding IT value into the Fundamental Tolerance Eq. (1), a rough estimation of working tolerances for each operation is obtained. Then by using the Stock Removal tolerance Eq. (5), an estimation of stock removal tolerances can be arrived. With the assumption that tolerance of a stock removal should not be larger than the nominal amount of the stock removal, the stock removal dimension can be assigned in the following ways:

a. Add a positive quantity to each stock removal tolerance. The amount added is based on the process capability of each operation, or
b. Multiply each stock removal tolerance by a figure greater than one. The chosen figure is different for each operation and reflects the process capability of the corresponding process.

Having assigned the stock removal dimensions, the combined SRBP dimension matrix Eq. (6) can be solved by using the Gauss-Jordan Elimination technique [1]. The solution is a set of working dimensions for the process plan.

4. Linear Programming (LP) Model for Working Tolerances

The tolerance assignment problem for tolerance charting can be summarized as to make each tolerance as large as possible subject to the BP specifications and process capabilities. This can be expressed as an LP model [5]:

$$\text{Maximize} \sum_{k=1}^{m} wd_k \tag{8}$$

$$\text{s.t.} \sum_{k \in DCSR_j} wd_k \leq sr_j \quad (j = 1, \ldots, m-n+1) \tag{9}$$

$$\sum_{k \in DCBP_i} wd_k \leq bp_i \quad (i = 1, \ldots, n-1) \tag{10}$$

$$wd_k \geq LPC_k \quad (k = 1, \ldots, m) \tag{11}$$

where
$DCSR_j$ = regular component set of dimension chain of the j-th stock removal,
$DCBP_i$ = regular component set of dimension chain of the i-th BP dimension,
LPC_k = minimum economic tolerance for the k-th working dimension.

The objective function Eq. (8) is to maximize the cumulative tolerance of each working dimension, which corresponds to assigning each tolerance to be as large as possible. Constraint set Eq. (9) reflects the dimension chains of stock removals. Constraint set Eq. (10) is derived from the BP dimension chains to meet the design requirements. Constraint set Eq. (11) is required from minimum economic capability of each process sequence which is obtained by using the Fundamental Tolerance Eq. (1).

By combining constraint set (9) with (10), the above LP model can be rewritten as:

$$\text{Maximize} \sum_{k=1}^{m} wd_k \tag{12}$$

$$\text{s.t.} \sum_{k \in DC_l} wd_k \leq q_k \quad (l = 1, \ldots, m) \tag{13}$$

$$wd_k \geq LPC_k \quad (k = 1, \ldots, m) \tag{14}$$

where
DC_l = regular component set of dimension chain of stock removal or BP dimension,
q_k = minimum allowed tolerance of stock removal or tolerance of BP dimension.

It can be shown [5] that the above model can be simplified into in a matrix format as follows:

$$\text{Maximize } [1][x]$$

$$\text{subject to } [|P|][x] \leq [q] - [|P|][LPC]$$
$$[x] = [wd] - [LPC] \geq [0]$$

The reduced LP model can then be solved by the Revised Simplex algorithm to calculate the working tolerance matrix, [wd].

5. Implementation of Computer Aided Tolerance Charting System

Based on the above methodology a Pascal program is developed to implement a Computer Aided Tolerance Charting System. The program takes the process plan and the corresponding BP specifications of the part as inputs, then the functional and stock removal equations are generated automatically. Stock removals are then assigned based on the capabilities of the machining processes involved. Working dimensions are obtained by solving the functional and stock removal equations. Then an LP model is set up for calculating the working tolerances. The condition of non-existence of feasible solution is checked and adjustment made before the LP model is solved by the Revised Simplex method. The resulted tolerance chart of the engine sleeve taken as an example is generated by the Computer Aided Tolerance Charting System as shown in Figure 3.

6. Conclusion

The algorithms employed by this method are proven to be correct and effective in providing a preliminary answer to the tolerance allocation and control problems in many complicated process plans. This method also carries greater implications to industries endeavouring to cut down the production cost and time by rationalizing each process along the manufacturing route. However, owing to the complex and diverse nature of manufactured goods, a universal computerized tolerance control system has yet to emerge. Further research and development work is therefore necessary to improve this system, making it more efficient and responsive to tolerance allocation problem in wider spectrum of industries.

7. References

[1] Ji, P., 1993, "A Tree Approach for Tolerance Charting", International Journal of Production Research, Vol. 31, No. 5, pp. 1023-1033.
[2] Ke, M.; Ji, P., 1987, "Computer Aided Operational Dimensions Calculation", Third International Conference in Metal Cutting Non-Conventional Machining and Their Automation, May 13-15, Nanjing, China.
[3] Farmer, L.E.; Harris, A.G., 1984, "Change of Datum of the Dimension in Engineering Design Drawings", International Journal of Machine Tool Design and Research, Vol. 24, No. 4, pp. 267-275.
[4] He, J.R.; Lin, G.C.I., 1992, "Computerized Trace Method for Establishing Equations for Dimensions and Tolerances in Design and Manufacture", International Journal of Advanced Manufacturing Technology, Vol. 7, No. 4, pp. 210-217.
[5] Chow, Y.K., 1994, Computer Aided Tolerance Control in Process Planning of Precision Parts, MSc thesis, the Hong Kong Polytechnic University, 103 pages.

Computer Aided Facilities Planning

An Evaluation of a PC CAD-Based Facilities Planning Package
    C. E. H. Teo

Systolic Algorithm for Improving Efficiency of Facility Layout Algorithms
    Y. L. Qi, B. Sirinaovakul and K. Narue-domkul

An Analysis of Computer Aided Facility Layout Techniques
    B. Sirinaovakul and P. Thajchayapong

An Evaluation of a PC CAD-based Facilities Planning Package

Christopher E. H. TEO
Mechanical & Manufacturing Department
Singapore Polytechnic
Singapore

Abstract

This paper attempts to evaluate a PC CAD-based facilities planning package called Factory. Factory is a CAD-based facilities planning program that can calculate material handling costs and distances within a manufacturing system. It allows departments to be moved in a CAD drawing as well as accommodates changes in other manufacturing parameters. Developed in the United States by David Sly and his associates in 1988. Evaluation criteria used were based on Apple and Deisenroth (1972), as well as criteria gleaned from Driscoll and Sangi's survey (1987). A comparison was also made with five other CAFL packages using the above criteria.

Keywords: Facilities planning, Manufacturing productivity, Systematic Layout Planning (SLP).

1. Introduction

Although more than half a century old, facilities planning continues to be a popular subject of current research, publications and conferences. Being a composite of facilities location and facilities design, the objective of facilities planning is thus to plan a facility that achieves facilities location and design objectives. Improved manufacturing productivity is closely associated with effective facilities planning which incorporates both quantitative and qualitative goals.

Economic considerations force a constant reevaluation and reorganisation of existing systems, personnel, and equipment. The introduction of Flexible Manufacturing System (FMS) or Cellular Production and the implementation of Just-In-Time (JIT) method of production have meant a reduction in inventory on the line and an increase in manufacturing productivity for many firms. Improved manufacturing productivity has been linked very closely with effective facilities planning. It is becoming apparent that the factory of the future will offer minimal lead times, unit batches, high quality, continuous operation, and integrated operation.

More recently, efforts have been made to present facilities layout solutions with both qualitative and quantitative goals in mind. While the quantitative goal is normally associated with product flow and the transportation of the product, both of which are cost related ,the qualitative goal is about the closeness of one department with another for reasons of sharing common utilities, personnel, etc. The combination of both qualitative and quantitative approaches into a single formulation will incorporate information from both approaches. Most of these single formulations consider only one qualitative parameter and one quantitative parameter in its objective function and are solved using computer heuristics similar to those used for single approach problems.

Five CAFL bespoke software, MOCRAFT, LAYOUT, STORM, PLANET, and BLOCPLAN will be used to compare with Factory in terms of accuracy, versatility of software, power and limitations to solve layout problems, and user-friendliness of the software.

2. Literature Review

In 1961, Richard Murther presented a method for systematically developing a layout to solve a single goal objective; for instance, the locating of different departments or workcenters in a plant. The method, called *systematic layout planning* (SLP) [16], has been widely used since its introduction.

Since 1979, a number of multi-goal approaches to the plant layout problem have been proposed. One common characteristic of past multi-goal approaches is that they consider just two factors (or parameters) in the objective function formulation, the quantitative flow data from a from-to chart and the qualitative relationships from a relationship chart. Each multi-goal approach uses a different scheme to assign numerical values to the relationship codes (A,E,I,O,U,X) from the relationship chart. Rosenblatt (1979), Dutta and Sahu (1982), Fortenberry and Cox (1985), Rosenblatt and Sinuany-Stern (1986), and Urban (1987) all present quadratic assignment problem formulations of the multi-goal objective. Although the specific cost term in the objective function is unique in each case, all consider only two goals or factors in the objective function. The complex nature of the facility layout process has prompted the development of several expert system based methods that provide assistance in the layout process. Typically, these methods are able to evaluate different layout criteria, select equipment, set a layout type, process qualitative and quantitative data, and incorporate traditional layout methods, with the aim of developing a highly efficient layout.

Previous approaches in solving the problems include the CRAFT algorithm (Buffa et al. 1964), the first successful program. The algorithm uses a heuristic approach to interchange facility locations where the initial layout is required before an improved solution based on material flow data can be found. Alternative approaches, such as CORELAP (Lee and Moore 1967) and ALDEP (Seehof and Evans 1967) do not start from the initial layout but generate a layout from scratch. They locate the activities according to the rating of the highest relationship activity. PLANET (Apple and Deisenroth 1972) uses the interdepartmental flow data, adding the 'penalty' cost associated with separating departments[1]. Abdou and Dutta (1990) developed an expert system approach to define appropriate layouts of machining facilities under specific combinations of manufacturing and material handling systems[12,13,17].

However, these conventional computerised techniques do not consider enough possible outcomes in the computer program to arrive at the optimal solution, especially for the construction methods such as CORELAP and ALDEP. Although improved methods such as COL (Vollmann et al. 1968) and COFAD (Tompkins and Reed 1976) are presented to obtain a better solution, the methods are still very sensitive to the initial layout.

Furthermore, it is difficult for conventional computerised techniques to take into account the practical limitations of facility design problems such as intangible factors as well as human aspects of the layout. This is because any facility layout problem has multiple criteria in selecting the best layout, being specific to the domain and problem.

The computer-aided facilities layout techniques may be classified by the method of recording flows among departments and by the method of generating layout. Flows among the departments may be either quantitatively recorded from a from-to chart, or qualitatively recorded in a relationship chart. While CRAFT and COFAD require quantitative flow inputs, PLANET accepts either quantitative or qualitative flow inputs. However, CORELAP and ALDEP require qualitative input.

The earliest computer algorithm, CRAFT (Armour and Buffa 1963) is an improvement logic which takes an initial block diagram layout and modifies it to find better layouts. Most of the newer improvement algorithms are based on the CRAFT like COFAD (Tompkins and Reed 1976) in which FactoryFLOW (David Sly 1990)[18] is designed around. In contrast, construction algorithms build layouts by placing departments with large flows next to previously placed departments. Some of the well known construction algorithms include CORELAP (Lee and Moore 1967), ALDEP (Seehof and Evans 1967), RAMAI (Murther and McPherson 1970), PLANET (Apple and Deisenroth 1972), DISCON (Drezner 1980, 1987) and FLAC (Sciabin and Vergin 1985). The computer heuristics often incorporate practical constraints such as fixed department locations and varying areas.

There is considerable development effort in linking CAFL to new advanced procedures and to currently topical manufacturing systems such FMS and JIT. Development effort is also directed at improving the level of graphics available as part of the CAFL packages, as evident from current generation of CAFL packages like Spaceplan and Factory.

3. FACTORY

Factory is a CAD-based facilities planning program that can calculate material handling costs and distances within a manufacturing system; for instance, within a factory area. The package also allows users to easily move departments in a CAD drawing, change other manufacturing parameters such as production volumes, parts routings, material handling systems or material handling paths, and number of product types to be produced. The recalculation results of costs and material handling distances of the revised layout are then shown and graphic regenerations of the Product Flow Diagram, Composite Diagram, and Distance-Intensity Chart also take place and are updated on the computer screen.

At the time of initial development there was only one module, FactoryFLOW, but it has since developed two other modules named FactoryPLAN and FactoryCAD, to enhance the whole software package.

The two survey reviews by Driscoll and Sangi in 1987 on CAFL software packages have shown that most of the first generation line printer-orientated layout programs are still used today but have limited ability to tackle real world problems. This is because they are difficult to use, run in batch mode, require mainframe computers, and often have no graphics capabilities. The survey further revealed that the users are interested in more interactive features as well as better graphics, akin to those in CAD in the CAFL software. Furthermore, in terms of improvement, users have indicated a need for economic evaluation in support of technical evaluation. Thus, based on the results of the survey, as well as on the suggested trend that the future CAFL packages may likely be one using proprietary support software aimed at meeting applications need, the software package Factory was designed.

The software package is designed to operate within AutoCAD graphics environment, and the reasons cited [18] for the choice of this CAD platform were as follows:

♦ AutoCAD offers a degree of programmability
♦ AutoCAD can operate on a microcomputer platform and is widely used;
♦ AutoCAD has superior 2-D graphics user interface,and has data interchange capability.

3.1 Capacity of the software

The program is not limited by the number of departments, products, or parts that the user may use. The number of paths that can be analysed is limited only by the amount of RAM available and the length of time for the program to analysis the results.

A) FactoryCAD

This module customises AutoCAD for facilities drafting and management. It aims to ease drafting of the layout. It comes with a library of electrical and mechanical utility symbols, and machine tool symbols available such as shown in figure 1. Users may even create their own symbols and add on to the library.

Figure 1: Process flow symbol

B) FactoryPLAN

This program module is a tool for designing and analysing layouts based on their qualitative relationship. The user can use this program to organise the many qualitative relationships. which need to be considered when designing or redesigning a facilities layout. Its relationship diagram visually displays the relationship between various departments, and allows the user to create, manipulate, and score each of the department closeness desirability. Although not a drafting tool but a planning one, it provides an introductory level solution for designing the layout of a wide array of spaces, from financial and service-oriented industries to manufacturing and food industries.

C) FactoryFLOW

In this software module, production data is integrated with a layout drawing. The outcome of this integration is a series of reports, charts, and graphic displays. The idea behind these modes of output format is to allow the option of viewing spatial problems in a spatial medium rather than just in a series of words and numbers that is difficult to take in quickly and manipulate. Material flow can be visually described in FactoryFLOW. Manipulation of the model (or layout drawing) can be easily made through changes made to the layout drawing, routings, production volume, material handling systems, and so forth. The effects of the changes can then be visually displayed and varied through text reports generated.

Once the input data is available, there are 8 available outputs in the form of graphics, charts and reports.

4. Case Study

The model used here is the model adopted from Driscoll [9] with the objectives to evaluate the practical working features and the flexibility of the software.

The test case involves the relocation of an initial layout of 34 main workcenters and 5 dummy workcenters in an area shown in Figure 2.

4.1 Approach adopted to solve the test case

1. There are 7 stages to an accurate and successful Factory analysis. They are to determine the level of detail for the analysis, get the layout into AutoCAD, organise the input data, decide on naming conventions for the products, parts, and workcenters, enter Products and Parts data, run module software FactoryFLOW and, if qualitative aspects of the layout are important such as noise, run module software FactoryPLAN as well.

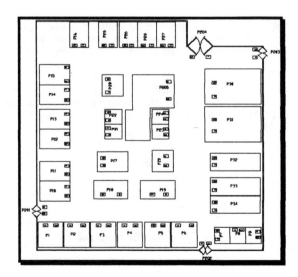

Figure 2: Test Case Original Layout (adapted from Driscoll)

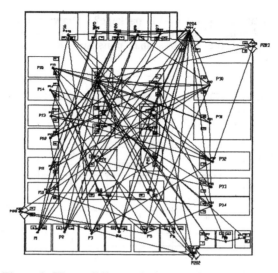

Figure 3: The euclidian path generated automatically.

Then cost is factored in to the euclidean path generated to produce the 'band' between work centres which is an indication of the cost of product flow. This is shown in Figure 4 below, and is known as the composite diagram. As with FactoryPLAN, the thicker the 'band' the more expensive the path is. After that the scores should be consulted so that the arrangement of the workcenters can be optimised quantitatively.

To minimise the cost, the following relayout plan was obtained.

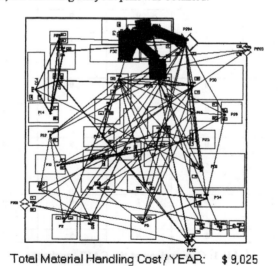

Total Material Handling Cost / YEAR: $ 9,025

Figure 4: The composite flow diagram (relayout euclidean path)

5. Comparison Study

However several questions need to be addressed: have the CAFL software developed so far been useful for practical applications, and are they user-friendly? Furthermore, does a commercially successful software have all or most of the ingredients that constitute a good CAFL package? Drawing on the findings in the survey by Driscoll and Sangi [7,8] and using them as the basis for comparing the 6 CAFL software, the findings of the features that constitute a good CAFL software are tabulated as shown in table 1.

Software \ Criteria	Factory	MOCRAFT	BLOCPLAN	LAY-OUT	STORM	PLANET
1. Acceptable manual	1	1	1	1	1	1
2. Graphics capability	1	0	1 (ILIMITED)	0	0	0
3. Accurate layout geometry	1	0	0	0	0	0
4. Interactive	1	0	0	0	0	0
5. Integration capability	1	0	0	0	0	0
6. PC base	1	1	1	1	1	1
7. Output analytical result printable	1	1	1	1	1	1
8. Plotable Graphics	1	0	0	0	0	0
9. Data input by keyboard or floppy disk	1	1	1	1	1	1
10. AI capabiility	1	0	0	0	0	0
11. Layout problem size (>= 30)	1 (NOT LIMITED)	0	0	0	0 (MAX 18)	1 (MAX 30)
12. Automated data collection	1	0	0	0	0	0
13. Capable of both Qualitative & Quantitative analysis	1	1	1	1	0	1
14. Financial evaluation	1	0	0	0	0	0
15. Possese construction & improvement	0	0	1	0	1	0
16 Able to deal with other shapes other than squares or rectangle	1	0	0	0	0	0 (SQUARE)
17 Able to deal multi-level layout	0	0	1	0	0	0
TOTAL	15	5	8	5	5	6

Table 1: Comparison of the CAFL packages

5.1 Criteria on Factory package (based on Apple and Deisenroth)

To compare Factory with the older CAFL software may not be a true judgement of the capability of the software to meet the requirement of fulfilling what is required in factory layout. The desirable requirements of CAFL software were put forward by Murther and McPherson and Apple and Deisenroth are used to examine the usefulness of CAFL software. Subjecting Factory to the criteria stated, the following conclusions can be made on the features and operating characteristics of Factory software:

1. The package is reliable, flexible, and able to deal with realistic problem representations.
2. Both qualitative and quantitative aspects of layout can be dealt with within the package by the interactive design approach of the package.
3. The use of interactive design approach creates a realistic layout and eliminates manual modification after a design has become accepted.
4. Graphical representation is detailed enough for a realistic layout area and workcentres.
5. The package has the ability to consider the location of any facility as fixed.
6. It is also able to deal within the confines and constraints of building restrictions and can evaluate realistic cost incurred by alternative layouts..
7. It can be used to evaluate detailed layout such as location of machines within a FMS system.
8. Very little human evaluation to layout solutions is required when using this software. Human evaluation occurs as part of the layout design process and is considered necessary only at the detailed layout design stage.
9. Any designer/user is able to tell at a glance the overlapping of facilities when presented in visual form; the package, however, is not able to detect this undesirable interrelationship.

6. Conclusions

The layout problem requires that both qualitative and quantitative factors be considered during the design stage, reflecting the need for an interactive method. As layout problems also involve physical constraints, apart from the qualitative and quantitative factors, a single evaluation criteria will not adequately reflect the problem requirements. Accurate physical representation of layout area and workcenters contribute significantly to the practical usefulness of the computer layout software, thus highlighting the value of using interactive graphics facilities. As an effective layout design is concerned with available material handling systems, the integration of material handling system with layout design would be beneficial.

The complexity of layout problems with volumes of calculations and information necessary for an industrial size problem, warrants the use of a computer. The computer software that is used should be versatile enough to calculate a wide variety of layout problems and be able to represent accurately facilities and layout areas, including details like internal obstructions, account for subjective factors, and allow the evaluation of alternative solutions

Based upon my evaluation of CAFL software, Factory$_{Release\ 3.2}$ has succeeded in incorporating several features to meet the shortcomings experienced in many CAFL software. Apart from having a good and accurate graphic representation of the layout, its product flow diagram enables the user to visualise the material handling costs within the plant, and recognise problem areas. Furthermore, the relative thickness of the paths visually and quickly draws the attention of the user to an area that requires layout alternatives. As all this can be done during the early stages of the design process, its usefulness especially to the management cannot be underestimated.

References

1. Apple, J., and Deisenroth, M., 1972, *A computerised plant layout analysis and evaluation technique.* Technical papers of AIIE 1972 Spring Conference, Norcross GA, 112-117.

2. Apple, J.M., *Plant Layout and Material Handling*, 3rd ed., Wiley, 1977.

3. *AutoCAD Reference Manual.* Autodesk, Inc., Switzerland

4. Block, T.E.,1979, *On the complexity of facilities layout problems*. Management Science, 25, 280-284.

5. Burkard, R.E., and Bonniger, T., 1983, *A heuristic for quadratic boolean problems with applications to quadratic assignment problems*. European Journal of Operations Research, 13, 374-386.

6. Das, S.K., *A facility layout method for flexible manufacturing system*, Int. J. Prod. Res., 1993, Vol. 31, No 2, 279-297.

7. Driscoll, J. And Sangi, N.A., *The development of Computer Aided Facilities Layout (CAFL) systems - International Survey report*, Dept. Ind. Studies, The University of Liverpool, 1986, 1-119.

8. Driscoll, J., and Sangi, N.A., *An international survey of Computer Aided Facilities Layout - Analysis of software.* 9th Int. J. Prod. Res. Aug. 1987.

9. Driscoll, J., *The Layout of Workcentres in a Job-shop situation*, Ph.D. Thesis, University of Aston, Birmingham, England, 1975.

10. Emmons, H., and others, eds., *STORM*, Holden-Day, Inc.,1989

11. *Factory software manuals - FactoryCAD, FactoryPLAN, FactoryFLOW manuals v3.2*, CIMTECHNOLOGIES CORPORATION, Ames, Iowa, USA., Copyright @ 1989, 1990, 1991, 1992, 1993.

12. Harmonosky, C.M., and Tothero, G.K., *A multi-factor plant layout methodology.* Int. J. Prod. Res., 1992, Vol.30, No 8. 1773-1789.

13. Lacksonen, T.A., and Enscore, E.E Jr., *Quadratic assignment algorithms for the dynamic layout problem*, Int. J. Prod. Res., March 1993.

14. Lilly, M.T., *Computer based design and simulation programs for manufacturing Facilities relayout problems*, Ph.D. Thesis, University of Liverpool, Liverpool, England, 1985.

15. Moore, J.M., *Computer Aided Facilities Design: An international survey*, Int. J. Prod. Res., 12 (1), 1974, 21-24.

16. Muther, R., *Practical Plant Layout*, McGraw-Hill, New York, 1956.

17. Sirinaovakul, B., and Thajchayapong, P., *A knowledge base to assist a heuristic search approach to facility layout*, Int. J. Prod. Res., 1994, Vol 32, No 1, 141-160.

18. Sly, D., *A method of industrial plant layout and material flow analysis in AutoCAD*, M.Sc. Dissertation, Iowa State University, Ames, Iowa, USA, 1990.

19. Teo, E.H.C., *An evaluation of a CAD based Facilities Planning package - Factory$_{Release\,3.2}$*, M.Sc. Dissertation, University of Surrey, Surrey, United Kingdom, 1994.

20. Tompkins, J.A., and White, J.A., *Facilities Planning*, Wiley & Sons, Inc. 1984.

SYSTOLIC ALGORITHM FOR IMPROVING EFFICIENCY OF FACILITY LAYOUT ALGORITHMS

Qi Yulu
Computer Science Program, Asian Institute of Technology,
Bangkok 10501, Thailand

Booncharoen Sirinaovakul
Computer Engineering Department, King Mngkut's Institute of Technology Thonburi,
Bangkok 10140, Thailand

Kanlaya Narue-domkul
Department of Mathematics, Mahidol University
Bangkok 10400, Thailand

ABSTRACT This paper proposes the systolic algorithm to improve the efficiency of the facility layout algorithms. The parallelism is created by simultaneously computing different total costs, different central points, different closeness weights and different distances. The systolic cell of all algorithms has been designed in the same architecture. The result has proved that this approach significantly decreases the processing time when compare to the uniprocessor facility layout algorithm. For some algorithms, the parallelism also produces the number of layout alternatives which mean that it increases the probability of getting the better solution.

Keywords: facility layout, systolic algorithm

1. INTRODUCTION

Systolic systems combine pipeline, array processing and multiprocessing to produce high-performance parallel computer system. There are many advantages of systolic architecture. One of the most important is multiple computations per memory access to ensure that it can speed up compute-bound computations without increasing I/O requirements. Other advantages are modular expandability, simple and regular data and control flows, using of simple and uniform cells, elimination of global broadcasting and fan-in, and (possibly) fast response time. Besides those advantages, a unique characteristic of the systolic approach is that as the number of cells expands the system cost and performance increase proportionally, provided that the size of the underlying problem is sufficiently large (KUNG, 1982). The suitable algorithms for implementation in systolic arrays can be found in many applications (FORTES et. al., 1983). Infact, most of the algorithms are computationally intensive and they require real-time system architecture for their implementations. These applications have two important characteristic sets. First, these applications require high throughput and large processing bandwidth possibly at the cost of increased response time. Second, these applications can be efficiently supported by algorithms that can be implemented on arrays consisting of a few types of simple processing elements (PEs); the array has simple controls and input/output ports in the peripheral PEs. Today, the VLSI technology allows us to design a special purpose chip or a general purpose chip to implement a systolic algorithm easily and efficiently.

This paper aims to improve an efficiency of the facility layout algorithm by using the systolic system. Among all facility layout algorithms proposed, four algorithms, Sirinaovakul and Thajchayapong, CLASS, ALDEP, CORELAP, and CRAFT, are selected as case studies. Sirinaovakul and Thajchayapong algorithm is a representative of construction layout using Exerted Energy technique. CLASS is an iterative improvement using Simulated Annealing technique. ALDEP is classical construction layouts whereas CRAFT is classical iterative improvement layout. These five representative algorithms are studied in details and the systolic systems are designed to improve the efficiency of these algorithms.

2. ANALYSIS OF FACILITY LAYOUT ALGORITHMS

By analyzing the selected facility layout algorithms, it is found that there can be classified into two groups based on the quantity of mathematics computation; computational approach and non-computational approach.

2.1 COMPUTATIONAL APPROACH

This approach needs a great deal of computations. It produces a number of layout alternatives and selects the best one to be the final layout based on some criteria. The steps for solving the problem of this strategy are; first, the algorithm generates layout alternatives and computes their criterion value. Then, it compares all values and selects the best alternative according to the setting criteria. Sirinaovakul & Thajchayapong (a so called latter as Siri&Thaj), CLASS and CRAFT are representatives of this strategy.

Siri&Thaj algorithm (SIRINAOVAKUL and THAJCHAYAPONG, 1994) designs a program for arrangement of facility areas to optimize the interrelationship between operating personnel, materials flow and information flow. The suitable designed layouts are selected under the constraint of minimizing interrelationship costs (a so-called later as exerted energy).

At each time of adding the new facility, the algorithm produces $(10 \times n)+6$ alternatives where n is a number of located facilities (rectangular facility only) and selects a minimum total exerted energy layout which needs $(10 \times n)+6$ computations for all possible generated alternatives.

Computerized LAyout Solutions using Simulated annealing (CLASS: JAJODIA et al., 1992) was designed based on simulated annealing, to overcome the dependence of the solution on the initial layout. The criterion of CLASS is to minimize the total material flow among facilities. The exchange of facilities, to produce a candidate layout, is considered under the total cost (E) of the exchange or under the acceptance probability. The acceptance probability is given by the value of $\exp(-\Delta E/T)$. Normally, ΔE is the cost increase. T (temparature) is a controlled value which is set to a very high value in the beginning of the process for accepting most of exchanges. Then T is gradually decreased so the cost increasing and the exchanges have less chance of being accepted. Ultimately, the T is reduced to a very low value so that only exchanges causing a cost reduction are accepted, and the algorithm converges to a low cost layout.

The basic concept of CLASS is to reduce the temperature in each iteration to converge to the minimum total cost layout. To meet this purpose, a number of trials for finding the appropriate layout candidate in each temperature has to be proceeded. In each trial, the acceptance of selected position depends on the total cost value. The generation technique is exchanging two facility locations. All possible layout alternatives are all combinations of exchanging two facility locations, $^nC_2$. Therefore, a number of computations in each temperature are $^nC_2$.

Computerized Relative Allocation of Facilities Technique (CRAFT: BUFFA, ARMOUR and VOLLMANN, 1964) interchanges facility locations of the initial layout to find the improved solutions based on material flow, successive interchanges leading to a least total cost layout. CRAFT uses the cost of material flow as a criterion for considering the best layout. The best design is the one that has the minimum total cost.

CRAFT does not guarantee the least-cost solution since all possible interchanges are not considered. The quality of final solution depends on the starting solution. Therefore, it is a common practice to specify a number of different initial layouts and try with all combinations of exchanges; then select the best final solution generated. For n input facilities with m initial layouts and pairwise exchange the at most possible number of layout is $m \times {}^nC_2$. We see that the number of alternatives is depended on the strategy of exchange and the interchanging condition.

These three algorithms compute the criterion value for all possible layout alternatives once they are generated. The number of computations depend on a number of layout alternatives. For large number of layout alternatives, the computational processing time can be reduced by doing the computations for each alternative in parallel. From the criterion formula shown in Table 1, distance, d_{ij}, is only a variable. The rest of them, w_{ij}, f_{ij} and d_{ij}, are parameter which have constant values for a certain period of time. By this manner, the systolic algorithm is suitable for improving this kind of algorithms because the systolic cell can store the loaded data inside the system for a sufficient length of time, so only one memory access is needed during whole computation time.

Algorithm	Criterion	Criterion formula
Siri&Thaj	Total exerted energy	$E = \sum_{i=1}^{n-1} \sum_{j=i+1}^{n} \frac{1}{2} w_{ij} (f_{ij} d_{ij})^2$
CLASS	Total material handling distance	$E = \sum_{i=1}^{n-1} \sum_{j=i+1}^{n} T_{ij} d_{ij}$
CRAFT	Total cost	$TC = \sum_{i=1}^{n} \sum_{j=1}^{n} f_{ij} d_{ij}$

Table 1 Criteria formula.

2.2 NON-COMPUTATIONAL APPROACH

The algorithm of this type does not deal with a difficult mathematics computation or with a great deal of computations. The example is ALDEP.

Automated Layout DEsign Program (ALDEP: SEEHOF and EVANS, 1967) designs the facility layout by selecting the first facility at random and placing the rest by considering both the activity relationship (REL) chart and the numerical weighted rating assigned to the closeness rating. The REL chart (MUTHER, 1973) defines the closeness rating of all pairwise combinations of relationship between patterns. The best layout design is the one which has the less score. The score of the layout is derived from the closeness rating.

ALDEP generates only one final layout at a time. To get more alternatives, user should repeat the layout generation process. A different final layout can be made by changing the first located facility. Therefore, a number of all possible alternatives can be determined by a number of input facilities. The efficiency of this algorithm can be improved by the systolic processing, to generate all possible layout alternatives in a processing time and compute all scores in parallel. In doing so, the probability of getting the best layout becomes higher.

3. DESIGNED SYSTOLIC ALGORITHM FOR FACILITY LAYOUT ALGORITHM

From analyzing the process of facility layout algorithms, it is found that their efficiencies can be improved by applying the systolic method to both computational and non-computational approach (NARUE-DOMKUL, 1994). Systolic method can decrease the computational processing time of the first group and increase the probability of getting the best layout for the second group.

3.1 DESIGNED SYSTOLIC ALGORITHM FOR COMPUTATIONAL APPROACH

The facility layout algorithm is considered to be a hard problem in the sense that it requires a great deal of computational time to obtain a good optimal solution. The computational time increases in exponential order when the number of facilities increases. Therefore, this paper proposed the improvement of processing time by using the systolic array processor. In doing so, the computational formulas in this algorithm have been redesigned to suit the systolic implementation as shown in Table 2.

Algorithm	Original criterion formula	Modified criterion formula
Siri&Thaj	$E = \sum_{i=1}^{n-1} \sum_{j=i+1}^{n} \frac{1}{2} w_{ii} (f_{ii}d_{ii})^2$	$E_k = \sum_{i=1}^{n-1} \sum_{j=i+1}^{n} c_{ii}D_{ii.k}$
CLASS	$E = \sum_{i=1}^{n-1} \sum_{j=i+1}^{n} T_{ii}d_{ii}$	$E_{lk} = \sum_{i=1}^{n-1} \sum_{j=i+1}^{n} T_{ii}d_{ii.lk}$
CRAFT	$TC = \sum_{i=1}^{n} \sum_{j=1}^{n} f_{ii}d_{ii}$	$TC_k = \sum_{i=1}^{n} \sum_{j=1}^{n} f_{ii}d_{ii.k}$

Table 2 modified formula for systolic algorithm.

The general form for computing the criterion value of these three algorithms is as follows:

$$out_k = \sum_{i=1}^{n-1} \sum_{j=i+1}^{n} in_{ij}d_{ij,k}$$

where out_k output number k
 in_{ij} characteristic value of between facility i and facility j
 $d_{ij,k}$ distance between facility i and facility j for layout alternative k
 n number of input facilities.

The systolic design for this computation is illustrated in Figure 1. In this case $d_{1j,k}, d_{2j,k}, d_{3j,k}, ...$ for each stream moves systolically to different cells. Each in_{ij} is preloaded and stays inside every cell. For each k, out_k is an output which is produced from the last cell on the right in the same cycle. The multiplication and addition are performed at all cells simultaneously.

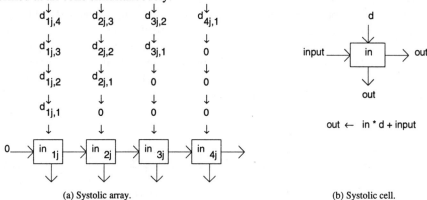

(a) Systolic array.   (b) Systolic cell.

Figure 1 Systolic design for energy calculation.

The input/output are defined as follows:

    Input: $d_{ij,k}$ distance between facility i and facility j for layout alternative k
                   in_{ij} characteristic value of between facility i and facility j
    where   n   number of facilities.
    Output: out_k criterion value of layout alternative number k.

In computing the criterion value of each algorithm, the distance calculation is required. To illustrate the distance computation, the Siri&Thaj algorithm is selected as an example. Let d_{ij} be a distance between central points of pattern i and pattern j and it is defined as:

$$d_{ij} = [(x_i-x_j)^2 + (y_i-y_j)^2]^{1/2}$$

where (x_i, y_i) and (x_j, y_j) are central points of pattern i and pattern j respectively.

In computing the energy E_{ij}, d_{ij}^2 is required. Therefore, d_{ij}^2 is computed directly to decrease the amount of computations. In doing so, d_{ij}^2 is denoted by D.

$$D = d_{ij}^2 = (x_i-x_j)^2 + (y_i-y_j)^2$$

The distance computation needs to know the central point of each located facility. By the layout generation of Siri&Thaj, the central point formula is shown in Table 3.

At point (x_2,y_1)	At point (x_2,y_2)	At point (x_1,y_2)	At point (x_1,y_1)
$(u_1,v_1)=(x_2+l,y_1+w)$	$(u_3,v_3)=(x_2+w,y_2-l)$	$(u_7,v_7)=(x_1+w,y_2+l)$	$(u_{13},v_{13})=(x-w,y_1+l)$
$(u_2,v_2)=(x_2+w,y_1+l)$	$(u_4,v_4)=(x_2+l,y_2-w)$	$(u_8,v_8)=(x_1+l,y_2+w)$	$(u_{14},v_{14})=(x_1-l,y_1+w)$
$(u_{15},v_{15})=(x_2-w,y_1-l)$	$(u_5,v_5)=(x_2-l,y_2+w)$	$(u_9,v_9)=(x_1-l,y_2-w)$	$(u_{15},v_{15})=(x_1+l,y_1-w)$
$(u_{16},v_{16})=(x_2-l,y_1-w)$	$(u_6,v_6)=(x_2-w,y_2+l)$	$(u_{10},v_{10})=(x_1-w,y_2-l)$	$(u_{16},v_{16})=(x_1+w,y_1-l)$

Table 3 formula for central point calculation.

Definitions of variables :

    (u_j,v_j) a central point coordinate of Pattern B at position j
    (x_1,y_1) a bottom-left rectangular coordinate of fixed Pattern A
    (x_2,y_2) a top-right rectangular coordinate of fixed Pattern A
    l   half length of to-be-located Pattern B
    w   half width of to-be-located Pattern B

We can have similar way to desing the systolic array and its cell definition for computing the distance and central point as we did it for energy caculation.

3.2 DESIGNED SYSTOLIC ALGORITHM FOR NON-COMPUTATIONAL APPROACH

While the layout generation is being processed, the generated table is also developed. This generated table shows the relative position of patterns in the designed layout. From the generated table, all possible combinations of adjacent patterns will be obtained. The layout score is defined by the twice of summation of all closeness ratings which associate with each of these combinations. To illustrate this computation, consider the sample problem involving 10 departments:

The combinations of adjacent patterns and the closeness ratings associate with these combinations are:

10-3,	10-4,	10-2,	10-1,	1-9,	1-2,	2-9,	2-3,	2-4,	4-3,	3-9,	9-6,	9-8,	9-7,	7-8,	8-5,
0,	1,	1,	16,	0,	0,	4,	0,	1,	4,	64,	0,	16,	64,	0,	4,

then the layout score = 2 * (0+1+1+16+0+0+4+0+1+4+64+0+16+64+0+4) = 350

The main computation of this algorithm is score computation. This computation has been written in the form which is suitable for systolic implementation as follow:

$$hscore_k = \sum_{i=1}^{n} c_{ki} \quad ; i \neq k$$

where $hscore_k$ half-score of layout candidate k
  c_{ki} closeness rating of combination i for layout candidate k
  N   number of input patterns.

The systolic design for this computation is similar with the energy caculation.

3.3 COMBINATION OF SYSTOLIC CELLS

From figure 2(a), the designed systolic cell for energy calculation has two inputs and two outputs. Figure 2(b) shows the systolic cells for central point finding which has one input and two inputs. Figure 2(c) is systolic cell for distance calculation which has two inputs and one output. Finally, Figure 2(d) shows the systolic cell for layout score calculation which has two inputs and one output. From these systolic cells a single systolic cell can be designed to combine all needs of these systolic cells for their inputs and output as shown in Figure 2(e).

4. CONCLUSION

This study proposes the systolic design for facility layout problem. Five algorithms have been selected as case studies. The systolic cell of these algorithms can be designed in one architecture. Each main computational part of each algorithm has proved that it can be done in parallel by using the proposed systolic design. Normally, the facility layout algorithms use the almost the same criteria in selecting the best layout. So, the main computational part for each algorithm is the computation of interrelationship cost which is including central point (or centroid) finding, distance calculation and energy (or total cost) calculation. Moreover, the selection of to-be-located pattern, which are including the closeness weight or closeness rating computation, is also important. Therefore, any facility layout algorithm which is based on such a computation algorithm or a computation algorithm of a similar formula even with different parameters can use these designs.

By using systolic algorithm, it not only reduces the processing time in Sirinaovakul and Thajchayapong, CRAFT and CLASS algorithm, but also produces a number of alternative layouts in ALDEP algorithm. As mentioned earlier, each processing time of ALDEP produces only one alternative layout. By using systolic algorithm, ALDEP produces more alternative layouts in each processing time.

This systolic algorithm is significantly faster than the operations required by algorithms performed on a sequential machine for a large number of input patterns. Since the nature of the systolic algorithm is the parallel processing, the improvement of the processing time depends on the number of the systolic cells. The more systolic cells there are, the faster the processing speed is. Unfortunately, the number of the systolic cells is a cost of the implementation. Therefore, the number of the parallel systolic cells has to be considered under the cost-effectiveness and the optimization of the processing time constraints. In the designed systolic algorithm, 8 processors are proposed. This number is estimated from the possibility of the average number of facilities to be laid out, so it can speed up the caculations 8 times then the normal algorithm.

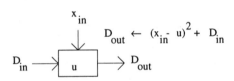

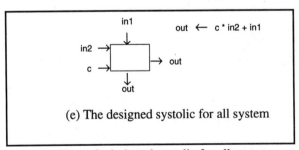

Figure 2 designed systolic for all system.

REFERENCES
1. BUFFA, E. S. ARMOUR, G.C. and VOLLMANN, T.E. (1964), Allocating Facilities with CRAFT. Harvard Business review, Vol. 42, pp. 136-157.
2. FORTES, J. A. B. FU, KING-SUN. and WAH, B. W. (1983), Systematic Design Approaches for Algorithmically Specified Systolic Arrays. In Milutinovic, V. M., editors, Computer Architecture, North-Holland.
3. JAJODIA, S. MINIS, I. HARHALAKIS, G. and PROTH, J. (1992), CLASS: Computerized Layout Solutions Using Simulated Annealing, International Journal of Production Research, Vol. 30, pp. 95-108.
4. KUNG, H.T. (1982), Why systolic Architecture. Computer, Vol. 15, pp. 37-46.
5. MUTHER, R. (1973), Systematic Layout Planning, Missouri: Management and Industrial research.
6. NARUE-DOMKUL, K (1994), A Systolic Algorithm for Improving a Computational Processing Time, Master's thesis, AIT, Bangkok.
7. SEEHOF, J.M. and EVANS, W.O. (1967), Automated Layout Design Program. Journal of Industrial Engineering, Vol. 18, pp.690-695.
8. SIRINAOVAKUL, B. and THAJCHAYAPONG, P. (1994), A Knowledge Base to Assist a Heuristic Search Approach to Facility Layout. International Journal of Production Research, Vol.32, pp. 141-160.

AN ANALYSIS OF COMPUTER AIDED FACILITY LAYOUT TECHNIQUES

BOONCHAROEN SIRINAOVAKUL
Faculty of Engineering, King Mongkut's Institute of Technology Ladkrabang
Ladkrabang, Bangkok 10520, Thailand.
E-mail: boon@cc.kmitt.ac.th
and
PAIRASH THAJCHAYAPONG
Faculty of Engineering, King Mongkut's Institute of Technology Ladkrabang
Ladkrabang, Bangkok 10520, Thailand.
E-mail pairash@nwg.nectec.or.th

ABSTRACT

This paper presents a new approach for classifying the facility layout techniques. The classification is based on the analysis of facility layout process as layout improvement, entire layout and partial layout model. This classification contributes two important concepts. First, it gives the idea of how the facility layout algorithm can be constructed; and second, it provides the idea of how the quality of facility layout algorithm can be improved. The previous researches of facility layout algorithms are also analyzed to support the classification concept.

1. Introduction

Computerized facility layout algorithms are characterized by one of the two distinct semantics modes; (1) they construct the layout by building up a solution from initial layout, or (2) they construct the layout by building up a solution from scratch. These semantic models are called *improvement algorithm* and *construction algorithm* respectively. In the later classification, two more models are added; hybrid and graph theory algorithm (Kusiak and Heragu 1987). The hybrid algorithm generates the layout from scratch and improves it by regeneration. The graph-theory algorithm generates the layout by employing graph-theory.

This paper presents a new method for classifying the facility layout algorithm which is done based on the analysis of facility layout process structure as *layout improvement*, *entire layout* and *partial layout* model. The presentation of the paper is divided into; analysis of computer aided facility layout techniques, facility placing order, facility allocation and alternative selection strategy. Few well-known algorithms are also presented as examples. The conclusion and comment are given in the conclusion remark section.

2. Analysis of Computer Aided Facility Layout Techniques

By analyzing the facility layout algorithm, it is found that there are three major processes; *facility placing order, facility allocation* and *alternative selection* process. The facility placing order determines the ordering of facilities to be located onto the layout area. The facility allocation process allocates facilities onto the layout area and, for some models, generates layout alternative(s). The alternative selection process selects the best layout from the generated alternative(s). However, not all algorithms consist all of these processes. Some algorithms have only two processes. From the different components of processes in constructing the layout algorithm, the facility layout can be classified as *layout improvement*, *entire layout* and *partial layout* model.

Table 1 shows the components of facility layout processes for each model.

Table 1 Analysis of facility layout model.

Layout model	Facility placing order	Facility allocation	Alternative selection
Layout improvement		O	O
Entire layout	O	O	
Partial layout	O	O	O

The *layout improvement* model uses *facility allocation process* to improve the existing layout by relocating the facilities to generate the new alternative layout. Then, the *alternative selection process* tests the newly generated alternative. If it gives the better cost, then it is selected as an improved layout and used as the initial layout for another improvement iteration. Otherwise the algorithm generates another alternative. The process is repeated until no further improvement can be made. The examples are CRAFT (Buffa, Armour and Vollmann 1964), and Moore's algorithm (Moore 1976).

The *entire layout* model selects and places a seed facility onto the given layout area. Then, another facility is selected by *facility placing order process*, one at a time, and placed by *facility allocation*. The processes are repeated until all facilities are located. The examples are the algorithms proposed by (Fortenberry and Cox 1985 Hassan and Hogg 1991).

In the *partial layout* model, the *facility placing order* selects a to-be-located facility and locates it to all possible locations, by *facility allocation*, to generate partial layout alternatives. Then, *alternative selection* selects the best partial layout alternative and adds another facility onto it by repeating the same processes until all facilities are located. The examples are CORELAP (Lee and Moore 1967) and Sirinaovakul and Thajchayapong's algorithm (Sirinaovakul and Thajchayapong 1994).

3. Facility Placing Order Strategy

The strategy is applied for *entire layout* and *partial layout* model. It considers the basis for the order in which the facilities enter the layout. The purpose of this strategy is to limit the problem space of the layout algorithm which means that the strategy reduces the search for a good layout effort. The followings are some possible strategies:

- Among all unallocated facilities, the next facility to-be-placed is selected at random. ALDEP (Seehof and Evans 1967) use this strategy.
- Among all unallocated facilities, the facility that has the highest weight relationship with the located facilities is selected. The examples are CRAFT and CORELAP.
- Among all unallocated facilities, the facility that has the highest weight relationship with the located facilities is selected. If there is no highest relationship weight occurred, the facility with highest area is selected. Sirinaovakul and Thajchayapong's algorithm employ this strategy.

4. Facility Allocation Strategy

According to the facility layout model, there are three strategies for allocating facilities onto the layout area. The facility allocations for layout improvement, entire

4.1 Layout Alternative Generation

The layout alternative generation is done by interchanging the facility location of the initial layout. Normally, the methods are designed for layout improvement model. CRAFT is the first model to present the way of generating the layout alternative. It divides the initial layout into an array of matrix. There are three required conditions for exchange; the facilities with common borders, the facilities with same area and the fixed out line of layout area. An exchange involving the greatest cost reduction is made and is used as a new layout. Figure 1 shows the pattern allocation of CRAFT algorithm.

```
A A B B C C C C C C      C C B B A A A A A A      A A B B D D D D D D
A A B B C C C C C C      C C B B A A A A A A      A A B B D D D D D D
A A B B C C C C C C      C C B B A A A A A A      A A B B D D D D D D
A A A A D D D D D D      C C C C D D D D D D      A A A A C C C C C C
A A A A D D D D D D      C C C C D D D D D D      A A A A C C C C C C
A A A A D D D D D D      C C C C D D D D D D      A A A A C C C C C C
       (a)                      (b)                      (c)
```

Figure 1 (a) initial layout (b) facilities A and C exchange by the same area condition and (c) C and D exchange by the common border condition.

Another technique used in layout alternative generation is graph theory layout which is also done by exchanging of facility location. Cutting Point Interchange Approach (Oten 1982) uses four operators to represent the slicing structure, "L" (left cut), "R" (right cut), "U" (up cut) and "B" (bottom cut).

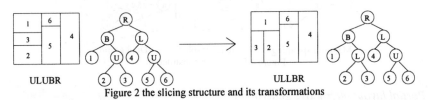

Figure 2 the slicing structure and its transformations

Figure 2 shows the slicing structure and its slicing tree and the structure transformation. The slicing structure can be represented by slicing tree in short form as 65U4L32U1BR or ULUBR. The representation of this form is similar to the way operators and operands are stated in post order arithmetic expressions. The algorithm relocates the facilities by changing the two opposite operators along each dimension (e.g. "L" and "R", "U" and "B"). The change of operator means that the facility location is relocated.

4.2 Pattern allocation

This approach generates the layout by locating a facility to the unique position which is considered from the relation between a to-be-located facility and located facilities. No alternative is generated for this approach. ALDEP places a facility onto grid array of matrix. It uses a vertical scan routine and places the facilities in the layout in a manner analogous to placement of strips of tape from left to right. The placement of this model must specify the length, width, and area requirements for each facility. Then, the unit area for each grid is calculated from the required area of all facilities. Figure 3 shows the example of facility allocation using ALDEP algorithm.

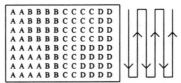

Figure 3 ALDEP facility allocation method.

Planar graph approach (Seppanen and Moore 1970) utilizes the properties of planar graph. The facility areas, in this algorithm, are considered as the vertices of the planar graph and then the layout is constructed by the dual graph created from this planar graph. The algorithm consists of three steps. First, the algorithm creates the maximal spanning tree of the facilities from the given relationship weights. Second, the high relationship weight edges are added to the maximal spanning tree to construct the maximal planar graph. Lastly, the dual graph is constructed from this maximal planar graph. This dual graph is considered to be the output of the facility design. The planar graph algorithm was later developed and called Detahedron Heuristic (Foulds and Robinson 1978 and Hassan and Hogg 1991). Figure 4 is the planar graph algorithm.

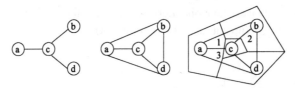

Maximal spanning tree   Planar graph   Dual graph
Figure 4 Dual graph construction

4.3 Partial layout alternative generation

This technique allocates a facility onto all possible locations. When a facility is placed, the alternatives are generated, called partial layout alternatives, by locating a to-be-located facility to all possible location. Sirinaovakul and Thajchayapong's algorithm is an example of this technique.

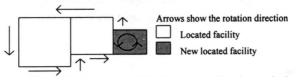

Figure 5 Sirinaovakul and Thajchayapong allocation method.

The process generates the partial layout alternative by locating a to-be-located facility onto a side of located facilities. This combined area is proposed as a first alternative. Next, a to-be-located facility is rotated around itself and placed so that its perpendicular side adjoins the same side of located facility and this combined area is also proposed as an alternative. Then, a to-be-located facility moves to another side of located facilities. The method repeats the above procedures until all possible combinations (or alternatives) of these facilities have been considered. Figure 5 shows the facility allocation method.

5. Alternative Selection

Normally, the alternative selection process is the heuristic search which is composed of heuristic search and objective function. The heuristic search searches for a good alternative and the objective function guides the search direction.

5.1 Heuristic Search

Three most popular heuristic searches are presented in this section.; hill climbing, beam search and simulated annealing.

Hill climbing method generates a set of alternatives and evaluates these alternative choices by using objective function. An alternative that appears to be the better solution is then chosen. No further reference to parent of other alternatives is retained. This process continues from node-to-node with previously expanded nodes being discarded. An example algorithm that uses hill climbing as a search process is CRAFT.

Beam search generates all possible alternatives for each level and the alternatives are evaluated by objective function. The n most promising alternatives are selected for further expansion. When n, called beam width, is an adjustable number. This process continues from level-to-level until no further alternatives can be generated. The examples of algorithm using beam search are AILAY (Shih, Enkawa and Itoh 1992) and Sirinaovakul and Thajchayapong algorithm.

Simulated annealing is a variation of hill climbing in which, at the beginning of the process, some downhill moves may be made. The idea is to do enough exploration of the whole space. Its computational process is patterned after the physical process of annealing. There is a probability that a transition to the higher energy state will occur. This probability is given by Eq. (1);

$$p = e^{-\Delta E/kT} \tag{1}$$

Where, ΔE is the positive change in the energy level, T is the temperature and k is Boltzmann's constant. In the algorithm, the moves that worsen the current solution by an amount, E, are accepted with probability p. T is a control parameter analogous to temperature in the annealing of physical systems. The examples of facility layout algorithm using simulated annealing are CLASS (Jajodia, Minis, Harhalakis and Proth 1992) and Tam algorithm (Tam 1992).

5.2 Objective Function

On searching for the best alternative, along the way the objective function calculates the costs in consideration of all alternatives. The following topics are some popular objective functions.

Total cost is defined as:

$$TR = \sum_{i=1}^{n} \sum_{j=1}^{n} f_{ij} d_{ij} \qquad (2)$$

Where n is the number of facilities, f_{ij} is the flow data between facilities i and j and d_{ij} is the distance between facilities i and j. The flow data are expressed in terms of the number of trips per time period. Therefore, TC, total cost, reflects the total material-handling cost required to move one unit of distance between combination of departments. The examples of algorithm using total cost are CRAFT and CLASS.

Total closeness rating uses closeness relationships for considering a pair of facilities for every combination. The vowel-letters are enumerated as fixed numerical values. Then the total cost is calculated by:

$$TCR = \sum_{i=1}^{n} \sum_{j=1}^{n} c_{ij} \qquad (3)$$

Where c_{ij} is the numerical value of the relationship weight rating between facilities i and j and n is the number of facilities. The examples of algorithm using total cost are ALDEP and CORELAP.

Multiple criteria defined by Fortenberry and Cox is

$$\min Z = \sum_{i=1}^{n} \sum_{j=1}^{n} \sum_{k=1}^{n} \sum_{l=1}^{n} a_{ijkl} b_{ijkl} X_{ij} X_{kl} \qquad (4)$$

subject to

$$\sum_{i=1}^{n} X_{ij} = 1, j = 1, 2, .. n \quad \text{and} \quad \sum_{j=1}^{n} X_{ij} = 1, i = 1, 2, ...n \qquad (5)$$

X_{ij} equals to 1 if facility i is assigned to location j or equals to 0 otherwise. The value of $a_{ijkl} = f_{ik} d_{jl}$. Where d_{jl} is a distance from location j to location l, f_{ik} is work flow from facility i to facility k, and b_{ijkl} (or r_{ik}) is vowel-value closeness weight of facility i and k.

There are two more alternative formulations. First, the formulation presented by (Rosenblatt 1979) is

$$\min Z = \sum_{i=1}^{n} \sum_{j=1}^{n} \sum_{k=1}^{n} \sum_{l=1}^{n} (a_2 a_{ijkl} - a_1 w_{ijkl}) X_{ij} X_{kl} \qquad (7)$$

Where a_1, a_2 and w are weight assigned to total material handling cost, total rating score and relationship cost respectively. The other formulation was proposed by (Dutta and Sahu 1982) is

$$\min C' = W_2 C - W_1 R \qquad (8)$$

Where W_1, W_2, C and R are weights assigned to closeness rating score, material handling cost, material handling objective function and closeness rating objective function.

Potential energy is defined as a potential field of two blocks that produces a force tending to separate them, when two blocks overlap. On the other hand, this force is tending to pull the two blocks closer when two connected blocks are separated. Without separation or overlap, blocks are said to be in equilibrium condition (Sirinaovakul and Thajchayapong 1994)

$$E_{ij} = 1/2*(w_{ij}*(d_{ij})^2) \quad (9)$$

Where w_{ij} is the closeness weight between facilities i and j, E_{ij} is the energy done by an external force in moving the facility from equilibrium position to d_{ij} which is obeying Hooke's law and d_{ij} is the length from the center of facility i to the center of facility j.

The alternative model of this approach is attractive force (Tam & Li 1991). The formulation is:

$$f_{ij} = 1/2*(w_{ij}*(d_{ij})^2) + \alpha \min\{0, d_{ij} - r_i - r_j\} \quad (10)$$

Where w_{ij}, d_{ij}, r_j, r_j and α are closeness weight, distance, radius of i, radius of j and coefficient of overlapping degree respectively.

6. Concluding Remarks

This paper has presented a new approach for analyzing the facility layout algorithms. Three major processes, facility placing order, facility allocation and alternative selection, are described. Based on these processes, the layout algorithms are classified as layout improvement, entire layout generation and partial layout generation algorithm. The results from this classification makes the researchers understand the facility layout programs structure and know the way to improve them. The comments for further research will be made on three points as follows:

1. The quality of the final layout can be measured by the number of generated alternatives in the algorithm. The more alternatives are generated, the better final layout can be carried out. On the other word, the higher number of alternatives provides the higher possibilities to result the better solution.

2. Efficiency of layout algorithms for each classification can be considered as;

The efficiency of layout improvement model depends on the number of iterations since the model generates only one alternative for each iteration. Therefore, more alternatives can be created by generating more iterations.

The final layout of entire layout model can be improved by changing the seed facility and the related parameters and then regenerating the final layout. This final layout can also be improved by using layout improvement algorithm.

The efficiency of the partial layout model depends on the alternative selection process since enough alternatives have already been generated. However, the problem of partial layout model is the speed of the processing time. The improvement of the model efficiency can be done by either improving the heuristic search performance or increasing the processing speed.

3. The processing speed of facility layout algorithm can be increased by systolic algorithm. The systolic algorithm not only reduces the processing time in facility layout

algorithm, but also increases the number of alternative layouts for the entire layout generation model (Narue-domkul 1994).

References

Buffa, E. S., Armour, G.C., and Vollmann, T.E., *Allocating facilities with CRAFT*, Harvard Business Review, **42**, (1964) 136-157.

Dutta N. K. and Sahu, S., *A multigoal heuristic for facilities design problem: MUGHAL*, International Journal of Production Research, **20**, (1982) 147-154.

Fortenberry, J. C. and Cox, J. F., *Multiple criteria approach to facilities layout problem*, International Journal of Production Research, **23**, (1985) 773-782.

Foulds, L. R. and Robinson, D. F., *Graph theoretic heuristics for the plant layout problem*, International Journal of Production Research, **16**, (1978) 27-37.

Hassan M. M. D. and Hogg, G. L., *On Constructing a Block Layout by Graph Theory*, International Journal of Production Research, **6**, (1991) 1263-1278.

Jajodia, S., Minis, I., Harhalakis, G. and Proth, J., *CLASS: Computerized Layout Solutions Using Simulated Annealing*, International Journal of Production Research, **30**, (1992) 95-108.

Narue-domkul, K., *A systolic algorithm for improving a computational processing time*, (Unpublished MS thesis: 1994), Computer Science Division, Asian Institute of Technology, Bangkok, Thailand.

Kusiak, A. and Heragu, S. S., *The facility layout problem*, European Journal of Operational Research, 29, (1987) 229-251.

Lee R. C. and Moore, J. M., *CORELAP-computerized relationship layout planning*, Industrial Engineering, **18**, (1967) 195-200.

Moore, J. M., *Facilities design with graph theory and strings*, Omega, **4**, (1976) 193-203.

Otten, R. H. J. M., *Automatic floorplan design,* Proc. 19th ACM/IEEE Design Automat. Conf., (1982) 261-267.

Rosenblatt, M. J., *The facilities layout problem: a multi-goal approach*, International Journal of Production Research, **17**, (1979) 323-332.

Seehof, J. M. and Evans, W. O., *Automated layout design program*, Journal of Industrial Engineering, **18**, (1967) 690-695.

Seppanen, J. and Moore, M. J., *Facilities planning with graph theory*, Management science, **17**, (1970) B242-B253.

Shih, L. C., Enkawa, T. and Itoh, K., *An AI-search technique-based layout planning method*, International Journal of Production Research, **30**, (1992) 2839-2855.

Sirinaovakul, B. and Thajchayapong, P., *A Knowledge Base to Assist a Heuristic Search Approach to Facility Layout,* International Journal of Production Research, **32**, (1994) 141-160.

Tam, K. Y. and Li, S. H., *A hierarchical approach to the facility layout problem,* International Journal of Production Research, **29**, (1991) 165-184.

Tam, K. Y., *A Simulated Annealing Algorithm for Allocating Space to Manufacturing Cells*, International Journal of Production Research, **30**, (1992) 63-87.

CAD DATA EXCHANGE

CAD DATA EXCHANGE

Visualizing STEP/EXPRESS Models Based on
Aggregation/Inheritance Hierarchies
   S. C. Hui, A. Goh and B. Song

Support for Overlapping Models in an Enterprise Data Model
   S. C. F. Chan, P. K. S. Tong and J. W. T. Lee

On the Implementation of EXPRESS Information Models onto
Versant OODB
   Q. Z. Yang and B. Song

A Reference Architecture for Information Sharing in Collaborative
Engineering Environments
   D. Domazet, D. Sng, F. N. Choong and S. Sum

PROSTEP- An Initiative of the Automotive Industry for Introducing
STEP into Industrial Applications
   M. Holland and D. Trippner

Visualizing STEP/EXPRESS Models Based on Aggregation/Inheritance Hierarchies

S.C. Hui*, A. Goh* and B. Song[†]

*Division of Computer Engineering, School of Applied Science
Nanyang Technological University, Singapore 2263
[†]GINTIC Institute of Manufacturing Technology
Nanyang Technological University, Singapore 2263
Email: asschui@ntuvax.ntu.ac.sg

ABSTRACT

STEP is a standard to support sharing and exchange of product design. EXPRESS is used to describe the product model within STEP. The EXPRESS models are difficult to design and validate. We present a visualization tool, called ExpressView, which helps EXPRESS designers to visualize and fine tune their models. The tool will aid the designer in two aspects: (i) to apply the concepts of inheritance to reduce the storage costs due to data duplication; (ii) to visualize the aggregation hierarchy to avoid difficult object retrievals. In this paper, we describe a STEP/EXPRESS environment on which the ExpressView is based. The graphical interface of the ExpressView is then presented. An example schema is used to illustrate the concepts of the ExpressView visualization tool.

1. Introduction

STEP (the STandard for the Exchange of Product model data) [1,2,3] is a standard to support sharing and exchange of product design. STEP is valuable for product databases, data exchange and concurrent engineering [4]. Part of the STEP is a data definition language called EXPRESS [5]. It is used to describe the product model. It not only defines the entities, but also specifies constraints on attributes, permissible operations on entities and so on. Users can define their product databases using EXPRESS. In view of the complexity of the products being handled, designing EXPRESS models can be difficult. The availability of design tools to assist product designers is important.

At present, no database systems support the EXPRESS model directly. However, the EXPRESS model could be translated and mapped onto an object-oriented database (ODBMS) or a relational database (RDBMS). This paper will focus on the design and development of a visualization tool to display the aggregation and inheritance hierarchy relationships of EXPRESS models. The visualization tool will help EXPRESS designers to visualize and fine tune his models. The tool will provide feedback to the designer so that they can decide if the EXPRESS model reflects his intentions accurately. If not, the model can be modified to capture the application semantics. Moreover, the tool will help the designer to apply the concepts of inheritance while viewing all the entities and their attributes/functions/constraints in the EXPRESS model. If the entity hierarchy is deep and narrow, then the EXPRESS model should be modified to reduce the depth in order to reduce the storage costs due to inheritance and data duplication. On the other hand, if too many complex objects are displayed, then an alternative representation of the EXPRESS model should be studied to avoid deep nesting of objects and thereby avoiding difficult retrievals.

This paper presents the visualization tool, known as ExpressView, to view the internal structure of STEP/EXPRESS models. This tool has been demonstrated to be very useful as a design aid for EXPRESS designers. Section 2 describes a STEP/EXPRESS environment. The visualization goals are discussed in Section 3. An EXPRESS schema example is given in Section 4. Section 5 presents the graphical interface of the ExpressView visualization tool. Section 6 discusses the implementation issues. Finally, a conclusion is given in Section 7.

2. STEP/EXPRESS

2.1 A STEP/EXPRESS Environment

There is a considerable overlap between EXPRESS features and object-oriented database (ODBMS) features. As the EXPRESS model is basically an object-oriented one, there is a much closer correspondence between the EXPRESS models and ODBMSs. The lack of constructs to model and manipulate complex data types in traditional database systems leads to poor support of CAD systems. It is anticipated that interactive querying into complex EXPRESS models places a heavy load on relational technology, resulting in deficiency in performance [6,7]. On the other hand, object-oriented systems promote rapid prototyping, high productivity, low rewrite [8]. Thus, one would expect the natural choice of database model for storing the EXPRESS model to be object-oriented, rather than relational or physical files.

In many C++ based ODBMSs, entities are mapped onto C++ classes, with attributes implemented as protected members. Inheritance is a property of C++ and therefore EXPRESS super and sub-typing are supported. Simple, aggregation, enumeration and defined data types are similarly supported. However, due to the richness of the EXPRESS language, some incompatibilities are likely to arise between EXPRESS and the target ODBMS. What this means is that there are certain features in EXPRESS which are not easily translated into the ODBMS. Since the translation of EXPRESS models into ODBMS is not straight forward, some translation tools [9,10] have been constructed to facilitate the translation process.

Figure 1 shows the EXPRESS model and its relationship with applications, STEP Data Access Interface (SDAI) [11], the translation tool and the visualization tool. Users can define their databases using EXPRESS. Visualization tool is provided to help the user to view the internal structure of the EXPRESS model. Once the EXPRESS model is defined, it is translated and mapped directly into an ODBMS or a RDBMS by the translation tool. SDAI is a proposal within STEP to provide a consistent set of access mechanisms regardless of the underlying data storage. Any application whose data is modeled in EXPRESS can utilize SDAI, thereby making the actual database transparent. The SDAI supports functions that are to be used by an application to access both data and metadata of the database. The user can manipulate the databases using SDAI and exchange data with other applications.

2.2 The EXPRESS Language

The EXPRESS language manual emphasizes that EXPRESS is not a programming language [5]. There is no EXPRESS compiler nor an EXPRESS executable code. The language itself resembles a conglomeration of different languages including ADA, C and SQL. As it is basically a data definition language, it does not contain any elements which support data manipulation in the form of input/output or data processing. It does, however, contain a rich set of constructs to enable an unambiguous object definition with constraints on the data to be clearly and concisely presented. These features make EXPRESS understandable by both humans and computers.

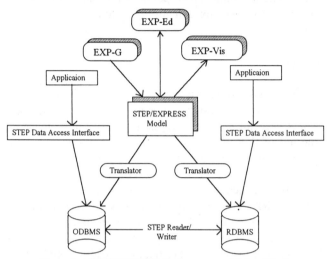

Figure 1. A STEP/EXPRESS Environment and Related Tools

EXPRESS enables a schema to be declared. The schema would consist of entities which are related objects of interest and data types to support the definitions of these entities. Encapsulated within the entities are attributes and constraints which restrict the values that these attributes may take. EXPRESS also supports functions, procedures and global rules.

Apart from the above constructs, two important hierarchy relationships, inheritance (supertypes and subtypes) and aggregation (entity data types), are also supported. Inheritance comes from the concept of generalization in semantic models. The classes are hierarchically organized. The generalization gathers the common properties of several classes (called subtypes) in a more general class (called supertype). The generalization expresses the "is a" relationship between a subtype and a supertype. Therefore, every instance of a subtype is also an instance of its supertype. The extension of a subtype is included in the extension of its supertype.

Each attribute of an entity is defined with a data type. Types are used to represent the domain of the attribute. The value of an attribute is an object owned by its domain. This value is the identity of the object if the domain is complex (when defined as entity data type) or it is the value of the object itself if the domain is atomic (when defined as simple type or aggregation data type). The first situation denotes the object as complex object. An example is given below to illustrate the above concept.

Example:

    ENTITY circle;
        center_point : POINT;
        radius : REAL;
END_ENTITY;
ENTITY point;
        x,y,z : REAL;
END_ENTITY;

The attribute *radius* of the entity is defined on the type *REAL* which is its domain, called atomic domain. The value of this attribute is the real value corresponding to the radius of the circle. The attribute *center_point* of the entity *line* is defined on *point* which is its domain, called complex domain. The value of this attribute is the object identifier of an object *point*.

3. Visualization goals

The visualization tool ExpressView can be used as a design tool for EXPRESS designers. It helps to display and design aggregation/inheritance hierarchy relationships of EXPRESS models. Structuring of supertypes/subtypes in inheritance hierarchy can be achieved with the aid of this tool. On the other hand, it can also help to design an EXPRESS model with better performance. One performance factor for EXPRESS database and thus the ODBMS is the storage overhead incurred by the objects and its indices. Another factor is related to the retrieval operations of objects from the database.

Logically, an EXPRESS database is arranged as a directed acyclic graph (DAG). The number of objects per entity and the distribution of objects over the DAG are contributed to the storage overhead. A simulation analysis carried out in [12] suggests that increasing the depth of the DAG rather than its breadth would bring higher storage costs due to inheritance and duplication of data. The study shows that the shape of the DAG has a significant impact on storage requirements. Issues addressed include the distribution of the objects and where the attribute values of the inherited object should be stored. To reduce storage requirements, it is recommended that they should be pushed as low as possible in the hierarchy (i.e. in the home entity) or the data should be contained as high as possible (i.e. in the entity where the object first appeared). All these issues need to be considered when designing an EXPRESS model. Thus, being able to view the entity hierarchy along with the entity attributes will enable a better design of the EXPRESS model.

Retrieval of objects from the EXPRESS database is a costly operation. It is especially true when the objects themselves are complex objects. In this case, the nesting of an object through the domains of its attributes immediately suggests that in order to fetch a complete complex object instance, the instance and all objects it references through its attributes must be recursively fetched. This means that to fetch one or more instances of an entity, the entity and all entities specified as non-primitive domains of the attributes of the entity must be recursively traversed. Such nesting of objects and entities is difficult to see from the EXPRESS code, whereas if it were displayed graphically, such nesting becomes obvious. On observing such nesting, the designer may need to study alternative representations of his complex objects to avoid unnecessary nesting of the object through the domains of its attributes.

4. An Example EXPRESS Schema

The example shown in Figure 2 describes the geometric objects extracted and modified from the EXPRESS/STEP Part 42 [13]. EXPRESS/STEP Part 42 is prepared by the Sub-committee 4 (Industrial data and global manufacturing programming languages) of ISO Technical Committee 184 (Industrial Automation Systems and Integration). This Part 42 specifies the integrated resources used for geometric and topological representation. Their primary application is to allow representation of the shape or geometric form of a product model. The geometric objects (gco_item) selected include point, cartesian_point, direction, curve, conic, circle, ellipse, hyperbola and parabola.

```
ENTITY geo_item
SUPERTYPE OF (ONEOF
    (point, direction, curve));
END_ENTITY;

ENTITY point
SUPERTYPE OF (cartesian_point)
SUBTYPE OF (geo_item);
END_ENTITY;

ENTITY cartesian_point
SUBTYPE OF (point);
    coordinates : LIST [2:3] OF REAL;
DERIVE
    dim : INTEGER :=
        count_dimensions (coordinates);
END_ENTITY;

ENTITY direction
SUBTYPE OF (geo_item);
    direction_ratios : LIST [2:3] OF REAL;
DERIVE
    dim : INTEGER :=
        count_dimensions (direction_ratios);
WHERE
    WR1 : vector_magnitude (SELF) > 0.0;
END_ENTITY;

ENTITY curve
SUPERTYPE OF (ONEOF (line, conic))
SUBTYPE OF (geo_item);
END_ENTITY;

ENTITY line
SUBTYPE OF (curve);
    pnt : cartesian_point;
    dir : direction;
END_ENTITY;

ENTITY conic
SUPERTYPE OF (ONEOF
    (circle, ellipse, hyperbola, parabola))
SUBTYPE OF (curve);
    position : REAL;
END_ENTITY;

ENTITY circle
SUBTYPE OF (conic);
    radius : REAL;
WHERE
    WR1 : radius > 0.0;
END_ENTITY;

ENTITY ellipse
SUBTYPE OF (conic);
    semi_axis_1 : REAL;
    semi_axis_2 : REAL;
WHERE
    WR1 : semi_axis_1 > 0.0;
    WR2 : semi_axis_2 > 0.0;
END_ENTITY;

ENTITY hyperbola
SUBTYPE OF (conic);
    semi_axis : REAL;
    semi_imag_axis : REAL;
WHERE
    WR1 : semi_axis > 0.0;
    WR2 : semi_imag_axis > 0.0;
END_ENTITY;

ENTITY parabola
SUBTYPE OF (conic);
    focal_dist : REAL;
WHERE
    WR1 : focal_dist <> 0.0;
END_ENTITY;
```

Figure 2. An Example EXPRESS Schema

5. Graphical Interface

In this section, we discuss the graphical interface of the ExpressView. The EXPRESS schema can be displayed as a hypergraph. The nodes correspond to the entities, the links are sets of entities gathered according to super/sub-typing and simple/complex entity data types. A link joins only two nodes. We refer to the links as inheritance links or aggregation links.

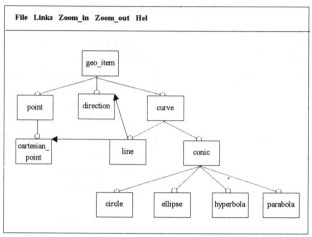

Figure 3. Graphical display for the example EXPRESS schema

Figure 3 shows the visualization of the EXPRESS schema from the example presented in Figure 2. The display presents entities as rectangles with the given entity names displayed within the rectangles. The entities are hierarchically organized. A supertype is connected to its subtypes through inheritance links. The inheritance link is denoted by a straight line with a circle ending at the subtype. In this schema, the *geo_item* entity is the supertype of entities *point*, *direction* and *curve*. The entity *point* is the supertype of entity *cartesian_point*. Similarly, the entity *curve* is the supertype of entities *line* and *conic*. And in turn, the entity *conic* is the supertype of entities *circle*, *ellipse*, *hyperbola* and *parabola*. In inheritance, the properties (attributes and functions) of a supertype are inherited by a subtype through the inheritance link.

The domain of an attribute of an entity can be either atomic (i.e. integer, string, etc.) or complex (that of another entity). For an entity whose attribute has a complex domain, the relationship of that particular entity to the complex domain is shown by the aggregation links (the attribute/domain links). The aggregation link is denoted by a line with an arrow from the entity containing the attribute to the complex domain's entity. For example in Figure 3, the entity *line* has two attributes, namely *pnt* and *dir*. The attribute *pnt* has the domain of entity *cartesian_point*. Similarly, the attribute *dir* has the domain of entity *direction*. The corresponding aggregation links are also illustrated in Figure 3.

From the main window of the graphical interface, five panel buttons are provided. They are "File", "Links", "Zoom_in", "Zoom_out" and "Help". The "File" button supports the opening and closing of EXPRESS files. The "Links" button allows the user to select the showing of the aggregation links, inheritance links or both. The "Zoom_in" and "Zoom_out" options enable the user to view either a portion or the whole of the schema on a single display.

Attributes, functions and constraints of the entities can be displayed in a separate window. When the particular entity is selected, a menu attached to the entity will appear. The menu contains options including "attributes", "functions", "constraints" and "all". It allows the user to select to show only attributes, functions or constraints or to display all of them on a separate window. In Figure 4, the entity *hyperbola* is chosen. On selecting the "all" option, a separate window appears with the attributes and constraints of the entity *hyperbola*. As shown, the attributes *semi_axis* and *semi_imag_axis* and the constraint rules WR1 and WR2 are displayed.

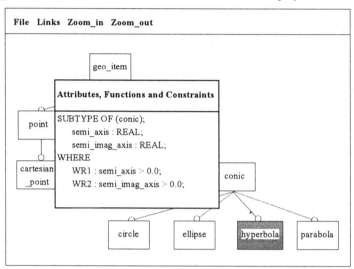

Figure 4. The example EXPRESS schema with the "all" option

6. Implementation Issues

The user interface is designed with an event-driven paradigm and conststructed using XView classes. In constructing the graphical display, the creation and traversal of a tree structure are needed. It is necessary to know the total number of displayed entities and all possible relationships among them beforehand, in order to create a well-proportioned tree structure. In particular, it is important to know at which level the tree is broadest, so that the distance between the entities can be calculated. In addition, the supertypes should be mapped onto the display first. Subsequently, the subtypes would use their supertypes' coordinates as references in plotting their own coordinates on the display.

In order to obtain all the necessary information mentioned above, two types of tree traversal algorithms are used: breadth-first and depth-first algorithms. The breadth-first algorithm is used to facilitate the finding of the level where the tree is broadest. This is done by keeping track of the current level and updating the count of the number of entities for that level every time when the tree structure is traversed from the higher level to the lower level. On the other hand, the depth-first algorithm is used to ensure that the higher levels of the supertype/subtypes hierarchy are always retrieved first before the lower ones. This allows the arrangement of the entities in such a way that the top level entities are always mapped onto the display first.

7. Conclusion

STEP has come a long way towards the realization of product data exchange and concurrent engineering. However, a crucial aspect of attaining wider acceptance by users is the availability of tools. Especially since the design of EXPRESS databases is a difficult task, design tools such as ExpressView can facilitate the design process.

ExpressView is a visualization tool which aids the EXPRESS database designers by displaying the aggregation and inheritance hierarchy relationships of EXPRESS models. Displaying aggregation relationships helps to avoid difficult retrievals of objects and displaying inheritance relationships provides information to reduce the storage costs due to data duplication. Thus, it enables EXPRESS designers to visualize and fine tune his models to achieve better performance.

The STEP/EXPRESS environment is being constructed using the Versant ODBMS and C++ running on Sun Sparc 2 workstations at Nanyang Technological University. The translator tool [10] which converts STEP/EXPRESS model into Versant ODBMS has been developed. The STEP Data Access Interface [14] has been implemented. Currently, the ExpressView visualization tool is being developed and will be linked with an EXPRESS text editor. Thus, while an EXPRESS schema is entered via the EXPRESS text editor, the ExpressView will be invoked automatically to display the underlying aggregation and inheritance links dynamically.

Acknowledgements

We wish to acknowledge the contributions of Brian Kho to the work in this project.

References

[1] Vergeest, J S M "CAD Surface Data Exchange using STEP," Computer-Aided Design, Vol 23, No 4, 1991, pp 269-281.

[2] Bloor, M S "CAD/CAM Product-data Exchange: The Next Step," Computer-Aided Design, Vol 23 No 12, 1991, pp 237-243.

[3] "Product data representation and Exchange - Part 11: The EXPRESS Language Reference Manual," Document ISO DIS 10303-11, National Institute of Standards and Technology, USA, 1992.

[4] Warthen, B.D. (Ed.) "What is STEP for ?," Product Data internat., Vol 3, No 6, 1992, pp 1-10.

[5] "Product data representation and exchange - Part 11: The EXPRESS Language Reference Manual," Document ISO DIS 10303-11, National Institute of Standards and Technology, U.S.A., 1992.

[6] Kemper, A and Wallrath, M "An Analysis of Geometric Modeling in Database System," ACM Computing Surveys, Vol 19, No 1, 1987, pp 49-91.

[7] Hardwick, M and Spooner, D.L. "Comparison of Some Data Models for Engineering Objects," IEEE Comp. Graphics and Appl., 7(3), 1987, pp 56-66.

[8] Kim W, Banergee, J, Chou H.T. and Garza, J.F. "Object-oriented database support for CAD," Computer-Aided Design, Vol 22, No 8, 1990, pp 469-479.

[9] Hardwick, M., "Implementing the PDES/STEP Specifications in an Object-Oriented Database," AUTOFACT '91, Chicago, Illinois, 1991.

[10] A.Goh, S.C.Hui, B.Song and F.Y.Wang, "A STEP/EXPRESS to Object-Oriented database Translator," submitted to the Journal of Integrated Computer-Aided Engineering.

[11] Fowler, J. (Ed.) "A proposal for the STEP data access interface specification," IGES/PDES Organisation, STEP Implementation Specifications Committee, January 1992.

[12] Willshire, M.J., "How spacey can they get? Space overhead for storage and indexing with object-oriented databases," Proceedings of the Seventh International Conference on Data Engineering, 1991.

[13] Subcommittee 4 of ISO Technical Committee, "Product Data Representation and Exchange - Part 42: Integrated Resources: Geometric & Topologic Representation," 1993.

[14] A.Goh, S.C.Hui, B.Song and F.Y.Wang, "A study of SDAI implementation on object-oriented databases," Computer Standards & Interfaces, 16, 1994, pp 33-43.

SUPPORT FOR OVERLAPPING MODELS IN AN ENTERPRISE DATA MODEL

STEPHEN C. F. CHAN
Department of Computing, Hong Kong Polytechnic University
Hung Hom, Kowloon, Hong Kong
E-mail: csschan@comp.polyu.edu.hk

and

PATRICK K. S. TONG
Hong Kong Polytechnic University

and

JOHN W. T. LEE
Hong Kong Polytechnic University

ABSTRACT

Both the bottom-up and top-down approaches for the design of an enterprise data model meet with many practical problems. Recently there is much interest in a compromise approach. Logical data models are first developed for individual application areas. Subsequently, these data models are integrated where they overlap. The international standard for manufacturing data exchange, STEP, is developed with this approach. This paper focuses on the support in STEP for such overlapping data models and discusses a method of implementation.

1. Introduction

The complete bottom-up approach to enterprise data modeling start with the modeling of low level data entities and attempts to integrate all the low level models into one. This approach requires much effort. Moreover, total integration is often unnecessary. The top-down approach start with a top-level, general model, and refines it into lower level models. This approach also encounters much difficulties because the amount of information involved in a real-life enterprise is so huge that it is very difficult even to cover all the major functions in a general model. During refinement of the data model, new types of information must be added and often leads of revision of the upper level models. And, in practice, many data models have already been developed for software development projects out of necessity. So starting from scratch again is just not practical. (Scheer 92) proposed a mixed approach which starts with the definition of analysis areas for which logical data models are designed separately, and which subsequently are integrated into the overall model.

In reality experts in individual application areas adopt data modeling methodologies, e.g., entity-relationship (Chen 76), IDEF-1X (see (Loomis 87)), NIAM (Verheijen 82), etc., and tools that are appropriate for that application area, or simply because these are what the experts are familiar with. Logical data models for these application areas need be integrated only where they overlap with one another, see Figure 1. Since these logical data models may be specified in diverse data modeling methodologies, one way to integrate them is to translate them into a common representation. Then the overlaps and inconsistencies can be analyzed and reconciled. This approach has been generally adopted in the development of STEP (ISO 93a) (Chan 93), where the common representation is the data modeling language EXPRESS (ISO 92) (Schenck 94). In this paper we discuss the support for overlapping logical data models, and discuss some of the software supports for implementation. EXPRESS/STEP can certainly be used to exchange standard-conforming data. Perhaps less obvious is the fact that EXPRESS can be used to define enterprise-specific data, and STEP-conforming data modeling and exchange tools can then be extended to integrate company-specific data into STEP-conforming data. Further, the methodology can be used to support the modeling and exchange of purely company-specific data. The advantage of such a approach is the benefits of a well-tested methodology and supporting tools, and the extendibility of the methodology.

2. A Common Data Model for Application Integration

Logical data models have been developed in many software projects for all kinds of purposes. These application models have been developed in a variety of methodologies and will continue to do so. For the purpose of developing an enterprise data model, however, it is necessary to have a common data model in order to compare and integrate these existing and evolving application models where necessary. This approach has been adopted by the International Standards Organization in the development of STEP, a set of standards for the modeling and exchange of product model data.

The common data modeling language employed in STEP, EXPRESS, supports the definition of complex data entity types, relationships, and constraints. Using EXPRESS, a number of logical data models have been developed. Related entities are grouped into schemas, and the definition of one entity often references another entity in a different schema. STEP defines an Application Protocol (AP) as the mechanism for specifying information requirements and for ensuring reliable data sharing in a specific application area. At the core of an application protocol is an Application Reference Model (ARM), defining the information requirements and constraints of the application using terminology and rules specific to the application. Hence an ARM may be defined as an entity-relationship model, IDEF-1X model, NIAM model, etc. The ARM is then mapped into the Application Interpreted Model (AIM), a logical data model (schema) defined in EXPRESS.

In a CAD system supporting a specific Application Protocol, the user will deal with concepts and terminology defined in the Application Reference Model. The data may be stored internally in a proprietary database, and coated with a STEP overcoat and appear to tbe conforming to the Application Intepreted Model. This way different systems can interact with the underlying data through STEP exchange files or the standard data access interface. For example, two CAD systems supporting the same Application Protocol will then be able to exchange data defined in the same AP freely, by one system outputting a STEP exchange file which is then read into the other system, by both of the accessing the same database through the same data access interface, etc. The relation between these models is illustrated in Figure 2. The essence is that ARM-based application systems must invoke the AIM-based standard data access interface to get AIM-typed data from a proprietary database.

An Application Protocol facilitates the sharing and exchange of data between two application programs in the same application area, e.g., two CAD systems used in the design phrase of a mechanical product. However, data must sometimes also be shared between application programs conforming to different Application Protocols. In practice, different application areas may have different concepts of similar data objects. For example, if a "hole" is interpreted as having certain attributes and constraints in one application (e.g., 3D mechanical design) but different constraints in another application (e.g., 2D printed circuit board design), then holes belonging to these two application areas are not identical and hence not interchangeable.

Some data may actually need be exchanged between different application areas. A circuit board layout created by a 2D circuit board design system may need to be enhanced by a 3D CAD system for the fabrication of the board, or used by a 3D CAD system to design brackets, connectors, and the sheet metal cage that houses the circuit board. In the case of the "hole" mentioned above. If the two kinds of hole are actually two different "views" of the same physical entity, then there may be a need to exchange data on instances of the hole between different application systems.

In order to facilitate the sharing of data between application programs conforming to different APs APs, common constructs between related APs are grouped into schemas called Application Interpreted Constructs (AICs) and referenced by APs as a complete unit (Chan 94). This step, however, involves detailed studies of both application areas and the needs for data sharing between them, and often involves lengthy negotiations between experts in the two application areas before the common constructs and their representations can be agreed upon.

For example, a CAD system supporting the Application Protocol for Mechanical Design Using Surface Representation may be used to create a complex design composed of many geometry data objects, by referring to concepts and terminology defined in the ARM of the Surface Representation AP. The design object is output as entities in the AIM schema of the Surface Representation AP and input into another CAD system which supports both the Surface Representation AP as well as the Configuration Controlled Design AP (ISO 93b), with common AIC schemas representing the common constructs between the two APs. One such AIC is the AIC: Geometrically Bounded Surface Shape Representation (ISO 94), which includes entity definitions for points, curves, surfaces, offset surfaces, retangular trimmed surfaces, curve bounded surfaces, boundary curves, etc. The designer is then able to enhance the original design with additional design information by employing concepts and terminologies defined in the Configuration Controlled Design AP. The result is a design objects composed of entities from both APs. In an organization supporting a central database, there may be a database supporting multiple application protocols, with each application program accessing the data according to its own application protocol, through a programming interface. This programming interface can be one conforming to the standard data access interface, SDAI being developed by the ISO (ISO 93d), see Figure 3.

The relation between the overlapping APs and the shared AIC is illustrated in the following sample EXPRESS code representing a skeleton of the related logical data models.

SCHEMA mech_desgn_w_surf_rep
    USE FROM aic_ geom_bounded _surf;
    ...
END_SCHEMA; -- mech_desgn_w_surf_rep

SCHEMA aic_geom_bounded_surf;
    ENTITY curve ...
    ENTITY surface ...
    ENTITY rectangular_trimmed_surface ...
    ENTITY curve_bounded_surface ...
    ENTITY boundary_curve ...
    ...
END_SCHEMA; -- aic_ geom_bounded _surf

SCHEMA config_control_schema
    USE FROM aic_ geom_bounded _surf;
    ...
END_SCHEMA; -- config_control_schema

The ISO has defined a number of application protocols and will certainly develop more. These typically cover the most common application areas. For any manufacturing enterprise, there are many more enterprise-specific applications that will not be covered by the standarized application protocols. The STEP methodology, however, can be applied to develop enterprise-specific application protocols and to integrate them into the set developed by the ISO. Further, the methodology and tools that are being developed to support it can be used to develop and implement enterprise data models that are totally independent from the STEP standard. In the following section we discuss some of the implementation methods and tools that are rapidly emerging.

3. Implementation Methods and Support Tools

3.1 Exchange File

A file format has been defined by the ISO for the ecoding and exchange of STEP data (ISO 93c). Each data file specifies the EXPRESS schema used, and contains instances of data entity types defined in the EXPRESS schema. The schema can be an Application Protocol (Application Interpreted Model) if data is being exchanged between two application programs in the same area. Alternatively, the schema may be an Application Interpreted Construct if data is being exchanged between two application programs conforming to two different Application Protocols sharing the common AIC. This file format, as well as the EXPRESS language and a number of basic schemas have already been made international standards.

3.2 Programming Interface

It is anticipated that more and more data will be shared between application programs by shared access to databases. Or application programs may input data according to one Application Protocol, enrich it, and output it according to another Application Protocol, as discussed briefly in the previous section, in each case going through a programming interface. The ISO is currently developing a standard (programming) data access interface to support data sharing in this manner. This standard interface between an application program and the database presents a standard view of the database to the application programs, irrespective of how the data in the database is structured. This view (which can be considered a virtual database) has a data dictionary, which is an in-memory version of the logical data model, or schema. The contents of the virtual database are all the data instances. Each data instance belongs to an entity definition in the data dictionary.

In order to support sharing between application programs conforming to different but overlapping Application Protocols, the shared entity definitions, embodied in the Application Interpreted Constructs as subschemas, must be stored in the data dictionary. This, however, may still not be enough, because an entity in the AIC may be subjected to different global rules in the different AIMs that it is used in. When and how should constraints (such as the global rules) on the contents of the virtual STEP database be validated and perhaps enforced is still largely an unresolved question.

There are two major levels of granularity of the data in the virtual database. In the simple case, data sharing is granulated at the instance level -- the contents part of the virtual database contains instances of entities, with reference to their respective entity definitions in the data dictionary. An entity instance can be assessed only by applications conforming to the Application Protocol that contains it. At a higher level, data sharing can also be granulated at the model level. Entity instances in the contents of the virtual database are grouped into models. These groupings may correspond to a complete design of a product, or some coherent subset of it. Each model contains instances of entities that must be defined within a single EXPRESS schema, and are associated with that schema and implicitly, all schemas that contain that schema, see Figure 4. If any of this owning schemas is an AP (AIM), then the model, or all instances within it, can be assessed by application programs that conform to that AP. If one of these owning schemas is an AIC, then the model can potentially be assessed by all APs that contain this AIC. In Figure 4, the arrows from Model to AP indicate the logical connection between data and its definition. For example, the entity instances in model A may be assessed only by application programs conforming to AP1, while entity instances in models B and C may be assessed by application programs conforming to AIC1, AP1 or AP2.

Figure 4 does not depict the possible interactions among the data instances themselves. In general, an instance (of an entity in a schema S) can reference an instance (of another entity) in a subschema (of S), but not the reverse. Hence, an instance in Model A can reference an instance in Models B, C & D. But an instance in Model B cannot reference an instance in Model A.

3.3 *Supporting Tools*

Many tool sets have been developed for the implementation of EXPRESS and STEP. One of the more popular ones is Data Probe (Sauder 93), currently supporting a Sun SPARCstation workstation. Its main advantage is that it is in the public domain and source code is available. The disadvantage is that there is very little documentation and, for obvious reasons, no support. It can be used to create a browser for examining and modifying an EXPRESS schema. At the heart of the Data Probe is an EXPRESS parser which generates class definitions in the C++ programming language as part of the Data Probe software system. These C++ class definitions can be used as part of the data structures needed to implement the standard data access interface in C++. Additional code that implement the access functions must also be written.

At this point the DataProbe translates each entity type definition into a C++ class. The cross-schema references, however, are not reflected in the C++ code. These can be done either by writing another (simpler) parser that specifically translates the schema references, or by modifying the EXPRESS parser in DataProbe itself. We are in the process of trying out both approaches to create the core data structures for implementing an interface into an object database (UNISQL and ObjectStore are being considered) which supports the sharing of data across application protocols.

4. Summary

The development of an enterprise data model involves the integration of logical data models developed in a variety of methodologies. EXPRESS is a data modeling language developed by the ISO for a common representation of these data models and their integration. The concept of a subschema grouping together the common portions of two overlapping schemas facilitates and sharing of data between application areas. Tools are beginning to appear to support practical implementations.

5. References

(Chan 93) S. Chan, et al, "Product Data Sharing with STEP", in P. Gu & A. Kusiak Eds., Concurrent Engineering: Methodology and Applications, Elsevier Sciences Publishers B.V., The Netherlands, 1993, pp. 277-298.

(Chan 94) S. Chan & P. Tong, "Manufacturing Data Sharing Through Integrated Data Models", The 10th Inter'l Conf. of CAD/CAM, Robotics & Factories of the Future, Ottawa, Canada, 21-24 Aug., 1994, pp.149-154.

(Chen 76) P. Chen, "The Entity-Relationship Model: Towards a Unified View of Data", ACM Trans. Database Sys., Vol. 1., No. 1, 1976.

(ISO 92) ISO/DIS 10303-11 Product Data Representation and Exchange - Part 11: The EXPRESS Language Reference Manual, ISO, 1992.

(ISO 93a) ISO DIS 10303-1 Product Data Representation and Exchange - Part 1: Overview & Fundamental Principles, ISO, 1993.

(ISO 93b) ISO/DIS 10303-203 Product Data Representation and Exchange - Part 203: Applicaton Protocol: Configuration Controlled Design, ISO, 1993.

(ISO 93c) ISO DIS 10303-21 Product Data Representation and Exchange - Part 21: Clear Text Encoding of the Exchange Structure, ISO, 1993.

(ISO 93d) ISO CD 10303-22 Product Data Representation and Exchange - Part 22: Standard Data Access Interface Specification, ISO TC 184/SC4/WG7 N350, 31 August, 1993.

(ISO 94) Application Interpreted Cosntruct: Manifold Surface Shape Representation, ISO TC184/SC4/WG4 N609a P6, ISO, 1994.

(Loomis 87) Mary Loomis, The Database Book, Macmillan, 1987.

(Sauder 93) David A. Sauder, Data Probe User's Guide, National PDES Testbed Report Series, NISTIR 5141, National Institute of Standards and Technology, March 1993.

(Scheer 92) August-Wilhelm Scheer and Alexander Hars, "Extending Data Modeling to cover the Whole Enterprise", Comm. of the ACM, Vol. 35, No. 9, Sept. 1992, pp. 166-172.
(Verheijen 82) G. Verheijen & J. VanBekkum, NIAM: An Information Analysis Method, in Information System Design Methodology, T. Olle, H. Sol, A. Verrijn-Stuart, Eds., North Holland, Amsterdam, 1982.
(Schenck 94) Schenck, D. & Wilson, P., Information Modeling: The EXPRESS Way, Oxford University Press, 1994.

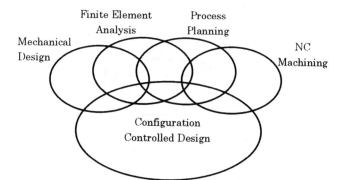

Figure 1. Overlapping Logical Data Models

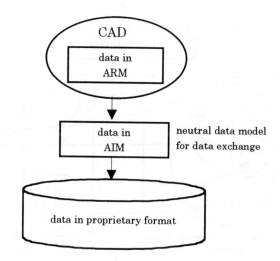

Figure 2. Relation between ARM, AIM and proprietary databases

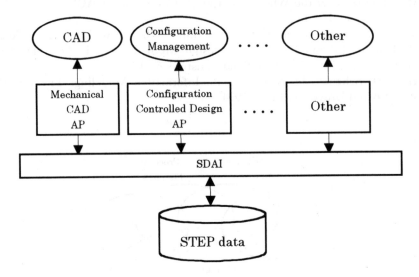

Figure 3. Multiple-application protocol access to the same data

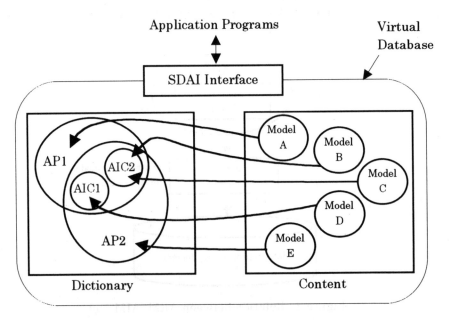

Figure 4. Virtual database for the programming interface

ON THE IMPLEMENTATION OF EXPRESS INFORMATION MODELS ONTO VERSANT OODB

YANG Qizhen and SONG Bin
Gintic Institute of Manufacturing Technology,
Nanyang Technological University
Singapore 2263
E-mail: gqyang@ntuvax.ntu.ac.sg

ABSTRACT

EXPRESS is a formal modeling language specified by STEP as a standard method of describing product model data. It provides facilities for defining data entities, relationships, constraints and functions for the representation of physical and functional characteristics of a product. However, EXPRESS does not contain elements which support execution. As a result, an EXPRESS information model needs to be implemented onto a programming environment or database system for exchange and sharing of product data defined by the model. Presented in this paper are approaches and techniques used for the development of a software prototype for implementing EXPRESS models on Versant object-oriented database. Issues on implementation and benchmark test are discussed.

Key Words: STEP/EXPRESS, Object-oriented database, Data exchange and sharing.

1. Introduction

The integration of computer aided systems is an important, but difficult, challenge that is faced by the industry. The core of achieving such an integration is to provide means enabling the exchange and sharing of product information among the systems. Over the years, a wide variety of approaches and efforts have been attempted. Among them, STEP (STandard for the Exchange of Product model data) is widely viewed as the most ambitious project [1-3] in promoting information exchange and sharing across different engineering environments.

STEP is to provide a complete and unambiguous representation for products over their life cycle. In scope are processes from the conceptualization, design, performance analysis, process planning, manufacturing to the maintenance of a product [4-8]. This means the integration of information covering geometry, material, tolerances, product structures and others necessary for the description of a product. In STEP, the method of achieving such a representation is by the use of a formal language, called EXPRESS, to describe product definitions.

An EXPRESS information model consists of one or more schemas. Each schema defines a set of entities and data types. An entity is defined by its attributes which together give the entity meaning to particular applications. A data type is declared to support the definitions of entities. Facilities are provided by the language to specify constraints on the instances of an entity attributes or global rules and general algorithms governing the use of the entities. EXPRESS also supports entity inheritance [9]. All these features enable product information defined in EXPRESS to be interpretable by software systems.

As a formal language, however, EXPRESS does not contain elements which support execution. Consequently, an EXPRESS information model needs to be implemented onto a database system for data sharing [7,10,11]. This implies that the features and capabilities of the database system should match that of the EXPRESS model to be implemented.

Among commercially available types of database systems, object-oriented database (OODB) systems are generally considered as most capable of managing engineering data which normally consists of complex objects and relationships. Versant is one of the most popularly used commercial OODBs in industries. It has programming interfaces for a number of executable languages, including C and C++ [12,13]. It also has database utilities for handling basic transactions between C++ class and Versant class definitions. These facilities make it an ideal platform for exploring automated establishment of a logical database schema from an EXPRESS model. The provision of such software tools is required for efficient implementation of STEP or STEP-based applications on databases.

This paper presents the considerations and issues in the development of a software prototype for automatic mapping of EXPRESS models to Versant database schemas. A STEP-based product data modeling approach is outlined and a benchmark test model developed for evaluating the developed software prototype is discussed.

2. Product Data Modeling and Implementation

Product development processes need all critical and relevant product information to be available for relevant engineering activities. A highly automated environment would require description of the information in a computer interpretable manner. Such a requirement demands the development of a methodology that enables integration and communication of engineering information. It also raises the question of how to represent and integrate complex product information like material behavior, shape geometry, surface characteristics, tolerances, structures or operational descriptions. Conceptual product data modeling by the STEP methodology offers one way to meet the challenge of engineering applications on product data integration and communication.

Using EXPRESS, STEP not only standardizes the description of engineering data for particular applications, but also provides a rich resources of integrated data definitions. This is evident from STEP's architecture of product models (see Fig.1). Each of the generic resource model defines a set of data for a particular discipline, e.g. geometry and topology, materials, dimensioning tolerances, etc. In addition, each application resource model provides data definitions necessary for an area of applications, e.g. draughting, finite element analysis, etc. An application protocol (AP) would draw definitions from both generic and application resources for the definition of information required by a particular application. The use of a standard set of resource data definitions for the description of application specific information facilitates the integration of APs in implementation. The same methodology can also be adopted in the development of information models for one's own applications.

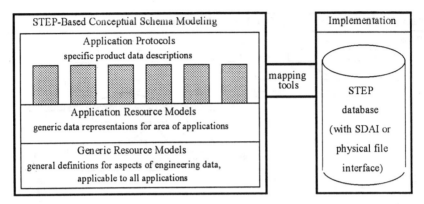

Figure 1: A STEP-based product data modeling and implementation

Regardless of actual applications, an information model at the application level must be implemented to arrive at an application system. In most of the applications, the system is an engineering database with a logical schema consistent with the information model. Such a database would enable data sharing among applications within the scope of the information model, facilitating teams in multi-disciplines to work cohesively in the development of a product.

Taking advantage of the computer interpretable nature of the EXPRESS language, the process of implementing an information model to its database can be handled automatically by the use of a software tool. Such a tool would perform syntactic and semantic analyses on the EXPRESS model to ensure that the model is error-free. It would then translate the EXPRESS constructs into their respective representation in a desired executable language that is acceptable by the targeted database. A further addition of necessary database functions would lead to the establishment of the database's logical schema through compiling.

In our practice, the target database is Versant OODB with a C++ programming interface. Accordingly, EXPRESS types, entities and semantics need to be migrated into C++ classes, and Versant data persistence and other database management functions can be added to the C++ classes to form a Versant schema. This schema defines the way how data is logically organized and accessed within the Versant OODB. It must therefore be an accurate representation of the original EXPRESS model.

When a database schema is set up from an EXPRESS model, a database schema viewer is required to check the database logical structures, the detailed messages on object state and behavior specified by a set of attributes and methods. In absence of more sophisticated tools, such a viewer is essential for the verification of the database setup.

3. Implementation Issues

Integrating the checking, translation, and viewing functions, a prototype system called MATEV (MAping Tool for Express to Versant) is developed to study the issues involved in the automatic processing of EXPRESS models for database implementation. One display of the system on the database object structures and relevant attributes generated from an EXPRESS benchmark test model is shown in Figure 2.

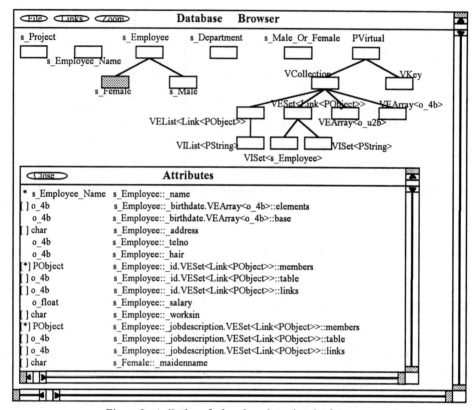

Figure 2: A display of a benchmark testing database

The implementation of MATEV used the following public domain software:
- NIST EXPRESS Toolkit (a library for building EXPRESS-related software)
- FEDEX (syntactic and semantic analyzer of EXPRESS)
- FEDEX-PLUS (EXPRESS to C++ translator)
- STEP Class Library (base classes supporting for EXPRESS to C++ translation)
- Flex lexical analyzer generator
- Bison parser generator
- GCC compiler

A SUN SPARCstaiton is used as the hardware platform for the implementation of the prototype MATEV developed in the current endeavor. Fig. 3 illustrates the processing flow of MATEV.

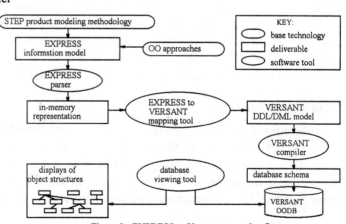

Figure 3: EXPRESS to Versant processing flow

MATEV is developed on the basis of FEDEX and FEDEX-PLUS software. The code is written in ANSI C and compiled using the GCC compiler. MATEV consists of two parts: the front end and the back end. The components used by both parts are illustrated in Figure 4.

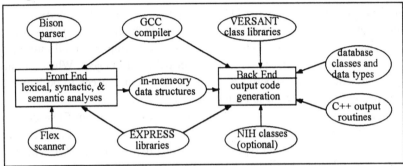

Figure 4: Components of MATEV prototype

The FEDEX is used as the front end in which its Flex scanner conducts the lexical, syntactic and semantic analyses. The EXPRESS conceptual models are processed by Bison parser and compiled into the EXPRESS Working Form which is a tightly linked set of in-memory data structures. The FEDEX-PLUS program is used as a component module of the back end by which the Working Form is translated into a Versant database schema. The entire back end consists of functions for accessing EXPRESS Working Form and header files generated by the front end, and standard output routines defined in FEDEX-PLUS.

Functionally, the back end is an object module which contains a single entry point traversing the given schema and writing the output format. The use of the Working Form accelerates the processing and encapsulates the internal representation of data structures from its formats of output. This allows MATEV to work for other database systems by simply adding additional format generators. The control flow for the back end of MATEV can be described as two levels of processing.

Level 1: Traversing schemas

After initialization, each schema element obtained from the linked_list of the front end is handed over to SCOPEprint() routine which controls the traversal of the instantiated Working Form. Relevant definitions of types, entities and variables in the schema together with the schema name are printed to an include file and a class source file. At the same process, C++ header files are also generated.

In SCOPEprint(), all entities defined in the schema are retrieved and implemented as instances of PClass of Versant in the source file by:
        Implement PClassObj<class_name>.
The TYPE SELECT is also implemented as an instance of PClass of Versant similar to entity elements. But the TYPE ENUMERATION is implemented differently using:
        Implement VPP_ENUM_DEFINED<enumeration_name>.

The Versant routine named SCHEMAprint() is also responsible for traversing the internal data structures. It directly calls SCOPEprint() to perform element printing.

Level 2: Primitive EXPRESS construct printing

Most of the output codes are produced by the routines for entity, type, and other concrete EXPRESS constructs using ENTITYprint(), TYPEprint() and VARprint() functions.

The Versant database uses the PObject class as the root of the Versant persistent class hierarchy. Deriving a user-defined class from PObject ensures persistency for instances of that class. As such, ENTITYprint() is used to implement all entities defined in EXPRESS as Versant persistent classes inherited from PObject. When an entity has more than one entity as its supertypes, multiple inheritance needs to be handled. This requirement can be tackled by a sub-function of ENTITYprint().

The function TYPEprint () is employed to generate the Versant Class definitions to represent the SELECT types of EXPRESS. It also provides Versant TypeDef definitions for EXPRESS user defined data types. TYPEprint() defines a SELECT type as a class with multiple pointers to the various types, allowing one pointer to be active at any one time. SELECT types are made to be inherited from PObject. ENUMERATION types and other user-defined types are not implemented as classes but as TYPEDEFs so that they do not need to be derived from PObject.

The back end processing for primitive EXPRESS constructs can be summarized as follows.
- Class definitions are printed in the include file with class inheritance relationships (if any) and class attributes. Declarations for constructors, destructors and copy constructors of each class are made in the file to meet the requirements by the Versant system. The access methods, such as get_attribute() and set_attribute(), for all attributes of each class as well as two specially defined database methods for listing and counting attributes in a class are also declared in this file.
- Corresponding to the classes defined in the include file, attribute access methods and class-level key methods, including constructors, destructors and copy constructors, are held in the class source file.

Arriving at the above two files, the EXPRESS product model is migrated into a Versant DDL/DML model ready to be compiled to generate a database schema.

4. Benchmark Testing and Other examples

A special designed EXPRESS schema, shown in Figure 5 in EXPRESS-G, is taken as a benchmark test model for examining the functionality of the MATEV prototype. Most of the EXPRESS elements are used in this schema, including some complex data types (see Fig. 5 and 6).

The testing criteria are that:
- all elements of EXPRESS should be accurately translated except constraint-related ones;
- after translation, all data structures and relationships between objects or attributes should remain correct and consistent; and
- the generated database schemas can be loaded into the Versant system to create STEP databases corresponding to the given EXPRESS models.

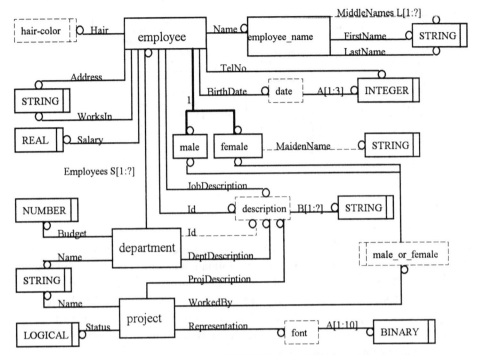

Figure 5: A complete entity-level EXPRESS-G model for benchmark testing

The result of the benchmark test indicates that the MATEV prototype satisfies all these criteria. The EXPRESS model is successfully processed by MATEV, a relevant DDL/DML model is generated and accepted by the Versant compiler, and the database is set up. Logically, the database setup from the benchmark test model should possess an object logical structure as shown in Figure 7. This is verified visually by the database schema displayed by a schema browser (see Fig. 2).

Other case studies are also conducted by using both the STEP generic resource schemas and self-defined EXPRESS schemas. MATEV gives satisfactory results for all these testing models.

```
ENTITY employee
  SUPERTYPE OF (ONEOF(male, female ));
    Name        : employee_name;      (* complex data *)
    BirthDate   : date;               (* user_defined type *)
    Address     : STRING;
    TelNo       : INTEGER;
    Hair        : hair_color;         (* enumeration type *)
    Id          : description;
    Salary      : REAL;
    WorhsIn     : STRING;
    JobDescription: description;
END_ENTITY;
```

Figure 6: An EXPRESS entity for modeling an "employee" object

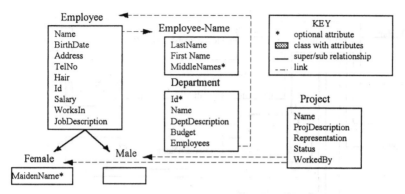

Figure 7: OO logical structure for benchmark testing

5. Conclusions

An important application for EXPRESS information models, either provided by STEP APs or user developed, is for sharing information among computer aided systems via databases. The use of suitable software tools for database implementations directly from an EXPRESS model would greatly enhance the efficiency. The MATEV prototype developed in this project has demonstrated the viability and capabilities of such an approach. Though targeted on Versant OODB, it can be conveniently expanded to process EXPRESS models for other databases.

Further studies are needed to enhance the MATEV prototype for handling constraints, implementing a subset of SDAI [14], and reading and writing STEP physical files.

References

1. A.Goh et al, *Computer Standards & Interfaces*, **16** (1994) p.33-43
2. M.S. Bloor & J.Owen, *Computer-Aided Design*, **23-4** (1991) p.237-243
3. B.D. Warthen, ed., *Product Data International*, **3-6** (1992) p.1-10
4. J.J.Shah & A.Mathew, *Computer-Aided Design*, **23-4** (1991) p.282-296
5. J.S.M.Vergeest, *Computer-Aided Design*, **23-4** (1991) p.269-281
6. L.H.Qiao et al, *Computers in Industry*, **21-1** (1993) p.11-22
7. Song Bin et al, *The development of a STEP-based OODB for standard parts of sheet metal dies*, Technical Paper #MS93-269 of AUTOFACT (1993)
8. C. Stark & M. Mitchell, *Development plan: application protocols for mechanical parts production*, PDES Testbed Report Series, NISTIR 4628 (1991)
9. ISO 10303-11: *The EXPRESS language reference manual*, (1993)
10. M. Hardwick et al, *Implementing the PDES/STEP specifications in an object-oriented database*, Technical Paper #MS91-429 of AUTOFACT (1991)
11. Song Bin et al, *A PDES/STEP-based relational database for PCB*, Technical Paper #MS92-384 of AUTOFACT (1992)
12. *System manual of Versant object database management system*, (1993)
13. *C++/Versant manual of Versant object database management system*, (1993)
14. ISO TC184/SC4/WG7, *Product data representation and exchange—Part 22: standard data access interface specifications*, (1992)

A REFERENCE ARCHITECTURE FOR INFORMATION SHARING IN COLLABORATIVE ENGINEERING ENVIRONMENTS

DRAGAN DOMAZET., DENNIS SNG
Gintic Institute of Manufacturing Technology, Nanyang Technological University
Nanyange Avenue, Singapore 2263, Republic of Singapore
E-mail: dragan@ntuix.ntu.ac.sg   dennis@ntuix.ntu.ac.sg

and

FOOK-NYEN CHOONG, STEPHEN SUM
Concurrent Engineering Unit, Siemens Pte Ltd
Singapore

ABSTRACT

In this paper, a reference architecture for information sharing is proposed. By using and combining its elements, various actual information infrastructures with different features can be derived. Also, basic requirements for product information sharing and management are specified. The proposed requirements and the reference architecture can help companies to classify their requirements, vendors to specify their products, and scientists to express their contributions and research areas.

Keywords: infromation sharing, collaborative engineering

1. Introduction

Heterogeneous computer environments and lack of standards prevent a smooth integration of different applications needed for achieving operational efficiency at the enterprise level. Many isolated and autonomous systems need to exchange or share data. Data sharing at an organizational level is needed when data integrity and consistency must be maintained at this level. Many solutions for information sharing or exchange in collaborative engineering environments have been proposed in the research community [1-15]. *Information exchange* implies that a copy of the source information as a package is handled from one source to another. If it is not in a format understandable by the receiver, it is translated into the correct format. *Information sharing* is achieved when information is maintained in only one logical location and is interactively accessed whenever needed, giving the appearance of one singular cohesive system. When a data item is changed, all copies of this data, regardless of their physical location, are automatically updated in an information sharing environment. Such a mechanism does not exist with information exchange and copies of the same data are not updated.

However, in practice companies face many problems when trying to integrate their "information islands" and usually they use a simple, but insufficient data exchange mechanism (such as IGES) or vendor specific data formats (such as DXF) for exchanging CAD data files. On the other hand, many systems for product data management are recently released offering product data integration and management. Users and prospective customers are sometimes confused trying to identify real features and information integration functions of all these systems, both on the market and in journals.

This paper proposes a generic reference architecture for information sharing and exchange that could help companes to classify their requirements, vendors to specify their products, and scientists to express their contributions and research directions. In Section 2 we specify two basic sets of requirements for information sharing in collaborative environments, one reflecting the *information content* that could be shared, and the other reflecting the *information management*. As product information are the basic integrating factor of a manufacturing company, our reference architecture, which is proposed in Section 3, is mainly related to product information sharing.

2. Product Information Requirements in Cooperative Product Development

For concurrent engineering, product life-cycle information are used in all organizational units. Two information aspects are of specific interest: product information content and product information management.

2.1 *Product Information Content Requirements*

1. Product models should support different, but consistent views and data representations, at different abstraction levels (Fig. 1)[1]. Product information should represent product structure, geometry, topology, form features, tolerances, notes, surface data, material data, version data, organizational data, design rationales, design history, etc., as specified with *integrated product information models*, such as STEP [16].

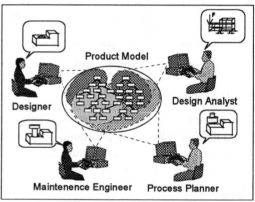

Fig.1 Different views of the product model

2. Besides data, *product related knowledge* is also needed, such as design functions, configuration rules and constraints, design change rules, and constraints reflecting concerns of all disciplines participating in the product development, manufacture, and maintenance. The knowledge part of a product model (reflecting multi-disciplinary concerns and providing for early involvement of different experts in a concurrent design process) is more stable then the data part that is growing rapidly with the progress of the product development (Fig. 2).

3. *Active product models* with derived data procedures, event-condition-action (ECA)

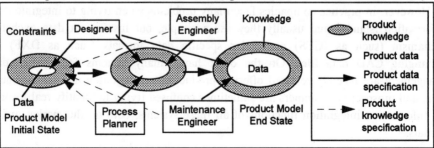

Fig.2 Product model life-cycle

rules, and automatic design change propagation rules are necessary to support dynamic design environments[2,3,4]. Product model objects should represent both product data and behavior using encapsulated programs, making product models "active".

2.2 Product Information Management Requirements

1. Both product information exchange and information sharing should be supported. Information exchange is mainly needed for supporting loosely coupled engineering activities. For tightly coupled activities and support of groupwork, information sharing is more appropriate (Fig.3) [5].
2. When *information sharing* is needed, all product information should be integrated, i.e., a data object may have multiple copies,

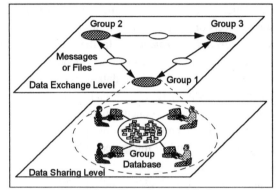

Fig.3 Information exchange and sharing

but they all should have only one logical address, and *its integrity and consistency must be maintained*. If a data object is distributed, an automatic update across the network must be provided when a change is specified. All usual DBMS functions must be provided globally, such as transaction and concurrency control, security, database recovery, etc.
3. To support product information exchange, a *message generation and handling mechanism* (such as message triggers) should be provided within the product model to reduce unnecessary user interaction and speed up the message exchange process.
4. Using object-oriented technology and the encapsulation of applications, legacy databases, and procedures, their representative objects, called *agents* can be created. Object distribution in *heterogeneous computer environments* should be provided (using CORBA-compliant software, for example).
5. Product information management should support *distributed architectures*, such as client/server architectures, and optimize the use of system resources in order to maximize information sharing/exchange efficiency.

3. A Reference Architecture for Product Information Sharing or Exchange

The information infrastructure to support collaborative product development depends on project specific requirements, as well as the generic requirements specified in the previous section which are required in all collaborative product development instances. We can identify at least three typical collaborative environments:
1) *Environmentss with low collaboration*: In this case, all applications are autonomous, they use local data models, and they exchange messages and data files irregularly. Data files are translated in accordance to internal data models and inter-application interaction is under the control of users. Data stored in different databases can be collected when needed for an application, but no global consistency is

maintained, i.e., modified data objects may have inconsistent copies.

2) *Environments with high collaboration:* In this case, applications in these systems are under control of an intelligent agent for partial or fully automatic collaboration and data translation (multi-agent systems). Agents can have different level of intelligence, and can provide different message and data exchange mechanisms [6-9]. Some of these systems can provide and manage global operations such as consistency of distributed data objects, recovery or concurrency. These DBMS-like operations are limited only to some objects or inter-object relations of global significance. [10,11]

3) *Integrated environments:* In these environemnts, all agents representing different applications use a common data model (directly or indirectly) and share product information using common data servers (for instance, implementing distributed or federated DBMSs)[12-15]. Usual DBMS data sharing functions, such as, data consistency, integrity, concurrent access and recovery are managed at the global level.

In reality, a combination of collaborative and integrated environments is needed to support both the loosely and tightly coupled activities. A hybrid environment should provide features of integrated environments for groupwork, where a tight cooperation of group members is needed, and features of a collaborative environments for supporting inter-group collaboration (where each group can be represented as a group agent) or inter-agent collaboration with high local autonomy (Fig. 3). The diagram in Figure 4 shows different features of these systems.

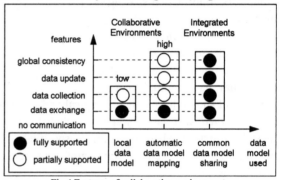

Fig.4 Features of collaborative environments

Because of the functional diversity of information systems that support collaborative product development, a reference architecture of their information infrastructure can help the classification of different implementations. Before presenting it, we shall identify its key elements.

1. Agents: An agent represents an intelligent application or a user. It interfaces to the rest of the information system. It can be represented as an object that encapsulates an application and has several sub-objects or modules (Fig. 5). The Communication Module handles the communication with other agents and elements of the system (exchange of messages). The Control Module knows what information and knowledge must be exchanged and what global operations must be executed. Depending on its control knowledge, an agent can broadcast its messages to all other agents, send to one or more specific agents, or send to a coordinator (Product

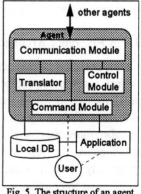

Fig. 5 The structure of an agent

Data Manager) that is responsible for message and data distribution. The Translator translates incoming data files from external to an internal data format. Some translators may also translate data from an agent's internal data model to a an external data model. The Command Module monitors local operations and executes actions needed to manage the collaboration with other agents by sending instructions to the Control Module. Interaction between the Command Module and a local application is highly specific to that application and its communication features.

2. File Servers: They store flat files and deliver them to applications upon request. They can be used only by applications that use the same data model.

3. Database Servers: These are database repositories of different entities (such as product, process, organization models, etc.) and provide concurrent data access, data consistency, integrity, security and recovery. Database management systems (DBMSs) can manage centralized or distributed databases. The data models they implement may be standard ones (such as STEP [16]), or custom design. In the first case, agents/applications in heterogeneous computer environments can directly access a database (using a standard access interface) and share data objects under the concurrency and transaction control of the DBMS[14,15]. In the second case, all agents/applications must use specific database access interfaces and translate data in accordance to their internal data models. Agents using a database server tightly cooperate and share database objects. This is convenient for highly cooperative groupworks, where all group members (individual agents) share the same data model.

4. Database Query System: It provides an uniform access to different database servers that use different data models. Using an SQL-like language, the system collects requested data and manage their transport to an agent. No data consistency or global data update is provided.

5. Federated Database: A Federated Database (FDB) is a collection of cooperating, but autonomous distributed and possibly heterogeneous DBMSs [11]. Shared data models have their local schema translated to a component schema using a common data model (Fig. 6). Parts of common schema (export schema) are accessible by specific group of agents. A federated schema determines export schema of different DBMSs that are under the control of a federated DBMS. Agents access data objects through external schema of federated schema (if they exist), providing them a specific view to data.

Agents using a FDB are under the impression that they all use the same DBMS (as a single logical database). A FDB provides all DBMS functions to its external users and coordinate all database operations among DBMSs that constitute the FDB. A global transaction management maintains data consistency while allowing concurrent updates across multiple databases. A FDB is used when it is necessary to maintain the global data

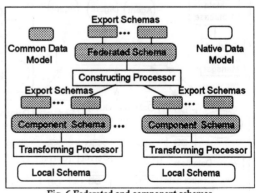

Fig. 6 Federated and component schemas

object consistency at the product level. Collaborating systems, using a Product Data Manager (explained later), cannot handle the consistency at this level as the corresponding control knowledge needed may be too complex. Instead of specifying an appropriate control knowledge in a case by case fashion (as in the case of a Product Data Manager), a FDB always maintains data consistency and perform other DBMS operations for all data accessed through its global, federated schema.

6. **Common Services:** They provide common services to all agents such as e-mail, printing, etc.

7. **Product Data Manager:** It organizes, accesses, and controls data related to products within a company. It also maintains relationships among product data, enforces the rules that describe data flow and processes, and performs notification and messaging functions. The Product Data Manager (PDM) does not interpret and manage product models as it only manages data file transport, translations and delivery to the end-users. Instead of storing data (as in database servers, a PDM stores pointers to a data repository and data control knowledge (such as rules for global operations) in its

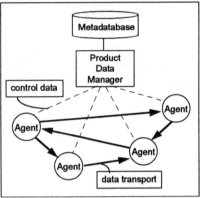

Fig. 7 Use of metadatabases

metadatabase (Fig. 7). Remote updates may be possible if special rules are specified for the update control. It can be a coordinator for inter-agent communication as well as a central manager and distributor for their data control knowledge [9,10]. It is suitable for the exchange of data, documents, images, and messages, and for the management of product configurations and design changes when no common product model (such as STEP) is in use.

8. **Data Bus:** The Data Bus provides data and message transport among agents and other elements of the information infrastructure (such as computer networks).

These eight elements of the proposed reference architecture (Fig.8) represent information sharing and exchange environments that support collaborative product development. Since the proposed reference architecture is generic, it can be instantiated to suit any information environment for collaborative product development by combining and deleting some of elements of this reference architecture. As these are logical elements, their number and physical implementation can be different from case to case.

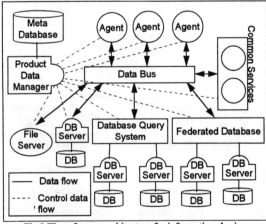

Fig.8 The reference architecture for information sharing

Information systems with a low degree of collaboration (Fig 9a) can be represented with the following elements: Agent, Data Bus, File Storage Common Services and optionally, Database Query System. Agents may use data files from the File Server, exchange data using the Data Bus and use printing and mailing services provided by the Common Services. If the Database Query System is available, the agents can collect data from various remote databases by making global query request

Information systems with a high degree of collaboration (Fig. 9b) normally use the following elements: an Agent, Data Bus, File Server, Common Services, Product Data Manager, and optionally Database Server and Database Query System. A multi-agent system sometimes may not use a PDM as a separate unit, as its functions are distributed among agents. In such cases they possess the necessary knowledge for data control to extract needed data from applications and manage the data transport among themselves. In contrast to this "peer-to-peer" approach, a more centralized multi-agent system implementation is also possible. A PDM then directly manages data transport between agents and maintains their data control knowledge stored locally (in agents control data repository). If the agents use database servers, then they can share data provided by these servers, using the usual DBMS functions. If a Database Query System is used, then they can collect data from remote databases using a uniform query operation. If a PDM has knowledge about global operations and the objects that perform global DBMS-like operations, then it could manage global consistency and recovery of these objects.

Integrated systems for product information sharing (Fig. 9c) use only logical elements that can manage and maintain both global and local data consistency, recovery and concurrent access control.

Fully integrated systems are difficult to realize in practice. Usually a hybrid system, consisting of both partially integrated and collaborating systems is a more realistic approach, especially in large organizations where global integration at the enterprise level may not be needed nor feasible. Many organizational units want to keep their autonomy, and so to share or exchange only a part of their data, as well as to maintain their global consistency. Also, many legacy systems are in use and their full integration is not easy. *A hybrid system* provides a collaborating environment for data exchange for collaboration between different organizations, departments and working groups. It supports their loosely coupled activities, as well as tightly coupled activities of workgroups thus providing information sharing found at integrated systems (Fig 3).

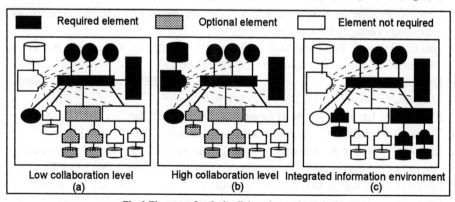

Fig. 9 Elements of typical collaborative environments

4. Conclusion

The reference architecture for information sharing/exchange in collaborative engineering environments proposed in this paper may be useful for classifying existing information infrastructures and for high-level design of new information systems. Three typical information infrastructures are derived from the proposed reference architecture, but many other specific architectures may be derived. By decomposing the proposed logical elements to finer functional and logical granularity, differences between various environments can be shown directly in the system architecture.

Integrated, consistent and active product models with encapsulated product knowledge are needed for integrated information systems. Data integrity and consistency should be maintained whenever product information sharing is needed. In a collaborative environment, data object consistency may be maintained for selected objects only if corresponding metaknowledge is available and mechanisms for update propagation are provided.

References

1. Krauze F.L., Kimura F., Kjellberg T., Lu S.C.-Y., *Annals of the CIRP*, **42/2** (1993) p. 695
2. Mattox, D., Smith K., Lu S.C.-Y., *Lecture Notes in Computer Science* **760**, (Springer-Verlag, 1993) p 379
3. Jasper H, *Proceedings of the Tenth International Conference on Data Engineering*, (IEEE Computer Society, 1994), p. 368
4. Cornelio, Navathe S., *Proceedings of the Ninth International Conference on Data Engineering*, (IEEE Computer Society, 1993), p. 100
5. Herman A.E., *Technical Report: KESRL-93-004*, Department of Mechanical and Industrial Engineering, University of Illinois at Urbana-Champaign (1994)
6. M.P. Papazoglou, S.C. Laufmann, T.K. Dellis, *International Journal of Intelligent and Cooperative Information Systems*, **1, No. 1** (1992), p. 169
7. Case M.P., *Ph.D. Thesis*, University of Illinois at Urbana-Champaign, (1994)
8. Tenenbaum J.M., Weber J.C, Gruber T.R, *Proceedings of the First International Conference on "Enterprise Integration Modelling"*, ed. C.J. Petrie, (MIT Press,) p. 356
9. Londono F, Cleetus K.J., Reddy Y.V, *Lecture Notes in Computer Science* **492**, (Springler-Verlag, 1991), p. 26
10. Hsu C., Babin G, Bouziane M.H., Cheung W., Rattner L., Rubenstein A., Yee L., *Journal of Intelligent Manufacturing*, **5** (1994), p. 333
11. Hsu C., Gerhardt L., Spooner D., Rubinstein A., *IEEE Transactions on Systems, Man, and Cybernetics*, **24**, No. 5, (1994), p. 828
12. Sheth A.P., Larson J.A., *ACM Computing Surveys*, **22**, No. 3, (1990),p. 183
13. Tiwari S., Howard H.C, *Engineering with Computers*, No. 10 (1994), p. 140
14. Manola F., Heiler S.,Georgakopoulos D., Hornick M., Brodie M., *International Journal of Intelligent and Cooperative Information Systems*, **1**, No.1 (1992), p. 5
15. Barhouti W.S., Kaiser G.E., *ACM Computing Surveys*, **23**, No. 3, 1991, p.269
16. Product Data Representation and Exchange standard - STEP - ISO 10303

ProSTEP - An Initiative of the Automotive Industry for introducing STEP into Industrial Application

Dr. M. Holland
ProSTEP Produktdatentechnologie GmbH
Darmstadt, 64293, Germany
Tel.: +49-6151-9287-0
Fax: +49-6151-9287-26
E-Mail: holland@prostep.darmstadt.gmd.de

and

D. Trippner
ProSTEP Produktdatentechnologie GmbH

ABSTRACT

A survey has shown that in Germany alone, car manufacturers and their 900 suppliers are using about 110 different CAD systems. This heterogeneous system environment and the current lack of sufficient interface formats necessitates expensive manual reworking of transmitted data. In the German automotive industry alone the resulting costs amount to about DM 100 million per year [Hand-93]. One way of improving the exchange of data is to use Product Data Technology (PDT) and correspondingly powerful interface formats like STEP. This paper outlines what Product Data Technology is and describes ProSTEP's approach to introduce PDT into industrial usage.

1. Product Data Technology

One essential factor of production in all design and manufacturing processes is the checking or control of information processing. In particular, the continuous availability and processing capability of product information described by means of CAD/CAM systems has a decisive effect on both the economy and flexibility, and therefore ultimately the competitiveness, of a company [Zimm-93]. The product life cycle comprises not only the generation of the product, in other words the fields of development, design, and fabrication, but also includes functions such as sales, maintenance and final disposal or recycling of a product. Product Data therefore describes all properties of a product and also includes the information required for its manufacture. Since this data involves a virtual description of the product, this description is termed "Product Model". Information defined by a Product Model is important for all divisions of a company as well as for customers and clients integrated in the process chains of designing and manufacturing a product. The technology which supports the use and exchange of Product Models is called Product Data Technology (PDT). Different computer aided systems are used in such a process chain, to give optimum support for each particular process step. But the different data formats of these systems handicap an efficient communication between the systems used in a process chain and therefore handicap simultaneous engineering.

2. Definition of the Problem

Especially at times of restructuring and cost-cutting in almost all sectors of industry, the ability of CAD/CAM systems to communicate must be forcefully promoted. The communication between systems is a prerequisite for measures which increase the efficiency. Such measures are e.g. simultaneous engineering or the support of closer partnerships between car manufacturers and suppliers with the aim to decrease the manufacturing activities at the car manufacturer side by subcontracting them to their suppliers. Communication in this context refers not only to the exchange of information between the different CAD/CAM systems. It also includes the ability to access, interpret and subsequently process the information across the boundaries of systems and companies throughout the entire computer-aided process in the development, design, fabrication, maintenance and disposal of products and production facilities.

The German automotive manufacturers and their approximately 900 suppliers use roughly 110 different CAD/CAM systems today. A large number of systems is used on the supplier side which all have to communicate with the corresponding systems at the manufacturer side. This situation shows very clearly the need for better communication between the different systems used in a process chain.

The agreements (interfaces) between the various systems that are necessary for the implementation of an efficient exchange of data must be equal to this task. The exchange of Product Model data is significant to companies for their internal and external data exchange between CA-Systems. The aim of internal exchange is to provide Information Technology (IT) links between the various departments of a company such as design, project planning, fabrication, assembly and quality assurance. External exchange is used for transmitting product information between manufacturers and suppliers.

The objective must therefore be
- independence of CAD/CAM systems by means of open selection of alternate systems according to purely functional and economic criteria;
- the ability to access, interpret and subsequently process Product Model data which is generated just once, throughout the product life cycle by means of data sharing on the basis of international Product Model standards;
- and not least, to reduce costs and delays incurred today during the exchange of data.

3. Product Model Data Interfaces

The above demands can only be met by open systems and appropriate interfaces. Open systems permit the configuration of a CAD/CAM solution according to functional and economic viewpoints. Since it is not practicable, in view of the high development costs, to develop a specific interface for every possible system configuration, it is wise to use a standardized interface for the exchange of Product Model data in order to minimize development costs. In addition, standardized interfaces ensure the return of investment in the area of information processing, despite the short life cycles of hardware and software and the adherence to the stringent legal requirements for product liability [Trip-93].

3.1. Existing solutions are inadequate

The interfaces used thus far, such as IGES, VDAIS, VDAFS or SET, do not meet the requirements of an interface for Product Model data exchange as described above. They are suitable for the transmission of product information in sub-areas of computer-integrated production in the form of technical drawings or simple geometric models [And-89]. More detailed information such as tolerance specifications, material properties, parts lists or even work planning information, however, can only be transmitted incompletely.
Several problems exist though:
- Neutral formats are limited to geometric data and drafting.
- The specification of these standards reflects the state-of-the-art of the late 70's and early 80's.
- Software has been developed by vendors and also by users without co-ordination.
- Users have to adopt incompatible solutions available on the market.
- High reworking and maintenance costs (in the German automotive industry alone about 100 million DM per year [Hand-93]) are the consequence.

A more dramatic situation than in CAD data exchange has arisen in Engineering Data Management. The large German automotive companies have developed their own systems for the management of CAD data, documents, bill-of-material data, etc. The mutual access to the data is limited or impossible. Therefore, data has to be input repeatedly and cost-intensive actions for conversions become necessary. The situation does not change if an EDMS bought on the market is being used. Each system has its specific logical data model and restricts the user to a certain system and vendor, thus making him dependent for a long period of time.
The incompatible native data formats are the limits of today's integration concepts.
The difficulties described above show the need for a system and company independent international standard for Product Model data exchange and management. The only alternative to today's situation is the new international standard STEP.

3.2. STEP

The failings of existing interfaces have been recognized by the International Standards Organization (ISO) and have led to the development of the STEP interface (Standard for the Exchange of Product Model Data). The objective of STEP is: "an unambiguous representation of computer interpretable product information through-out the life of a product" [ISO-10303-94]. Simultaneously with the standardization of the interface specification, a draft methodology was developed in order to eliminate the failings of existing standards, such as inadequate formal description and non-prescribed test criteria.

3.2.1. STEP structure

The methods and principles used for the development of STEP are reflected in its structure (figure 1). STEP will appear as a series of the 10303 standards which, in addition to the pure Product Model data, also define description methods, implementation methods and a conformance testing methodology and framework.

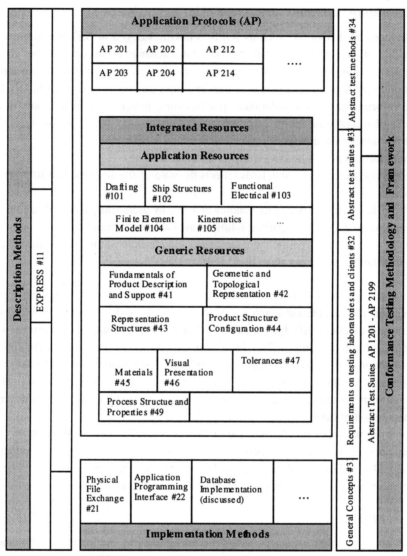

Figure 1: Architecture of STEP

3.2.2. Description methods

For the consistent, non-contradictory and semantically unambiguous description of the STEP Product Model, the EXPRESS formal description language and its EXPRESS-G graphical representation were defined. EXPRESS is not a programming language but a specification language with object-oriented characteristics for the formal, unambiguous description of the STEP Product Model in the form of information models. The Integrated Resources and Application Protocols of STEP are defined using this language.

3.2.3. Implementation methods

An information model described in EXPRESS can be compared to a conceptual schema description for databases. It thus remains independent of a specific implementation. It is, however, available in a form suitable for computer processing. The specification can therefore be simulated in different target implementations [And-92]. For the transfer of a model specified in EXPRESS into a special implementation form, various implementation methods are used. Until now, the following implementations are supported by standard implementation methods:
- Physical file exchange for the exchange of Product Data by means of a sequential file.
- An application programming interface for database access of Product Data.

3.2.4. Conformance testing methodology and framework

Basic principles and methods for verifying conformance were defined for the validation of STEP supporting software. These describe general conformance criteria and test procedures as well as test methods. An abstract test method is defined for each implementation method. Abstract test suites are developed accordingly for the Application Protocols.

3.2.5. Integrated Resources

The Integrated Resources constitute the core of STEP. These are information models that describe the Product Data independently of a special application in EXPRESS [ISO-10303-1-94]. They describe, for example shape and geometry, material property, shape variation tolerance and product structure configuration data of a product. It is also possible to define parameters and rules for the visual presentation of Product Data. In addition, a process model has been defined for the representation of the logical sequence of process activities and their parameters (e. g. cutting speed).

3.2.6. Application Protocols

It is not practical to implement the entire scope of STEP for industrial use. On the one hand, not all elements defined in the Integrated Resources are required for the support of a special application. On the other hand, a complete STEP implementation demands inordinately high development costs. For this reason, Application Protocols (AP) are standardized by ISO. APs describe the extract from the Integrated Resources that is required for a specific application. At present, 27 Application Protocol drafts are in the process of being standardized. Two APs have already reached the level of International Standard.

4. Application Protocol 214 Automotive Design

In collaboration between the German Automotive Association (VDA), ProSTEP and other international partners, the application protocol "Core Data for Automotive Mechanical Design Processes" is currently being developed as ISO standard 10303-214.

4.1. Scope of performance

AP 214 describes the product and production facilities data of development process chains in the automotive and mechanical industry. It includes geometric and non-geometric data of parts, assemblies and production facilities for mechanical parts and assemblies in automobile manufacture. AP 214 comprises the following classes of Product Data:
- Product Structure Configuration
- Geometric and Topological Representation (Wireframe, B-REP and CSG)
- Shape Variation Tolerance
- Part and Assembly descriptions by explicit drawings
- Kinematics
- Finite Element Analysis
- Material property
- Surface condition

4.2. Benefits

AP 214 represents a comprehensive information model for various process chains. Since it describes all relevant data, it can offer both manufacturers and suppliers considerable strategic benefits. These benefits, that can be transferred to almost all STEP applications, include:
- system independence,
- data consistency across the entire product life cycle,
- redundance free Product Data description,
- higher quality of the Product Data due to unambiguous definitions,
- possibilities of long-term archiving.

These advantages become very apparent on considering that about 100 different CAD/CAM systems are in use by German car manufacturers alone.

4.3. Current status

The application reference model has been checked in detail on examples taken from industrial practice in Germany, Japan and France. AP 214 will go out for international CD Ballot in June 1995.

5. Industrial launch of STEP

But strategies alone are not sufficient. The fast introduction together with a high quality of the used software is essential to industrial users of STEP. The lesson learned from IGES is that a standard can only be introduced successfully through a strong co-operation between system users and vendors. Only compatible STEP implementations through a close co-operation of the various vendors can assure a broad industrial usage of STEP. The user of STEP also needs to know which AP he needs in order to support his applications and he wants to use compatible software products. But the user does not want to deal with to many details of STEP. He wants to use STEP in order to improve his business processes.

5.1. The ProSTEP Center

The idea of the ProSTEP center is to support the co-operation of the CA-vendors in order to develop compatible products, to help the CA-users introducing STEP and to communicate their requirements to the standard development. It consists of two legal units: The ProSTEP Association and the ProSTEP Company (GmbH).

5.2. ProSTEP's standardization approach

Developing and implementing an international standard requires a strong co-operation with international projects and standardization organizations particularly with ISO and the associations of the French, American, Japanese and German automotive industry. A mutual exchange of information has been initiated with the PDT centers in the US and Europe including PDES, GOSET and CADDETC. With PDES, Inc. a Memorandum of Understanding was signed. ProSTEP collects and focuses the requirements especially from the association members and presents them to the vendors and standardization bodies. It also supports Vendors in implementing the standard and opens the market for STEP based products. ProSTEP and vendors are business partners! The user should only buy compatible solutions at the vendors. Also services and additional products to introduce this solutions into practice are offered. It is ProSTEP's objective to accompany the standardization process with prototype implementations. This will ensure firstly the fast availability of AP 214 based processors and secondly a feedback to the standardization process. A central activity of the Association is therefore the moderation of the Round Table of STEP processor developers. The Round Table brings together system vendors in order to assure the development of compatible STEP processors by the participating software vendors. Today about 17 AP 214 based processors are under development by vendors participating at the Round Table.

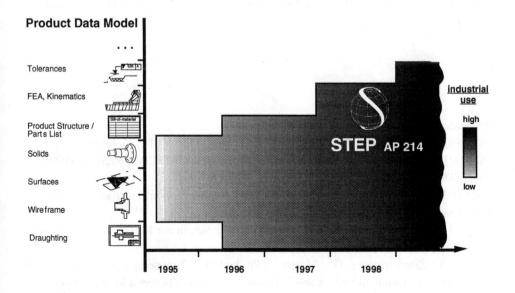

Figure 2: Stepwise implementation of AP214

The functionality of the processors is defined by the user requirements mainly coming from the association members and will grow in parallel with the standardization process (figure 2).

AP 214 is the first Application Protocol STEP covering a whole process chain in mechanical engineering. AP 214 contains overlapping areas to several other APs. These APs either have already been standardized or are currently being developed. It is the strategic goal to ensure the interoperability between implementations based on overlapping APs. Therefore, ProSTEP is an active player to address this topic at ISO. With AP 212 (Electrotechnical design) a second AP is under development by ProSTEP, with the scope to support electrotechnical applications.

6. Summary

The initial release of STEP has been published. First STEP based products are available on the market. STEP has become a critical success factor for business reengineering.
That means
- the development of company-wide data models shall be based on these standardized concepts and methods;
- data modelling has to be in accordance with STEP application protocols;
- migration strategies to open system architecture will only be successful if they take STEP into account.

ProSTEP's mission is to help its customers introducing this new technology in their business processes. Initiated by the German industry it has become an international movement. ProSTEP has representatives in the US, the UK and China. As of right now (May 1995), the Association has 115 member companies. They are not only from Germany but also from Austria, Canada, France, Netherlands, Spain , Sweden, Switzerland and UK. The fastest and easiest way to benefit from ProSTEP is to join the Association.

References
[And-89]: Anderl, R.; Das Produktmodellkonzept von STEP; VDI-Z 12.1989
[And-92]: Anderl, R.; STEP-Schritte zum Produktmodell; CAD-CAM Report 8.1992
[Hand-93]: Handelsblatt; German Newspaper; 6.10.1993
[ISO-10303-1-94]: ISO-10303-1; Product Data Representation and Exchange-Part 1: Overview and Fundamental Principles; ISO/IEC Schweiz (1993)
[Trip-93]: Machner, B.; Trippner, D.; ProSTEP: Der Schritt zur Datenintegration; CAD-CAM Report Nr. 5; Mai 1993
[Zimm-93]: Zimmermeyer, G.; Mastering Product Data Technology as a Fundamental Contribution to Competitiveness and Efficient Cooperation between Manufacturers and their Suppliers; VDI Berichte Nr. 1096; VDI-Verlag; Düsseldorf; 1993

FEM/FEA

FEM/FEA

Numerical Simulation of Melt Flow Behavior in Injection Molding
  J. F. Bao, P. F. Shao and Y. M. Jin

The Influence of Tool Geometry on Blanking Characteristics
  C. M. Choy and R. Balendra

FE Analysis of Pre-Stressed Press Frames
  R. Balendra, H. Ou, X. Lu and K. Chodnikiewicz

Numerical Simulation of Melt Flow Behavior In Injection Molding

BAO JIAFU, SHAO PENGFEI, JIN YONGMEI

Department of Precision Machinery and Precision Instrumentation
University of Science and Technology of China
Hefei, Anhui, P.R.China

ABSTRACT

Mathematical model of the filling process in a thin cavity is established by using finite difference method, combined with utilization of a number of reasonable assumptions and appropriate boundary conditions. A computational simulation program of melt flow behavior in a thin rectangular cavity is developed on microcomputer, proving the model and the program acceptable.

Introduction

Injection molding is a very important industrial process for the manufacturing of plastic parts. The injection molding cycle is composed of three stages: filling, packing and cooling. Filling stage is of importance because it significantly affects the part quality. If the melt flow behavior is to be predicted, the part quality can well be controlled and improved. Now the computer aided design can provide mold designers a good tool. The model and the program in this paper provide a way to the numerical simulation of melt flow in injection molding.

Mathematical Model

The model of the polymer melt flow inside a cavity can be represented by the following equations of mass, momentum and energy.

Continuity:
$$\frac{d\rho}{dt} = -\rho \, \text{div} \, \mathbf{V} \tag{1}$$

Momentum:
$$\rho \frac{d\mathbf{V}}{dt} = -\text{grad} \, P - \text{div} \, \tau + \rho \mathbf{g} \tag{2}$$

Energy:
$$\rho C_v \frac{dT}{dt} = [\nabla \cdot q] - T[\frac{\partial P}{\partial T}] \rho [\nabla \cdot \mathbf{V}] - [\tau \cdot \nabla \mathbf{V}] \tag{3}$$

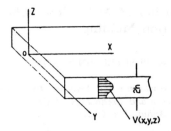

Fig.1 Schematic representation of melt filling in the mold

Assume that z coordinate along the thickness, x,y coordinates along the flow plan.

Since the resulting system of differential equations is rather complex, a variety of physical assumptions are employed to simplify the governing equtions without significant loss of generality of the treatment. The assumptions are listed below.

- The melt flow is two dimensional
- The fluid motion is laminar
- The fluid is incompressible and isothermal
- The melt flow is isothermal two-dimensional Newtonian fluid flow
- Body forces and inertic forces are neglected. The effects of gate and elasticity are neglected too.

With these assumptions, the above equations take on the following forms.

Continuity:
$$\frac{\partial V_x}{\partial x} + \frac{\partial V_y}{\partial y} = 0 \tag{4}$$

Motion:
$$\frac{\partial P}{\partial x} = \mu \frac{\partial^2 V_x}{\partial z^2} \tag{5-A}$$

$$\frac{\partial P}{\partial y} = \mu \frac{\partial^2 V_y}{\partial z^2} \tag{5-B}$$

$$\frac{\partial P}{\partial z} = 0 \tag{5-C}$$

Upon double integrating Eq(5) with boundary conditions
$$V_x = V_y = 0 \quad \text{at} \quad z = \pm h \tag{6}$$

An expression for the velocity field is obtained as following:

$$V_x = \frac{Z^2 - h^2}{2\mu} \frac{\partial P}{\partial x} \tag{7-A}$$

$$V_y = \frac{Z^2 - h^2}{2\mu} \frac{\partial P}{\partial y} \tag{7-B}$$

From Eq(4) and Eq(7), the following governing equation is obtained:

$$\frac{\partial^2 P}{\partial x^2} + \frac{\partial^2 P}{\partial y^2} = 0 \qquad (8)$$

In the above equations, μ is fluid viscosity, h is semi-thichness of the cavity, V_x and V_y are velocity components, P is pressure.

Numerical Calculation

A square mesh is adopted upon the two-dimension flow field.($\triangle x = \triangle y$) With utilization of five point finite difference method, Eq(8) becomes

$$-4P(i,j)+P(i+1,j)+P(i-1,j)+P(i,j+1)+P(i,j-1)=0 \qquad (9)$$

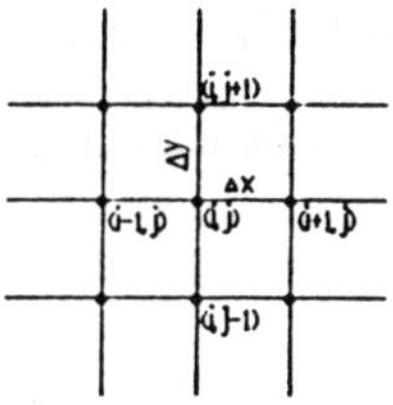

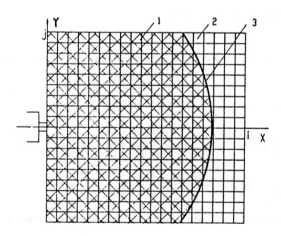

Fig.2 Mesh configuration of finite difference method.

Fig.3 Schematic representation of the flow pattern during the filling.

1.filled unit 2.unfilled unit 3.front position

evaluating equation (9) for each node of the filled unit, we obtain the equations set for pressure of each node with boundary conditions,

$$\frac{\partial P}{\partial n} = 0 \quad \text{along the side wall} \tag{10}$$

where n is a line segment normal to the boundary,

and
$$P=0 \quad \text{along the melt front} \tag{11}$$
$$P=P_o \quad \text{along tip of gate} \tag{12}$$

We can deduce an equation of matrix form,

$$[A][P]=[B] \tag{13}$$

where A is matrix of coefficients of pressure, B is array of constant term, P is array of model point prssure.

By solving equations set (13), the pressure of each node of the filled units is gotten. From Eq(7) the mean velocity is expressed as:

$$Vx = \frac{h^2}{3\mu} \frac{\partial P}{\partial x} \tag{14-A}$$

$$Vy = \frac{h^2}{3\mu} \frac{\partial P}{\partial y} \tag{14-B}$$

then maximum shear rate can be expressed by

$$\dot{\gamma} = 3 \sqrt{V_x^2 + V_y^2} / h \tag{15}$$

when $P(i,j)$ is known, the mean velocity and maximum shear rate of each node is obtained as

$$Vx(i,j) = \frac{h^2}{6\mu \Delta x} [P(i-1,j) - P(i+1,j)] \tag{16-A}$$

$$Vy(i,j) = \frac{h^2}{6\mu \Delta y} [P(i,j-1) - P(i,j+1)] \tag{16-B}$$

$$\dot{\gamma} = 3 \sqrt{[Vx(i,j)]^2 + [Vy(i,j)]^2} / h \tag{17}$$

Eqations set(14) and Eq(16) and Eq(17) are used to analyse melt flow process in a thin cavity.

System Structure of the Program

The flowchart of the computational program is shown in Fig.4.

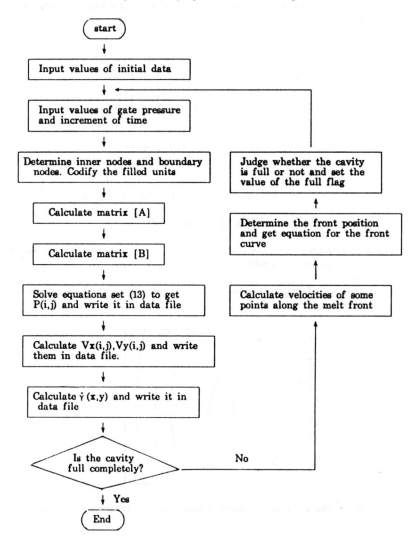

Fig.4 Flowchart of computational program

Illustration Example

The rectangular cavity dimension is 45mm×47.5mm×2mm. Since the cavity is symmetric, we need only analyse the upper section. The upper section is devided into 171 units with 200 nodes, as shown in Fig.5.

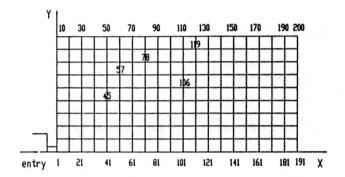

Fig.5 Mesh configuration of the rectangular cavity

Assuming that gate pressure is constant, and letting P be 100KPa and μ be 10 Ns/m^2. The pressures and velocities of points which are not just on the nodes can be calculated by linear interpolation.

By the program, melt front profiles, pressures, and velocities at different time are obtained and shown in Fig.6 and Table 1.

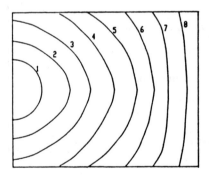

Fig.6 The shape and position of melt front at different time

Table 1 Pressures, velocities and shear rates of some nodes at the time when melt front is at the seventh position.

node number	21	45	57	78	81	106	119	131
pressure(KPa)	55.35	34.88	29.41	23.09	22.62	14.21	9.83	5.55
velocity (cm/s) Vx	60.61	17.16	14.53	15.91	20.09	17.77	18.87	16.74
velocity (cm/s) Vy	0	8.95	4.61	1.98	0	1.97	1.16	0
shear rate(1/s)	909.1	290.3	228.7	240.5	601.4	268.1	283.6	251.0

References

1. Tang Zhiyu, *Application of Rheology in Engineering Design of Plastics Die and Mold*, (National defense Industry Press,1991).
2. Wang Xingtian, *Injection molding Technology* (National Chemical Industry Press,1989).
3. Lu Jinfu, Guan Zhi, *Numerical Solution to Practical Differential Equations*, (Tsinghua University Press, 1986).
4. Sung Kuk Soh and Chin Jui Chang, *Polym, Eng.Sci.* Vol.26,No.12(1986) 893.

THE INFLUENCE OF TOOL GEOMETRY ON BLANKING CHARACTERISTICS

C.M. CHOY
Dept. of Design, Manufacture & Engineering Management
University of Strathclyde, Glasgow G1 1XJ, UK
E-mail: cler16@strath.ac.uk

and

R. BALENDRA
Dept. of Design, Manufacture & Engineering Management
University of Strathclyde, Glasgow G1 1XJ, UK

ABSTRACT

Little is known about the effect of tool geometry on blanking operation; particularly on fracture initiation, burr height and blanking force. The blanking process was modelled using Finite Element method; a plane-strain condition and the Von-Mises yield criterion were assumed to prevail. Simulation was conducted for seven different punch-die clearances and four different punch radii. Simulation with these conditions produced results which resembled previous experimental results. The models enabled an understanding of the stress and strain distribution in the shearing zone. Increases in punch radius disperses the work hardening zone, delays fracture initiation and does not increase the maximum blanking force.

Keywords: Blanking, Finite Element Analysis.

1. Introduction

Blanking is a constrained shearing operation which involves elastic and plastic deformation, and fracture. The process of shearing and the form of the sheared surface (the product quality) are mainly influenced by three factors; these are the tool geometry, the shearing rate and the properties of the work material; investigations have shown a comprehensive understanding of the above factors[1-5]. However, literature on blanking research was based on well-defined punch and die conditions. Little is known about the effect of wear on the blanking tools, particularly the influence on the blank quality (sheared surface and burr height), the nature of crack initiation and propagation, the maximum blanking force and energy consumed. Research [6-8] only reviewed the effect of different levels of wear on punch on product burr height and burnished area. Recent experimental work[9] clearly defined the relationship between different level of punch radius, product quality, autograph characteristic and the maximum blanking force. However, no attempt has been made to conduct theoretical and numerical analysis in this particular aspect.

The objective of this research is to use Finite Element Analysis to obtain a clearer understanding of the effect of tool geometry changes (punch radius) on the product quality and the autograph characteristics. A detail study of the distribution of stress and strain near the punch and die shearing edges, crack initiation and propagation, maximum blanking force and burr height of the blanked component was conducted. The FE results were compared with the previous experimental results[9].

2. Materials And Procedures

2.1. FE Modelling

ABAQUS FE code[10] was used for blanking simulation. PATRAN was used for pre & post-processing. Since the 3-D plastic analysis of material deformation is time-consuming and the punch-die clearance(c/t) is usually very small in relation to the blank diameter, the plane-strain model was used to enable an approximate simulation of the crack initiation and propagation during blanking. Figure 1 shows the two-dimensional blanking model. As the constrained shearing operation is symmetrical about the centreline, only one half of the model (region ABCD) needs to be analysed (Fig. 2). The region near the punch and die shearing edges, where cracks would be initiated, was modelled with a mesh containing 400 axisymmetric quadrilateral elements (Fig. 3). The punch and the die were modelled as rigid bodies. Interface elements were used to model the contact and friction between the sheet and these rigid bodies. The friction coefficient (μ) between tools and the work material was assumed to be 0.25.

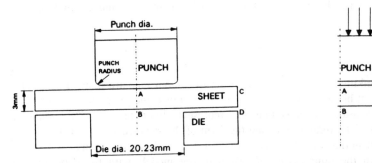

Fig. 1 2-D Blanking Model

Fig. 2 Blanking Model Used for FEA

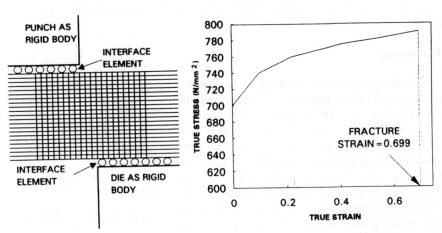

Fig. 3 Constrained Shearing Zone

Fig. 4 True Stress/Strain Curve

2.2. Crack Initiation Criterion

Material used for the experiments were 3mm thick 0.16% carbon steel. Young's Modules is 2×10^5 N/mm^2 and the Poisson's ratio is 0.3. The true stress and true strain curve (Fig. 4) of the material was obtained from tensile tests. It is assumed that the material is isotropic and yield occurred as prescribed by Von-Mises yield criterion. As crack initiation in constrained shear is a local phenomenon, it is postulated that a crack is initiated when the effective strain (ϵ) of the sheet, near the punch and die shearing region, reaches the fracture strain (ϵ_f) of the material (Fig. 4).

2.3. Procedure

Experiments with seven punch-die clearances (c/t), 1%, 7%, 10%, 12%, 14%, 16% and 18%, and four punch shearing edge conditions, R=0mm, R=0.2mm, R=0.4mm and R=0.6mm, was conducted. A total of twenty-eight blanking models were generated and analysed. The incremental value of the punch penetration was 0.04mm. When the effective strain (ϵ) reached the fracture strain (ϵ_f), crack was initiated and the analysis was terminated.

3. Result

Eight sets of blanking models with effective strain contour map are shown Fig. 6. The first four sets represent blanking with 1% punch-die clearance and punch radii of R=0mm to R=0.6mm (Fig. 6 a-d). The next four sets represent blanking with 7% punch-die clearance (Fig 6 e-h). The figures illustrate the form of the strained zone and the plastic deformation of the material prior to crack initiation. Fig. 7 shows the comparison of the crack initiation from the punch edge of two blanking models with the experimental results[9]. Three sets of autograph pertaining to 1%, 7% and 18% punch-die clearance are shown in Fig. 8. Each set of autograph consists of four curves which represent the effect of different punch radii on blanking characteristics. Fig. 9 and Fig. 10 summarise the effect of the punch radii on the maximum blanking force and the punch displacement prior to crack initiation respectively.

4. Discussions

In blanking operations, wear takes place on the punch shearing edge, punch free surface and the punch face (Fig. 5). The blanking models used did not take into account the wear on the punch free surface and the punch face into consideration; wear on the punch was represented by different radius on the shearing edge. As the main aim was to study the effect of punch radii on the sheared surface and maximum blanking force required for blanking, quality characteristics of blanking, such as doming, dishing and roll over were not evaluated.

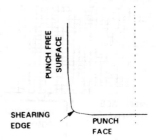

Fig. 5 Wear on Punch

4.1. Crack Initiation

Fig. 6a to 6h clearly show that crack initiation and propagation are not only dependent on the punch-die clearance but was also sensitive to the punch radius. The blanking models in Fig. 6a and 6e show that, when tools without radii are used, stress and strain distributions are limited to a narrow area near the punch and die shearing edges. The cracks are initiated and propagated simultaneously from the punch and die shearing edges when blanking with a 1% punch-die clearance (Fig. 6a). However, when blanking with a 7% punch-die clearance (Fig. 6e), the crack is initiated form the punch shearing and then followed by the fracture near the die shearing edge. When punch radii were introduced into the blanking models, the work hardening zone near the punch shearing edge was dispersed; this agrees with experimental results[9]. The punch radius promotes more plastic deformation and also suppresses the initiation of cracks[9,11]. Fig. 6d and 6h show that effective strain (ϵ) values of the material near the punch shearing edge reduce substantially when punch radii are increased from 0.2mm to 0.6mm. Thus, when the punch penetrates further into the material, the crack is initiated and propagates from the die edge when the effective strain (ϵ) value has reached the fracture strain value (ϵ_f). The models (Fig. 7a and 7c) show that the crack is initiated form the free surface of the punch. The predicted crack initiation and burr height form the blanking models (Fig. 7a and 7c) are in agreement with the experimental results[9] (Fig. 7b and 7d).

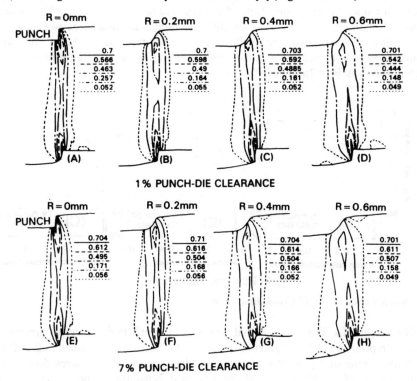

Fig. 6 Blanking Models with Effective Strain Contour Map

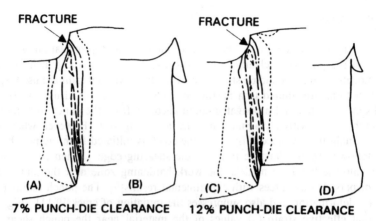

Fig. 7 Comparison of Blanking Models with Experimental Results[9]

4.2. Maximum Blanking Force

Fig. 9 shows that maximum blanking force does not bear an exact relationship to punch radius. However, it was observed that the maximum blanking force is a maximum when the punch radius R=0.2mm. The graph also shows that further increases of punch radius will reduce the maximum punch force required to effect blanking. This is the case for all the blanking models for which different punch-die clearances were considered. This result contradicts earlier results[8]. However, the result is similar to the previous research[9] which showed that the maximum blanking force reduced for increases in the punch radius from R=0.2mm to R=0.6mm.

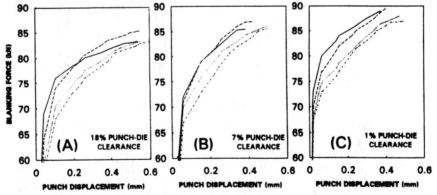

Fig. 8 Blanking Autographs

4.3. Punch Penetration At Crack Initiation

The autographs (Fig. 8) show that increases in punch radius requires greater punch penetration prior to crack initiation of the blank. As discussed, the punch radius disperses the strained zone and promotes more plastic deformation, this requires deeper punch penetration to effect blanking and metallurgical separation. The graph in Fig. 10 summarises the punch displacement required to effect blanking with respect to various punch radii and punch-die clearances.

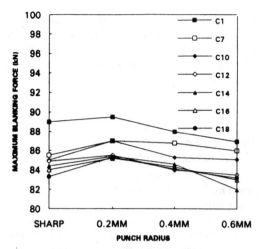

Fig. 9 Max. Blanking Force

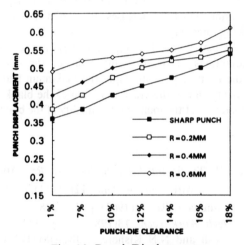

Fig. 10 Punch Displacement

4.4. Blanking Process Control

In the sheetmetal working industries, monitoring of the blanking process, the tools condition and the quality of blanked component, is achieved by monitoring the maximum blanking force. If the maximum blanking force is outside a defined range, the process is defined to be out of control. However, investigation shows that maximum blanking force does not necessarily increase with the bigger punch radii. Fig. 9 shows that maximum blanking force required by a punch with sharp edge and a punch with R=0.4mm are almost identical. Thus, monitoring of the maximum blanking force may not be a suitable method of defining the conditions of the tools or the quality of the blanks. Fig. 8 shows that punch radius modifies the form of the autograph, the maximum blanking force and the punch displacement prior to crack initiation of the blank, monitoring of these three variables may be required in order to enable the determination the punch radius and the burr height of the blank. This could be achieved by mapping the whole or the partial autograph during the process[12].

5. Conclusions

The following conclusions are drawn:-
1. Increases in punch radius requires greater punch penetration prior to crack initiation and metallurgical separation.
2. Increases in punch radius disperses the strained zone, delays crack initiation from the punch shearing edge.
3. Cracks initiate and propagate simultaneously from the punch and the die shearing edges when 1% punch-die clearance is used. When 7% punch-die clearance is used, crack is initiated form the punch shearing edge and then followed by the shear at the die shearing edge.
4. The blanking force, required to effect blanking, is a maximum when the punch radius is R=0.2mm. Further increases in punch radius will decrease the maximum blanking force required.
5. Monitoring of the maximum blanking force may not be a suitable method of defining the conditions of the blanking tools or the quality of the blanks. Monitoring of the form of the autograph, the maximum blanking force and the punch displacement prior to the crack initiation of the blank may be required to determine the condition of the tools and the quality of the blank.

6. References

1. R. Balendra and F.W. Travis, *Int. J. Mach. Tool Des. Res.*, **10**(1970), p. 249-271.
2. T.M. Chang and H.W. Swift, *J. Ins of Metal*, **128**(1951), p. 393-414.
3. M. Masuda and T. Jimma, CIRP, **11**(1962), p. 224-228.
4. P.B. Popat, *J. of Mech. Work. Tech.*, **18**(1989), p. 269-282.
5. P.B. Popat, A. Ghosh and N.N Kishore, 12th AIMTDR Conference, (1986), p.423-427.
6. E. Sondershaus, *Mit. Dtsch. Forschungsges. Blechverarb. u. Oberflachenbehand.*, **19**(1968), p. 142-156.
7. K. Buchmann, *Werkstattstechnik*, **53**(1963), p. 128-134.
8. T.W. Timmerbeil, *Werkstattstech. u. Maschinenbau*, 46(1956), p. 58-66.
9. C.M. Choy and R. Balendra, *Proceeding of the 10th National Conference on Manufacturing Research*, (Loughborough, UK,1994), p. 582-586.
10. Hibbitt, Karlsson & Sorensen Inc. 1992, ABAQUS User Manual.
11. A Metallographic Study of the Mechanism of Constrained Shear, *Production Engineering Research Association*, 93(1961).
12. C.M. Choy and R. Balendra, *Proceeding of the 10th National Conference on Computer-Aided Production Engineering*, (Palermo, Italy,1994), p. 30-38.

FE ANALYSIS OF PRE-STRESSED PRESS FRAMES

R. Balendra, H. Ou, X. Lu
DMEM, University of Strathclyde, UK
E-mail: hengan@zephyr.dmem.strath.ac.uk
and
K. Chodnikiewicz
Warsaw University of Technology, POLAND

ABSTRACT

This paper deals with FE modelling and analysis of pre-stressed press frames. The behaviour of a simplified pre-stressed frame was simulated and an iterative FE solution is proposed for determining the initial pre-stressing force in the tie-rods. Results derived from the FE simulation show that, in spite of a high pre-stressed force, the operational load may cause a partial gap between parts of the frame. The approach can be used to calculate the stresses, strains, contact pressures and pre-stressing force of frames of other types of machines.

1. Introduction

The accuracy of components formed by the deformation of materials depends, to a large extent, on the vertical (axial) stiffness C of the press; the stiffness may be defined as follows:

$$C = \frac{C_f C_d}{C_f + C_d} \qquad (1)$$

where C_f is the vertical stiffness of the press frame and C_d is the vertical stiffness of the driving system. In the case of stroke-restricted presses, C_f influences the accuracy of the component directly; the dimension affected would be the height of the component [1,2].

Press designers use computational methods which enable the prediction of the stiffness of press frames [3-5]. Large-capacity presses are designed, as a rule, with closed pre-stressed frames. Frames of this type comprise at least the following members: bed with columns (referred to as lower frame), cross-head and two tie-rods. Often, the bed is constructed as a member which is independent of the columns. Such frames are pre-stressed during assembly.

Traditional methods of calculating the pre-stressing force which have been used for several decades [1,6] and are based on the assumption that the stiffness of the tie-rods and the columns are known. In such cases, it is possible to plot the elastic characteristics of both the tie-rods and the columns as shown in Fig.1. These trends intersect at a point P which represents the force equilibrium in the assembly. If this assembled frame were subjected to a forming force F, the tie-rods would be loaded with a force $F_R > F_P$ whereas the frame would be subjected to a force $F_F < F_P$. Normally, it is a requirement that the interface between the two members should remain in compression under the forming force. According to the traditional method, the pre-stressing force F_P should fulfill the relationship $F_P = kF_N$, in which F_N is the nominal forming force (the nominal force capacity of the press) and the coefficient k is a function of the nature of the load on the press; the following values have been recommended for this constant [5,7]: $k \geq 1.07$ for presses which are subjected to axial loads only and $k \approx 1.4$ for presses which are subjected to off-set, axial loads.

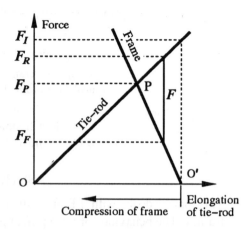

Fig.1 Loading diagram for a prestressed frame.

The stiffness of a press frame, made up of several members, may only be defined using nonlinear force-deflection relationships because of the contact interactions. In such cases, the traditional method [3,7] which relies on an accurate definition of the elastic characteristics of both, the tie-rod and the prestressed structure, is not reliable. For this reason, the behaviour of a simplified pre-stressed frame was examined using FE simulation; the research defines a general procedure for FE simulation of pre-stressed metal-forming presses which is made up of several members. An automatic FE model generator was developed using the PCL programming tools contained in PATRAN and the FE simulation of the deflection of the frame was performed using ABAQUS together with the standard interfaces for data exchange between the FE model and the solution. An interface was developed to define the contact elements between contact members of the press. An iterative FE solution is proposed for determining the initial pre-stressing force in the tie-rods.

Results derived from the FE simulation show that, in spite of a high pre-stressed force, the external load causes a partial gap between parts of the frame. The approach may be used to compute the pre-stressing force, stresses, strains and contact pressures for other types of presses.

2. FE Simulation of Pre-stressed Press Frames

2.1 Simplified Pre-stressed Frame

A simplified frame was used as an illustration and is defined with the assumption that its structure is symmetrical and that the external load (the forming force) acts vertically along the axis of symmetry (Fig.2). It is therefore, possible to

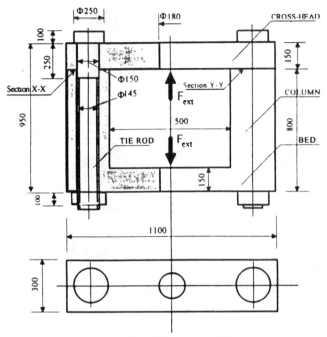

Fig. 2 Prestressed frame

consider half the structure for the FE simulation. Three components are considered in the FE model, (1) cross head, (2) lower frame (bed and column) and (3) tie-rod. As the deflection and stress-flux of the whole structure is the main concern and not the detailed stress distribution at the interface of the tie-rod and the nut, these members are treated as an integral component in which the local stress concentration is neglected. However, both parts of the frame and the tie-rod are treated as individual components which are in contact with each other on specific surfaces. The cross-head is in contact with the lower frame along the surfaces X-X and Y-Y in Fig.2.

2.2 Procedure for FE Simulation

The majority of available CAD software and FE codes cannot provide complete support for the analysis of complex structures; in some cases, the

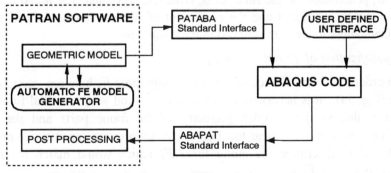

Fig.3 Schematic diagram of FE analysis procedure

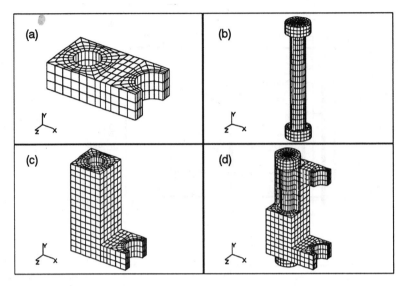

Fig.4 FE mesh of the components of the press, a) cross head, b) tie-rod, c) lower frame, d) assembly press with a revealing show of inside connection.

structure is too complex to model while in others, the CAD and FE software are insufficient, in capacity terms. In such circumstances a specific computational procedure would have to be adopted and developed.

A schematic diagram of the FE analysis procedure is shown in Fig.3. PATRAN [8] and ABAQUS [9] were used for pre/post processing and FE computation respectively. Data exchange between PATRAN and ABAQUS was effected using standard interfaces PATABA and ABAPAT. An automatic FE model generator for press frames was developed using the PCL programming tool contained in PATRAN; this auxiliary generator requires only a limited number of input parameters for the generation of the FE model of a press frame.

Fig.4 shows the mesh for individual members and for the assembled frame. An interface was also developed for the application of the contact element, GAP, at the contact surfaces. The GAP element in ABAQUS enables simulation of deflection and stress along contacting surfaces by allowing sliding and separation to occur; penetration of the surfaces is, however, prevented. The initial clearances between contact surfaces and the coefficient of friction can be defined as input data. Coulomb friction was assumed in the simulation.

2.3. Qualification of Initial Force F_I

In order to achieve the required pre-stressing force F_P between sections of the frame (Fig. 5a), it is necessary to elongate the tie-rod with an initial force of F_I. However, due to the complex geometry of the frame parts and the contact deflection which is a nonlinear function of contact pressure, an iterative procedure is required to determine the initial force F_I which would match the required pre-stressing force F_P.

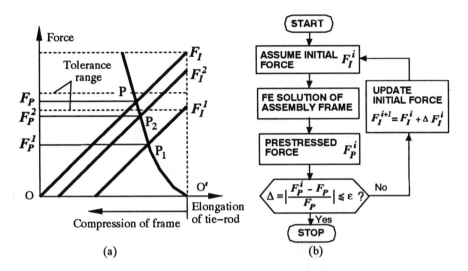

Fig.5 Determination of initial force, a) Loading diagram, b) Iteration procedure.

The procedure is initiated by assuming an initial force F_I^i, where i is the iteration superscript; applying this initial force on the tie-rod, the corresponding pre-stressing force F_P^i can be calculated using FE simulation. The subsequent estimate of the initial force can be determined by

$$F_I^{i+1} = F_I^i + \Delta F_I^i \qquad (2)$$

where Δ^i is defined as the relative error of the pre-stressing force,

$$\Delta = (F_P^i - F_P)/F_P \qquad (3)$$

When the error falls within the defined tolerance, $|\Delta| < \varepsilon$, the iteration terminates. At this stage, the press is under the previously defined prestressing condition without an external force. The flowchart of this iterative procedure is shown in Fig.5b. Subsequently, the external force can be applied on the prestressed press to simulate the deflection responses under the loading conditions.

3. Results

FE simulation was used to evaluate the deflection behaviour of the frame shown in Fig.2. It is assumed that the external force on the press is $F_N = 4000kN$. A high value of prestressing force was assumed, $F_P = 2 \times F_N = 8000kN$; the corresponding initial force in one tie-rod was calculated from the FE simulation $F_I = 5200kN$.

The deflection of the pre-stressed frame is shown in Fig.6. The total deflection of the frame was $2.23mm$; the deflection of the cross head was $1.36mm$, while that of the lower frame was $0.874mm$. The contact surfaces between the cross head and the lower frame partly separate as shown in Fig.6d. The maximum opening of the gap was $0.995mm$.

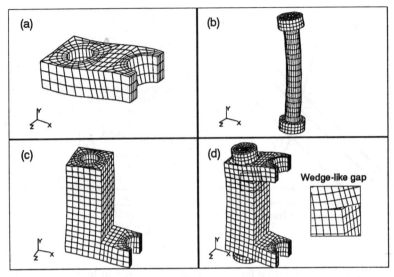

Fig.6 Deflection of components and the whole press under the external force.

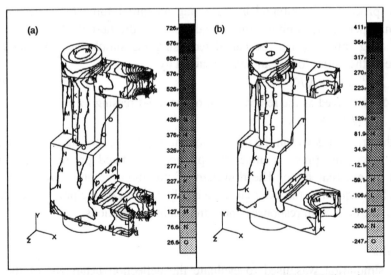

Fig.7 Stress contour plot, a) Mises stress, b) Stress along Y direction.

The Mises equivalent stress contours and the axial stress contours along the direction of the external load are shown in Fig.7. The maximum Mises stress in the structure was $726MPa$ on the central part of upper frame. The maximum contact stress at the joint interface has a value of $247MPa$; at this section, the tensile stress in the tie-rod was $441MPa$.

4. Conclusions

The implementation of the FE modelling and the analysis of the pre-stressed press frames enable the following conclusions:

(a). The automatic FE model generator which was developed can be used to simulate the behaviour of a pre-stressed press frame; the simulation enables the definition of the extent of the separation at the joint interface, during forming operations. The use of this generator increases the efficiency of the simulation procedure.

(b). An iterative FE solution which has been implemented can be used for determining the initial force for pre-stressing of frames.

(c). The wedged-shaped separation which occurs between contact members of the frame cannot be predicted by traditional mechanics of materials considerations. FE analysis enables the simulation of the incidence of such separation.

References

1. Lange, K., Handbook of Metal-forming. McGraw-Hill Book Company. New York (1985).
2. Zhang, H. G. and Dean, T., Computer modelling of tool loads and press/tool deflections in multi-stage forging, Int J Mach and Manf, Vol 35, No1 (1995).
3. Makelt, H., Mechanische Pressen. Munchen, Hanser (1961).
4. Hupfer, P., Berechnung von Pressesgestellen in Zweistander-und Saulenbauart. Maschinenbautechnik 14, Heft 7 (1965).
5. Lanskoj, E. N. and Banketov, A. N., Computational Method of Mechanical Presses (in Russian), Moscow (1966) .
6. Shigley, J. E. and Mischke, C. R., Mechanical Engineering Design. McGraw-Hill, New York (1989).
7. Bauketov, A. N. and Lauskoj, J. E., Metal-Forming Machinery (in Russian). Moscow (1982).
8. PATRAN Plus user Manual, PDA Engineering (1990).
9. ABAQUS User Manual, Hibbitt, Karlsson & Sorensen Inc. (1992).

MANUFACTURING PLANNING AND CONTROL

MANUFACTURING PLANNING & CONTROL

Cellular Group Technology

An Improved Assignment Model for Group Technology Application
    A. K. Agrawal and Abhinav

An Agile Line Balancing Procedure for Versatile Market
    H. Katayama and M. Tanaka

Simulated Annealing Approach to Group Technology
    S. M. Sharma and N. Viswanadham

Fuzzy Set Based Machine-Cell Formation in Cellular Manufacturing
    C. W. Leem and J. J. G. Chen

AN IMPROVED ASSIGNMENT MODEL FOR GROUP TECHNOLOGY APPLICATION

ANIL KUMAR AGRAWAL
Mechanical Engineering Department, Institute of Technology, Banaras Hindu University
Varanasi - 221 005, INDIA

and

ABHINAV
Mechanical Engineer (Jr.), Geological Survey of India
Calcutta - 700 016, INDIA

ABSTRACT

For increasing the productivity of mid-volume and mid-variety batch manufacturing system, emphasis is given to the division of the system into a number of subsystems each containing a set of machines desiring to process independent set of parts having similar manufacturing requirement. The process of division based on the simple concept of exploiting processing similarity of parts is one such application of Group Technology. Of a number of approaches that have been proposed by the researchers for this purpose, is one the assignment model due to Srinivasan et al. [8]. This paper discusses some of the weaknesses of this model and suggests certain modifications for improvement. Justification of modifications is established by solving a number of example problems taken from the related literature.

KEYWORDS: Group Technology, Similarity Coefficient and Assignment Technique.

1. Introduction

Group Technology (GT) is a philosophy applied for organizing, planning and control of industrial systems, viewed generally as a set of interrelated entities as parts and machines with independent attributes, to achieve a common goal of producing useful products and/or services. The approaches for GT application are normally based on exploitation of similarity or commonness amongst the various parts and machines to form groups or clusters. The basic idea behind GT is to decompose a manufacturing system into subsystems. Parts are classified into part families and machine into machine cells based on their similarity such that every member of the part family undergoes almost all its processing in the assigned machine cell. For this objective of group formation a number of approaches have been developed over time. Details of these approaches can be found in the work of King and Nakornchai [14], and Ballakur and Steudel [1].

A majority of GT cell formation approaches make use of some kind of similarity defined either for a pair of parts, or for a pair of machines, or for a pair of a part and a machine. Among these probably the oldest one is the Jaccard's Similarity Coefficient (JSC) first used by McAuley [12] for GT application. Other popular measure is Simple

Matching Similarity Coefficient (MSC) used by Kusiak [2], Waghodekar and Sahu [21] and many others. The superiority of JSC over MSC has been illustrated by Chandrasekharan and Rajagopalan [20], and also by Shanker and Agrawal [17]. The similarity coefficient measures simply identify the closeness and then some kind of procedure is applied to form the groups. These procedures can be purely heuristic or based on optimization models. The heuristics based on similarity coefficient have been developed by several researchers such as Mosier [6], Mosier and Taube [7], and Seifoddini and Wolfe [11] to mention a few. Methods formulating the GT problem as a mathematical programming problem are p-median formulation [2], graph theoretic formulations [5,15,24] besides several others. Some of these methods are improvement over the earlier proposed methodologies. The Average Linkage Clustering (ALC) methodology due to Seifoddini [9] is an improvement over Single Linkage Clustering [12] methodology. Similarly, to remove the deficiencies of p-median formulation, a heuristic approach has been suggested by Srinivasan et al. [8] that uses assignment technique for initial group formation.

The assignment technique based model for GT cell formation [8] uses assignment technique on similarity coefficient matrix to identify groups of machines and/or parts. In this assignment method, one part (machine) is assigned to only one of the other parts (machines) with which it is expected to have maximum similarity while maximizing the total similarity of such assignments. In this kind of assignment where no two parts can be assigned to one single part, chances are that some of the parts may not be assigned to parts with which they have maximum similarity. However, a good number of parts may have maximum similarity with a single part and thus need to be assigned to that part. Thus the assignment method goes against the fundamental concept of natural alignment of like parts for the formation of the groups. The second weakness in the model is due to the use of MSC similarity measure that compared to JSC has, as mentioned earlier, lower discriminating power.

The purpose of the present work is to bring in improvement in the assignment model by modifying those steps of their stepwise procedure that are unnecessary and contrary to GT principle of yielding natural groups. The efforts have been to make the procedure more efficient to result better grouping solution. Section 2 contains the details of the modified assignment model. The steps of this proposed model are further clarified by taking an example problem in Section 3. Comparative performance of the two assignment models, original and proposed, has been studied in section 4. The last section contains conclusions.

2. The Modified Assignment Model

The proposed model starts for assignment with first row and makes an assignment at that cell of the matrix where it finds the corresponding entry to be the maximum in that row. In case there are more than one such entry, then assignment is made at the first entry from the left. This is continued for all the rows. This will clearly show as which machine (part) represented in the matrix by a row is to be grouped with which machine (part) represented in the matrix by a column. A column machine (part) together with those machines (parts) that are assigned to it represents a separate machine-cell (part-families) formed at this stage.

In the proposed method of assignment, one may encounter situations where machine-cells (part-families) may overlap as some of the machines (parts), after assignment process as discussed above, may be in more than one group thus formed. To resolve this problem, the concepts used in the Average Linkage Clustering (ALC) methodology of Seifoddini [9] is used. A machine that overlaps over two or more machine-cells is assigned to that machine-cell with which the average of its similarity with the cell members is the maximum. Similarly overlapping parts are assigned. Rest of the steps of the model are the same except the refinement method which is described below.

2.1. Group Refinement

In the grouping results from the assignment models, one may find that a part is not in a group where it can find its maximum processing requirements met. Similarly, a machine may not be in a group where it is maximally required. Such a grouping means more number of exceptional elements and thus more of intercell movements. However, a better grouping means least number of intercell movements.

In view of the above, the machines and parts that are not properly assigned are reallocated to their proper cells and families, respectively. First all the machines and then all the parts are checked. This process continues until there is no more reassignment. Then similar approach is followed to reduce the voids subject to the condition that there is no increase in the number of exceptional elements. This results in the minimum number of exceptional elements for the total groups formed, and for that number of exceptional elements least number of voids.

2.2. The Stepwise Procedure

The proposed assignment model consists of the following steps.

Step 1: Compute the Jaccard's Similarity Coefficient between all the pairs of machines and construct matrix S of these coefficients.

Step 2: Use matrix S thus obtained to assign the machines to those with which they are maximally similar.

Step 3: Identify the groups from the assignments made above in step 2.

Step 4: Merge a machine-cell into another machine-cell if it is a subset of the other. Repeat this process until no further merger of machine-cells is possible.

Step 5: In case a machine is in two or more groups, then allow it to stay in that machine-cell with which its average similarity is maximum. Ties can be broken arbitrarily.

Step 6: For each such machine-cell formed, list the parts that visit them.

Step 7: Now scan the list of parts visiting each machine-cell. Wherever the part-family corresponding to a cell is a subset of another, merge the two machine-cells together. Repeat this process until no further merger of machine-cells is possible.

Step 8: If the part-families are disjoint, stop.

Step 9: Repeat steps 1 to 5 for parts to determine part-families.

Step 10: Assign part-family l to machine-cell k such that $Q = \sum_{i \in \{m_k\}} \sum_{j \in \{n_l\}} a_{ij}$ is maximum.
Notation $\{m_k\}$ denotes the set of machines in cell k and $\{n_l\}$ the set of parts in family l, and a_{ij} is one if machine i is required by part j and zero otherwise. Repeat this procedure to assign all the part-families. Ties can be broken arbitrarily.

Step 11: If there is any machine group k that has no part-family assigned to it, merge it with a group l such that $Q = \sum_{i \in \{m_k\}} \sum_{j \in \{n_l\}} a_{ij}$ is maximum. Repeat this step till all such machine groups are assigned. Ties can be broken arbitrarily.

Step 12: Refine the groups thus formed.

Step 13: Merge two groups and their part-families if the number of voids created by the merger is not more than the number of exceptional elements eliminated by the merger. In case of merger go to step 11, else stop.

3. An Example Problem

For a better understanding of the proposed model, consider an example problem shown in Fig. 1. The similarity coefficient matrix of JSC for machines is shown in Fig. 2. The places where assignments are made are marked by '*'.

PART

	1	2	3	4	5	6	7	8	9	10
M 1	1	1			1					
A 2	1			1						
C 3										1
H 4	1	1		1						
I 5			1		1					
N 6	1					1	1	1		
E 7		1	1		1	1		1		

MACHINE

	1	2	3	4	5	6	7
M 1	0.00	0.00	0.00	0.20	0.00	0.00	0.33*
A 2	0.00	0.00	0.25*	0.20	0.25	0.14	0.00
C 3	0.00	0.25*	0.00	0.00	0.00	0.17	0.00
H 4	0.20	0.20	0.00	0.00	0.00	0.14	0.33*
I 5	0.00	0.25*	0.00	0.00	0.00	0.17	0.00
N 6	0.00	0.14	0.17	0.14	0.17	0.00	0.25*
E 7	0.33*	0.00	0.00	0.33	0.00	0.25	0.00

Fig. 1: Machine Part Incidence Matrix.   Fig. 2: Matrix of Similarity Coefficients for Machines.

These assignments lead to four machine-cells: {1,7}, {2,3,5}, {3,2} and {1,4,6,7}. Since machine-cell {1,7} is a subset of machine-cell {1,4,6,7}, these two are merged together and so the machine-cells {3,2} and {2,3,5}. Finally we get two machine-cells:

        Machine-cell # 1 = {1,4,6,7}
and   Machine-cell # 2 = {2,3,5}.

Corresponding to these two machine-cells, the two respective part-families are:

        Part-family # 1 = {1,2,3,4,6,7,8,9,10}
and   Part-family # 2 = {1,5,8,10,11}.

Since the part-families are not disjoint due to parts 1, 8 and 10, step 9 of the procedure is followed to determine part-families. The similarity coefficient matrix for parts is formed and assignments are made at appropriate places as shown in Fig. 3 marked by '*'.

From these assignments, seven part-families are formed. They are: {3,6,7}, {1,4,8,9,10}, {5,11}, {6,3}, {7,2}, {8,5} and {9,4}. Since part-family (6,3) is a subset of part-family {3,6,7}, they are merged together and so part-families {9,4} and {1,4,8,9,10}. The following part-families are obtained: {3,6,7}, {1,4,8,9,10}, {5,11}, {2,7} and {5,8}. Since parts 5, 7 and 8 are overlapping each over two part-families, the concepts of average linkage clustering (ALC) is to be used to assign these parts to their proper part-families. Using concepts of ALC, parts 5, 7 and 8 are assigned to families {5,11}, {2,7}, and {1,4,8,9,10}, respectively. Finally, the following part-families are obtained: Part-family # 1 = {3,6}, Part-family # 2 = {2,7}, Part-family # 3 = {5,11} and Part-family # 4 = {1,4,8,9,10}.

PART

		1	2	3	4	5	6	7	8	9	10	11
	1	0.00	0.00	0.20	0.25*	0.25	0.25	0.00	0.25	0.25	0.25	0.25
	2	0.00	0.00	0.33	0.00	0.00	0.00	0.50*	0.00	0.00	0.00	0.00
	3	0.20	0.33	0.00	0.25	0.00	0.67*	0.67	0.00	0.25	0.00	0.00
P	4	0.25	0.00	0.25	0.00	0.00	0.33	0.33	0.33	1.00*	0.33	0.00
A	5	0.25	0.00	0.00	0.00	0.00	0.00	0.00	0.33*	0.00	0.00	0.33
R	6	0.25	0.00	0.67*	0.33	0.00	0.00	0.33	0.00	0.33	0.00	0.00
T	7	0.00	0.50	0.67*	0.33	0.00	0.33	0.00	0.00	0.33	0.00	0.00
	8	0.25	0.00	0.00	0.33*	0.33	0.00	0.00	0.00	0.33	0.33	0.00
	9	0.25	0.00	0.25	1.00*	0.00	0.33	0.33	0.33	0.00	0.33	0.00
	10	0.25	0.00	0.00	0.33*	0.00	0.00	0.00	0.33	0.33	0.00	0.33
	11	0.25	0.00	0.00	0.00	0.33*	0.00	0.00	0.00	0.00	0.33	0.00

Fig. 3: Matrix of Similarity Coefficients for Parts.

These four part-families are assigned to the two machine-cells based on the similarities between them (steps 10 and 11). The details of part-families' assignment to the machine-cells are given below.

Group No.	Machines in cell	Parts in family
1	{2,3,5}	{5,11}
2	{1,4,6,7}	{1,2,3,4,6,7,8,9,10}

At this stage the number of exceptional elements is 3 and of voids is 22. Group refinement does not help to reduce the number of exceptional elements, but the number of voids to 20 by assigning parts 8 and 10 now to the part-family first from the second.

The details of exceptional elements and voids for this grouping can be had from the rearranged part-machine matrix as shown in Fig. 4.

4. Performance Evaluation and Discussion

The performance of the modified assignment method (MAM) in comparison with the one (to be referred as SAM) presented by

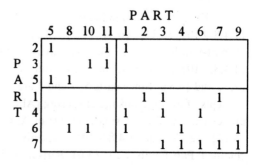

Fig. 4: The Rearranged Machine-Component Matrix.

Srinivasan et al. [8] has been evaluated for 28 different grouping problems available from the literature. The grouping details for these problems are summarized in Table 1.

Table 1: Grouping Results for Different Methods.

Problem No.	Matrix Size	SAM				MAM				Reference No.
		G	E	V	E+V	G	E	V	E+V	
1	5 X 7	2	2	4	6	2	2	3	5	[14]
2	10 X 5	2	0	6	6	2	0	6	6	[23]
3	7 X 11	2	3	18	21	2	3	20	23	[3]
4	8 X 12	4	10	1	11	4	10	1	11	[11]
5	10 X 10	3	0	10	10	3	0	10	10	[7]
6	12 X 10	3	3	7	10	3	3	7	10	[12]
7	12 X 12	4	3	3	6	4	3	3	6	[11]
8	15 X 10	3	1	5	6	3	1	5	6	[1]
9	8 X 20	3	9	0	9	3	9	0	9	[19]
10	10 X 20	4	1	1	2	4	1	1	2	[16]
11	11 X 22	3	15	31	46	3	10	16	26	[10]
12	14 X 24	5	9	15	24	4	2	26	28	[22]
13	14 X 24	4	2	27	29	5	9	14	23	[18]
14	23 X 16	5	37	58	95	3	22	62	84	[25]
15	20 X 20	4	53	66	119	5	55	51	106	[7]
16	24 X 18	5	35	35	70	5	24	40	64	[4]
17	23 X 20	8	62	12	74	5	38	47	85	[15]
18	16 X 30	4	22	26	48	5	21	23	44	[8]
19	16 X 43	6	45	50	95	5	29	53	82	[13]
20	20 X 35	4	6	41	47	4	5	39	44	[4]
21	27 X 27	4	76	91	167	3	38	181	219	[25]
22	24 X 40	7	10	9	19	7	10	9	19	[20]
23	24 X 40	7	24	22	46	7	22	21	43	[20]
24	24 X 40	8	49	41	90	8	46	38	84	[20]
25	24 X 40	10	62	53	115	10	56	33	89	[20]
26	24 X 40	5	47	185	232	8	60	32	92	[20]
27	30 X 41	7	27	124	151	10	30	36	66	[5]
28	30 X 50	11	45	41	86	12	46	20	66	[18]

For comparing the quality of groups, the best solution is taken to be the one in which the sum of exceptional elements (E) and voids (V) is the minimum. This criterion is the same as used by Srinivasan et al. [8]. The total number of groups is denoted by G.

It can be said, in general, that more number of groups will lead to more number of exceptional elements and fewer of voids, and vice-versa. Further, in situations where the number of groups is not changed, decrease in the number of exceptional elements may bring increase in the number of voids and vice-versa.

Cases in which the numbers of groups are the same, comparison is easy to make. For 9 problems out of such 17 cases, the solution obtained is the same. In one case (Problem no. 1) where G and E are the same, V is comparatively less and so (E+V); whereas in one other case (Problem no. 3) though G and E are the same, V is comparatively high and so (E+V). In cellular manufacturing the intercell movements are more of a problem. Thus, due to the same value of E for the same value of G, it can be said that both SAM and MAM have given almost the same results. In other five cases (Problem nos. 11, 20, 23, 24 and 25), for MAM not only E is less but V is also less and so (E+V). It clearly exhibits the superiority of MAM over SAM. In remaining one case (Problem no. 16), V is more due to less E. However, (E+V) is less. It is mainly due to the fact that the group refinement scheme has more priority for minimizing the number of exceptional elements than for voids for the reasons already discussed before.

Out of all the remaining problems, only in three cases (Problem nos. 12, 17 and 21), the sum (E+V) is found to be more. In these cases, decrease in G brings decrease in E with simultaneous increase in V as expected. However, in other two cases (Problem nos. 14 and 19), still with this expected behaviour (E+V) is found to be less. In other cases where G is more, (E+V) is still less. The result for problem no. 18 perfectly shows the superiority of MAM over SAM where even though G is more, E and V both are less and so (E+V).

On solving these problems on computer, it was observed that MAM takes comparatively more CPU time on computer mainly due to group refinement. In addition, it was felt that instead of breaking the tie arbitrarily while assigning part-families and machine-cells to each other in steps 10 and 11 of MAM, use of some other relevant criterion such as least increase in the number of voids may further improve the grouping results.

5. Conclusion

In the present work, certain weaknesses of the assignment model of Srinivasan et al. [8] for group technology application have been removed. Also, a new refinement procedure has been added to it. The superiority due to the proposed modifications and additions in their model in usually giving an improved solution has been established through various numerical examples. The proposed method is found usually to take comparatively more CPU time on computer for solving the problems.

6. References

[1] A. Ballakur and H.J. Steudel, *Int. J. of Prod. Res.* **25** (1987) 639.

[2] A. Kusiak, *Int. J. of Prod. Res.* **25** (1987) 561.
[3] A. Kusiak and W.S. Chow, *J. of Mfg. Syst.* **6** (1987) 117.
[4] A.S. Carrie, *Int. J. of Prod. Res.* **11** (1973) 399.
[5] A. Vannelli and K.R. Kumar, *Int. J. of Prod. Res.* **24** (1986) 387.
[6] C.T. Mosier, *Int. J. of Prod. Res.* **27** (1989) 1811.
[7] C.T. Mosier and L. Taube, *Omega* **13** (1985) 577.
[8] G. Srinivasan, T.T. Narendran and B. Mahadevan, *Int. J. of Prod. Res.* **28** (1990) 145.
[9] H. Seifoddini, *Computers Ind. Engng.* **16** (1989) 419.
[10] H. Seifoddini, *Computers Ind. Engng.* **17** (1989) 609.
[11] H. Seifoddini and P.M. Wolfe, *I.I.E. Trans.* **19** (1986) 266.
[12] J. McAuley, *Prod. Engr.* **5** (1972) 53.
[13] J.R. King, *Int. J. of Prod. Res.* **18** (1980) 213.
[14] J.R. King and V. Nakornchai, *Int. J. of Prod. Res.* **20** (1982) 117.
[15] K.R. Kumar and A. Vannelli, *Int. J. of Prod. Res.* **25** (1987) 1715.
[16] K.S. Badrinarayana, *Formation of Part-Families and Machine-Cells Using Interval Data* (Unpublished Masters Thesis, R.E.C. Calicut, INDIA, 1987).
[17] K. Shanker and A.K. Agrawal, *Proc. of the 2nd Int. Conf. on Comp. Integrated Mfg.* **2 (1993) 690.**
[18] L. Stanfel, *Engg. Costs and Prod. Econ.* **9** (1985) 73.
[19] M.P. Chandrasekharan and R. Rajagopalan, *Int. J. of Prod. Res.* **24** (1986) 1221.
[20] M.P. Chandrasekharan and R. Rajagopalan, *Int. J. of Prod. Res.* **27** (1989) 1035.
[21] P. Waghodekar and S. Sahu, *Int. J. of Prod. Res.* **22** (1984) 937.
[22] R.G. Askin and S.P. Subramanian, *Int. J. of Prod. Res.* 25 (1987) 101.
[23] R. Pannerselvam and K.N. Balasubramanian, *Engg. Costs and Prod. Econ.* **9** (1985) 125.
[24] R. Rajagopalan and J.L. Batra, *Int. J. of Prod. Res.* **13** (1975) 567.
[25] W.T. McCormick (J.), P.J. Scweitzer and T.W. White, *Opns. Res.* **20** (1972) 993.

AN AGILE LINE BALANCING PROCEDURE FOR VERSATILE MARKET

HIROSHI KATAYAMA and MASAYUKI TANAKA

Department of Industrial Engineering and Management,
School of Science and Engineering, Waseda University.
Ohkubo 3-4-1, Shinjuku Ward, Tokyo, Japan.(postal code 169)

ABSTRACT

This paper proposes an agile line balancing procedure of multi-item switching line production system in the versatile market environment. The procedure is composed of database generation of production formation for each product (Step 1) and design of balanced multi-product switching production formation (Step 2). The kernel of Step 1 is conventional single-product line balancing algorithm and various balanced production formations of each item are generated for various cycle time. Problem of Step 2 is formulated in terms of quadratic programming model for stable operation resource (number of workers) utilisation between periods of each item production as well as higher resource utilisation rate. Performance of proposed procedure is evaluated through simulation experiments based on the actual data from the case factory, and results indicate that effective formation is expected to be obtained quickly under practical scene.

Keywords : Agile Manufacturing, Line Balancing, Production Formation

1. Introduction

Recent years, based on diversification of customer needs, shorter product life cycle, higher competitive environment in the global market, principal demand feature is summarised as its versatility. To cope with such unstable market structure, many leading manufacturing enterprises are trying to reform their production systems supported by conventional design and implementation procedures with new production technologies. The class of production activities with such technologies is called agile manufacturing[2], and researches on them are becoming vital these years.

On the other hand, computer based peripheral control, network and software technologies have been advanced and amount of successful application in manufacturing businesses are accomplished in terms of computer integrated manufacturing (CIM), which discharges powerful support roles to manage large- scale, complex production activities.

Looking at the field of line production system management, efforts for capability improvement on the product variety have been paid by increase of product mixture rate, set up time reduction *etc.* so as to alive in the competitive market. Various efforts for constructing CIM have been performed also, as it is widely understood to be a strong weapon to cope with such tuff problem, *i.e.* construction of relevant agile manufacturing technologies. However, current performance of such courageous efforts is still inadequate because of ambiguity increase in the recent market. Understanding these background, this paper deals with an agile management procedure of production formation for the popular efficient manufacturing scheme, *i.e.* multi-product switching line production system.

2. Proposed Procedure

2-1. General Flow of the Procedure

The proposed prototype procedure is constructed through investigation of actual modification procedures of line production system managed in several production management division of factories, and composed of two principal steps, *i.e.* database generation of balanced production formation for each product (Step 1) and design of balanced multi-product switching production formation (Step 2). Main structure and its components, their input and output information are illustrated in Figure 1.

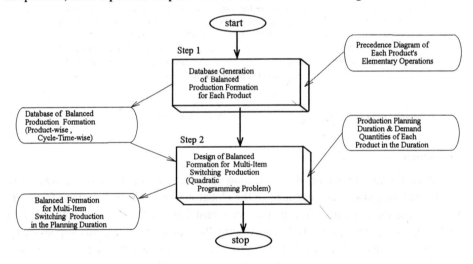

Figure 1. General Flow of the Proposed Procedure

2-2. Database Generation (Step 1)

This step is the preparatory phase of the entire procedure equipped by the conventional single- product line balancing algorithm, and this kernel technology is to generate and maintain various balanced production formations of each item for various cycle time. This database is referred by Step 2 as management knowledge.

The input/output information of the step and outline of the procedure are summarised as follows.
(1) Input Information : precedence diagram of each product's elementary operations
(2) Output Information : database of balanced production formation of single- product line production system (product- wise, cycle- time- wise)

(3) Procedure of the Step : The conventional algorithm used in the single-product line production system design is formulated as follows[3].

< Mathematical Formulation >

Objective Function : Min. ϕ_i

$$\phi_i = \frac{T_i}{J_i \cdot C_i} \tag{1}$$

Where,
- ϕ_i : balancing efficiency of considered product i
- J_i : number of stations for considered product i
- C_i : cycle time for considered product i

$$C_i = \frac{P_i}{Q_i}$$

- T_i : total operation time of considered product i defined by the sum of OT_{ij}

$$T_i = \sum_{j=1}^{J_i} OT_{ij}$$

- OT_{ij} : operation time of station j for considered product i
- Q_i : demand quantity of considered product i in the planning duration H
- P_i : considered length of production period for product i

Constraints :
1) All the precedence relations among elementary operations should be satisfied.
2) Each OT_{ij} must not exceed the cycle time of regarding product i.

$$OT_{ij} \leq C_i \quad \forall j = 1,2,\cdots,J \tag{2}$$

This problem formulated above can be solved by branch and bound method, in which lower bound function is defined as Eq.(3).

Lower Bound of Objective Function for Product i :

$$LB_i = r_i \cdot C_i + q_i \cdot C_i \tag{3}$$

$$q_i = \left\lceil \left\{ T_i - \sum_{j=1}^{r} OT_{ij} \right\} \middle/ C_i \right\rceil \tag{4}$$

So, optimistic estimation of balancing efficiency $\hat{\phi}_i$ is in Eq.(5).

$$\hat{\phi}_i = T_i \Big/ LB_i \tag{5}$$

Where,
- LB_i : the function expressing minimum possible capacity which is necessary to maintain for the production duration
- r_i : number of stations that elementary operations are assigned already
- q_i : minimum possible station number required for assigning the elements remaining at the current node
- $\lceil x \rceil$: minimum integer greater than or equal to x
- $\hat{\phi}_i$: estimated maximum balancing efficiency

2-3. Design Procedure of Balanced Multi-Product Switching Production Formation

This step is the main phase of the entire procedure for desirable multi-product switching production formation equipped by quadratic programming model, which realises stable operation resource (number of workers) allocation as well as higher average resource utilisation under given demand quantities of each product item in the specified production duration.

The input/output of the step and the mathematical representation of the contents are summarised as follows.

(1) Input Information
   ①length of production duration and demand quantities of each product item
   ②database of balanced production formation (product- wise, cycle- time- wise)
(2) Output Information
   multi- product switching production formation realising stable operation, reasonable resource (number of workers) allocation and higher average resource utilisation during the production duration
(3) Procedure of the Step
   For considered multi- product production formation design, the following four step procedure including quadratic programming sub-procedure, which is for stable operation resource (number of workers) utilisation between periods of specified item production as well as higher resource utilisation rate is developed.
   ①initial value setting of target number of workers ($\overline{m}$) in the entire production duration H
   ②quadratic programming problem solving
      Following mathematical model could be solved, e.g. by "HYPER LINGO"[1] generalised non- linear programming package.
< Mathematical Formulation >
   Objective Function : Min. f

$$f = \sum_{i \in I} \sum_{j \in J_i} \left(m_{ij} x_{ij} - \overline{m}\right)^2 \quad (6)$$

Constraints :

$$\underline{\eta} \leq \sum_{j \in J_i} \eta_{ij} x_{ij} \leq \overline{\eta} \,, \quad i \in I \quad (7)$$

$$x_{ij} = 0 \text{ or } 1 \quad (8)$$

$$\sum_{j \in J_i} x_{ij} = 1 \,, \quad i \in I \quad (9)$$

$$\underline{w}H \leq \sum_{i \in I} \sum_{j \in J_i} C_{ij} Q_i x_{ij} \leq \overline{w}H \quad (10)$$

Where,
   I :set of production items
   J_i :set of alternatives of balanced production formation for item i
   $\overline{m}$:target number of workers occupied in the production duration H
   m_{ij} :necessary number of workers for the j- th alternative formation of item i
   η_{ij} :balancing efficiency of the j- th alternative of item i
   $\underline{\eta}$:admissible minimum balancing efficiency
   $\overline{\eta}$:admissible maximum balancing efficiency
   C_{ij} :cycle time of the j- th alternative formation of item i
   Q_i :demand quantity of product item i during the production duration H

H : available production time during the entire production duration
$\underline{w}$: admissible minimum utilisation rate of the production system
$\overline{w}$: admissible maximum utilisation rate of the production system
x_{ij} : 0- 1 decision variables indicating adoption of the j- th alternative of item i
    $x_{ij} = 1$ if the alternative is adopted
    $x_{ij} = 0$ if the alternative is not adopted

③ Is the new f value less than the f value calculated for the last $\overline{m}$?
  Go to Step ④   if true
  Terminate the procedure   if false
④ Increase $\overline{m}$ by 1 and return to the Step ①.

3. Case Studies

3-1. Assumptions/Conditions

Simulation experiments are executed under the following assumptions/conditions to validate the feasibility and effectiveness of the proposed procedure.

(1) Precondition
  Number of Product Item : 4
  Cycle Time : several possible values for each item set by experienced data
  Demand Quantities : several possible values for each item set by experienced data

(2) Feature of Each Product
  Item 1 : product item that spend typical moderate operation- hours
  Item 2 : product item that spend the least operation- hours (the easiest product)
  Item 3 : product item whose process includes operations of basic item 1 and a few additional operations
  Item 4 : product item that spend the biggest operation- hours (the most difficult to produce)

3-2. Precedence Diagram of Each Product's Elementary Operations

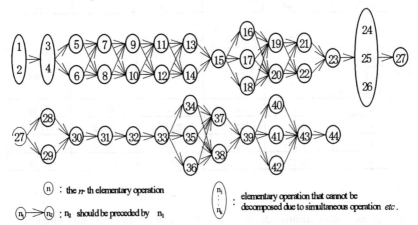

Figure 2. Example of Precedence Diagram (Item 1)

An example of precedence diagram of elementary operations necessary to produce specific product item, which is collected and summarised through factory observation, is illustrated in Figure 2.

3-3. Line Balancing Data for Both Each Item & Cycle Time

Database of balanced production formation for single-product line production system obtained in the procedure 2-2 is illustrated in terms of matrix data as following example table, where each elementary operation is assigned to proper station to fill each cycle time, and balancing efficiency of each case is summarised in the separated table.

Table 1-1. Example Database of Balanced Formation (Item 1)

Cycle Time / Station Number	125.0	150.0	155.2	175.0	200.0
1	1,2	1,2	1 - 4	1 - 4	1 - 5
2	3 - 6	3 - 6	5 - 8	5 - 10	6 - 11
3	7 - 10	7 - 10	9 - 12	11 - 13	12 - 15
4	11,12	11 - 13	13 - 15	14 - 18	16 - 22
5	13,14	14 - 17	16 - 20	19 - 23	23
6	15 - 18	18 - 22	21 - 23	----	----
7	19 - 22	23	----	----	----
8	23	----	----	----	----
9	The Process Operated by Parallel Facilities(PFs) [2-5 PFs]				
10	27,28	27 - 29	27 - 29	27 - 29	27 - 29
11	29,30	30,31	30,31	30,31	30,31
12	31,32	32,33	32,33	32,33	32 - 34
13	33,34	34 - 36	34 - 36	34 - 36	35 - 39
14	35 - 37	37 - 40	37 - 40	37 - 42	40 - 44
15	38 - 40	41 - 44	41 - 44	43,44	----
16	41 - 44	----	----	----	----

Table 1-2. Balancing Efficiency (η) [%]

Cycle Time / Number of Parallel Facilities	125.0	150.0	155.2	175.0	200.0
2	----	----	----	81.44	77.20
3	82.35	77.20	79.59	75.62	71.26
4	78.01	72.66	74.61	70.58	66.17
5	74.11	68.62	70.22	66.17	61.76

3-4. Production Planning Duration and Demand Quantities

The production planning duration and the demand quantity data of each product investigated in this paper are as follows.

(1) Length of Production Planning Duration: one week (= 240000 DM)
(2) Demand Quantities of Each Product: Two fictional cases and one specific case based on the actual demand in recent month are studied as examples.

Table 2. Demand of Each Product Item (Pieces)

Case	No. 1	No. 2	"Actual"
Item 1	300	100	735
Item 2	300	900	128
Item 3	300	100	75
Item 4	300	100	225

3-5. Computational Results

Production formations of the investigated cases and the criteria as well as related variables obtained through the procedure 2-3 are summarised in Table 3, and the performance of the current formation of the "Actual" case is in Table 4.

Table 3. Variables and Criteria of Examined Formation

	Case	No. 1	No. 2	"Actual"
Criteria & Variables				
	$\overline{m}$	14	11	14
	f	0	5	0
	g	81.99	83.33	81.38
Item 1	M	14	11	14
	PFs	2	2	2
	C	163.0	209.5	163.0
	η	81.19	80.40	81.19
Item 2	M	14	11	14
	PFs	0	0	0
	C	108.6	131.9	108.6
	η	80.35	84.20	80.35
Item 3	M	14	13	14
	PFs	2	2	
	C	178.5	194.0	178.5
	η	84.55	83.77	84.55
Item 4	M	14	10	14
	PFs	3	2	3
	C	199.0	278.6	199.0
	η	81.26	81.26	81.26

Table 4. Variables and Criteria of Current Formation

Item	M_i	C_i	η	Q_i	f	g [%]
1	16	155.2	79.7	735		
2	13	155.2	65.6	128	13	77.72
3	18	155.2	80.2	75		
4	16	155.2	75.9	225		

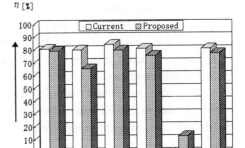

Figure 3. Performance Comparison between Current and Revised Formation

In the proposed algorithm, stability of number of necessary workers is considered as the first important criterion and calculation is performed to aim minimum variance of number of workers (f) during the production duration. However, following criterion g, i.e. working rate of total capacity (total man-hour being maintained during the production planning duration), is also evaluated for checking validity of the alternatives.

$$g = \frac{\sum_i M_i Q_i C_i \eta_{i\bullet}}{\sum_i M_i Q_i C_i} \times 100 \quad [\%] \tag{11}$$

Where,
M_i : number of workers for producing item i
$\eta_{i^\bullet}$: balancing efficiency of item i

From Table 3, results of the fictional case 1 to 2 indicate that proposed procedure is not only feasible but also able to qualify, as f and g values are sufficiently small and over 80% respectively. In addition, such effectiveness is much clear, as "Actual" case prepared based on the actual demand in recent month also indicates the certainty of balancing efficiency improvement in the practical scene. The quantitative difference between current and proposed procedure, and the advantage of the latter could be understood clearly by Figure 3.

4. Concluding Remarks

In this paper, an agile balancing procedure of multi- product switching line production system is proposed and performance is evaluated through simulation experiments based on the actual data from the case factory. Noteworthy results obtained from this study could be summarised as follows.

(1) Relevant agile procedure of production formation generation for multi- product line production system is developed, which consists of two principal steps, generation of balanced production formation database and multi- product production formation design, and guarantees minimum fluctuation of necessary work force as well as higher rate of production system utilisation.

(2) Standing on the basis of several case calculations, performance of the proposed procedure is considered to be satisfactory. Particularly, the formation result obtained by using the actual demand data of the case factory is certified to be relevant by the collaborated company.

It is, also, recognised that relevancy of the formation database is considered to be significant for successful results, and it is expected that further improvement of the criteria values is realised through enrichment of the database.

Acknowledgement

We, the authors, sincerely acknowledge to be supported by Waseda University Grant of Special Projects '93-'94 to complete this paper.

References

1. K. Cunnigham and L. Schrage, *LINGO User's Manual* (LINDO SYSTEM INC., 1988).
2. L. Gerhardt, *Adaptive / Agile Integrated Manufacturing Enterprises (AIME) - An Evolving CIM Strategy for the 21 Century* (Key Note Lecture in the 2nd ICCIM, Singapore, 1993).
3. R. Muramatsu, *Basic Production Management* (Kunimoto Publishing Co. Ltd., Tokyo, Japan, 1979), p.169-219.

SIMULATED ANNEALING APPROACH TO GROUP TECHNOLOGY

Shashi M Sharma
Department of Computer Science and Automation,
Indian Institute of Science, Bangalore 560018, India
E-mail: sms@bhaskara.csa.iisc.ernet.in

and

N Viswanadham
Department of Computer Science and Automation,
Indian Institute of Science, Bangalore 560018, India
E-mail: vishu@csa.iisc.ernet.in

ABSTRACT

In this paper, we use the simulated annealing approach to solve the group technology(GT) problem. It is well known that this problem belongs to the class of NP complete nonlinear integer programming problems. Several heuristic approaches have been reported in the literature but these do not work for the large scale problems. We report in this paper, a simulated annealing based approach that is very effective for solving large scale GT problems in terms of CPU time and the quality of the solution.

1. Introduction

Group technology (GT) problem involves clustering of machines that are dedicated to the manufacture of part families in one physical location to form the cells. GT is the central concept behind cellular manufacturing systems. The benefits of cellular manufacturing systems include

- Reduced set-up time and work in process inventory.

- Increased design and manufacturing efficiency.

- Reduced non-value adding flow of parts resulting in improvements of manufacturing lead time and increased throughput.

In order to successfully implement GT, the problem of machine grouping has to be solved together with the machine cell layout problem. In machine grouping problem one has to determine the composition of the machine cells and corresponding part families. The facility layout problem on the other hand determine the arrangement of the machine cells in the factory such that the transportation time of material and parts handled by material handling systems is minimized. In this paper, we consider the GT problem only.

The rest of the paper is organized as follows: In section 2, we formulate the generalized part family problem as integer quadratic programming. In section 3, we introduce simulated annealing algorithm and techniques to generate a cell feasible solution. In section 4, we apply simulated annealing method to various cell formation examples from the literature. In the last section we discuss our computational experience and compare our method with various techniques available in the literature.

2. Problem Formulation

Consider a manufacturing system consisting of n machines and produces m parts. The objective is to group n machines in to a distinct number, say B cells with the constraint that each machine resides in exactly one cell.

Let x_{ik} be a binary valued variable which takes a value 1 if i'th machine is in k'th cell and 0 otherwise. Let p_q be the production requirement of part q in a given period. The machines requirement for the various parts is represented by the matrix $A = [a_{iq}]$ matrix, where a_{iq} is the processing time for part q in the i'th machine. If machine i is not required for processing part q, a_{iq} is assumed to be zero. Let T_i be the total available time on machine i in a given period of time.

One has to satisfy several engineering constraints while performing the cell grouping. Sometimes it may be necessary to have two machines of different type in the cell to enable smooth production of the required parts and yet other times it is necessary that two different machines are not placed in the same cell. To take care of such constraints. we define the relationship indicator matrix $R = [r_{ij}]$. If $r_{ij} = 1$ then the machines i and j are to be placed in the same cells. $r_{ij} = 0$ indicates that their is no special adjacency relationship between machines i and j. $r_{ij} = -1$ on the other hand, indicates that machines i and j are not be placed in the same cell. In our discussion, we will limit ourselves to two values of r_{ij}, 0 and -1. The case $r_{ij} = 1$ can be converted to the standard formulation by clubbing the machine i and j as a composite machine.

We define $W = [w_{ij}]$, as a $n \times n$, symmetric matrix which is derived from the A matrix as follows

$$w_{ij} = \begin{cases} \sum_{q=1}^{m}(a_{iq} \wedge a_{jq})p_q & i \neq j \\ 0 & i = j \end{cases} \quad (1)$$

wherer $\wedge$ denote the boolean AND operation defined by

$$a_{iq} \wedge a_{jq} = \begin{cases} 1 & \text{if } a_{iq} = 1 \text{ and } a_{jq} = 1 \\ 0 & \text{otherwise} \end{cases}$$

Thus w_{ij} represents the cumulative part flow between machines i and j. Naturally to minimize the inter cellular movement, the pair of machines with high value of w_{ij} are to be grouped in the same cell. One can define w_{ij} in many other ways as appropriate to the problem. The only restriction being that the resultant W matrix should be symmetric, in which case, the resultant cell formation will be the one which will have minimal interaction among the cells in terms of the w_{ij} metric.

We further define $V = [v_{iq}]$ the $(m \times n)$ matrix and $M = [m_{ji}]$ as an $(B \times m)$ average cell load matrix with components

$$v_{iq} = \frac{a_{iq} \times p_q}{T_i}$$

and

$$m_{jq} = \frac{\sum_{i=1}^{n} x_{ij} v_{iq}}{\sum_{i=1}^{n} x_{ij}}$$

The element v_{iq} is the fraction of the workload induced by part q on machine i, m_{jq} is the total load in cell j induced by part q and $\sum_{i=1}^{n} x_{ij}$ is the total number of machines in cell j.

With the above notation, we define the objective function Φ_1 as

$$\Phi_1(X) = \sum_{i=1}^{n} \sum_{j>i}^{n} \sum_{k=1}^{B} x_{ik}(1 - x_{jk})w_{ij} \qquad (2)$$

In order to ensure that the average workload in each cell is distributed uniformly, we define the objective function Φ_2 as

$$\Phi_2(X) = \sum_{i=1}^{n} \sum_{j=1}^{B} x_{ij} \sum_{q=1}^{m} (v_{iq} - m_{jq})^2 \qquad (3)$$

Now we formulate a number of optimization problems essentially minimizing $\Phi_1(X) + \gamma \Phi_2(X)$ subject to certain feasibility and engineering constraints. The constant $\gamma > 0$ is used to scale the two objective functions appropriately. We consider fixed cell size and bounded cell size problem below

Case 1 Fixed Cell Size Problem

In this formulation the sizes of the cells(number of machines in the cell) are fixed. Let c_i be the size of the i'th cell. Then the optimization problem

$$\text{Minimize } \Phi_1(X) + \gamma \Phi_2(X)$$

subject to

$$x_{ik} = \{0,1\}, \quad \forall i = 1, \cdots n; \ k = 1, \cdots B \qquad (4)$$

$$\sum_{k=1}^{B} x_{ik} = 1 \qquad \text{for } i = 1, \cdots, n \qquad (5)$$

$$\sum_{i=1}^{n} x_{ik} = c_i \qquad \text{for } k = 1, \cdots, B \qquad (6)$$

$$r_{ij} x_{ik} x_{jk} \leq 0 \qquad \forall i \neq j \qquad (7)$$

gives the solution to the GT problem. The constraint (5) ensures that the machine i resides exactly in one cell. The second constraint (6) ensures that there are exactly c_k machines in k'th cell. The third constraints (7) imposes relationship constraint among machines i and j as dictated by R.

Case 2 Bounded Cell Size Problem

The cell size in this case is not fixed but has a lower bound $\bar{c}_k$ and upper bound $\bar{C}_k$. The GT problem can be formulated as a minimization problem:

$$\text{Minimize } \Phi_1(X) + \gamma \Phi_2(X)$$

subject to

$$x_{ik} = \{0,1\}, \quad \forall i = 1,\cdots n; \; k = 1,\cdots B \qquad (8)$$

$$\sum_{k=1}^{B} x_{ik} = 1 \qquad \text{for } i = 1,\cdots,n \qquad (9)$$

$$\bar{c}_k \leq \sum_{i=1}^{n} x_{ik} \leq \bar{C}_i \qquad \text{for } k = 1,\cdots,B \qquad (10)$$

$$r_{ij} x_{ik} x_{jk} \leq 0 \qquad \forall i \neq j \qquad (11)$$

The constraint (9) and (11) are similar to case 1. The constraint (10) in this case enforces the upper and lower bounds on the cells sizes.

3. Simulated Annealing Algorithm

Simulated annealing approach (SA) is an algorithmic approach for the solution of optimization problems. The name of the algorithm derives from an analogy between solving optimization problems and annealing of solids as proposed by Metropolis et al [1]. Computational experiments with this algorithm on a variety of notorious combinatorial optimization problem are reported by Korst [2]. A description of the simulated annealing algorithm appears in Figure 1.

A simulated annealing algorithm, has two kind of parameters

- Problem specific
- Generic

The problem specific parameters are

1. **Configuration** The configuration of the system is represented by the binary valued variables $x_{ij} : i = 1,\cdots,n; j = 1\cdots,B$

2. **Neighborhood of the configuration** The neighborhood of configuration S is the set of configurations in which the system can move from S with non-zero probability. For our problem the neighborhood of configuration S is the set $\mathcal{S}$ of configuration which are generated by swapping the cell of two machine at random or picking up one machine at random and putting it to some other cell.

Step 1.Get an initial configuration.
Step 2.Get an initial temperature $T > 0$.
Step 3.
    3.1   For $1 \leq i \leq L$
        3.1.1 Pick any random neighbor $\hat{S}$ of S.
        3.1.2 Let Δ = cost($\hat{S}$) - cost(S).
        3.3.3 If $\Delta > 0$ then $S = \hat{S}$
        3.1.4 If $\Delta \geq 0$ then set $S = \hat{S}$ with probability $\exp(-\Delta/T)$
    3.2   Set $T = \alpha.T$
Step 4 Return S

Figure 1: The Simulated Annealing Algorithm

3. **Cost of the configuration** The cost of the configuration is the measure of the goodness of the configuration. We have already defined an appropriate cost function for our problem in the previous section.

4. **Initial Configuration** The initial configuration is chosen picking up a random configuration from the set of all possible feasible configurations.

The generic parameters of the simulated annealing are

- Initial value of the temperature T_0.

- Decreasing factor α .

- Stopping criterion.

- Length of the iteration L, at each value of the temperature.

Starting from an initial configuration, the SA generates at random a new configuration in the neighborhood of the original configuration. The change Δ, difference in the cost of configurations between the current and the previous state, is calculated and if the change represents a reduction in the value of the objective function then the transition to the new configuration is accepted. If the change represents an increase in the objective function, then the transition to the new configuration is accepted with a specified probability. The acceptance probability function usually takes the form $\exp(-\Delta/T)$ where T is the control parameter. This mechanism enables the SA algorithm to avoid becoming trapped in a local minimum during its search for the global minimum.

The initial temperature T_0 is chosen such that a cost increasing transition occurs with probability $P_0 > .9$. The length of the iteration is chosen in the order of the size of the cardinality of the neighborhood. The temperature is decremented using the following rule

$$T_{k+1} = \alpha T_k \tag{12}$$

Prob-lem #	Size	Optimal Cost	
		Venugopal's SA	Linear SA
1	5 × 8	0.56000	0.56000
2	9 × 9	0.19417	0.19417
3	9 × 10	0.46667	0.46667
4	8 × 14	0.62833	0.62833
5	11 × 16	1.41167	1.41660
6	10 × 20	1.47500	1.47500
7	14 × 20	1.64500	1.64500
8	15 × 30	7.31600	7.31600
9	7 × 11	0.47580	0.45592
10	8 × 20	0.73922	0.73922
11	12 × 19	0.52999	0.52999
12	20 × 20	1.79239	1.78528

Table 1: Comparative analysis with Venugopal's SA algorithm

where typical value of α lies between 0.85 to 0.95.

The last detailed that must be filled is the stopping critireon which will determine when the execution is terminated. The algorithm terminates either when the optimal value is reached or when there is no improvement in the optimal value for a number of temperature reduction stages. In this paper, we have used a fast annealing algorithm by incorporating incremental calculation of cost function.

4. Computational Results

In order to demonstrate the efficiency of the method developed here, we solved large number of problems with number of machines ranging from 5–20 and parts ranging from 8–25. We present these results below.

Example 1

In this example we will compare our method with the method of Venugopal and Narendran [4]. The authors have used the simulated annealing algorithm to solve GT problem and our method differs with theirs in the calculation of Δ and the cooling schedule. We have used linear cooling schedule while theirs is exponential cooling schedule. We will take their examples as benchmark problems for the purpose of comparative analysis.

Their are 12 examples ranging from 5 parts, 8 machines to 20 parts and 20 machines. In the absence of the additional data we assume W, R matrix to be identically zero.

We have conducted extensive investigation on these examples and our method is superior to Venugopal's [4] in terms of the quality of the solution which are equal or better for all the examples.(see Table 1)

	# Cell	Size of Cell	α	T_0	L	Cost	CPU Time
1	2	12,31	0.95	20	50	13	1.782
2	2	21,22	0.95	20	50	23	1.717
3	2	21,22	0.95	80	50	22	6.815
4	3	14,14,15	0.95	20	50	33	1.516
5	3	14,14,15	0.95	150	50	32	13.250

Table 2: Computational result for Example 2 for the Case 1

	# Cell	$\bar{c}_i$	C_i	α	T_0	L	Cost	CPU Time
1	2	10	35	0.95	20	50	13	1.433
2	2	20	30	0.95	20	50	21	1.467
3	3	10	30	0.95	20	50	29	1.300
4	3	13	15	0.95	20	50	32	1.333
5	4	9	15	0.95	150	50	42	1.516
6	4	9	15	0.95	80	50	41	5.400

Table 3: Computational result for Example 2 for the Case 2

Example 2

In order to compare the performance of annealing algorithm with the heuristics method used for group technology we applied our method to the example of Ravi Kumar et al. The weight matrix(say $\tilde{W}$) as given in Ravi Kumar et al[3] corresponds to the adjacency matrix for a bipartite graph. The corresponding W matrix is obtained as follows

$$W = \begin{bmatrix} 0 & \tilde{W} \\ \tilde{W}^T & 0 \end{bmatrix}$$

The matrices V, R are assumed to be zero matrices, i.e., $\Phi_2(X) = 0$, and there is no constraint on the machine placement.

The result of our algorithm and the parameters for the GT problem are given in Table 3 for Case 1 and in Table 4 for Case 2. The annealing parameters are also listed in the corresponding tables. In their paper, Ravi et. al.[3], report an optimal cost of 13 which we also have obtained for the 2 cell problem for the following configuration

Cell 1 = { 5,7,8,9,13,14,17,25,30,33,34 }
Cell 2 = { 1,2,3,4,6,10,11,12,15,16,18,19,20,21,22,23,
24,26,27,28,29,31,32,35,36,37,38,40,41,42,43}

It is evident from the tables that the algorithm gives good result for the fast cooling schedule and there is little improvement in the optimal value with the increase in T_0 and L. It can be seen that in general that algorithm for Case 1 runs faster than the algorithm for Case 2 because implementation of Rule 2 is faster than Rule 1.

The time reproted here is for SPARC IPC and the code was written in C++ and using gnu compiler gcc-3.1.2.

5. Conclusion

In this paper we developed a methodology to solve the GT problem by using simulated annealing algorithm. We modelled the GT problem as quadratic integer optimization problem with linear inequality constraints. We have shown that the formulation is equivalent to multi-graph partitioning problem with constraints. We report that the annealing algorithm gives superior result even for the faster cooling schedule.

Our method can be easily extended to accomdate all the other variants of GT available in the literature. Our method is superior when applied to some large scale problems, however the preprocessing takes time of the order $O(Bn^2)$. It may be possible to obtain further computational saving in this regard using by using special data structure in the future.

References

1. N. Metropolis, A.W. Rosenbluth, M.N. Rosenbluth, A.H. Teller, *Equation of state calculations by fast computing machines*, J. Chem. Phys. 21 (6), 1087-1092, 1953.
2. E. Aarts and J. Korst, *Simulated annealing and boltzmann machine: A stochastic approach to combinatorial optimization with neural computing*, John Wiley, New York, 1989.
3. D.S. Jhonson, C.R. Aargon, L.A. McGeoch and C. Schevon, Optimization by Simulated Annealing : An Experimental Evaluation. Part I, Graph Partitioning. Operation Research,Vol-37, 865-892, 1989.
4. K. Ravi Kumar and Anthony Vannelli, *Efficient algortihms for grouping component-processor families*, Tech. Report RC 10636 (# 47695) 7/20/84, IBM Research Divison, 1984.
5. V. Venugopal and T. T. Narendran, *Cell formation in manufacturing systems through simulated annealing: An experimental evaluation*, European Journal of Operation Research **63** (1992), 409–422.
6. J. R. King, *Machine-component grouping in production flow analysis:An approach using a rank order clustering algorithm*, International Journal of Production Research **18** (1980), 213–219.
7. J. R. King and V. Nakornchai, *Machine-component group formation in group technology:Review and extension*, International Journal of Production Research **20** (1982), no. 2, 117–133.

FUZZY SET BASED MACHINE-CELL FORMATION IN CELLULAR MANUFACTURING

Choon-Woo Leem

LG-PRC, LG-Electronics Inc., 19-1, Cheongho-Ri, jinwuy-Myun, Pyungtaek-Gun Kyunggi-Do, 451-860 Korea

and

Jacob Jen-Gwo Chen

Department of Industrial Engineering, University of Houston, 4800 Calhoun Road, Houston, Texas, USA

E-mail: jgchen@uh.edu

ABSTRACT

In this article, a fuzzy set based machine-cell formation algorithm for cellular manufacturing is presented. The fuzzy logic is employed to express the degree of appropriateness when alternative machines are specified to process a part shape. For machine grouping, the similarity coefficient based approach is used. The algorithm produces efficient machine cells and part families which maximize the similarity values. A numerical example is given to illustrate this approach.

1. Introduction

Cellular manufacturing (CM) has been recognized as the second generation of Group Technology (GT) (Wang and Li 1991). In the first generation of GT, only part families were formed. In a part family, similar geometric shapes or parts requiring similar manufacturing processes are included without physically changing the machine layout. In the second generation of GT, the part families and machine cells are formed concurrently. By physically forming machines into cells, the materials-handling distance is greatly reduced (Wang and Li 1991).

A large number of methods have been developed to form part families and machine cells. The assumption of conventional methods is that a part visits only one machine for a particular shape, which is expressed by "0" or "1." The "1" element means that the part visits the corresponding machine for processing a particular shape. The element is "0" if the part is not assigned to the corresponding machine. The binary machine-part incidence matrix is used to express binary values. Then, machine cells are formed by manipulating the machine-part incidence matrix in a manner such that the grouping of all similar objects is possible. The basis for the implementation of cellular manufacturing is the development of the initial matrix into a subsystem or block diagram.

However, in practice, a part shape can be processed by alternative machines with different levels of efficiency and processing times. It requires the specification of the degree of appropriateness for alternative machines. In addition, since machine tools can perform more than one operation, a part can be routed through the same machine for different operations (Sundaram and Doshi 1992). The specification of alternative machines for an operation can be useful in choosing the second machine in case the first machine breaks down. The nonbinary machine-part incidence matrix based on the fuzzy theory approach is used to represent the relationships between machines and parts. The relationships can be expressed by membership values, which is the degree of appropriateness that a part will visit the corresponding machine.

2. Fuzzy Set Based Approach

Suppose that there are m machines and p parts to be grouped into c cells. The binary machine-part incidence matrix is shown in Table 1:

Table 1 Binary Machine-part Incidence Matrix

Machine	Part				
	P1	P2	P3	...	Pp
M1	μ_{11}			...	μ_{1p}
M2	*μ_{21}	μ_{22}	μ_{23}	...	μ_{2p}
M3				...	μ_{3p}
.	.	.	.	...	.
Mm	μ_{m1}	μ_{m2}	μ_{m3}	...	

$$*\mu_{ij} = \begin{cases} 1 & \text{if part } j \text{ is processed by machine } i, \\ 0 & \text{otherwise.} \end{cases}$$

The notation μ_{ij} denotes the relationship between machine i and part j. Because of the limitations of the binary matrix approach explained in the previous section, the nonbinary matrix approach is presented. The nonbinary matrix scheme offers a unique approach to the cell formation problem since it has more flexibility in grouping and clustering than the binary logic approach (Li, et al. 1988). The nonbinary machine-part incidence matrix is shown in Table 2.

Table 2 Nonbinary Machine-part Incidence Matrix

Machine	Part				
	P1	P2	P3	...	Pp
M1				...	μ_{1p}
M2	μ_{21}	μ_{22}	μ_{23}	...	μ_{2p}
M3				...	μ_{3p}
.	.	.	.	...	.
Mm	μ_{m1}	μ_{m2}	μ_{m3}	...	

The constraints are such that

a) $0 \leq \mu_{ik} \leq 1$ for $i = 1,\ldots,m$ and $k = 1,\ldots,p$ and (1)
b) $\sum_p \mu_{ik} > 0$ for $i = 1,\ldots,m$. (2)

The constraint (1) ensures that the membership value μ_{ik} is restricted not to a binary value [0, 1] but to a value {0, 1}. The constraint (2) requires that a part shape be processed by more than one machine.

3. Proposed Machine-cell Formation Algorithm

3.1 Similarity Coefficient

The similarity coefficient approach is a well-known methodology in cellular manufacturing because it is the most efficient in forming machine cells (Seifoddini and Wolfe 1986). In the case of binary machine-part incidence matrix, the similarity coefficients defined in SLCA (McAuley 1972) and ALCA (Seifoddini and Wolfe, 1986) are commonly used.

In the nonbinary machine-part incidence matrix, Li et al. (1988) proposed different types of the similarity coefficient. In this research, a similarity coefficient (S_{ij}) for machines i and j is defined as

$$S_{ij} = \frac{\sum_{k=1}^{p}(\mu_{ik} \wedge \mu_{jk})}{\sum_{k=1}^{p}(\mu_{ik} \vee \mu_{jk})} \quad \text{for } i=1,\ldots,m, j=1,\ldots,m, \text{ and } i \neq j. \tag{3}$$

The similarity coefficient in equation (3) reflects the proportion of degree for parts visiting M_i and M_j. The values of the similarity are standardized such that the value near one is more desirable to form M_i and M_j into the same cell. If all the elements μ_{ij} are identical, the similarity coefficient indicates one. Also, if all the elements are inverse, the similarity coefficient is zero. The value near zero means comparatively unimportant. The values of S_{ij} have the following properties:

a) $0 \leq S_{ij} \leq 1$ for $i \neq j$, (4)
b) $S_{ij} = S_{ji}$, and (5)
c) $S_{ii} = 1$. (6)

Pairwise similarity can be arranged in a matrix form as follows:

$$S = \begin{vmatrix} 1 & S_{12} & \cdots & S_{1p} \\ S_{21} & 1 & \cdots & S_{2p} \\ \vdots & & & \\ S_{m1} & S_{m2} & \cdots & 1 \end{vmatrix}. \tag{7}$$

3.2 Machine-chaining Problem

The machine-chaining problem can arise when the machines are assigned in cells improperly. Seifoddini and Wolfe (1986) and Chow and Hawaleshka (1992) have introduced efficient algorithms to overcome such a problem. The key for the machine-chaining problem is to regard all steps of the procedure as continuous ones. If the machines M_i and M_j have maximal similarity value, then M_i and M_j are formed into the same cell and elements in M_i and M_j are transformed into a single unit. The transformation of elements in M_i and M_j into a new machine unit, $M_{(i,j)}$, is as follows:

$$M_{(i,j),k} = (\mu_{ik} \vee \mu_{jk}) \quad \text{for } k = 1,\ldots,p. \tag{8}$$

Since the output of the previous step is the input to the next step, such a continuous re-evaluation is repeated until the desired number of cells are clustered. For instance, there are five parts and the membership values for M_i and M_j are shown in equations (9) and (10). Then the new elements for $M_{(i,j)}$ in equation (11) are transformed as follows:

$$M_i = (0.1, 0.3, 0.0, 0.0, 0.7), \tag{9}$$
$$M_j = (0.3, 0.1, 0.1, 0.0, 0.0), \text{ and} \tag{10}$$

$$M_{(i,j)} = (0.3, 0.3, 0.1, 0.0, 0.0, 0.7). \tag{11}$$

The machine-chaining problem should be considered in the manufacturing-cell design stage. When similar machines are formed in a cell, the density of a cell can be increased while those of other cells can be decreased. It may result in decreasing overall machine utilization in machine cell. However, the cost of intercellular and intracellular movements can be reduced when this problem is considered. Therefore, before a manufacturing cell is designed, the objective of the system should be identified.

3.3 Performance Measuring in Cell Formation

After grouping the machine cells and part families, the configuration and the performance measures of the cell should be considered throughout the design process to provide the information for the selection of best operational procedure. Numerous performance measures have been proposed (Kamrani and Parsaei 1993), such as, minimum cost of intercellular movements, Total bond energy/measure of effectiveness. There can be other factors to consider during the evaluation of the cell. It is necessary to select a proper measure to satisfy the system objectives.

3.3.1 Machine Utilization Measure

The machine utilization (MU) measure is the density of machines being utilized in each cell. It is expressed as

$$MU = \frac{\text{total number of 1's in each cell}}{\sum_{k=1}^{c} m_k p_k}, \tag{12}$$

where m_k : the number of machines in machine cell-k,
 p_k : the number of parts in part family-k, and
 c : the total number of cells.

The value of MU will be large when each part in a cell uses more machines. Therefore, a large value of MU is preferred.

3.3.2 The Number of Intercellular Movement Measure

The definition of the number of intercellular movement (NIM) measure is the total number of nonclustered elements with a value of 1 in the solution. Since intercellular movement may lead to the materials-handling problem, it is desirable to minimize the number of intercellular movements. This measure can be used when there is a cluster of more than one cell.

3.3.3 Bond Energy Measure

The bond energy (BE) measure (McCormick et al., 1972) is to maximize the number of elements having a value of 1 close to each other. The definition of BE is expressed as

$$BE = \frac{1}{2} \sum \sum a_{i,j}(a_{i,j+1} + a_{i,j-1} + a_{i+1,j} + a_{i-1,j}). \tag{13}$$

The higher the value of BE, the better the solution. However, it is not a good measure because the same solution with different arrangement of rows and columns may give different values of measures (Kusiak and Cheng 1991).

3.3.4 Group Efficiency Measure

The group efficiency measure (Chandrasekharan and Ralagopalan 1986) is the weighted average between the number of exceptional parts and machine utilization. The group efficiency is defined in equation (14). A higher GE measure leads to a higher grouping measure.

$$GE = \alpha m_a + (1-\alpha) m_b, \qquad (14)$$

where $m_a = MU$,

$$m_b = 1 - \frac{\text{number of exceptional parts}}{mp - \sum_{k=1}^{c} m_k p_k}, \qquad (15)$$

a: weight factor,
c : the total number of cells,
m : the total number of machines, and
p : the total number of parts.

In the case of the nonbinary machine-part incidence matrix, it is not necessary to cluster the nonzero elements because alternative machines are available for processing a shape. The primary measure is to maximize the density of each cell, i.e., rearrange the majority of large nonzero elements into the cells. On the other hand, Lee et al. (1991) used the number of exceptional elements (NEE) and the sum of the exceptional element values (SEV) to compare their algorithm with others. Other performance measures can be considered for cell evaluation.

3.4 Machine-cell Formation Procedure

The proposed procedure consists of three basic rules. The first rule is to calculate the similarity values among the machines using equation (3) and to construct a similarity matrix. The second rule is to form the cell including the machines having maximal similarity values. The final rule is to transform the elements of selected M_i and M_j into the new machine unit using equation (8). The combination of these three rules provides the machine cells and part families. The procedure is described as follows:

Step 0) Initialization
Set the Current Number of Cells (CNC) to be m and the desired numbers of cells to be c. Compute the similarity values of the given machine-part incidence matrix.

Step 1) Cell Formation
Find the machines M_i^* and M_j^* that have the maximal similarity coefficient, and include these two machines into the same cell $M_{(i^*,j^*)}$.

Step 2) Transformation
Transform the elements of the machines M_i^* and M_j^*, and rearrange the machine-part incidence matrix.

Step 3) Similarity Values Calculation
Update the similarity values from the rearranged machine-part incidence matrix and reduce the CNC to one unit.

Step 4) Evaluation
Check the CNC. If the CNC $> c$, go to Step 1, otherwise repeat it until the CNC $= c$.

4. Conclusion

The manufacturing cell design module specifies the machines and allocates parts to the machines. The algorithm presented in this article is based on fuzzy clustering algorithm to express the degrees of appropriateness for machines in processing an operation. It is possible that an operation can be processed by more than one machine, and the machines that can process a particular operation are called alternative machines. Specification of alternative machines for an operation can be useful to choose the second machine in case of the first machines broken down or busy. Second, the similarity coefficient based machine-cell formation algorithm is used to show the degree of similarity between machines or parts. A similarity coefficient is defined for calculating the similarity value between machines. Third, the machine-chaining problem is considered to prevent the improper assignment of machines in cells.

Although the proposed algorithm produces better solutions compared to other methods, it is only concerned with machine cell formation. That is, a second ancillary procedure is required to identify part families. Further research should be conducted to develop an extended fuzzy logic algorithm, where machine cells and part families are formed simultaneously.

5. References

1. M.P. Chandrasekharan and R. Rajagopalan, An ideal seed non-hierarchical clustering algorithm for group technology, *International Journal of Production Research*, **11** (1986) 835-850.
2. W.S. Chow and O. Hawaleshka, An efficient algorithm for solving the machine chaining problem in cellular manufacturing, *Computers and Industrial Engineering*, **22** (1992) 95-100.
3. A.K. Kamrani and H.R. Parsaei, A survey of design methods for manufacturing cells, in *Proceedings of the 15th Conference in Computers and Industrial Engineering*, 1993, p. 499-502.
4. A. Kusiak and C. Cheng, Group technology: analysis of selected models and algorithms, *Design, Analysis, and Control of Manufacturing Cells*, ASME, **PED-Vol. 53**(1991), 99-114.
5. S. Lee, C. Zhang and H.P. Wang, Fuzzy set-based procedures for machine cell formation, *Design, alysis, and Control of Manufacturing Cells*, ASME, **PED-Vol. 53**(1991), p. 31-45.

6. J. Li, Z. Ding, and W. Lei, Fuzzy cluster analysis and fuzzy pattern recognition methods for formation of part families, in *16th North American Manufacturing Research Conference (NAMAC) Proceedings, Society of Manufacturing Engineers*, 1988, p.558-563.
7. J. McAuley, Machine grouping for efficient production, *The Production Engineer*, **51**(1972), 53-57.
8. W.T. McCormick, P.J. Schweitzer and T.W. White, Problem decomposition and data reorganization by a clustering technique, *Operations Research*, **52** (1972), 993-1009.
9. H. Seifoddini and P.M. Wolfe, Application of the similarity coefficient method in group technology, *IIE Transactions*, **19**(1986) 271-277.
10. R.M. Sundaram and K. Doshi, Formation of part families to design cells with alternative routing considerations, *Proceedings of the 14th Annual Conference on Computers and Industrial Engineering*, **23**(1992), 59-62.
11. H.P. Wang and J.K. Li, *Computer-Aided Process Planning* (Elsevier, Amsterdam, 1991).

Planning

An Object-Oriented Bill-of-Materials System for Dynamic
Product Management
    A. J. C. Trappey, T. K. Peng and H. D. Lin

Application of MRP in a Local Manufacturing Company — A Case Study
    N. Bin Mohd. Yusof and C. L. Ngeow

MRP II as a Base of Different Production Management Techniques
    E. L. J. Bohez and M. A. A. Hasin

A Decision Support Framework for PWB Assembly
    Y. Y. Su and K. Srihari

AN OBJECT-ORIENTED BILL-OF-MATERIALS SYSTEM FOR DYNAMIC PRODUCT MANAGEMENT

Amy J. C. Trappey, Ting-Kou Peng and Hsuei-Dian Lin
Department of Industrial Engineering
National Tsing Hua University
Hsinchu, Taiwan, R.O.C.
Telephone: 886-35-715131 Ext. 3968, Fax: 886-35-722685
E-mail: trappey@ie.nthu.edu.tw

ABSTRACT

The Bill-of-materials (BOM) represents key information in a manufacturing system. Most of the information relevant to products and their components is described in this document. The information is critically related to data from product design, manufacturing planning, production, procurement, inventory control, and accounting within the organization. A conventional BOM structure, which only manages data in a stand-along relational database management style, cannot satisfy the needs requested by all departments within a company. A BOM should relate and communicate with other domains of information system. Thus, a newly designed BOM system uses the object-oriented programming (OOP) concept to dynamically represent product information. The object-oriented BOM (OOBOM) system can capture and encapsulate data that will simultaneously satisfy the information needs of various departments. Furthermore, the system provides several interfaces for easy data access and manipulation. This product management approach guarantees the consistency and efficiency of BOM data representation.

Keywords: Bill-of-material, Computer-integrated manufacturing, Object-oriented programming, Engineering data management system.

1. Introduction

The Bill-of-Materials (BOM) is an information entity that is widely used in a manufacturing system. Due to the various combinations of information requested by departments, BOM must store most of the information related to the products, their components and their productions. As shown in Figure 1, a well-defined BOM should include product definition, manufacturing instructions, engineering changes control, service parts support, liability or warranty protection, order entry facility, costing, and pricing (Mather, 1987). Traditionally, the BOM data is stored in a relational database using a record-based information model. The record-based information model describes the structure of the data in a record structure that stores attributes, but not the behaviors or operations of the data or related to the data. As discussed by Andleigh and Gretzinger (1992), there are limitations to relational databases. The limitations include that: (1) the record structure is too rigid; (2) the data and their description are maintained separately; (3) the data description is hard to change; (4) behavioral relations are not described. These characteristics restrict the performance of a relational BOM database. The limitation is more obvious when it is considered to be a company-wide information system in a CIM environment.

Thus, a new way of storing and managing BOM data is essential to overcome the limitation of the traditional relational BOM database. Chung and Fisher (1992) proposed the basic concepts of using an object-oriented database (OODB) to store BOM information. The concept of constructing the structure of BOM with OOP methodology is addressed under an existing OODB management system, ORION. Chen (1993) used the object-oriented methodology to model BOM, which is capable of representing product components and assembly sequences. He also addresses the problems of the traditional BOM structure, such as the lack of complete description of products and the difficulty in realizing component configurations.

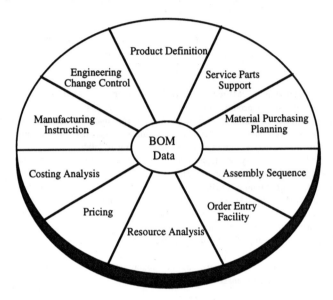

Figure 1   Major BOM data types and their functions.

In our research, a unique object-oriented representation for structuring a BOM system, called OOBOM, is discussed. We describe how a product structure is uniquely modeled and how a product, its sub-assemblies, and its components are linked in a hierarchical tree form. The differences between the OOBOM representation and the other BOM models using object-oriented representation are also discussed.

2. The Object-Oriented BOM Structure

In the OOBOM system, there are many classes that serve as templates for the data entities. These data store the BOM relevant data. The **Part** class and its sub-classes, **MfgPart** and **ProcPart**, are the major templates for encapsulating the BOM relevant data in a manufacturing system. Moreover, there are sub-classes of **MfgPart** created to represent data related to a manufactured part or product, such as **Routing** and **Packaging**. **Routing** specifies the process sheet of a part and **Packaging** specifies the packaging container of a manufactured product. **ProcCo**, a sub-class of **ProcPart**, records the information relevant to the supplier of a procured part. These data objects aggregate to the part's object as properties of a part, sub-assembly, or final product.

2.1. The Part, MfgPart and ProcPart Classes

In a manufacturing system, the final product is composed of components and sub-assemblies. Among these components and sub-assemblies, some are made within the manufacturing system, and some are not. Therefore, a meta-class of the BOM structure named **Part** is defined, which is an abstraction of components, sub-assemblies, and products. Due to the characteristics differences between the made (-in-house) parts and procured (-from-others) parts, there are two sub-classes representing these types of parts. One of the classes is **MfgPart** that abstracts the made parts and the other is **ProcPart** that represents the procured parts. The most elementary object of a BOM structure is generated by these two classes. Different characteristics of parts can be further classified into sub-classes of **MfgPart** or **ProcPart**. For example, a turnbuckle belongs to a part collection which is different from bolts. If the turnbuckles are made in-house, it is reasonable to define a sub-class of the **MfgPart** class representing the turnbuckle collection. However, the implementation and application of a BOM system would not be practical if the lowest-level object is defined this way as in (Chung and Fischer, 1992; Chen, 1993). The system operator of a BOM system will have to update the system's class definitions whenever a new part comes in. This approach lacks security concerns and is not user friendly. It is difficult and ambiguous to find a criterion to determine where the lowest level should be. For instance, caps of pens can be sub-classified as caps "with pocket-holder" and "without pocket-holder." Caps can also be sub-classified as the round-top type and the square-top type. Under such circumstance, defining the lowest-level object will be very confusing to the system engineers.

The definition of an object (e.g., a bolt) is encapsulated into the "instant variables" of its belonging class (LaLonde and Pugh, 1990). For instance, **Part** has instant variables such as partName, partNo, parents, type and attributes. The **MfgPart** class has all instant variables inherited from its super-class, **Part**, and the additional instant variables such as children (sub-assemblies), consumableTool, packaging, routing, and cadLink. The names of the instance variables generally self-explain what the variables represent. The **attributes** instance variable is a **Dictionary** data type (LaLonde and Pugh, 1990) and stores key-value data pairs. For example, if a turnbuckle has a parameter named "diameter" and valued 10, the list "(diameter 10)" is described as a pair data in the **attributes** variable. Thus, **attributes** can be used to describe objects of different collections.

2.2. The Routing, ProcCo and Packaging Classes

ProcCo is a class that represents the definition of a supplier. Regarding the information related to a company, each instance of this class should at least store the name, the telephone number, and the address of the company. Thus, the **ProcCo** class is defined with three instance variables, i.e., coName (the company name), coAddr (the company address) and coTel (the company phone number). Additional variables can be added for specific applications. The **Routing** class encapsulates information such as the operation number, the operation description and the material required. In a manufacturing system, the manufacturing process sheet of a part is recorded in more detail. Therefore, this system creates the definition of the **Routing** class as a sub-class of **MfgPart** to fit the information needed. **Packaging** describes how a part is packaged when it is sold or stored. The definition of the **Packaging** class includes two instance variables, i.e., "container" to describe the packaging container and "size" to describe the packaging size for the calculation of storage space and packaging material.

2.3. The BOM Data Storage

The OOBOM system has numerous data entities representing objects of BOM, such as parts and their associating attributes. While a data entity is a logical representation of an object, a table or a record is a physical representation of an object. The system has a facility that manages and stores these data entities, both in logical and physical representation. In an object-oriented information system, the data management is under the control of the data dictionary. The data dictionary stores the logical relationship among data entities and the relationships between an object and its attributes. In addition, the data dictionary manages the relationships in a physical database among tables, as well as between a table and its columns (Andleigh and Gretzinger, 1992; Tonshoff and Dittmer, 1990).

When an instance of a part relevant class, such as 'mfgPart' (an instance of the **MfgPart** class), is created from the OOBOM system, the data dictionary manages and stores it so that the system can access it. The **BomRecord** class serves as the data dictionary. It is defined in Smalltalk as the sub-class of **Model**. **BomRecord** is defined with a set of global variables (or called class variables). These variables are LibPartDic, LibraryCategory, PartRecord, Product Category and ProductCategoryDic. LibPartDic is to record all existing standard parts in the categorized sets. LibraryCategory is a variable that lists all part categories. PartRecord lists all existing parts and all part definitions are encapsulated in PartRecord by the part names. ProductCategory stores the categories of the final products. Finally, ProductCategoryDic lists all final products organized by product categories.

In these class variables, the relationships and manipulations of the objects can be handled adequately as well as transparently. In general, the partNo of a part data entity seems to be the appropriate choice as an index for accessing a data object. However, since part name presents a more recognizable semantic meaning, it could facilitate the interaction between the system and the users. Thus, part objects with a duplicate part name are not allowed and part names are used to identify objects and locate their definitions.

Due to the inheritance characteristic of OOP, every sub-class can inherit attributes and instructions of its super-class. In the hierarchical structure, the classes for transaction objects are all the sub-classes of the **BomRecord** class. The hierarchy of these classes and their instance variables (listed in the parentheses) are represented as:

Object ()
    Model ('dependents')
        BomRecord ('model')
            BOMPrint ('selProduct' 'selType' 'maxLength')
            NodeTree ('buttonSel' 'onPart' 'myWindow' 'myComposite' 'compositeSet' 'lineComposite' 'howManyNode')
            PartLibrary ('libSel' 'libSet' 'partSel' 'partSet')

PartWindow ('dealPart' 'itemSel' 'itemSet' 'valueSet' 'valueSel' 'parentSel' 'childSel')
ProductBrowser ('categorySelectOn' 'finalProductSelectOn' 'finalProductSet' 'textMode')

Therefore, each sub-class of the **BomRecord** class can access the data dictionary of the OOBOM system. Whenever a transaction occurs, the transaction object can manipulate the adequate data via the inheritance relationship. Since **BomRecord** is a sub-class of **Model**, it can define many views as its dependents (i.e., the Model-View-Controller Paradigm of Smalltalk programming) (LaLonde and Pugh, 1990). The dependents (virtual models) of the **BomRecord** model are generated from the following classes: BOMPrint, NodeTree, PartLibrary, PartWindow and ProductBrowser. The BomRecord model handles these virtual models, while each virtual model handles its dependents (views: the transaction windows). Therefore, these transaction windows can self-update, whenever a related data change occurs. For example, the child-list (sub-assembly list) of a part is modified by a **PartWindow**, then this 'changed' message for the child-list will pass to the **PartWindow** virtual model and in turn this virtual model will pass it to the **BomRecord** model. Further, **BomRecord** will notify its dependents with the changed message, and then the dependents will self-update with the changed data if the data is related. For instance, the **BomTree** window may need to update the tree branches when a part is added into the assembly structure.

3. The Transaction Model

The transaction model of the OOBOM system describes the interface between BOM data and data management. The behavior of a part object can be specified and controlled by the methods of its class. In regards to a database system, the most important task is to access the value of an object's instance variable, i.e., the information query of a part. Because an assembled part is composed of components, there should be methods to perform the linkage functions between parts, sub-assemblies, and products. Each assembled part should link its "children," the component parts, in a single level BOM form and then iteratively construct the multi-level BOM. Calculations, such as the cost estimate of an assembled part, can provide more complete information for specific departmental operations.

Furthermore, there are classes in the OOBOM system which manipulate the transaction management via the Model-View-Controller (MVC) paradigm of Smalltalk. They are **BOMPrint, NodeTree, PartLibrary, PartWindow** and **ProductBrowser**. Each uses its own window interface to perform the related manipulation. The relationship between interfaces is shown in Figure 2. **PartWindow** supports the basic operations on part data, such as part creation, data modification and part query. This window is invoked by all the other windows to support part data entry and editing. **ProductBrowser** provides users with functions to browse the final products in the BOM database. As shown in Figure 3, these data are organized in product categories for a better product management. The users can start from this window to select the final product or its sub-part to perform relational links. It can invoke a BOM tree to define or edit a product structure. An indented BOM structure can be shown in the window interface to show the assembly structure of a final product. **PartLibrary** manages all standard parts by grouping them in categories (Figure 4). For example, pencil_assy, body, clip and barrel are included in a pencil-intrinsics category. Therefore, the creation of a part is always started from this interface for entry to part database and completed via **PartWindow** for detail part definition. Besides, the BOM tree (**BomTree**) uses the categories of **PartLibrary** to organize the search of sub-assemblies for a product. As shown in Figures 5, a **BomTree** shows the assembly structure of a product. In addition to the graphical structure, each node of a **BomTree** encapsulates the data of the corresponding part. The user can select a node in the structure to perform operation relevant to the represented part data. The **BomTree** operations may invoke **PartWindow** to modify the part data, invoke **PartLibrary** to search for a sub-part, or modify its own structure by adding and deleting nodes. These actions result in the update of relevant information.

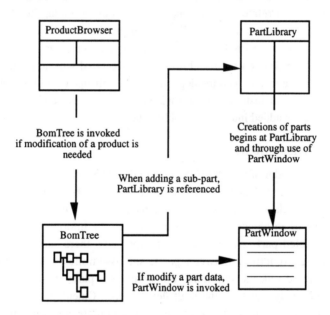

Figure 2. Relationships between data management tools in the OOBOM system

4. Conclusion

This research develops a methodology for structuring BOM based on an object oriented paradigm. Components, sub-assemblies and products are encapsulated into the OOP objects. With the benefit of OOP characteristics, such as abstraction, inheritance, encapsulation and polymorphism, objects can be easily linked and manipulated in a hierarchical BOM structure with user-friendly interfaces. A prototype of the OOBOM system is implemented in the Smalltalk environment. The OOBOM system architecture can be further implemented into an Object-Oriented DBMS for an engineering data management system (EDMS) development. Further more, the OOBOM system can be integrated with a company's CAD/CAM system. The preliminary solution for integration between OOBOM and CATIA CAD/CAM system is studied (Trappey and Lin, 1994). Our current efforts are in the areas of EDMS and CAD/CAM/BOM integration.

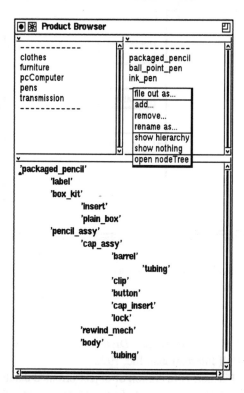

Figure 3. The ProductBrowser, its three sub-windows, and its activated PopUpMenu for updating the product database.

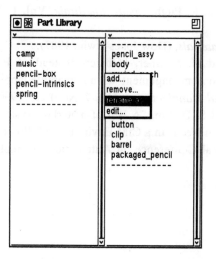

Figure 4. The Part Library Window and the activated PopUpMenu.

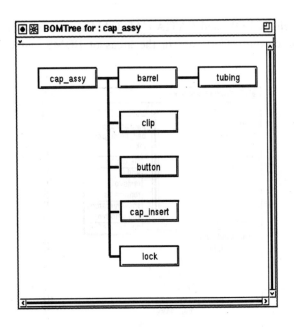

Figure 5. A BOM tree example: the added 'barrel' and the modified structure for cap_assy after the 'add' operation.

5. References

1. P. K. Andleigh and M. R. Gretzinger, *Distributed Object Oriented Data-System Design*, (Prentice-Hall International, Inc, 1992).
2. Y. Chung and G. W. Fischer, "Illustration of object-oriented databases for the structure of a bill of materials," *Computers in Industry*, **19**(1992), pp. 257-270.
3. W. S. Chen, "An Object Oriented Approach for modeling Bill-of-materials," *Proceedings of the CIIE National Conference*, 1993, pp. 188-193.
4. W. R. LaLonde and J. R. Pugh, *Inside Smalltalk*, **Vol. 1, 2**. (Prentice-Hall, Inc, 1990).
5. H. Mather, *Bills of Materials*, (Dow Jones-Irwin, 1987).
6. H. K. Tonshoff and H. Dittmer, "Object- instead of Function-Oriented data management for tool management as an example application," *Robotics & Computer-Integrated Manufacturing*, **Vol. 7**. No. 1 /2, (1990) pp. 133-141.
7. A. J. C. Trappey and S. D. Lin, "A Bill-of-materials System Designed to Provide Dynamic Information Needs in a CIM Environment," *Proceedings, Automation'94*, The 3rd International Conference on Automation Technology, **Vol 1** (1994) pp. 169-175.

APPLICATION OF MRP IN A LOCAL MANUFACTURING COMPANY - A CASE STUDY

NOORDIN BIN MOHD. YUSOF

and

C. L. NGEOW

Universiti Teknologi Malaysia
Locked Bag 791, 80990 Johor Bahru, Malaysia

ABSTRACT

Closed-Loop Materials Requirement Planning (MRP), a major tool in assisting manufacturers to become competitive, is not widely used in Malaysia. Among the reasons for this are lack of funds, lack of expertise on MRP, lack of knowledge on its significance and perception of not needing it. This paper presents a case study that investigates the implementation of MRP in a local, medium sized and discrete product manufacturer. Among the issues covered are the identification of the critical success factors for implementing MRP, the evaluation of MRP utilisation status, performance measurement, benefits realised, problems faced and strength and weakness analysis. In view of the above, specific recommendations are made to improve the MRP implementation. It is anticipated that this study will guide non-users of MRP to a better understanding of the issues involved in implementing MRP and be encouraged to consider it.

Keywords: MRP, Manufacturing

1. Introduction

Manufacturing is a dominant sector of the Malaysian economy (Noordin 1993). It requires the use of the latest developments in information, control, communication and processing technology together with developments in organisation and management to meet the more exacting needs of more discerning clients for quality products (Noordin and Shariff 1993). One of the tools available to assist manufacturers is Manufacturing Resource Planning (MRP II).

MRP II is an explicit and formal manufacturing information system that integrates marketing, finance and operations planning. It coordinates the above mentioned activities to assure the feasibility and consistency of the business plan. An important module under MRP II is the Closed-Loop Materials Requirement Planning (MRP) which is the focus of this study. An overview of the MRP system is given in Figure 1.

The introduction of MRP, which is designed to handle ordering and scheduling of dependent demand items, into an organisation has relieved much of the burden and problems in planning and scheduling faced by manufacturers of discrete products. Successful implementation of MRP system will enable manufacturers to, amongst others, exercise better inventory and raw material control, react faster to market and production changes, and improve communication and information sharing within and between departments (Yeo and Ong 1988, Aziz 1991).

A more comprehensive listing of factors ensuring the successful implementation of MRP is identified and summarized as follows:
- Develoment of software 'tailored' to company's operation requirement,
- Top management commitment,
- Total involvement,
- Sufficient training,
- Money available for viable project,
- Right consultation,
- Well structured Bill of Materials (BOM) and part number system,
- Well structured Standard Operating Procedure (SOP) and good coordination in all related departments,
- Proper material documentation,
- Commitment to performance evaluation,
- Low employee turnover and
- Materials suppliers' support.

These factors are identified based on literature review and investigation in the company where the case study is conducted through discussion with company's top management, results of a questionnaire developed on MRP implementation and self observation on the existing system.

2. MRP Implementation in Malaysia

Implementation of an MRP system requires a substantial and continuous investment in terms of both money and effort, involving an overhaul of the companies' planning, monitoring, controlling and information management activities. However potential users are not aware of the benefits or are unconvinced that such benefits can be realised easily. This can be seen from, among others, a survey of 72 companies chosen randomly from all over Peninsular Malaysia (Aziz 1991) which revealed the following:
- The percentage of companies using MRP is small, ie., 27%,
- A computerised MRP system requires high initial as well as training and maintenance cost,
- The three most problematic obstacles towards successful MRP implementation were as follows:
    * lack of company expertise on MRP,
    * lack of MRP communication in company and
    * lack of support from personnel in marketing and production.
- Some users enjoy benefits from the MRP systems and
- Among the reasons for not implementing MRP are lack of funds, lack of expertise on MRP, do not know its significance or do not need it.

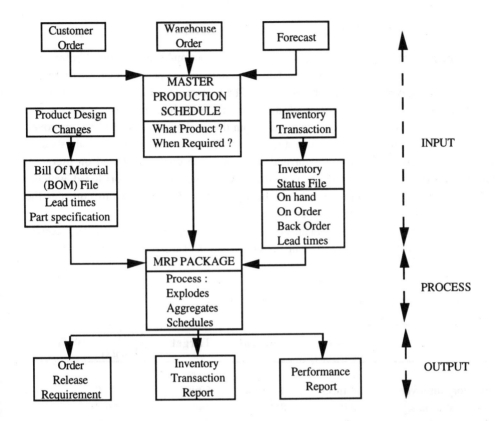

Figure 1. MRP System Overview : Input - Process - Output

3. Case Study

This case study was conducted to further understand the issues and challenges faced by a local manufacturing company in implementing MRP. It documents an example of the implementation issues and a measure of the benefits to be realised. It is hoped that this study, and further cases, will be able to provide a guide to manufacturers considering implementing MRP, as well as to initiate a collection of actual cases for use by consultants and academicians to further study MRP implementation in Malaysia.

The company chosen for this case study is a subsidiary of a multinational corporation, medium sized and involved in the manufacture of integrated electrical equipment and appliances. Computerised MRP II was implemented in the company in 1991. We shall name the company ABC.

3.1. MRP Utilisation Status

The Wight's MRP checklist (Magad and Amos 1989) was used to determine the utilisation status and the result of the evaluation indicates that ABC is a class B MRP user (Class B : 70 - 90 points). This implies that ABC has achieved good result, i.e., is using the system to schedule works primarily in manufacturing and materials management (MRP).

3.2. Performance Measurement

The performance of the company in relevant functions was monitored before and after the implementation of the MRP system. The data for the performance measurement was provided by ABC's production controller. The performance measurement was solely based on various tangible factors such as inventory reduction, inventory accuracy etc. Intangible factors such as confidence gained from the implementation of MRP, improved operator's morale etc., were however not measured. The target values were established by consultation between the investigator and the production controller based on literature review on MRP systems whereby the percentage of accuracy for the factors listed, with the exception of inventory turnover, should be more than 95% in order to consider the implementation of the MRP system effective.

A comparison of the performance measures before implementation (Before), current - based on 1992 data (Current) and company's expectation (Target) is given in Table 1.

Table 1
Performance measurement of MRP implementation

Factor	Before	Current	Target	$\dfrac{\text{Current}}{\text{Target}} \times 100\%$
Inventory turnover	4 turns	4.5 turns	6 turns	75.0
Inventory accuracy	82%	85%	98%	86.7
Master schedule	89%	90%	95%	94.7
Purchase schedule	80%	88%	95%	92.6
Capacity plan	75%	87%	95%	91.6
Production plan	84%	85%	95%	89.5
Delivery schedule	80%	85%	95%	89.5

3.3. Benefits Realised

The benefits ABC realised from the MRP implementation were as follows:
- More efficient production and materials requrement planning,
- Better inventory and raw material control,
- More efficient and effective data processing,
- Reduced clerical supports and
- Improved communication.

3.4. Problems faced

Problems on both data and operational aspects encountered by ABC were as follows:
- Inaccurate inventory data,
- Slow response to engineering change,
- Long lead time for imported components,

- Variation in forecast order,
- Short lead time for confirmed orders,
- Long system run time and
- Lack of performance analysis.

3.5. Strength and weakness analysis

A strength and weakness analysis pertaining to factors affecting MRP implementation, as given in section 1.0, was carried out. The results of the analysis were as follows:

Company's strength
- Top management commitment,
- Money is available for viable project,
- Well structured BOM and part number system,
- Sufficient training,
- Right consultation,
- Low employee turnover,
- Total involvement and
- Materials supplier's support.

Company's weakness

- Incorrect software selection,
- No standard operating procedure,
- Improper material documentation and
- Lack commitment to performance evaluation.

3.6. Discussion and recommendation

The results obtained from the evaluation of ABC's MRP utilisation status is commendable as most of the MRP users in USA are also class B MRP user (Apics 1989). The result of the performance measurement indicates that improvements can be seen after the implementation of MRP. However, the achievements on the whole are not as good as could have been. There is still room for further improvement.

It can be inferred that the problems faced by the company and the weakness identified by the strength and weakness analysis have influenced the performance measurement. In order to improve the implementation of the MRP system at ABC, several actions were suggested to overcome the above weaknesses. The following recommendations were made:-

3.6.1. Development of a Standard Operating Produre (SOP)

A SOP for Materials Requirement Planning was developed for ABC. This may serve as the starting point for the development of other SOP's. The SOP contains the purpose, scope, responsibility and procedures in carrying out the materials requirement planning activities.

3.6.2. Development of good materials documentation.

A modified materials documentation system was proposed to ABC to help reduce the problems faced. In the proposed system, every document has a unique function which eliminates confusion arising from a 'multi-purpose' document. It also has a well-diagrammed document flow which will better serve the task of tracing and control. The modified system uses existing documents, modified documents and entrely new documents. Table 2 shows the proposed materials documentation whilst Figure 2 shows the proposed document flow chart.

3.6.3. Improved performance evaluation system

It was proposed that ABC changes its performance evaluation from being manual and irregular to one which is computerised and periodic. The computer system available should be utilised for this purpose whereby standards and formula for quantitative measurement can be programmed. Examples of formulae for performance measurement can be obtained from (Shah 1988). The performance report generated can then be compared with the target set to identify actions needed for improvement.

3.6.4. Development of a software evaluation checklist

The importance of selecting a correct MRP software has been emphasised. The availability of numerous MRP softwares have made the task of selecting the correct software more difficult. A software evaluation checklist has been developed based on literature review and consultation with ABC's production controller to assist would-be users in this respect. This however is not discussed in this paper.

4. Conclusion

It has been determined that successful implementation of MRP in a local manufacturing company improved some measures that can assist it to become competitive. Successful implementation, however requires consideration of the critical success factors. The issues and challenges faced by this company in implementing MRP were reviewed and recommendations were made to improve its implementation. It is hoped that by documenting this case study, non-users of MRP would have a better understanding of the issues involved in implementing MRP and be encouraged to consider it.

Acknowledgement

Sincere thanks are due to fellow project supervisor, Mr. Ahmad Zaidi Bahari from the Faculty of Management and Human Resource Development, for his input in this project and to company ABC for their willingness and cooperation in allowing us to study

Table 2
Proposed Materials Documentation

Code	Defination	Purpose	Originator
GRN	Goods Received Note	Raised when items are received	Receiving-Store
GRS	Goods Return To Supplier	To return excess, rejected or wrong items to supplier	Receiving-Store
RRI	Rejected Finished Goods Received Inward	Used for all rejected finished goods return by customer	Receiving-Store
RRP	Rejected Finished Goods Released To Production	To sent rejected finished goods for rework	Raw Materials Store (RM Store)
RRS	Rejected Finished Goods Return To Store	To transfer reworked finished goods from line to finished goods store	Production
RRO	Rejected Finished Goods Returned Outward	To returned reworked finished goods to customer	Finished Goods Store (FG Store)
MIV	Materials Issue Voucher	To issue needed materials for production	RM Store
MRL	Materials Return From Line	Wrong, reworkable or excess parts return to raw materials store	Production
ITN	Internal Transfer Note	To transfer semi-finished or sub-assembled parts to raw materials store	Production
ITF	Internal Transfer Of Parts For Fitting	To send parts to FG store through which parts are send to supplier for fitting	RM Store
MSA	Materials Stock Adjustment	To adjust stock after physical audit	RM Store
SCN	Scrap Note	To transfer scrap items to RM store for disposal	QC-Production
WOF	Write Off	To dispose rejected goods or parts	RM Store
FGN	Finished Goods Note	To transfer finished goods from line to finished goods store	Production
FGW	Finished Goods Rework	Finished goods return to production for rework	FG Store
FGR	Finished Goods Return	To return reworked finished goods to finished goods store	Production
DOC	Deliver Order For Customer	To deliver finished goods to customer	FG Store
DOS	Deliver Order For Sample	To deliver sample (finished goods) to headquarters or sister company for testing	FG Store
DMA	Deliver Order For Outgoing Materials Advice	To deliver parts to other supplier for fitting	FG Store

their MRP implementation. Special thanks are also due to Dr. Mohd. Shariff b. Nabi Baksh for his valuable suggestions in improving the original paper.

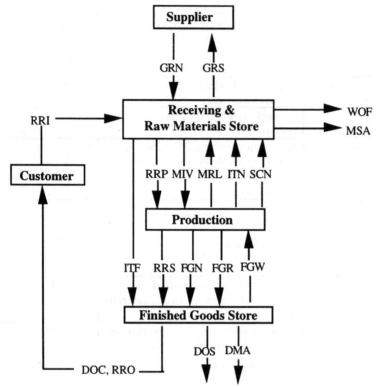

Figure 2. Proposed document flow chart

References

1. APICS, Implementation of MRP, Production and Inventory Management Journal, 70 - 75, 3rd Quarter, 1989.
2. Aziz A. Hamid, s et. al., Computerised MRP in manufacturing companies in Malaysia, Int. Journal of Production Economics, 25; 73 - 79, 1991.
3. E. L. Magad and J. M. Amos, Total Materials Management, Van Nostrand Reinhold, New York, 208, 1989.
4. Noordin M. Yusof, The use of CAD/CAM system in manufacturing, CAD/CAM seminar for small and medium sized industries, 1993.
5. Noordin M. Yusof and M. Shariff Nabi Baksh, Trends in Modern Manufacture : The UTM perspective, Seminar on CMM in Advanced Manufacturing and Quality Control, 1993.
6. Shah N.M., An integrated concept of materials management, Tata McGraw-Hill Pub. Co., New Delhi, 457 - 458, 1988.
7. K.T. Yeo and N.S. Ong, Applications of MRP in local industries - A user survey, Vision, vol. 4/ No.2, 5- 7, 1988.

MRPII AS A BASE OF DIFFERENT PRODUCTION MANAGEMENT TECHNIQUES

ERIK L. J. BOHEZ

Manufacturinng Systems Engineering Program
Asian Institute of Technology
G.P.O. Box 2754, Bangkok 10501, Thailand.

and

M. AHSAN AKHTAR HASIN

Manufacturing Systems Engineering Program
Asian Institute of Technology, Bangkok, Thailand
E-mail: hasin@emailhost.ait.ac.th

ABSTRACT

Over the years, the stochastic and complex characteristics of shop-floor activities and inventory replenishment policies have led to the development of different philosophies, like Manufacturing Resource Planning (MRPII), Just-In-Time (JIT), Optimized Production Technology (OPT), Leitstand, Gantt Chart, some hybrid technologies, etc. But none of them can solve all the problems in any manufacturing environment. So, naturally a question arises - Which manufacturing and management policy should be adopted ? There is no general answer to this question. The solution depends on the type of production environment and related conditions. But assessment of different alternatives under different environments can help the managers to select one or more (hybrid) technologies for their industry.

Keywords: MRPII, JIT, OPT.

1. Introduction

Over the last couples of decades, the industries have been testing different operations management techniques for competitive advantages. MRPII, JIT, OPT, FMS (Flexible Manufacturing System), Leitstand, Disaster, etc. invaded production planning and control in quick succession. The new techniques overcame the limitations of the older techniques, still with limitations in different areas.

Different types of manufacturing industries have some unique features of their own.

So, their MPS (Master Production Scheduling) strategies, requirements calculation, materials flow and shop floor scheduling techniques are partially or totally different. That's why the same production management techniques can not be applied to each of them. Without detailed study of the company's manufacturing environment, it is not possible to show the way out from the planning and controlling problems.

2. Manufacturing Resource Planning (MRPII)

MRPII is a planning tool to prepare purchasing and manufacturing orders based on company data. The orders show which part, how much, and when to purchase or manufacture.

If we consider MRPII as a planning tool, then MRPII is not a database system. Rather it depends on a database system in order to generate plans. In that case, the definition of MRPII should be - "MRP is a formal planning system to prepare manufacturing and purchasing orders based on a Database Management System (DBMS) with central data repository and distributed processing capability". Distributed processing capability gives the power to individual departments to update their data and information separately and without any interference. So, this planning tool as a whole has two parts (Fig.1.) : 1) A Data Repository with DBMS (part a), 2) A planning part with planning algorithms (part b).

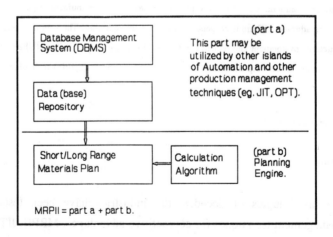

Fig.1 MRPII as a Planning Tool.

If the CIM implementation is database oriented, then it is definite that MRPII may take the role of central hub. The role of MRPII in the integrated planning paradigm has been defined by several authors (Chua, T.J., et al. 1990/91). The database part (part a of Fig. 1) will store and supply data to different islands. Also there will be direct data communication among the islands.

2.1. What Can Not Be Done in MRPII

MRPII, no doubt, is a very good planning tool, though it can not cover some major functions of manufacturing. These pitfalls have raised several questions regarding MRPII's wide acceptability. The following are those manufacturing functions:

1) Capacity Scheduling : Only infinite Capacity Scheduling is possible at the present moment. Finite scheduling is done mostly by trial and error method. A Gantt chart may be used as a capacity scheduling tool. Some third party software and systems are available for this finite scheduling.
2) Inventory : MRPII attempts to reduce stock-outs to zero. It does not try to reduce inventory to zero (Aggarwal 1985).
3) Quality : There is no quality control algorithm or rules incorporated in MRPII system. It never tries to reduce waste or scrap (Compare JIT !), rather it just keeps track of scraps (Mozeson 1991).
4) Shop floor Scheduling : It follows a simple backward scheduling rule by offsetting setup and other times to calculate the order starting date. This scheduling rule is never an optimal one. Also sequencing of operations and/or jobs are never done in this system. Some third party systems and software are available for Shop Floor scheduling.
5) Lot Sizing : It attempts to find out an optimal lot size part by part (i.e. one item at a time). But requirements of parts for an end item are not independent of each other; they are tied by the BOM. So, optimizing lot sizes for one item at a time may be considered as a suboptimizing process. Better results can be obtained by looking at more items and products at a time (Kamenetzky 1985).
6) Time Consideration : Setup time, queue time, operation time, etc. are assumed to follow a Uniform Distribution function. It does not follow actual statistical distribution functions. Lead times are stochastic variables with 90% of the time spent in queueing.

3. Just-In-Time Philosophy

Just in Time is a manufacturing philosophy with a very simple goal, i.e. produce the required items, at the required quality and in the required quantities, at the precise time they are required. This system is mostly used in a repetitive manufacturing environment. There are some distinct advantages of JIT over other systems.

This is the first methodology which says that the product should be designed for ease of manufacturing and assembly. So the process plan should be considered first, then the product should be designed to make the production process an easy one. But in case of unstable suppliers and circumstances, JIT can not work properly. It also needs a highly trained and devoted workforce.

4. Optimized Production Technology (OPT)

The OPT is a scheduling tool which operates well at the shop floor. Here, the leadtimes are not inputed, but are outputed from calculation. It is similar to MRPII scheduling in one sense that it also uses a simple offsetting rule for forward and backward scheduling with special emphasis on bottlenecks. This is the first philosophy which says that the batch size should not be a fixed one, rather the transfer batch should be different from the process batch. It is similar to JIT in the sense that it tries to reduce the setup times with special attention at the bottleneck. OPT preschedules everything like a simulation tool. So the users can know precisely what will happen each hour of each day in advance. In contrast, since Kanban pulls material through the system only when needed, it can not give any pre-idea.

5. Other Techniques

Because of limitations in all of the above techniques, more and more new ideas and concepts are emerging with special emphasis on Artificial Intelligence, knowledge-based Expert Systems and Decision Support Systems (Chong 1985).

5.1. Leitstand

This is quite a new planning tool with powerful planning algorithms and graphic scheduling capability (Van Landeghem 1993). Orders may be scheduled selecting any priority rule or simply manually rearranging the tasks one by one. Some of the Leitstand software have already incorporated Knowledge-based scheduling techniques. This system also needs a computer-based database to define the Bill of Material and Routing as the MRPII system.

6. Are There Boundary Walls Between The Techniques ?

Implementation of one technique does not mean that the company has completed its optimization task (Bermudez 1991). Combination of the techniques are possible because the systems cover some common structural characteristics with complementary gaps. These complementary gaps may be covered using another one. So the answer to the above question is clear. There are no boundary walls around the systems.

MRPII and JIT are specially considered to be two production planning & controlling techniques which are totally overlapping. MRPII may be considered to be the first step to attain manufacturing goals through functional optimization, and JIT may be the next step to eliminate waste, ensure better quality and further enhance the optimal point towards manufacturing excellence. Many of the US companies have implemented JIT starting their journey from MRPII.

On-hand inventory, %scrap in the shop floor, lot sizing, lead time, etc. are the elements of MRPII which are not optimized in this technique. The JIT technique can optimize these. So JIT can serve to fill-in the gaps of MRPII. Also the MRPII database and some calculation algorithms, like BOM explosion with low-level coding, forecasting, accounting & financial integration, RCCP (Rough-Cut Capacity Planning), etc. can serve JIT to enhance its functionality. The shop floor control may be left to JIT.

The OPT system tries to reduce lead time and inventory level similar to JIT with special attention at the bottleneck resource. It makes use of a data repository like MRPII. The main focus of attention of OPT is to produce a better shop floor schedule than MRPII does. In that sense, OPT can replace the shop floor scheduling methodology of MRPII taking the advantages of other philosophies and the data repository of the MRPII system. The BUILDNET module can serve this integration to extract information, like BOM, routings, inventory, machine capacity, etc. from MRPII database (Jacobs 1984, Wheatley 1989). So no boundary walls exist between MRPII and OPT, rather a clear combined approach may be adopted to attain a better result.

7. Application of Production Management Techniques In the Asian Countries

Though different production management techniques have been proved to be a useful tool to improve productivity, there are several problems in most of the asian countries to implement them. There are several reasons for JIT implementation failure. But MRPII implementation should be successful only if the managment tries to understand that it is a "System" to be "Implemented", not a simple "Package" to be "Installed" only. The major problems are identified below:

1) JIT Implementation: As explained before, JIT technique aims at producing right

amount of products at the right moment. Since this is a pull-type production technique, it should also procure materials in the right amount at the right moment. Because of several problems encountered in these countries, the market is totally unstable. It is really difficult to procure good quality materials at the right moment. So JIT procurement will not be successful down stream. The present structure of management is a barrier against Quality-Circle concepts. The management of the industries never wants to involve the downstream staff in the decision making process. Also the management has lack of knowledge on these topics.

In Thailand, our survey has seen that the local industries have a bureaucratic management structure. Most of the industries have some unnecessary levels of management, and official papers flow through several unnecessary steps. Ultimately, it takes some days to reach a final decision. Some of the industries in Thailand have tried to implement JIT, but because of the above reasons, they were never successful.

2) MRPII Implementation: The above factors are also applicable in case of MRPII. The lack of knowledge is the main barrier against successful implementation.

In Thailand, most of the industries have a misconcept about MRPII. They think this as a simple package, which needs only some commands to run the software, as word-processors or a drafting package. They do not think that MRPII is a system which needs to be "Implemented" and "Managed". They actually wanted to "Install" (not "Implement") the package. In one multi-national company, they said that they want to "Install" the package in one PC (inside a computer room) which also contains AutoCAD drafting software. There is no network at all.

There are around one hundred industries in Thailand who have bought some MRPII or similar systems. But they are mostly using these systems as a simple database package, to store and retrieve data. But for that they don't need to invest so much money, because a simple database package, like FOXPRO, or DBASE IV, could be sufficient. Some industries in Thailand (and the vendors of MRPII system) claims that they are "Using" the system successfully (?). But they never generate their plans from the system. They only store and retrieve data.

Actually, the industries do not know even their own requirements. We know that one reputed international shoe making company in Thailand has their own Mainframe computer to run only one accounting software, for which one 286-based desktop computer is sufficient. There are at least one hundred companies in Thailand who own

Mini-computers for running some simple packages. But their requirements show that a PC-Local area network could be more than sufficient.

In Singapore, there are several researches going on under the Government sponsorship. Because, the industries in these countries do not have so strong foundation that they may conduct their own research to find out their requirements. They also can not develop their own systems (Computer-aided Systems), like CAD-CAPP-MRPII, because of several limitations. In Thailand, and some similar countries, their is no government support or sponsorship to conduct research on problems and prospects of different techniques to fit in their environment. There should be government sponsored massive education on these Computer-Aided technologies in conjunction with the reputed Educational Institutes.

7.1. Implementation of Different Systems:

The approach of implementaion of different systems should not be the same in all countries. Also the selection of techniques should also be different depending upon the market characteristics. In the USA, it is possible to implement JIT starting from MRPII (Fig. 2.b) because of stable market condition. Materials may be available in the right amount at the right moment, responsibility may be delegated to the downstream staff for quality assurance. But in most of the asian countries, it is practically impossible to implement JIT starting from MRPII implementation. For better scheduling, the industries may lead towards implementing OPT or other third party scheduling software (Fig. 2.a). The implementation approach of hybrid systems are shown below:

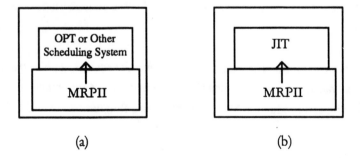

Fig 2. Hybrid system selection for different market situation.

8. Conclusion

This paper gave an overview of computer-based production management techniques where database is common to all. All these systems need a database. So a good step to start is to build the database first with any database software. Interface of this database with other software is a must which should be available with good production management packages.

8. References

1. Aggarwal, Sumer C., *MRP, JIT, OPT, FMS ? making sense of production operations systems*, (Harvard Business Review, **Sept-Oct**, 1985), p. 8-13.

2. Bermudez, John., *Using MRP system to implement JIT in continuous improvement effort*, (Industrial Engineering, **November** 1991), p. 37-40.

3. Chong, Philip S., *System combines Manufacturing Resource Planning and Decision Support approach*, (Industrial Engineering, **October**, 1985), p. 72-81.

4. Jacobs, F. Robert., *OPT uncovered: Many production planning and scheduling concepts can be applied with or without the software*, (Industrial Engineering, **October** 1984), p. 32-41.

5. Chua, T. J., Ho, N. C., and Yeo. K. T., *Integrating CAD & MRPII systems*, (A report on a research project conducted at GINTIC Institute of CIM, Nanyang Technological Institute, Singapore. Reference: Annual report, 1990/91), p. 18.

6. Kamenetzky, Rocardo D., *Successful MRPII implementation can be complemented by smart scheduling, sequencing systems*, (Industrial Engineering, **October** 1985), p. 44-52.

7. Van Landeghem, R., *The Role of 'Leitstand' schedulers within the integrated planning paradigm*, (Proceedings of International Conference on Industrial Engineering and Production Management, Mons, Belgium, 2-4 June, 1993), p. 677-686.

8. Mozeson, Mark H., *What your MRPII systems cannot do*, (Industrial Engineering, **December** 1991), p. 20-24.

9. Wheatley, Malcolm, *OPTimising production's potential*, (International Journal of Operations & Production Management), **Vol. 9**, No. 2, 1989, p. 38--44.

A DECISION SUPPORT FRAMEWORK FOR PWB ASSEMBLY

Yann-Yean Su
Doctoral Candidate

Department of Systems Science and Industrial Engineering
State University of New York, Binghamton, New York 13902-6000

and

K. Srihari, Ph.D.
Associate Professor

Department of Systems Science and Industrial Engineering
State University of New York, Binghamton, New York 13902-6000
E-mail: srihari@bingvmb.cc.binghamton.edu

ABSTRACT

A prototype decision support framework that can assist in generating a manufacturing plan for surface mount PWB assembly is presented in this paper. This framework assists in decision making from design through manufacturing process planning. It combines the use of expert systems, artificial neural networks and fuzzy theory techniques that work in a mutually supportive manner to identify near optimal solution(s) for facets of the surface mount PWB manufacturing planning problem.

1.0 Introduction to Research

Market driven functional needs in consumer electronics have resulted in a continual increase in the density of components on Printed Wiring Boards (PWBs) along with a concomitant decrease in component lead pitch, and often smaller volumes. This has resulted in an increased use of Surface Mount Technology (SMT) in PWB assembly. Manufacturing problems in the surface mount PWB domain are typically complex and varied in nature often mandating the need for decision support in problem solving. Facets of shopfloor control such as process planning, scheduling, sequencing, and line balancing are not either purely serial or hierarchical in surface mount PWB assembly. The dynamic nature of this domain is an important factor that needs to be considered when developing an automated manufacturing planning system.

Advances in computer science and improvements in computer hardware and software have, in recent years, resulted in the increased maturity and practicality of concepts related to Artificial Intelligence (AI). AI has been applied in a wide variety of areas (expert systems, natural language processing, robotics, etc). It has been recognized as a technique that has significant potential for problem solving in today's manufacturing environment [Turban 1992]. Expert systems (a facet of AI) have been shown to be successful commercial products as well as interesting research tools. The knowledge used by an expert system to perform these tasks, in general, has been extracted from human experts in the domain and encoded in a formal language [Rolston 1988]. Expert systems tend to be specialists focusing on a narrow set of problems. Their knowledge tends to be both theoretical and practical having been perfected through experience [Srihari and Westby 1992]. Expert systems are often limited by the (lack of) ability to deal with uncertain information. However, Artificial Neural Networks (ANNs) and fuzzy theory provide a novel approach and an advanced technology to deal with the weaknesses of expert systems. ANNs are biologically inspired, i.e., they are composed of elements that perform in a manner that is analogous to the most elementary functions of the biological neuron [Wasserman 1989]. Often, ANNs require fewer assumptions and less precise information about the system modeled than some of the more traditional techniques [Padgett and Roppel 1992].

This paper addresses the use of a combination of expert systems, ANNs, and fuzzy theory procedures that work in a mutually supportive manner for manufacturing problem solving within the surface mount Printed Wiring Board (PWB) assembly domain. The combination of these techniques will help provide a solution mechanism with features which exceed those that any one technique (expert systems, ANNs, and fuzzy theory) can provide [Caudill 1991].

2.0 Problem Statement and Research Objective

The assembly of PWBs has several characteristics that distinguish it from other forms of manufacturing. Electronic assemblies often contain an unusually large number and variety of individual parts. PWB assembly is highly dependent on the equipment used to perform the operations in order to generate the sequence of operations. Planning for PWB assembly requires a relatively simple Computer-Aided Design (CAD) interface. Many PWB assembly facilities need to produce multiple products while Work-In-Process (WIP) inventory investment needs to be reduced by processing smaller batch sizes. Many current processes, however, are dependent on the set-up for each type of PWB product and are inflexible in product routing and change-over capability. These obstacles must be overcome if batch size reduction and desired flexibility are to be achieved [Taylor and Graves 1990]. The PWB manufacturer also must deal with growing pressures caused by product proliferation, shorter life cycles, quality concerns, responsiveness requirements, and cost. It is becoming increasingly important to optimize the PWB assembly processes to achieve high levels of utilization without causing excessive WIP [McGinnis, Ammons, Carlyle, Cranmer, Depuy, Ellis, Tovey and Xu 1992].

A proper understanding of the complex process related inter-relationships is critical to achieve effective process control and for the realization of high yields in surface mount PWB assembly. Most decisions in SMT process and manufacturing control require the consideration of several variables [Srihari and Westby 1992]. The procedures followed for assembling electronic components on PWBs is fairly rigid. Numerous parameters are involved in each process of electronic assembly. Therefore, choosing the appropriate parameters, processes and equipment suited for today's application while being flexible enough to be changed for tomorrow's new technology can be a bewildering task. The dynamic nature of the PWB assembly environment is an important factor to consider when developing an automated process planning system. To develop a system that can be used in a changing environment, the sources of these changes must be identified. The interactions between process planning, scheduling, sequencing, and line balancing need to be understood and managed. While several expert systems have been developed for manufacturing process control in the PWB assembly domain, full control and management of information flow in these systems has not yet been achieved. This is mainly because of data inconsistencies and the lack of established functional relationships among these different systems.

The objective of this research was to investigate the use of ANNs, fuzzy theory and expert systems as decision support tools that work together within a comprehensive problem solving framework for the surface mount PWB assembly domain. The sub-objectives of this

research were to:
1. Understand the use of expert systems, ANNs, and fuzzy theory as they apply to PWB assembly.
2. Study the PWB assembly domain to identify possible applications of the technologies mentioned above in solutions to specific problems.
3. Design and develop a prototype updateable decision support system to assist the manufacturing engineer in process and systems related decisions within the surface mount PWB assembly domain.

3.0 Methodology

Production planning and manufacturing activity in PWB assembly are extremely distinct from conventional machining processes. The process sequence, the functions and properties of machines, the features of materials, and the parameters to be selected are drastically dissimilar. A general dynamic production planning and production manufacturing system is an ideal mechanism to preserve expertise. It provides valuable and useful tools for decision support in this domain. The prototype system described in this research can present responses quickly to dynamically changing circumstances in surface mount PWB assembly. This research uses concepts related to:
* PWB assembly using Surface Mount Technology (SMT).
* Complex problem solving techniques including knowledge based or expert systems, neural networks, fuzzy theory, and Object Oriented Programming (OOP).

Combinations of solution methodologies have been used in this research. Solution mechanisms have been chosen based upon their appropriateness for specific problems and environments. AI based technologies such as expert systems, ANNs and fuzzy theory are three dominant methodologies for specific SMT issues. ANNs can be embedded within expert systems and vice versa. Expert systems handle rule-based data and many also provide user interfaces. The ANNs receive input through the expert system interface and perform estimation and correlation. Rule-based decision making may overlay the net output fed to the expert system by a neural network. A neural net could analyze data, and a rule-based system could then select among several discrete responses based on this analysis. The uncertainty within the domain should be molded into the knowledge base to make the system a more representative model of the real world. Fuzzy theory helps to provide answers to uncertainty management issues in expert systems.

This research uses C++ under the Windows operating environment. The system is integrated with CAD drawings which serve as its input. A process plan is a part of the output. This prototype system was designed in a modular fashion, where each module was developed and designed independently. OOP techniques were extensively used in software development. The use of C++ and Windows programming made the system more user-friendly through graphical user interface screens while facilitating change and modification.

3.1. System Related Assumption and Constraints

Assumptions and constraints were used to limit the scope of this research. The PWB assembly facility considered in this research can assemble single sided and double sided surface mount PWBs. It has multiple surface mount PWB assembly lines. However, some machines such as a hot bar bonder are shared by two lines. The three process steps, namely, solder paste printing, surface mount component placement, and solder paste reflow soldering are the primary procedures used in surface mount PWB assembly. The machines considered in this research are used for stencil printing, surface mount component placement, solder reflow, and for hot bar bonding. The assumptions used can be classified as job related, machine related, process related, control related, and system related assumptions.

3.2. Manufacturing Tasks Considered

Inputs to the prototype decision support framework described in this research are first reviewed by a 'Design For Manufacturing' (or DFM) system with respect to their manufacturability and cost. During DFM review, a macro-level manufacturing analysis is executed. The 'finalized' PWB design is then used to establish the process and production related parameters for a specific PWB batch. Input information derived from the PWB's design includes relevant parameters from the circuit diagram, board dimensions, component types, required package density, expected thermal and electrical requirements, required testing conditions, and cleaning specifications. The CAD file and other supplementary information is used by the personal computer based decision support system to perform the following functions: process planning, scheduling, sequencing, and line balancing. The process planning system is realtime and generative. The prototype system determines process parameters for each process in the assembly. The system's outputs are the relevant manufacturing sequence and the processing instructions for each step in the process.

Process planning helps to determine the sequence and the process details associated with the individual processing and assembly operations needed to produce and manufacture a product. It bridges the gap between design and manufacturing. Scheduling determines when and what resources are used for the jobs (or work orders) to be accomplished in the factory. The scheduling function needs to consider the other jobs which are competing for the same resources (machines, tools, materials, personnel, etc.) often with limited availability [Sarin and Salgame 1990]. This task is difficult since typically there are numerous jobs on the PWB production floor at any given time, and their routings depend upon the availability of required resources and the relative importance of time and cost. The dynamic variation in factory status is an important factor that influences the scheduling of jobs [Sarin and Salgame 1990]. The availability of workers, tools and machines in surface mount PWB manufacturing can change unpredictably over time thereby requiring the jobs to be rescheduled. The sequencing phase of this research selects the (near) optimal sequence for a specific batch of PWBs and dispatches working orders for a set of requirements. The system attempts to optimize production throughput, maximize machine utilization, and reduce flow time and WIP. Line balancing in the surface mount PWB batch production environment requires the arrangement of individual processing and assembly tasks at the machines so that the total time required at each machine is approximately the same. The PWBs move successively at an almost consistent rate through a sequence of surface mount processes. Segments of the PWB assembly task may be performed at different machines.

3.3 System Design and Architecture

The AI based generative dynamic surface mount PWB planning system develops manufacturing process plans while considering the dynamic status of the shop floor. The system is composed of several modules such as the process planning module, scheduling module, sequencing module, line balancing module, explanation facility module, uncertainty management module, knowledge base module, and the parameter module. The selection of the processes and operations and establishing their scheduling and sequence is accomplished by the process planning module, scheduling module, sequencing module, and the line balancing module. These modules perform forward and backward chaining inferences by utilizing the information available in the knowledge base to generate a process plan.

In the process planning procedure, the system will consider the design of the PWB, the required batch quantity, production control constraints, and production system construction in identifying the line (and therefore the machine) to be used. Due to the setup time needed for manufacturing a PWB batch and the comparatively short amount of time it takes to actually assemble the PWB, it is not advisable to change lines after the line has been setup. This segment of the system uses individual modules to deduce the manufacturing process parameters associated with individual machines. These modules include the solder paste printing module, component placement module, reflow soldering module, and the TAB bonder module. The solder paste printing module identifies stencil printer set up parameters. The component placement module determines the sequence in which components need to be placed at the component placement machines in a line. The reflow (profile identification) module helps the user identify the thermal profile that can be used (along with an acceptable range) for the pre-heating, flux activation, reflow spike, and cooling zones of the oven. The TAB bonder module provides the user with machine set-up information. These (near) optimal parameters are identified by using expert systems and/or ANNs (using a backpropagation approach). The prototype system provides in-depth output based on known or proven rules for each process step.

A major emphasis of this research has been to deal with the uncertain nature of the surface mount PWB production on a realtime basis. Uncertainty can be considered as the lack of adequate information to make decisions. Uncertainty may prevent systems from making the best decision and may even cause a bad decision. This mandates the use of a method of uncertainty management to maintain system consistency. The prototype system can proceed with individual machines being 'up' or 'down' on the floor. Any change in the status of any machine at a later point in time makes the system revise the information stored earlier, and maintain the correct and realtime shop floor status. The uncertainty management module requires a set of rules, an inference mechanism and a method of uncertainty management. Shop floor status is tracked using a realtime database. This allows the system to check shop floor status at any time, and take appropriate process planning and routing decisions. A change in shop floor status will activate the passive truth maintenance system that removes any inconsistencies in the system.

The knowledge base contains the information necessary for understanding, formulating, and solving problems. It includes two basic elements, namely, facts and rules. The system allows users to create a new knowledge base or modify an existing knowledge base when knowledge/information is expanded or changed. It also helps maintain correct rule syntax, and performs consistency checks on the updated knowledge base. A hypertext based help facility is provided for the user to access information in a nonlinear fashion by following a train of thought.

A flowchart that depicts the working of the proposed prototype decision support framework for surface mount PWB assembly is shown in Figure 1. Initially, the CAD drawing is input to the DFM module. The DFM module then evaluates and checks the PWB's design to review whether the design is acceptable from a manufacturing point of view. If the CAD drawing is not acceptable after DFM review, it will need to be modified. If the design is acceptable, other relevant knowledge bases are accessed. They provide the supplementary inputs required by the prototype system to execute manufacturing tasks. These manufacturing tasks include process planning, scheduling, sequencing and line balancing. Next, manufacturing planning information for the specific surface mount PWB batch is identified. If the solution is acceptable, the system presents the manufacturing planning and process parameters to the manufacturing engineer to enable the production of the PWB batch and related machine set-up. If the manufacturing plan generated by the system is not accepted, then an alternate plan(s) will need to be generated.

3.4 Information Used and Outputs

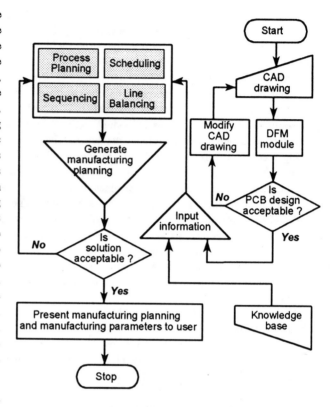

FIGURE 1: THE SYSTEM FLOWCHART

The primary input to the system is through a CAD drawing. Supplementary information is obtained through a consultation between the end user and the system, and from the relevant knowledge bases. The input information used includes: PWB number, assembly batch size, product description and dimensions, PWB design and configuration details, the type and quantity of components to be placed, component dimensional information, component location and orientation on the boards, type of solder paste (composition) and flux used, type of cleaning equipments and the cleaning solvent used, substrate material and properties of the boards, and machine specifications and capabilities.

The output from the prototype system includes the part manufacturing process sequence(s), the machines selected, alternate routes that are possible with an indication of the applicable penalty for choosing the alternate route(s), and the process instructions for each step and the process parameters. The outputs provided to the user includes: manufacturing routing, manufacturing scheduling, the suggested process sequence for each PWB, the estimated throughput time for a particular batch, rerouting information, machine setup parameters, processing time(s), sequence of required machine operations, machines up/down status, machine loading, and machine breakdown time.

4.0 Conclusion

The design and development of an integrated surface mount PWB production planning and manufacturing system requires an abstract understanding of the relationship between factors such as PWB design, production system construction, capacity, production control needs, and manufacturing process knowledge. The decision support framework discussed above helps establish a dynamic generative process planning system to help process engineers in surface mount PWB assembly. This system reduces the time spent from design to actual manufacture in a product's life cycle. Design and manufacturing would perform their tasks in an integrated and synergistic fashion for the purpose of achieving total quality design in a timely manner, and above all a lower cost. It is expected that the use and application of expert systems, ANNs, and fuzzy theory can be established in the manufacturing and design community as design or manufacturing aids to ill-defined and diverse problems in surface mount PWB assembly.

References

1. Turban, E., *Expert Systems and Applied Artificial Intelligence* (Macmillan Publishing Company, New York, 1992).
2. Rolston, D. W., *Principles of Artificial Intelligence and Expert Systems Development* (McGraw Hill, New York, 1988).
3. Srihari, K., and Westby, G., *'Decision Support for PWB Assembly Using Knowledge Based Expert Systems'*, Proceedings - NEPCON West 1992, Anaheim, California, Vol. **2** (1992), pp. 545 - 554.
4. Wasserman, P. D., *Neural Computing: Theory and Practice* (Van Nostrand Reinhold, New York, 1989).
5. Padgett, M. L., and Roppel, T. A., *'Neural Networks and Simulation: Modeling for Applications'*, Simulation, Vol. **58**, No. **5** (1992), pp. 295 - 305.
6. Caudill, M., *'Expert Networks'*, Byte, Vol. **16** (1991), pp. 108 - 116.
7. Taylor, D., and Graves, R. J., *'An Examination of Routing Flexibility for Small Batch Assembly of Printed Circuit Boards'*, International Journal of Production Research, Vol. **28**, No. **11** (1990), pp. 2117 - 2135.
8. McGinnis, L. F., Ammons, J. C., Carlyle, M., Cranmer, L., Depuy, G. W., Ellis, K. P., Tovey, C. A., and Xu, H., *'Automated Process Planning for Printed Circuit Card Assembly'*, IIE Transactions, Vol. **24**, No. **4** (1992), pp. 18 - 30.
9. Sarin, S. C., and Salgame, R. R., *'Development of a Knowledge-based System for Dynamic Scheduling'*, International Journal of Production Research, Vol. **28**, No. **8**, 1990, pp. 1499 - 1512.

Scheduling

A Scheduling Strategy For Efficient Operation of PCB Assembly Line
    Y. H. Lee and D. H. Kim

Modeling the Effect of Hot Lots in Semiconductor Manufacturing Systems
    Y. Narahari and L. M. Khan

Integrating Intelligent Job-Scheduling into a Real-World Production-Scheduling System
    K. Kurbel and A. Ruppel

A Scheduling Algorithm for Dynamic Job Scheduling
    Y. X. Zhang, S. Di, H. Cheng and K. F. Cheng

Scheduling Utilizing Market Models
    H. H. Adelsberger, W. Conen and R. Krukis

Developing Knowledge-Based System for Calibration Scheduling
    F. T. S. Chan

Disjunctive Constraints for Manufacturing Scheduling: Principles and Extensions
    P. Baptiste and C. Le Pape

Intelligent Simulation-Based Scheduling of Work Cells
    H. S. Tan and R. de Souza

A SCHEDULING STRATEGY FOR EFFICIENT OPERATION OF PCB ASSEMBLY LINE

YOUNG-HAE LEE and DUCK-HAN KIM
Department of Industrial Engineering
Hanyang University
Seoul, 133-791 Korea
LEEYH@KRHYUCC1. bitnet

ABSTRACT

Printed circuit board assembly line is characterized by very long set-up times. In this paper, we examine the scheduling methods that may significantly reduce the set-ups. And the greedy sequence dependent scheduling method is proposed which is based on component commonality among PCB types. Using the typical traditional scheduling method as a benchmark, three methods are compared in terms of three performance measures: line throughput, average work in process inventory level and implementation complexity. Guidelines for selecting the most appropriate method for a given production environment are proposed.

Keywords: scheduling; PCB assembly.

1. Introduction

The electronics industry today is well-developed, world-wide industry, in which a large variety of products are produced. In america, the industry is valued at $200 billion-a-year in 1992. The printed circuit board(PCB) is the 'brains' of the electronic products, and their cost plays a major role in determining the competitiveness of an electronic firm. This paper focuses on the reduction of production costs via the efficient scheduling methods.

A PCB consists of two major parts: the raw board and the electronic components assembled on it. Generally, the PCB assembly line is a flowshop, in which the sequence of operations is predetermined by the technological constraints. Batch production is common, and market pressures often require electronic manufacturers to produce a large variety of small-lot, customized PCBs. There have been some researches to reduce the cost and to increase the throughput of small-lot PCB assembly lines focusing on the time-consuming set-up operation (Tali F. Carmon, Oded Z. Maimon and Ezey M. Dal-El 1989, P. Cunningham and J. Browne 1986, Ezey M. Dar-El and Oded Maimon 1988, Oded Maimon and Avraham Shtub 1991, Oded Z. Maimon, Ezey M. Dar-El, Tali F. and Carmon 1993).

The traditional serial production method used in the assembly of the electronic components on PCBs requires that new set-up of all the components be done each time the PCB type is changed. This procedure results in extended set-up times, since components that are common to several PCB types are set up several times. We examine three scheduling methods, the grouped set-up (GSU) method, the sequence dependent scheduling (SDS) method and the greedy sequence dependent scheduling (GSDS) method.

The GSU method was recently introduced (Tali F. Carmon, Oded Z. Maimon and Ezey M. Dal-El 1989). In an attempt to overcome the disadvantages of the GSU method, the SDS method is suggested as an alternative scheduling method that sacrifices some of the reduction in set-up time but has some advantageous qualities for some type of a production system. Although the idea of sequence dependent scheduling has been used in the past to reduce set-up time, the SDS concept is applied here in an inherently different manufacturing enviroment. In an attempt to overcome the disadvantages of the GSU method and the SDS method, the GSDS method is suggested as an alternative scheduling method that gets some advantage of reduction in set-up time and WIP inventory level. The industrial data is used to evaluate the above methods. The manufacturing enviroment considered in this paper is a flowshop which consists of two machines which can be modeled using the mixed integer programming (Oded Maimon and Avraham Shtub 1991).

2. Scheduling Methods

2.1 Traditional Method

The traditional scheduling method is the simplest scheduling method which can be used for process control. The groups of PCBs (from the short term production plans) are produced sequentially, and so are the lots within each group. When a PCB type is changed, the machine is shut down, and all the components are unloaded from the machine. The components for the next PCB type are then loaded. It is technically possible to leave common components on the machines, but this has not been done in practice since it makes the process control complicate. It also requires checking for component commonality among the PCBs and adapting the computer program for assembly to the machines

Since the set-ups in the traditional method are independent of the job sequence, it is possible to implement a simple scheduling rule like the Johnson's rule for two machines to 'minimize' the makespan of the group(Ronald G. Askin and Charles R. 1993). Note that the set-up required for each PCB type in the traditional method is the full set-up of all the component types, and the result may not be a minimal makespan.

2.2 GSU Method

The GSU approach has been recently introduced(Tali F. Carmon, Oded Z. Maimon and Ezey M. Dal-El 1989). First, the common components (i.e., components that are shared among two or more PCB types in the group) are set up on the machine and are assembled for all PCBs. We refer to this stage as the common set-up and assembly. In the next stage, referred to the residual components set-up and assembly of the remaining components on each PCB type are seperated and performed sequentially. The method results in high throughput but also high work-in-process (WIP) inventory level. The analysis of its performance is presented in the later sections.

2.3 SDS method

The SDS approach has been recently introduced and has been used in metal-processing(Tali F. Carmon, Oded Z. Maimon and Ezey M. Dal-El 1989, C. C. Gallagher and W. A. Knight 1986). The SDS approach applied the component commonality concept differently. The idea is that PCB types should be sequenced in order that PCB may have the largest number of common components with the following PCB. The ultimate goal is to minimize the number of component changes required during the operation.

Variation of the SDS method have been used in other applications, especially in the metal processing industry, where the common resources are tools and parts. In other applications, similar parts requiring common resources like pallets are scheduled separately in order to reduce idle times. Focusing on the electronics industry, heuristic methods have been developed for sequencing reels of components for PCB assembly (P. Cunningham and J. Browne 1986, Yhaya Fathi and Javad Taheri 1989).

2.4 GSDS Method

In this method, groups of component type are formed as long as the number of bins or reels allows. After that, we seperate PCBs with the groups. Using the advantages of the SDS and the greedy methods, we can save more set-up times(M.L. Brandeau and C. A. Billington 1991, Ezey M. Dar-El and Oded Maimon 1988). Also using the Bayes's rule, more set-up times are saved (Edward J. Dudewicz and Satya N. Mishra 1988).

The GSDS method consists of two stages. First stage determines the sequence of PCB types to maximize the commonality between the present PCB type and the following PCB type. Comparing to the SDS method, the heuristic algorithm is used for solving shortest Hamiltonian path problem. Second stage groups PCBs as long as the number of bins or reels allows. This method has resulted in high throughput and low WIP inventory level. A schematic presentation of GSDS is shown Figure 2.1, where the shaded area represents the set-up time saved when using this method. One group is PCB 1 and 2, another group is PCB n-1 and n.

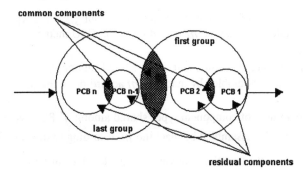

Figure 2.1 The GSDS method.

3. Criteria for Comparing Scheduling Methods

The GSU, the SDS, and the GSDS methods improve the performance of the PCB assembly line in different ways. In this section, we compare the four scheduling methods; the GSU, the SDS, the GSDS and traditional scheduling method, and define the conditions under which each method is superior to the others. Specially, we will focus on the GSDS method that is superior to the others. There are three performance measures that are the most important in a PCB assembly: the line throughput which determines the revenues, the average WIP inventory level which is the major production cost, and the implementation complexity which is usually a neglected factor in scheduling theory, but can be crucial from the practical point of view.

3.1 Line Throughput

When the machines assembles the PCBs continuously, the assembly line throughput (D) is inversely proportional to the machine occupation time (T) of PCB. That is,

$$D = 1/T \tag{3.1}$$

When the line consists of two machines, the machine occupation time is like the equation (3.2)

$$T = \max(T_1, T_2) \tag{3.2}$$

The machine occupation time consists of the set-up time and the assembly time. Since the assembly time is constant for all scheduling methods, the only variable affecting the machine occupation time is the total set-up time requied(Oded Z. Maimon, Ezey M. Dar-El and Tali F. Carmon 1993).

Notations are defined as follows:

i : The PCB type index, $i = 1, 2, \ldots, n$

m : The machine index, $m = 1, 2, \ldots, M$.

s : The average set-up time (including loading and unloading) required for a single component type.

$C_{i,m}$: The number of component types assembled by machine m on PCB type i.

$C^*_{k,m}$: The number of component types shared among k PCB types in the group assembled by machine m according to the GSU method.

$C'_{h,m}$: The number of component types shared among h PCB types, which are sequentially assembled by machine m according to the SDS method.

$C^\#_{f,m}$: The number of component types shared among f PCB types, which are sequentially assembled by machine m according to the GSDS method.

The set-up time for the traditional method is calculated as follows:

$$S_{trad} = s \sum_{m=1}^{M} \sum_{i=1}^{n} C_{i,m} \tag{3.3}$$

The set-up time for the GSU method is

$$S_{GSU} = s \sum_{m=1}^{M} \sum_{i=1}^{n} C_{i,m} - s \sum_{m=1}^{M} \sum_{k=2}^{n} (k-1) C^*_{k,m} \tag{3.4}$$

the set-up time for the SDS method is

$$S_{SDS} = s \sum_{m=1}^{M} \sum_{i=1}^{n} C_{i,m} - s \sum_{m=1}^{M} \sum_{h=2}^{n} (h-1) C'_{h,m} \tag{3.5}$$

and the set-up time for the GSDS method is

$$S_{GSDS} = s \sum_{m=1}^{M} \sum_{i=1}^{n} C_{i,m} - s \sum_{m=1}^{M} \sum_{f=2}^{n} (f-1) C^\#_{f,m} \tag{3.6}$$

The set-up time for the GSU and the GSDS methods depends on the number of common components among the groups. The set-up time for the SDS and the GSDS methods also depends on the distribution of common components among the PCB types. The saving of set-up time in the SDS and GSDS method must be less than or equal to that of the GSU method, since in the latter method, each component is set-up only once, whereas in the SDS and GSDS method, each component is set-up at least once depending on the PCB sequence. To caculate the line throughput let us denote by

T_m : The occupation time of machine m, which is the sum of the set-ups and the assembly times of all the PCBs in the group on machine m.

$Ct_{i,m}$: The total number of components assembled on PCB type i by machine m (allowing for multiple components of the same type).

p : The assembly time of a single component on each machine.

N_i : The lot size of PCB type i.

$P_{i,m}$: The assembly time of a lot of PCBs type i on machine m, as follows:

$$P_{i,m} = pN_iCt_{i,m} \tag{3.7}$$

The occupation time of machine m is given by

$$T_m^{trad} = p\sum_{i=1}^{n} N_iCt_{i,m} + s\sum_{i=1}^{n} C_{i,m} \tag{3.8}$$

$$T_m^{GSU} = p\sum_{i=1}^{n} N_iCt_{i,m} + s\sum_{i=1}^{n} C_{i,m} - s\sum_{k=2}^{n}(k-1)C_{k,m}^* \tag{3.9}$$

$$T_m^{SDS} = p\sum_{i=1}^{n} N_iCt_{i,m} + s\sum_{i=1}^{n} C_{i,m} - s\sum_{h=2}^{n}(h-1)C_{h,m}' \tag{3.10}$$

$$T_m^{GSDS} = p\sum_{i=1}^{n} N_iCt_{i,m} + s\sum_{i=1}^{n} C_{i,m} - s\sum_{f=2}^{n}(f-1)C_{f,m}^{\#} \tag{3.11}$$

3.2 Average WIP Level

Under the GSU method, the WIP of each PCB is constant as long as the assembly of the residuals on Machine 2 is begun, and it is reduced in the every step until the last lot of PCBs in the group is completed. The introduction of new group to machine 1 radically increases the average WIP level while machine 2 is working the previous one. With the traditional, the SDS and the GSDS methods, the WIP function is a reduction step function - each step representing the completion of a lot of PCBs on Machine 2. In the three methods a new group is introduced to Machine 1 after Machine 2 completes the last lot in the previous group (the time in which two groups are in the line concurrently is assumed to be negligible for these methods hereafter). With only one group at a time in the assembly line, the average WIP level for the GSDS, the SDS and traditional methods is much lower than the average WIP level for the GSU method.

Calculating the average WIP level for all methods and particularly for the GSU method can be remarkably simplified by approximating the WIP function which is the monotonic decreasing function.

$$I_{GSU} = WG_t \tag{3.12}$$

$$I_{trad} = I_{SDS} = I_{GSDS} = 0.5\sum_{i=1}^{n} W_i = 0.5WG_t \tag{3.13}$$

where

I : The average WIP in the line.

W_i : The average WIP of PCB type i.

WG_t : The average WIP of a group.

4. Evaluations

The data was taken from a particular electronics firm(Oded Maimon and Avraham Shtub 1991). The eight PCB types are the most frequently assembled PCBs in this company. 20 PCBs of each type are usually assembled every week, on the average. Assembly rate is taken as 3600 components per hour, and assembly time is given in minute. The average set-up time is 1 minute per component type. Other relevant data is given in Tables 4.1. A Gantt chart demonstrating the scheduling methods is shown in Figure 4.1.

Since Machine 1 is substantially dominant in this case, the optimal schedule for Machine 1 is chosen as the SDS sequence. This sequence, 1-3-5-2-7-8-6-4 is the optimal solution determined by a branch and bound method.

Table 4.1 PCB data.

PCB type i	1	2	3	4	5	6	7	8	total
No. of comp. types for M1	29	20	41	19	19	19	17	13	177
Total no. of comp. for M1	63	40	198	190	42	41	29	13	616
No. of comp. types for M2	14	10	6	9	8	9	10	12	78
Total no. of comp. for M2	86	62	111	187	43	42	41	30	602
Assembly time of PCB i in M1	1.05	0.7	3.3	3.17	0.7	0.68	0.48	0.22	
Assembly time of lot i in M1	21	13.3	66	63.3	14	13.7	9.7	4.3	205.3
Assembly time of PCB i in M2	1.43	1.03	1.85	3.12	0.72	0.7	0.68	0.5	
Assembly time of lot i in M2	28.7	20.7	37	62.3	14.3	14	13.7	10	200.7

Compared with traditional method in this example, the GSU and the GSDS methods are saved the set-up time 65 / 255 = 25%, respectively. And the SDS method is saved 42 / 255 = 6%

Machine 1 dominates the assembly line performance. Using the GSU and the GSDS methods, it is possible to save (382.3 - 342.3) / 382.3 = 10.4% in Machine 1 occupation time, and therefore, increase the line throughput from the base throughput obtained by the traditional method by 10.4%. Using the SDS method, it is possible to increase the throughput by 6.8%. Using the data, we get the WIP vector as

$$W = (W_1, W_2, W_3, W_4, W_5, W_6, W_7, W_8) = (149, 102, 309, 377, 85, 83, 70, 43)$$

$WG_t = 1218$ components,

$I_{GSU} = 1218$ components,

$I_{trad} = I_{SDS} = I_{GSDS} = 609$ components.

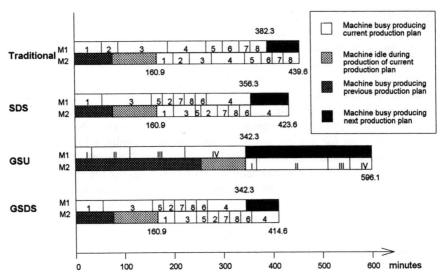

Figure 4.1 A Gantt chart for the scheduling methods.

In particularly, when the number of component types is 200 and the number of PCB types is 5, the result is shown in Table 4.2.

Table 4.2 The result of experiment, when the number of component types is 200.

# of PCB	# of comp.	50		60		70	
		set-up	WIP	set-up	WIP	set-up	WIP
5	Traditional	116.31	62.5	260.36	150.0	378.30	275.5
	GSU	67.98	125.0	127.64	300.0	161.05	475.0
	SDS	71.76	62.5	143.02	150.0	188.73	275.5
	GSDS	67.98	62.5	127.64	150.0	161.05	275.5

# of PCB	# of comp.	80		90		avg. % of decrease	
		set-up	WIP	set-up	WIP	set-up	WIP
5	Traditional	535.51	387.5	682.72	575.0	-	-
	GSU	183.52	775.0	195.40	1150.0	62.7	-100.0
	SDS	202.83	387.5	239.24	575.0	57.1	0.0
	GSDS	183.52	387.5	195.40	575.0	62.7	0.0

5. Discussion and Conclusions

The scheduling methods for the assembly of PCBs have been presented, which outperform the traditional method in terms of reduced set-up times and higher throughput. The GSU and GSDS methods were shown to perform better than the SDS method in terms of throughput, whereas the SDS and GSDS methods were better than the GSU method in terms of the average WIP level.

The decision as to which method to be used for a specific production enviroment depends on several considerations. Generally, the traditional method should be used when the number of common components is small, since there is no point in complicating the process control for a small saving in set-up time. The GSU should be chosen when the PCB assembly line is a bottleneck operation in the production process, and any increase in its throughput is significant to the whole plant. The necessary conditions for implementing the GSU method require a large number of common components and high level of process control. The SDS method is superior to the others when the common components are distributed among the PCB types such that the difference in set-up time between the GSDS and SDS methods is small. The operational simplicity and the reduced WIP levels of the SDS and the GSDS methods over the GSU method are typically significant. It could be concluded that the GSDS method is superior to the other methods. Therefore we suggest that the GSDS method is the most appropriate method for PCB assembly line.

6. References

1. Ronald G. Askin and Charles R. Standridge, *Modeling and Analysis of Manufacturing Systems* (John Wiley & Sons 1993).
2. M. L. Brandeau and C. A. Billington, "Design of manufacturing cells : operation assignment in printed circuit board manufacturing", *Journal of Intelligent Manufacturing*, 2 (1991) 95-106.
3. Tali F. Carmon, Oded Z. Maimon and Ezey M. Dal-El, "Group set-up for printed circuit board assembly", *International Journal of Production Research*, 27 (1989) 1795-1810.
4. P. Cunningham and J. Browne, "LISP-based heuristic scheduler for automatic insertion in electronics assembly", *International Journal of Production Research*, 24 (1986) 1395-1408.
5. Edward J. Dudewicz and Satya N. Mishra, *Modern Mathematical Statistics* (John Wiley & Sons 1988).
6. Ezey M. Dar-El and Oded Maimon, "Proposed Scheduling Methods for Printed Circuit Board Assembly", *Annals of the CIRP*, 37 (1988) 13-15.
7. Yhaya Fathi and Javad Taheri, "A mathematical model for loading the sequencers in a printed circuit pack manufacturing enviroment", *International Journal of Production Research*, 27 (1989) 1305-1316.
8. C. C. Gallagher and W. A. Knight, *Group Technology Production Methods in Manufacture*, (Ellis Horwood Limited 1986).
9. Oded Maimon and Avraham Shtub, "Grouping Methods for printed circuit board assembly", *International Journal of Production Research*, 29 (1991) 1379-1390,
10. Oded Z. Maimon, Ezey M. Dar-El and Tali F. Carmon, "Set-up saving schemes for printed circuit boards assembly", *European Journal of Operational Research* 70, (1993) 177-190.

MODELING THE EFFECT OF HOT LOTS IN SEMICONDUCTOR MANUFACTURING SYSTEMS

Y. Narahari
Department of Computer Science and Automation
Indian Institute of Science, Bangalore, 560 012, INDIA
E-mail: hari@chanakya.csa.iisc.ernet.in

and

L. M. Khan
Department of Computer Science and Automation
Indian Institute of Science, Bangalore, 560 012, INDIA
E-mail: mohd@chanakya.csa.iisc.ernet.in

ABSTRACT

The presence of hot lots or high priority jobs in semiconductor manufacturing systems is known to significantly affect the cycle time and throughput of the regular lots since the hot lots get priority at all stages of processing. In this paper, we present an efficient analytical model based on re-entrant lines and an efficient, approximate analysis methodology for this model, for predicting the performance of a semiconductor manufacturing line in the presence of hot lots. The proposed method explicitly models scheduling policies and can be used for rapid performance analysis. Using the analytical method and also simulation, we show the severe effects hot lots can have on the performance characteristics of regular lots.

1. Introduction

In this paper, we model semiconductor manufacturing systems using *re-entrant lines* [1] and study the effect of hot lots or high priority jobs through an approximate analysis of the re-entrant line model using mean value analysis (MVA) [2,3]. The MVA-based method facilitates explicit modeling of buffer priority based scheduling policies used in re-entrant lines [1] and is computationally much more efficient than simulation, the traditional technique that has been employed in the relevant literature. We provide numerical results obtained using the analytical method and also simulation, to study the effect of hot lots on performance characteristics such as mean cycle time, variance of cycle time, and mean throughput rate, of regular lots and hot lots.

1.1. Modeling Semiconductor Manufacturing Systems

In semiconductor manufacturing systems, wafer fabrication constitutes the most important step. Wafer fabrication involves a large, complex sequence of processing steps. An important feature of wafer fabrication processing is *re-entrancy*, which

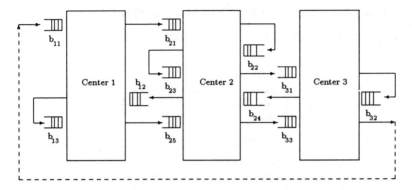

Figure 1: A Re-entrant line with 3 stations and 11 buffers

refers to multiple visits by a wafer lot to the same processing center at various times. *Re-entrant lines* [1] constitute an appropriate queueing model for wafer fabrication lines. Figure 1 shows a typical re-entrant line, with three processing centers – center 1, center 2, and center 3, and 11 buffers.

The route that a wafer lot follows during its processing is also shown in Figure 1. Note that each wafer lot visits the same center several times during its processing. The buffers $b_{11}, b_{12}, ...$, represent physical or logical buffers that contain wafer lots waiting for processing.

For the purposes of this paper, let us assume that there are m service centers, denoted by the set $\{1, 2, \ldots, m\}$. Processing center i has n_i logical or physical buffers, $b_{i1}, b_{i2}, \ldots, b_{in_i}$. For $j \in \{1, 2, \ldots, n_i\}$, the buffer b_{ij} contains parts visiting center i for the jth time. A wafer lot visits these buffers in a given deterministic sequence.

1.2. Scheduling in Re-entrant Lines

The scheduling in re-entrant line becomes interesting because several parts at different stages of processing may be in contention with one another for service at the same machine. Several researchers have focused on the issue of scheduling in re-entrant lines [1,4,5,6]. Wein [5] has investigated the effect of dispatching policies, in a detailed simulation study. Srivatsan, Bai, and Gershwin [6] have developed a software testbed for experimenting with hierarchical scheduling algorithms. Distributed scheduling policies based on buffer priorities and due dates have been formulated and investigated by Kumar [1], and Lu, Ramaswamy, and Kumar [4]. These authors have investigated buffer priority policies such as: FBFS (First Buffer First Serve), LBFS (Last Buffer First Serve), and FCFS (First Come First Serve). They have also investigated due-date based policies, such as EDD (Earliest Due Date first) and LS (Least

Slack first), and *fluctuation smoothing* scheduling policies.

1.3. Hot lots in Semiconductor Fabs

In the semiconductor manufacturing terminology, hot lots refer to the class of jobs with the highest priority, often introduced into the system purely out of marketing and business considerations [9]. Hot lots often lead to irregular flow of parts and can drastically alter the cycle time and throughput of regular jobs. From a modeling point of view, hot lots have the same route and the same processing times at all processing centers as the regular lots but get priority over regular lots everywhere. Ehteshami, Petrakian, and Shabe [9] have carried out a simulation study to understand the impact of hot lots on the cycle time of regular lots in the system. The results show that, as the proportion of hot lots in the work-in-process increases, both the average cycle time and the corresponding standard deviation for all other lots increase rather drastically. Their study only considers the fixed work-in-process policy as the input release policy and FCFS as the dispatching policy, and does not account for any other dispatching policy. It is of much importance and practical interest to study the effects of hot lots under specific scheduling policies. This study attempts to address this problem.

1.4. Organization of the Paper

In this paper we first propose, in Section 2, an efficient method for approximate analysis of re-entrant lines in the presence of hot lots. The method is based on mean value analysis (MVA) [2,3]. An important object of the analysis method is the ability to model scheduling policies such as LBFS, FBFS, and FCFS. an efficient implementation of the MVA recursions. We next present in Section 3, detailed results obtained, through the proposed method and simulation, and study the effect of hot lots on the performance of regular lots.

2. An Approximate Analysis Methodology

MVA yields expressions for mean values of performance measures such as steady-state queue lengths, delays, and throughput rates. Since re-entrant lines with buffer priority scheduling policies and hot lots do not belong to the class of product-form networks [1,7], MVA is not exact and so the MVA formulation we propose is only an approximate analysis method.

2.1. Assumptions and Notation

We shall illustrate the formulation of MVA equations by assuming the LBFS scheduling policy. We assume that each processing center has exactly one machine and that the processing time of a regular lot (and also a hot lot) visiting center i on its

j^{th} visit is an independent exponentially distributed random variable with rate μ_{ij}. In the LBFS scheduling policy, parts visiting a center i for the jth time get priority over parts visiting this center for the rth time where $r = 1, \ldots, j-1$. For example in center 2 of Figure 1, parts in buffer b_{25} would get priority over parts in buffers b_{21}, b_{22}, b_{23}, and b_{24}. We assume that hot lots get non-preemptive priority over regular lots at all processing centers.

To apply MVA, we have to assume that the re-entrant line is a closed queueing network. Let N be the total number of regular lots and H, the total number of hot lots in the system. We shall use the following indices: i denotes a processing center; j denotes a buffer at a given processing center; k denotes a current regular lot population and has the range, $1, \ldots, N$, and h denotes a current hot lot population and has the range, $1, \ldots, H$. Let stage (i, j) correspond to the waiting or the processing of regular lots or hot lots visiting center i for the jth time. Let the performance measures of the network be denoted as follows.

$L_{ij}(k, h)$: expected number of regular lots in stage (i, j) when the network has k regular lots and h hot lots.

$L'_{ij}(k, h)$: expected number of hot lots in stage (i, j) when network has k regular lots and h hot lots.

$W_{ij}(k, h)$: mean steady-state delay for regular lots in stage (i, j) (mean waiting time in buffer b_{ij} + mean processing time).

$W'_{ij}(k, h)$: mean steady-state delay for hot lots in stage (i, j) (mean waiting time in buffer b'_{ij} + mean processing time).

$\lambda(k, h)$: mean-steady state throughput rate of regular lots when the network has k regular lots and h hot lots.

$\lambda'(k, h)$: mean-steady state throughput rate of hot lots when the network has k regular lots and h hot lots.

If $W(k, h)$ and $W'(k, h)$ denote the mean total delay (mean cycle time) in the entire network, we immediately have

$$W(k, h) = \sum_{i=1}^{m} \sum_{j=1}^{n_i} W_{ij}(k, h) \qquad (1)$$

$$W'(k, h) = \sum_{i=1}^{m} \sum_{j=1}^{n_i} W'_{ij}(k, h) \qquad (2)$$

Using MVA, we compute $W(N, H)$, $W'(N, H)$, $\lambda(N, H)$ and $\lambda'(N, H)$ in a recursive way.

2.2. Performance Measures of Hot Lots

We compute $W'(N, H)$ recursively in terms of $W'(N, H-1), W'(N, H-2), \cdots, W'(N, 0)$. Consider $W'_{ij}(N, h)$ where $h \in \{1, 2, \ldots, H\}$. This is the mean total delay of hot lots in stage (i,j) when there are N regular lots and h hot lots in the entire network and can be viewed as the sum of four terms, call them term 1, term 2, term 3 and term 4. Term 1 is the mean delay that an arriving hot lot would experience if it finds a regular lot being processed at the time of arrival. This would be equal to the product of probability of an arriving hot lot finding a regular lot undergoing processing and the mean remaining processing time of a regular lot. Term 2 accounts for the mean total time required for processing all hot lots resident — at the time of arrival of the distinguished hot lot into the buffer b'_{ij} — in higher priority buffers, and the hot lots in the buffer b'_{ij} waiting ahead of it (FCFS is followed for the parts in a given buffer). Term 3 gives the mean total time required to process all higher priority hot lots that arrive at the center i during the waiting time of the distinguished hot lot in buffer b'_{ij}. Finally, term 4 is the mean processing time, $(1/\mu_{ij})$, of the distinguished hot lot itself.

The details of computation of Terms 1, 2, and 3 are presented by Narahari and Khan [8].

2.3. Performance Measures of Regular Lots

Note that the processing of a regular lot is delayed until all hot lots and higher priority regular lots are processed. Here we shall first consider the computation of $W_{ij}(k,H)$ where $k \in \{1,2,\ldots,N\}$. The delay W_{ij} can be viewed as the sum of five terms, say, Term 1, Term 2, Term 3, Term 4, and Term 5. Term 1 is the time required for processing all hot lots present at service center i when a typical regular lot arrives into buffer b_{ij}. Term 2 is the mean total time needed to process all hot lots that arrive at processing center i during the waiting of the distinguished regular lot in the buffer b_{ij}. Term 3 represents the mean total time required for processing all regular lots that are resident — at the time of arrival of the distinguished regular lot into buffer b_{ij} — in higher priority buffers and the regular lots in b_{ij} waiting ahead of it. Term 4 is the mean total time required to process all higher priority regular lots that arrive at processing center i during the waiting of the distinguished regular lot in buffer b_{ij}. The computation of this is similar to that of term 3 of hot lots. Finally Term 5 represents the mean processing time of the distinguished regular lot and is equal to $(1/\mu_{ij})$.

3. Numerical results

We present the results for a two station, four buffer system, shown in Figure 2. In this system, only four buffer priority policies are possible :

1. LBFS at station 1 and LBFS at station 2 (Policy 1)

2. LBFS at station 1 and FBFS at station 2 (POlicy 2)

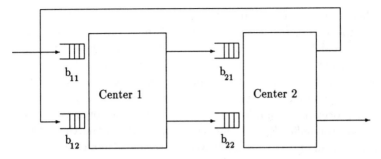

Figure 2: A re-entrant line with two stations and four buffers

3. FBFS at station 1 and LBFS at station 2 (Policy 3)

4. FBFS at station 1 and FBFS at station 2 (Policy 4)

Let
$$\frac{1}{\mu_{11}} = \frac{1}{\mu_{12}} = \frac{1}{\mu_{21}} = \frac{1}{\mu_{22}} = 1$$

Assuming a constant regular lot population of 50 and varying the hot lot population from 0 to 10, we computed the mean cycle time (MCT) and the mean throughput rate (TR), in the steady state, of regular lots and hot lots, using the proposed analysis method and also simulation. Table 1 shows these values, assuming LBFS policy at both the stations. We find that the presence of hot lots brings down significantly the throughput rate of regular lots and consequently the mean cycle time of the regular lots rises quite dramatically. From this table, we also see a close agreement between the values obtained using the analytical and simulation

The MCT and TR values were obtained for the other three policies also and were found to be virtually the same for all four scheduling policies. This is because of the closed nature of the network. The mean steady state delay at individual buffers will however be different for different scheduling policies. Table 2 shows the delay of regular lots at buffers b_{11}, b_{12}, b_{21}, and b_{22}, as predicted by our analysis method, for two different scheduling policies : Policy 1 (LBFS at station 1 and LBFS at station 2); Policy 2 (LBFS at station 1 and FBFS at station 2).

Since the overall MCT and TR values for the policies were found to be virtually the same, we computed the standard deviation of cycle times as a function of the hot lot population. This enables to capture the variability of performance offered by various buffer priority policies and assumes importance in the light of desirability of fluctuation smoothing [4]. Here different scheduling policies yeild different trends. In all the cases however, the cycle time of the regular lots shows a very high standard deviation. Table 3 shows these results. The LBFS policy applied to both the stations is found to yield the lowest values of the standard deviation. also, in all the cases, it is found that the coefficients of variation increase with the hot lot populations.

Population		MCT-Reg		MCT-Hot		TR-Reg		TR-Hot	
Reg	Hot	SIM	MVA	SIM	MVA	SIM	MVA	SIM	MVA
50	0	101.95	102.97	0	0	0.49029	0.48558	0	0
50	1	137.37	141.93	7.89	7.88	0.36384	0.35228	0.12669	0.12683
50	2	189.22	197.83	8.83	8.92	0.26412	0.25275	0.22657	0.22417
50	3	263.13	270.63	9.97	10.27	0.18986	0.18475	0.30099	0.29201
50	4	355.64	355.36	11.41	11.89	0.14044	0.14070	0.35053	0.33651
50	5	461.45	477.42	13.05	13.66	0.10826	0.11175	0.38303	0.36596
50	6	576.21	545.47	14.82	15.52	0.08666	0.09166	0.40477	0.38657
50	7	699.74	649.29	16.66	17.42	0.07136	0.07701	0.42029	0.40174
50	8	825.56	785.74	18.54	19.35	0.06043	0.06590	0.43138	0.41335
50	9	960.55	873.77	20.46	21.30	0.05192	0.05722	0.43993	0.42253
50	10	1093.83	994.31	22.40	23.26	0.04558	0.05029	0.44646	0.42995

Table 1: Analytical and simulation results for Example 1 (Figure 2)

Population		Delay at b_{11}		Delay at b_{12}		Delay at b_{21}		Delay at b_{22}	
Reg	Hot	Policy 1	Policy 2	Policy 1	Policy 2	Policy 1	Policy 2	Policy 1	Policy 2
50	0	47.80	49.01	3.18	2.81	47.80	3.54	3.18	46.77
50	1	64.40	66.30	4.49	3.86	64.40	5.21	4.49	62.23
50	2	87.50	91.32	7.09	5.82	87.50	8.30	7.09	84.51
50	3	119.89	125.14	11.73	9.48	119.89	14.02	11.73	114.91
50	4	158.57	165.58	19.30	15.41	158.57	22.66	19.30	150.67
50	5	200.92	211.45	29.78	23.67	200.92	34.50	29.78	191.29
50	6	245.40	259.77	42.67	33.91	245.40	48.80	42.67	232.05
50	7	292.00	311.02	57.90	47.45	292.00	66.26	57.90	272.44
50	8	338.03	359.20	74.99	63.19	338.03	87.14	74.99	315.65
50	9	385.85	416.38	94.79	79.66	385.85	108.62	94.79	362.96
50	10	430.51	466.99	116.77	96.60	430.51	130.56	116.77	399.74

Table 2: Mean delays at individual buffers for various scheduling policies

Population		Policy 1		Policy 2		Policy 3		Policy 4	
Reg	Hot	SD-Reg	SD-Hot	SD-Reg	SD-Hot	SD-Reg	SD-Hot	SD-Reg	SD-Hot
50	0	10.11	0	19.56	0	10.11	0	10.84	0
50	1	14.39	2.79	29.03	2.79	14.35	2.80	15.79	2.80
50	2	23.01	3.04	45.21	3.06	22.61	3.05	26.02	3.09
50	3	36.62	3.27	70.86	3.34	36.20	3.31	42.62	3.37
50	4	56.04	3.52	102.49	3.73	55.26	3.53	65.08	3.69
50	5	79.65	3.76	143.10	4.19	79.51	3.80	94.34	4.03
50	6	106.81	4.02	187.31	4.72	109.09	4.03	131.26	4.37
50	7	136.19	4.28	234.11	5.29	141.14	4.28	170.09	4.70
50	8	162.76	4.45	280.24	5.81	182.14	4.50	205.82	5.02
50	9	199.95	4.68	339.20	6.33	217.38	4.74	255.96	5.31
50	10	237.16	4.88	382.43	6.90	255.03	4.95	299.34	5.54

Table 3: Standard deviation of cycle times of regular and hot lots

4. Conclusions

The analytical method, based on MVA approximation, presented in this paper has been found to be efficient and quite accurate in predicting the performance of semiconductor manufacturing lines in the presence of hot lots. The advantages of the analytical method can be summarized as follows.

1. The method is much faster (by almost three to four orders of magnitude in observed experiments) than detailed simulation.

2. The method yields, as an attractive by-product, the performance measures for all intermediate populations of regular lots and hot lots.

3. Many scheduling policies, such as FBFS, LBFS, FCFS, and any fixed priority policy can be easily analyzed using the method.

5. References

1. P. R. Kumar. Re-Entrant lines. *Queueing Systems: Theory and Applications*, 13:87–110, 1993.
2. M. Reiser and S. S. Lavenberg. Mean value analysis of closed multichain queueing networks. *Journal of the ACM*, 27(2):313–322, April 1980.
3. R. Suri and R.R. Hildebrant. Modeling flexible manufacturing systems using mean value analysis. *Journal of Manufacturing Systems*, 3(1):27–38, 1984.
4. S.H. Lu, Deepa Ramaswamy, and P.R. Kumar. Efficient scheduling policies to reduce mean and variance of cycle-time in semiconductor manufacturing plants. *IEEE Transactions on Semiconductor Manufacturing*, 7(3):374–388, August 1994.
5. L. M. Wein. Scheduling semiconductor wafer fabrication. *IEEE Transactions on Semiconductor Manufacturing*, 1(3):115–130, August 1988.
6. N. Srivatsan, S. X. Bai, and S. B. Gershwin. Hierarchical real-time integrated scheduling of a semiconductor fabrication facility. Technical report, Laboratory for Manufacturing and Productivity, Massachusetts Institute of Technology, 1992.
7. Y. Narahari and L.M. Khan. Performance analysis of scheduling policies in re-entrant manufacturing systems. *Computers and Operations Research*, To appear: 1995.
8. Y. Narahari and L.M. Khan. Modeling The effect of hot lots in semiconductor manufacturing systems. Technical report, Department of Computer Science and Automation, Indian Institute of Science, September 1994.
9. B. Ehteshami, R. G. Petrakian, and P. M.Shabe. Trade-offs in cycle time management: Hot lots. *IEEE Transactions on Semiconductor Manufacturing*, 5(2):101–106, May 1992.

INTEGRATING INTELLIGENT JOB-SCHEDULING INTO A REAL-WORLD PRODUCTION-SCHEDULING SYSTEM

KARL KURBEL, ANDREAS RUPPEL

Institute of Business Informatics, University of Muenster,
Grevener Strasse 91, D-48159 Muenster, Germany
E-mail: kurbel@uni-muenster.de

ABSTRACT

The paper addresses the problem of scheduling production orders (jobs). First an approach based on simulated annealing and Hopfield nets is described. Since performance was unsatisfactory for real-world applications, we changed problem representation and tuned the scheduling method, dropping features of the Hopfield net and retaining simulated annealing. Both computing time and solution quality were significantly improved. The scheduling method was then integrated into a software system for short-term production planning and control ("electronic leitstand"). The paper describes how real-world requirements are met and how the scheduling method interacts with the leitstand's database and graphical representation of schedules.

Keywords: Scheduling, simulated annealing, electronic leitstand.

1 Approaches to Production Planning and Scheduling

Software systems for production planning, scheduling, and control have been available since the late sixties. They were called MRP (Material Requirements Planning), MRP II (Manufacturing Resource Planning), PPC (Production Planning and Control), or PPS (Production Planning and Scheduling) systems. Their focus is on long and medium-range planning. For short-term issues including scheduling, dedicated systems called *electronic leitstands* were introduced in the late eighties. Since they originated from Germany, the term 'leitstand' (= production control post) has been adopted in English publications before (Adelsberger and Kanet 1991, Kurbel 1993).

The problem of scheduling production orders (jobs) has been an issue in theory and practice for many years. It is well understood, but it is of exponential complexity in time. Optimizing algorithms based on complete enumeration or branching-and-bounding fail when applied to large real-world problems. Various heuristics and dispatching rules (e.g. shortest operation time, first come first serve, cost over time, etc.) have been proposed to overcome this shortcoming. They terminate in polynomial time, but since the problem is NP-hard, they cannot guarantee optimal solutions.

This paper presents a scheduling method based on principles from simulated annealing (SA), Hopfield networks (Hopfield and Tank 1985) and Boltzmann machines (Aarts and Korst 1987). It is heuristic, too, but it comes very close to optimal solutions. Moreover, it is fast and easily adaptable to different objective functions.

2 Scheduling with Simulated Annealing and Energy Function Motivated by Hopfield Nets

2.1 Initial Model: Mapping the Scheduling Problem to a Hopfield Net

Our first neural-net approach to scheduling was based on a simplified model of the problem domain. It included:

- A set of jobs, each one consisting of several operations with precedence relations among them.
- Machine groups, each one composed of several machines able to perform the same operations.
- Closed production, i.e. all jobs and operations to be considered within a period are available when scheduling starts with an empty schedule.
- Non-preemptive scheduling (Drexl 1991, Talbot 1982), i.e. an operation must not be interrupted by another one.
- Continuous time scale, i.e. 24 hours working time, no breaks, no holidays.

Sequencing now means that each *operation* has to be placed on a certain *machine* at a certain *position*. A three-dimensional representation is straightforward. Binary neurons are employed to construct a three-dimensional grid where the activation of neuron u_{ijk} becomes 1 *iff* operation i is to be dispatched on machine j as the k-th one on that machine. The outputs of all neurons determine the state of the net. They represent a certain schedule which, however, may be inconsistent. In each step of the algorithm, a neuron is determined, its state is complemented ("flip"), and the quality of the new state is measured with respect to the energy function. A flip in our model is equivalent to an operation being placed at or removed from a particular position. When *simulated annealing* is applied to the net, the new state is accepted with a certain probability that depends on the change ΔE of the energy function and on an external parameter T ("temperature"). T is a monotonically decreasing function of time.

The *objective function* of the scheduling problem (e.g. minimize order-flow time) can easily be incorporated into the energy function. It is very difficult, however, to treat the *constraints* (e.g. sequencing restrictions) adequately. In a typical Hopfield-net application, the values of the weights associated with neuron links would be determined according to domain-dependent circumstances, and in some way represent problem constraints. The energy would then be computed from the state of the net and from the weights. For the job-scheduling problem, however, finding the right weights in advance proved to be impossible. Therefore, we treated domain-dependent constraints by means of penalty terms within the energy function.

The purpose of penalties is to prevent infeasible solutions. Basically, four types of inconsistencies might occur. For each type, a different penalty is assigned to the respective weights. The four illegal cases are:

1. An operation is scheduled twice.
2. Two operations are scheduled at the same position of the same machine.
3. A precedence relation between two operations is violated.
4. An operation is assigned to the wrong machine.

Apart from the penalties, the most important term of the energy function is the objective function. It is used to evaluate the quality of the respective states of the net. Before the objective function can be computed, the Hopfield net has to be transformed into a schedule for which times can be calculated. The energy function is minimal if the net is in a consistent state and the implied schedule is optimal with respect to the objective function.

Although the Hopfield-based approach to scheduling works, there are several drawbacks. First, solution quality strongly depends on how well the values of the penalty parameters are chosen. Second, even if optimal parameter values were available, the process would not necessarily terminate with a consistent solution. Third, a major shortcoming is that the energy function has to be recomputed after each flip, implying that all operations must be dispatched and the schedule must be evaluated first. For a problem of 109 operations to be placed on 27 machines, it took about one second to compute the energy function. This is far too long when hundreds or thousands of flips have to be evaluated during execution. Therefore, the model had to be redesigned.

2.2 Tuning the Model for Real-world Application

In the improved model, the number of dimensions was reduced, and a different flip algorithm was employed to decrease computation time and to avoid inconsistent solutions altogether. A new dispatching procedure was also developed. It uses knowledge about which machines are permitted for which kind of operations, and which machines belong to which groups. Operations are now assigned to machine groups instead of single machines. Allocating a particular machine is no longer part of the Hopfield net but of the dispatching heuristic.

Operations are kept in an ordered linear list. The list contains only *feasible* sequences. The scheduling model thus looses the two dimensions of machines and positions. A one-dimensional representation remains. Weights do not have to be considered now, because the violations they are supposed to prevent cannot occur any more. Instead, feasibility of the list is maintained by the scheduling procedure. The energy function is no longer a function of the weights; the only remaining component is the problem-dependent optimization criterion (order-flow time).

Operations can now be treated one after the other. They are represented in a list u_1, u_2, ..., u_n, where u_i is a single operation. The list is said to be consistent *iff*

$$\forall \ i, j: ((u_i \text{ has to precede } u_j) \Rightarrow i < j).$$

The scheduling algorithm makes use of the dispatching algorithm to create schedules and of the shifting algorithm to improve schedules.

Dispatching Algorithm

Dispatching means inserting an operation into a schedule. In figure 1, operations u_1, u_2, u_5, and u_{11} belong to the same job. They have to be performed in the sequence indicated by numbers 1 to 4. The third operation (u_5), for example, may be placed anywhere between the second and fourth one without violating sequence restrictions.

Shifting Algorithm

In analogy to a flip in the Hopfield net, the algorithm will select an arbitrary operation and determine the permissible shifting range (between the predecessor and the successor of the operation). The operation will then be inserted at a new position determined randomly. All operations between the old and the new position have to be shifted by one position into the appropriate direction. In this way, any consistent sequence of operations will remain consistent after any number of shifts.

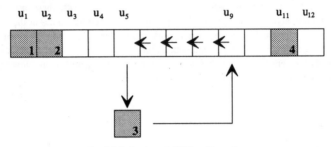

Fig. 1: Mechanism of Shifting Operations

Scheduling Algorithm

The scheduling algorithm is now basically an extension of simulated annealing. A schedule is created by applying the dispatching algorithm to all operations of the list and then improved by shifting. The algorithm proceeds as follows:

1. Place all operations of all jobs in a consistent starting list $u_1 \dots u_n$.
2. Create a schedule from the starting list using the dispatching algorithm. Compute the value of the energy function E_1.
3. Change the sequence according to the shifting algorithm.
4. Create a new schedule from the list using the dispatching algorithm. Compute the new value E_2 of the energy function. Compute the difference $\Delta E = E_2 - E_1$.
5. Accept the new sequence with probability $p = \dfrac{1}{1 + e^{\frac{-\Delta E}{T}}}$.
6. Decrease T logarithmically in time, i.e. $T = \dfrac{T_0}{\log(t)}$. Increment time t.
7. Test termination criterion; if true, terminate with the best sequence found so far, otherwise continue with step 3.

The termination criterion is met if T drops below a given bound (e.g. 0.5 degrees) or if the number of sequence changes rejected exceeds a certain limit (e.g. four times the number of operations in the list).

3 Real-world Scheduling Using an Electronic Leitstand

3.1 Electronic Leitstand Systems

An *electronic leitstand system* is a dedicated computer-aided graphical decision-support tool for short-term production planning and control (Kurbel 1993, Kurbel, Schneider and Singh 1995). Most leitstands are capable of creating and modifying schedules automatically, but they also assist the user in performing those operations by hand.

The user interface is primarily a graphical one. Schedules are represented by bar charts, similar to Gantt diagrams. Machines are plotted on the vertical axis and time units on the horizontal axis, respectively. Figure 2 shows a screen dump. Rectangles represent operations of jobs. Colors (or shades of grey) indicate different status informations. The characters within the rectangles are job and operation numbers. The window at the bottom shows part of the pool of operations waiting to be scheduled.

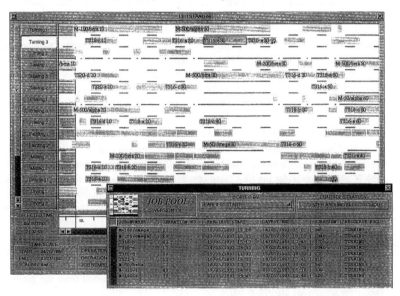

Fig. 2: Schedule Representation in an Electronic Leitstand

3.2 Scheduling Requirements

Real-world scheduling is far more complex than the simple models discussed above. Some requirements will be addressed now.

First, PPS systems as well as leitstand systems keep their data in *databases*. In particular, jobs, operations, production dates, technological sequences of operations, capacity utilization, and other data used by scheduling algorithms are stored by means of a database management system. This means that the simulated-annealing algorithm somehow has to be given access to the database or has to be provided with the relevant portions in some other way.

Second, if the user wishes to recalculate a schedule or to determine a new one from a different pool of operations, the respective information has to be read from the user interface (which is part of the leitstand) and given to the scheduling algorithm. After execution of the algorithm, results must be stored in the database and visualized on the leitstand screen again.

Third, the premise of continuous manufacturing time (24 hours per day, no breaks, no holidays) cannot be maintained. *Plant calendars* have to be introduced so that working days actually available for production can be modelled in more detail. Since calendars depend on the country, on the state, on contracts with labor unions, etc., more than one calendar may have to be considered at the same time. Furthermore, many companies work in several shifts per day. Sometimes shifts only apply to particular machine groups, workshops, etc. To allow maximum flexibility and accurateness, different *shift models* have to be included, in addition to plant calendars. By using both schemes together, actual working periods can be calculated with respect to individual machines.

When shift models and plant calendars are introduced it is obvious that an operation may be interrupted by a *break* (night, weekend, etc.) and continued afterwards. Total elapsed time for an interrupted operation has to be increased by the duration of the break.

4 Integrating Simulated-annealing Based Scheduling into a Leitstand

Both the scheduling method and an electronic leitstand were developed at the University of Muenster, Germany. The leitstand *ooL* (*o*bject-*o*riented *L*eitstand) runs on a NeXTCube station under the NeXTStep operating system. It was developed in Objective C. Data are kept in a relational database managed by the Sybase RDBMS. Data access is achieved by embedded SQL scripts. Source programs for scheduling were written in ANSI-C on a different computer first and then ported to the NeXT hardware.

Operation and order data are now taken from the leitstand database. As data have to be read and written hundreds or thousands of times during execution, performance would be very slow it each access were realized by a separate database operation. Therefore, an internal representation of jobs and operations is constructed from the database before the scheduling algorithm starts.

Figure 3 shows how the leitstand and simulated annealing interact. Triggered by a user request, the leitstand asks the SA component to create a schedule for a given set of operations on a given set of machines within a certain period. All relevant data are then extracted from the database by the *SA Request* module. This module generates appropriate SQL statements and builds up the internal representation.

When scheduling is finished database entries are modified by the *SA Result* module. In particular, starting and ending dates of operations, and the machines the operations were assigned to, are essential data updating the database. To present the results to the user, the leitstand then has to read the new data from the database and transform them into a visual representation on screen.

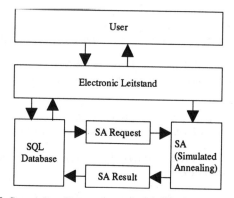

Fig. 3: Cooperation of Leitstand and Simulated-annealing Component

In order to meet real-world requirements as outlined above, our model of chapter 2.2 had to be extended in several ways. Plant calendars, shift models, and breaks required implementation changes. Breaks are now treated the same way the leitstand treats them: Whenever an operation is interrupted by a break, elapsed time will be extended by the lenght of the break. The idea is illustrated in figure 2. For example, the operation in the middle of row 5 (M500/beta30) seems to be a rather long one. The only reason is, however, that the operation cannot be finished before the weekend. The weekend break is indicated by the narrow bar below the operation rectangle. Part of the processing time is left for the following week. Extending elapsed time in this way ensures that resources are allocated by operations as long as they are really needed; no other job could have been scheduled on that machine prior to the end of the interrupted one.

5 Conclusions and Outlook

In computational tests, the scheduling method described in this paper performed quite well. It is fast and reaches the optimal solutions, or comes close to them, in most cases. Test results have been presented in detail elsewhere (Kurbel 1993).

In recent work, we improved the scheduling method by introducing the population concept of genetic algorithms and parallelizing the hybrid method. The simulated-annealing mechanism was combined with a genetic algorithm following an approach by Mahfoud and Goldberg (Mahfoud and Goldberg 1992). The resulting algorithm, called *PRSA* (Parallel Recombinative Simulated Annealing), was parallelized and implemented on a parallel computer (transputer system Parsytec MultiCluster/2 with 32 transputer nodes Inmos 805/25). Solutions calculated by one node are distributed to other nodes and vice versa.

Current tests indicate that if PRSA parameters (e.g. migration strategy, migration rate, genetic selection strategy, number of nodes involved) are tuned carefully results are better than those obtained by "simple" simulated annealing (Kurbel, Schneider and Singh 1995). Parameter settings will be tested and tuned in our further work. Modifying the annealing mechanism (e.g. using mean field annealing (Bilbo et al. 1989)) will be subject of investigations, too.

References

E.H.L. Aarts and H.M.Korst, *Boltzmann Machines and Their Applications*, in *PARLE (Parallel architectures and languages Europe)* (Eindhoven, 1987), pp. 34 - 50.

H.H. Adelsberger and J.J. Kanet, *The Leitstand - A New Tool for Computer-integrated Manufacturing*, in *Production and Inventory Management Journal* **32** (1991) 1, pp. 43 - 48.

G. Bilbo et al., *Optimization by Mean Field Annealing*, in *Advances in Neural Information Processing Systems 1*, ed. D.S. Touretzky (San Mateo, 1989), pp. 91 - 98.

A. Drexl, *Scheduling of Project Networks by Job Assignment*, in *Management Science* **37** (1991) 12, pp. 1590 - 1602.

G.E. Hinton and T.J. Sejnowski, *Learning and Relearning in Boltzmann Machines*, in *Parallel Distributed Processing, Vol. 1*, eds. D.E. Rumelhart and J.L. McClelland (Cambridge, MA, 1986), pp. 282 - 317.

J.J. Hopfield and D.W. Tank, *"Neural" Computation of Decisions in Optimization Problems*, in *Biological Cybernetics* **52** (1985), pp. 141 - 152.

K. Kurbel, *Production Scheduling in a Leitstand System Using a Neural-net Approach*, in *Artificial Intelligence Technology - Applications and Management*, eds. E. Balagurusamy and B. Sushila (New Delhi, New York et al., 1993), pp. 297 - 305.

K. Kurbel, B. Schneider and K. Singh, *Parallelization of Hybrid Simulated Annealing and Genetic Algorithm for Short-term Production Scheduling*, in *Proceedings of International Conference on Intelligent, Knowledge and Integration for Manufacturing*, Nanjing, China, March 28-31, 1995 (to appear).

W.S. Mahfoud and D.E. Goldberg, *Parallel recombinative simulating annealing: a genetic algorithm*, in *Technical Report No. 92002* (University of Illinois, 1992).

Y. Takefuji, *Neural Network Parallel Computing* (Boston, Dordrecht, 1992), pp. 37 - 50.

F. Talbot, *Resource-constrained Project Scheduling with Time-resource: The Nonpreemptive Tradeoffs Case*, in *Management Science* **28** (1982) 10, pp. 1197 - 1210.

A Scheduling Algorithm for DYNAmic Job Scheduling

Zhang YaoXue
Di Shuo Cheng Hua Cheng KangFu
Dept. of Computer Science & Technology, Tsinghua University
Beijing, China
E-mail: zyx@dcs.tsinghua.edu.cn

ABSTRACT

This paper reports a scheduling algorithm for practical job scheduling in the low-volume/high-variety manufacturing environment. This algorithm takes into account of the influence of many factors such as machine setup times, cell changes, replacement machines and load balancing between machines. Numerical testing example are taken from a real-manufacturing factory of Japan, and high-quality results are efficiently generated.

1. Introduction

This paper reports a scheduling algorithm for dynamic job scheduling, which can dynamically assign the workers, the products to different types of cells(an assigned machine group which can have different types) with satisfying due dates and maximizing machine utilization. The algorithm takes into account of influence of cell changes, machine setup times, replacement machines, load balancing, personnel calendar and overwork times. Therefore, this algorithm can be used in the low-volume/high-variety manufacturing environment. We implement this algorithm in the client/server architecture and different users can share the database server, so that the system becomes more powerful.

2. Scheduling Problem

To let our scheduling algorithm can be used in the low-volume/high-variety manufacturing environment, we take the following factors into account: order information, process design information, shop floor personnel information, cell and machine information, production time and set up times, load balancing information, and actual production progress information.

2.1 Order information

The order information contains due dates, number of the products in an order, and the names of the products. We assume that the orders are all divided into different lots by a process designer of the manufacturing company. A lot is an unit of jobs to be scheduled in our algorithm. An order from a customer may includes several lots or a lot may consist of several different orders for the same product. The due date is one of the most important factors which are usually considered in common scheduling algorithms. However, we use assigned due date and basic due date in our system to let the algorithm more practical.

2.2 Process design information

This information includes the parameters of the machines used in each process, the replacement machines, the process flows and the process priorities to manufacture each lot. Here, we assume that the designer gives the process flows for each lot and the machine types for manufacturing each process.

2.3 Shop floor personnel information

Shop floor personnel information includes the personnel group names, the numbers of the personnel in each group, the calendar of the personnel groups, the calendar of every personnel, and the working time horizon of each personnel. To satisfy due dates of the orders, personnel is sometimes asked to work overtimes. Consequently, the working time horizon of the personnel is considered as variable in our algorithm. However, the length of the working overtime is given by the foreman as one of the input.

2.4 Cell and machine information

To manufacture low-volume/high-variety products, we compose the different machines into cells. A cell is a set of machines with different types or a line to complete one or more than one manufacturing process. We consider three types of basic cells. One of them consists of independent machines which can process different lots in different machines but only complete one process for every given lot(see Fig. 1.a). Another one consists of machines which continually complete several processes but only for an assigned lot, and any machine of the cell does not do the same process with other machines in the same cell(see Fig. 1.b). The last one is similar to the above second one, but a part of the machines of the cell can do the work on the same process(see Fig. 1.c). These three types of cells are shown in Fig.1. The other types of the cells can be obtained by composing of the above three basic types.

We assume that any machine can be the element of several different cells, but only one cell including such a machine is active at any assigned time. Moreover, we assume that any machine in our system must belong to a cell.

2.5 Production time and set up time

The production time of each process is always different in the case of low-volume/high-variety manufacturing, though the used machine is the same one. The production time $P_{Ti}(m)$ of a lot in process i with machine m by:

$$P_{Ti}(m) = t_{lot}$$

The setup time varies with the tooling, materials, and the products types. The setup time $S_{Tij}(m)$ of a lot in process i with machine m to next process j is given as follows:

$$S_{Tij}(m) = S_{tool} + S_{mat}$$

Here, S_{tool} is the time to set up tooling for the lot, which depends on the tooling required and the current tooling on the machine; and S_{mat} is the time to change materials, e.g., change the back-up board in the manufacturing of flexible printed circuit.

Consequently, we have that the machine time T_i for the production of a lot in process i is given by:

$$T_i = P_{Ti}(m) + S_{Tij}(m).$$

According to this formula, we can calculate the time for the production of any lot in the above three basic types of cells.

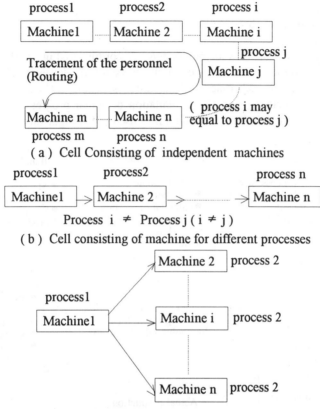

Fig. 1 Three types of basic Cells

2.6 Load balancing information

In low-volume/high-variety manufacturing, the load balancing between machines must be considered to obtain the maximizing machine utilization, specially, in the case of machine breakdowns or unsatisfying due dates, the replacement machines or cells should be used.

2.7 Actual production progress

The actual data from shop floor are input by using data scanners or by human hand, the information includes the status of every lot which has been produced or not, the status of the cells and the personnel in the shop floors, and the status of tools and materials.

It is obvious from above discussion that the dynamic scheduling for low-volume/high-variety manufacturing is a very complicate problem. At present, there is no such a scheduling system working at the client/server environment which can be used in the practical shop floor scheduling to our knowledge. The objective of our scheduling heuristics is to give a practical solution to this problem with the satisfaction of due dates, the maximizing machine utilization and the rapidly dynamic rescheduling.

3. The Scheduling Algorithm

Fig. 2 shows the data flow of our system. According to Fig. 2, the inputs of the algorithm are the information discussed in above section II. However, the outputs of the algorithm are products schedule, cell(machine) schedule, cell(machine) utilization, postpone or delay analysis, and the presentation of actual progress of the production. While, Fig. 3 gives the relations of the data files used in our algorithm.

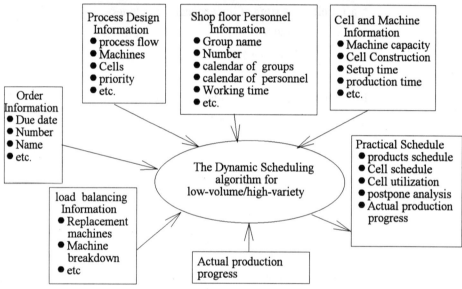

Fig. 2 The data flow of the DYNAS3

3.1 The principle of the scheduling algorithm

Our algorithm is a knowledge-based step adjusting one. It mainly includes the following parts: the calculation of last starting time of each lot in each process, the calculation of available resources, assignment of the lots to be scheduled to the cells and the workers according to heuristics rules, and the adjustment of the scheduling result.

3.1.1 The calculation of last starting time

The scheduling algorithm firstly selects the scheduling objects, i.e., some lots, from the plan file according to the experimental maximum production ability of a shop; and then calculates the last starting time(LS) of every process for every selected lot according to the given due date or assigned due date of the lot. The LS of a selected lot in a process is the time that the production must start to satisfy the due date. We can get the LS_i of a lot in the process i by using the following formula:

$$LS_i = LS_{i+1} - S_{Ti\ i+1} - P_{Ti}$$

Here, the LS_{i+1} is the last starting time of the lot in next process. If the next process is the last one, then LS_{i+1} is the due date. Moreover, the $S_{Ti\ i+1}$ is the setup time from process i to process $i+1$ and the P_{Ti} is the production time of the lot in process i. We must calculate the setup time and the production time to obtain the LS at first.

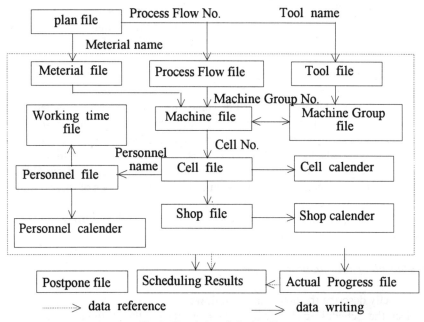

Fig. 3 The relation of the data files

3.1.2 The calculation of available resources

Available resources include machines, cells, personnel and material. The material status can be obtained from plan file, the types of machines or cells can be obtained from cell file, while the personnel information can be obtained from personnel calendar file. The calculation is mapping of the personnel working time, cell available time and etc.. This calculation is very complicate because there are too many constraint conditions on the machines, the cells, and the personnel working times. For example, the type of workers can be divided into the formal employers and the part time. Moreover, the workers should be divided into different classes according to their knowledge, experience and some other conditions. We use different data files to classify them.

3.1.3 Assignment of the lots

When we have finished the above calculations, we assign the given lots to the correct usable cells process by process, until the final process of every given lot has been scheduled. Over 60 rules are used for this assignment in our algorithm. For example, the following rules are a part of the policy for deciding the priority of each lot:

    a. The lots with earlier due date have the higher priorities.

    b. If the user assigns a lot with a higher priority than others, then the lot has a higher priority.

c. To those lots which have the same due dates and no user-assigned priorities, we give the following formula to calculate the priorities:

$$P_i = a_1 * A + a_2 * B + a_3 * C + a_4$$

Here, the a_1, a_2, a_3, a_4 are constants given by the user, and A represents the number of processes to be passed from the current process until the final process, B is the production time including the setup time from the current process until the final process, C is the waiting time until the current time. According to this formula, a lot with longer waiting time, longer production time and more processes will get higher priority.

3.1.4 Adjustment of the scheduling result

If the assignment of the given lots brings some due dates not to be satisfied or with low machine utilization, our algorithm automatically adjusts the assignment to get good machine utilization and to satisfy the due dates. Two methods are considered. One is usage of replacement cells, and another one is to ask the users to do overtime work. The selection of replacement cells is based on two conditions, i.e., a set of candidates of replacement machines or cells to a given lot, and the utilization rates of these candidates. The algorithm selects a candidate with lowest utilization rate and no loading in the current time as the replacement one. If there is no such a candidate, then the overtime work is needed. Otherwise, the delay will occur.

3.2 The scheduling algorithm

We can briefly describe the algorithm as follows:

Step1: Select the scheduling objects from the plan file and the postpone queues according to the due dates.
Step2: Check the materials and tools status required by the selected lots. If the required materials or tools are not ready, then the corresponding lots are added into the postpone queues and repeat to Step1.
Step3: Decide the process flows of the selected lots with materials and tools ready.
Step4: For each process of every selected lot, calculate the setup time and the LS time.
Step5: For each machine, calculate the active times of the cells which belong to it and the required workers for operating these cells.
Step6: Assign the given lots to the corresponding usable cells by the LS time of each corresponding process. If overability occurred at any cell, then adjust the overload to other cells.
Step7: If a lot can not be assigned to satisfy its due date, then the algorithm searches a replacement cell and go to Step4. If the replacement cell can not be found, then ask overtime work or add this lot into the postpone queues.
Step8: Decide the scheduling and show the scheduling results.

4. Implementation of the Scheduling Algorithm

We have implemented the above scheduling algorithm in the BC++ language and the Btrieve database running on Netware 3.1.1 network operating system and MS-Windows. The database server is the PC486 with 200MB hard disk, and the clients are the PC386 with 120MB hard disks. The system has been implemented for practical usage with two objectives in mind: user-friendliness and high executive speed.

The system is mainly constructed with 6 modules as shown in Fig. 4. These modules are: User interface module, Scheduling module, Analysis module, two Communication modules in the client and the server, and the database management module.

The scheduling module performs scheduling according to the heuristics given in previous sections, the analysis module analyzes the scheduling results and the postpone reasons. While the Communication modules provides us with the client/server architecture. Moreover, the database management module provides us with the operation methods to the Btrieve database. It is possible to install the Btrieve database and the management module into the client. However, the user interface module provides us with the user friendly graphical interface.

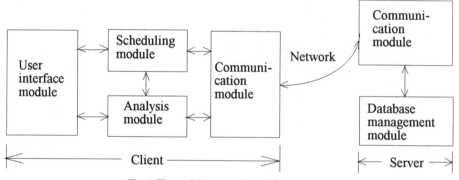

Fig. 4 The module construction of the DYNAS3

5. Evaluation

The system we developed has been applied in a Japanese manufacturing factory which has over one thousand machines with different types, over one thousand workers, and over ten shop floors. The factory which produces the Flexible printed Circuit manufacturing is a typical example in low-volume/high-variety. We have used our system in the scheduling of the exposure shop and the laminate shop of the factory. These two shops both have over one hundred machines and over one hundred workers. Using our system, the foreman can automatically schedule the shop production activities of 3~5 days with less than 1600 lots(a lot always includes 50~200 sheet parts). The executive time for scheduling in a 386/33 client is near to 2~5 minutes and the machine utilization can averagely arrive 90% in general cases. Moreover, comparing with the case of performing the schedule in hand, the system provides us with less postpone of the due date of the lots, because the machine utilization is increased and the workers are used more reasonable by the system.

6. Conclusion

This paper reports a scheduling algorithm for practical usage in low-volume/high-variety manufacturing. This algorithm has been implemented in client/server architecture and the influence of many factors on the scheduling process such as set up times, worker overtime working, cell constructions, and replacement machines has been considered. The system has been used to the practical shop floor scheduling in a manufacturing factory of Japan and the scheduling results have provided with much better machine utilization and less postpone then the scheduling in hand.

7. Acknowledgments

The authors are thankful to the system department of MEKTRON Company of Japan for their helpful discussion and commits. Many of rules used in the system are also abstracted from their experiences.

8. References

1. Lee H.S.Luong, "A database System for job scheduling in plastic injection moduling", IKDME, HongKong, PP 31~36, 1994.
2. D.J.Hoitomt and etal., "A practical approach to job-shop scheduling problems", IEEE Trans. on Robotics and Automation. Vol.9, No.1, PP 1~11, Feb.1993.
3. J.Ahn and etal., "Scheduling with alternative operations", IEEE Trans. on Robotics and Automation, Vol.9 No.3, PP 297~303, June 1993.
4. Zhang YaoXue and etal., "A dynamic scheduling system for job scheduling in low-volume/high-variety manufacturing" to be presented, CAPES 95, Beijing, China, May 1995.

SCHEDULING UTILIZING MARKET MODELS

HEIMO H. ADELSBERGER, WOLFRAM CONEN
Wirtschaftsinformatik der Produktionsunternehmen,
Universität Gesamthochschule Essen, 45117 Essen, Germany
E-mail conen@wi-inf.uni-essen.de

and

RITA KRUKIS
FZI - Forschungszentrum Informatik Karlsruhe,
Haid-und-Neu-Strasse 10-14, 76131 Karlsruhe, Germany

ABSTRACT

This paper presents an approach to job shop scheduling based on fictive markets. Various, often conflicting goals and interests influence the scheduling process in pratice. Markets seem to offer instruments to balance such interests. In the approach described below, Job and Resource agents act as buyers and sellers of allocations in a fictive market, their budget and cost calculations reflecting the organizational environment of the scheduling process. A scheduler (and other members of the organization interested in the resulting schedule) may influence the market in various ways (defining a regulatory framework, setting of budgets, etc.). An introductory market-model is described and some scheduling results are presented. Extensions to the model (integration of cost accounting, "intelligent" agents, interactive scheduling) are suggested and the used modeling and distributed implementation techniques are briefly mentioned.

1. Introduction

Scheduling allocates resources over time to perform a collection of tasks. A (large) number of methods for solving job shop problems have been proposed in the last decades, among them such different approaches as branch-and-bound and genetic algorithms. Nevertheless, job shop scheduling in practice requires often more than solving a mathematically formulated abstract scheduling problem. A multitude of controllable and uncontrollable (viewed from within the manufacturing environment) parameters leads to a framework of interests, uncertainties, and conditions influencing the (already complex) scheduling task. Thus, the integration of scheduling tools into a coherent view of the underlying organization and a balanced system of permanent planning and control seems to be difficult or even impossible.

Our approach to solve this problem starts with modeling the changing scheduling environment, guided by identifying sources of influences and interests, e.g., the goals of senior management (e.g., to maximize profit, to fulfil market requirements, to satisfy customer needs), manufacturing manager (e.g., to distribute capacity utilization equally, to reach a stable manufacturing situation), jobs (e.g., to be produced in time), and resources (e.g., to produce efficiently).[a] We call the observed sources agents. Each agent needs specific informations to follow his interests—they

---

[a]Here, jobs and resources are viewed as natural chrystalization points of interests and goals of humans in the organization and, to simplify the situation, we allow them to *act* as a surrogate for those instances.

can be obtained by communicating with other agents or by observing his environment. The agent may follow strategic considerations and may change his strategy over time. Each agent needs to generate plans, evaluate the current situation against his plan, decide about the next action(s) necessary to implement his plan, act, and check the situation again.

The most "simple" and successful form of integrating different interests and still reaching an overall goal of maximizing welfare are markets. This paper demonstrates the utilization of market-models to solve job shop problems: At a first stage we did only select job and resource agents as market members. All other agents can be integrated into the market-model(s) by viewing them

- as agents influencing the general setting/regulatory framework of the market
- as sources of information for the market members (which probably have to be paid)
- as interest groups directly influencing the behaviour of the market members (which probably will pay the market members for representing their interests)

The job agents act as buyers of resource allocations. Considering a job shop problem without alternatives in the process plan of the job routing, the resources can act as monopolists on the market, this induces the necessity of controlling their behaviour through the definition of appropriated market mechanisms. Alternative routings lead to a oligopol/polypol situation for the resource agents.

Determining the budget of the job agents is problematic—this can be done due to external priorities given to jobs or priorities determined through priority rules or by analyzing former and expected cash-flows connected with each job (which is a major problem itself). We define the primary goal of the *job agents* simply as: meet the due date! The underlying cost model which allows the job to reason about prices they are willing to pay for certain allocations is an important instrument to influence the scheduling process. The cost-model should (ideally) be derived from the interests of agents, e.g., stemming from the different management levels they belong to.[b]

The *resource agents* try to maximize their benefits. Relevant cost parameters result from sequence-dependend costs, processing costs (regular, overtime), maintainance considerations etc. The goals of the agents and informations flowing in a pre-trading phase between market members or other agents allow the market-members to generate an "ideal" (and local) plan (e.g., resource agent: an optimal sequence of jobs). Each agent should be able to select a strategy for his behavior on the market depending on the rules for the transactions on the market.

2. A Simple Model

To demonstrate the mapping of a job shop situation to a market a simple

---

[b]This derivation process is dynamic again, e.g., the substitution of a marketing manager/agent could lead to different goals (e.g., the importance of due dates increases) and therefore to a different evaluation of cost-relevant parameters. For our studies we decided to simplify this aspect by selecting a fixed cost-model.

model of exchange processes between non-intelligent agents is given. Extensions to the model will be discussed in subsequent sections.

2.1. Jobs, Resources, and a Broker

The following assumptions are part of the simple model (most of them may be discarded, compare Section 3)

- Resources: one machine per machine type; no concurrent processing of jobs on a machine; no maintenance or failure disruptions

- Jobs: date of job arrivals known (static); each job is processed at each machine at most once; no concurrent processing of a job on different machines; no pre-emption; processing time is known and constant; the jobs are mutually independent

To be produced in time is set to be the elementary need of the jobs. This induces the following ordinal preference structure[c]

1. The job exactly meets the due date.

2. The job is produced too early (negative lateness, earliness).

3. The job is produced too late (positive lateness, tardiness).

A broker regulates the exchange process by means of complete price differentiation. Result is a pareto-optimal allocation. According to Gutenberg (Gutenberg 1971), a linear cost function do often suffice to describe the resource situation.[d]

$$C(x) = cx + d \qquad (1)$$

where $VC(x) = cx$ are the variable costs, $FC = d$ the fixed costs, $UC(x) = C(x)/x$ the costs per unit, and $C'(x) = dk/dx = c$ are the marginal costs. The following equilibrium condition holds for this market (Feuerstein 1993): Price $P(x)=UC(x)$.

The demand of a single buyer depends on the due date dd, the current date d, the price of the pre-period P_{d-1}, the individual valuation of the importance of due delivery δ, and on the budget weight β he might use. This is subsummed in the following equation:

$$P_d = (P_{d-1} + (d - dd) * \delta) * \beta \qquad (2)$$

δ reflects the preference structure related to earliness and tardiness, in a more general setting this could be exponential[e]. α is the gradient of the branch describing earliness, $\alpha + \gamma$ is used for tardiness.

$$\delta = \begin{Bmatrix} \alpha+\gamma & \text{for } dd<d \\ \alpha & \text{for } dd\geq d \end{Bmatrix}, \ \gamma \geq 0 \ \wedge \ \alpha \geq 0 \qquad (3)$$

---

[c]The ordering follows an example in (Davis and Kanet 1993), it defines a non-regular cost measure, so most of the standard scheduling procedures are not applicable!

[d]The linearity of the cost function used is no hard constraint, some cost considerations are given in Subsection 3.5.

[e]Linear functions are used here for comprehensiveness.

The buyers are willing to pay a relatively higher price to prevent tardiness than to prevent earliness penalties (as, for example, inventory costs).[f] The broker is informed about pre-period prices, and actual cost and price functions. The following steps lead to allocations:

- The broker calculates the price the buyer is willing to pay.

- The broker allocates the demanded quantities according to the ordering of prices as long as the resource has free capacity and as long as the bids are higher than the unit costs (this may lead to problems when allocating the first demands because of the U–shape of the cost-per-unit function).

- An average price is calculate as a proviso for the next period. If no allocation took place the last pre-period price is used again.

- If a demand could not be satisfied the job has to re-apply for allocation in the next period.

2.2. Scheduling

To demonstrate the possibilities the above model offers, some experimental data are given.

Setting: 20 Jobs, 1 resource, cost function $c(t) = 3.5t + 500$ below standard capacity (435 min.),[g] $c(t) = 6,5t + 500$ above standard capacity. It is not allowed to start processing a new job in overtime—it's only allowed to finish work-in-process. Each job has to pay the bidded price (complete price differentiation). A job is characterized by the parameters given in Tab. 1 (which are set to identic values for each job in this introductory example).

Table 1. Parameter settings for exp. 1

Id	UnitPrice	* Dur.	=Price	α	β	γ	dd
1	4.5	100	450	0.5	1	0.1	07/11/94
...	...	...	...	...	...	...	...
20	4.5	100	450	0.5	1	0.1	07/11/94

The pre-period price necessary to calculate the price in the first period was set to 5. The allocation process starts on the 10th of July 1994. The allocation results are documented in Tab. 2. Variations in the parameter setting were introduced in subsequent experiments. The second experiment distributes the demand (again 20 jobs) over 4 due dates. Everything else was left unchanged. The third experiment complements the second in varying the due delivery parameter, such that the price a job is willing to pay for preventing tardiness is increased and the difference between early and timely produced jobs is diminished ($\alpha = 0.2, \gamma = 0.5$). Compared to experiment 2, the allocation efficiency criteria show better results. Experiment 4 shows the consequence of increasing all budget parameters (20 %).[h]

---

[f]This is due to the preference structure given above and reflects the partly uncertainty about consequences of late delivery.
[g]t = time in "minutes", parameters selected according to an example in (Gutenberg 1971)
[h]what suggest that the ratio between cost per unit and the budget available per unit is relevant (the effect stops when increasing the budget above a certain bound)

Further experiments demonstrated the relevance of time and budget parameter for the control of the scheduling process. The simple model was also used to schedule a three-machines problem. The results show the applicability of even the simple model in more complex circumstances.[i]

Table 2. Results of experiments 1-4

E	Cap.	FT	Av.tard.	#early	#late	Av.price	Av.cost	Profit	Revenue
1	76%	3.95	1.95	0	16	7.28	5.75	3060	14560
2	76%	4	0.9	0	16	6.62	5	3240	13240
3	91%	3	-0.1	4	2	5.065	4.75	630	10130
4	91%	2.85	-0.25	6	1	8.298	6.25	4096.38	1.66e+04

2.3. Some Remarks concerning the Simple Model

This simple model has one main drawback:[j] it is very difficult to interpret the terms profit and revenue. This is due to the fact that the budgeting of the jobs is done in a "relative" manner—the job's are more or less prioritized, the real value connected with the jobs is not considered. The resource agents use real cost information to calculate their offers—however, this is only of limited use since the job agents evaluate the price they are willing to pay relative to this offer. Possible extension of the model towards integration of cost and value aspects connected to the jobs are considered in Section 3.5.

3. Model Extensions

Most of the constraints used as preconditions for the simple model can be relaxed. This, and possible extensions to the model are to be discussed next.

3.1. Dynamic Environment

The static structure of the production environment and the static model of job arrivals imposed non-essential constraints on the problem. They may be relaxed as follows.

- Due to their mutual independence, jobs may enter and leave the market without disturbing the allocation process (no reservations are made; no constraints on the jobs sequence are present).

- Given a process structure without alternative routings, removing resources would simply result in a break-down of the production process for all the jobs with the resource laying on their production path. The allocation process itself (as every resource determines it's schedule completely independend of other resources in a sense that jobs cannot choose between alternative resources)

---

[i] A complete description of the experiments and the results can be obtained from the authors.

[j] In fact, there is another one: An equilibrium price is not determined before all allocations take place (no tatoonement as necessary for walrasian equilibria). This might prevent the process from computing a system's equilibrium (ensuring an optimal assignment of allocations). See any textbook on micro economics for detailed informations. Our work extending the simple model towards more complex equilibria processes will be described in a subsequent paper. Interested readers should contact the authors.

is purely local, and a break-down or addition of a resource affects the other resources only in the way described above: Jobs will leave the markets for all resources on their production path (break-down) or new jobs will enter the markets for all resources on their production path (including the new resource).[k]

3.2. Priority Rules

A comparison of the results of applying the simple model given above to job shop scheduling problems and solving the same tasks with a variety of priority rules is currently in progress. In designing the simulation experiments we follow closely the recommendations given in Ramasesh's paper (Ramasesh 1990). We expect that it is possible to emulate the behavior of certain priority rules (as MOD, see (Kanet 1983)) with appropriate parameter settings. The motivation behind this experiments is that this model and it's extension may serve as a framework for developing an easy-to-use decision support tool offering access to a variety of scheduling strategies via an unified instrument: markets.[l]

3.3. Interactive Model Execution/Simulation

In further experiments, the user of the model was allowed to change the budget of single jobs or the allocation criterion.[m] The execution of an allocation process stops and interaction is requested whenever idle times are expected. We call this *interaction on exception*. This experiments showed that the possibility to interactively adapt the behavior of the model in case of irregularities is essential for the simulation of the trading process to successfully match the scheduler's needs. This will be true until the agents acting on the market will be intelligent enough to handle such situations as flexible and effective as a human scheduler would do.

3.4. Intelligent Agents and Adaptation to Environment

Certain environments obey certain conditions and constraints. In order to allow an adaptation of the market system to it's environment, the participating agents (job and resource agents as well as other influencing agents) have to be able to change their behavior in time. In a certain situation, certain expectations regarding the future development of the market lead to a certain decision about the strategy to follow. Expectations and predictions could be disappointed in the course of time. This disappointment should initiate a process of learning and, as a consequence, of behavioral adaptation through the change/development of strategies.[n] Techniques and models of case-based reasoning could help to model an appropriate form of behavior. The use of "intelligent" strategic decisions in our models is still very limited, switching between strategies is only used to handle

---

[k] If alternative routings are possible, the situation changes. This will be discussed in a follow-up paper.
[l] For more details on decision support see Subsection 3.6.
[m] average or marginal costs.
[n] See (Katona 1946) for a discussion of the role of expectation and experience for behavior.

well-recognizable "ill- situations", for example, situations, were deadlocks occur (if reservations are allowed).[o]

3.5. Active Budgeting and Integration of Cost Accounting

Budgeting plays a central role for the success of the market model. Success in this context means: the allocation decisions lead to "good" plans, i.e. plans that reflect the cost, time, and quality constraints the production process should obey and are thus acceptable to people at shop floor and managment levels. One way to reach this is to integrate the market model into the cost accounting and budget planning system of the firm. Additionally that would open up the possibility to use absolute budgets instead of the budget parameter used in Section 2. If the jobs budget reflects revenues and costs, scheduling decisions can be rationalized, and, more important, evaluated in a manner understandable and agreeable by all members of the underlying organisation: in terms of money.

The idea behind using absolute budgets is to create a standardized instrument every agent (human or not) influencing the scheduling process may use to express his interests. The budgeting process is called *active* since the budgets are

- actively used to influence and control processes. This could be scheduling processes at shop floor level as well as product development processes or other processes using resources, leading to expenses, and inducing revenues.

- are actively pre-determined by agents of higher hierarchical levels to control the behavior of subordinate agents.

Ideally, active budgeting should be embedded in a market-like organisation of the organization. This seems to be impossible (inspite of the advantages markets exhibit over other organizational forms (Malone 1988 or Miller and Drexler 1988) due to high overhead costs connected with a market-oriented organization of internal business processes (see, for example, (Coase 1937)). Nevertheless, this have to be re-thought again, since information systems could reduce the overhead costs to an irrelevant amount, especially if one takes vulnerability costs into account?[p]

Active budgeting of agents, embedded into a cost and budget planning system, together with a market-driven implementation of business processes may serve as an instrument allowing to integrate planning and control processes with cost and finance considerations and to connect direct and indirect production processes in a coherent manner.

3.6. Decision Support

Market-driven planning processes might not only be useful at shop floor level, but also for tactical and strategic production planning tasks as well as for other pro-

---

[o]A detailed example of the consequences of applying different behavioral strategies for job agents is given in a technical report, prepared by Christian Klee as a master thesis (in German) — it can be obtained from the authors.
[p]Compare (Malone 1988)—vulnerability costs are assumed to be proportional to expected costs due to failures of task processors and mangers.

cesses where partially exclusive interests and conflicting goals are to be considered, e.g., concurrent engineering processes. It seems difficult to answer a question like "Is it possible to implement an organization based on planning processes utilizing markets" in general. A few arguments for such an organizational form are given above. We think that the mechanisms of active budgeting and market-driven processes could be used on aggregate levels of information and planning horizons supporting the integration of different planning levels and functional branches within a firm. Further studies on this topic have to be performed.

3.7. Modeling, Implementation, and Execution

Modeling for the first experiments was based upon OMT with a subsequent implementation in C++. Since the suggested market models exhibit an inherent parallelism, we switched our modeling to an PVM–extended version of Charles Lakos' Object Petri Nets (LOOPN)(Lakos 1991)[q]

4. Conclusion

At least did the modeling (and the application of the model) itself lead to a better understanding of real-world planning processes and the micro and macro environment in which the planning process takes place. We think that scheduling with the help of market models might improve the integration of planning and costing/budgeting, might open up a way to re-structure exchange and allocation processes in a firm, and might still deliver reasonably good schedules.

Current research activies are mentioned in Section 3. More detailed results can be obtained from the authors and the on-going work and further experiences with the model(s) will be documented at the conference.

5. References

1. K.R. Baker and J.J. Kanet, *Journal of Operations Management* **4** (1983), 1.
2. R. H. Coase, *The Nature of the Firm*, in Economica, New Series **IV** (1937)
3. J.S. Davis and J.J. Kanet, *Single Machine Scheduling with Early and Tardy Completion Costs*, Naval Research Logistics (1993)
4. Erich Gutenberg, *Grundlagen der Betriebswirtschaftslehre, Bd. 2 - Die Produktion, 17th Edition* (Springer 1971)
5. Feuerstein, *Wirtschaftswissenschaftliches Studium* **22** (1993), pp. 286-290
6. G. Katona, *Am. Ec. Rev.* **36** (1946), pp. 44-62
7. C.A. Lakos, *LOOPN user manual*, Technical Report, University of Tasmania
8. Thomas W. Malone, in *Readings in Distributed Artifical Intelligence* eds. A.H. Bond and L. Gasser (Morgan Kaufmann Publishers, Inc., San Mateo, 1988)
9. M.S Miller and K.E. Drexler, in *The Ecology of Computation*, ed. B.A.Huberman, (North-Holland 1988), pp. 133-176
10. R Ramasesh, in *OMEGA - Int. J. of Mgmt Sci.* **18** (1990), pp. 43-57

---

[q]A version of our software will (hopefully) be available to public in June '95. Please, contact one of the authors.

DEVELOPING KNOWLEDGE-BASED SYSTEM FOR CALIBRATION SCHEDULING

FELIX T.S. CHAN
School of Manufacturing and Mechanical Engineering, University of South Australia
Ingle Farm, PO Box 1, The Levels, SA 5098, AUSTRALIA.
e-mail: metsc@levels.unisa.edu.au.

ABSTRACT

This paper proposes an expert scheduling system to arrange daily calibration in a semiconductors company. In the existing system of the company, the scheduling of re-calibration time is dependent on the equipment capability, it has not taken into account the availability of resource and work load forecasting. There are a lot of problems existed in the system, such as lengthening the equipment idle time, competition of using calibration standards and uneven work load distribution hence result in poor quality. This paper discusses how the Expert Calibration Scheduling System (ECSS) works and solves the existing problems.

Keywords: Scheduling, Knowledge-based system, Calibration.

1. Introduction

Calibration is preventive maintenance procedure, by which ensure the accuracy of an equipment or gage within its tolerance (usually is the published specification by manufacturer). Calibration is performed by comparing an unknown accuracy of a being calibrated equipment with a known accuracy equipment, we call it as Standards, and make adjustment once its accuracy out of its specification.

A semiconductors manufacturing company wants their products achieve a well recognized quality level, such as certification to conform ISO9000, calibration of their equipment is a basic requirement. In the company, there is an in-house calibration laboratory to serve their equipment. There are nine technical staff in the laboratory to serve about 3000 sets equipment. On average, there is more than 70 sets of equipment to be calibrated every week.

In the existing calibration system, an equipment to be re-calibrated totally depends on the equipment capability and has no consideration of whether having sufficient resources, and how much work load within the scheduling period. Sometimes this will cause a lot of troubles such as shortage of man power, competition of using calibration equipment etc. In this paper, an Expert Calibration Scheduling System (ECSS) will be proposed to resolve the existing problems in calibration arrangement.

2. Problem Background

Calibration of in-house equipment is routine maintenance procedure. Unlike other job orders, the calibration date is assigned by the calibration recall system (i.e. the system is developed to determine when the equipment should be re-calibrated), the work load is deterministic and can be forecasted. In the existing recall system, the next calibration due-date is merely determined by the equipment capability. If the history of the recorded calibration data of an equipment shows a good stability and high accuracy, then the next calibration interval can be lengthened. On the other hand, if the performance of an equipment is deteriorated in a short period, the next calibration interval will be shortened.

When the scheduled calibration date is over due in the following week, the concerned user will be notified to deliver their equipment to the laboratory for calibration. Once the equipment have been calibrated, a next due-dated will be assigned. Then the cycle will be repeated again.

This recall system causes a lot of troubles for equipment users and the laboratory staff. Because there is no consideration of the availability of resources when assigning the next calibration date of each equipment. Although there are thousands of equipment, the commonly used calibration standards are about 30 sets. If some equipment are recalled at the same time, and suppose they are all calibrated using the same standards, then problem of competition of standards will be existing. Hence the queuing time and eventually the cycle time will be lengthened. It has stressed here that when the scheduled calibration date is due, no matter how urgent work is that equipment being engaged, it has to be released for calibration. Otherwise they violate the company policy for using any over-due equipment. The situation is even worse, if the equipment is being used in the middle of production (not all the equipment is used for manufacturing, some are used in R & D or product evaluation), the company may lose thousands US dollars a day if stopping the equipment during production.

Solving these problems may not need a large amount of resource, because it seems to be a resource management problem, an effective scheduling system is helpful to improve the situation.

3. Analysis Of Existing System

In order to find out some possible solutions to solve the problems, the process flow and capability of the existing system are analyzed. By reviewing the process flow, the bottleneck and some non-productive, non-value added processes can be identified. We could evaluate whether the existing resources are able to cope with the workload, hence determine how much additional resources are needed.

3.1 Process Flow

Calibration recall system is a semi-automatic system. The heart of the system is an Equipment Management Database which stores most of the equipment calibration information, such as due-dates and last calibration dates, serve times, equipment user information, and used by which department (this data is important because the equipment is used in different ways according to the responsibility of the department) etc. At the beginning of each week, the recall system listed out those equipment which will be over-due in the following week. The users of those equipment will be notified to deliver their equipment on or before the due-date to the laboratory for re-calibration. The laboratory supervisor then arranges the duty manually according to the job list. Basically, the assignment is scheduled according to an individual's experience and the relative work load of each sub-ordinate. After calibration, the next re-calibration time (next due date) is assigned according to the historical performance. Then the calibration cycle will be repeated.

Ideally, each responsible person will calibrate his/her assigned equipment according to the schedule. However, the supervisor has not taken into account the availability of those required calibration resources and the tardiness of equipment. At present, problems existed in the system include:-

1) Shortage and surplus of man power
   Because the existing scheduling system did not consider the work load in the forecast period, sometimes it will cause uneven distribution of work loading. In some period, there is shortage of man power while surplus of work force in the other period. During busy period, the staff would have to work over-time to complete the scheduled calibration items on time.
2) Absence of standards
   Standards also require periodic calibration. Some standards are calibrated by outside agents, such as government laboratory and equipment manufacturers (usually in US). A typical cycle time is more than one month. During this period, the equipment requiring those standards for calibration cannot be performed. It has to wait for the standards to be released.
3) Competition of using standards
   Although there are thousands of equipment, the commonly used calibration standards are about 30 sets. If some equipment are recalled at the same time, suppose they are all calibrated using the same standards, then there is competition of standards. Because the equipment are waiting in a queue, the cycle time will be lengthened and the original schedule will be disturbed.
4) Interruption of schedule due to late equipment
   Due to the tardiness of delivered equipments, the standards are sometimes idled for a long period. On the other hand, the late arriving equipment will interrupt the normal schedule.
5) Ineffective scheduling method
   Normally, efficiency and quality of work improve as experience cumulates. It is also true for calibration. According to the historical data, typically, the time needed for calibrating an equipment by an inexperienced personnel is about three times more than that by an experienced person, because the learning curve effect prevails. Although, the job assignment has already taken this factor into account, this is purely dependent on the supervisor to remember which person is familiar with which particular type of equipment. This is not an effective and accurate way to schedule the jobs. Although, calibration is a skilful job, anyone with reasonable technical background can handle the job efficiently provided that sufficient training is available. However, if the job is manually scheduled, the supervisor would likely disregard the opportunity of providing the sufficient training for his/her subordinates to familiar with their inexperienced items.
6) Lack of communication with the equipment users
   Although a schedule for equipment re-calibration has been determined, there is no consideration of whether the user can actually release the equipment on time. In the absence of a proper channel to communicate with them, they usually do not release their equipment on time without prior notification. Obviously, this will interrupt the normal planned schedule. In fact, the user very often violates the company policy for using over-due equipment.
7) Interruption of manufacturing process
   If the equipment is due to be calibrated, no matter how urgent it is in use, it has to be released for calibration. Otherwise it will violate the company policy for using over-due equipment. The situation is even worse if the equipment is engaged in a manufacturing

process. It may lose thousands of US dollars a day caused by removing the equipment during production.

3.2 Capacity Analysis

To analyse the problem which may be caused by insufficient resources or ineffective allocation of resources, two important figures are studied. First is the average resource (calibration standards) utilization rate, the other is a direct comparison between the total required calibration time and the normal available working hours. These two figures indicate whether the supplied resources are adequate to provide sufficient service.

Figure 1 shows the utilization time for some most intensive used standards. Assume 8 working hours per day, all the standards are under utilized. It can be interpreted that, if the calibration time can be effectively organized, there will be sufficient time for standards to be allocated for calibrating each equipment.

Figure 2 shows the comparison between total required calibration time and the normal available working hours for each month. Besides there are other duty of the laboratory staff, it should have sufficient man power to cope with the requirement. In some period, the man power is in shortage (e.g. Feb.) but other is in surplus (e.g. Jan.), it is due to the uneven distribution of workload. If the required calibration time is greater than the normal working hours, then over time is needed to complete all the scheduled job.

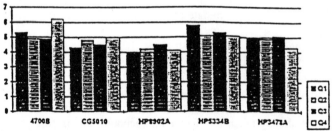

Figure 1. Average daily utilization in each quarter for the five mostly intensive used calibration standards in 1994

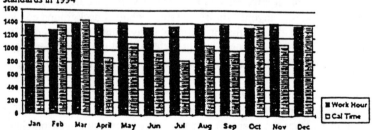

Figure 2. Comparison between the total required calibration time and the normal available working time in 1994

4. Proposed Solution

By analysing the process flow, we can observe the following problems:
1) Assign the next calibration date without consider the availability of resource in that period.

2) Do not monitor the progress of the equipment usage, sometimes the laboratory supervisor is informed in the last minute that the scheduled equipment is not ready for calibration.

An Expert Calibration Scheduling System (ECSS) is proposed to resolve the problems. It consists of three schedulers, i.e. a Coast Scheduler, a Main Scheduler and a Real Time Scheduler. They are arranged in hierarchical structures. The Coast Scheduler, is the highest level scheduling system which formulates the plan by considering only the resource in a macro-view. Then the schedule will be refined to a detailed plan by the Main Scheduler. The final stage is the Real Time Scheduler, it provides real time schedule modification function to cope with the minor change of resource and some unexpected events.

The three schedulers are working in different time frame, the Coast Scheduler is allocating working load in few months to a year from the present. The Main Scheduler is used to arrange the jobs in more detail, it only forecasts the work load and resource in the near future. So that it has more details and accurate information to produce the schedule. The Real Time Scheduler is operated in real time, it can be used to cope with any interruptions.

4.1 Coarse Scheduler

When an equipment is calibrated, according to its capability, the next calibration due-date will be assigned (usually few months to two years later). In the existing system, the due-date is the latest date to re-calibrate the equipment, we define this as 'Real Due-date'. Beyond this Real Due-date, the accuracy of the equipment would likely to drift out from its specification.

The function of the Coarse Scheduler(see Fig. 3) is to assign an appropriate period for re-calibrate an equipment. By taking into account the required calibration standards may not be available (the standards may be sent to outside calibration laboratory), it searches for a period from the 'Real Due-date' backward in such a way that the work load on that period is relatively low, but this period should not be too far away from the 'Real Due-date' (typically a few weeks before). Then the system will look into more details of resource availability in that period, called 'Floating Due-date'. There are two major parameters to be considered in determining the 'Floating Due-date':-

1) It is the utilization rate of the required standards used for the calibration. The utilization rate indicates how busy of the standards being used in the period. If the rate is very high, most likely the use of the standards is very busy and the Main Scheduler cannot find an appropriate period for using it. Thus the Coarse Scheduler will allocate another calibration period on which the rate is relatively low, i.e. the due-date will be floated to another period.
2) The second parameter is man power adequacy. Although the studied period is far away from the present, some long term plan may cover this period, such as the planning for people on trip, and employ additional staff etc. So that the scheduler can avoid allocating too much work load in the period of shortage of man power.

The scheduler determines a coarse work load allocation, and stores it in the database. It is not a finalized schedule, it can be adjusted if there are major changes of resource. For instance, if there is a new additional calibration standard arrived in a few months later, it may resolve the congestion of using the single standard in certain period. The scheduler will regenerate the coarse work load planning hence develop a more balanced schedule.

The Coarse Schedule produce a job list, it states which equipment should be calibrated in certain period. A detail schedule will be worked out by the Main Scheduler. Fig 4 shows an example of the job list.

4.2 Main Scheduler

A detailed work plan of calibration is assigned by the Main Scheduler. It extracts the job list of the following weeks from the Coarse Work Plan Database. Because the working schedule will be used for the near future, it has a clear picture of the available resources, such as who and when taking annual leave or attending training class. The Main Scheduler will work out a detailed calibration schedule according to the job list and available resource. As there are many factors affecting the final schedule, it is difficult to construct an mathematical model. The traditional scheduling techniques such as linear and dynamic programming is insufficient and inefficient to perform the task[1]. Knowledge-based techniques use intuition and robust heuristic, which embedded the experience and knowledge of scheduling, thus it can avoid the explosion of possible combination of arrangement and quickly reduce the search space to approach a satisfactory solution[2].

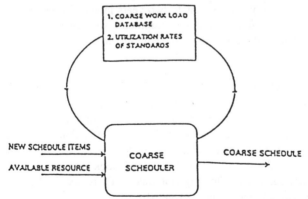

Figure 3. Coarse Scheduling Process

WORK WEEK	EQUIPMENT TO BE CALIBRATED
41, 42	MMDG-123X
	TMID-012X
	ZMIS-120X
	ZMIS-019X
	TMOV-023X
	.
43, 44	CAVE-013X
	ENPH-130H
	MMDG-340X
	TMRD-032X
	WTSD-048H
	.

Figure 4. Example of coast schedule

The scheduling process in the Main Scheduler can be divided into two stages: an initial assignment stage, and iterative improvement stage. Figure 5 shows an example of a possible schedule worked out by the Main Scheduler. Initial assignment stage fills the empty grid (initial stage of build up a schedule) with tokens (duty), and allows violation of some of the constraints. Iterative improvement stage tries to resolve the violated constraints by sweeping and shifting entries in the grid. The goodness of the schedule is measured by the degree of violated constraints.

Selection assignment of filling the grid with tokens is in accordance with the scheduling rules which is coded as forward chaining rules. Some of the scheduling rules can be stated as follows:

IF technician X calibrated the type of the equipment AND X has time THEN assign the equipment to X for calibration
IF required calibration standards are available at time T AND technician X is free for calibration THEN assign the calibration at time T.

All constraints are checked in the iterative improvement stage. If a constraint is violated, all feasible swapping and shifting actions that may resolve that constraints are enumerated. Usually, some constraints are more difficult to fulfil thus the harder constraints will be attempted to resolve first then the easier one. The process stops if all constraints are resolved or no further improvement can be made.

Once the calibration schedule is completed, the system will generate an equipment calibration notice to inform the user to deliver their equipment on the scheduled date. If the user cannot release their equipment on time, he/she can request another favourable time. The Main Scheduler can re-adjust the schedule by repeating the scheduling process until a compromise solution exist. Figure 6 shows the block diagram of the Main Scheduler.

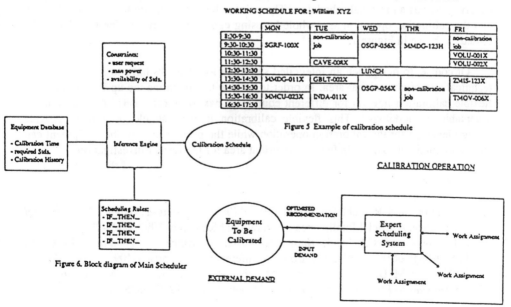

Figure 5 Example of calibration schedule

Figure 6. Block diagram of Main Scheduler

Figure 7 Recommended Integrated System

4.3 Real Time Scheduler

Similar to the Main Scheduler, it is a knowledge-based system, but equipped with different rules. The main function of this scheduler is to fine tune the calibration work plan in order to cope with some minor change conditions. It responses to the real time resource and job queuing for calibration. If there is an additional job, the scheduler attempts to searching for a free room from the grid or sweeping some items to accommodate the change. However, if there is a major change that affects the whole defined schedule, or some changes cannot be handled by the Real Time Scheduler, then the change request will transfer back to the Main Scheduler for jobs rearrangement, as it can arrange the job in a longer time frame and have high flexibility to make a 'better' schedule.

5. Conclusion

This paper states the methodology of developing a knowledge-based scheduling system, but, this is only one of the proposal stages, there may have some fine tunes and adjustments before it is finalized. However, the paper drafts out the skeleton of the scheduling system.

According to the research work, there are two main causes for the inefficiency of the calibration operation. Firstly, the works are poorly scheduled within the calibration operation. Secondly, there is a mis-alignment between the external demand or loading and the calibration operation. The proposed expert system is feasible to solve the first problem. But more important, the second issue remains a major bottle-neck in the calibration operation. If the demand or loading is extremely uneven, the calibration works can never be performed efficiently no matter how good the expert system can schedule the works.

Therefore, in order to solve the whole problem, it is not adequate by only using an expert system as an internal scheduler. The point is how to integrate the external demand and the calibration operation as a whole by using certain mechanism. The loop has to be closed as shown in Figure 7.

Of course, to build such an integrated system, a lot of works have to be done in both technical and organisational area. In order to be able to streamline the up front demand, a flexible calibration interval for different equipment has to be established according to an acceptable methodology. This flexible calibration interval should be able to provide enough leverage for the calibration operation while the performance of the equipment can still be maintained. This is in fact an important area for the further investigation.

6. References

1. H. Walters and R. Schtakle, *From OR to Knowledge-based System: An Industrial Experience*, Artificial Intelligence in operational Research, 1992, pp23-29.
2. K.P. Chow and C.K. Hui, *Knowledge-based Approach to Scheduling Problems*, Proceedings of IEEE Asian Electronics Conference, 1987, pp404-409.
3. M. Zweben, M. Deale and R. Gargan, *Anytime Rescheduling*, Workshop on Innovative Approaches to Planning, Scheduling and Control, Nov. 1990, pp251-259.

DISJUNCTIVE CONSTRAINTS FOR MANUFACTURING SCHEDULING: PRINCIPLES AND EXTENSIONS

PHILIPPE BAPTISTE and CLAUDE LE PAPE
ILOG S.A., 2 Avenue Gallieni, BP 85
F-94253 Gentilly Cedex FRANCE
E-mail: baptiste@ilog.fr lepape@ilog.fr
Url: http://www.ilog.fr or http://www.ilog.com

ABSTRACT

Disjunctive constraints are widely used to ensure that the time intervals over which two activities require the same resource do not overlap in time. Two types of extensions of disjunctive constraints are proposed: (1) extensions allowing the representation of more complex constraints including activities that may or may not require the resource, "state resources" to represent activities which may use resources only under specific conditions, and a mechanism to represent setup times between activities; (2) extensions of the disjunctive constraint propagation algorithm to deduce more precise time-bounds. These extensions are integrated in ILOG SCHEDULE, a C++ library for constraint-based scheduling.

1. Introduction

Scheduling is the process of assigning activities to resources in time. Basically, the three main things to consider when building a scheduling system are:

- **The complexity of the scheduling problem.** Most scheduling problems are known to be NP-hard (Garey and Johnson 1979). In practice, this means that one must design robust approximate algorithms, to generate appropriate (possibly optimal but often sub-optimal) solutions in a bounded amount of time.

- **The specificity of the problems to address.** Different manufacturing environments induce different scheduling constraints, some of which may be very specific to the problem under consideration.

- **The integration with the overall manufacturing system.** A scheduling system must get its data from the information system globally in use in the factory, and must return its results (i.e., the constructed schedule) for factory-floor execution.

Scheduling problems are very different one from the other:

- First, three broad families of scheduling problems can be distinguished depending on the degrees of freedom in positioning resource supply and resource demand intervals in time. In pure *scheduling* problems (e.g., job-shop machine scheduling), the capacity of each resource is defined over a number of time intervals and the problem consists of positioning resource-demanding activities over time, without ever exceeding the available capacity. In pure *resource allocation* problems (e.g., allocation of personnel to planes or trains), the demand for each resource is known beforehand and the problem consists of allocating resources in time to guarantee that the supply always equals or exceeds the demand. In *joint* scheduling and resource allocation problems, degrees of freedom exist for deciding both which activities to perform and when, and which resources to make available for these activities.

- Different environments are subjected to different constraints which more or less contribute to the complexity of the problem. For example, a factory scheduling problem may involve only machines as resources, while another may also require the consideration of the abilities of human operators.

- The size of a scheduling problem may vary from a few dozens activities to thousands of activities. For complexity reasons, algorithms that work well for the small problems may not be applicable to the bigger problems.

- Depending on the environment, the suitable response time for the construction of a schedule may vary from a few microseconds to a few days. Also, it may be necessary to incrementally modify the schedule, either as a response to environmental changes, or because it is more appropriate for a human to "make the decisions."

In response to this important variety and variability of scheduling problems, ILOG initiated the development of ILOG SCHEDULE, a C++ library enabling the representation of scheduling constraints in terms of *resources* and *activities*. SCHEDULE (Le Pape 1994a) is itself based on SOLVER, the generic software tool for object-oriented constraint programming developed and marketed by ILOG (Puget 1994). This enables the user of SCHEDULE to benefit from the functionalities of SOLVER to develop specific types of constraints and implement specific problem-solving procedures, in response to the requirements of each particular application.

The interest of constraint-based programming lies in using constraints to reduce the computational effort needed to solve combinatorial problems. Constraints are used not only to test the validity of a solution, as in conventional programming languages, but also in a constructive mode to deduce new constraints and rapidly detect inconsistencies. For example, from $x < y$ and $x > 8$, we deduce, if x and y denote integers, that the value of y is at least 10. If later we add the constraint $y \leq 9$, a contradiction is immediately detected. Without propagation, the "$y \leq 9$" test could not be performed before the instantiation of y: no contradiction would be detected at this stage of the problem-solving process.

SCHEDULE includes three categories of predefined constraints:

- **Temporal constraints.** Users may link any two activities together by any type of precedence constraint (A starts after $start(B)$, A starts after $end(B)$, A ends after $start(B)$, A ends after $end(B)$). In addition, minimum and maximum delays between activities can be imposed. When only temporal constraints are considered, constraint propagation suffices to determine whether a set of temporal constraints is consistent and to compute the earliest and latest start and end times of activities (Le Pape 1988).

- **Capacity constraints.** SCHEDULE offers many different ways to express that a resource is available in finite amounts over time: *unary resources* to represent resources of capacity one (like a specific person); *volumetric resources* to represent resource pools of many, non-differentiated resources (like a group of people with the same capabilities); *state resources* to represent situations where an activity uses a resource only under specific conditions (like waiting for an appropriate oven temperature). In addition, the capacity of a resource can be constrained either at each time unit (number of people available each day) or over given time periods (number of people-days available over one week).

- **Resource utilization constraints.** An activity may require, consume, provide and produce resources, in an amount represented either as a constant or as a constrained variable. This allows the representation of the case where the duration of the activity varies with the amount of resources assigned to the activity. The propagation of resource utilization constraints results in adjusting the earliest and latest start and end times (time-bounds) of activities.

The problem of determining whether a set of resource utilization constraints is consistent with given time-bounds is NP-hard. The propagation of resource utilization constraints is consequently incomplete. This means that it is usually necessary to explore the search space to generate a solution to the scheduling problem under consideration. Constraint propagation is useful in this process as it allows the pruning of many impossible decisions. Three constraint propagation methods are available in the current version of SCHEDULE (version 1.1).

- The first method relies on a generic mechanism allowing the definition of time-tables as discrete arrays or sequential lists of constrained variables. This mechanism is presented in details in (Le Pape 1994a). It applies to all types of resources: unary, volumetric and state resources.

- The second method applies to unary resources and state resources. It consists of posting a generic "disjunctive" constraint to ensure that the time intervals over which two activities require a unary resource (or require different states of a state resource) cannot overlap in time. For instance, if a resource is required (or provided) by two activities throughout two time intervals $[s_i\ e_i)$ and $[s_j\ e_j)$, the disjunctive constraint states that either e_i is less than or equal to s_j, or e_j is less than or equal to s_i. Such a disjunctive representation has been used in the scheduling domain for years (Erschler 1976) (Carlier 1984) (Le Pape 1988). It is a priori more time-consuming but often results in more precise time-bounds than the propagation of the corresponding time-table constraints. Section 2 presents the basic principles used to propagate such constraints. Section 3 presents three extensions made in SCHEDULE to (1) allow activities that may or may not require the resource, (2) apply disjunctive constraint propagation to state resources, and (3) provide a mechanism to deal with sequence-dependent setup times between activities.

- The third method extends the disjunctive constraint propagation algorithm to deduce more precise time-bounds for each activity. Section 4 presents this extension.

2. Propagation of Disjunctive Constraints for Scheduling

The most basic disjunctive constraint states that two activities A and B that require the same unary resource R cannot overlap in time: either A precedes B or B precedes A. This can be stated as follows:

$$[end(A) \leq start(B)] \text{ or } [end(B) \leq start(A)]$$

In this formula, $start(A)$, $end(A)$, $start(B)$ and $end(B)$ denote constrained variables. This means that the values of $start(A)$, $end(A)$, $start(B)$ and $end(B)$ are initially unknown. Constraint propagation consists in reducing the set of possible values for these variables: whenever the minimal possible value of $end(A)$ (earliest possible end time of A) exceeds the maximal possible value of $start(B)$ (latest possible start time of B), A cannot precede B; hence B must precede A; the timebounds of A and B can consequently be updated with respect to the new temporal constraint $[end(B) \leq start(A)]$. Similarly, when the earliest possible end time of B exceeds the latest possible start time of A, B cannot precede A. When neither of the two activities can precede the other, a contradiction is detected.

Notice that the disjunctive formulation above does not necessarily imply the explicit creation of a disjunctive constraint for each pair of activities. In ILOG SCHEDULE, a unique global constraint is created for the overall set $\{A_1 ... A_n\}$ of activities that require a given unary resource R. The propagation of this constraint is equivalent to applying the process described above to the $n(n-1)/2$ pairs of activities $\{A_i\ A_j\}$. Three extensions are described in the next section.

3. Extended Functionalities

3.1. Optional Activities and Resource Alternatives

In manufacturing scheduling, it is often the case that several resources can be used to perform the same activity. These resources can be strictly equivalent (e.g., n identical machines), in which case a discrete resource is used. They can also be different (e.g., have different operating speeds), in which case it is absolutely necessary to determine which resource is going to be used for which activity.

Some activities can also be optional if, for example, the production of parts can be either done "in house," or sub-contracted to another company.

In both of these cases, it is necessary to represent the situation in which an activity may or may not require a resource. This can be done in two ways:

- The fact that an activity may or may not require a resource can be stated by attaching a Boolean variable $demand$ to the resource requirement constraint. In a given solution, $demand$ is either 0 or 1. When $demand$ is 1, it means that the resource is used for executing the activity. When $demand$ is 0, it means that the resource is not used for executing the activity. A choice between m resources can consequently be implemented by using m Boolean variables, and stating that exactly one of these Boolean variables must be 1.

- A fully optional activity can be represented by an activity the duration of which can be either 0 or the time actually necessary to execute it. When the duration is 0, the activity does not really require the resource.

The disjunctive constraint can easily be modified to represent these two cases. For any two activities A and B that require the resource R, the disjunctive formula to satisfy is:

$$[end(A) \leq start(B)] \text{ or } [end(B) \leq start(A)]$$
$$\text{or } [duration(A) = 0] \text{ or } [duration(B) = 0]$$
$$\text{or } [demand(A) = 0] \text{ or } [demand(B) = 0]$$

The propagation of this constraint consists in imposing one of its six disjuncts when the five other disjuncts are proven false.

3.2. State Resources

A state resource is a resource (a priori of infinite capacity) which can operate in different states. Each activity may require the resource to remain in a particular state throughout its execution. Consequently, two activities which require distinct states cannot overlap. The formula to satisfy is:

$$[end(A) \leq start(B)] \text{ or } [end(B) \leq start(A)]$$
$$\text{or } [duration(A) = 0] \text{ or } [duration(B) = 0]$$
$$\text{or } [state(A) = state(B)]$$

Notice that the states required by A and B can be variables. This allows, for example, the representation of the situation where a batch of products can be "cooked" at different temperatures, with the duration of the activity or the quality of products varying with the chosen temperature.

SCHEDULE also allows the representation of constraints such that: (1) activity A requires resource R in any state different from a given state s (the state may change during the execution of A, but can never be s); (2) activity A requires resource R in any state that belongs to a given set of states $\{s_1 \ldots s_n\}$ (the state may change during the execution of A, but must remain in the given set of states); (3) activity A requires resource R in any state that does not belong to a given set of states $\{s_1 \ldots s_n\}$ (the state may change during the execution of A, but can never belong to the given set of states). The disjunctive constraint can be generalized to handle constraints of types (1) and (3) when the number of possible states for the resource is infinite. When the number of possible states is finite, or when constraints of type (2) are used, the time-table mechanism (Le Pape 1994a) must be used. Nevertheless, it may still be interesting to propagate the redundant disjunctive constraint in these cases, as the disjunctive constraint may deduce more precise time-bounds.

3.3. Transition Times Between Activities

Another extension allows the definition of "transition times" between any two activities that require the same unary or state resource. Given two activities A and

B, the "transition time" between A and B is an amount of time that must elapse between the end of A and the beginning of B when A precedes B. Transition times typically exist in factories where tools must be set up prior to the execution of a manufacturing operation: the setup time varies not only with the operation for which the setup is made, but also with the preceding operation.

The disjunctive constraint can easily be modified to take transition times into account. If $tt(A, B)$ denotes the transition time between A and B, the formula to satisfy is:

$$[end(A) + tt(A, B) \leq start(B)] \text{ or } [end(B) + tt(B, A) \leq start(A)]$$
$$\text{or } [duration(A) = 0] \text{ or } [duration(B) = 0]$$
$$\text{or } [demand(A) = 0] \text{ or } [demand(B) = 0]$$

The tt function is defined by the user of SCHEDULE with respect to the particular needs of his or her specific application. The default tt function always returns 0. Hence, if the user does not redefine tt, the disjunctive constraint behaves as if there were no setup times between activities. The main advantage of this implementation is that it does not necessitate the creation of a matrix containing $tt(A, B)$ for every A and every B. For example, if A and B are painting operations and if the setup time depends only on the colors used for A and B, $tt(A, B)$ could be defined by a matrix the dimensions of which are the number of colors — a number potentially much smaller than the number of activities.

Transition times can also apply to state resources, assuming that the transition time between two activities that require the same state is 0.

4. Extended Propagation

The previous section has shown how disjunctive constraints can be extended to represent a variety of manufacturing scheduling situations. This section presents an extension of the propagation process: the meaning of the disjunctive constraint remains the one developed in Section 2; but the constraint propagation algorithm is improved to allow the deduction of more precise time-bounds. In the context of a particular scheduling application, more CPU time may be spent propagating the constraints, but this extra propagation may result in a better exploration of the search space and, consequently, in a drastic improvement of the overall CPU time.

The extended propagation algorithm still allows the consideration of activities that may or may not require the resource, and of sequence-dependent transition times between activities. In fact, the extension of the propagation process does not interfere with the extensions of functionalities proposed in Section 3, thereby allowing the combination of both types of extensions in the same propagation algorithm.

The extended propagation algorithm works as follows: rather than just looking at pairs of activities $\{A\ B\}$, the algorithm compares the temporal characteristics of A to those of a set of activities $\Omega = \{B_1 \ldots B_n\}$. Let est_A denote the earliest possible start time of A, let_A denote the latest possible end time of A, and p_A denote the smallest possible duration of A. Let est_Ω denote the smallest of the earliest possible start times of the activities in Ω, let_Ω denote the greatest of the latest possible end times of the activities in Ω, and p_Ω denote the sum of the smallest possible durations of the activities in Ω. The following rules apply:

$$let_{\Omega \cup \{A\}} - est_\Omega < p_A + p_\Omega \Rightarrow A \text{ is before all activities in } \Omega$$
$$let_\Omega - est_{\Omega \cup \{A\}} < p_A + p_\Omega \Rightarrow A \text{ is after all activities in } \Omega$$

New time-bounds can consequently be deduced. When A is before all activities in Ω, the end time of A is at most $let_\Omega - p_\Omega$. When A is after all activities in Ω, the start time of A is at least $est_\Omega + p_\Omega$.

The technique which consists in applying these rules is known as the *edge-finding* technique. Notice that if n activities require the resource, there are potentially $O(n*2^n)$ pairs $\{A\ \Omega\}$ to consider. (Carlier and Pinson 1990) presents an algorithm that performs all of the possible time-bound adjustments in $O(n^2)$. A variant of this algorithm, described in (Nuijten et al 1993), is used in SCHEDULE. (Carlier and Pinson 1994) presents a variant running in $O(n * log(n))$ that we have not tested yet.

The following table provides results obtained on ten 10x10 instances of the job-shop scheduling problem used by (Applegate and Cook 1991). In this table, BT and CPU denote the total number of backtracks and CPU time needed to find an optimal solution and prove its optimality. BT(pr) and CPU(pr) denote the number of backtracks and CPU time needed for the proof of optimality. CPU(one) denotes the CPU time needed to find one solution. This includes the time needed for stating the problem and performing initial constraint propagation. Finally, TM denotes the total amount of memory used to represent and solve the problem (in kilobytes). CPU times are expressed in seconds on a 100MHz 486DX4 PC under SOLARIS. The SCHEDULE program performs better or about as well as the specific procedure of Applegate and Cook on five problems in terms of CPU time, and on seven problems in terms of the number of backtracks needed to solve the problem. Other results are available in (Baptiste 1994) and (Le Pape 1994b).

Instance	CPU(one)	BT	CPU	BT(pr)	CPU(pr)	TM
MT10	.4	69758	1498.9	7792	179.5	140
ABZ5	.5	17636	322.6	5145	92.5	136
ABZ6	.4	898	22.4	291	6.8	136
LA19	.4	21910	424.1	5618	109.6	140
LA20	.4	74452	1207.2	22567	353.8	136
ORB1	.5	13944	318.2	5382	121.4	144
ORB2	.4	114715	2838.4	30519	742.8	140
ORB3	.5	190117	4485.5	25809	620.7	140
ORB4	.4	64652	1601.5	22443	561.5	140
ORB5	.5	11629	241.3	3755	76.9	140

5. Conclusion

Disjunctive constraints have been used by many researchers and practitioners (e.g., (Erschler 1976) (Carlier 1984) (Le Pape 1988)) to represent activities competing for the same unary resource. Two types of extensions have been proposed in this paper, to allow the representation of more complex constraints and to deduce more precise time-bounds for each activity.

These extensions are available in ILOG SCHEDULE, a C++ library aimed at simplifying the representation and the resolution of industrial scheduling problems. The integration of these extensions within the same constraint programming tool provides a strong basis for the development of scheduling applications that fit the constraints of complex manufacturing environments.

6. References

1. D. Applegate and W. Cook, "A Computational Study of the Job-Shop Scheduling Problem," *ORSA Journal on Computing* **3** (1991) 149-156.
2. P. Baptiste, *Constraint-Based Scheduling: Two Extensions* (MSc Thesis, University of Strathclyde, 1994).
3. J. Carlier, *Problèmes d'ordonnancement à contraintes de ressources : algorithmes et complexité* (Thèse de Doctorat d'Etat, University Paris VI, 1984).
4. J. Carlier and E. Pinson, "A Practical Use of Jackson's Preemptive Schedule for Solving the Job-Shop Problem," *Annals of Operations Research* **26** (1990) 269-287.
5. J. Carlier and E. Pinson, "Adjustment of Heads and Tails for the Job-Shop Problem," *European Journal of Operational Research* **78** (1994) 146-161.
6. J. Erschler, *Analyse sous contraintes et aide à la décision pour certains problèmes d'ordonnancement* (Thèse de Doctorat d'Etat, University Paul Sabatier, 1976).
7. M. R. Garey and D. S. Johnson, *Computers and Intractability: A Guide to the Theory of NP-Completeness* (W. H. Freeman and Company, 1979).
8. C. Le Pape, *Des systèmes d'ordonnancement flexibles et opportunistes* (PhD Thesis, University Paris XI, 1988).
9. C. Le Pape, "Implementation of Resource Constraints in ILOG SCHEDULE: A Library for the Development of Constraint-Based Scheduling Systems," *Intelligent Systems Engineering* **3** (1994a) 55-66.
10. C. Le Pape, "Using a Constraint-Based Scheduling Library to Solve a Specific Scheduling Problem," in *Proceedings of the AAAI-SIGMAN Workshop on Artificial Intelligence Approaches to Modelling and Scheduling Manufacturing Processes* (TAI, New Orleans, Louisiana, 1994b).
11. W. P. M. Nuijten, E. H. L. Aarts, D. A. A. van Erp Taalman Kip and K. M. van Hee, *Job-Shop Scheduling by Constraint Satisfaction* (Computing Science Note, Eindhoven University of Technology, 1993).
12. J.-F. Puget, *A C++ Implementation of CLP* (Technical Report, ILOG S.A., 1994).

INTELLIGENT SIMULATION-BASED SCHEDULING OF WORKCELLS

TAN HOCK SOON
SPL Worldgroup (S) Pte Ltd
410 North Bridge Road
#03-00 Cosmic Insurance Building
Singapore 0718

and

ROBERT DE SOUZA
School of Mechanical and Production Engineering
Nanyang Technological University
Nanyang Avenue
Singapore 2263

ABSTRACT

Manual scheduling will have difficulties in reacting on time to changes on the shop floor. It also is not capable of analysing the goodness of the schedule because the scheduler is not capable of accurately predicting the outcome of the schedule. In addition, a human scheduler cannot generate operation sequences in any great detail given the short time he needs to come up with a schedule.

The use of simulation for detailed production fills an important void in the management of the factory floor. Through experimentations, it will enable the engineer to be able to predict a good schedule for his work orders. Unfortunately, there are instances when he will not have time to experiment. Hence, a faster system is necessary. This paper presents an approach to build such a system combining a neural network to predict rules for scheduling and a schedule generator using a simulation based commercial product.

Keywords: Scheduling, Neural Networks, Simulation

1. Introduction

"What is Scheduling and why is it difficult?" Scheduling is one of the most important functions in a manufacturing firm. It is the allocation of available production resources over time to meet some set of performance criteria. Typically the scheduling problem involves a set of jobs to be completed, where each job comprises a set of operations to be performed [Rodammer and White 1989]. Operations require machines and material resources and must be performed according to some feasible technological sequence. Schedules are influenced by such diverse factors as job priorities, due-date requirements, release dates, cost restrictions, production levels, lot-size restrictions, machine availabilities, machine capabilities, operation precedences, resource requirements, and resource availabilities. Performance criteria typically involve trade-offs between holding inventory for the task, frequent production changeovers, satisfaction of production-level requirements, and satisfaction of due-dates.

Developing a schedule involves designating the resources needed to execute each operation of the process routing plan and assigning the times at which each operation in the routing will start and finish execution.

Except in very special cases, there are no good(polynomially-bound) algorithms for scheduling problems, despite decades of intense activity [Buchanan 1990]. In other words, it is in the Non-deterministic Polynomial (NP) Complete Domain, where a large number of possibilities exists in which job operations can be sequenced and does not lend itself to satisfactory mathematical analysis and solution.

Apart from the intrinsic complexity of the scheduling task, there are difficulties associated with the very definition of the problem. The quality of a schedule is a function of several measures, some with an easily-achieved quantification, but others may be ill-defined and have at best subjective or surrogate quantification. Thus on-time delivery performance and levels of work in progress(WIP) may be fairly readily measured, but other concepts such as customer satisfaction or schedule robustness will not be so readily admit of quantification. Even with a collection of appropriate metrics, the multi-criteria feature leads to issues involving judgements about trade-offs among the measures.

A third source of difficulty is the recognition that scheduling takes place in a dynamic environment, and the definition of the problem is continually shifting. The solution proposed by a scheduling system may become irrelevant, at least in part, due to events deviating from plan. Thus work may be lost through unplanned scrap, resources may be lost through breakdown, operation activity times may deviate from normal, new work may be introduced to be expedited, and work originally planned may be withdrawn through cancellation of an order. An ideal scheduling system should not only be able to function in a predictive manner, but also react effectively to unscheduled events. This will almost mean that the scheduling system should be able to address the new information in an incremental manner, and avoid a complete re-schedule unless the changes in the domain are drastic [Buchanan 1990].

2. On Simulation and Neural Networks

Most manufacturing systems are too complex to allow realistic models to be evaluated analytically [Rodammer and White 1989]. As an alternative, simulation based scheduling can provide an effective tool for shop floor scheduling while requiring few assumptions. The schedules generated are based on an accurate, realistic model of the production facility. It can allow various dispatch rules, or decisions regarding the system to be tried out and selected based on performance results. Simulation based scheduling has gained acceptance by both researchers and practitioners [Grant and Clapp 1988]. Within the last ten years, high level simulation languages and systems have been developed and specifically tailored for manufacturing [Koh, Ho and de Souza 1993]. This is partly due to its ability to conduct powerful experimentation with reproducible results.

The main disadvantage is that it does not inherently optimise, and the computation burden can be significant for large and complex models. It has also been proved to be inadequate in modelling complex systems which include some element of human decision making [O'Keefe 1990]. A good example is the scheduling problem mentioned in this paper which involves scheduling of job orders and job steps within a particular order bearing in mind the constraints of machines, workers, due dates, utilization, etc. A "good" solution can be obtained only after experimenting and running different alternatives with different rules in the model.

Neural networks have shown good promises for solving combinatorial optimization and constraint satisfaction problems such as shop floor scheduling. Lee [Lee 1991] has successfully used back propagation networks to solve job shop scheduling problems. Zhou et. al. [Zhou, Cherkassky, Baldwin and Olson 1991] have presented an enhanced approach which can handle medium sized problems. Lim et. al. [Lim, Khaw and Lim 1993] have applied hybrid neural networks to schedule Group Technology (GT) manufacturing cells.

Extensive research has shown that the capabilities of artificial neural networks for automatic learning, association, generalisation and pattern recognition through their ability to provide non-linear transformations to model highly complex functions. In addition, the approach does not require the strong underlining assumptions on the structure of the data required by many traditional techniques. The complex interactions of the dynamic considerations at the shop floor are therefore candidates for investigating the possibility of using neural networks to learn and perform dynamic scheduling for a shop floor. The disadvantage of neural networks is that "knowledge" in the network is not easily available to the user.

Based on the above observations, an approach to schedule a manufacturing cell was conceived using both simulation and neural methods.

3. Intelligent Schedule Generation

The scheduler is envisaged to be a module by itself, sitting between the upper hierarchies of the CIM architecture and the shopfloor control module as shown in figure 1.

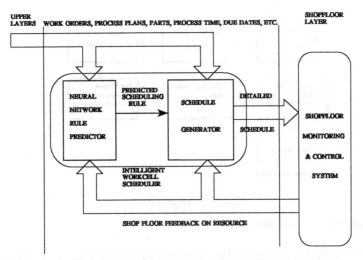

Figure 1: The Two Modules of the Intelligent Workcell Scheduler

The scheduler itself is made up of two modules, the *schedule generator* and the *neural network rule predictor*. The scheduling rule to be used is obtained from a given range: FIFO, LIFO, Earliest Order Due Date, Shortest Time for Current Job Step, Longest Time for Current Job Step, Least Number of Remaining Job Steps, Least Static Slack, Least Dynamic Slack. To obtain a "good" schedule, the neural network rule predictor is employed using feedbacks from the shop floor as well as work order information. The scheduling rule predicted by the neural network is then used by the schedule generator to create the workcell dispatch list. The workcell environment used in this model is based

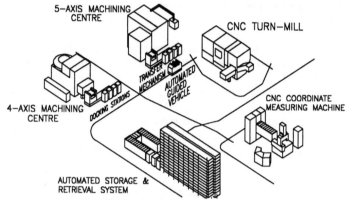

Figure 2: Layout of F.M.C.

on the Flexible Manufacturing Cell (FMC) located in the German-Singapore Institute's CIM Centre. The workcell consists of three CNC machines, a coordinate measuring machine (CMM), an Automated Storage and Retrieval System and an Automated Guided Vehicle (AGV). The layout of the cell is as shown in figure 2. Figure 3 shows the systematic method used in constructing the scheduler.

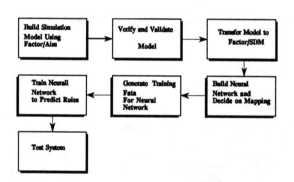

Figure 3: Approach

3.1 Building the Simulation Model

A commercial product called FACTOR developed by Pritsker Corporation was used to develop the schedule generator. Firstly, a simulation model of the workcell was created in FACTOR/AIM (Analyzer for Improving Manufacturing), a simulation tool.

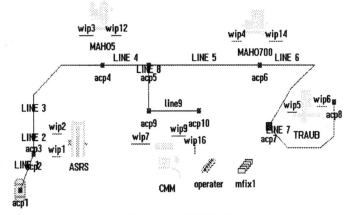

Figure 4: AIM Model

3.2 Verify and Validate

The simulation model was then verified and validated using the following techniques:
i)   Review by Experts
ii)   Sensitivity Analysis
iii) Comparing Model vs system performance and
iv)   Animation, as recommended by Sargent [Sargent 1992] and Banks [Banks 1994].

3.3 Port Model to Scheduler Module

The model was then ported over to the scheduler module, FACTOR/SDM (Schedule Development Module). This module is able to generate finite schedules for the machines in the workcell. See figures 5 and 6.

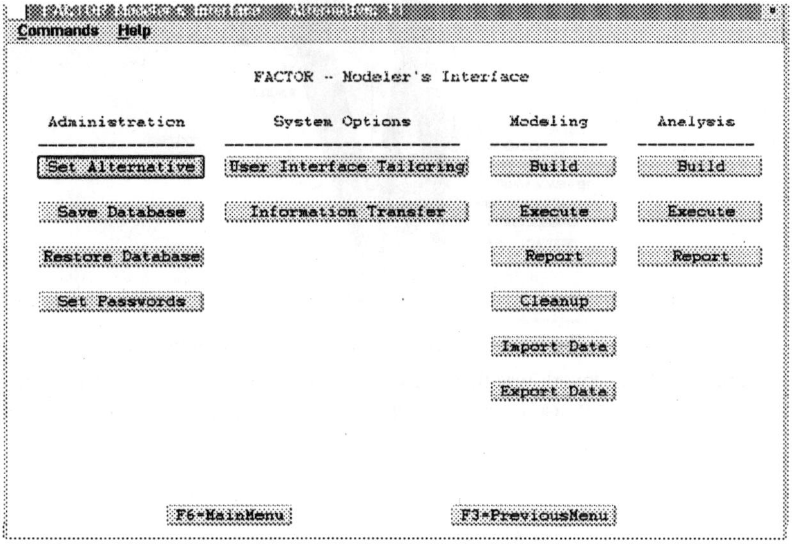

Figure 5: SDM Modeller's Interface

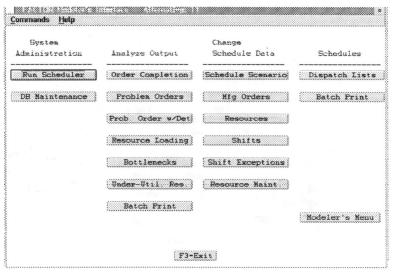

Figure 6: SDM Scheduler's Interface

3.4 Build Neural Network & Decide on Mapping

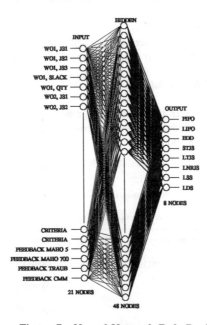

Figure 7: Neural Network Rule Predictor

A backpropagation network was then constructed to map the constraints of the resources, job sequence, process times, machine occupancy, due dates, etc. to a scheduling rule that gives the best performance measure. The methodology for constructing the neural networks were taken from Lawrence and Andriola [Lawrence and Andriola 1992]. The network architecture is as shown in figure 7. It has 21 input nodes, 48 nodes at the hidden layer, and 8 output nodes.

3.5 Generate Training Files for Neural Network

Using FACTOR/SDM, different alternatives (each using a different scheduling rule) of the scenario were simulated. The reports obtained were then analyzed and the information needed for neural network training extracted.

3.6 Train and Test the Neural Network to Predict Rules

The network was then trained to select rules giving the best performance measure/criteria. Performance measures or criteria used in this scheduler include:

    a. Total Flow Time b. Order Lateness
    c. Order Tardiness d. Resource Utilisation

Once the network is trained, its predictions are to be tested against the simulation model or the real system. Finally, interfaces between the neural network rule predictor, the schedule generator, the shopfloor module and the upper layers of the CIM hierarchy will have to be built.

4. CONCLUSION

We have described an approach for creating a simulation and neural network based scheduler. The scheduler will take outputs from higher level production control (eg. MRPII systems) and generate a detailed plan for sequencing and dispatching the orders. The system is also able to quickly re-schedule orders in the shopfloor should the need arise because of its link to the shopfloor control module. Besides that, the system has self learning capabilities due to the inherent nature of neural networks.

REFERENCES:

(Banks, J. 1994), *Pitfalls in the Simulation Process,* **Proceedings of Conference on New Directions in Simulation for Manufacturing and Communications 1994,** pp. 57-63.

(Buchanan, J.T. 1990), *Current and Future Scheduling Systems,* **Conference Proceedings of CAD/CAM '90,** pp. 286-294.

(Chan, D.Y. 1989), *Design of a Scheduling System For a Flexible Manufacturing Cell,* **PhD. Dissertation, Arizona State University, U.S.A., 1989.**

(Grant, H. and Clapp, C. 1988), *Making Production Scheduling More Efficient Helps Control Manufacturing Costs and Improve Productivity,* **IE'88, June 1988,** pp. 54-62.

(Koh, K.H., Ho, N.C. and De Souza, R. 1993), *Current Approaches To Job-Shop Scheduling*, **Proceedings of the 2nd International Conference on Computer Integrated Manufacturing - ICCIM '93, Vol. 1,** pp. 681-687.

(Lawrence, J. and Andriola, P. 1992), *Three-Step Method Evaluates Neural Networks for Your Application*, **EDNAsia, December 1992,** pp. 47-60.

(Lee, W.H. 1991), *Application of Artificial Intelligence Methodology in Job-Shop Scheduling: A Neural Network Approach*, **M.Eng., Thesis, National University of Singapore, Singapore.**

(Lim, B.S. ,Khaw, J. and Lim, L. 1993), *Hybrid Neural Networks for Scheduling in a Group Technology Manufacturing Cell*, **Proceedings of the 2nd International Conference on Computer Integrated Manufacturing -ICCIM '93, Vol. 2,** pp. 828-835.

Neural Computing: A Technology Handbook for Professional II/Plus and Neural Works Explorer, Technology Publishing Group, NeuralWare Inc., Pittsburgh, 1993.

(O'Keefe, R.M. 1990), *The Role of Artificial Intelligence in Discrete-Event Simulation*, **Artificial Intelligence, Simulation, and Modelling,** Widman, L.E., Loparo, K.A. and Nielsen, N.R. (Eds), John Wiley and Sons, pp. 359-379.

(Rodammer, F.A. and White, K.P. 1989), *A Recent Survey of Production Scheduling*, **IEEE Transactions on Systems, Man and Cybernetics, Vol. 18, No. 6,** pp. 841-851.

(Sargent, R.G. 1992), *Validation and Verification of Simulation Models*, **Proceedings of the Winter Simulation Conference, 1992,** pp. 104-114.

(Zhou, D.N., Cherkassky, V., Baldwin, T.R. and Olson, D.E. 1991), *A Neural Network Approach to Job-Shop Scheduling*, **IEEE Transactions on Neural Networks, Vol. 2, No. 1, January,** pp. 175-179.

Control

Intelligent Object Networks — The Solution for Tomorrow Manufacturing Control Systems
   J. Gausemeier, G. Gehnen and K. H. Gerdes

Object-Oriented Integration of Distributed Flexible Manufacturing Systems
   S. K. Cha

Intelligent Object Networks -
The Solution For Tomorrows Manufacturing Control Systems

Prof. Dr.-Ing. J. Gausemeier, Dipl.-Ing. G. Gehnen, Dipl.-Ing. K.-H. Gerdes

Heinz Nixdorf Institut, Universität-GH Paderborn
Pohlweg 47-49, D-33098 Paderborn, Germany
E-mail:gehnen@hni.uni-paderborn.de, gerdes@hni.uni-paderborn.de

ABSTRACT

Today's manufacturing control systems (MCS) do not meet the requirements of market driven manufacturing processes. An increased flexibility of the MCS is necessary for adaption to changing production conditions during a system's lifetime of about 10 to 15 years.

The MCS described in this paper is based on a generic view of object orientation. It is composed mainly of intelligent objects and a global manager. Compared with conventional MCS this concept has important advantages. The flexibility of the system referring to changes in the manufacturing system is enhanced significantly. The system can also be easily distributed on parallel computers for speed-up and adaption to the required performance.

1 Introduction

The manufacturing industry worldwide faces strong competitors in the fight for market shares. The manufacturing processes must be flexible and fulfil customer orders with regard to time, quality and cost. Thus, the flexible automation is state of the art in manufacturing technology [HS90]. While the material flow has reached a high level of flexible automation, the adequate high potential of information processing has not been realized yet. In most cases an inflexible control system is used which has to be completely redesigned when manufacturing processes are changed [Sch92]. Additionally, the introduction, customizing, and maintenance of MCS is too expensive today.

In this paper an object oriented approach for the development of MCS is discussed. Based on this approach, a new concept of an object oriented MCS will be presented. This MCS is based on intelligent objects, representing existing devices of the plant. The key concept is the partition and distribution of system elements to reach the flexibility goals and to meet the performance requirements of high automatized manufacturing processes. The concept has been implemented at the Heinz Nixdorf Institut. The results gained confirm the effectiveness of the object oriented approach.

2 Starting Position

2.1 The Present Level of Technology

Flexible manufacturing systems (FMS) consist of stand-alone machines, conveyers, manipulators and flexible production cells. Information processing in a FMS takes place on three hierarchically structured levels (see Figure 1), the machine control level, the cell control level, and the production management level. On the machine control level the machines and all transport functions done by conveyer systems are directly controlled by NC- and programmable controllers. Interactions between the equipment within manufacturing and assembly cells are controlled by cell computers. The MCS has to coordinate the machine controllers and cell computers. Its task is to schedule, control, and monitor the orders. A production planning system supplies orders with fixed delivery dates.

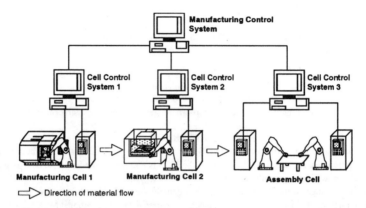

Figure 1: Structure of information processing in flexible manufacturing systems

During the job processing, accidental events (e.g. disturbances) may change the workload and the capacity of the manufacturing system. Whenever such an event occurs control strategies will be applied to achieve the disposition targets despite of the changed conditions. This causes differences in time scheduling; the actual scheduling order must be changed accordingly to the scheduling targets.

2.2 Research and Development Goals

The software of today's MCS consists of modules, for example the sequence control module, the scheduling module, or the maintenance planning module, etc. The structure of the manufacturing system, the scheduling and the control strategies are represented in a fixed and plant-specific manner. There are no analogies between this structure and the real structure of the manufacturing system. With these systems, flexibility to changes can hardly be realized. Consequently, the following objectives result:

Expansion flexibility: The MCS has to be flexible for expansion because the requirements of the flexible manufacturing system can change during the phase of introduction and the later operating phase, which can take 10 or more years. Today's MCS are not flexible enough to handle the changes during this period. Future systems have to allow the adaption to changes more easily and handable by the MCS-user.

Communication structure: A coordinated work of different manufacturing devices requires close communication between them, for example to avoid collisions in a common workspace. This communication is done today by exchanging binary signals between the PLCs of the machines. Thus, as far as the information structure is concerned, the machines are embedded in their environment in a fixed way. If the structure of communication between the machines is changed, the machine's PLC interconnections have to be changed as well. In future systems hard-wired binary signals no longer are used for the communication on the PLC level. Instead of this, the communication takes place on the cell and manufacturing control computer level, respecting the real time requirements. Modifications in the communication structure will be much easier with this architecture.

Efficient Rescheduling: Rescheduling is necessary, if disturbances occur [MB91].

These disturbances may change the capacity of the production system. The reaction time on changed capacities has to be minimized. Since processing data is sampled in real-time, the time needed to run the scheduling operation determines the reaction time of the MCS. The control system has to generate and evaluate a number of scheduling alternatives. The alternatives have to be evaluated depending on the system's workload and the scheduling targets respecting all necessary resources. Future control systems have to reduce the needed time for generating and evaluating adequate schedules. They have to select the most efficient alternative in real time.

3 An Object Oriented Approach for the Development of MCS

3.1 Distributed Systems on Object Oriented Base

Construction principles of the nature are often very useful examples for engineering solutions. The variety of life forms, realized by using always the similar basic elements can be an efficient way to build up flexibe MCS. Complex organisms are based on organs, controlled by a network, the nervous system. The organs themselves are made up of different, specialized cells. However, every cell of an organism has the same basic functionality. The organs have specialized functions and they operate autonomously. The entire functionality of an organism is established by coordinating the organs via the nervous system.

The same principle can be used to develop MCS, constructing them from small functional elements and connecting them to a network. Functional elements with similar operations, able to operate almost autonomously, must be found. Chosing real world objects in the manufacturing (MO= manufacturing object) is the best way to design these functional elements. A MO can be a single robot or a machine tool, but also a complete manufacturing line, depending on the required flexibility. For completing an order, every MO has to be scheduled and the execution has to be directed. Because there are many different influences which can obstruct or stop the production process, a thorough observation of processing is necessary. Every order on its way through the manufacturing system passing the different MO has to be controlled. Therefore, every MO has to obtain methods for scheduling, controlling, and monitoring.

3.2 Intelligent Objects as a Base of New Manufacturing Control Systems

Intelligent objects are models of all objects existing in the production reality, which have to be scheduled, controlled, and monitored. These objects are called intelligent because they expand the abilities of the real objects.

For example the abilities of a machine are expanded with the help of the master, who schedules the orders of the machine, the toolsetter, who prepares the machine, and the operator, who controls and troubleshoots the machine in case of disturbance.

The required "intelligence" of the people assigned with these tasks, has to be reproduced by intelligent objects (IO) in the MCS with the help of algorithms and rules. Objects which are designed in this way complement each other in their functionality and can be compared with the organs in the analogy of nature, as shown above. But in the same way the organs must be coordinated by a nervous system, the control system needs a coordinating instance, called global manager. The tasks of the global manager can be summarized as central jobs. Such jobs are the communication with the MRP-system and the user, setting the guidelines of the control strategy and solving deadlock situations among the IO.

The introduced "intelligent objects" (IO) are supersets of objects for special use of manufacturing control. They are objects on a functional higher level based on the principles of object oriented programming[Gra91], [Hel91], [R+92], [R+91]. Thus, inheritance and the class libraries allow an easy adaption of the MCS in every individual case.

4 The Concept of an Object Oriented Manufacturing Control System

4.1 The Architecture of the Communicating Intelligent Objects

As mentioned above, the IO structure follows the object oriented principle. They have local data and methods which act on this data. The data are encapsulated so that they can only be acessed by messages transmitted to the methods. The internal structure mirrors the functional division into three tasks. Therefore, every IO has methods for scheduling, controlling, and monitoring.

The MCS uses a standardized protocol for the communication between all components. The communication between the IO and the machines are done by a common fieldbus interface objects or by machine specific interfaces for machines which have no fieldbus support. The unified message protocol between the objects is vital for the communication in the MCS because communication is a basic principle of the presented conception of a MCS. The IO communicate with each other and with the global manager for coordinating tasks. The coordinating tasks are used to synchronize the scheduling or controlling actions.

4.2 Functions of Intelligent Objects

4.2.1 Scheduling

The IO schedules the production device which it represents in the MCS, but in communication with the global manager and all other related IO. It receives from the global manager all information regarding manufacturing jobs, including all parameters for the execution. The global manager can also give a scheduling strategy as guideline to the IO.

To do the scheduling, the IO has a set of several different scheduling algorithms. To perform all partial jobs according to the correct work flow order a communication between all involved IO is necessary. The global manager provides all jobs to suitable groups of manufacturing objects. All suitable production devices compete for the job. All IO have a unique grading mechanism in which several different parameters of interest are calculated. The one with the highest score among all competitors will obtain the job. This means, that the selected IO is the best choice in regard to the relevant parameters. Normally, the jobs are provided by the MRP-system in a cyclic way. In case of an eminent disturbance of the production device, causing the IO to give back its jobs, an acyclic distribution round is started. Now all interested IO are informed about the jobs and a new grading cycle begins. Thus, the necessary rescheduling actions are done without any delay.

After scheduling the first job, the following technological group is able to start its scheduling based on the given guidelines and the schedule of the predecessor. The presented mechanism allows a parallel processing in most parts of the program. This leads to an increased velocity of scheduling compared with a sequential working process.

4.2.2 Controlling

The results of the scheduling are placed into a time-schedule which exist for every IO. The control operations in the IO have to guaranty a coordinated and undisturbed realization of the planned schedule. The IO for a machine tool for example has to coordinate the material support and the tool support by generating jobs for the tool adjustment and the material supporting system. These jobs have to be generated soon enough before the end of the preceding job. The supply with material and tools can be handled with robots or with manual work.

In the MCS workers are also represented by intelligent objects, called 'personal assistents' (PAS). PAS are connected to PDA-terminals. The workers get their jobs from the PDA terminal in the same way a tool management systems for example would do.

The completion of the supporting jobs is reported to the IO. Adjustment data for tools are sent over the over the IO of the toolsetting device to the IO of the machine tool.

In cases of disturbance, which is the normal case for manufacturing, the IO handles the most likely cases with a set of methods and strategies. Only if the object cannot operate this disturbance on its own, the operator is informed by dialogue functions of the MCS. The operator will get all the information and diagnocstics about the failure and maybe also means to eliminate them, as far as the MCS can provide them.

4.2.3 Monitoring

The monitoring functions help to automate the control functions. They are realized by an immediate comparison of incoming process data with predetermined limits, for example, technological limits, limits given by the operator, and information of the orders, etc. The observation of the number of the produced pieces could be used as an example. A method informs the operator when the ordered number of pieces has been reached. It is also possible to implement methods which reduce the piece counter when a scrapped piece has been detected and starts all necessary operations to maintain the ordered number of pieces (e.g., ordering of new material, correcting the time schedule).

4.3 The Architecture of the Manufacturing Control System

An overview of the MCS is given in figure 3. The system is arranged around a softwarebus, which handles the communication. The system architecture consists of the following parts:

MRP-interface: The MRP-system releases manufacturing orders for the MCS and receives feedback information of the manufacturing process from the MCS. The MRP-interface allows the adaption to different MRP-systems without any substantial changes of the MCS.

Man-machine-interface (MMI): The MMI-module encapsulates the user interface from all other modules of the MCS. This allows an exchange of the graphical user interface without conflicting other system's functions.

MAP/Fieldbus-interface: All equipment of the manufacturing is ideally connected via fieldbus or MAP to the MCS. This interface enables the connection of different fieldbus systems to the MCS without changing other parts of the system.

Functional tools: The blocks in the upper part of the MCS represent the functional tools. They implement central functions, which can not be distributed to the IO. Examples for functional tools are: the online simulation for the graphical preview of the manufacturing situation, the configuration tool for the configuration of network and IO, the maintenance management for the administration of maintenance orders, the tool management and the DNC for the administration of tools as well as NC-programs and the global manager. The global manager divides orders from the MRP into technology specific orders and adds transportation and toolsetting orders on its own. The technology specific orders are later submitted to the assigned technology specific manufacturing group (technology group). The technology groups are build up from IO by configuration. These IO represent production resources with equal capabilities referring to the manufacturing process. All members of a specific technology group receive the relevant information about the partial orders.

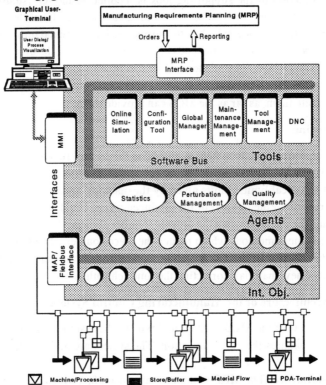

Figure 2: Architecture of the Manufacturing Control System

Agents: Agents have to fullfil specific tasks dedicated to several IO. In figure 3 they are represented as ellipses. Examples for agents are the perturbation management, the system log (statistics) and the quality management.

Intelligent Objects: The IO are represented as circles in figure 3. The functionality of the IO has been described above.

The presented MCS system is structured decentrally. The hardware of the MCS consists of an unlimited number of linked computer nodes in a flat hierarchy (for example bus topology). This allows a high availability of the total computing power and a scaling of the computing power to the given performance requirements. A high availability of the entire production system can be attained by the decentral structure together with the high degree of autonomy of all the MCS components realized through the IO.

5 Current Level of Realization and Experiences

5.1 Testing Environment

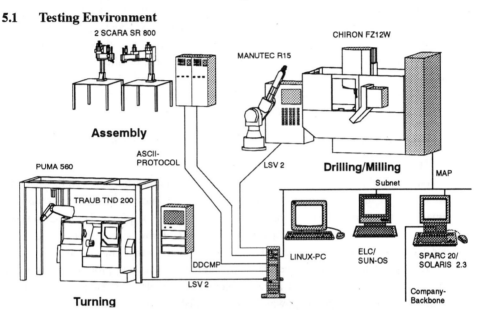

Figure 3: The Testing Environment

The conception of a new MCS has been implemented at the Heinz Nixdorf Institut. The basic manufacturing system consists of a NC turning machine, a flexible manufacturing cell, and an assembly cell. The material flow is organized by a material transfer system. The simulation of a much bigger plant is realized by using dummy-objects representing virtual machines. The material flow and the machine supply is handled by robots. A set of different products can be produced. The equipment is controlled by a control system, organized in only two levels. The machine controllers form the lower level. The MCS forms the upper level and it coordinates the machine controllers, and performs scheduling and monitoring tasks

5.2 Current Level of Realization

The following level of realization is achieved, described by these aspects:
- Each device in the testing environment is represented in the MCS as an IO.
- As functional tools the global manager, the MMI, the configuration tool and the DNC-module are implemented.
- Control strategies are implemented in the global manager.
- Communication between the production devices is established in the MCS. The machines are no longer interconnected on the PLC level.
- The devices are linked to their specific IO. The operations for transfer and management of the NC-programs for machines and robots are realized.

5.3 Design Process of a Manufacturing Control System

The effort to design a MCS is reduced by using intelligent objects and object libraries. The system is designed by defining new objects or selecting objects from libraries.

The system design is transparent and free of redundancies because of the object oriented structure and an exactly defined communication structure between the objects. The communication is managed by the MCS and not between the NC- and programmable controllers on the PLC level as realized in conventional systems. Thus, the effort of the interconnection on controller level is omitted. Additionally a more detailed communication in the MCS can be realized. The objects can exchange, for example, binary signals or information about the actual joint-positions of a robot. Working in a common workspace is more efficient and the machine's waiting times are reduced.

5.4 Expanding Flexibility of a Manufacturing Control System

The expanding flexibility is increased by the object oriented structure because of the encapsulation of objects and the transfer of the intercommunication between the machine controllers to the manufacturing control level. Thus, no interconnections between programmable controllers, as used in conventional systems, are required. The modular construction allows the exchange of objects, if the manufacturing system is changed. New objects can be linked to the existing system without affecting other objects adversely.

The complete system can be scaled by adding or removing processors. By using cheap standard processors, it is possible to implement a scalable system at low cost. If there is a demand for more computing power, there is no need to invest in a new system, but only in additional processors and maybe additional mass storage [CD90].

6 Literature

[CD90] Coulouris, G. F., Dollimore, J.: *Distributed Systems - Concepts and Design*, Reading, 1990.

[GGGL94] Gausemeier,J.; Gehnen, G.; Gerdes, K.-H.; Leschka, St.: *Cell control by Intelligent Objects- A New Dimension of Production Control Systems*. Proc. of the IEEE/RSJ/GI Int. Conf. on Intelligent Robots and Systems - IROS '94, München, 12.-16. 9.1994

[Gra91] Graham, I.: *Object Oriented Methods*. Addison Wesley Publishing Comp., 1991.

[HS90] Hars, A.; Scheer, A. W.: *Entwicklungsstand von Leitständen*, VDI-Z 123, Verein Deutscher Ingenieure, 1990.

[Hel91] Held, G.: *Objektorientierte Systementwicklung: Modellierung und Realisierung komplexer Systeme*. Siemens Aktiengesellschaft, Berlin und München, 1991.

[MB91] Milberg, J.; Burger, C.: *Produktionsregelung als Erweiterung der Produktionsplanung und Steuerung*, CIM Management 2/91.

[R+92] Robinson, P. J. et al.: *Object Oriented Design*. Chapman&Hall, London, 1992.

[R+91] Rumbaugh, J. et al.: *Object Oriented Modeling and Design*. Prentice Hall, Englewood Cliffs, 1991.

[Sch92] Schmidt, G.: *Informationsmanagement in der Fertigung*. Proceedings of the FLS '92, Berlin, 1992.

OBJECT-ORIENTED INTEGRATION
OF
DISTRIBUTED FLEXIBLE MANUFACTURING SYSTEMS

Sang K. Cha
School of Electrical Engineering
and Engineering Research Center for Advanced Control & Instrumentation
Seoul National University
San 56-1, Shinrim-dong, Kwanak-ku, Seoul 151-742, Korea
E.Mail: chask@kdb.snu.ac.kr

Abstract

The flexible manufacturing system(FMS) is a distributed network of heterogeneous programmable manufacturing machinery, such as assembly lines and numerically controlled machines. Despite these interconnected, programmable hardware elements, the success of building a truly flexible manufacturing system has been limited so far by the lack of flexibility in its control software layer. In integrating heterogeneous machinery, many existing FMS control software systems structurally depend on specific machinery and job scheduling strategies, thus difficult to incorporate the new development in FMS organization and operational requirements.

In searching for an open architecture for the distributed FMS control software system, this paper presents an object-oriented FMS data model. Among others, it represents each physical cluster of related machinery (called flexible manufacturing cell), as an object. To facilitate the integration of heterogeneous physical cells, such cell objects share a common protocol of interacting with the main control process through the inheritance from the abstract cell class. Other related physical and abstract entities in FMS are also modeled as objects, with their similarity and difference captured in inheritance hierarchies. To verify experimentally the proposed approach, a prototype FMS control software system named FREE(Fms Runtime Executive Environment) has been implemented on top of a commercial object-oriented database system.

Keywords: Flexible Manufacturing System Control Software, Object-Oriented Integration, Object-Oriented Database System

1. INTRODUCTION

FMS(Flexible Manufacturing System) is a distributed network of heterogeneous programmable manufacturing machinery, such as assembly lines, NC(numerically controlled) machines, and AGV(automatically guided vehicles)s[1]. Typically these physical FMS elements are hierarchically organized into cells performing high-level tasks such as assembly, machining, and material transport. For the autonomous operation of an FMS, a control system is needed that interconnects these individual cells as a single integrated system and coordinates the overall shop floor activities. However, building an FMS control system, especially its software is difficult due to the complexity of the underlying FMS organization and the heterogeneous behavior of its elements. Ad hoc approaches to building FMS control software systems are likely to ruin the intended flexibility of FMS because of the difficulty of evolving with the frquent changes in the underlying FMS.

This paper is concerned with defining an open architecture for the distributed FMS control software systems. To cope with the complexity of integrating heterogeneous elements, it proposes an object-oriented FMS data model, in which each physical cell is represented by an object encapsulating the state and behavior of the physical cell. Such objects, instantiated from their corresponding classes, share a common protocol of interacting with the main control process, and this sharing is achieved through the inheritance from the abstract cell class at the root of the hierarchy. This approach of heterogeneous cell integration simpifies the job of the main control process, and localizes the necessary changes in the control system software as changes occur in physical cells.

In addition to cells, the model includes various object classes for other related physical and abstract FMS entities, and binary relationships among these objects. Following the object-oriented principle[2],[3],[4], these object classes are structured into inheritance hierarchies by their commonality and difference. This hierarchy makes it easy to add new cell classes and to extend existing class definitions.

A prototype FMS control system executive has been implemented based on the proposed FMS data model, and is described in the paper. Named FREE(Fms Run-time Executive Environment), this system stores the modeled FMS objects and relationships in the object-oriented database system, and interacts with physical cells over a local area network.

This paper is organized as follows. Section 2 discusses the general organization of FMS and presents the object-oriented FMS integration model. Section 3 presents major FMS object classes

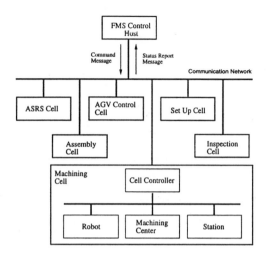

Figure 1: An exemplary distributed FMS

and the relationships among them. Section 4 discusses the architecture and implementation environment of FREE. Section 5 concludes this paper with the summary and the on-going research.

2. OBJECT-ORIENTED FMS INTEGRATION MODEL

An FMS is organized with multiple cells performing various high-level manufacturing tasks such as assembly, material transport, and machining. Figure 1 shows such an FMS organization made up with six cells: assembly, machining, set up, inspection, ASRS, and AGV control Cells. All of these cells are equipped with a cell controller, as shown in the case of the machining cell, and are connected to the control host over the communication network.

The assembly cell, equipped with robots and conveyor belts, assembles parts fed in from its input buffer stations and produces the assembled results through its output buffer station. The parts and the finished products are carried in groups on the units called pallets. They are either placed on pallets by human workers at the set up cell, or produced by processing raw materials at the machining cell or by assembling lower-level parts at the assembly cell. If needed, the inspection cell, sorts out the parts that fall out of the predefined tolerance range. The AGV control cell dispatches AGVs to transport pallets from one cell to another. The ASRS(Automatic Storage and Retrieval System) cell temporarily stores pallets carrying parts, finished products,

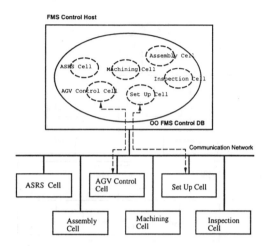

Figure 2: Object-Oriented Model for Integrated Control of FMS

and tools.

The control host dispatches high-level commands to the cells and keeps track of their status. The control system software integrates the various FMS cells as a single whole system and schedules the tasks for individual cells to perform. It is desirable that such a software is flexibly structured so that it can easily incorporates changes in the underlying FMS organization and its operation strategies. Ad hoc approaches to the control system software are likely to make it difficult to change as the FMS evolves.

To provide the control system software with the structural flexibility, this paper introduces an object-oriented FMS integration model. Its essence is to represent each physical cell explicitly as a virtual cell object in the control system software, as shown in Figure 2. Encapsulating the state and behavior of the corresponding physical cell, such an object acts as the agent of the corresponding physical cell. This encapsulation localizes the changes necessary in the control system software as the physical FMS undergoes changes. If changes occur in the FMS, only the affected cells need to be changed in the software.

By the object-oriented principle, the state and behavior of FMS objects are captured in their class definitions. Various cell classes are defined by the functions that cells perform: Assembly, Machining, ASRS, etc. These class definitions share some commonality, such as the data and functions for interaction with the physical cell. This commonality is captured by the inheritance hierarchy in the object-oriented modeling. The Cell class is defined as the most general class,

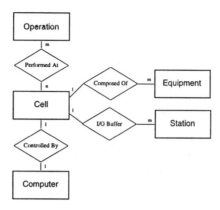

Figure 3: Cell Class and Some of Its Relationships

capturing the data and functions that are shared by all of its specialized classes. This structuring of commonality in the inheritance hierarchy also helps to localize the changes in the control system software if the changes occurring in the FMS affect all physical cells.

Not only cells but also other interesting entities, operations, and processes in manufacturing are modeled as objects by the object-oriented principle. These objects are connected to related objects by the binary relationship. Figure 3 shows the Cell class and its relationships to four other object classes: Operation, (Control) Computer, Equipment, Station. The cardinality information is attached to the links representing relationships. Even the scheduling function of the control system software is modeled as an object so that we can build the hierarchy of various scheduler classes. Sharing the common interface of the Scheduler class, its specialized classes implement different scheduling strategies. Multiple scheduler objects may reside and be switched in a single control system software.

3. FMS OBJECT SCHEMA

A collection of FMS object class definitions form the FMS object schema. These classes roughly fall into three groups:

1. Classes on physical FMS organization: Cell, and other classes representing physically existing objects such as Control Computer, Equipment, Buffer Station.

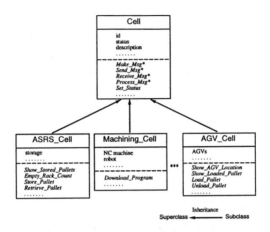

Figure 4: Cell Class Inheritance Hierarchy

2. Classes on FMS operation: Operation (and Operation Sequence), Part (and Reference Part), Task, Scheduler.

3. Classes on communication with the physical cell: Message, Message Handler

Figure 4 shows the Cell class hierarchy. Each box representing a class is divided into three areas for displaying its name, data members, and methods(member functions). The methods marked with * in the non-leaf class represent virtual methods which define only their signatures. Their actual implementations are defined by the lower-level classes. Virtual methods capture the functions which are conceptually identical among the specialized classes but whose actual implementations differ. The data members and methods defined in the Cell class represent the common data and functionality of its specialized classes. Additional data members and methods are added to individual specialized classes. For example, the ASRS_Cell class defines the data member "storage" for keeping track of stored pallets, and methods for commanding the physical ASRS.

Figure 5 shows classes on operations and parts, and the important relationships among them. The Op_Seq and Ref_Part classes model that a sequence of operations is needed to make a part. Other objects and relationships model that one of operations in an operation sequence is performed at cells, and parts and operations know which operation to perform next. Figure 6 shows the scheduler hierarchy. The scheduler class implements the methods for generating the task object, which holds information on what to do next in the FMS, such as moving a specific

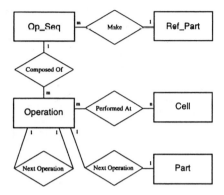

Figure 5: Classes on Operations and Parts and Some of Their Relationships

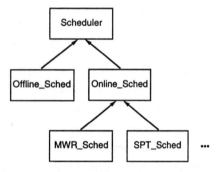

Figure 6: Scheduler Class

pallet to a specific buffer station.

The communication between the control host and the physical cell needs to be logged on both sides so that the FMS control system can be restored to the consistent state after the crash. Figure 7 shows the Message class hierarchy. Each message is identified by a serial number unique per cell object. Incoming_Msg knows from which cell it came, and Outgoing_Msg knows to which cell to send.

4. FREE IMPLEMENTATION

To verify the proposed object-oriented integration model, a prototype FMS control system executive called FREE has been implemented, and tested in connection with the actual model FMS

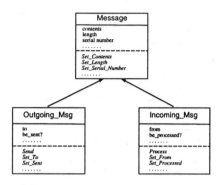

Figure 7: Message Class

plant at the ASRI(Automation and Systems Research Institute) of Seoul National University, which has the same configuration as shown in Figure 1. A commercial object-oriented database system Objectivity/DB[5] has been used to represent the FMS object schema. Figure 9 exemplifies the Objectivity/DB representation of the Cell class and its specialized class ASRS_Cell.

Figure 8 shows the architecture of FREE on the control host. A SUN Sparc workstation is used as the control host, and is connected to the physical cells over the 10 Mbps ethernet. The FREE system spawns three types of processes on the control host. For the ease of FMS scalability, a communication process is created for interaction with each physical cell. Monitoring processes are created as needed for monitoring a whole or part of FMS. The main control process provides a simple event processing loop to these processes.

5. SUMMARY

This paper has proposed an object-oriented FMS data model in dealing with the problem of flexibly structuring the FMS control system software. Each physical cell is represented by an object encapsulating its state and behavior. Other important FMS entities, operations, and processes are also modeled as objects, and the connections among them are captured by the binary relationships. The concept of inheritance hierarchy has been explored to facilitate the addition and update of FMS object class definitions.

To verify the proposed model, an object-oriented FMS control system executive has been implemented and tested with a physical model FMS plant. The ongoing research addresses extending the proposed model to incorporate manufacturing software activities such as CAD/CAM.

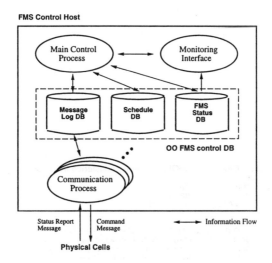

Figure 8: FREE Architecture

The research for dealing with event processing at a higher-level is also pursued.

REFERENCES

[1] W. W. Luggen, *Flexible Manufacturing Cells and Systems.* Prentice-Hall, 1991.

[2] F. Bancilhon, "Object-oriented database system," in *Proc. ACM SIGACT SIGMOD Symposium on Principles of Database Systems*, Mar. 1988.

[3] D. H. H. Ingalls, "The smalltalk-76 programming system design and implementation," in *Conference Record of the Fifth Annual ACM Symposium on Principles of Programming Languages.*

[4] B. Stroustrup, *The C++ Programming Language.* Addison-Wesley, second ed., 1991.

[5] Objectivity, Inc., *Objectivity/DB System Overview, ver. 2.1*, 1993.

```
class Cell: public ooObj{ // Inherit persistence from ooObj
protected:
    char          id[IDSIZE];
    int           status;
    ooVArray(char)    description;

    // User-defined relationships
    ooHandle(Computer) Controlled_By <-> CellA;
    ooHandle(Station) IOBuffer[] <-> CellA;
    ooHandle(Equipment) Composed_Of[] <-> CellA;
    ooHandle(Operation) Performed_At[] <-> CellA;

Public:
    // Public methods
    int Status(void);
    ooStatus Set_Satus(int stat);   or oocError
    ooStatus Show_Description(void);

    virtual ooStatus Make_Message(ooHandle(Outgoing_Msg) &messageH);
    virtual ooStatus Send_Message(ooHandle(Outgoing_Msg) &messageH);
    virtual ooStatus Receive_Message(ooHandle(Incoming_Msg) &messageH);
    virtual ooStatus Process_Message(ooHandle(Incoming_Msg) &messageH);
    . . . . . . .
};


class ASRS_Cell: public Cell{
protected:
    ooHandle(Storage) StorageA <-> ASRS_CellA;

Public:
    void Show_Stored_Pallets(void);
    int Empty_Rack_Count(void);
    ooStatus Store_Pallet(ooHandle(Pallet) &palletH);
    ooStatus Retrieve_Pallet(ooHandle(Pallet) &palletH, char* name);

    ooStatus Make_Message(ooHandle(Outgoing_Msg) &messageH);
    ooStatus Send_Message(ooHandle(Outgoing_Msg) &messageH);
    ooStatus Receive_Message(ooHandle(Incoming_Msg) &messageH);
    ooStatus Process_Message(ooHandle(Incoming_Msg) &messageH);
    . . . . . . .
};
```

Figure 9: An example of integrated FMS control database schema

FMS

A Method for Describing Operations Sequences in Flexible Manufacturing Systems
   G. Jones and M. S. Sodhi

Planning for Modular Fixtures in Flexible Manufacturing Systems
   P. C. Pandey and P. Ngamvinijsakul

A Method For Describing Operations Sequences In Flexible Manufacturing Systems

Gregory B. Jones
Navy Undersea Warfare Center
Newport, RI 02841, USA
E-mail: jones@cs.uri.edu

and

Manbir S. Sodhi
Industrial and Manufacturing Engineering
Kingston, RI 02881, USA
E-mail: sodhi@egr.uri.edu

ABSTRACT

Flexible Manufacturing Systems offer several benefits in producing diverse part mixes at low medium volumes as compared to conventional systems. Some of these benefits come from the various flexibilities offered in such systems. Among the principal strengths of FMSs is the flexibility they offer in routing parts through the system, thus allowing rapid response to dynamic disruptions. This flexibility results from the ability to load several copies of tools necessary to perform an operation on different machines, thus providing alternate routes through the system. For real-time control of such systems, every time a machine completes the processing requirements of an operation of a part assigned to it, it may be necessary to consider all routing options available for the next operation. In this paper a scheme for describing the process plan to encapsulate the routing flexibility is described. It is shown that this scheme is sufficient to describe all processing sequences possible for processing of discrete parts. Several examples illustrate the use of this scheme.

1. Introduction

A Flexible Manufacturing System (FMS) is a configuration of machine tools, usually linked together by an automated material handling system, intended to produce a variety of parts in low/medium production volumes. The machine tools are usually general purpose, programmable machine tools, each capable of a variety of operations and are often equipped with automatic tool changing devices [1]. All machines may be set up identically, or they may be set up so that each performs a different set operations. While such systems are very versatile, careful planning is required to fully exploit their capabilities [2]. Studies ([3]) show that the availability of alternative processing routes in FMSs increases throughput and reduces the vulnerability of these systems to breakdowns. However, to realize these advantages, scheduling and dispatching should be capable of dynamic adjustment - which in turn requires information of all feasible processing options.

A scheme for describing operations plans for parts to be produced in FMSs is detailed in this paper. The aim of this scheme is to provide a representation of processing plans that can be readily accessed for real-time control. Also, procedures for updating the plan as operations are performed are specified. This paper is organized in the following manner - §2 details the scheme for describing processing sequences. §3 describes the operations required to access and maintain this list. §4 describes the use of this scheme for describing system capabilities. Examples of the use of this scheme are provided in §5, and conclusions follow in §6.

2. A Scheme for Description of Process Plans

One of the principal flexibilities offered in FMS systems is operation flexibility whereby the same part can be produced by different processing sequences [4]. It can be argued that this flexibility is available in conventional manufacturing systems as well – however, the use of automatic tool changing capabilities facilitates the use of alternative operations by greatly reducing the setup time and operator intervention associated with tool changes in conventional systems. Greater routing flexibility allows easier control of flexible manufacturing systems, and permits a quicker response to disruptions ([4], [5], [6]). However, to take advantage of this routing flexibility, it is necessary to maintain representation of the processing plans for the parts in a manner that is easy to scrutinize for alternative operations when necessary. While it is possible to use tables to store the alternative processes for each primary operation, since this is essentially a lattice representation, a compact representation of all possible sequences, which correspond to a tree structure, may not be possible.

Process plans for machining parts typically involve sequences of operations with the following characteristics:

1. Un-interruptable sequences of operations consisting of sequences in which all operations must be completed immediately upon initiation of the first operation of the sequence.

2. Preemptable sequences, which may be interrupted before completion and resumed from the point of interruption later.

3. Ordered sequences where operations must be executed in the given order.

4. Sequences of operations which can be performed in any order.

5. Alternative sequences which describe optional means of processing for achieving the same end-result.

In the proposed method, the following notation is used to describe each of the above sequences:

1. Operations enclosed in square brackets [ ], indicate a sequential ordering of operations. These operations are to be performed in the given order. However, the sequence can be interrupted by another sequence, and can be processed to completion later.

2. Operations enclosed in parenthesis or round brackets ( ), indicate random or no special ordering for enclosed operations. These operations may be performed in any order.

3. Operations enclosed in curly braces, { }, indicate non-interruptable or non-preemptable sequences. Once an operation of this sequence is initiated, all operations in the list must be performed before undertaking any other operations from the list.

4. Operations enclosed in angle brackets, ⟨ ⟩, represent alternate sequences that can be performed to achieve the same end result. Only one of the sequences of those enclosed within these braces must be performed.

As a convenience in recalling this notation, the authors have found it useful to associate the type of bracket with the nature of operations represented. Thus, Round brackets are associated with Random operation sequences, Square brackets with Sequential lists of operations, Angle brackets with Alternative operations and cUrly braces with Un-interruptable sequences of operations.

2.1. Operation Sequences

Any sequence of operations is denoted as a list enclosed in one of the above types of braces. The elements of the list may either be single operations or other lists. Operations may be nested to any level as long as all open and closed parenthesis match. For the purpose of clarity, all operations are denoted by alpha-numeric characters and delimited by spaces. A requirement of this scheme is that an operation appear only once in a sequence – if an operation is to be performed more then once, then it must be given additional identifiers.

Using the above notation, the list ([1 2] [3 4]) specifies a sequence of four operations such that operation 2 cannot be performed until after operation 1, and 4 cannot be performed until after 3. The legal processing sequences corresponding to the plan given above are: 1-3-2-4, 1-3-4-2, 1-2-3-4, 3-1-2-4, 3-1-4-2 and 3-4-1-2. Another example is ({ 1 2} [3 4]) . The legal sequence corresponding to this are: 1-2-3-4, 3-1-2-4, or 3-4-1-2. Note that 2 always immediately follows 1. Finally, consider ⟨ [1 2] 3 ⟩ . The possible sequences resulting from this are 1-2 or 3.

Although integers have been used in the above illustrations, letters or keywords can just as easily be used with a more sophisticated parser. Other information related to the part and operations such as processing time, or setup time for the operation, can easily be included with the operation designator, as shall be illustrated later in the paper.

It is easy to prove that this language can express any possible ordering of operation sequences. To see this, note that any ordering of operations can be described by a tree where each node represents an operation. If all operations where required and all sequences were permissible, then there would be **n!** possible traversals of the tree from the start node to any end node (where n is the number of operations). When all sequences are not permissible, we are left with a truncated tree where only some of the **n!** traversals are possible. However, each permissible traversal (up to **n!**) can be expressed as an alternate sequence of operations (enclosed in ⟨ ⟩). Hence, this language is obviously capable of describing any sequence of operations.

While the proof that the language is complete is quite simple, in practice the exclusive use of alternate sequences is unsatisfactory. Obviously, the number of possible sequence can be quite large (up to **n!**) and the enumeration of all alternative processing sequences may lead to large and cumbersome representations for even simple process plans. Fortunately, the use of additional operators such as (, [, { , } ,] and) results in a compact representation, yet preserves the flexibility of alternative operation selection which would be lost if only one of the multitude of alternative sequences was committed to.

3. Operation List Processing

As mentioned earlier, the process plan for a part is a list of operations enclosed in brackets. This list can be maintained as a string (as in our implementation) or table of tokens. In either case, the following operations are required to process the list:

1. Given an operations list, report the operations which are ready to be performed immediately.

2. Given an operations list, modify the list by deleting a ready operation selected to be performed.

Here, it is assumed that error checking for syntax, semantics, and incorrect precedences has been performed by suitable pre-processing of the list. Scanning the list for operations to be performed consists of a recursive search through the list. An operation is ready if it is any operation enclosed by ( ) or ⟨ ⟩ , or it is the first operation enclosed by [ ] or { } . This rule applies recursively to the nested list in which the operation appears as well as all higher level lists.

Deleting operations that have been performed is slightly more complicated. For operations enclosed in ( ) and ⟨ ⟩ , the operation is simply deleted. After the deletion, any empty parenthesis or brackets are also deleted.

When an operation is selected from a non-preemptable list, { } , then the next operation selected must be the next operation in the sequence. One way to implement this is to delete the operation then rewrite the list to place the non-preemptable sequence on the outside of the list as a sequence [ ] . For example, take the list of operations: (1 2 { 3 4 5 }) . Ready operations are 1 2 and 3. If 3 is selected then rewrite the list as: [4 5 (1 2)] . This rewriting must be performed recursively from the outside (upper level) brackets in to the lower level.

When an operation is selected from a alternate list, ⟨ ⟩ , then all other alternatives must be deleted in addition to the selected operation. This deletion is performed recursively from the lowest level out to the highest level. For example, consider ⟨ (1 2) [⟨ 3 4 ⟩ 5 6] ⟩ . Operations 1, 2, 3 and 4 are ready. The legal operation sequences for processing are : 1–2, 2–1, 3–5–6, 4–5–6. If 1 is selected, [⟨ 3 4 ⟩ 5 6] must be deleted, leaving 2. Similarly, selection of 2 first leaves 1. If 3 is selected, 4 and (1 2) are deleted in that order, leaving [5 6] . In the same manner, selection of 4 leads to the sequential deletion of 3 and (1 2) , again leaving [5 6] . Thus, operation sequences that would otherwise be quite tedious to represent can be tersely represented using this syntax.

4. Machine Operations List

The notation for specifying the ordering of operations on parts can be easily extended to describe machine capabilities as well. In this case, the list of operations would refer to the operations that the machines in the system are capable of performing. As an example, consider a machine with three tools that can be accessed in any order. The capabilities of this machine are specified by all sequences generated from the list (1 2 3) . Similarly, an indexing table with three stations can be described by { 4 5 6 } . The capabilities of a "shop" consisting of these two machines can be described by ((1 2 3) { 4 5 6}) . Note that unlike the use of this syntax for part operations, subsequences are permissible here. As with part operations, the machine operation can be a label other than an integer and can encode specific information related to that operation on that machine, such as operation cost. In addition, if a machine has a current "setup" which differs from the list of all possible operations for the machine; that setup list with lower costs associated with each operation could also be maintained. In scheduling parts for machines, it is desirable to select machines already setup for the necessary operation. Scheduling parts to machines is a complex problem and well beyond the scope of this paper. However, the parts operations list and the machine tool lists provide a means to match part to machine. In fact, finding the machine with tools available that matches a part sequence suggests a method of selection to minimize the transfer of parts between machines. Such a search would require the recursive comparison of the ready lists for machine and part and selecting an operation in common.

5. Examples:

The use of the above scheme for representation of process plans is illustrated in this section. The part shown in Figure 1 is a milled workpiece with a through hole and a threaded hole in it.

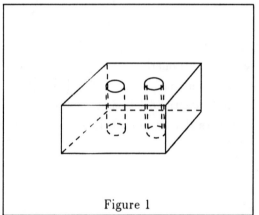

Figure 1

The list of operations required to produce this part is:
A Mill face **D** Drill Hole (1)
B Mill Sides **E** Drill Hole (2)
C Mill Base **F** Tap Hole

Assuming the base must be machined first to provide a reference, the legal sequences of operations possible for processing this part are all contained in the following list:

$$[\; C \; (\; A \; B \;) \; (\; D \; \{ \; E \; F \; \} \;) \;] \; .$$

Now, consider a part produced primarily by turning operations, as shown in Figure 2.

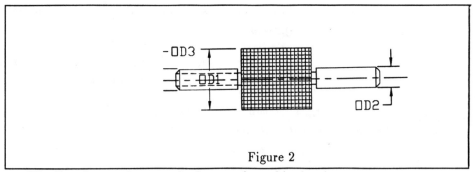

Figure 2

The list of operations required to produce this part is:

A Turn OD 1	**D** Chamfer 1	**G** Groove 2
B Turn OD 2	**E** Chamfer 2	**H** Knurl
C Turn OD 3	**F** Groove 1	**I** Thread

Using the method of description of this paper, all sequences of operations for processing this part can be represented by the following list:

$$[A (\{ B (D F) \} \{ C (E G) I \} H)]$$

It is easy to extend the notation to include additional information in this syntax. For example, information about the time required for an operation can be included by incorporating it as a suffix to the operation token. Thus, if operation **A** above takes 2 minutes to perform, the representation can be **A:2.00**. Setup times can be specified in a similar manner.

In an industrial setting, this language has successfully been used to describe the operations needed to machine components for valves and the representation has provided an easy interface for simulation studies of the system.

6. Conclusion

The specification of part operations is necessary for scheduling and simulation of FMS systems. The language presented here provides a means of specifying any possible sequence of operations and with the appropriate extension, also provides the convenience of precedence ordering for sequences that are best described by precedence.

This language will be especially useful in flexible manufacturing systems, where parts are dynamically routed to machines based on dispatching rules that take tool loading into consideration.

7. References

[1] Lenz, J. E., " Changing Tool Configurations to Change Production Capacity", *Industrial Engineering*, (1993):28-29.

[2] Talavage J., and Hannam, R., *Flexible Manufacturing Sytems in Practice: Applications, Design and Simulation*, Marcel Decker, New York, (1988).

[3] Stecke, K. E., and Solberg, J. "Loading and Control Policies for a Flexible Manufacturing System", *International Journal of Production Research*, (1982):481-490.

[4] Sodhi M.S., Agnetis A. and Askin R.G., "Tool loading strategies for flexible manufacturing systems", *International Journal of Flexible Manufacturing Systems*, Vol. 6, No. 4, pp. 287–310(1994).

[5] Benjaafar S., "Models for Performance Evaluation of Flexibility in Manufacturing Systems", *International Journal of Production Research*, (1994):1383-1402.

[6] O' Grady, P. J. and Menon U. A., "A Concise Review of Flexible Manufacturing Systems and FMS Literature", *Computers in Industry*, (1986): 155-167.

PLANNING FOR MODULAR FIXTURES IN FLEXIBLE MANUFACTURING SYSTEMS

Prof. P. C. Pandey
and
Pairoj Ngamvinijsakul
Industrial Engineering/ Manufacturing Systems
School of Advanced Technology
Asian Institute of Technology
P. O. Box 2754
Bangkok 10501, Thailand.
Email No : pcpandey@rccsun.ait.ac.th

Abstract : This paper describes the development of a software for checking the availability of modular fixturing elements for use in flexible manufacturing systems. Based on production data initial allocation of the elements is made and its feasibility checked by comparing the attainable performance with the desired system performance. For processing different part types initial allocation of the fixtures has been made and its performance in terms of attainable production has been evaluated by modeling the flexible manufacturing system as a closed queuing network. The software contains a number of databases which have been organized to facilitate data retrieval and storage for efficient working. Working of the system has been demonstrated by an illustrative example.

Keywords : Flexible manufacturing systems (FMS), modular fixtures, closed queuing network (CQN).

1. Introduction

Flexible manufacturing system (FMS) is seen as the most viable solution for problems faced in manufacturing of large product variety in small/medium sized batches. A truly flexible system will be comprised of a number of highly flexible subsystems with the overall system flexibility dependent upon its most inflexible subsystem. Flexible processing machines and flexible part handling systems are two of the most important constituents of FMS. Machining center is a typical example of flexible processing machine. Flexible part handling systems can range from simple pick and place devices to intelligent robots/AGVs. Using standardized pallets with custom built fixtures, provides part handling flexibility. These however require large setup times and fixture inventory. Use of modular fixtures helps in achieving clamping/ location flexibility but in some cases their setting up may prove to be tedious and time consuming. Modular fixtures also suffer from loss of repeatability between batches. Thus, problems concerning flexibility, location and clamping of the workpieces remain unresolved to a certain extent.

Use of modular fixtures involves extensive planning for procurement and stocking of adequate quantities of fixturing elements so as to cater to the needs of a variety of parts. This study focuses on the development of an algorithm for checking the availability of and generating a list of modular fixturing elements to satisfy a specified production plan. A methodology for optimum allocation of fixture elements required for the processing of different part types has also been developed. The utility of the system has been tested through an illustrative example.

2. Flexible Fixtures

Fixtures used for rotational and non-rotational parts are classified as :
- Traditional or dedicated fixtures
- Modular/universal fixtures
- Flexible mechanical fixtures
- NC fixturing machines

Dedicated fixtures are designed to locate, support and clamp one type of workpiece only and are unsuitable for small batch production and large part variety. Modular fixtures (MF) use standard, reusable and interchangeable elements analogous to a building block set and, are capable of locating and clamping a large part variety. These can be assembled/ disassembled in relatively shorter periods of time. Modular fixture elements are more expensive compared to elements used in conventional fixtures because of superior construction materials, and dimensional tolerances. Although, modular fixtures offer flexibility of location and clamping but may suffer from poor repeatability on account of possible tolerance stack up. Each time a modular fixture design is constructed tool proving runs may be necessary.

2.1. Design of Modular Fixtures

Modular fixtures are often designed by trial and error and do not require a formal design methodology. One of the advantages of modular fixtures is that they can be constructed at the last minute after the part, to be processed, becomes available. Lewis (4) provides a comprehensive discussion on the advantages of modular fixtures. Siong, Imato et.al. (7) have described a systematic procedure for the design of modular fixtures. Design requirements of the fixtures, for use in FMS, have been discussed by Luggen (5). Intelligent, knowledge based systems, for fixture design, has been mentioned in (1).

Majority of the research, on modular fixtures, deals with design and related problems and little or no work has been done on the allocation and stocking policy for flexible fixture elements. Inadequate allocation of fixturing elements may result into inefficient production, poor machine utilization, job tardiness etc. Fixture element location plan should be based on product demand requirements, processing times, operation sequence, number of machines etc.

This paper describes a system for checking the availability of modular fixture elements to meet a certain FMS production requirements. Based on the part through-put data and product mix, the system determines the optimum allocation of modular fixturing elements, and generates a list of the elements that should be available/procured. The utility of the system has been demonstrated through an example.

3. System Modeling

Schematic diagram of a flexible manufacturing system is shown in Fig. 1. All the incoming and outgoing parts pass through a single fixturing/defixturing station. Incoming parts are first loaded into fixtures which are next mounted on to the pallets for processing. It is assumed that each part type requires a specific type of fixture and the parts are released for processing after fixturing and palletizing. In the event of their non-availability the parts must wait. It is also assumed that sufficient numbers of parts are always available and the fixtures are never idle. The number of different part types circulating within the system are however, limited by fixture availability/allocation. When the output demand for a particular part type has been met, the fixture under use is released and dismantled.

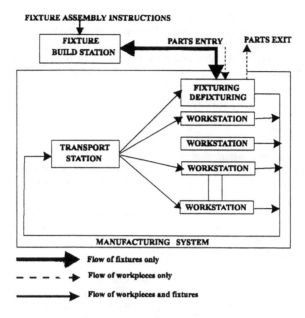

Figure 1 : Schematic diagram of a FMS.

3.1 Fixture Planning

The number of fixtures circulating within a FMS controls the part mix ratio. Thus, it is essential to specify the number of fixtures for each part type to satisfy the demand. Other parameters to be considered are processing times, routing probability, number of processing stations, etc. The methodology used for the determination of optimal number of fixtures is summarized in Fig. 2.

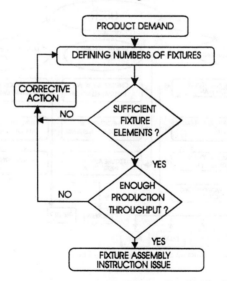

Figure 2 : Procedure for optimal fixture planning in FMS using modular fixtures

After appropriate number of fixtures, to process different part types, have been allocated the user must check that the required numbers of fixture elements can be made available. If sufficient numbers of the elements are not available steps can be taken to purchase additional numbers of elements or the fixtures be redesigned to reduce the demand for fixture elements that are critical (7).

It may also be possible that after the allocation of appropriate number of fixtures desired production throughput can not be achieved. This may be on account of excessively large job waiting time due to poor fixture design, bottleneck operations or inadequate distribution of fixtures amongst the job types to be processed.

3.2 System Design

The overall system design to meet the objectives of this study has been summarized in Fig. 3. The system has been organized at two levels viz., data processing and data base levels. Three modules of the system at the data processing level manipulate, compare and compute the necessary data. (Fig 4) Whereas a number of data files have been created at the data base level. Some of the important data files are; product data file, fixture assembly data file, processing data file, fixture element data file, work station data file, assembly instruction data file, information preparation data file, result data file etc.

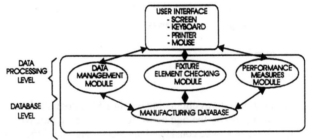

Figure 3 : Overall system design

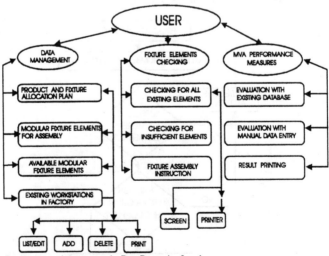

Figure 4 : Components and structure at the Data Processing Level

3.2.1 System Implementation

From the computation point of view, the system can be divided into two sections. The first section deals with data storage and manipulation. Fox-pro 2.0 has been used as data base software. The second section deals with mathematical computation and is provided with calculation capability and has memory allocation for the programming language. For this Turbo C version 2.0 has been employed. The system has been implemented on IBM-PC or compatible with microprocessor model 286 or upper.

3.3 Program Structure

The program has been broken down into 35 units. One of the units contains the main program where as, the remaining 34 unit are the subprograms. All the units have been developed with Fox-Pro 2.0 except one unit for computation which uses Turbo C-2.0. The program source code can be obtained from the authors whereas, the system details can be found in reference (6).

4. Case Study

Processing of a hypothetical product with three different versions, using FMS with 6 work stations and modular fixtures has been considered. The fixtures makes use of VENLIC block jig elements (3). Necessary data for processing of the jobs is provided in Table 1

Optimum allocation of fixtures is given in Table 2 whereas Table 3 gives the fixture element shortage list . Initial allocation of fixtures in Table 2 has been based on the average time (Tr) for part type r (without queue) (eqn 1) and the part mix ratio.

$$T_r = \sum_{m=1}^{M} \frac{t_{r,m} \cdot f_{r,m}}{L_m} \qquad (1)$$

where $t_{r,m}$ = Processing time of product r at the station m.

$f_{r,m}$ = Routing probability of r station m.

L_m = Number of facilities available at the station m.

Table 1: Part Processing Data

Product Name/No	Station No	Processing Time (min)	Routing Probability
Hook (EX11)	S1	2.0	2.0
(throughput desired 6/hr)	S2	1.5	4.0
	S3	5.0	1.0
	S4	4.0	1.0
	S6	1.5	1.0
Bracket (EX12)	S1	1.5	2.0
(desired throughput 6/hr))	S2	1.5	5.0
	S3	6.0	1.0
	S4	4.0	1.0
	S5	5.0	1.0
	S6	2.0	1.0
Housing (EX20)	S1	2.5	2.0
(desired throughput 6/hr)	S2	1.5	5.0
	S3	4.0	1.0
	S4	10.0	1.0
	S5	3.0	1.0
	S6	3.0	1.0

S1: Fixturing/defixturing (2 nos.) S2: Transporter (3 nos;AGVs)
S3: CNC milling (2 nos.) S4: Machining center (2 nos)
S5: CNC grinder (1 nos) S6: Inspection (1 nos)

Table 2: Allocation of fixtures

Iteration	Product Type	Allocated no of Fixtures	Production pcs/h from CAN-Q
1	EX.11	4	6.223
	EX.12	6	6.044
	EX.20	7	5.871*
2	EX.11	4	5.849*
	EX.12	7	6.419
	EX.20	8	6.114
3	EX.11	5	6.602
	EX.12	7	6.079
	EX.20	9	6.226

*throughput requirements not met

Table 3: Availability of fixture elements (corresponding to iteration 3)

Element Name	Nos. Available	Nos. Required	Net in Stock
Hook clamp	8	9	-1
Double tapered strap	8	9	-1
Support blocks	22	29	-2
Rack block	8	9	-1
Square adapter	8	9	-1

5. Conclusions

A system for checking the availability of modular fixture elements for use in FMS has been developed and implemented on an IBM Compatible PC. The software capabilities are :
- Storage and Management of appropriate manufacturing data.
- Manufacturing system performance evaluation based on closed queuing network modeling.
- Determining the number of fixture element for each work type to satisfy the production requirements and generating a list of fixture elements that be procured/modified.
- Providing fixture assembly instructions.

The present system uses relational data base structure this however, limits the system flexibility. Work is being undertaken to use object oriented data base structure to achieve flexibility in system design.

6. Reference

1. Darvishi, A. R. And Gill, K. F., *Expert System Rules for Fixture design*, IJPR, **28/10**, (1990), p.1901-1920.
2. Gandhi, M. V. And Thompson, B. S., *Phase Change Fixturing for Flexible Manufacturing Systems*, Jour. Manu. Systems, **4/1**, (1985), 29-38.
3. Imao Corporation, *VENLIC Bloc Jig System; Fixture Elements*, Catalog, **PRT-9010-10-600** Imao Corp., Japan.
4. Llewis, G. (1983), *Modular Fixturing Systems*, Proc. 2nd Int. Conf. On FMS, (1983), IFS (publications Ltd), London.
5. Luggen, W. W. (1991), *Flexible Manufacturing Cells and Systems*, Prentice-Hall Int. Inc. (1991), p.323-345.
6. Ngamvinijsakul Pairoj, *Planning for Modular Fixtures in Flexible Manufacturing Systems*, M.Engg. Thesis (1993), Industrial Engineering and Management, AIT, Bangkok.
7. Siong, L. B., Imato, T. Et.al., *Integrated Modular Fixture Design*, Pricing and Inventory Control Expert System, IJPR, **30/9**, p.2019-2044.

8. Thompson, B. S., Flexible Fixturing-*A Current Frontier in the Evolution of Flexible Manufacturing Cells*, **ASME Paper # 84-WA/Prod-16**, (1984).
9. Tuffentsammer, K. *Automated Loading of Machining System and Automatic Clamping of Workpieces*, Annals of CIRP, (1981), **30/2**, p.553-558.

Inventory

Ordering Alternatives in JIT Production Systems
   K. Takahashi and N. Nakamura

Manufacturing System Performance Evaluation with Stock Profiles
   M. Aldanondo and B. Archimede

An Integrative Model for Automatic Warehousing Systems
   E. Eben-Chaime and N. Pliskin

ORDERING ALTERNATIVES IN JIT PRODUCTION SYSTEMS

KATSUHIKO TAKAHASHI

*Dept. of Industrial & Systems Engineering, Hiroshima University,
4-1, Kagamiyama 1 chome, Higashi-Hiroshima, Hiroshima 739, Japan*
E-mail: takahasi@pel.sys.hiroshima-u.ac.jp

and

NOBUTO NAKAMURA

*Dept. of Industrial & Systems Engineering, Hiroshima University,
4-1, Kagamiyama 1 chome, Higashi-Hiroshima, Hiroshima 739, Japan*
E-mail: nakamura@pel.sys.hiroshima-u.ac.jp

ABSTRACT

As ordering alternatives in Just-in-Time (JIT) production systems, this paper deals with the pure Kanban system, the well-known Kanban system, and the concurrent ordering system. Mathematical models of the systems are developed, and the performances of the models are analyzed and compared with each other by means of simulation experiments.

1. Introduction

As a system of production planning and inventory control for multi-stage production inventory systems, Just-in-Time (JIT) production system has been developed and researched from various view points. In the JIT production systems, the Kanban system controls the order release for each production process and transportation process. The mechanism and performance of the system are investigated by a lot of researchers and from various view points. However, the system is only an ordering alternative in JIT production systems. Based on the difference of the information for order release, the ordering alternatives, that is, some variations of order release can be proposed.

After introduction, this paper describes the information for order release in JIT production systems. Also, the ordering alternatives based on the difference of the information utilized are pointed out. A mathematical model of the ordering system for each ordering alternative is developed, and the performances of the models are analyzed and compared with each other by means of simulation experiments. Finally, this paper summarizes the findings obtained in this paper as conclusions.

2. Information for Order Release and Ordering Alternatives

In JIT production systems, orders are released without utilizing the forecasted information. Only the actual results relating demand and supply are utilized in releasing orders. The information about the actual results relating demand and supply can be classified into three as follows;
 1. Demand arrival from the succeeding stage (or market),
 2. Parts supply from the preceding stage,
 3. Process (production or transportation) completion at the stage.

The difference of the information utilized in releasing orders leads to some variations of order release, that is, ordering alternatives.

Table 1: Information for order release in each ordering alternative

Information for Order Release	Pure Kanban	Well-known Kanban	Concurrent
Demand arrival	O	O	O
Parts arrival	O	O	
Process completion	O		

O : the information is considered in order release

The Kanban system, as an ordering system in JIT production systems, releases an order to a stage when the Kanban arrives from the succeeding stage (this means demand arrival from the succeeding stage), the parts are supplied from the preceding stage, and the orders previously released to the stage have been processed. That is, in the Kanban system, the orders will be released only when the three kinds of information shows that the process at the stage can be started at once. Also, at the time, the Kanban is removed from the part and the information of order is transmitted to the preceding stages.

Therefore, in the Kanban system, even if the demand is arrived at a certain stage, the start of process will be delayed as long as the work-in-process inventory runs out because of the delay of parts supply, or as long as the orders previously released are processed at the stage. Additionally, as the order release is sequentially transmitted to the preceding stages in the Kanban system, the delay at a certain stage means the delay at all the preceding stages.

The Kanban system described above differs from the Kanban system that has been studied by a lot of researchers [1][2][3][6][8][9], except for Spearman [7] (Spearman's model is equivalent to the Kanban system described above). We call the former the pure Kanban system and the latter the well-known Kanban system hereafter in this paper. These two kinds of Kanban systems can be regarded as ordering alternatives in JIT production systems.

In the well-known Kanban system, the orders are released when the demand arrives at the stage from the succeeding stage and the parts necessary to process the order are supplied from the preceding stage. The third information, that is, the process completion at the stage, is not considered in releasing orders in the well-known Kanban system. Therefore, the well-known Kanban system can be considered as a system that the first and the second information is utilized in releasing orders.

The other ordering alternative is the concurrent ordering system. The concurrent ordering system is developed by Izumi and Takahashi [2]. Also, it is modified by Takahashi et al. [8]. In the concurrent ordering system, the orders are released to all the stages concurrently when the product demand arrives at the production system. The parts supply from the preceding stages and the process completion at the stage are not considered. That is, the concurrent ordering system is a system that only the first information is considered in releasing orders.

Table 1 shows the information for order release and ordering alternatives.

3. Modeling JIT Production Systems

In this section, a mathematical model of the ordering system is developed for each ordering alternative in JIT production systems.

3.1. Assumptions

We define the following assumptions for the system considered in this paper.
1. The production system deals with a standard product that can be made to stock.
2. The product is produced through a serial production system with N stages. Each stage has a production process, and it is called 1st, 2nd, ..., or N-th stage accordingly as the process proceeds.
3. The production time at each production stage distributes stochastically.
4. The transportation process between the $(n-1)$st and the n-th production stages is called the n-th transportation stage.
5. Each production stage has two inventory points, the inventory points before and after production stage. The initial inventories at the inventory points before and after the n-th production stage, $S_B^{(n)}$ and $S_A^{(n)}$ are given and constant.
6. The backorder of product demand can be allowed.

3.2. Pure Kanban System

In the Kanban system, Kanbans are attached onto the buffer stocks at each inventory point. In the dual-card Kanban system, the Kanbans are classified into the production-ordering Kanban and the withdrawal Kanban. The withdrawal Kanbans are attached onto the buffer stocks at the inventory point before each production stage. When the part stocked before a production stage is consumed for production, the withdrawal Kanban is removed from the part and used to order the transportation of the part to the inventory point after the immediately preceding production stage. On the other hand, the production-ordering Kanbans are attached onto the buffer stocks at the inventory point after each production stage. When the product or part stocked after a production stage is consumed for demand or transportation, the production-ordering Kanban is removed from the product or part and used to order the production of the product or part to the production stage. By sequentially connecting the orders of the withdrawal Kanban and the production-ordering Kanban, the consumption of product becomes a trigger to produce at each production stage and to transport at each transportation stage.

By any of the Kanbans, the orders are released when the part is consumed. This means when the demand arrives from the succeeding stage, the parts for the order are supplied from the preceding stage, and the process for the order is available at the stage.

We define the times when the i-th order is released for the n-th production stage and the n-th transportation stage as $OP_i^{(n)}$ and $OT_i^{(n)}$, respectively. And, they can be formulated as follows;

$$OP_i^{(n)} = \max\{OT_i^{(n+1)}, P_{i-S_A^{(n)}}^{(n)}, T_{i-1}^{(n+1)}\}, \qquad (n = 1, 2, \ldots, N-1) \quad (1)$$

$$OP_i^{(N)} = \max\{D_i, P_{i-S_A^{(N)}}^{(N)}\}, \qquad (2)$$

$$OT_i^{(n)} = \max\{OP_i^{(n)}, T_{i-S_B^{(n)}}^{(n)}, P_{i-1}^{(n)}\}, \qquad (n = 1, 2, \ldots, N) \quad (3)$$

where D_i means the time when the i-th demand of product arrives. Let $P_i^{(n)}$ be the time when the i-th production is completed at the n-th production stage, and $T_i^{(n)}$ the time when the i-th transportation is completed at the n-th transportation stage. Then, they are formulated as follows;

$$P_i^{(n)} = \max\{OP_i^{(n)}, T_{i-S_B^{(n)}}^{(n)}, P_{i-1}^{(n)}\} + p_i^{(n)}, \qquad (n = 1, 2, \ldots, N) \quad (4)$$

$$T_i^{(1)} = \max\{OT_i^{(1)}, T_{i-1}^{(1)}\} + t_i^{(1)}, \qquad (5)$$

$$T_i^{(n)} = \max\{OT_i^{(n)}, P_{i-S_A^{(n-1)}}^{(n-1)}, T_{i-1}^{(n)}\} + t_i^{(n)}, \qquad (n = 2, 3, \ldots, N) \quad (6)$$

where $p_i^{(n)}$ means the i-th production time at the n-th production stage, and $t_i^{(n)}$ the i-th transportation time at the n-th transportation stage.

3.3. Well-known Kanban System

A lot of researchers have developed various models of the well-known Kanban system. In the models, the orders are released not on the basis of all the three kinds of information but on the basis of only two kinds of information.

Mitra and Mitrani [3] and Tayur [9] developed mathematical models of the well-known Kanban system. In their models, the times when the orders of production and transportation are released are formulated, and they are determined by the maximum of the time of the demand arrival from the succeeding stage and that of the parts supply from the preceding stage. The process completion at the stage does not considered in their models.

Huang et al. [1] and Sarker [6] developed the queueing network models of the well-known Kanban system by means of Q-GERT and SLAM II. In their queueing network models, the order at a certain stage is released at the time when the information flow of the Kanban from the succeeding stage and that of the part from the preceding stage are assembled, and the information of order release is transmitted to the preceding stage. That is, their queueing network models are equivalent to the mathematical models developed by Mitra and Mitrani [3] and Tayur [9], and only two kinds of information, that is, the demand arrival from the succeeding stage and the parts supply from the preceding stage are considered in releasing orders in their models.

By any approach to modeling the well-known Kanban system, the orders are released at the time when the demand from the succeeding stage arrives and the parts for the order are supplied from the preceding stage.

Therefore, in the well-known Kanban system, the times when the i-th order is released for the n-th production stage and the n-th transportation stage, $OP_i^{(n)}$ and $OT_i^{(n)}$ can be formulated as follows;

$$OP_i^{(n)} = \max\{OT_i^{(n+1)}, P_{i-S_A^{(n)}}^{(n)}\}, \qquad (n = 1, 2, \ldots, N-1) \quad (7)$$

$$OP_i^{(N)} = \max\{D_i, P_{i-S_A^{(N)}}^{(N)}\}, \qquad (8)$$

$$OT_i^{(n)} = \max\{OP_i^{(n)}, T_{i-S_B^{(n)}}^{(n)}\}. \qquad (n = 1, 2, \ldots, N) \quad (9)$$

Also, in the well-known Kanban system, $P_i^{(n)}$ and $T_i^{(n)}$ are formulated as the same equations shown in the pure Kanban system (Eq. (4)–(6)).

3.4. Concurrent Ordering System

As another ordering alternative in JIT production systems, the concurrent ordering system [8] can be pointed out. In the concurrent ordering system, an order is released immediately after the demand arrives from the preceding stage, and it means immediately after the product demand arrives at the production system. As a result, in the concurrent ordering system, only one kind of information, that is, the demand arrival from the succeeding stage is considered in releasing orders.

Therefore, in the concurrent ordering system, the times when the i-th order is released for the n-th production stage and the n-th transportation stage, $OP_i^{(n)}$ and $OT_i^{(n)}$ can be formulated as follows;

$$OP_i^{(n)} = D_i, \qquad (n = 1, 2, \ldots, N) \qquad (10)$$
$$OT_i^{(n)} = D_i. \qquad (n = 1, 2, \ldots, N) \qquad (11)$$

Also, in the concurrent ordering system, $P_i^{(n)}$ and $T_i^{(n)}$ are formulated as the same equations shown in the pure Kanban system (Eq. (4)–(6)).

4. Analyzing and Comparing the Performances

By means of simulation experiments, the performances of the ordering alternatives are analyzed and compared with each other. Also, the trade-off relationship between the performance measures is analyzed in this section.

4.1. Experimental Conditions

The experimental conditions in the simulation performed are as follows.
1. The inter-arrival time of product demand follows an exponential distribution with the parameter $\lambda=1$.
2. The number of stages, $N=5$.
3. The production time at each production stage follows an identical independent gamma distribution with mean μ_p and variance σ_p^2 [4].
4. The initial inventories are equivalent among the inventory points ($S_B^{(n)} = S_A^{(n)} = s$).
5. Simulation run-length is 100,000 including the warm-up run of the length 500.

4.2. Performance Measures

In this paper, the following performance measures are considered.
1. The mean waiting time of product demand :

$$wt = \lim_{I \to \infty} \left[\sum_{i=1}^{I} \max\{0, P_{i-S_A^{(N)}}^{(N)} - D_i\}/I \right] \qquad (12)$$

2. The mean of work-in-process (WIP) inventory :

$$wip_X^{(n)} = S_X^{(n)} + \lim_{\tau \to \infty} \left[\sum_i W_{X,i}^{(n)}/\tau \right] \qquad (X = A, B, n = 1, 2, \ldots, N) \qquad (13)$$

Where $W_{B,i}^{(n)}$ and $W_{A,i}^{(n)}$ are the waiting times of the i-th demand before the time τ at the inventory points before and after the n-th production stage, respectively. They are formulated as follows.

$$W_{B,i}^{(n)} = \min\{P_i^{(n)} - p_i^{(n)}, \tau\} - \min\{T_i^{(n)}, \tau\} \qquad (n = 1, 2, \ldots, N) \qquad (14)$$
$$W_{A,i}^{(n)} = \min\{T_i^{(n+1)} - t_i^{(n+1)}, \tau\} - \min\{P_i^{(n)}, \tau\} \qquad (n = 1, 2, \ldots, N-1) \qquad (15)$$
$$W_{A,i}^{(N)} = \min\{\max\{D_i, P_{i-S_A^{(N)}}^{(N)}\}, \tau\} - \min\{P_i^{(N)}, \tau\} \qquad (16)$$

4.3. SLAM II Network Models

Simulation experiments are available to estimate the performance of each ordering alternative. For the simulation experiments, a SLAM II [5] network model of pure Kanban system, which is equivalent to the mathematical model of the system (Eq. (1)-(6)), is developed as shown in Fig. 1. Also, a SLAM II network model of well-known Kanban system and that of concurrent ordering system have been developed by Takahashi et al. [8].

4.4. Trade-off between WIP inventory and Waiting Time

Generally, the waiting time of product demand decreases by increasing the WIP inventory, and we can easily understand that there is the trade-off relationship between both of the performance measures. Also, the trade-off relationship is different among the ordering alternatives. Therefore, in order to evaluate each ordering alternative, the trade-off relationship between the performance measures must be analyzed.

In analyzing the trade-off relationship, a total measure for the WIP inventories must be introduced. In this paper, the following total measure is considered;

$$TWIP = \sum_{n=1}^{N} \alpha^{(n)}(wip_B^{(n)} + wip_A^{(n)}), \qquad (17)$$

where $\alpha^{(n)}$ is the weighting parameter for the n-th stage.

By simulating the models developed above under various levels of the initial inventory and estimating the performance measures for each ordering alternative, Fig. 2 shows the trade-off relationship between the total measure of WIP inventory and the mean waiting time of product demand for each ordering alternative.

With Fig. 2, the followings can be pointed out.

1. The mean waiting time of the concurrent ordering system is shorter than those of the other systems under the condition that the total measure of WIP inventory is much less than a certain level (about 52 in the case of Fig. 2).
2. Under the other condition of WIP inventory, the mean waiting time of the pure Kanban system is shorter than those of the other systems.
3. As a result, it can be claimed that the pure Kanban system or the concurrent ordering system is an effective system among the systems considered in this paper.

5. Conclusions

In order to develop an effective ordering system, this paper considered the information for order release and three ordering alternatives, that is, the pure Kanban system, the well-known Kanban system, and the concurrent ordering system. With simulation experiments for the models developed in this paper, the performances of the ordering alternatives were analyzed and compared with each other. As a result, the followings were obtained in this paper.

1. Three kinds of information can be utilized in releasing orders in JIT production systems, and three ordering alternatives based on the difference of the information utilized are pointed out.

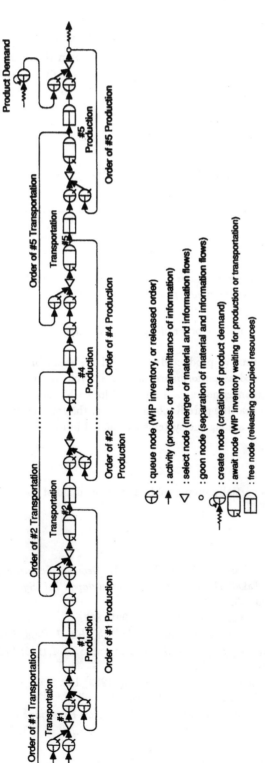

Figure 1: A SLAM II network model of pure Kanban system

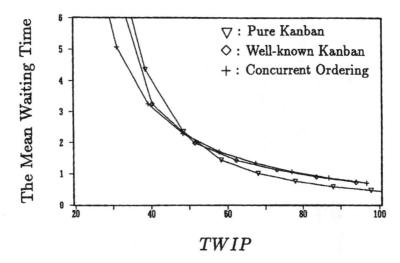

Figure 2: The trade-off relationship between the total measure of WIP inventory and the mean waiting time of product demand ($\mu_p=0.8$, $\sigma_p^2=0.8$, $\alpha^{(n)}=1$, $n=1, 2, \ldots, 5$).

2. A mathematical model of the ordering system is developed for each ordering alternatives.
3. The mean waiting time of product demand and the mean of work-in-process inventory are considered as performance measures. And, the trade-off relationship between the performance measures is analyzed for each ordering alternative. As a result, it is clarified that the pure Kanban system or the concurrent ordering system is an effective system among the systems considered in this paper.

Acknowledgement

The work in this paper is partially supported by the grant-in-aid of the Japanese Ministry of Education in 1994–1995.

References

1. P. Y. Huang, L. P. Rees, and B. W. Taylor III, *Decision Sciences*, **14** (1983) 326.
2. M. Izumi and K. Takahashi, in *Proc. 2nd China-Japan Int. Sympo. Indust. Manage.*, Beijing (1993) 53.
3. D. Mitra and I. Mitrani, *Oper. Res.*, **39** (1991) 807.
4. K. Muralidhar, S. R. Swanseth, and R. L. Wilson, *Int. J. Prod. Sys.*, **30** (1992) 1.
5. A. A. B. Pritsker, *Introduction to Simulation and SLAM II, 3rd ed.* (John Wiley & Sons, New York, 1986).
6. B. R. Sarker, *Computers & Ind. Engng.*, **16** (1989) 127.
7. M. L. Spearman, *Oper. Res.*, **40** (1992) 948.
8. K. Takahashi, N. Nakamura, and K. Ohashi, in *New Directions in Simulations for Manufacturing and Communications*, eds. Morito, S., et al. (ORSJ, Tokyo, 1994) 105.
9. S. R. Tayur, *Manage. Sci.*, **39** (1993) 1347.

MANUFACTURING SYSTEM PERFORMANCE EVALUATION WITH STOCK PROFILES

M. ALDANONDO B. ARCHIMÈDE

Ecole Nationale d'Ingénieurs de Tarbes
Laboratoire Génie de Production
Avenue d'Azereix BP 16- 65016 Tarbes Cedex FRANCE
Phone : (33) 62 44 27 21. Fax : (33) 62 44 27 27

ABSTRACT

Indicators definition is essential for manufacturing system performance evaluation. When the relevant decision system is hierarchical, so must be indicators. We propose in this communication an indicator : the stock profile, which fits multi level decision making for time relevant aspects.

Keywords : manufacturing decision system, information system, manufacturing database, raw manufacturing data, indicators

1 - Introduction

We are concerned with manufacturing system where production is launched with small batches according to a known demand. In this case decision is generally hierarchised as explained in the surveys of Jones[1] and Doumeingts [2]. Relevant hierarchical information system problems have been less studied; Scheer[3], Winter[4], Aldanondo[5] have provided some elements for information architecture modelisation. Our communication deals with indicators or how information collected at the physical level can be used for hierarchical decision making.

An industrial study of indicator utilization proposed by Gelders & all [6] shows the importance of indicators and identifies their relevant domains: time related aspects, quality and cost. It is explained by Aldanondo & all [7] how these three aspects can fit a database design for data collected at the physical level while keeping a multi level decision making possibilities. Our purpose is to show that the stock profile : single level indicator presented by Archimède & all [8], can be extended to multi-level decision system for some time relevant aspects.

Our communication is therefore divided in three parts. A global approach of the main indicator parameters is firstly described to locate our goal accurately. Then, stock profiles are defined and potential uses explained. Finally a model of stock profiles able to deal with multi-level decision making is proposed.

2 - Elements and global approach for indicators definition

The aim of this part is to present the main aspects of indicators and raw manufacturing data management. Raw manufacturing data correspond to the information that can be collected from the activities effectively completed in the shop floor. This information is then more or less aggregated and proved to the different decision centers.

When dealing with hierarchical decision system and in order to propose to each decision activity a correct set of data, the management of manufacturing data must take into account three aspects :
- the decision level of the activity,
- the nature of information which will be required for decision making,
- the basic entities which support the information.

These three aspects are presented and a resulting conceptual frame is defined allowing us to situate our approach. More details are presented in aldanondo & all (7).

2.1 - Decision level and information organization

Raw manufacturing data is collected at the physical system level of the manufacturing workshop. Therefore they feed :
- the lower level of the decision system, with an instantaneous image of the workshop behavior, in order to control in real time the progress of the work and to determine which activities have to be launched,
- upper decision levels, with workshop behavior information on longer time horizons, these global views are generally obtained through the construction or updating of different indicators.

Two database configurations can be designed in order to support these two aspects. The first one assumes that each decision level has its own database while the second one is based on an unique ground level raw manufacturing database on which rely decision making of any decision level. lOur presentation is single data base oriented.

2.2 - Information nature

Main production objectives deal with time relevant aspects, quality and cost. So three types of data can be identified.

Time relevant information corresponds to events related to beginning and ending dates of production activities. From these events it is always possible to get the duration of a state or activity. They may concern for example, at a lower decision level : an operation effective duration, a set-up state occupancy ratio for a machine; and for upper levels : a working state occupancy ratio for a manufacturing cell, a manufacturing lead time for a customer order or a stock level.

Quality information can characterized both product and process and can be set for different abstraction level. Product quality data may characterize for example at a lower level a scrap quantity for an operation or part measurements, for upper levels a scrap ratio correlated with a raw material supplier. Process quality data is more and more used especially for quality certification purpose (ISO...) and can correspond with : SPC control charts or resource capability.

Cost indicators result from a mix between the costs of material and resource and data of the two above natures. For instance, the cost of a part (or batch, or customer order) may be calculated from the cost of material and resource combined respectively with the total used material (including scrap) for this part, (quality data), and the duration of the resource utilization, (time relevant data).

2.3 - Information supporting entities

Manufacturing data are supported by two kind of entities : the product and the process. Data linked with the product according to the abstraction levels can characterize physical part, manufacturing order and customer order.... The three natures : time aspects (beginning and ending dates...), quality (amount of scrap...) and cost (raw material prices...) show the indicators combination possibilities. Process data, corresponding with physical resource, as machine, cell, workshop... shows the abstraction level concern. The resource influences on time aspect, quality and cost are easily understandable.

2.4 - Conceptual frame and proposed approach situation

As shown in figure 1, indicators deal with three basic aspects. On this schema, the stock profile indicator can be located as a multi-level indicator, dealing mainly with products which provides information related to time aspects (cost can be indirectly taken into account).

In order to avoid detailed explanations we assume that the database supporting the raw manufacturing information is able to provide us detailed operation information as : beginning and ending dates, quantity of parts and part references.

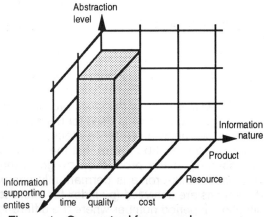

Figure 1 - Conceptual frame and approach location

3 - Stock-profiles
3.1 - Basic element

The stock profile is a diagram which represents the inventory level between a destination resource B and an origin resource A. Two kinds of stock profile can be defined :
- the partial stock profile which characterizes a single reference, with two attributes time and load,
- the global stock profile aggregates the partial stock profiles, with three attributes time, load and a color corresponding to the reference mix, therefore the global load is the sum of the partial load. In the following developments the term stock profile will be used for global stock profile.

These two kinds of profile are represented in Figure 2 where it is assumed that only ref1 and ref2 are present between the two resources. The load comes from resource A. The load-level indicates the amount of work that the resource B is going to undertake. The global stock profile color represent the reference mix on which the resource B will operate.

With respect to the indicator problematic, the aim of the stock-profiles is to reflect various transfer situations between resources. For instance :

(a) At t1, end of some processing of resource A
(b) Between t1 and t2, the intermediate stock remains constant
(d) At t3, resource B starts to work,

Note that the orientation of a stock-profile is important. A->B ≠ B->A .

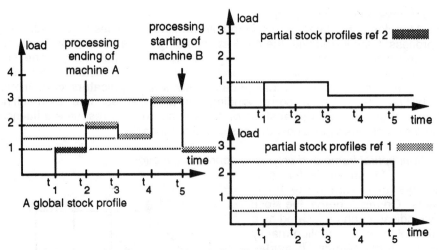
Figure 2 - Stock-Profile diagrams

3.2 - Basic utilization of the stock-profile between two resources
The stock-profile is used for behavior understanding of resources A and B. For this, we need a reference stock-profile which reflects the given production program. We must be aware that the theoretical performances of the resources, as assumed in the production program may be different in the reality. This reference stock-profile is permanently compared with the effective profile and differences are analyzed (p : index for reference production program, r: real actual situation, i: period number where the difference is observed).

The various types of difference causes are identified by comparing the reference stock profile and the effective stock profile (Figure 3). For this analysis, we consider the stock profile as a suite of period L_i where the load level is constant. The difference between the two diagrams may occur with the length L of the waiting time (period during the stock remains constant) and with the load level N.

if	if		The perturbation-causes may be :
$L_p^i < L_r^i$	$N_p^{i+1} > N_p^i$	⌐‾‾ p ⌐...r	-resource failure of the uphill unit A -under-evaluation of the processing time of A
$L_p^i < L_r^i$	$N_p^{i+1} < N_p^i$	‾‾⌐...r ⌐_ p	-resource failure of the downstream unit B -procurement delay at B because of conveying problems
$L_p^i > L_r^i$	$N_r^{i+1} > N_r^i$	⌐...r ⌐_ p	-over-evaluation of the processing time at A
$L_p^i > L_r^i$	$N_r^{i+1} < N_r^i$	‾‾⌐ p ⌐...r	- the transfer between A and B is too fast

Figure 3. Comparison of load diagrams

3.3 - Elements of stock-profile for a workshop

When dealing with a workshop regrouping single resources it is interesting to define some aggregated stock profiles. For each resource, it can be defined :
- Ress-Prod : generated by one resource, supplying all the resources,
- Ress-Cons : consumed by one resource, provided by all the resources,
- n : the number of the resources of the workshop.

The figure 4 shows relations between various stock-Profiles between the resource-situations :
(a) n*n detailed profiles for n resources, between to single resources,
(b1) n Ress-prod profiles, one for each resource, between each resource and the totality of the n resources
(b2) n Ress-cons profiles, one for each resource, between each resource and the totality of the n resources
(c) workshop profile of the workshop which results of the aggregations of all the Ress-prod or Ress-cons profiles. These aggregations provide the same result.

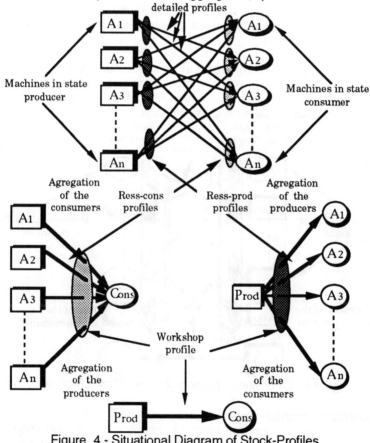

Figure 4 - Situational Diagram of Stock-Profiles

3.4 - Workshop stock-profile utilization

As the total number of detailed profiles is n*n, the different aggregated profiles allows to reduce the number of profile analysis in order to understand the behavior of the facility. The analysis is divided in three steps :

Step 1 - Evaluation of the tendency. The "tendency" information is provided by the analysis of the workshop profile. An uphill tendency (some resources produce a load too great or too small) indicates that the perturbation is on the "prod" side while a downstream tendency (some resources consume a load too great or to small) indicates that the perturbation is on the "cons" side.

Step 2 - Determination of the "first resource". At this step, we compare the reference and the perturbed Ress-cons and Ress-prod profiles. When the tendency is "PROD", we identify the perturbed producer as the first resource. When the tendency is "CONS" we identify the perturbed consumer as the first resource. This need n profile analysis.

Step 3 - Determination of the "second resource". Now, we compare the detailed stock-profiles. Note that, from step 2, only one producer or consumer is identified. Therefore, at step 3, depending on whether this is a producer or a consumer, we only compare one set of n corresponding consumers or n corresponding producers.

The search of the critical point requires therefore : 1 comparison at step 1, n comparisons at step2 and step3. We therefore reduce a problem of complexity $O(n^2)$ to a complexity of $O(2n+1)$. Once the faulty stock profile is found, it is possible to analyze the partial stock profiles in order to find the reference concerned by the problem.

4 - Stock-profiles in a hierarchical manufacturing system

We are now going to show, how stock profile can be used with hierarchical system. Therefore a model which fits hierarchical system is defined firstly, then the utilization of stock profile for analysis is done.

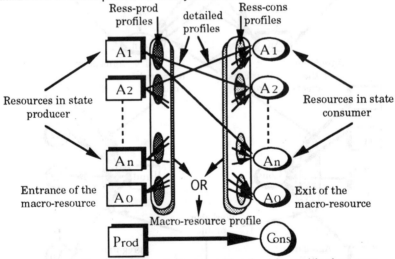

Figure 5 - Definition of stock profiles for hierarchical system

4.1 - Model for hierarchical system

Stock profile has been presented in chapter 3 on a workshop containing single resources. The problem is to define a model valid for manufacturing systems where the resources are aggregated according to the location criterion, this can provide for example : machine/cell/workshop/factory system architecture. Our goal is to have a vertical consistency of the different profiles which allows various analysis relevant to each level. The model of figure 5 can be used for any abstraction level. The differences with the basic model are :
- the definitions are given for a macro-resource containing resources, which can correspond according to the level with cell/machine, workshop/cell or factory/cell,

- the definition of an entrance and an exit of the macro-resource which are considered like machine with null duration,
- detailed profiles, Ress-prod and Ress-cons profiles are the same. Note that entrance and exit have respectively only a Ress-prod and Ress-cons profile.

4.2 - Stock profile utilization with a hierarchical system

For this section we explain how the analysis at a high level of abstraction (level 2) can be refined at an immediate lower level (level 1). This analysis which can be performed recursively with any number of levels is as follow :
Step 1 - Level 2 Macro-resource profile analysis is explained in the next table.

level 2 Macro-resource profile	Analysis to be performed	Results	Problem identification or further analysis requirements
uphill tendancy ⌐‾‾ p --- r	For each level 2 resource Ress-prod analysis	No tendancy found	Entrance delay
		Uphill tedancy found	Level 2 resource or level 1 Macro-resource generation delay - CASE 1
uphill tendancy ,--- r ⌐__ p	For each level 2 resource Ress-prod analysis	No tendancy found	Unpredicted or earlier entrance
		Uphill tedancy found	Level 2 resource or level 1 Macro-resource erlier generation - CASE 2
downhill tendancy ‾‾⌐ --- r ⌐__ p	For each level 2 resource Ress-cons analysis	No tendancy found	Exit delay
		Downhill tedancy found	Level 2 resource or level 1 Macro-resource consumtion delay - CASE 3
downhill tendancy ‾‾⌐ p ,--- r	For each level 2 resource Ress-cons analysis	No tendancy found	Unpredicted or earlier exit
		Downhill tedancy found	Level 2 resource or level 1 Macro-resource earlier consumtion - CASE 4

Thus the level 2 resource or level 1 Macro-resource generating a problem is identified or a load movement delay at the entrance or exit of the level 2 macro-resource can be understood. In the four cases (1,2,3 and 4) it is always possible with the analysis of detailed level2 profiles to find the "second resource" as explained in section 3.4.

Step 2 - Going from level 2 to level 1 profiles.

Therefore the analysis of the level 1 suspected Macro-resource suspected must be undertaken. Going from level 2 to level 1 logically reverse the profiles (a level 2 uphill or downhill tendancy corresponds respectively with a level 1 down hill and uphill tendancy) as shown in the following table.

	CASE 1	CASE 2	CASE 3	CASE 4
level 2 resource profile	uphill tendancy ⌐‾‾ p --- r	uphill tendancy ,--- r ⌐__ p	downhill tendancy ‾‾⌐ --- r ⌐__ p	downhill tendancy ‾‾⌐ p ,--- r
level 1 Macro-resource profile	downhill tendancy ‾‾⌐ --- r ⌐__ p	downhill tendancy ‾‾⌐ p ,--- r	uphill tendancy ⌐‾‾ p --- r	uphill tendancy ,--- r ⌐__ p

This operation allows to have the level 1 Macro-resource profile, and the analysis of this Macro-resource is the same as the one described in step 1 for level 2.

5 - Conclusions

Using the previous method, it is therefore possible to analyze the different stock-profiles of manufacturing system with a large number of abstraction levels with consistence.

Generally when abstraction is higher the information update gets less frequent. Typical updating period can be estimated for example : cell/day, workshop/week and factory/month. This allows to have different level of perturbation reactivity in control, and therefore to allow autonomy to each control level. For example a machine generation delay on Wednesday can be resume on Thursday, the cell level will analyze that problem while the workshop level updated on Friday will not know this minor problem. When a level detects something wrong, the origin of the perturbation can be very simply identified and normally the immediate lower level can tell the reason of such problem.

These various profiles can be directly built from the raw manufacturing database with the beginning and ending dates of the operations collected on the physical system. Furthermore, their calculations can be entirely independent. These stock profiles can be linked with cost of materials if the color attribute representing the reference type is taken into account. In this case cost indicator for the work in progress can be built for any macro-resource of any level. These different analysis have assumed a single perturbation, some work need to be done in the case of multi-perturbation occurrence.

References

1. Jones A Saleh A., 1990 : "A multi level/multi layer architecture for intelligent shop floor control", *International Journal of CIM,* (Taylor & Francis), **vol 3 n°1**.
2. Doumeingts G., 1990 : "Design and Specification Method for Production Systems", *CIM : integration aspects*, (Ed Teknea)
3. Scheer A. 1989 : *Enterprise Wide Data Modeling*, (Springer & Verlag).
4. Winter R.E.,1991 : "On the utilization of an active, integrated database for the vertical integration of production planning and control", *IFIP CAPE 91* (North-Holland).
5. Aldanondo M. Mercé C. 1991 : "Hierarchical data model for scheduling and monitoring in manufacturing systems", *IFIP CAPE 91,* (North-Holland).
6. Gelders.L Mannaerts.P Maes.J, 1994 : "Manufacturing strategy, performance indicators and improvement programs", *International Journal of Production Research*, **Vol 32 n°4**, (Taylor & Francis).
7. Aldanondo M, Breuil D. 1993 : "Raw manufacturing data for hierarchical decision system. Identification and potential uses", *Journal of Decision Systems*, **Vol 2 n°3-4**, (Ed Hermes).
8. Archimède B,Pun L,Bérard C, Doumeingts G,1993 : "Flow-Profiles and Potential Graphs based FMS dynamical control", *Control Eng. Practice*, **vol 1 n°1**.

AN INTEGRATIVE MODEL FOR AUTOMATIC WAREHOUSING SYSTEMS

Moshe Eben-Chaime

and

Nava Pliskin
Department of Industrial Engineering and Management
Ben-Gurion University of the Negev
P.O. Box 653, 84105 Be'er Sheva, Israel
Phone: 972-7-472206/3   Fax: 972-7-280776
E-mail: even@bgumail.bgu.ac.il

ABSTRACT

Automatic warehouses, for the most part, have been studied in isolation. This paper departs from this approach by proposing an integrative model in which warehouse activities interact with other functions of the total system. The model has been simulated under three modes of operation: single command, dual command (DC), and hybrid. Simulation results confirm that the length of stay of unit loads outside the warehouse affects performance and also suggest that, under the DC mode, the warehouse may lose stability. Hence, the hybrid mode is proposed as an alternative. The hybrid mode outperforms the DC mode on most performance measures, except that the length of the storage queue at high throughput levels is longer. Finally, saving opportunities, via a reduction in the number of storage/retrieval machines, are discussed.

1. Introduction

An automatic warehousing (AW) system consists of racks, storage/retrieval (S/R) machines, input/output (I/O) station(s), and computerized control devices. The racks are paired back to back with aisles between the pairs. The S/R machines are cranes that travel in the aisles and move objects between the racks and the I/O stations. Each S/R machine can move horizontally and vertically at the same time and can access the front (pick) face of the racks on both sides of the aisle. An S/R machine is either dedicated to a single aisle, or can move between aisles. The cranes operate under the control of a computerized system in one of two modes of operation. These operation modes are either based on a single command (SC) cycle, during which a single storage or retrieval operation is performed, or on a dual command (DC) cycle, during which both a storage operation and a retrieval operation are performed between two consecutive visits to I/O stations.

Operations management is concerned with the sequencing of storage and retrieval requests, and the matching of both types of requests in DC cycles. Bozer and White (1984) proposed the "standardization and approximation" approach, while assuming randomized storage and FIFO sequencing. They offered to standardize (or normalize) the pick face of the racks and to use continuous approximations for developing general expressions of the expected travel times for single and dual command cycles in AS/RS. This work inspired the investigation of alternative sequencing policies by Han et al. (1987), who developed the efficient "nearest neighbor" (NN) heuristic rule.

Expected travel time was the only concern in these studies, which applied the NN rule to block sequencing. The travel time is the major component of the service time in AW systems. The other components are the pick and deposit (P/D) times which are constant and independent of the operation mode, sequencing rule, and the like. Travel time was considered because often, in queuing systems, the mean service time is highly correlated with the mean response time which measures the service level of the system. The response time is the time elapsed since a request for service is issued until it is completed, including both the waiting time in the queue and service time. Eben-Chaime (1992) showed that block sequencing can be hazardous in terms of response times and queue lengths. As an alternative he proposed a dynamic application of the NN rule for dispatching, an alternative which was shown to maintain response times and queue lengths at the levels of FIFO sequencing, while significantly reducing crane travel times.

Previous operations management studies considered the storage function in isolation, ignoring its relationships with, and dependencies on, the other functions of the total system. While this might be acceptable for the distribution center studied by Seidmann (1988), the isolation premise does not hold for the more general case, where storage sub-systems interact with various other subsystems such as manufacturing. Some studies assumed FIFO service and did not account for the economic benefits which may result from improved operation management policies, such as fewer cranes and a smaller storage area. Recently, it was noted that "what makes AS/RS most justifiable is when you integrate it into an entire manufacturing system scheme" (1993).

This paper presents an integrative systematic model that considers the relationships between storage and other functions of the total system. The proposed model, in Section 2, provides a conceptual framework for investigating, by simulation, the operational characteristics of the system under different circumstances. The simulation environment is also described in Section 2. Queue behavior and the effects of sequencing rules are studied in the third section while Section 4 is devoted to a comparison of various operations modes. It is noteworthy that although the paper focuses on operations management, the model also lends itself to the study of storage policies and assignment and system design.

2. The Model and Simulation Environment

Many automated warehouses (AWs) function on the basis of unit loads (UL) such as bins, drawers, or pallets, which are used to store items, rather than on the basis of single items. When performing a retrieval operation, an S/R machine pulls a UL from a rack and delivers it to an I/O station. When performing a storage operation, a UL is taken from an I/O station and is deposited in an empty rack slot. A retrieval (storage) of a UL does not necessarily imply that items are retrieved (stored). A UL, for instance, may be retrieved in order to add to its contents, i.e., store item(s). Similarly, ULs are often re-stored after some items are removed, i.e., retrieved from them. Hence, storage requests are not generated independently but as a result of prior retrievals. Further, warehouse activities are triggered by other system functions that generate UL retrieval requests. These observations imply that there are two separate queue types, one for storage and one for retrieval, and more than a single queue of each type may exist. There are two principle differences between the two queue types. First, the queues of retrieval requests are lists stored in the computer memory while ULs are physically waiting to be stored in the storage queues. Second, an exact location of a UL is specified for each retrieval, while under the assumption of randomized storage, ULs can be stored in any empty slot in the warehouse. The addresses of the empty slots are stored in another list in the computer memory.

The integrative model for AWs proposed below consists of the system, the racks, the S/R machines, the storage and retrieval queues, and the list of empty slot addresses. The operation cycles are illustrated graphically in Figure 1 where the cause-and-effect relationships between both service types are clarified by the directed arrows. The length of time a UL spends out of the warehouse is determined by activities performed on the items in the UL by other system functions. Since this time segment may influence the warehouse operation, it must be incorporated into the analysis.

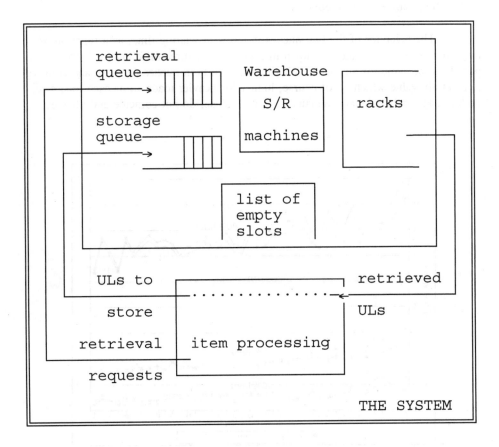

Figure 1: The Integrative Model for Automatic Warehouses

The integrative model has been simulated on an 80486 platform to study an AS/RS consisting of a pair of racks, an S/R machine, and a single I/O station. The simulation was programmed using both Turbo-Pascal 6.0™ and Paradox™ in a complementary manner. For the sake of generality, the continuous approximation and rack standardization of (Bozer and White, 1984) is adopted in the simulation. This approach allows analysis in terms of the "standard time unit", T, and the "shape factor", b, of the rack, disregarding physical attributes, size and structure of the rack and speeds of the S/R machines. Our simulation, of a pair of squared in time racks and an S/R machine dedicated to serve them, assumes randomized storage and a single I/O station at the lower-left corner of the racks. The inputs to the simulator include the following: 1) the distribution of the inter-arrival times of retrieval requests; 2) the distribution of the length of stay of ULs outside the warehouse; 3) the standard time unit T (squared in time racks, b=1, were assumed); 4) the number of empty slots; 5) the P/D times; 6) the length of the simulation in terms of completed service cycles; and 7) the mode of operation.

3. The Behavior of the Queues

The behavior of the two queues in terms of waiting time, as a function of mean item-processing time, is shown in Figure 2, for three different load sizes. The pattern is the same for all loads. Waiting times in both queues decrease and converge to a certain value which is, of course, higher for heavier loads. For each load size, both storage and retrieval time curves, though close, do not coincide and intersect once.

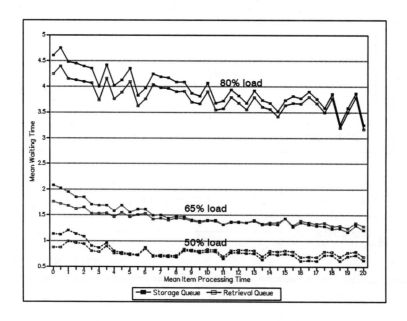

Figure 2: Queue Performance - Sc Mode

The influence of the sequencing rule on warehouse performance is studied via a comparison between FIFO and SPT. The simulation confirms that throughput rates are equal under both rules for the same load size. Figure 3 displays the effect of the load size on waiting time in the storage queue and the retrieval queue, the retrieval queue is hardly affected by the sequencing rule, in contrast to the remarkable effect on the storage queue.

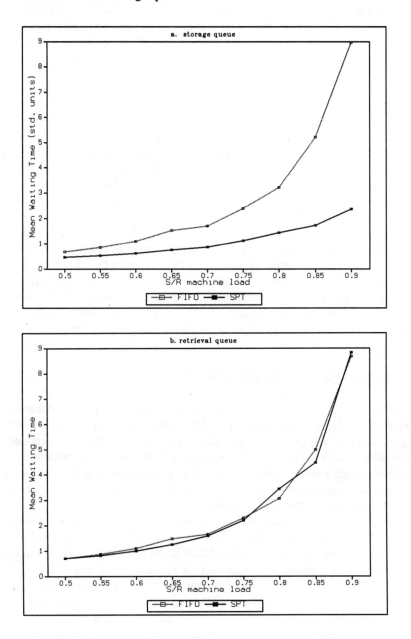

Figure 3: Sequencing Rules Comparison - SC Mode

Next, DC cycles are performed. The use of DC cycles increases the dependency between both service types. The patterns for FIFO sequencing are presented in Figure 4. Similar patterns are observed under NN dispatching except that, as expected, the means of the waiting time are shorter under the NN rule. The length of stay outside the warehouse, the item processing time, seems to have a vital influence on both queues, as can be observed in parts (a) of Figure 4. For longer mean processing time the waiting time in the storage queue decreases in parallel to a growth of the waiting times in the retrieval queue. The load, controlled primarily by the arrival rate, has a similar effect, part (b) of Figure 4. It is noteworthy and disturbing that for heavy loads and long processing time (e.g., 80% load and 8 time units) the system collapses and the simulation is aborted. The trends are switched with respect to the number of empty slots, as shown in Figure 4(c). In order to apply the NN rule, there should be a number of empty slots in the racks. This number remains fixed since, during each cycle, one slot is filled up by storing a UL in it, while another slot is emptied when its UL is retrieved. The larger the number of empty slots, the larger is the cycle time reduction and the consequent throughput increase obtained under the NN rule. Clearly, the cycle time is reduced at the expense of much longer waiting time in the storage queue. Longer waiting time implies longer queues, creating a need for a larger space to hold the actual ULs that are waiting in the storage queue.

The advantage of the DC mode over the SC mode is the resultant reduction in travel time per transaction, in the order of 40%! However, the sensitivity of warehouse performance (in terms of waiting time) under pure DC operation mode and the resulting system collapse motivated us to search for a better alternative. A hybrid mode of operation is proposed in an attempt to maintain stability while taking advantage of the short travel times under the DC mode. Performance levels under the proposed hybrid mode of operation are analyzed and compared with the DC mode, in the next section.

4. Comparative Analysis of the Hybrid Mode

Under the hybrid mode, DC cycles are performed whenever possible. Otherwise, SC cycles are performed. The S/R machine halts only when both queues are empty. Both the FIFO and the NN rules can be used to sequence operations. The motivation to this mode is to achieve stability in the warehouse queuing system. The average waiting time in both retrieval and storage queues is plotted in Figure 5 against mean item processing time, for load size of 50%. Clearly, the performance under the hybrid mode is more stable than under the DC mode.

Performance levels under the hybrid mode and the DC mode are compared in Table 1 as a function of the mean inter-arrival time between retrieval requests. The mean inter-arrival time of retrieval requests is the reciprocal of the average number of retrievals completed per time unit, or half the throughput of the AS/RS studied here. The other half consists of storage operations, since each retrieval is eventually accompanied with a storage operation. Apparently, utilization of the S/R machine is superior under the hybrid mode. For most activity levels, the consequence is a reduction in the average length of the storage queue as well as in the maximum number of ULs waiting. Only for the shortest inter-arrival time, the

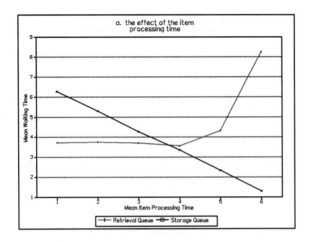
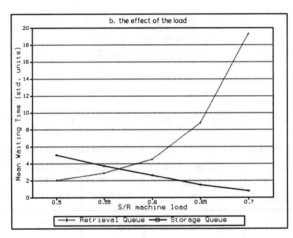
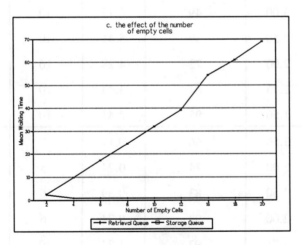

Figure 4: Queue Performance - DC Mode

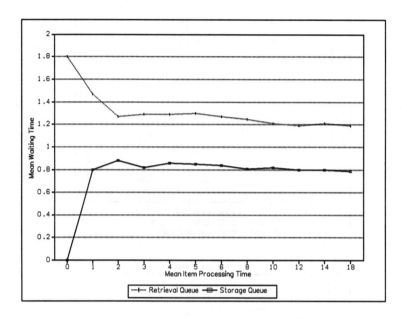

Figure 5: Queue Performance - Hybrid Mode

Table 1: Performance Comparison of the DC and Hybrid Modes

Operation mode	Inter-arrival time	S/R machine utilization %	Storage queue length		Retrieval response time
			average	maximum	
DC	3.50	42	6.72	10	2.04
	3.00	49	6.17	10	2.24
	2.50	59	5.41	10	2.58
	2.00	73	4.28	10	3.55
	1.75	84	3.43	10	5.32
	1.50	95	2.41	10	19.30
Hybrid	3.50	66	0.17	5	1.96
	3.00	74	0.24	5	2.13
	2.50	83	0.39	6	2.50
	2.00	92	0.74	6	3.58
	1.75	97	1.24	10	5.09
	1.50	99	3.01	15	13.00

highest throughput, the storage queue under the hybrid mode (bottom row of Table 1) is longer than under the DC mode (sixth row of the Table 1). Note, however, the remarkable difference between the respective means of the response time and the gap between these two means and the other means. At very high activity levels, evidently, the performance levels are very sensitive and any improvement in one dimension comes at the expense of other dimensions. Response time impacts the operation of the system to which the warehouse provides services. Since service is the main function of the warehouse, service level, and hence response time, should be a major factor in evaluating warehouse performance.

5. Conclusions

"By all means, buffer is the operative word when AS/RS in manufacturing is being discussed. ... In this case the AS/RS is not a storage device; rather it is a computer-controlled production machine" (Knill, Schwind and Witt, 1993). Based on this distinction, the integrative model presented here considers the AW as a function that provides UL storage services to other functions of the total system which process UL items. A computer program that was developed to simulate the integrative model, has been used to study the operational implications of these distinctions. Indeed, the results shed light on the performance of automatic warehousing systems. Activities of other functions, represented in the simulation by item processing delays, affect the activity and the performance of the warehouse. As the simulation demonstrates, the effects of these and other factors, such as sequencing rules and the number of empty slots, are interrelated.

The result of the most practical significance is the sensitivity to heavy loads under the DC mode of operation, which might even cause a failure of the warehouse. The hybrid operation mode is proposed as an alternative, under which the utilization of the S/R machine is increased. Consequently this mode outperforms the DC mode at most activity levels. Only the storage queues under the hybrid mode are longer and only when very high throughput is required. Even then, stability of the system and reduced response time, which are obtained under the hybrid mode, outweigh this growth in storage queue lengths.

6. References

Y. A. Bozer and J.A. White: Travel Time Models for Automated Storage and Retrieval Systems, *IIE Transactions* **16** (1984) 329-338.
M. Eben-Chaime: Operations Sequencing in Automated Warehousing Systems, *Int. J. Production Research* **30** (1992) 2401-2409.
M.H. Han, L.F. McGinnis, J.S. Shieh and J.A. White: On Sequencing Retrievals in Automated Storage/Retrieval Systems, *IIE Transactions* **19** (1987) 56-66.
B. Knill, G. Schwind and C. Witt: AS/RS: Full Partner in Manufacturing, *Material Handling Eng.* **48** (1993) 43-52.
A. Seidmann: Intelligent Control Schemes for Automated Storage and Retrieval Systems, *Int. J. Production Research* **26** (1988) 931-952.

Tools & Fixture Planning

Planning and Control of Tool and Cutter Grinding in Manufacturing Systems
    G. Petuelli and U. Muller

A Computerized Tool Planning and Scheduling System
    K. N. Krishnaswamy, B. G. Raghavendra and D. Sampath

Application of Neural Network in Tool Selection Problem
    P. C. Pandey and S. Pal

PLANNING AND CONTROL OF TOOL AND CUTTER GRINDING IN MANUFACTURING SYSTEMS

PROF. DR.-ING. G. PETUELLI
Universität GH Paderborn Abt. Soest
FB 12 Maschinenbau - Automatisierungstechnik
Steingraben 21
59494 Soest

and

DIPL.- ING. U. MÜLLER
Gerhard-Mercator-Universität GH Duisburg
FB 7 Werkzeugmaschinen
Lotharstr. 65
47048 Duisburg

ABSTRACT

The Event-Oriented Tool-Management was created to reduce the downtime of the machines which is caused by problems in their tooling. This software tool is supposed to enable the production planning department to take into account and to optimize any activity necessary to supply the machines with the correct tools at the point and time they are needed. The system enables the planning and control of the reconditioning and resharpening of the metal cutting tools and the appropriate tooling of the machine. By aid of this system the productivity of the machines will be increased and the tooling costs will be reduced.

1 Introduction

Most of the efforts to increase the productivity of metal cutting machines in production engineering are concentrated onto the improvement of the machining technology, the automation of the processes, the optimisation of the production schedule as well as the sequence of the workpieces to be machined.

Regarding the tools needed to enable the machining it can be found that there are lists generated indicating the appropriate tools, but there is no information available, e.g. about the status of the tools and when they will be used next. Consequently quite often the machining must be stopped due to the improper tooling of the machines. Different evaluations showed that about 53% of all interruptions of metal cutting processes are caused by the poor tooling mainly due to the supply with improper tools. Improper tools means e. g. that the cutting edges were damaged, the quality of the tool resharpening was poor or even the incorrect type of tool was supplied (Wiendahl and Ullmann 1988 [1]). By this kind of problems the availability of the machines is reduced by up to 9 % of the total production time. In total the set-up time including the exchange of tools for whatever reason reduces the machine uptime by about 18% (Milberg and Ebner 1994 [2]).

Thus it becomes obvious that special attention must be given to any activity related to the flow and application of tools, especially when taking into account the large number of tools and the money involved for the purchase, the reconditioning and resharpening of the tools. As can be found in figure 1 about 46.5 million US $ respectively 25.6 US $ were spend in 1992 by two different automotive companies for the purchase and resharpening of tools.

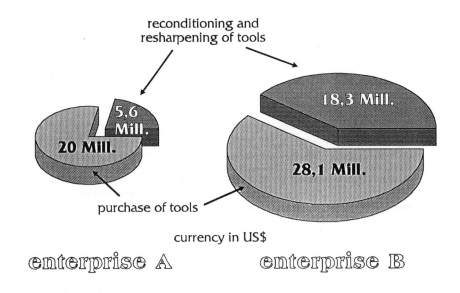

Figure 1: Costs of tooling in automotive industry in 1992.

Furthermore it generally can be found that about 40% of the total costs of machining centres are due to their tooling are used for the tooling of machining centres. Therefore it is essential to develop powerful software tools enabling the integration of the metal cutting tools into the production planning system as well as their circulation to and from the machines, the tool shop and the tool room. It could be shown using simulation techniques that the application of this kind of planning system assures the purchase of the different kind of tools to the number necessary to maintain the production at the highest level possible (Petuelli and Müller 1994 [3]).

On the other hand keeping in stock just one replacement of the individual tools results in an increase of production cost per workpiece by about 200 US $. This was found by simulation of the machining of parts in the paper-machine industry. The workpieces are machined in a flexible manufacturing system consisting of two machining centres and the appropriate material handling system. Based on this knowledge the Event-Oriented Tool-Management (Petuelli and Müller 1994 [4]) was created. The major parts and the performance of this system under development will be discussed

2 Structure of the Event-Oriented Tool-Management

The basic structure of the Event-Oriented Tool-Management is predefined by the different components of production systems such as machines, workpieces and their material flow, the circulation of tools to and from the machine shop to the tool shop and the storage as well as the handling systems. Contrary to the standard production planning and control systems these above mentioned components are taken care of and integrated into the system by the appropriate exchange of information (fig. 2).

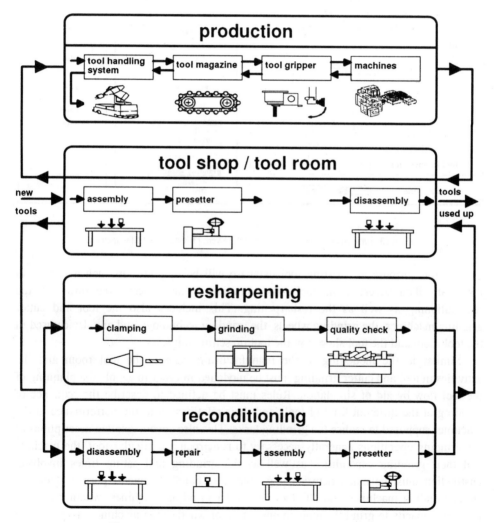

Figure 2: Material flow to enable Event-Oriented Tool-Management

Since the circulation of the tools is considered as being important with respect to the productivity of the machines it must be described in detail in order to enable the planning of the reconditioning and resharpening in conjunction with the tooling of the machines. Reconditioning in this case means maintenance, repair and exchange of indexable inserts and assembly of the tools. Thus the information about the status of the different tools must be integrated into the production planning and the other major departments of the enterprise (fig. 3).

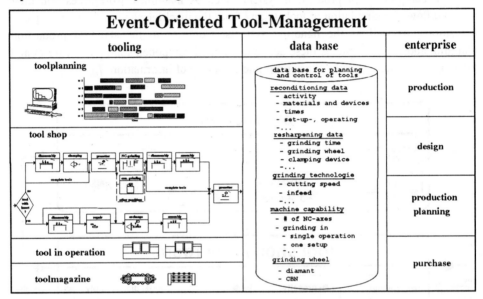

Figure 3: Performance and components of the Event-Oriented Tool-Management

For this purpose an existing tool database will be upgraded in such a way as to insert any data covering the tools, the planning of their usage, the flow and the reconditioning as well as the resharpening. This includes also the tool and cutter grinding machines, the grinding wheels, the grinding technology and any device used in the tool room and the tool shop, e.g. the tool presetting units.

Similar to the modelling of the manufacturing systems the tool room and the activities in tool and cutter grinding must be modelled in order to enable the planning of the tool flow by aid of simulation. Rules must be defined to describe the sequence of grinding of the different kind of tools and their relationship to the performance to the machines supposed to realise the grinding tasks. This also covers detailed descriptions of the grinding operations generally necessary to resharpen a particular tool, the machine and their grinding capabilities as well as the grinding and set-up times involved. Distinction must be made between, e.g. manually handled general purpose tool and cutter grinding machines and NC tool and cutter grinding machines, whether they do have the capacity to grind the tools in one set-up or whether the machine set-up must be changed. Similar information about the reconditioning have to be collected and integrated into the data base.

On the other hand the handling of the tools and the procedure of the clamping are important information which influence the time required to replace worn-out tools respectively indexable inserts or to prepare the tooling for the machining of a different kind of workpiece. The different parts of the planning and the control system will be developed in a way such as to enable the integration of the components sequentially according to the needs of the companies, thus the Event-Oriented Tool-Management will be of modular design.

3 Planning of the Tooling of the Machines

Today's production planning systems do optimize the sequence of the machining of the workpieces according to the capacity of the factory taking into account the needs of the assembly respectively the delivery date of the products scheduled. Based on the sequence of the workpieces lists are created giving information about the tooling of the machines which are provided for the manufacturing of the parts (fig. 4)

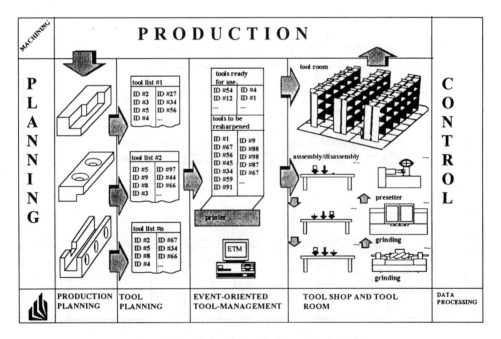

Figure 4: Description of the method to prepare the tooling

The creation of these catalogues is based on the actual tooling of the individual machine and the tooling required to manufacture the next batch of workpieces. No provision is made to take into account the status of the tools actually stored in the magazine of the machine, e.g. no information is given about their residual tool life or whether different tools will be used lateron and should be left at the machine. Furthermore nowadays there is no information given about the status of tools which were sent to the tool shop or the tool room. This especially relates to tools that must be reconditioned or resharpened. Due to the lack of integration the status of the tools circulating in the tool room and the tool shop is unknown. Thus quite often new tools are purchased in order to ensure the manufacturing which consequently result in an enormous stock of tools. This situation will change after the realisation of Event-Oriented Tool-Management. According to the information flow shown in figure 4 the status of the tools will be taken care of and different procedures initiated in order to fulfil the requirements of the production. Based on the production program in total the status and the availability of the tools necessary to machine the workpieces will be checked and catalogues will be created which contain the ID-numbers of the tools ready for use and those to be reconditioned or resharpened. At this point and time it is of utmost importance to assign priorities according to which the tools should be handled in the tool room. On the other hand based on the priorities the reconditioning and resharpening will be scheduled. For this purpose the individual tool room must be modelled in detail as shown in figure 5.

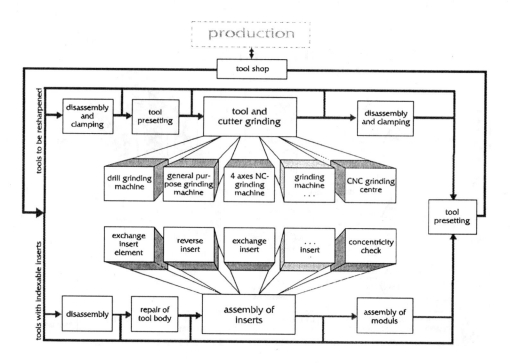

Figure 5: Modelling of the tool room (Petuelli and Müller 1994 [5])

When describing and planning the flow of the tools first of all distinction must be made between tools to be resharpened and tools with indexable inserts. To simulate the flow sheet of the individual tools the different activities must be taken into account. This also includes the different machines and devices and in many cases the possibility to handle the tools in several ways, e.g. to regrind a tool in one set-up or in several set-ups using different special purpose machines. Any way the quality of the regrinding and the time consumption to make the tool ready for use may differ to a large amount.

It becomes obvious that by using the simulation technique to schedule the activities to be done in the tool room in order to fulfil the requirements of the production, information will be provided at which point and time the tools will be available necessary to machine the workpieces in the sequence given by the production program. Thus the production planning department may check and decide whether the production program can be fulfilled or what kind of activity must be started to reach the target and to meet the schedule. This may also include the rescheduling of the production program.

4 Conclusion

The structure of the Event-Oriented Tool-Management is described which enables the planning and control of metal cutting manufacturing systems taking into account the handling, the resharpening and the reconditioning of tools. This kind of integration is necessary in order to assure the tooling of the metal cutting machines. By this measure the productivity and the number of tools and thus the costs will be reduced. For this purpose means are developed to describe the activities to be done in the tool room and thus to schedule and optimize the sequence of reconditioning and resharpening of tools with respect to the workpieces to be manufactured. The software tool Event-Oriented Tool-Management does not only provides the capability of modelling existing manufacturing system but also to set-up new manufacturing systems including their tooling to find the best solution possible for the machine selection as well as their tooling.

5 References

[1] Wiendahl, H.-P. and Ullmann, W.: „Anforderungen an die Bereitstellungsplanung von Werkzeugen in der Fertigungssteuerung", Werkstatt und Betrieb **121** (1988), **Nr. 2**, pp. 133-13

[2] Milberg, J. and Ebner, C.: „Verfügbarkeit von Werkzeugmaschinen" VDI-Z **136, Nr. 1/2**, pp. 31-35, 1994

[3] Petuelli, G. and Müller, U.: „Ereignisorientiertes Tool- Management", VDI- Spezial Werkzeuge, pp. 72-.80, 1994

[4] Petuelli, G. and Müller, U.: „Simulation zur Planung und Steuerung von Werkzeugversorgungssystemen", Fortschritte in der Simulationstechnik **Vol 8**, pp 339-353, 1994

[5] Petuelli, G. and Müller, U.: „Anwendung der Simulation zur Planung der Werkzeugaufbereitung und Werkzeuginstandsetzung", Fortschritte in der Simulationstechnik **Vol 9**, pp 573-578, 1994

A COMPUTERIZED TOOL PLANNING AND SCHEDULING SYSTEM

K. N. KRISHNASWAMY [*]   B. G. RAGHAVENDRA [†]   D. SAMPATH [‡]

Indian Institute of Science, Bangalore - 560 012, India
E-mail: sampath@vidya.iisc.ernet.in

Abstract

There have been significant advances in scheduling techniques in production scheduling theory, but few might have attempted to use them in practice. This paper deals with computerizing the planning and scheduling of tool orders in an in-house tool room of an engineering firm. The system successfully adopted the well known Critical Ratio(CR) method for scheduling on each of the work centres. The software developed is a user-friendly and menu-driven package and caters to tool planning and scheduling of more than 7000 tools used by the firm. The concepts and the techniques used can easily adaptable to in-house tool rooms of any manufacturing industries.

1 Introduction

This paper deals with the production planning and scheduling system for an in-house tool room of FASTSUN, a medium sized engineering firm. FASTSUN manufactures a variety of high precision, high strength pressed mechanical components mostly for the automotive industry. No major assembly operation is involved in the manufacturing of these components. The characteristics of the production items, which, by and large are against specific orders, call for intricate tooling, all of which is met by the in-house tool room. FASTSUN operates in a highly volatile environment in which the products and the order size for each product, keep changing dynamically without any predictable pattern thereby introducing a very high degree of uncertainty on tool demand.

There have been significant advances in scheduling techniques in production scheduling theory, but the use of these in practice has been very difficult to assess (Graves 1981). Though a large number of static and dynamic heuristic scheduling rules have been developed (Gere 1966, Panwalkar and Iskander 1977) few appear to have been successfully used in practice. Critical Ratio(CR) scheduling (Monks and Joseph 1988) seems to be one of the most widely used dynamic rule in computerized scheduling systems. Putnam etal (1971) and Berry and Rao (1975) have dealt with comparative experimental studies on the implementation of this rule. This paper

---
[*]Department of management Studies
[†]Department of Management Studies
[‡]Supercomputer Education and Research Centre

presents the features of a computerized system developed for effective planning and scheduling of tool production using the CR technique, which offers some interesting features. The development of the computerized system was carried out in two phases. The first phase was a detailed analysis of the existing manual system to identify areas for development of effective procedures for planning and control. The second phase was the detailed system design, development and implementation of the associated computer software system.

2 Analysis of the Existing System

The tools required by FASTSUN are complex and exacting from the point of view of precision and dimensional accuracy. Generally, the tools are classified as basic element tools and assembly tools. Common assembly tools are interchangeable among the operations of different products and different operations, whereas specific assembly tools are used for a particular operation. When a tool assembly is required, matching quantities of associated tool elements are ordered.

The demand on the services of the tool room can be broadly grouped into the following categories : (i) Manufacture of tools for regular production items, (ii) Manufacture of developmental tools for new products, (iii) Rework or Repair of improper or worn-out tools, and (iv) Service jobs for other departments in the industry.

During any period (month), the load on the tool room in terms of the machine hours required for processing the jobs vary considerably both among the four categories listed above as well as within each category. Preliminary studies revealed that on an average, approximately 40-45% of scheduled time is spent on regular tool production items, 10-20% for service and rework and 40-45% for developmental tools for new products. The following paragraphs describes in detail the nature of the above four categories of job.

Regular Tools: The starting point for the production planning of regular tools is the three month rolling market forecast. Approximately 4-5 days before the beginning of each month a firm forecast is prepared by marketing department based on committed customer orders. Simultaneously it prepares tentative forecasts for the subsequent two months.

Based on a specific month's firm forecast of regular products, the Production Planning and Control (PPC) department develops a **production plan** for manufacture. While doing so it tacitly takes into account the existing stock of products and the ideal batch sizes for the production of these products. Simultaneously the PPC translates these production targets into requirements of tool assemblies and tool elements using estimated (standard) tool life. This datewise tool elements requirement list is then passed on to the tool room for manufacture and supply with specific work-order numbers. Using estimates of time required for the manufacture and the work load, the tool room planners decide the commitment dates for the completion of each tool element so required, purely based on subjective judgement and experience of the tool room scheduler.

Developmental Tools: In the case of developmental tools, the Production Engineering Department (PED) develops appropriate designs of the tools required. The tool drawings and work-orders are then released to the tool room. The scheduling of work-order is done on a subjective basis depending on production of other tools and priorities. Sometimes these developmental tools go through a redesign process through a Design Change Notice (DCN) and may result in rework. After a developmental production run in the production shop a tool design is finalized. The lead time for productionizing a developmental tool varies from 3 to 6 months.

The developmental tools constitute a major hurdle to the effective planning and scheduling of the tool orders, because it is difficult to estimate in advance the types of materials and machines required for processing and machine times due to high degree of uncertainties. These tools often cause an imbalance of loads on the tool room machines. They also necessitate unplanned procurement of tool materials requiring 3 to 10 weeks of lead time.

Service and Rework: The frequency and arrival of service jobs at the tool room was purely random. The time required to process each job and its priority varied depending on the type and criticality of jobs in the production process. As a result, the load on tool room machines was highly unpredictable.

Each tool service or rework job is accompanied by a service slip which is used as a work-order. These work-orders are scheduled subjectively on the machines taking into consideration the urgency and the amount of work involved.

Loading Schedule: The loading schedule on the individual machines was being made on a daily basis considering the commitment dates for different category of jobs. The whole process was tedious as more than 7000 speciality tools each requiring processing at 8 to 15 work centres with differing work sequences were being produced. Further, decision making was based on subjective judgements and trial and error often leading to non-availability of tools at the right time resulting in production interruptions.

3 Observations from the System Analysis

As a result of the system analysis conducted in the first phase of the work, the following observations were made:

(i) Changes by marketing department in the quantities of products requirements were incorporated in the production plan during the month. Tentative forecasts for the second and third month were not known to the tool room at the beginning of the current month.

(ii) The tool requirements provided by PPC at the beginning of each month almost always led to a *crisis management* situation.

(iii) Considerable variability existed in machine operation times and routings.

(iv) Service and developmental tool work-orders were being scheduled purely on judgement and pressure from the production shops.

(v) Reliable data on tool life was not available. Actual tool life varied anywhere from 50% to 200% of the estimated life.

(vi) Systematic inventory control of finished tool stock either in the production shop or in the tool room was not being practised.

(vii) Subcontracting facilities were available for some operations of a few tools and was used extensively as the subcontract costs were generally lower than the in-house cost. However, the lead times for subcontract jobs were relatively high and varied.

4 Framework for System Design

Considering all the aspects discussed earlier it was decided to develop an overall system consisting of two sub-systems; namely planning system and scheduling system.

Planning System: A system to generate tool production requirements for each production plan period giving planned quantities of tool items to be ordered on the tool room well in advance of the main production plan period. It computes the tool requirements taking into consideration the tool inventories available. It also makes a rough cut capacity plan indicating machinewise the total extent of subcontracting hours that the planner should plan if there is overload, or the extent of additional orders to be scheduled if there is underload. The output of this planning system is the number of tools(batches) of various tools planned for production during the month, cumulative quantities required for the rolling planning horizon of 3 months and the overload or underload indicated for the planning horizon. The system is intended for operation once every month.

Scheduling System: A scheduling system for the tool room gives daily work-order loading on each machining centre, their sequence and weekly hours of subcontracting (machine-centrewise, operationwise) during the tool production scheduling period. It generates the work-orders for the tool batches indicated in the output of the planning system for the first month. It also indicates work centres (including subcontractors) on which they have to be loaded, estimates processing times, computes the CRs and prioritizes them. An important output of the scheduling system is the list of work-orders, in the order of priority to be loaded on each work centre, and work-order status report including completed orders. To incorporate flexibility into the system, provision is made for manual overriding of priority, so that special and emergent situations can be easily handled by management judgement.

5 A Computerized Planning and Scheduling System

As per the systems and procedures developed to meet the requirements a computerized Tool Planning and Scheduling system was developed. The software was made user-friendly and menu drive. It basically consists of three modules; (i) Data Interface system, (ii) Planning system, and (ii) Scheduling system.

Data Interface System: The Data Interface system facilitates the easy access and operation of the system. At the core, a number of database files, some of which are one-time "static" files while others are "dynamic". The static data files basically consisted of four categories: products, tools, machines and work centres and the last one for linking tool production operation on various machines and work centres. The dynamic data files relates to the production plans and inventory of tools. Some of the important data files among these are briefly described as follows:

(a) A Product Operations Data File (POPN) consisting of product code, product description, operations sequence, operation code, operation description, tool code, quantity of tools required for the product, etc. This is prepared for every product, both in the regular and developmental product categories.

(b) A Tool Master File (TMF) consisting of Tool code, Tool description, standard batch size, quantity in stock, desirable minimum and maximum inventories, tool type, estimated tool life, etc. An additional Assembly Master (AM) file consisting of all the assembly elements, element tool code and quantities required also forms part of this system.

(c) A Machine Centre database (MC) giving details of machine (work) centres, operation code, machine code, machine description, available machine hours in-house and with subcontractors, utilization factor and service factor.

(d) A Tool Operations Master (TOM) providing all the information on the operations involved in the manufacture of the tool and auxiliary tools required, if any. This mainly consists of tool code, operation sequence number, operation code, machine centre, set-up time, operation time, auxiliary tool requirement, sub-contractors/in-house facility, etc,.

The database file for the production plan consists of the product code, quantity of products required in each one of the three months, and related information.

Figure 1: Net Tool Requirements for all Products

Tool Code	Tool Description	Batch Qty	Current Invent.	Month of March.. Tools Required			Priority
				P1	P2	P3	
C10904	Anvil	2	0.6	1.2	2.0	4.0	High
C27005	Die Insert	2	1.5	2.0	3.5	4.8	High
C15128	Auto Top Ring	4	2.5	3.0	7.2	15.0	High
..	..	..	..	..	..	..	..
C28046	Shuttle Spacer	1	1.2	1.0	1.5	1.8	Low
C28079	Guide Plate	1	2.4	0	1.4	2.0	Low
..	..	..	..	..	..	..	..

Planning System: Given the firm or committed production plan for one period and tentative forecasts for the subsequent two periods alongwith the initial inventory levels, the planning system generates the following outputs:

(a) Estimated tool requirements for the three months, reported as net tool requirements in actual quantities as well as in terms of appropriate batch quantities.

This is given in terms of both the absolute monthly quantities and in terms of cumulated quantities, cumulated over the three months. It also gives a priority indicator for the tool manufacture in the first month as "high" or "low' according to whether the tool is required in the current period production or it is for building up the inventory for future use. The tool requirements are given both productwise and toolwise. A sample of one of the planning system report for a case example is shown in Figure 1.

(b) An overload/underload report giving machine groupwise report on the available hours and required hours in each one of the three periods. The hierarchical loading pattern of the machines are taken into consideration in this report which is further classified by in-house and subcontract machines. The output for a case example is shown in Figure 2.

Figure 2. Work Centrewise Overload/Underload report

M/c Code	Available M/c Hrs.	Required M/c Hrs.	Month of March..		
			Overload	Underload	Inhouse/Subcontract
11	128.0	103.6	–	24.4	I
16	64.0	106.6	42.6	–	I
43	53.1	240.0	186.9	–	I
71	96.0	11.6	–	84.4	S
..	..	..	..	..	..

Scheduling System: Taking the output of the planning system for the first month's tool requirements as the basic input, the scheduling system has a number of programs which will perform the following functions:

(a) Generation of work-order numbers for the planned tools;

(b) Provision of a user facility to change the in-house operation to sub-contract or vice-versa based on information from the planning system, so as to balance the machine loads;

(c) Hierarchical loading of machine groups;

(d) Preparation of overload/underload report based on rescheduled loads;

(e) Computation of the throughput times for each operation of each tool taking the setup and operation times and lead time between operations;

(f) Estimation of due hours and due date for each operation of the tools;

(g) Computation of the CR and development of a priority loading schedule for each work-order operation on a machine group;

(h) Sorting of sub-contract jobs according to sequence of operations between stages;

(i) Generation of reports on work-orderwise, machine groupwise for priority loading schedules for both in-house and sub-contract, bunching of sub-contract jobs between stage codes for manual allocation to sub-contractors, work-order status reports, auxiliary and main tool work-order linkages, elements and assembly tool work-order linkages; and

(j) Provision for introducing new work-orders for tools; manual change of the batch quantity in case of breakage of tools during the manufacturing process or rejection after inspection and for changing the CRs for forcing priorities.

Figure 3: Machine Centrewise Work-Order Report

Machine Centre: 22   Location: In-house   Month of March..

Pri. No	Work Order No	Tool Code	Tool Desc.	Qty	CR	D.Date	Expected C. date
1	8CNO38	C15128	A.T Ring	3	1.0441	21.3.93	29.3.93
2	8CN009	C15109	P. Sleeve	1	1.7352	31.3.93	29.3.93
3	8CN06	C15904	Hard Plate	1	1.8024	31.3.93	30.3.93
..	..	..	..	..	..	..	..

One of the output of a case example is shown in Figure 3.

6 Computational Aspects of the System

In this section, we describe some of the computations involved in both the planning and scheduling system.

Calculation of Tool Requirement: The product operation file contains data related to an operation of the product and the tools required for that operation. Tool requirements are calculated as follows for each operation of the product.

(a) If it is an element tool, Quantity of tool required$(Q_{TE}) = \frac{Q_P \times N_{TE}}{L_{TE}}$

Where
$Q_P \longrightarrow$ Product quantity;
$N_{TE} \longrightarrow$ Element tool quantity per set for product operation, and
$L_{TE} \longrightarrow$ Tool life

(b) If it is an assembly tool,
(i) Quantity of assembly tool required$(Q_{TA}) = \frac{Q_P \times N_{TA}}{Min(L_{TA})}$

Where
$N_{TA} \longrightarrow$ Assembly tool quantity per set for product operation, and
$L_{TA} \longrightarrow$ Minimum Tool life among all elements of the assembly tool

For the elements of the tool assembly,
(ii) Quantity of tool required$(Q_{TE}) = \frac{Q_P \times N_{TA} \times N_{ATE}}{Min(L_{TA})}$

Where
$N_{ATE} \longrightarrow$ Element tool quantity per set of tool assembly

(c) If it is a common assembly,

Quantity of tool required $(Q_{CTA}) = I_{min} - Q_S$

Where

$I_{min} \longrightarrow$ Minimum inventory level for common assembly, and

$Q_s \longrightarrow$ Quantity in stock

For the elements of the common assembly,

(i) Quantity of tool required $(Q_{TE}) = Q_{CTA} \times N_{CTE}$

Where

$Q_{CTE} \longrightarrow$ Quantity of element tool required per set of common assembly

For the subassembly of a common assembly

(ii) Quantity of tool required $(Q_{TSA}) = Q_{CTA} \times N_{CATE}$

Where $N_{CATE} \longrightarrow$ Quantity of subassembly required per set of common assembly

For the elements of the subassembly

(iii) Quantity of tool required $(Q_{TES}) = Q_{CTA} \times N_{CATE} \times N_{ATE}$

Where

$N_{ATE} \longrightarrow$ Quantity per set of subassembly

After the calculation of tool requirements for the above types of tools, let the tool required be Q_{TR}. Based on Q_{TR}, quantity of tool to be ordered is calculated as follows:

If $(Q_{TR} + Q_S) < Q_{max}$

Order Quantity $= Q_{TR}$

Otherwise

Order Quantity $= Q_{max} - Q_S$

This will ensure building up of stock to the desired level over a period of time.

Underload/Overload Calculation: For all the machines under a group, first the available hours are recorded and then the hours required for manufacturing the tools are calculated from the planned tool quantities. The calculations are done as follows for each of the machine group.

(1) Net hours available for each M/c group (HA) = Hours available × Utilization factor × (1 - service load factor)

(2) Net Hours required for each M/c group (HR) = (batches × set-up times) + (batches × batch quantity × operation time)

(3) If HA > HR then indicate underload
    else if HA < HR indicate overload.

Hierarchical Loading of Machine Groups: The practice at FASTSUN is to load and group the machines in a hierarchical way. Therefore if a work-order operation can be performed by more than one machine or machine group, the planning system loads the operation on an appropriate machining centre and updates the new machining centre code in the work-order. A hierarchy of these (the highest, the next highest, the least in terms of capability) is identified and stored for this purpose. If the most suited machine is overloaded then the hierarchy is scanned and the next higher capability machine having underload (with underload greater than or equal to hours required for the work-order operation) is found, the work-order operation is loaded on it and the availability is updated accordingly.

Calculation of Critical Ratio(CR): Critical Ratio scheduling technique is a priority scheduling technique that has been found effective for advance scheduling as well as for current review and revision of existing schedules. In contrast to the static scheduling rules like Smallest Operation Time, Least Slack, etc,. CR scheduling rule is dynamic. It facilitates constant updating of priorities and is quite a simple system. It develops a comparative index for each job, and with a computer system, it can be updated daily to provide relatively close and timely control. The design of the CR is such that priority is given to those jobs which most urgently require the work (machine time) in order to be completed on schedule. As a job tends to get farther behind schedule its CR becomes lower (the jobs with lower CRs get precedence over others) and is accorded higher priority.

The factors which controls the calculation of the CR are the expected completion date and the due date. The CR is calculated as a ratio of :

$$\frac{\text{Estimated completion date} - \text{Today's date}}{\text{Estimated completion date} - \text{Due date}}$$

While the numerator reflects the current total load on tool room facilities, the denominator reflects the tool room's commitment to the work-order.

7   Some Special features of the System

Tool Inventory buildup: The system provides for building up stocks of tools within specified maximum limits. This is a feature suited to the present condition of tool supply which is inadequate. However, simple changes in the system can alter this to keep inventories to minimum limits when required.

Changing the mix of Subcontract and In-house loads: The system has built-in flexibility to alter the rate of overall tool production through decisions interfacing the planning and scheduling system, i.e., increasing subcontracting to boost up production. In the scheduling system when difficulties are encountered, provisions are made to switch from in-house operation to subcontracting and vice versa in order to adjust short term variations in loads and speed up production of tools.

Flexibility to absorb service and development loads: The system provides for service and developmental work-orders enabling manual overriding of CRs to hasten a job through in the scheduling phase and providing a prorata capacity buffer in the planning system.

8 Implementation of the System

The entire software development evolved out of an indepth analysis of the existing system and procedures and several meetings and discussions between the authors and FASTSUN representatives over a ten month period. Starting with the feasibility report as the basis, the functional specifications of the system were the result of many of the suggestions based on specific requirements of the firm. The functioning of the planning and scheduling systems were first demonstrated to the firm's representatives who were satisfied with the outputs and operation of flexibilities provided in the system. Further, during the initial phases of implementation when the database was being built using part of the information, schedules were successfully developed by using the software on computer system which were checked manually for physical loading on the machines. It is believed that such a simple-rule based planning and scheduling system would be useful in many engineering firms in which job-shop type scheduling environment prevails.

9 Acknowledgements

The authors wish to place on record the excellent cooperation and rapport provided by the management and staff of the company in developing and implementing this software system. Thanks are also due to an anonymous referee for helpful suggestions to improve an earlier version of the paper.

References

[1] W. L. Berry and V. Rao, *Critical Ratio Scheduling and Experimental Analysis*, Management Science, 22 (1975), p. 192-201.

[2] W. S. Gere, *Heuristics in Job Shop Scheduling*, Management Science, 13 (1966), p. 167-190.

[3] Graves, C. Stephen, *A Review of Production Scheduling*, Operations Research, 29-4 (1981), p. 676-697.

[4] Monks, G. Joseph, *Operations Management - Theory and Problems*, 4th EDITION (McGraw Hill, 1988).

[5] S. S. Panwalker, W. Iskander, *A Survey of Scheduling Rules*, Operations Research, 25-1 (1977), p. 45-61.

[6] A. O. Putnam, R. Everdell, G. H. Dorman, R. R. Cronan, and L. H. Lindgren, *Updating Critical Ratio and Slack Time Priority Scheduling Rules*, Production Inventory Management, 12 (1971), p. 51-72.

APPLICATION OF NEURAL NETWORK IN TOOL SELECTION PROBLEMS

Prof. P.C. Pandey

and

Suparna Pal
Email No: pcpandey@rccsun.ait.ac.th
Asian Institute Of Technology, School Of Advanced Technology, Manufacturing Systems Engineering Division, GPO BOX 2754, Bangkok 10501, Thailand.

Abstract: Limiting the size of tool inventory in flexible manufacturing systems, using shared tools, has been a well known policy. However, much work has not been done towards the development of a suitable algorithm to yield for the optimum number of tools for processing a set of workpieces on a machining center. This paper presents an integrated neural net-simulated annealing algorithm for determination of the optimum number of tools for a flexible CNC machine tool processing a set of jobs over a particular time period. The convergence behavior of the algorithm has also been studied through simulation.

Keywords: Neural Networks, Simulated annealing, Multicriteria optimization.

1. Introduction

The overall flexibility in flexible manufacturing systems (FMS) can be constrained by the availability of palletized fixtures and the available tool storage capability. In FMS operations one of the strategies for limiting the size of tool inventory, is to employ tools that can be shared by more than one workpiece/operation during the fixed production period. However, the shared tools may not perform optimally for all types of operations for which they may be suitable. Moreover it may be possible to recommend more than one tool for a particular operation from the available set of tools in the tool library. Thus, problem of optimizing the number of tools, for a set of jobs to achieve optimum overall performance, is quite complex. One possible solution approach, to this problem, would be to enumerate the operational cost of each tool for all possible operations and select the one with minimum cost. This however is not practical and can not ensure global optimality.

The problem of tool allocation and workpiece assignment with the objective of minimizing the processing time, can be treated as 0-1 mixed integer programming problem [3]. The model however does not consider constraints e.g., availability and limited capacity of tool magazine and suffers from computational complexity. Available heuristics [1, 5] for tool selection problems can be applied under specified conditions only and are not valid for all types of situations. Nee and Poo (7) have developed a neural network based solution to the tool selection problem, with the objective of minimizing the number of tools required. Their work however, provides little information concerning the use and development of the algorithm.

In this work the authors have considered the tool selection as a problem of multicriteria optimization and have proposed an integrated neural-net simulated annealing algorithm for optimization purpose. The behavior of the algorithm, particularly its convergence, has also been studied.

2. Mathematical model formulation

Let us consider a set of jobs being processed on a CNC machine equipped with automatic tool changer. Each job requires a set of tools which are placed in a finite capacity tool magazine before the start of the operation. If a particular tool, required for processing, is not available in the tool magazine it has to be transported

from a secondary storage and loaded into the tool magazine. This would lead to tool switching. A good tool management policy should aim at minimizing the number of tool switches during the operation. The following general assumptions have been made.
1. Only one tool change occurs at a time.
2. Each tool occupies one slot in the tool magazine.
3. Any combination of tools can be loaded in the magazine.
4. One operation is performed at a time.

For a machining situation where m operations are performed by n tools, the tool allocation problem can be modeled as,

Minimize: $$Z = \sum_{i=1}^{n} \sum_{j=1}^{m} x_{ij} c_{ij} + \sum_{i=1}^{m} \Omega \sum_{j=1}^{n} x_{ij}$$

Subject to: $$\sum_{j=1}^{m} x_{ij} = 1 \quad \text{for } i=1 \text{ to } n$$

Where $x_{ij} = 1$ when j th tool is used for ith operation
$= 0$ otherwise.

and $$c_{ij} = \frac{t_{ij}}{T_j} \times P_j \tag{1}$$

The first term in the objective function (Z) refers to the total tool related cost whereas, the second term pertains to the number of tools required. P_j is the unit cost of the tool, T_j the tool life and t_{ij} is the time for which the tool j was operational during processing of job i.

Efforts to solve the optimization problem in eqn 1, using standard packages e.g., LINDO and XA proved to be infeasible. Furthermore, possibility of employing exhaustive search was ruled out in view of large search space.

3. Hopfield and Tank TSP model

For the solution of certain class of optimization problems neural networks have proved to be superior particularly as regards computational time and speed of convergence. In Hopfield and Tank model for traveling salesman problems (TSP) [6, 8], the network state is defined by a binary variable x_{ij}. [$x_{ij}=1$ when the neuron is on whereas $x_{ij}=0$ when neuron is off] The energy of the network defines the objective function whereas the state of the network is controlled by the states of the neurons. During the iterative process of state transition the network stabilizes at the lowest energy level and this corresponds to a state yielding the optimal value of Z.

4. Modified Hopfield and Tank model

To adapt Hopfield and Tank model for tool optimization problems, the energy function to be used has to be modified so as to map the objective function and the constraints. The neuron state x_{ij} assumes a value of 1 if the operation i is served by the tool j, 0 otherwise.

The 2nd energy term, in the TSP problem, which restricts the assignment of one city only once, has been deleted as one tool can serve more than one operation. Third energy term in TSP problem has been adopted to restrict the number of operations. The energy E can be expressed as,

$$E = \frac{A}{2}\sum_{x=1}^{n}\sum_{i=1}^{m}\sum_{j=1}^{m}x_{ij}x_{ji} + \frac{B}{2}(\sum_{x=1}^{n}\sum_{i=1}^{m}x_{ij} - n)^2 + \frac{C}{2}\sum_{j=1}^{m}\sum_{i=1}^{n}\Omega x_{ij} \quad (2)$$

Where A and B and C are constants, to be chosen heuristically.

To achieve minimum overall number of tools an energy term $\frac{C}{2}\sum_{j=1}^{m}\sum_{i=1}^{n}\Omega x_{ij}$ has been added in eqn (2) which is responsible for the successive oring of the matrix columns and addition of the results. Change in energy ∂E due to a change in the neuron state x_{ij} is given by,

$$\partial E = [A\sum_{l=1}^{m}x_{il} + B(x_{ij}\sum_{k=1}^{n}\sum_{l=1}^{m}x_{kl}x_{lk} - x_{ij}n) + C(1-\sum_{k \neq i}^{n}\Omega)]\partial x_{ij} \quad (3)$$

∂x_{ij} is the change in the neuron output due to change in state. $\partial x_{ij} = 1$ if the neuron is turned on and -1 if turned off. The general form of energy function can be written as:

$$\partial E = -[\sum_{k=1}^{n}\sum_{l=1}^{m}w_{kl,ij}x_{kl}]\partial x_{ij} \quad (4)$$

Combining eqns (3) and (4) the function describing the synaptic strength is obtained as:

$$W_{kl,ij} = -Ax_{ij}x_{kl} - B\delta_{ik}(1-\delta_{jl}) \text{ where } \delta_{xy} = \{^{0 \text{ if } x \neq y}_{1 \text{ if } x = y} \quad (5)$$

As Hopfield network is a non-adaptive network, the weights for A, B, C are computed initially. These do not change through the state transition process.

5. Integration of Simulated Annealing

The most common drawback [9] of Hopfield network modeling is that one may get trapped into a local minima. To overcome this difficulty simulated annealing is integrated with the network modeling. To achieve this, during each state transition the value of Δ is calculated as per eqn (6) and a decision regarding the move to the next state is taken according to the algorithm given in Fig 1.

$$\Delta = \Delta C + \Delta E \quad (6)$$

if($\Delta < 0$)
        accept the move
else
{

$$\text{if}(P_{ij} > \text{acceptance probability})$$
$$\text{accept the move}$$
$$\text{else}$$
$$\text{reject the move.}$$
$$\}$$

where $P_{ij} = [\dfrac{1}{1+\exp(\dfrac{\Delta}{T})}]$ is the acceptance probability of the neuron.

Fig 1: Move criterion in simulated annealing.

In the above scheme the probability of acceptance is kept high to ensure that for large positive value of Δ, a move will be rejected. The complete algorithm is given below.

6. The Integrated Algorithm

The following algorithm has been developed for integrating the neural network and simulated annealing in C++ (Fig 2).

```
void run_modified_hopfield(void)
{
   Set_the_synaptic_weights( );

  Energy_term1 = calculate_energ_term1( );
  Energy_term2 = calculate_energy_term2( );
  Energy_term3 = calculate_energy_term3( );

  clock_t start, end;
  start = clock( );

  while(Energy_term1!=0 || Energy_term2 !=0 || Energy_term3 > ((D/2)*number_Of_Tools))
  {
          valid_flag=FALSE;
          iteration_count=0;

          for(i=0;i<number_Of_operation;++i)
          {
          for(j=0;j<number_Of_Tools;++j)
            state[i][j]=neighbour[i][j];
          }
          number_of_neighbour = calculate_number_of_neighbour( );
          cost = calculate_Total_Cost();
          Energy_term1 = calculate_energy_term1( );
          Energy_term2 = calculate_energy_term2( );
          Energy_term3 = calculate_energy_term3( );

          nm=9;
       do
       {
Again: iteration_count++;
     option = random(2);
     operation_count++;

     option=0;

     if(nm != 0)
     {
     if(operation_count%nm==0)
           option=1;
     }

     if(iteration_count>=number_of_neighbour)
           break;
     for(i=0;i<number_Of_operation;++i)
```

```
             {
                     for(j=0;j<number_Of_Tools;++j)
                       neighbour[i][j]=state[i][j];
                     }

                     xco = random(number_Of_operation);
                     yco = random(no_Of_Tools);
     if((valid=validCheck(xco,yco))==TRUE)
     {
             if(option==0)
             {
               neighbour[xco][yco]=0;
             }
             else
              {
               neighbour[xco][yco]=1;
              }
     }
     else
             {
             goto Again;

             }

     if(option == 0)
       del_energy = calculate_del_energy(xco,yco,-1);
     else
       del_energy = calculate_del_energy(xco,yco,1);

     del_cost = cost - calculate_TotalCost_neighbour();
     del = del_energy + del_cost;
     if(del < 0)
     {
             valid_move++;
             valid_flag = TRUE;
     }
     else if(probability(del,temp)>=0.45)
     {
             valid_move++;
             valid_flag = TRUE;
     }
     else
             {
                Again;
             }
     }while( valid_flag==FALSE);

     temperature = cooling_rate*temperature;
     t_count++;
 }
 end = clock();
 cpu_time=(end - start) / CLK_TCK;

}
```
Fig 2: The C++ code of the Integrated Algorithm.

In the outer loop validity of the constraints or the feasibility of the particular state is checked along with total number of tools. In the inner loop all the neighbors of that state are evaluated. If the transition from a state to a neighbor is accepted according to the simulated annealing scheme the neighbor is considered as the next state. Similarly, in the next iteration the neighbors of the new state are evaluated. This state transition process continues until the network stabilizes to a feasible point where all the constraints are satisfied. Each move from a state to a neighbor is generated randomly. As the move generation scheme is random it may also generate invalid moves. If during a move a neuron, initially in state 0, is turned on (state 1) then such a move can be considered as invalid. Validity of every upward move is checked by comparing the state value of the neuron, responsible for the state transition, with its corresponding value in the initial feasible matrix. The neighbor[ ][ ] and state[ ][ ] are the datastructure to hold the neuron states and the states of the neighbors.

During upward move the value of Δ increases whereas in a downward move it decreases. For a faster convergence a controlled upward move would be beneficial. This can be achieved by selecting a suitable value of turbulence factor. A turbulence factor of $\varpi = 1/3$ implies that out of 3 moves one upward move will be generated.

7. Results and discussion

The simulation of the network has been achieved in Borland C++ using a 80486 machine under DOS environment. A sample of the results is given Fig 3. Whereas the effect of employing different values of the turbulence on the results can be seen in Table 1 and Fig. 4.

NO	Turbulence Factor	Initial Cost ($)	Minimum Cost ($)	Initial No. Of Tools	Minimum No Of Tools	CPU Time (Secs)
1	1/4	152.77	20.69	7	5	215.43
2	1/7	152.77	44.948	7	4	152.14
3	1/11	152.77	54.56	7	4	146.56
4	Random	152.77	48.43	7	7	173.18

Tab 1: Effect of turbulence factor on the results obtained.

Operations ⇒

	0	1	1	0	0	0	0
T	0	1	0	1	0	0	1
o	1	0	1	0	0	0	1
o	0	0	0	1	0	1	0
l	0	1	1	0	1	0	1
s	0	1	1	0	1	0	1
⇓	1	1	0	1	0	0	0
	0	0	0	1	0	1	1
	1	0	0	0	0	1	1

0	0	1	0	0	0	0
0	0	0	0	0	0	1
1	0	0	0	0	0	0
0	0	0	0	0	1	0
0	1	0	0	0	0	0
0	1	0	0	0	0	0
0	1	0	0	0	0	0
0	0	0	0	0	1	0
0	0	0	0	0	0	1

(a) The initial feasible matrix (b) The output matrix; Turbulence 1/5

0	0	1	0	0	0	0
0	0	0	0	0	0	1
0	0	1	0	0	0	0
0	0	0	1	0	0	0
0	0	0	0	0	0	1
0	0	0	0	1	1	0
0	0	0	1	0	0	0
0	0	0	1	0	0	0
0	0	0	0	0	0	1

100	9.1	5.2	100	100	100	100
100	5.4	100	8.0	100	100	100
6.3	100	7.4	100	100	100	100
100	100	100	6.7	100	8.1	9.4
100	6.1	2.8	100	3.4	100	4.8
100	2.1	4.8	100	4.8	100	100
6.0	5.2	100	5.4	100	100	5.6
100	3.7	6.8	100	3.9	7.9	100
2.3	100	100	8.5	9.4	100	5.7

(c) The output matrix; Turbulence 1/7   (d) Cutting time matrix

Fig 3 : The initial feasible matrix and output matrices for different turbulence factors.

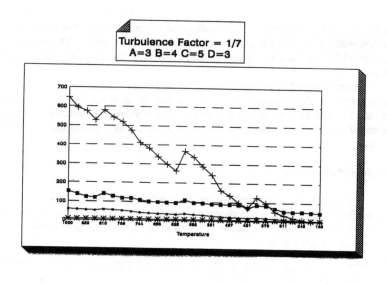

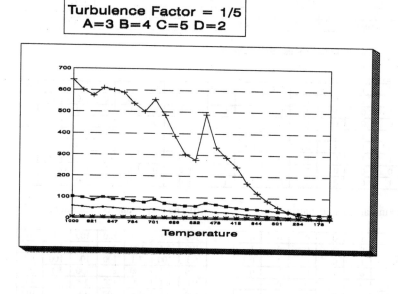

Fig 4 : Temperature Vs Cost and Energy for different Turbulence Factor.

The rate of cooling used for the simulated annealing, in the present study, was chosen heuristically. It has been found that slow cooling rate enhances the possibility of convergence to the global minima.

The input data to neural network was generated by a tool selection expert system developed in TURBO PROLOG. The parameters A, B, C, D in the Hopfield and Tank model were chosen in the same manner as the cooling rate. However, for selecting suitable values of these parameters out of a wide range of the data it may be possible to develop a suitable genetic algorithm to determine their fitness values. After evaluating the fitness values the weak datasets can be replaced by child dataset born due to cross over between two parent datasets with high fitness values. This process can be continued iteratively until the stopping criteria is met. More work however, would be necessary in this direction.

8. Conclusions

A mathematical model of the tool selection and allocation problem has been formulated and the difficulties in solving the model are indicated. Use of neural network to solve the combinatorial optimization problem by carrying out necessary modifications to suit the problem requirements, has been demonstrated. A simulated annealing scheme has been integrated with the neural network to overcome the problem of being trapped at one of the local minima and to arrive a near optimal solution.

The value of turbulence factor to be used during the computation has been shown to affect the convergence of the results and the computational effort required. To help in the selection of other network parameters, a possible selection procedure has been suggested. This forms a topic of further research.

9. References

1. Bard, J. F, *A heuristic for minimizing the number of tool switches on a flexible machine*, IIE Transactions, **Vol. 20**, No 4, pp. 382-391, (1988).
2. Buzacott, J. A and Shantikumar, J. G, *A survey of Flexible Manufacturing Systems*, Decision Sciences, **Vol 8**, pp. 28-55, (1980).
3. Chakravarty, A. K and Shtub, *A Selecting parts and loading Flexible Manufacturing Systems*, Proceedings of the first ORSA/TIMS FMS Conference, Ann Arbor, MI, (1984).
4. Chan, B. W. M, *Tool management for Flexible Manufacturing, International Journal Of CIM*, **Vol 5**, pp. 255-265, (1992).
5. Follonier, J. P, *Minimization of the number of tool switches on a flexible machine*, Belgian Journal of Operations Research, Statistics and Computer Sc, **Vol 34(1)**, pp 55-72, (1994).
6. Hopfield, J. J and Tank, D. W, *Neural Computation of decision in optimization problems*, Biol. Cybernet, **Vol 52**, pp. 141-152, (1985).
7. Nee, A. Y. C, Poo, A. N and Shan, X. H, *An AI/Neural network based solution for cutting tool selection*, Computer-Aided production Engineering, Elsevier, pp. 251-260, (1991).
8. Ramanujan, J and Sadayappan, P, *Optimization by Neural Networks*, Proc. IEEE Second Int. Neural Neywork Conf. **Vol II**, pp.325-332. (1988).
9. Yue, T. W and Fu, L. C, *Ineffectiveness in solving combinatorial optimization problems using Hopfield network: a new perspective from a aliasing effect*, proc. Int. Joint. Conf. Neural Networks, San Diago, Calif, **Vol III**, (1990).

Petri-Net Applications

An Integrated Object-Oriented Petri Net Approach for Developing the Manufacturing Control Systems
    L. C. Wang

An Integrated MMS and Colour Petri-Net Model for the Distributed Control of Flexible Assembly Systems
    W. H. R. Yeung and P. R. Moore

AN INTEGRATED OBJECT-ORIENTED PETRI NET APPROACH FOR DEVELOPING THE MANUFACTURING CONTROL SYSTEMS

Li-Chih Wang, Ph.D.
Department of Industrial Engineering, Box 5-985, Tunghai University,
Taichung, Taiwan 40704, R.O.C.
E-mail: wanglc@ie.thu.edu.tw

ABSTRACT

This paper presents a new manufacturing control software development methodology called IOPN (Integrated Object-oriented Petri Net paradigm). The IOPN consists of four phases: (1) static analysis, (2) dynamic analysis, (3) integration, and (4) implementation. The IOPN not only possesses the characteristics of the object-oriented method and the Petri net approach, but also provides a straightforward mapping between the high-level specification of the control logic and the code of the control software, the time required for developing a new control software or modifying an existing control software to accommondate to the operational changes will be greatly reduced.

Keywords : objected-oriented method, Petri net, manufacturing control.

1. Introduction

Development of a flexible and reliable control system for flexible manufacturing systems has long been a bottleneck. Traditionally, the sequence of operations/activities performed by the manufacturing systems is usually controlled by the programmable logic controller (PLC) which employs either ladder logic diagram or Boolean equations to specify its input and output procedures. The effort required for programming becomes quite cumbersome when either the system becomes more complex or the control logic changes. Currently, Petri nets and its variations, such as colored Petri nets (CPNs), were widely applied to develop either shop floor control model or FMS control software due to the characteristics of graphical representation and of the mathematical analysis of deadlock, liveness, boundedness, and reachability of the control logic; but a Petri net-based control model is usually suitable for the specification of sequential control of a manufacturing system, and is very difficult to include control decision/knowledge (e.g., scheduling knowledge and dispatching policy) in the control logic, this drawback will limit the application of Petri nets in the development of real-time control software. Besides, the development of a Petri net-based control model is highly system dependent and does not have the properties, such as modular programming style and the higher degree of flexibility, reusability and maintenanability, which are commonly required by the modern manufacturing control models.

Chaar et al. (1993) reviews a number of recent methods (e.g., rule-based expert system, operations research, object-oriented method, programmable logic controller, Petri nets) for developing manufacturing control software and concludes that these approaches support only a very restricted part of the planning, scheduling, simulating, and monitoring activities and an integrated approach is desired in order to enhance the control software performance. The objective of this paper is to develop a new manufacturing control software development methodology, called IOPN (Integrated Object-oriented Petri Net), which consists of four phases: static analysis, dynamic analysis, integration, and implementation.

2. Object-oriented Petri net (OPN)

In an OPN model, a system, S, is composed of mutually communicating physical objects and their interconnection relations [similar to the OPNets developed by Lee and Park (1993)]. Mathematically, a system may be defined as:
$S = (O, R)$
in which
O = a set of physical objects in the system = $\{O_i, i=1, 2, ..., I\}$

R = a set of message passing relations among physical objects
= $\{R_{ij}, i,j =1, 2, ..., I; i \neq j\}$

The dynamic behavior of a physical object is characterized by its internal behavior transitions and external message passings. The former is represented by the state places and activity transitions, the latter is achieved through a number of message places which are responsible for sending and receiving messages (i.e., tokens) among distinct objects. In order to simplify the modeling and analysis process, the concept of colored Petri net (CPN) is employed; each state place (message place) is associated with a set of colors that defines the types of tokens (messages) that may reside in that state place (pass through that message place). Similarly, a transition will make its response according to the associated colors of the input tokens (messages). Mathematically, an object O_i ($i=1, 2, ..., I$) may be defined as follow:
$O_i = (SP_i, AT_i, F_i, IM_i, OM_i, C)$
in which
SP_i = a finite set of state places for object O_i
AT_i = a finite set of activity transitions for object O_i
F_i = the input and output relationships between activity transitions and state/message places
IM_i (OM_i) = a finite set of input (output) message places for object O_i
C = the color set associates with SP_i, AT_i, IM_i, and OM_i of object O_i

In an OPN model, the communication between distinct objects (e.g., message sending object O_i and message receiving object O_j) is specified by the interconnection relations, R_{ij} ($i \neq j$), which may be defined as a set of three-tuple (OM_i, g_{ij}, IM_j), where g_{ij} is a special type of transition called *gate*. In a manufacturing system, *when* to perform the internal activities/methods, AT_j, of a physical object O_j is determined by the timing that its corresponding input message place, IM_j, receives messages from other objects (i.e., a token is in IM_j). Then, the sequence of performing certain specific activities/methods is determined by F_j which specifies the input/output relations between SP_j, AT_j, IM_j, and OM_j in the OPN of object O_j. Input message place, IM_j, can be triggered only when both the output message place of sender object, say OM_i, has tokens and the connection gate g_{ij}, specified in the interconnection relation R_{ij}, is fired. The decision of whether to fire a connection gate is made by the control/decison objects according to the suitable decision rule inferenced from the decision knowledge base. Readers who are intrested in the analysis of an OPN model may refer to Wang (1995).

3. The integrated object-oriented Petri net (IOPN)

In order to increase the reusability, modularity and maintainability of the code of a control software which allows the scheduling decision/knowledge to be included, the control hierarchy of an object-oriented Petri net manufacturing control system, shown in Figure 1, is divided into three levels: (1) physical level, (2) control level, and (3) decision level. Physical level realizes the physical production activities of shop floor equipments and is responsible for the reporting of the shop floor status information to the control level. Control level follows the control decisions (e.g., scheduling rule, dispatching policy) suggested by the decision level to control and coordinate the concurrent production activities of the physical level. Decision level is responsible for the determination of the suitable control decisions or strategies according to the current manufacturing system status. Object classes in the control hierarchy are classified into physical, information, and control/decision object classes (similar to the classification of Mize, 1992).

The IOPN paradigm consists of four phases: static analysis, dynamic analysis, integration, and implementation. The static analysis phase is to identify all possible object classes required by the manufacturing control software and to describe the static relationships (e.g., inheritance, aggregation, and association) among the object classes. The dynamic analysis phase will construct an OPN control model to model and analyze the dynamic behaviour/control logic of the proposed manufacturing control system. After the control logic of the OPN model is validated, results of the static analysis must be integrated with the OPN model to allow the scheduling/dispatching knowledge be included in the OPN model. Finally, the specification of the complete OPN model is transformed into the generalized production rules for a rule-based control system.

3.1 Static analysis

In the static analysis phase, a *requirement analysis* must first be performed so that the functional or nonfunctional (e.g., interface and hardware constraints) requirements of a control software may be clearified. Results of the requirement analysis may be represented as either in data flow diagram (DFD) or in requirement statements (Wang *et al.*, 1994). Following the concept of the control hierarchy shown in Figure 1, we need to identify all possible object classes forming the control software and construct the *object model* which describes the attributes and methods of each object class and the relationships among all related object classes (Wang, 1995).

3.2 Dynamic analysis

Since all the control activities in an object-oriented control system are derived from the OPN model, the quality of the OPN model will greatly affect the degree of successfulness of a control system. Procedure for developing the OPN model for a manufacturing control system may be briefly described as follows:

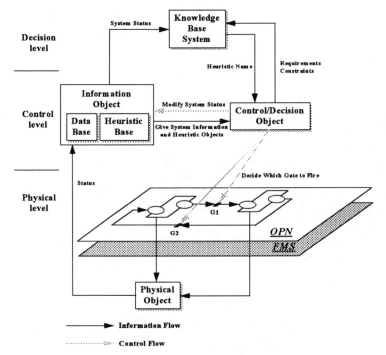

Figure 1. The control hierarchy of an object-oriented Petri net manufacturing control system.

Step 1: *Construct the state transition diagram (STD) for each physical object identified in the static analysis phase.* By identifying all the states and the events/actions which change these states for each physical object, the STD for that object may be constructed.

Step 2: *Use the structured analysis and design technique (SADT) to analyze the dynamic behavior of the production activities for all physical objects in the manufacturing system.* This step may be further divided into the following sub-steps:

(1) *Construct the functional model for each physical object.* By employing the characteristics of the top-down analysis and functional decomposition possessed by the SADT, a functional model will specify the input/output relationships, in terms of the information flow and the material flow, between the distinct production activities performed by an object. The mechanisms (or resources) used to perform an activity should also be specified in the functional model.

(2) *Construct the control flow model for each physical object.* Based on the functional model, a control flow model may be constructed by adding the control signals or the constraints which are required by performing each production activity.

Step 3: *Combine the STD and the control flow model obtained in steps 1 and 2 to construct the corresponding OPN for each physical object.* This step may be further divided into the following sub-steps:

(1) *For any production activity in a control flow model obtained in step 2 has two or more mechanisms, decompose that activity into the more detailed sub-activities so that each sub-activity has only one mechanism.*
(2) *Construct the complete control flow model for each physical object by incorporating the STD obtained in step 1 and the modified control flow model obtained in step 3(1).*
(3) *Transform the complete control flow model of each physical object into its corresponding OPN.* The following transformation rules are employed:
  Transformation rule 1. Transform each activity into a transition.
  Transformation rule 2. Add a state place between any two connected activities.
  Transformation rule 3. For any activity with some external input information/materials, add an input message place to connect each input with this activity. If both the input information and the input materials of an activity come from the same external object, only one input message place is required.
  Transformation rule 4. For any activity with some output information/materials sent to the external objects, add an output message place to connect this activity with each output. If both the output information and the output materials of an activity will be sent to the same external object, only one output message place is required.
Step 4: *For each physical object, add the syntax of colored Petri net in the OPN obtained in step 3.*
Step 5: *Construct the message passing net to obtain the complete OPN model for the manufacturing control system.* The message passing net is constructed by connecting all the related OPNs obtained in step 3 through gates since the input/output message places in the OPN of each physical object have already specified the type of messages (e.g., information or material) needed to be received from or sent to the external related objects.

After the complete OPN model for a manufacturing control system is constructed, the control logic specified by the OPN model must be analyzed to make sure that the control logic of the OPN model will not deadlock.

3.3 Integration

Once the control logic of the OPN model is validated, the control/decision objects and the information objects at the control level must be incorporated in the OPN model at the physical level to control the flow of messages (either information or material) inside every OPN or among the distinct OPNs. This is accomplished by the message flow diagram (MFD; Adiga, 1993).

3.4 Implementation

The last phase in IOPN is to implement the manufacturing control software according to the specification of the OPN control model. The specification of a complete OPN model is easily transformed into the generalized production rules for a rule-based control system (see Wang 1995 for the illustration).

4. An application of the IOPN

In this section the integrated object-oriented Petri net paradigm is demonstrated to develop a control system for a flexible manufacturing cell which is composed of one buffer, one CNC lathe, one CNC milling machine, and one robot. The buffer contains either the parts waiting to be processed or the processed parts. The robot will load (unload) the milling machine/lathe, which is feeded by the buffer, whenever a part needs to be milled/turned (is processed). The manufacturing cell processes parts in a batch production manner, the sequence of processes for a part depends on its routing.

We start with the static analysis for both the specification of the functional and nonfunctional requirements and the identification of all possible object classes for the control software. There are four types of physical object classes: robot, machine, part, and buffer in the control system, the dynamic analysis starts from constructing the STD [Step 1], the functional model [Step 2(1)], and the control flow model [Step 2(2)] for each physical object, then the complete control flow model for each physical object is constructed [Steps 3(1) and 3(2)]. The OPN for each physical object is then obtained by applying transformation rules 1, 2, 3, and 4 [Step 3(3)]. The complete OPN model for each physical object and for the control system is illustrated in Figures 2, and 3, respectively.

After the dynamic behavior of the complete OPN model is analyzed and shows that the control logic of the control system specified by the OPN model will not deadlock, any activity transition/gate which needs the control/decision objects or information objects to determine the necessary message passings or decision makings must first be identified, then the corresponding MFD is constructed. For example, selecting the suitable machine (lathe or milling machine), commanding the selected machine to begin setting up, and returning a completion signal are the three activities involved in firing activity transition T21 in the OPN of object machine, a 'machine selection' MFD should be used to determine the most suitable machine to perform activity T21. Finally, the code of the manufacturing control system may be immediately transformed from the complete OPN model. For example, the code of T21 in macro rule block 'lathe' ,when *CELLworks* is employed as the control platform, is:

> "when message "MP21" received in Mill_MP21
> if P21=="1" then do...
> let P21="0"
> let P22="1"
> endif."

5. Conclusion

In this paper, a new manufacturing control software development methodoloy, called IOPN (Integrated Object-oriented Petri Net paradigm for manufacturing control systems), is introduced and applied to develop a control system for a flexible manufacturing cell. The object-oriented Petri net (OPN) control model plays the key role in the IOPN paradigm. Results of the static analysis and dynamic analysis for the proposed control system may be directly transformed into the OPN control model, the scheduling decision or control knowledge can then be included in the OPN control model after the control logic represented by it was analyzed. Finally, the specification of the OPN control model may be transformed into the code of the control system in terms of the generalized production rules or the specific rules (e.g., rule blocks) if any control platform is used (e.g., *CELLworks*).

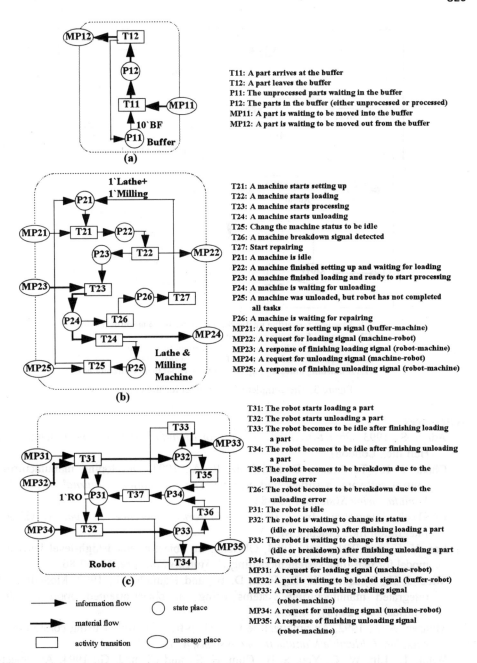

Figure 2. The complete OPN for object (a) buffer, (b) lathe/milling machine, and (c) robot.

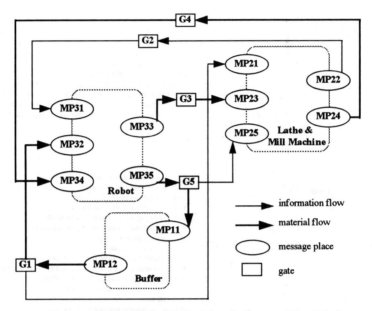

G1: A request for setting up (loading) from buffer to machine (robot)
G2: A request for loading sent from machine to robot
G3: A finishing loading signal sent from robot to machine
G4: A request for unloading sent from machine to robot
G5: A finishing unloading signal sent from robot to machine and buffer

Figure 3. The complete OPN model for the control system.

References

Adiga, S., 1993, *Object-oriented Software for Manufacturing Systems* (Chapman & Hall Inc., London).

Chaar, J. K., Teichroew, D. and Volz, R. A., 1993, Developing manufacturing control software: a survey and critique. *The International Journal of Flexible Manufacturing Systems*, **5**, 53-88.

FASTech Integration Inc., 1991, *CELLworks reference manual* (FASTech Integration Inc., Lincoln, M.A.).

Lee, Y. K. and Park, S. J., 1993, OPNets : an object-oriented high-level Petri net model for real-time system modelling. *J. Systems Software*, **20**, 69-86.

Mize, J. H., Bhuskute, D. B., Pratt, D. B., and Kamath, M., 1992, Modelling of integrated manufacturing systems using an object-oriented approach. *IIE Transactions*, **24** (3), 14-26.

Mohsen, A. J., 1992, An architecture for a shop-floor controller using colored Petri nets. *Int. J. Flexible Manufacturing Systems*, **4**, 159-181.

Wang, L., Lin, W. T., You, S. B., Chiu, H. S., and Chen, J. G., 1994, An object-oriented analysis method for the manufacturing information systems. *Journal of the Chinese Institute of Industrial Engineers*, **11** (3).

Wang, L., 1995, The development of an object-oriented Petri net cell control model. *International Journal of Advanced Manufacturing Technology* (accepted, in press).

AN INTEGRATED MMS AND COLOUR PETRI-NET MODEL FOR THE DISTRIBUTED CONTROL OF FLEXIBLE ASSEMBLY SYSTEMS

Yeung, W.H.R.
Department of Manufacturing Engineering, City University of Hong Kong,
83 Tak Chee Avenue, Hong Kong,
E-mail: MERICKYY@CITYU.EDU.HK

and

Moore, P.R.,
Mechatronics Research Group, School of Engineering & Manufacture,
DeMontfort University, LeiCester, The GateWay LeiCester LE1 9BH, UK.

ABSTRACT

In this paper, a reference system architecture is presented which utilizes Colour Petri-Nets (CPN) and the Manufacturing Message Specification (MMS) model for the control of a flexible assembly system which consists of multiple robots, conveyor systems and sensory devices. To provide co-ordination amongst robots and other devices, a CPN system is designed which provides a graphical editor for the configuration of a Petri-Net which represents the assembly processes and a special run-time executive is employed to co-ordinate the resources and resolve conflicts and deadlock. The first part of the paper presents an overview of the object oriented based design concept using MMS in distributed FAS control. In the second part, a CPN model is presented, which acts as a cell controller keep tracking of the realtime status of the devices and despatches commands to the corresponding devices accordingly. A Fieldbus sensory I/O network was used to monitor the Input/Output (I/O) points of the conveyor system and other digital devices and synchronise it with the CPN controller.

Introduction To MMS-Based Control of Flexible Assembly Systems

MMS [1] is an application protocol in the Opens System Interconnection (OSI) seven layer model. It is designed to support messaging communications amongst programmable devices in a CIM environment. In MMS, a server is a system which has physical control over a set of real devices. The other components of the communication network will be a client who sends requests to the server to perform a task. Figure 1 shows their master and slave relationship. MMS describes the behaviour of the communicating systems using an abstract object known as a Virtual Manufacturing Device (**VMD**). Real physical devices, such as robots and conveyor lines are mapped to a whole or part of the VMD through the concept of capabilities. Since the behaviour of the VMDs are standardised and the same sematics can be extracted from the information contained in the MMS protocol data unit, it ensures that all MMS-based devices can communicate amongst themsevles regardless of the proprietary communication format imposed by different vendors. The use of MMS can thus greatly enhance the inter-operability of manufacturing devices in a CIM environment.

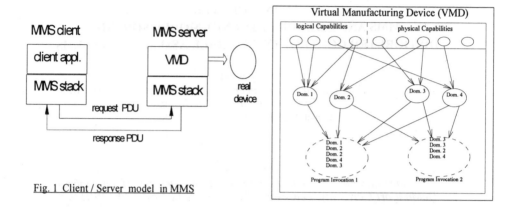

Fig. 1 Client / Server model in MMS

Fig. 2 Relationship between VMD, Domain, Program Invocation and Capabilities

A revolutionary approach in MMS is the definition of Domain and Program Invocation which provide flexibility in the control of dynamic systems. A Domain is a subset of the VMD which includes the required Capabilities set and a unique domain content for a specific task. More than one Domain can co-exist and they can be created statically or dynamically according to the requirements of a specific assembly task[2]. The execution sequence of the tasks, or domains is defined by Program Invocation. A Program Invocation contains a set of Domains and when it starts, the real device will act according to the domain information contained in the invocation. Figure 2 shows the relationship between VMD, Domain and Program Invocation.

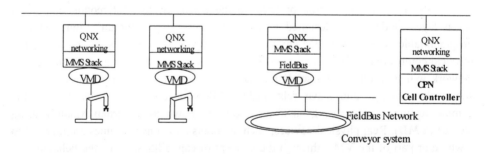

Fig. 3 Logical setup of the MMS-based integrated assembly system

In the Flexible Assembly System described here[3], a VMD is formed for each robot which acts as the MMS server waiting for assembly task requests from the remote cell controller. Each request from the controller is a specific assembly related operation. All these operations are constructed as autonomous domains and are made known to the robot servers at the system initialisation phase. Therefore, when the cell controller asks a specific robot to perform an operation on a component, it in effect sends a "Start

Program Invocation" request to the remote robot server which in turn acts on the real device according to the static domain content predefined in the system. In addition to robot servers, a distributed FieldBus [4] sensory I/O controller is also employed and a minimial set of MMS services are available which turn the FieldBus controller into an MMS-based I/O server responsible for maintaining the realtime state of the digital I/O of the assembly system as well as the co-ordination of the materials handling system. Figure 3 illustrates the logical setup of the MMS-based integrated assembly system. Since the main topic of this paper is the CPN control, no further detail of the Fieldbus controller will be presented here.

The co-ordination of the assembly process and the sequencing and sharing of the resources and prevention of possible system deadlock is controlled by a Coloured Petri-Net model. It is described in more details in the following section.

Introduction To The Coloured Petri-Net (CPN)

Since its introduction in 1960 by Adam Petri, the Petri Net has gained world-wide recognition for its capability to model and simulate dynamic systems which have asynchronous and concurrent events with resource constraints. During the past decade, a number of variances have been developed and entities such as time, colours and distribution functions, etc.,[5] were added to enrich the modeling power of Petri-Nets. In this paper, Colour Petri-Net (CPN) with timed-transitions are employed to model and control the operation of a Flexible Assembly System (FAS) capable of manufacturing multiple products simultaneously.

A CPN is a five-tuple (P,T,C,IN,OUT) where
$P = \{P_1, \ldots P_n\}$ is a set of n places
$T = \{T_1, \ldots T_m\}$ is a set of m transitions
C(p) and C(t) are the sets of colors associated with place $p \in P$ and transition $t \in T$
IN(p,t): $C(p) \times C(t) \longrightarrow N$ is an input function that defines directed arcs from input place(p) to transition (t).
OUT(p,t): $C(p) \times C(t) \longrightarrow N$ is an output function that defines directed arcs from transition (t) to output place (p).

A transition t_j is said to be enabled with respect to a color C_k if
$M(p_i) >= IN(p,t_j/C_k) \ \forall \ P_i \in {}^\circ t_j$ where ${}^\circ t_j$ denotes the input places of t_j
and $M(p_i)$ is the marking of p_i.
After the transition tj is fired, the new marking M' of the CPN will be
$M'(p_i) = M(p_i) + OUT(p,t_j/C_k) - IN(p,t_j/C_k) \ \forall \ P_i$

In the domain of FMS/FAS, some typical examples of colour sets will be component types to be manufactured $\{C_1, \ldots, C_n\}$ or the available machines in the system $\{m_1, \ldots, m_n\}$. But in general, a colour can be represented by a n-tuple $Ck = \{c_{k1}, \ldots, c_{kn}\}$ which conveys a complex piece of information.

Recently, there has been extensive research in applying PN for the modeling of FMS/FAS. Most of them have emphasized the modeling, simulation and analysis for performance evaluation[6], deadlock prevention[7][8] or resources conflict resolution[9]. Few have attempted to apply PN in the realtime control of manufacturing systems. An example of applying PN in realtime control is given in [7]. This paper presents a methodology and model for applying timed-transition CPN in the distributed control of a Flexible Assembly System.

Virtual and Real CPN Models for Deadlock Prevention

As shown in figure 4 below, the integrated CPN controller consists of two main modules, the Virtual CPN and the Real CPN (**VCPN** and **RCPN**). VCPN is used to provide a predictor scheme, similar to [7], to simulate and check for system deadlock for the incoming order before it is fed to the RCPN for the actual control of the FAS. The VCPN, when it executes, which is triggered by the Firing Controller of the RCPN, takes the first three components from the input schedule and performs the deadlock analysis simulation. To enhance the accuracy of the simulation result, the real state of the system (Marking $M(p_i)$ $\forall P_i$) is copied from the RCPN before each check. If deadlock is found, the VCPN will delay the dispatch of the components to RCPN. The VCPN in this system uses the same control code as the RCPN except that it is not connected to the real manufacturing devices and a pre-defined operation time is assigned to each of the transitions of the CPN model. The VCPN also operates at a time-slice of about 10:1 to the RCPN to guarantee that enough simulation runs are performed.

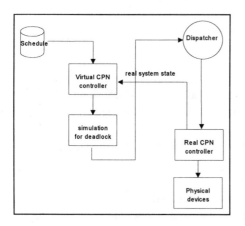
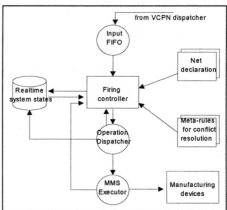

Fig. 4 The VCPN/RCPN interaction for deadlock prevention

Fig.5 the RCPN control architecture

RCPN Control Architecture

Figure 5 depicts the architecture of the RCPN controller. The deadlock free order (in batches of three components) is dispatched from the VCPN into a FIFO queue. The **Firing Controller** takes overall responsibility to coordinate the operation of the CPN model. According to the current status of the model, it searches and compiles a list of conflict free to-be-fired transitions and passes these on to the **Operation Dispatcher**. The Operation Dispatcher distinguishes between two kind of transitions, the immediate transition T_I and the timed transition T_T. The T_I are those used for system state change and hence there is no time defined in the transitions. The timed-transitions T_T always represents some physical operation which takes a finite operation time. Both T_I and T_T are pre-defined in the CPN model. The Operation Dispatcher thus parses the to-be-fired transition list one by one. If the transition is a T_I, it immediately modifies the marking of the CPN and the Firing Controller progresses to the next iteration. However, if the transition is a T_T, it determines the actual operations to be performed from the colour token and constructs the action list. The action list is then passed to the MMS Executor which operates in parallel and independently to the rest of the system. The MMS Executor, the details of which can be found in [3], operates here as a MMS client which creates the client concurrent processes sending requests to the remote MMS servers to action the assembly operations and waits for the completion of these operations. Upon completion of a remote operation, it signals the Firing Controller to trigger the corresponding transition and it in turn updates the marking of the CPN and initiates next iteration. When the Firing Controller starts its next iteration, it fetches a new component from the FIFO into the system for execution by adding a new component token to the first input place of the CPN model and signal the VCPN to start its next deadlock simulation.

A Case Study Flexible Assembly System

Figure 6 below shows the physical setup of an experimental assembly system, which consists of four industrial robots, a programmable conveyor line and four work stations. Robot 2 (R2) in this system is mounted on a linear slideway and services two different work stations. At each work station, a number of assembly operations can be performed. The component are transported on standard pallets which run through the conveyor line.

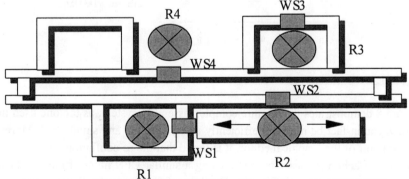

Fig. 6 The Experimental Setup of the MMS-based Flexible Assembly CEll(MFACE)

To demonstrate the CPN model of this system and the operation of the integrated controller, three product types (P1,P2,P3) are fed into the system to be processed simultaneously. Table 1 below illustrates a typical system configuration detailing the operations that can be performed at each work station while Table 2 shows the corresponding process plan of each product type. Figure 8 shows the complete CPN model for this product mix. The event conditions E_i adjacent to the transitions are employed for the remote sensor I/O controller to signal to the RCPN the arrival of a component to work station and thus synchronise to the real physical state of the system.

Table 1 System Capabilities

Operation	WS1	WS2	WS3	WS4
R1	OP1			
R2	OP1,OP2	OP3		
R3			OP4	
R4				OP5

Table 2 Process Plan for each product

	Operation Sequence		
P1	OP1 →	OP2 →	OP4
P2	OP1 →	OP3 →	OP5
P3	OP3 →	OP4 →	OP5

1. P1 reaches WS1, signalled by Field Bus, OP1 starts operated
2. OP1 completed on P1, signalled by robot server R2
3. OP2 completed on P1, signalled by robot server R2 WS1 released by Field Bus.
4. P3 reaches WS1, signalled by Field Bus.
5. no operation on P3 at WS1.
6. P1 reaches WS2, signalled by Field Bus.
7. no operation on P1 at WS2.
8. P3 reaches WS2, signalled by Field Bus, OP3 starts operated
9. OP3 completed on P3, signalled by robot server R2 WS2 released by Field Bus.

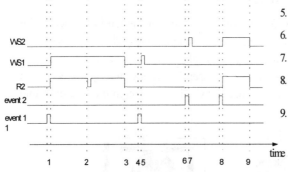

Fig.7 A sample of state changes on the CPN model

Figure 8 demonstrates that a quite complex assembly system, as the one used in the case study, can in fact be modeled efficiently using a simple and small CPN. Moreover, with the support of the graphical editor, it is found that the re-configuration of assembly systems can be achieved easily by just loading a similar CPN model from the database and performs some minor modifications. Database of the standard cell configurations can also be created in the future and stored in the CPN as some form of high level objects to further improve the system configuration cycle.

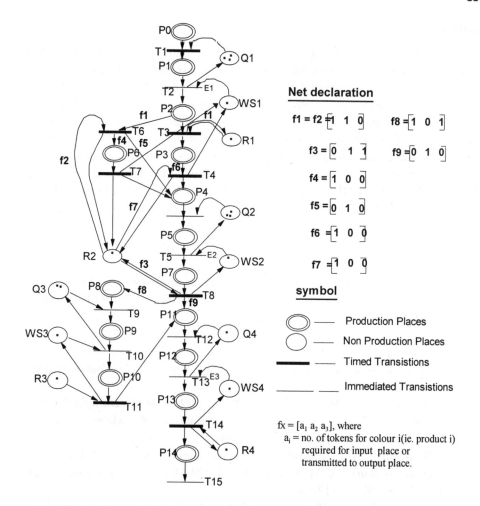

Fig. 8 The Coloured Petri-Net model for the Flexible assembly system

Figure 7 illustrates a sample of state changes on the CPN model when component P1 and P3 are fed into the system. When P1 is first fed into the system, the Firing Controller (FC) detects that T1 was enabled and it was a T_T. The Operation Dispatcher constructs an action list for the MMS Executor which in turn sends a MMS write request to the remote FieldBus controller to open the input gate of the system. P1 then travels to WS1 and MMS Executor responds immediately and the FC continues to operate. It determines that T2 is enabled, however, event 1 (E1) was not true and hence FC is blocked for another cycle. When P1 reaches WS1 (1), the FieldBus controller is alerted and triggers the MMS executor which in turn sets event 1 to be true. Transition T2 is then fired. As Robot R2 has a higher priority than R1(defined by meta-rules), thus T6, followed by T7 are fired and OP1 and OP2 are performed on P1 by R2(2,3). P3 then enters the system. However, because of the input function (f1), it does not stop at WS1 (4,5) and passes on to WS2 where T8 is fired and operation OP3 performed.

Conclusion

In this paper, a control architecture is outlined which utilizes a Coloured Pertri-Net model to implement a cell controller for Flexible Assembly Systems which can simultaneously manufacture multiple products. A window-based CPN editor has been created which runs under on the QNX multiple-users operating system. The editor allow users to generate the CPN model graphically and implemented the desired control algorithms as well as to facilitate the re-configuration of the systems. The CPN model has been used successfully to simulate and control the Flexible Assembly Cell outlined. To further the research in the control of Flexible Assembly/Manufacturing Systems, we will explore the study of MMS in particular the area of 'Virtual Manufacturing Cells' and intelligent task scheduling. Moreover, the modern control concept like heterogenous or hotonic control are also the interested topic in the future research.

References

1. ISO. *"Industrial Automation Systems - Manufacturing Message Specification, Part 1: Service definition. Part 2: Protocol specification"*. **ISO/IEC 9506-1 and 2.** 1990.
2. Mo J.P.T., & Wang Y. *"Integrated Robot Control Using Manufacturing Message Specification Protocol Based on NetBIOS"*. Proc. 12th Worl Congress, Int. Federation of Automation Control, Sydney, Australia, 1993, **Vol.6**, pp. 165-168.
3. Yeung, W.H.R., & Mo, P.T.J., *"A MMS-based Open Architecture for Flexible Assembly Cell"*, Prco., Int. Conference on Data and Knowlege systems for Manufacturing and Engineering", 1994, **Vol.2**, pp. 478-484.
4. C.D. Roeder, *"The economic & technical implications of FieldBus in the chemical industry"*, CIM, Proc. of the 8th CIM-Europe Annual conf., 27-29 May, 1992, UK. pp. 117-126.
5. R. David, & H. Alla, *"Petri Nets & Grafcet, Tools for modelling discrete event systems"*, Prentice Hall, 1992.
6. S.R. Menon, & P.M. Ferreira, *"Analysis of Colored Petri Net based Models for coordination control of Flexible Manufacturing Systems."*
7. N.Viswanadham, Y. Narahari, & T. Johnson, *"Deadlock Prevention and Deadlock Avoidance in Flexible Manufacturing Systems Using Petri Net Models"*, IEEE trans. Robotics and Automation, **Vol.6**, No.6, Dec., 1990, pp. 713-723.
8. Z. Banaszak, & B. Krogh, *"Deadlock Avoidance in Flexible Manufacturing Systems with Concurrently Competing Process Flows"*, IEEE trans. Robotics and Automation, **Vol. 6**, No.6, Dec., 1990, pp. 724-734.
9. S. Son, B.K. Choi, T.U. Park, & H.L. Kwang, *"Resolution of multiple conflicts in the Petri-net model of FAS"*, Int.J. Computer Integrated Manufacturing, 1991, **Vol.4**, No. 4, pp. 253-262.